D0890086

Methods in Enzymology

Volume 225
GUIDE TO TECHNIQUES IN
MOUSE DEVELOPMENT

METHODS IN ENZYMOLOGY

EDITORS-IN-CHIEF

John N. Abelson   Melvin I. Simon

DIVISION OF BIOLOGY
CALIFORNIA INSTITUTE OF TECHNOLOGY
PASADENA, CALIFORNIA

FOUNDING EDITORS

Sidney P. Colowick and Nathan O. Kaplan

Table of Contents

Section I. General Methodology

Section II. Germ Cells and Embryos

MAR 1 1994

Section V. Gene Expression: Reporter Genes

Section VI. Gene Expression: Proteins

Section VII. Gene Expression: Methylation

Section VIII. Gene Identification

Section IX. Nuclear Transplantation

Section X. Transgenic Animals: Pronuclear Injection

Section XI. Transgenic Animals: Embryonic Stem Cells and Gene Targeting

Section XII. Lineage Analysis

Contributors to Volume 225

Article numbers are in parentheses following the names of contributors.
Affiliations listed are current.

SUSAN J. ABBONDANZO (49), *Department of Cell and Developmental Biology, Roche Institute of Molecular Biology, Roche Research Center, Nutley, New Jersey 07110*

MAYI Y. ARCELLANA-PANLILIO (18), *Department of Medical Biochemistry, University of Calgary Health Sciences Center, Calgary, Alberta, Canada T2N 4N1*

CHRISTOPHER P. AUSTIN (56), *Department of Genetics, Harvard Medical School, Boston, Massachusetts 02115*

SHEILA C. BARTON (44), *Wellcome/CRC Institute of Cancer and Developmental Biology, University of Cambridge, Cambridge CB2 1QR, England*

ANTHONY R. BELLVÉ (6, 7), *Departments of Anatomy and Cell Biology and Urology, and Center for Reproductive Sciences, College of Physicians and Surgeons, Columbia University, New York, New York 10032*

JOHN D. BIGGERS (9), *Department of Cellular and Molecular Physiology, Laboratory of Human Reproduction and Reproductive Biology, Harvard Medical School, Boston, Massachusetts 02115*

JEFFREY D. BLEIL (14), *Department of Molecular Biology, The Scripps Research Institute, La Jolla, California 92037*

CLAIRE BONNEROT (27, 28), *Unité de Biologie Moléculaire du Développement, Unité Associée 1148 du Centre National de la Recherche Scientifique, 75724 Paris Cedex 15, France*

ALLAN BRADLEY (51), *Institute for Molecular Genetics, Baylor College of Medicine, Houston, Texas 77030*

GERARD BRADY (36), *Molecular Pharmacology, School of Biological Sciences, University of Manchester, Manchester M13 9PT, England*

PASCALE BRIAND (27), *Laboratoire de Génétique et Pathologie Expérimentales, INSERM, Institut Cochin de Génétique Moléculaire, 75014 Paris, France*

MIA BUEHR (4), *Institut for Molekylær Biologi, Århus Universitet, DK-8000 Arhus C, Denmark*

QIU PING CAO (19), *Cell Biology Group, Worcester Foundation for Experimental Biology, Shrewsbury, Massachusetts 01545*

RICHARD A. CARDULLO (8), *Department of Biology, University of California, Riverside, Riverside, California 92521*

CONSTANCE L. CEPKO (56), *Department of Genetics, Harvard Medical School, Boston, Massachusetts 02115*

KERRY B. CLEGG (15), *Developmental Biology Laboratory, Veterans Administration Medical Center, Sepulveda, California 91343*

RONALD A. CONLON (23), *Samuel Lunenfeld Research Institute, Mount Sinai Hospital, Toronto, Ontario, Canada M5G 1X5*

JULIE E. COOKE (3), *Wellcome/CRC Institute, Cambridge University, Cambridge CB2 1QR, England*

ROGER D. COX (38), *Genome Analysis Laboratory, Imperial Cancer Research Fund, London WC2A 3PX, England*

WILLIAM R. CRAIN (19), *McLaughlin Research Institute for Biomedical Sciences, Great Falls, Montana, 59401*

ANN C. DAVIS (51), *Institute of Molecular Genetics, Baylor College of Medicine, Houston, Texas 77030*

CLAYTUS A. DAVIS (31), *Division of Biology, California Institute of Technology, Pasadena, California 91125*

JULIE A. DeLOIA (35), *Department of Physiology, University of Pittsburgh, Magee-Womens Hospital, Pittsburgh, Pennsylvania 15213*

MELVIN L. DePAMPHILIS (25, 26, 30), *Department of Cell and Developmental Biology, Roche Institute of Molecular Biology, Roche Research Center, Nutley, New Jersey 07110*

JOHN J. EPPIG (5), *The Jackson Laboratory, Bar Harbor, Maine 04609*

DONNA M. FEKETE (56), *Department of Biology, Boston College, Chestnut Hill, Massachusetts 02167*

CHARLES FFRENCH-CONSTANT (3), *Wellcome/CRC Institute, Cambridge University, Cambridge CB2 1QR, England*

P. A. FLECKNELL (2), *Comparative Biology Centre, University of Newcastle upon Tyne, Medical School, Newcastle upon Tyne NE2 4HH, England*

HARVEY M. FLORMAN (8), *Worcester Foundation for Experimental Biology, Shrewsbury, Massachusetts 01545*

LYNN R. FRASER (13), *Anatomy and Human Biology Group, Biomedical Sciences Division, King's College London, University of London, Strand, London WC2R 2LS, England*

GLENN FRIEDRICH (41), *Fred Hutchinson Cancer Research Center, Program in Molecular Medicine, Seattle, Washington 91809.*

INDER GADI (49), *Department of Genetics, Roche Biomedical Laboratories, Inc., Raritan, New Jersey 08869*

JAMES I. GARRELS (29), *Cold Spring Harbor Laboratory, Cold Spring Harbor, New York 11724*

MICHELLE F. GAUDETTE (19), *Cell Biology Group, Worcester Foundation for Experimental Biology, Shrewsbury, Massachusetts 01545*

BRIAN J. GAVIN (39), *Department of Receptor Mechanisms, Sandoz Research Institute, Sandoz Pharmaceuticals Corporation, East Hanover, New Jersey 07936*

MAUREEN GENDRON-MAGUIRE (48), *Department of Cell and Developmental Biology, Roche Institute of Molecular Biology, Roche Research Center, Nutley, New Jersey 07110*

ISABELLE GODIN (3), *Institut d'Embryologie, Centre National de Recherche Scientifique et Collège de France, Nogent-sur-Marne 94736, France*

JON W. GORDON (12, 45), *Department of Obstetrics, Gynecology, and Reproductive Science, Mount Sinai School of Medicine, New York, New York 10029*

MARK GRANT (33), *St. Edmund's College, Cambridge CB3, England*

THOMAS GRIDLEY (48), *Department of Cell and Developmental Biology, Roche Institute of Molecular Biology, Roche Research Center, Nutley, New Jersey 07110*

JANET HEASMAN (3), *Wellcome/CRC Institute, Cambridge University, Cambridge CB2 1QR, England*

BERNHARD G. HERRMANN (23), *Max-Planck Institut für Entwicklungsbiologie, Abteilung Biochemie, D-7400 Tübingen, Germany*

DAVID P. HILL (40), *Division of Molecular and Developmental Biology, Samuel Lunenfeld Research Institute, Mount Sinai Hospital, Toronto, Ontario, Canada M5G 1X5*

JOAQUIN HUARTE (21), *Institute of Histology and Embryology, University of Geneva Medical School, CH-1211 Geneva 4, Switzerland*

NORMAN N. ISCOVE (36), *Ontario Cancer Institute, Toronto, Ontario, Canada M4X 1K9*

JOEL JESSEE (35), *Life Technologies, Inc., Gaithersburg, Maryland 20877*

DABNEY JOHNSON (35), *Biology Division, Oak Ridge National Laboratory, Oak Ridge, Tennessee 37831*

ROSS A. KINLOCH (17), *Department of Cell and Developmental Biology, Roche Institute of Molecular Biology, Roche Research Center, Nutley, New Jersey 07110*

BARBARA B. KNOWLES (35), *The Jackson Laboratory, Bar Harbor, Maine 04609*

FRANK KÖNTGEN (52), *Cellular Immunology Unit, The Walter and Eliza Hall Institute of Medical Research, The Royal Melbourne Hospital, Victoria 3050, Australia*

ZOIA LARIN (37, 38), *Department of Biochemistry, University of Oxford, Oxford OX1 3QU, England*

KEITH E. LATHAM (29, 43), *Fels Institute for Cancer Research and Molecular Biology, Temple University School of Medicine, Philadelphia, Pennsylvania 19140*

JOEL A. LAWITTS (9), *Transgenic Facility, Beth Israel Hospital, Boston, Massachusetts 02115*

HANS LEHRACH (37, 38), *Genome Analysis Laboratory, Imperial Cancer Research Fund, London WC2A 3PX, England*

SADHAN MAJUMDER (26), *Department of Cell and Developmental Biology, Roche Institute of Molecular Biology, Roche Research Center, Nutley, New Jersey 07110*

JEFFREY R. MANN (46, 47), *Division of Biology, Beckman Research Institute of the City of Hope, Duarte, California 91010*

YORDANKA S. MARTINOVA (7), *Institute of Cell Biology and Morphology, Bulgarian Academy of Sciences, 1113 Sofia, Bulgaria*

ANNE MCLAREN (4), *MRC Mammalian Development Unit, Wolfson House, University College London, London NW1 2HE, England*

K. J. MCLAUGHLIN (55), *Division of Biology, Beckman Research Institute of the City of Hope, Duarte, California 91010*

ANDRE P. MCMAHON (39, 46), *Department of Cell and Developmental Biology, Roche Institute of Molecular Biology, Roche Research Center, Nutley, New Jersey 07110*

SEBASTIAN MEIER-EWERT (37, 38), *Genome Analysis Laboratory, Imperial Cancer Research Fund, London WC2A 3PX, England*

MIRIAM MIRANDA (25, 26), *Department of Cell and Developmental Biology, Roche Institute of Molecular Biology, Roche Research Center, Nutley, New Jersey 07110*

ANTHONY P. MONACO (37, 38), *Human Genetics Laboratory, Imperial Cancer Research Fund, Institute of Molecular Medicine, John Radcliffe Hospital, Oxford OX3 9DU, England*

MARILYN MONK (33), *Institute of Child Health, London W1C1N 1EH, England*

JEAN-FRANÇOIS NICOLAS (27, 28), *Unité de Biologie Moléculaire du Développement, Unité Associée 1148 du Centre National de la Recherche Scientifique, 75724 Paris Cedex 15, France*

M. ANGELA NIETO (22), *MRC Laboratory of Eukaryotic Molecular Genetics, National Institute for Medical Research, London NW7 1AA, England*

G. P. PFEIFER (34), *Department of Biology, Beckman Research Institute of the City of Hope, Duarte, California 91010*

LAJOS PIKÓ (15, 16), *Developmental Biology Laboratory, Veterans Admnistration Medical Center, Sepulveda, California 91343*

RAMIRO RAMÍREZ-SOLIS (51), *Institute for Molecular Genetics, Baylor College of Medicine, Houston, Texas 77030*

WILLIAM G. RICHARDS (21), *Department of Pharmacology, State University of New York at Stony Brook, Stony Brook, New York 11794*

A. D. RIGGS (20, 34), *Department of Biology, Beckman Research Institute of the City of Hope, Duarte, California 91010*

RICHARD J. ROLLER (17), *Department of Cell and Developmental Biology, Roche Institute of Molecular Biology, Roche Research Center, Nutley, New Jersey 07110*

NADIA ROSENTHAL (24), *Cardiovascular Research Center, Massachusetts General Hospital East, Charlestown, Massachusetts 02129*

MARK ROSS (38), *Genome Analysis Laboratory, Imperial Cancer Research Fund, London WC2A 3PX, England*

JAY L. ROTHSTEIN (35), *Departments of Microbiology/Immunology and Otolaryngology, Jefferson Cancer Institute, Thomas Jefferson University, Philadelphia, Pennsylvania 19107*

ELIZABETH F. RYDER (56), *Department of Genetics, Harvard Medical School, Boston, Massachusetts 02115*

FERNANDO J. SALLÉS (21), *Department of Pharmacology, State University of New York at Stony Brook, Stony Brook, New York 11794*

DAVID SASSOON (24), *Department of Biochemistry, Boston University School of Medicine, Boston, Massachusetts 02118*

BRIAN SAUER (53), *Biotechnology, Du Pont Merck Pharmaceutical Company, Wilmington, Delaware 19880*

GERALD SCHATTEN (32), *Department of Zoology, University of Wisconsin, Madison, Wisconsin 53706*

GILBERT A. SCHULTZ (18), *Department of Medical Biochemistry, University of Calgary Health Sciences Center, Calgary, Alberta, Canada T2N 4N1*

RACHEL D. SHEPPARD (42), *Genetic Therapy, Inc., Immunology Group, Gaithersburg, Maryland 20878*

LEE M. SILVER (1, 42), *Department of Molecular Biology, Princeton University, Princeton, New Jersey 08544*

CALVIN SIMERLY (32), *Department of Zoology, University of Wisconsin, Madison, Wisconsin 53706*

J. SINGER-SAM (20, 34), *Department of Biology, Beckman Research Institute of the City of Hope, Duarte, California 91010*

JACEK SKOWRONSKI (35), *Cold Spring Harbor Laboratory, Cold Spring Harbor, New York 11724*

DAVOR SOLTER (29, 35, 43), *Max-Planck Institut für Immunbiologie, D-7800 Freiburg-Zähringen, Germany*

PHILIPPE SORIANO (41), *Fred Hutchinson Cancer Research Center, Program in Molecular Medicine, Seattle, Washington 91809*

COLIN L. STEWART (49, 50, 52), *Department of Cell and Developmental Biology, Roche Institute of Molecular Biology, Roche Research Center, Nutley, New Jersey 07110*

SIDNEY STRICKLAND (21), *Department of Pharmacology, State University of New York at Stony Brook, Stony Brook, New York 11794*

KARIN STURM (10), *Embryology Unit, Children's Medical Research Institute, University of Sydney, Wentworthville NSW 2145, Australia*

M. AZIM SURANI (44), *Wellcome/CRC Institute of Cancer and Developmental Biology, University of Cambridge, Cambridge CB2 1QR, England*

PATRICK P. L. TAM (10, 11), *Embryology Unit, Children's Medical Research Institute, University of Sydney, Wentworthville NSW 2145, Australia*

KENT D. TAYLOR (16), *Developmental Biology Laboratory, Veterans Administration Medical Center, Sepulveda, California 91343*

EVELYN E. TELFER (5), *Institute of Ecology and Resource Management, University of Edinburgh, School of Agriculture, Edinburgh EH9 3JG, Scotland*

JEAN-DOMINIQUE VASSALLI (21), *Institute of Histology and Embryology, University of Geneva Medical School, CH-1211 Geneva 4, Switzerland*

MURIAL VERNET (27), *Laboratoire de Génétique et Pathologie Expérimentales, INSERM, Institut Cochin de Génétique Moléculaire, 75014 Paris, France*

CHRISTOPHER WALSH (56), *Department of Neurology, Beth Israel Hospital, Boston, Massachusetts 02115*

PAUL M. WASSARMAN (17), *Department of Cell and Developmental Biology, Roche Institute of Molecular Biology, Roche Research Center, Nutley, New Jersey 07110*

MARIA WIEKOWSKI (26, 30), *Department of Molecular Biology, Schering-Plough Corporation, Kenilworth, New Jersey 07033*

MICHAEL V. WILES (54), *Basel Institute for Immunology, CH-4005, Basel, Switzerland*

DAVID G. WILKINSON (22), *MRC Laboratory of Eukaryotic Molecular Genetics, National Institute for Medical Research, London NW7 1AA, England*

WOLFGANG WURST (40), *Division of Molecular and Developmental Biology, Samuel Lunenfeld Research Institute, Mount Sinai Hospital, Toronto, Ontario, Canada M5G 1X5*

CHRISTOPHER C. WYLIE (3), *Wellcome/CRC Institute, Cambridge University, Cambridge, CB2 1QR, England*

WENXIN ZHENG (7), *Department of Pathology, New York Hospital Cornell Medical Center, New York, New York 10021*

MAURIZIO ZUCCOTTI (33), *Dipartimento Biologia Animale and Centro Studio per L'istochimica del CNR, University of Pavia, Pavia 27100, Italy*

Preface

Publication of a *Guide to Techniques in Mouse Development* is timely in view of the already enormous and rapidly growing interest in the mouse as an experimental organism. Perhaps nowhere is the impact of technology on developmental biology seen so clearly as in the case of research on mouse development. The recently acquired ability to add specifically engineered genes to the mouse genome by the production of transgenic animals, as well as to remove ("knockout") specific genes from the mouse genome by homologous recombination in embryonic stem (ES) cells, has virtually revolutionized the field. Genetic manipulations that not so long ago were feasible only with nonmammals, such as fruit flies and worms, are now performed routinely with mice. A plethora of new methodology is available that can be applied to classical questions involving cellular behavior during mouse development; such questions can now be addressed at the level of individual genes, messenger RNAs, and proteins.

Our purpose in assembling this volume is to create a source of state-of-the-art experimental approaches in mouse development useful at the laboratory bench to a diverse group of investigators. The aim is to provide investigators with reliable experimental protocols and recipes that are described in sufficient detail by leaders in the field. Although technology in this area is changing rapidly, it is likely that much of the *Guide* will remain relevant for many years to come.

We extend our thanks to the authors for their contributions as well as for their cooperation and patience during the preparation of the volume. Also, we are grateful to our colleagues Tom Gridley, Andy McMahon, and Colin Stewart who provided good advice throughout this long venture.

PAUL M. WASSARMAN
MELVIN L. DEPAMPHILIS

METHODS IN ENZYMOLOGY

VOLUME 181. RNA Processing (Part B: Specific Methods)
Edited by JAMES E. DAHLBERG AND JOHN N. ABELSON

VOLUME 182. Guide to Protein Purification
Edited by MURRAY P. DEUTSCHER

VOLUME 183. Molecular Evolution: Computer Analysis of Protein and Nucleic Acid Sequences
Edited by RUSSELL F. DOOLITTLE

VOLUME 184. Avidin–Biotin Technology
Edited by MEIR WILCHEK AND EDWARD A. BAYER

VOLUME 185. Gene Expression Technology
Edited by DAVID V. GOEDDEL

VOLUME 186. Oxygen Radicals in Biological Systems (Part B: Oxygen Radicals and Antioxidants)
Edited by LESTER PACKER AND ALEXANDER N. GLAZER

VOLUME 187. Arachidonate Related Lipid Mediators
Edited by ROBERT C. MURPHY AND FRANK A. FITZPATRICK

VOLUME 188. Hydrocarbons and Methylotrophy
Edited by MARY E. LIDSTROM

VOLUME 189. Retinoids (Part A: Molecular and Metabolic Aspects)
Edited by LESTER PACKER

VOLUME 190. Retinoids (Part B: Cell Differentiation and Clinical Applications)
Edited by LESTER PACKER

VOLUME 191. Biomembranes (Part V: Cellular and Subcellular Transport: Epithelial Cells)
Edited by SIDNEY FLEISCHER AND BECCA FLEISCHER

VOLUME 192. Biomembranes (Part W: Cellular and Subcellular Transport: Epithelial Cells)
Edited by SIDNEY FLEISCHER AND BECCA FLEISCHER

VOLUME 193. Mass Spectrometry
Edited by JAMES A. MCCLOSKEY

VOLUME 194. Guide to Yeast Genetics and Molecular Biology
Edited by CHRISTINE GUTHRIE AND GERALD R. FINK

VOLUME 195. Adenylyl Cyclase, G Proteins, and Guanylyl Cyclase
Edited by ROGER A. JOHNSON AND JACKIE D. CORBIN

VOLUME 196. Molecular Motors and the Cytoskeleton
Edited by RICHARD B. VALLEE

VOLUME 197. Phospholipases
Edited by EDWARD A. DENNIS

VOLUME 198. Peptide Growth Factors (Part C)
Edited by DAVID BARNES, J. P. MATHER, AND GORDON H. SATO

Methods in Enzymology

Volume 225

Guide to Techniques in Mouse Development

EDITED BY

Paul M. Wassarman

DEPARTMENT OF CELL AND DEVELOPMENTAL BIOLOGY
ROCHE INSTITUTE OF MOLECULAR BIOLOGY
ROCHE RESEARCH CENTER
NUTLEY, NEW JERSEY

Melvin L. DePamphilis

DEPARTMENT OF CELL AND DEVELOPMENTAL BIOLOGY
ROCHE INSTITUTE OF MOLECULAR BIOLOGY
ROCHE RESEARCH CENTER
NUTLEY, NEW JERSEY

ACADEMIC PRESS, INC.
A Division of Harcourt Brace & Company
San Diego   New York   Boston   London   Sydney   Tokyo   Toronto

QP601
,C733
vol. 225

Front cover photo, paper edition only: Shown are an unfertilized mouse egg with sperm bound to the zona pellucida and chimeric mice produced by embryonic stem (ES) cell injection.

This book is printed on acid-free paper. ∞

Copyright © 1993 by ACADEMIC PRESS, INC.
All Rights Reserved.
No part of this publication may be reproduced or transmitted in any form or by any means, electronic or mechanical, including photocopy, recording, or any information storage and retrieval system, without permission in writing from the publisher.

Academic Press, Inc.
1250 Sixth Avenue, San Diego, California 92101-4311

United Kingdom Edition published by
Academic Press Limited
24–28 Oval Road, London NW1 7DX

International Standard Serial Number: 0076-6879

International Standard Book Number: 0-12-182126-9 (Hardcover)

International Standard Book Number: 0-12-736450-1 (Papercover)

PRINTED IN THE UNITED STATES OF AMERICA
93 94 95 96 97 98   EB   9 8 7 6 5 4 3 2 1

Hints on Using the *Guide to Techniques in Mouse Development*

The methods described have been limited to the most current ones available for studying the molecular biology of mouse development. They are presented in fifty-six chapters that have been organized into twelve sections. Since it is not possible to discuss in a single volume everything one might need to know to carry out experiments in mouse development, the *Guide* focuses in depth on a limited number of methodologies, providing more than one description of the same method in several instances. In many cases, the same experimental approach is described by more than one contributor. This provides the reader with the opportunity to identify important variations in a particular approach. Each variation has advantages and disadvantages, depending on how the method is applied and the type of data one wishes to extract. The authors contribute their own personal experiences, observations, and helpful hints on dealing with technical problems likely to be encountered. Moreover, some problems addressed by one author may not be addressed by another. Considering the relative complexity of carrying out experiments in molecular biology on mice, it is not surprising that differences exist among the various protocols and recipes for each method. Table I provides a directory that cross-references general subjects addressed in more than one chapter.

Background information, descriptions of certain basic methodology used in studies on mouse development, and compendia of useful information about mouse molecular biology that may not be described in detail in this volume are presented in several other monographs (see Table II and the list of Recommended Monographs) as well as in the periodical *Mouse Genome*. The manual by Hogan *et al.* (1986) is particularly well suited as a companion to this *Guide*. Some general information about mice that may prove useful is summarized in Table III. A time-course for the sequence of embryological events in mouse development is summarized in Table IV and Fig. 1, with additional information provided in the various chapters.

TABLE I
GENERAL SUBJECTS

Subject	Chapters
Animal care	1, 2
Isolation and culture of primordial germ cells	3, 4, 10
In vitro fertilization	12–14
Preparation of sperm	6–8, 12
Preparation of eggs	5–8, 12
Preparation of embryos	9, 12
Analysis of messenger RNA	15–20
In situ hybridization	22–24
Microinjection	21, 25–28, 32, 45, 46
Nuclear transplantation	43, 44
Retroviruses	41, 56
Embryonic stem cells	40, 49–52, 54
Gene libraries	35–39
DNA methylation	33, 34
Histochemistry	11, 31, 32
Transgenic animals	40, 45–55

TABLE II
BACKGROUND INFORMATION

Topic	References
Manipulation of germ cells and embryos	Rafferty (1970); Hogan *et al.* (1986); Monk (1987); Rugh (1990); Copp and Cockroft (1990)
Isolation and use of embryonic stem cells	Robertson (1987)
Morphology of embryogenesis	Theiler (1989); Kaufman (1992)
Postimplantation development	Ciba Foundation Symposium No. 165 (1992)
Chromosome linkage map	*Mouse Genome* **90**(1), 1–75 (1992)
Known genes	*Mouse Genome* **90**(2), 112–199 (1992)
Origins and characteristics of inbred strains	*Mouse Genome* **90**(3), 231–352 (1992)
DNA clones, probes, and molecular markers	*Mouse Genome* **90**(4), 457–598 (1992)

TABLE III
CHARACTERISTICS OF MICE[a]

Characteristic	Value
Life span	~1.5 years
Gestation	~20 days
Estrous cycle	~4 days
Breeding age	
Male	~60 days
No. sperm/ejaculation	~5 × 10^7
Female	~55 days
No. eggs ovulated	
Natural	~10
Superovulation	~30 (<15 to >50)
Breeding life	
Male	~1.5 years
Female	~8 litters
Litter size	~9 pups
Weaning age	~19 days
Weight	
Birth	~1.25 g
Adult male	~35 g
Adult female	~30 g
Blood volume (adult)	~2 ml
Urine volume (daily)	~2 ml
Food consumption (daily)	~4.5 g
Water consumption (daily)	~6 ml
Chromosomes	
Diploid number	~40
Total DNA	~6 pg
Genes	~10^5
Recombination units	~1.6 × 10^3 cM

[a] The values listed are only a guide, since many of these characteristics are dependent on mouse strain, as well as other factors.

TABLE IV
TIMETABLE FOR MOUSE DEVELOPMENT

Developmental stage	Hours post-hCG injection[a]
Two-cell	~48
Four-cell	~60
Eight-cell (compacted)	~72
Morula	~84
Blastula (expanded)	~110
Blastula (hatched)	~120

Days of gestation	Hours post-hCG injection
Implantation	~5
Egg cyclinder	~6
Amnion	~7
Primitive streak	~7
Neural plate	~7.5
Germ cells	~7.5
Gastrulation	~7–10
Somites	
~8–10 pairs	~8.5
~15–20 pairs	~9.5
~20–30 pairs	~10
~40–44 pairs	~11
Organogenesis	~8
Heartbeat	~9
Genital ridge	~11
Embryo length	
~6 mm	~11
~13 mm	~15
~18 mm	~17
~25 mm	~19
Birth	~19

[a] hCG, Human chorionic gonadotropin.

Early Mouse Development

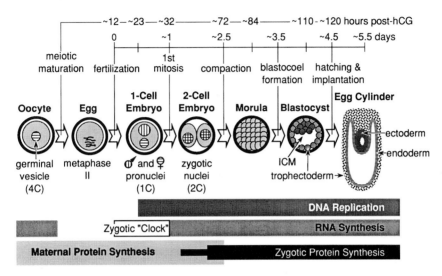

FIG. 1. Outline of preimplantation mouse development. Oocytes and embryos are surrounded by the zona pellucida (dark ring), and contain either a maternal nucleus (solid circle), paternal pronucleus (open circle), or zygotic nuclei (shaded circles) as indicated along with their nuclear ploidy (*N* value). Bars indicate at which stages maternal (solid), and zygotic (shaded) DNA replication, DNA transcription, and protein synthesis occur. Hours post-hCG indicate approximate times after injection of human chorionic gonadotropin.

Recommended Monographs

C. R. Austin and R. V. Short, "Reproduction in Mammals." Cambridge Univ. Press, Cambridge, 1982.

M. Balls and A. E. Wild, "The Early Development of Mammals." Cambridge Univ. Press, Cambridge, 1975.

K. Burki, "Experimental Embryology of the Mouse." Karger, Basel, 1986.

Ciba Foundation Symposium No. 165. "Postimplantation Development in the Mouse." Wiley, New York, 1992.

A. J. Copp and D. L. Cockroft, "Postimplantation Mammalian Embryos: A Practical Approach." IRL Press, Oxford, 1990.

F. Costantini, and R. Jaenisch. "Genetic Manipulation of the Early Mammalian Embryo." Cold Spring Harbor Laboratory, Cold Spring Harbor, 1985.

J. C. Daniel, "Methods in Mammalian Reproduction." W. H. Freeman/Academic Press, New York, 1971/1978.

H. L. Foster, J. D. Small, and J. G. Fox. "The Mouse in Biomedical Research." Academic Press, New York, 1983.

E. L. Green, "Biology of the Laboratory Mouse," Dover, New York, 1975.

R. B. L. Gwatkin, "Manipulation of Mammalian Development." Plenum Press, New York, 1986.

B. Hogan, F. Costantini, and E. Lacy, "Manipulating the Mouse Embryo: A Laboratory Manual." Cold Spring Harbor Laboratory, Cold Spring Harbor, 1986.

M. Johnson, "Development in Mammals." Elsevier/North Holland, Amsterdam, 1977.

M. H. Kaufman, "Early Mammalian Development: Parthenogenetic Studies," Cambridge Univ. Press, Cambridge, 1983.

M. H. Kaufman "An Atlas of Mouse Development." Academic Press, San Diego, 1992.

M. F. Lyon, and A. G. Searle, "Genetic Variants and Strains of the Laboratory Mouse." Oxford Univ. Press, Oxford, 1989.

A. McLaren, "Mammalian Chimeras." Cambridge Univ. Press, Cambridge, 1976.

M. Monk, "Mammalian Development: A Practical Approach." IRL Press, Oxford, 1987.

K. A. Rafferty, "Methods in Experimental Embryology of the Mouse." The Johns Hopkins Univ. Press, Baltimore, 1970.

E. J. Robertson, "Teratocarcinomas and Embryonic Stem Cells: A Practical Approach." IRL Press, Oxford, 1987.

Rossant J. and R. A. Pedersen, "Experimental Approaches to Mammalian Embryonic Development." Cambridge Univ. Press, Cambridge, 1986.

R. Rugh, "The Mouse: Its Reproduction and Development." Oxford Univ. Press, Oxford, 1990.

M. I. Sherman, "Concepts in Mammalian Development." MIT Press, Cambridge, Massachusetts, 1977.

K. Theiler, "The House Mouse: Atlas of Embryonic Development." Springer-Verlag, New York, 1989.

Section I

General Methodology

[1] Recordkeeping and Database Analysis of Breeding Colonies

By Lee M. Silver

General Strategies for Recordkeeping

Requirements

A breeding mouse colony differs significantly from a static one in the type and complexity of information that is generated. In a nonbreeding colony, there are only the animals and the results obtained from experiments on each one. In a breeding colony, there are animals, matings, and litters, with specific connections among various members of each of these data classes. Classical genetic analysis is based on the transmission of information between generations, and, as a consequence, the network of associations among individual components of the colony is often as important as the components in and of themselves.

An ideal recordkeeping system would allow one to keep track of (1) individual animals, their ancestors, siblings, and descendants; (2) experimental material (tissues and DNA samples) obtained from such animals; (3) matings between animals; (4) litters born to such matings, and the animals derived from such litters used in experiments or to set up the next generation of matings. Ideally, one would like to maintain records in a format that readily allows one to determine the relationship, if any, that exists between any two or more components of the colony, past or present.

Based on these general requirements, two different systems for recordkeeping have been developed by mouse geneticists over the last 60 years. The "mating unit" system centers on the mating pair as the primary unit for recordkeeping. The "animal/litter" system treats each animal and litter as a separate entity. As discussed below, there are advantages and disadvantages to each. In a later section, I describe a computer software version of the animal/litter system that can be implemented on a personal computer.

Mating Unit System

With the mating unit system, each mating unit is assigned a unique number and is given an individual record. When recordkeeping is carried out with a notebook and pencil, each mating pair is assigned a page in

METHODS IN ENZYMOLOGY, VOL. 225
Copyright © 1993 by Academic Press, Inc.
All rights of reproduction in any form reserved.

the book. The cage that holds the mating pair can be identified with a simple card on which the record number is indicated; this provides immediate access to the corresponding page in the record book.

When litters are born, they are recorded within the mating record. Each litter is normally given one line on which the following information is recorded: (1) a number indicating whether it is the first, second, third, or a subsequent litter born to the particular mating pair; (2) the date of birth; (3) the number of pups; and (4) other information of importance to the investigator. At a later time, individual mice within a litter can be identified (and recorded, if necessary, on further lines within the record page) by the mating number along with a secondary simple number or letter combination to distinguish siblings from each other. For example, the fourth pup in the third litter born to mating unit 7371 could be numbered 7371-3d, where "3" indicates the litter and "d" indicates the pup within it. This system provides for the individual numbering of animals in a manner that immediately allows one to identify siblings.

At the outset, parental numbers are incorporated into each mating record, and since these are linked implicitly to the litters from which they come, it becomes possible to trace a complete pedigree back from any starting individual. It also becomes possible to trace pedigrees forward if, as a matter of course, one cross-references all new matings within the litter records from which the parents derive. For example, if one sets up a new mating assigned the number 8765 with female 5678-2e and male 5543-1c, the number 8765 would be inscribed on appropriate lines in records 5678 and 5543.

There are several important advantages to a recordkeeping system based on the mating unit: (1) only a single set of primary record numbers is required; (2) one can easily keep track of the reproductive history of each mating pair; and (3) information on siblings is readily viewed within a single location. The major disadvantage to this recordkeeping system is that, for most investigators, it is impossible to determine ahead of time how much space will be required ultimately for any one record. One mating may yield no litters, while another may be highly prolific and require more record space than was originally set aside. A second disadvantage comes into play in those colonies where mating units are not infrequently taken apart and re-formed with new combinations of animals. In this situation, where the mating unit is not sacrosanct, the animal/litter system described below is more amenable for recordkeeping.

Animal/Litter System

In a second system developed originally by the geneticist L. C. Dunn, there are two primary units for recordkeeping—the individual animal and the individual litter. Each breeding animal is assigned a unique number (at the time of weaning) that is associated with an individual record occupying one line across one or two facing pages within an "animal record book." Each animal record contains the numbers of both parents, and through these numbers it is possible to trace back pedigrees. Each litter that is born is also assigned a unique number with an individual one-line record in a separate "litter record book." Both animal numbers and litter numbers are normally assigned in sequence.

A third independent set of numbers are those assigned to individual cages. Cage numbers can be assigned in a systematic manner so that related matings are in cages with related numbers. For example, different matings that derive from the same founder of a particular transgenic line may be placed in cages numbered from 2311 to 2319. A second set of matings that carry the same transgene from a different founder could be placed into cages numbered 2321 to 2329, and so on. Thus, the cages between 2300 and 2399 would all have animals that carried the same transgene; however, different sets of ten would be used for different founder lines. For matings of animals with a second transgene, one might choose to use the cages numbered 2400 to 2499. This type of numbering allows one to classify cages, which represent matings, in a hierarchical manner. Although at any point in time, every cage in the colony will have a different number, once a particular cage is eliminated, its number can be reassigned to a new mating. Cage cards from dismantled matings are saved in numerical order.

When a litter is born, the litter record is initiated with an identifying number, the birth date, the numbers of the parents, the number of the cage in which the litter was born, and any other important information. In addition, the litter number is inscribed on the cage card (which may or may not have additional information about the mating pair). When an animal is weaned from a litter for participation in the breeding program, an animal record is initiated. The most important information in the animal record is the number of the litter from which it came, the cage that it goes into, and the date of that move. The cage number is particularly important in allowing one to trace pedigrees forward from any individual at a future date. If an animal is moved from one cage to another at some later date, this can easily be added to the record.

With three unrelated systems of number assignment and the need for extensive cross-referencing, the animal/litter system is complex, and implementation on paper is labor intensive. However, it does provide the investigator with additional power for analysis. For example, by choosing cage numbers wisely and saving cage cards in numerical order, it becomes possible to go back at any point in the future and look at all of the litters born to a particular category of matings over any period of time. With the mating unit system, this can only be accomplished by using different record books, or different sections of a book, for different categories of matings. However, with a complex breeding program, it is often difficult to predict how much space a particular category of matings is likely to subsume over an extended period of time. This problem could be solved with the use of a loose-leaf book in which pages (representing mating units) can be added without limits to any particular section.

Another difference between the mating unit system and the animal/litter system is the ease with which it is possible to keep track of animals that are moved from one mating unit to another. The mating unit system is most effective for colonies where "animals are mated for life." The animal/litter system is effective for colonies of this type as well, but it is also amenable to those where animals are frequently switched from one mate to another.

Electronic Mouse Colony Recordkeeping System

Overview of Program

The animal/litter system for recordkeeping has been incorporated into a more extensive computer software package that greatly simplifies data entry, with automatic cross-referencing and built-in error checking. This software package is called the "Animal House Manager" or AMAN and can be licensed for use by Princeton University as described at the end of this chapter. AMAN is a specialized database program that allows users to record and retrieve information on animals, litters, tissues, DNA samples, and restriction digests generated from one or more breeding mouse colonies. Data are entered through a series of queries and answers. With automatic cross-referencing, the same information never has to be entered more than one time. Hard copy printouts can be obtained for cage cards, individual records, or sets of records uncovered through searches for positive or negative matches to particular parameters. Search protocols are highly versatile; for example, it is possible to print out a cage-ordered list of litters that are old enough for weaning or a list of live mice ordered according to birth cage.

AMAN provides investigators with the ability to maintain control over a complex breeding program with instant access to each record, current and past. AMAN can store 100,000 records in each of four files for (1) animals, (2) litters, (3) DNA/tissue samples, and (4) restriction digests. In the sections that follow, a detailed description is provided for the utilization of various components of this software package. Principles of data entry are described first, followed by protocols for data retrieval.

Entry of Founder Animals into Database

All animals that are brought into the colony from elsewhere are considered "founders." All animals that are born in the colony are considered "colony offspring." The parents of all colony offspring (who can be either founders or other colony offspring) must have been entered previously into the AMAN database. Obviously, there must be an original set of founders to get the colony going. Additional founders can be added at any later point, whenever animals are brought in from elsewhere to breed with each other, prior founders, or any colony offspring.

On choosing the "new animal entry" from the menu, AMAN will ask a series of questions concerning the animal; in most cases, an answer is optional. However, the mechanism by which AMAN identifies founder animals is through the nonoptional answer to the question regarding litter number or supplier. For all founding animals, one should designate a supplier with a nonnumerical answer of up to five characters. The default value for the weaning date is the date on which the animal is entered; the weaning date, in the case of founding animals, can be used to designate the receiving date. Birth date is optional.

The six primary animal description fields (with their symbols in the database) are sex (SEX), coat color (COAT), phenotype (PHNO), strain (STRN), genotype (GENO), and generation (GENR). These fields are considered primary because the information within each will show up in all abbreviated, two-line descriptions of animals. These abbreviated descriptions are used in the output from search routines and on cage cards. Each of these fields has a different length (visible on the screen). Only the sex field is nonoptional and restricted in terms of the information that can be placed in it: you must enter either M for male, F for female, or ? for unknown. Entry into the other fields is optional and unrestricted. For example, you may decide to use one of these fields to enter a form of information that does not correspond to the actual field designation. However, it is critical to follow two rules for data entry. First, always enter the same type of information in the same field; this is essential because the search routines must be provided with a particular field name in which to look for a particular piece of data. Second, always use the same exact symbol or phrase to describe the same information. For example, to record a white belly spot, you might choose to enter "wh-bsp." You can switch between lowercase and uppercase letters, but do not sometimes use "wh bsp" or "whbsp." If you find at a later date that you have accidentally used two different symbols for the same characteristic in different sets of records, you can use the String substitution routine (from the main menu) to change one symbol into the other in all records where it has been used.

After the primary description fields, you will be asked to enter the initial cage that the animal will be placed into. You must enter something in this field. All active cages in the colony must have a name that begins with a digit (0–9) but which can be followed by up to three digits or letters. AMAN recognizes a name that begins with a digit as an indication that the animal is alive in a cage with the full name. If the name begins with a letter, AMAN recognizes this as an indication that the animal is no longer in the colony by virtue of death or by its being exported to another investigator or colony.

It is extremely useful to use cage names as a means for organizing the colony into a hierarchy of subcolonies. For example, if you have three kinds of experiments under way, you might choose cages numbered from 1000 to 1999 for one, 2000 to 2999 for the second, and 3000 to 3999 for the third. Within the second experiment, you might have multiple crosses set up, or you might be maintaining multiple transgenic lines. You could divide the 2000–2999 cages into ten sections (2000–2099, 2100–2199, etc.) and then further divide the 2100–2199 into ten more sections (2100–2109, etc.). Unlike animal numbers, which are used only once, cage numbers only last as long as the animals within them are alive, and they can be used over and over again. All of the routines for retrieving data allow one to identify rapidly animals that are present in any subset of cages, and lists of animals can be sorted according to both cage of residence and cage of birth.

After entering the cage that the animal is placed into, you will be prompted for the date that the move took place. The date fields for all cage moves allow entry of only a month and day. This should not be a problem for most mouse colonies since experimental mice are usually not maintained for more than 1 year. The remaining fields are all optional. The request for further information will place any entered data (up to 78 characters) into the information 1 field (INF1). This field is useful for sentence-like descriptions of unique characteristics or any other data.

Certain investigators, especially those involved in transgenic work, may use the same inbred or F_1 strains over and over as recipients for transgenic embryos or as the mothers and fathers that give rise to the embryos to be injected. In such cases, it is possible to save time, effort, and space by designating certain animal records as generics. For example, record #10 could represent a generic B6 female, and the number 10 could be used many times to represent a different founding mother for many different lines.

Once any animals have been entered into the database, you can print cage cards. Just go to the "animal submenu" and then to the "search submenu" within it, and choose the "cage card" option.

Entry of Colony Offspring into Database

General. There are two general approaches to recordkeeping with a breeding animal colony, and both can be followed with AMAN. The first approach is the more inclusive one, and it should be followed whenever an investigator wants to keep close track of animal breeding from birth to death. With this approach, each litter that comes out of a mating is given a separate record in the litter file at the time of birth. When juveniles reach weaning age and some are to be used for subsequent matings, the litter record will serve as a template to facilitate their entry into individual animal records. Litter numbers and animal numbers are automatically cross-referenced, in both directions, within the database. By maintaining litter records, it is possible to keep track of the complete breeding history associated with each mating group.

In some cases, it may not be necessary to maintain such detailed records of breeding when all the investigator wants to do is keep track of the pedigrees along a particular line. In these cases, it is possible to skip the litter entry step and enter weaned animals directly into the animal file. AMAN allows the investigator to choose both approaches within the same database. The two approaches are discussed separately in the following two sections.

Entering New Animals Directly. The quickest way to maintain breeding data is to bypass the litter entry protocol and enter information directly only on those colony offspring that are of particular interest and/or will be used in further matings. The easiest means to accomplish this task is to generate a new record for each animal at the time of weaning. AMAN provides time-saving routines for quickly entering common information on each animal within the same litter.

To begin data entry on animals from a litter that is being weaned, choose the animal submenu and then the New record entry option. When prompted to enter a litter number or supplier, just enter 0 (zero). You will then be prompted to enter the mother's cage number and birth date (an answer to both questions is optional but often very useful). Next, you will be prompted to enter the record numbers for the parents. You must enter at least one mother and one father. The additional parent fields are useful for several different purposes. In some cases, there may be more than one potential mother and/or father in the cage of birth. In other cases, investigators working with transgenic animals may wish to use the Mom 2 or Mom 3 field to record the foster mother, and the Dad 2 field to record the stud male used to induce pseudopregnancy in the foster mother.

Entering information into the primary description fields can be facilitated by the ability to copy information from either the first mother, the first father, or the previous animal record. At any point in the entry of primary description data, if the field in view and all remaining primary fields are identical to those of the previously entered animal, type ** to copy over all of this information.

As before, you must enter the description of the cage that the animal will be placed into, and the date that this is done. A carriage return at the date question will automatically enter today's date. Additional questions will follow that are optionally answered.

At the completion of the record, AMAN will ask whether or not you wish to enter another record from the same litter. If you answer yes to this question, AMAN will automatically copy over information from the just-entered record into the new record, including the birth date, mother's cage, and the record numbers of the parents. You can begin directly with primary description information, and if any of this is the same as Mom 1, Dad 1, or the previously entered animal, the use of *m, *d, *, or ** in any of these fields will copy it over.

Entering Litters as Prerequisite to Animal Entry. Although entering litter information requires additional time and effort, it often pays back in terms of providing a more complete history of a breeding colony. To enter a record on a new litter, choose the litter submenu and the appropriate option therein. Follow the questions as they are asked. The parents of all litters must be entered into the database before it is possible to record little information.

When litters are entered into the database, it is useful to print out litter tags that can be taped onto the cages in which the litters reside. This is accomplished by choosing the "search" option from the litter submenu, then narrowing the search accordingly and printing in the "short" format. Abbreviated descriptions of each litter will be printed that can be taped directly onto cage cards.

When litters are routinely entered into the database, it becomes possible to print out a list of only those litters that are old enough to be weaned by using either the "search" or "weaning list" options from the litter submenu. (The search option provides more flexibility in the choice of various parameters.) With the search option, you can choose to list all litters that have reached a certain age and are still with their mothers (considered alive by AMAN).

When the time comes to record individual animals within a litter (usually at the time of weaning), choose the animal submenu again and the new entry option. When the litter number or supplier is requested, enter the litter number. AMAN will then automatically retrieve information from the litter record to place into the animal record (birth date, mother's cage, parents' numbers). You will then be prompted to enter the primary description of the animal as described above. If you decide that the animal entry is correct, you will save the record. AMAN will then change the STATUS field of the associated LITTER record to indicate that the litter is no longer "alive" with its mother. The status field will now hold the number of the first animal weaned from that litter.

If you decide to eliminate a litter, or if a litter dies, without any individual animals from it having been recorded, you must indicate this to AMAN. Go into the litter submenu, and choose the "status change or adding information option." Record whatever you wish in the information, but be sure to include a *k if the litter was killed or a *d if the litter died. This will cause AMAN to change the status of the litter to KILLED or DEAD, respectively. It is important to carry out this protocol in order to keep the database up to date.

General Error-Checking Routines. Through the entry of breeding information, you will notice the various error-checking routines that AMAN employs. You will only be allowed to enter males as fathers and females as mothers. If the cage in which a litter was born does not match that of the parents, you will receive a message to that effect. Likewise, if a parent was not in the right cage when conception was likely to have taken place (21 days before the birth date for mice), you will receive another message. If you choose a litter number from which animals were previously weaned, you will be informed and asked if you intend to wean additional animals from this litter. Other error-checking routines are in operation during all data information entry routines.

Parental Descriptions. When you look at a litter record or animal record, you will see the numbers of the parents as well as their primary descriptions. The primary descriptions of the parents do not exist within these records; rather, each time that you look at a record, AMAN goes back to the parents' records directly to retrieve information for display. Thus, if you change the primary description in a parent's record, the next time that you view a record of any offspring, you will see the changed information. An important consequence of this fact is that you cannot change parental description information within the records of offspring, although you can change parental numbers. If you change parental numbers with the editor, be careful to put in a number of an animal that actually exists in the colony.

Further Data Entry

Editing Records. There is a general editor available within each of the submenus as well as a special means for editing particular features. The general editor allows you to move around each record with the arrow keys and change or modify any field. In an improvement from earlier versions of the AMAN program, the up, down, right, and left arrow keys actually function as they should, making it much easier to move to specific fields. Another improvement allows you to bring up the current information in a particular field for modification. This makes it much easier to simply change a character or two at the end of a long phrase.

The general editor does not have special error-checking routines. You can type any letters, numbers, or characters into any field. However, if, for example, you place letters into a field for a parental number, the next time that you try to view that record, AMAN will crash. So, be careful. This is why it pays to back up your files often.

There are also several protocols for special editing. For example, to add information to an animal record and/or to record a cage change, it is much more efficient to use the "moving animal" routine. To change the status or to add information to a litter record, it is much more efficient to use the "status change" option. If you change certain information in sequential sets of DNA/tissue samples or restriction digests, use the special routines in each of the corresponding submenus.

Moving Animals from One Cage to Another or Out of Colony. Choose the appropriate option from the animal submenu for the task of moving animals. You can move animals up to eight times. AMAN keeps track of the latest cage that the animal is in, as well as all previous residences. When animals die, are sacrified, or are given to another investigator, they are "moved" into an alphabetic icon for each. If you have additional information to add to a record, you can do it within the context of this protocol.

Supplemental Date Information. In some cases, it may be useful to be able to record a date, of some kind, in a series of different animal records. The supplemental date field is available for this purpose. To record rapidly the same date in a series of animal records, choose the appropriate option from the animal submenu, enter the date (if other than "today"), and then list the animal numbers. The date will be recorded automatically in each of these records.

DNA/Tissue and Digest Records. AMAN provides the user with the ability to maintain records on tissue samples and DNA obtained from mice and litters. Choose the DNA/Tissue submenu and the new sample entry option. You will be prompted through a series of questions to enter the sample. Again, you can use most of the various fields to record any type of information that you wish. On saving a record in this file, AMAN will automatically mark the INFormation field of the animal or litter record with the number of the DNA/tissue sample for cross-reference. In the case of animals, AMAN will also indicate that the animal has been sacrified.

The last file maintained by AMAN is for records of DNA digests. You can use this file to maintain information on restriction digests of DNA samples recorded in the DNA file. These two files are also automatically cross-referenced. The various features of both the DNA and digest sub-menus should be self-explanatory.

String Substitution. If you wish to change a set of characters or a word in the same field of a number of records, use the string substitution option from the main menu. Choose the file and field, and the "character string" that you wish to change. You must choose a new "character string" of the same length. (The space bar produces a space which is considered a character.) Thus, you could change "Th-34 + " into "Th 34 " (you must hit the space bar for that last space.)

Retrieving Data and Printed Lists

General. There are numerous means for retrieving the data entered in AMAN. It is possible to scroll through a list of records on screen in numerical order with the scroll option in each submenu. For the animal and litter files, this option gives only an abbreviated two- or three-line version of each record. To see individual animal and litter records in their entirety, use the view/edit option. To print any screen full of information, be sure to start up with this option when you first begin, and then use the shift-PrtSc combination.

All printing occurs through the COM1 port. A printer should be hooked up directly to this port. If you wish to print through a networked printer, be sure to purchase a license for the network version of this program.

All printing of lists of records occurs through the search/print option. If you just wish to print all of the records between two numbers, indicate these numbers as answers to the appropriate questions, and then press return for all of the further search questions. You can print either directly to a printer or to a file (of your own naming) that can be opened later by a word processor for printing or manipulation.

When the colony grows very large, it becomes important to limit the search as much as possible. AMAN keeps track of the oldest living animal and the oldest litter still with its mother. Thus, if you have 20,000 animals, but only those with numbers in the 19,000–20,000 range are still alive, when you limit your search to "live animals," AMAN will only search through the last 1000 records. If you do not limit your search to live animals and do not place limits on animal numbers, AMAN will search through the entire set of 20,000 animals, which can take a very long time.

Searches through both the animal and litter records can be limited according to a number of different parameters (which appear as questions), and, in both cases, it is possible to print out results in a number of different ways. In an improvement from previous versions of the program, it is possible in both the animal and litter search routines to limit the search specifically to cages between any two numbers.

Cage Card Searches and Printed Lists of Animals. If you would just like to see a list of all of the animals currently alive in the colony according to cage number, with males listed above females in each cage, choose the "current cage" option from the search submenu with the animal submenu. This option can also provide a complete list of cage numbers alone. Within each cage, all animals will be ordered according to sex, with males first and females next.

The cage card option in the same submenu allows the printing of cage cards in a 3 by 5 inch format (again with animals ordered according to sex). If you would just like to print up cage cards for a new set of matings put together on a single date, you can set the data parameter accordingly.

Another useful option is printing according to cage of birth. This option is very useful if you wean animals that are not used directly for mating. It is possible to set aside a set of cage numbers for stocks of females, combining females from many different stocks together in the same "storage cages" according to age, for example. (In an improvement from previous versions of the program, you can choose any sets of cages between any pairs of numbers for both the current cage and the mother's cage.) Then to prepare matings, you can print out a list of all the storage animals according to the cage of birth. If you have divided up the cage numbers properly at the outset, different sets of breeding cages will represent different sets of genotypes or experiments.

It is also useful to generate two lists of the same set of animals, according to current residence as well as cage of birth. This allows you to get a sense of the relatedness of different matings.

Finally, there is the general search option which allows you to find and list animals (or litters or DNA samples or restriction digests) according to the presence or absence of any "string of characters" in a defined field. To search for the absence of a string or phrase, proceed it with a *. You can choose to make the search case-sensitive (upper- and lowercase letters are distinguishable). An example of the use of this protocol follows.

Suppose you are breeding a line of mice that are segregating two transgenes Tg427 and Tg551. When animals are first weaned and recorded, you do not know if they have either transgene and hence you input a genotype of "Tg427?,Tg551?" You then clip their tails and test the DNA to see if the mice carry either transgene: if they carry Tg427 but not Tg551, for example, you change the genotype to "Tg427 + ,Tg551 − " and so on. Now, you want to print out several lists of mice from this complex line. First, to print out all mice that derive from this line, you might search for "427" in the GENO field. This list would include all 427?,427 + , and

427 − animals irrespective of their Tg551 genotype. To identify all animals that had not yet been tested for Tg427, you would search for "427?"; to identify all mice that *had been tested* for Tg427, you would search for "427" and *not* Tg427? (which would be entered into the second search field as "*427?"). This search would give only Tg427+ and Tg427− animals, but not Tg427? animals. Finally, you could search for any combination of positive and negative string sets. For example, to identify all animals that were positive for Tg427 and negative or unknown for Tg551, you would input the following search strings for the GENO field (in any order): first "427+", second "551", third "*551−". All these searches could also be limited in a number of other parameters, such as whether live animals only are considered or whether only mice between two particular cage numbers are considered.

Offspring of Particular Parents. It is possible to list all of the litters or animals born to a particular parent by using the general search routine in either the litter or animal submenu. Just choose a particular parental field (MOM1, MOM2, MOM3, DAD1, or DAD2) or all fields of a particular type (MOM* or DAD*) and then put in the parental number that you would like to search for. This search can obviously be combined with other parameters as described above. In an improvement from earlier versions, the search routine looks at the parental fields in a different way from other fields. So if you search for parent number 42, animals numbers 342 and 421 will not show up as positive.

Searches through List of Litters. General searches through the list of litters can be conducted according to the same general principles just described for animal searches. In addition, you can limit the search to litters having a certain minimum age and/or a certain maximum age. Lists of litters can be printed either according to litter number or according to cage number.

Hardware and Licensing Information

The Animal House Manager (AMAN) can be licensed for use through Princeton University (Princeton, NJ 08544-1014). AMAN will run on all IBM-compatible computers under the DOS operating system. AMAN will also run on Macintosh computers within the context of a DOS emulator program called SoftPC which is available from distributors of Macintosh software. SoftPC is produced by Insignia Solutions Inc., 254 San Geronimo Way, Sunnyvale, CA 94086. The latest version of the program is compatible with data files generated under all previous versions (provided under other names including "Princeton Mouse Recordkeeping Program"). To receive further information, contact Dr. Lee M. Silver, Department of Molecular Biology, Princeton University, Princeton, NJ 08544-1014 (FAX: 1-609-258-3345).

[2] Anesthesia and Perioperative Care

By P. A. FLECKNELL

Introduction

Providing safe and effective anesthesia for laboratory mice is essential to ensure that they experience no unnecessary pain and distress during experimental procedures. In addition, the anesthetic regimen selected should be considered a significant factor within the overall protocol of the study. All of the currently available anesthetics have some side effects that potentially could frustrate certain types of experiments. Careful selection of the anesthetic regimen, so that agents with particular side effects are avoided, can minimize some of these interactions. The selection process is not easy, but a careful assessment of the range of anesthetic regimens that are available, and their particular physiological and pharmacological effects, can help to minimize the interactions between the anesthetic and a particular animal model. It is important to realize that this type of assessment is not undertaken by all research groups. Simply selecting an anesthetic regimen described in publications dealing with the same model will not necessarily assure that the most appropriate technique is used. Whichever anesthetic regimen is selected, it is important that it provides humane restraint, which will usually require loss of consciousness, a sufficient degree of analgesia to prevent the animal from feeling pain during the procedure, and a relaxation of muscle tone so that surgery can be carried out quickly and efficiently.

Several practical points should be considered when selecting a method of anesthesia. If volatile anesthetics are used, the agent chosen should be nonirritant, and the delivery system should be designed to avoid contact between the animal and the liquid anesthetic. If an injectable anesthetic is to be used, intramuscular administration should be selected only if a very small volume of anesthetic (<0.05 ml) is to be injected. The muscle mass in a mouse is extremely small, and the leg muscles can easily be disrupted by intramuscular injection of large volumes of drug. This will be painful for the animal, and it also results in unpredictable absorption of the anesthetic. Whichever method of anesthesia is selected, it is important that the animal is handled carefully, so that any stress that might be caused by restraint and movement prior to anesthesia is minimized.

Whichever anesthetic regimen is selected, it is of critical importance that high standards of intraoperative and postoperative care are main-

Copyright © 1993 by Academic Press, Inc.
All rights of reproduction in any form reserved.

tained. Poor perioperative care can result in unnecessary stress, prolonged recovery from anesthesia, and an increase in anesthetic mortality rates. This is undesirable because of concern for animal welfare, but it also reflects extremely poor scientific standards. Most researchers set out to produce an animal model that is carefully defined, with close control of all experimental variables. Poor anesthetic practice can increase stress, pain, fear, and distress, and these represent uncontrolled variables that can adversely affect an experiment. A mouse that develops severe hypothermia, hypovolemia, acidosis, and hypoxia and fails to eat and drink postoperatively can hardly be considered a good animal model.

Selection of Anesthetic Regimen

During the initial stages of selecting an anesthetic regimen, a choice may be made between inhalational and injectable agents. Injectable anesthetics are easy to administer, requiring only a syringe and needle and the necessary expertise to carry out a simple injection. Inhalational anesthetics can be administered using simple delivery systems such as an "ether jar," but this method of anesthesia is anachronistic and has virtually nothing to recommend it except that it is inexpensive. It is preferable to deliver volatile anesthetics into an induction chamber from an anesthetic machine. Both onset of anesthesia and recovery are rapid when using volatile anesthetics, whereas recovery following the administration of injectable agents can be very prolonged. The main reason for this prolonged recovery arises because of the route of administration. Most injectable anesthetics are administered to mice by the intraperitoneal route. Absorption into the circulation is slow compared to intravenous administration, and the production of anesthesia requires a large total quantity of drug to be administered.

A second problem arises when injectable anesthetics are administered by the intraperitoneal route. In large animals and in humans, injectable anesthetics are usually administered by intravenous injection, and the required dose is evaluated as the drug is delivered. Once a satisfactory depth of anesthesia has been attained, no further anesthetic need be administered. This adjustment of the required dose is not possible when it is administered as a single intraperitoneal, intramuscular, or subcutaneous injection. Although this might not seem to represent a problem, it is important to appreciate the very large variation in drug responses which exist between mice of different strains, sex, and housing environments. For example, the duration of unconsciousness following a standard dose

TABLE I

ANESTHETICS AND OTHER COMPOUNDS FOR USE IN MICE[a]

Compound	Dose rate	Comments	Duration of anesthesia	Sleep time[b]
Acepromazine	2.5 mg/kg	Moderate sedation	—	—
Alphaxalone/alphadolone	10–15 mg/kg i.v.	Surgical anesthesia only if i.v.	5 min	10 min
Atropine	0.05 mg/kg	Give to reduce salivation	—	—
Chloral hydrate	370–400 mg/kg	—	60–120 min	2–3 hr
Diazepam	5 mg/kg	Moderate sedation	—	—
Hypnorm (fentanyl/ fluanisone)	0.4 ml/kg i.m.	Surgical analgesia, immobilization, poor muscle relaxation	20 min	60 min
Hypnorm (fentanyl/ fluanisone) + diazepam	0.3 ml/kg i.m., 5 mg/kg	Surgical anesthesia	45–60 min	2–4 hr
Hypnorm (fentanyl/ fluanisone) + midazolam[c]	10 ml/kg	Surgical anesthesia	45–60 min	2–4 hr
Innovar Vet (fentanyl/ droperidol)	0.5 mg/kg i.m.	Analgesia, immobilization, poor muscle relaxation	20–30 min	1–2 hr
Ketamine	100–200 mg/kg i.m.	Sedation, immobilization	20–30 min	2 hr
Ketamine/acepromazine	100 mg/kg, 5 mg/kg	Light anesthesia	20–30 min	2 hr
Ketamine/diazepam	100 mg/kg, 5 mg/kg	Light anesthesia	20–30 min	2 hr
Ketamine/medetomidine	75 mg/kg, 1.0 mg/kg	Light to moderate surgical anesthesia	20–30 min	2–3 hr

of pentobarbitone varies 3-fold in different strains of mice.[1] This implies that a dose of anesthetic which would produce surgical anesthesia in one strain of mouse may be ineffective in a second strain, and yet may represent a lethal overdose in a third. This has three important practical implications in selecting an injectable anesthetic. The regimen chosen should, if possible, include an agent with as wide a safety margin as possible. When using a particular anesthetic for the first time, or when changing the supplier, strain, or sex of mice used in a study, anesthetize one or two animals and carefully observe their response, before anesthetizing large numbers of animals. Finally, consider the possibility of administering

[1] D. P. Lovell, *Lab. Anim.* **20,** 85 (1986).

TABLE I (*continued*)

Compound	Dose rate	Comments	Duration of anesthesia	Sleep time[b]
Ketamine/xylazine	100 mg/kg, 10 mg/kg	Surgical anesthesia	20–30 min	2–4 hr
Methohexitone	10 mg/kg i.v.	Surgical anesthesia only if i.v.	5 min	10 min
Metomidate/fentanyl	60 mg/kg, 0.06 mg/kg	Surgical anesthesia	60 min	120 min
Pentobarbitone	45 mg/kg	Narrow safety margin	15–60 min	2–4 hr
Propofol	26 mg/kg i.v.	Surgical anesthesia only if i.v.	5 min	10 min
Thiopentone	30 mg/kg i.v.	Surgical anesthesia only if i.v.	10 min	15 min
Tiletamine/Zolezepam	40 mg/kg (of commercial preparation)	Light to moderate surgical anesthesia	15–25 min	1–2 hr
Tribromoethanol	125–300 mg/kg	Surgical anesthesia, but postoperative mortality possible	15–60 min	1–2 hr
Xylazine	10 mg/kg	Sedative, some analgesia	—	—

[a] Considerable variation in response between different strains of mice can be anticipated. Always undertake a pilot study when changing to a new anesthetic regime. All dose rates are for intraperitoneal injection unless otherwise stated.

[b] Sleep time is duration of loss of righting reflex.

[c] Mixture of one part Hypnorm, two parts water for injection, and one part midazolam.

anesthetics by intravenous injection. This enables easy adjustment of the anesthetic dose and also results in rapid recovery times (Table I).

Inhalational Anesthetics

Apparatus

It is possible to place volatile anesthetic on a cotton wool swab, to place the swab in a glass jar, and to drop in the mouse. This technique has been widely used for ether anesthesia. It does not enable a controlled and reproducible depth of anesthesia to be attained, however, and scavenging of waste anesthetic gas is difficult. There is a significant risk that the animal may come into direct contact with liquid anesthetic, and the system cannot be used safely with modern anesthetics such as halothane

TABLE II

MINIMUM ALVEOLAR CONCENTRATION AND CONCENTRATIONS
OF INHALATIONAL ANESTHETIC AGENTS FOR USE IN MICE[a]

| Anesthetic | MAC_{50}[b] | Concentration (%) | |
		Induction	Maintenance
Enflurane	1.95	3–5	0.5–2
Ether	3.2	10–20	4–5
Halothane	0.95	3–4	1–2
Isoflurane	1.34	3.5–4.5	1.5–3
Methoxyflurane	0.22	3.5	0.4–1

[a] R. I. Mazze, S. A. Rice, and J. M. Baden, *Anesthesiology* **62,** 339 (1985); C. J. Green, "Animal Anaesthesia." Laboratory Animals Ltd., London, 1979.
[b] MAC_{50}, minimum alveolar concentration.

and isoflurane. For these reasons an anesthetic machine should be obtained, together with an induction chamber. The chamber should be transparent so that the animal can be observed during induction; it should be easy to clean and have both an inlet for delivery of fresh gas and an outlet to allow easy and effective removal of waste anesthetics.

Induction and Maintenance of Anesthesia

The mouse should be placed in the anesthetic chamber, gas scavenging equipment should be activated, and anesthetic vapor delivered at the appropriate concentration (Table II). The mouse will become ataxic and lose its righting reflex. If a potent anesthetic such as halothane or isoflurane is used, surgical anesthesia will be attained after a further 30–60 sec. The mouse should be removed from the chamber, and a brief (<30 sec) procedure can be undertaken immediately. If a longer period of anesthesia is required, the mouse should be maintained on a face mask connected to the anesthetic chamber. A suitable design that enables continued removal of waste anesthetic gas has been described by Hunter *et al.*[2] and is available commercially (International Market Supply, Dane Mill, Broadhurst Lane, Congleton, Cheshire, UK). When maintaining anesthesia, the concentration of anesthetic should be reduced from that used for induction, as indicated in Table II.

If the mouse is to be maintained for prolonged periods of anesthesia, or if blood gas parameters must be controlled, then assisted ventilation

[2] S. C. Hunter, J. B. Glen, and C. J. Butcher, *Lab. Anim.* **18,** 42 (1984).

will be required. Mice can be intubated using purpose-made endotracheal tubes and a specially constructed laryngoscope.[3] Alternatively, if the animal is not required to recover from anesthesia, then a tracheostomy can be performed. The mouse can then be connected to a suitable ventilator (e.g., Harvard rodent ventilator, Harvard Biosciences, 3900 Birch Street, Commerce Park, Newport Beach, CA 92660). A ventilation rate of 60–100 breaths per minute and a tidal volume of 0.15 ml/10 g body weight are usually required to maintain adequate respiratory function.

Agents Available

Halothane. Halothane is a potent and effective anesthetic; it causes a dose-dependent depression of the cardiovascular and respiratory system, but this rarely results in clinical problems in healthy mice. Halothane undergoes extensive hepatic metabolism resulting in microsomal enzyme induction, but this is likely to be significant only after prolonged periods of anesthesia.[4,5]

Isoflurane. Isoflurane resembles halothane in providing effective and safe anesthesia, but both induction and recovery are even more rapid. Isoflurane also causes circulatory and respiratory depression, but this should not be of clinical significance. The anesthetic undergoes virtually no biotransformation,[6] and so it may be particularly suited to studies involving drug metabolism.

Methoxyflurane. Induction of anesthesia and recovery are slower when using methoxyflurane in comparison with halothane or isoflurane. This has some advantages for the less experienced researcher, as it allows more time for the assessment of the depth of anesthesia. Methoxyflurane has a relatively high boiling point and thus vaporizes less readily than halothane. This enables the compound to be used safely in simple anesthetic chambers as a replacement for ether. Unlike ether, it is nonflammable and nonirritant, but waste anesthetic vapor must be effectively scavenged to prevent any possible risk to human health. Methoxyflurane undergoes some metabolism resulting in inorganic fluoride ion release, which can cause renal damage.[7] This is likely to be significant only after long periods of anesthesia (>2–3 hr), but care should be taken if other potentially nephrotoxic agents are administered simultaneously.

[3] D. L. Costa, J. R. Lehmann, W. M. Harold, and R. T. Drew, *Lab. Anim. Sci.* **36,** 256 (1986).
[4] B. R. Brown and A. M. Sagalyn, *Anesthesiology* **40,** 152 (1974).
[5] H. W. Linde and M. L. Berman, *Anesth. Analg. (N.Y.)* **50,** 656 (1971).
[6] E. I. Eger, *Anesthesiology* **55,** 559 (1981).
[7] W. J. Murray and P. J. Fleming, *Anesthesiology* **37,** 620 (1972).

Injectable Anesthetics

Apparatus

A 23- or 25-gauge needle should be used for administration of injectable agents by the intraperitoneal, intramuscular, or subcutaneous routes. The volume of anesthetic required is usually small, and in some instances it may be preferable to dilute the commercially available product to help ensure accurate dosing. Acceptable volumes for injection by the intraperitoneal and subcutaneous route range from 0.1 to 0.5 ml for an adult mouse. Volumes for intramuscular injection should not exceed 0.05 ml, to minimize muscle trauma and to ensure rapid absorption. Intravenous injection can be made into the lateral tail vein, using a 25- or 26-gauge needle. Injection is made easier if the mouse has first been warmed by placing it in an incubator maintained at around 30°. This results in vasodilation of the tail veins. Volumes for injection should be in the range of 0.05–0.1 ml, administered using a tuberculin syringe or disposable insulin syringe.

Short-Acting Agents

Propofol. Propofol is an anesthetic that produces about 2–3 min of surgical anesthesia when administered by intravenous injection (20–26 mg/kg i.v.).[8] Recovery is smooth and rapid, and administration of repeated successive doses of anesthetic does not unduly prolong the recovery period. Propofol can cause a short period of respiratory depression immediately after administration, but this rarely causes significant problems. Prolonged anesthesia can be produced by continuous infusion of propofol (2–3 mg/kg/min).

Alphaxalone/Alphadolone. The steroid anesthetic alphaxalone/alphadolone produces short periods of anesthesia (<5 min) after intravenous injection (10–15 mg/kg i.v.). Effects after intraperitoneal or intramuscular administration are unpredictable, and these routes are not recommended. Recovery from anesthesia is rapid. Prolonged periods of anesthesia can be produced by continuous intravenous infusion of alphaxalone/alphadolone (0.25–0.75 mg/kg/min).[9]

Methohexitone. The short-acting barbiturate methohexitone produces about 5 min of anesthesia following intravenous injection (6–10 mg/kg i.v.). The drug produces moderate cardiovascular and respiratory depression, and anesthesia can be prolonged by one or two additional injections

---

[8] J. B. Glen, *Br. J. Anaesth.* **52,** 731 (1980).
[9] C. J. Green, M. J. Halsey, S. Precious, and B. Wardley-Smith, *Lab. Anim.* **12,** 85 (1978).

without unduly delaying recovery. Effects after intraperitoneal injection are unpredictable, and this route is not recommended. Recovery after intravenous administration is rapid but may be associated with involuntary excitement.

Thiopentone. Like methohexitone, thiopentone produces a short period of anesthesia with rapid recovery, but it must be administered by the intravenous route (30–40 mg/kg i.v.). Thiopentone causes moderate respiratory and cardiovascular depression. Recovery is rapid, but not as smooth as with propofol. This agent is ineffective when administered by the intraperitoneal or intramuscular route.

Medium Duration Anesthesia

Several agents produce anesthesia of medium duration (up to 1 hr). Combinations of agents are often employed.

Ketamine. When ketamine is administered as the sole anesthetic agent (100–200 mg/kg i.m. or i.p.), insufficient analgesia is provided to allow even superficial surgery. Even at high doses, some mice fail to lose their righting reflex or may have persistent spontaneous movements during the period of sedation. When combined with other compounds, however, ketamine can provide light to medium planes of anesthesia.

Ketamine/xylazine (100 mg/kg + 10 mg/kg i.p.) is the most widely used ketamine combination. Xylazine is an α_2-adrenergic agonist that has sedative and analgesic properties. The newer α_2-agonist medetomidine may be used as an alternative to xylazine, as this compound is reported to be more specific in its actions.[10] In practice, the effects of both ketamine/ xylazine and ketamine/medetomidine (75 mg/kg + 1.0 mg/kg i.p.) are similar. Respiration is depressed, a moderate hypotension is produced, and mice develop hyperglycemia and diuresis. High doses of the mixture may be required to produce surgical planes of anesthesia. At the higher doses, recovery can be prolonged. It is strongly recommended that anesthesia be partially reversed by administration of the α_2-adrenoreceptor antagonist atipamezole (Antisedan, SmithKline Beecham). This compound completely reverses the effects of xylazine or medetomidine, and since ketamine alone has only sedative effects in mice, a partial recovery from anesthesia is produced. Atipamezole is a highly specific antagonist with no significant side effects, unlike older agonists such as yohimbine.[11]

An unusual side effect that has been noted in mice anesthetized with

---

[10] R. Virtanen, *Acta Vet. Scand. Suppl.* **85,** 29 (1989).
[11] N. S. Lipman, P. A. Phillips, and C. E. Newcomer, *Lab. Anim. Sci.* **37,** 474 (1987).

ketamine/xylazine is the production of acute temporary cataracts in some animals.[12] The lens opacity reverses following recovery from anesthesia.

As an alternative to xylazine, ketamine can be combined with acepromazine, a phenothiazine tranquilizer, or midazolam, a water-soluble benzodiazepine. Neither of these compounds has any analgesic action, so, not surprisingly, only light anesthesia is produced. High doses of these mixtures (ketamine/acepromazine, 100 mg/kg + 2.5 mg/kg i.p.; ketamine/midazolam, 100 mg/kg + 5 mg/kg i.p.) can produce surgical planes of anesthesia in some strains of mice but this can be accompanied by marked respiratory depression. A benzodiazepine antagonist, flumazenil, is available commercially (Anexate, Roche) and can be used to reverse some of the effects of ketamine/midazolam.

Neuroleptanalgesics. Several commercial preparations are available that combine a potent opioid analgesic with a tranquilizer. Innovar Vet (fentanyl/droperidol; 0.65 ml/kg i.m.) and Hypnorm (fentanyl/fluanisone; 0.33 ml/kg i.p.) are the most widely available preparations. Although similar in basic pharmacology, the effects of these combinations vary considerably. When the agents are used as the sole anesthetic, mice are immobilized, and profound analgesia is produced. This is accompanied by muscle rigidity and pronounced respiratory depression. These undesirable effects can be overcome in some instances by reducing the dose of the neuroleptanalgesic and incorporating a benzodiazepine in the anesthetic regimen. The combination of Hypnorm (fentanyl/fluanisone) and midazolam has gained widespread popularity as an anesthetic regimen in mice[13] since it produces good analgesia and muscle relaxation [10 ml/kg i.p. of a mixture of 2 parts water for injection, 1 part Hypnorm, and 1 part midazolam (5 mg/ml)]. There are no reports of the use of Innovar Vet/midazolam combinations.

The combination of a hypnotic agent (metomidate) and fentanyl, a potent opioid analgesic, produces stable, surgical anesthesia for 60–70 min.[14] The recommended dose rate is 60 mg/kg metomidate plus 0.06 mg/kg fentanyl, and administration is by the subcutaneous route.

Chloral hydrate (370–400 mg/kg i.p.) produces 45–60 min of light surgical anesthesia following intraperitoneal injection. The depth of anesthesia produced varies considerably between different strains of mice, and in some strains a depth of anesthesia sufficient to allow major surgery is attained. Chloral hydrate has been reported to be particularly useful when

[12] L. Calderone, P. Grimes, and M. Shalev, *Exp. Eye Res.* **42,** 331 (1986).
[13] P. A. Flecknell and M. Mitchell, *Lab. Anim.* **18,** 143 (1984).
[14] C. J. Green, J. Knight, S. Precious, and S. Simpkin, *Lab. Anim.* **15,** 171 (1981).

studying central nervous system (CNS) function, as it may have fewer depressant effects on neuronal function than other anesthetics.

Tribromoethanol (125–300 mg/kg i.p.) can be used to produce medium planes of surgical anesthesia in mice.[15] If it is to be used for recovery anesthesia, it is essential to prepare a fresh solution on each occasion that it is to be administered, since decomposition of the material can result in peritonitis, gut disorders, and death of the animal.[16,17] Administration of tribromoethanol on a subsequent occasion can result in peritonitis and death, even when a freshly prepared solution is used.[18] It is also recommended that dilute solutions of the compound be administered (1.25%, v/v) as higher concentrations of tribromoethanol seem to be more frequently associated with high postanesthetic mortality. Because of the risk of adverse effects following anesthesia, alternative anesthetics should be used whenever possible.

Pentobarbitone, a barbiturate, has been one of the most widely used injectable anesthetics for laboratory mice. It has the advantage that a single intraperitoneal injection (45 mg/kg) can be given to produce light surgical anesthesia. It has a narrow safety margin, however, and until an appropriate dose for a specific strain, sex, and age of mouse has been established, an unacceptably high mortality rate is likely to be encountered if surgical anesthesia is produced. Pentobarbitone produces severe cardiovascular and respiratory depression. Recovery is prolonged, and no specific antagonist is available.

Recommended Techniques

As discussed earlier, the choice of a particular anesthetic regimen will be influenced by the overall objectives of the experiment. If not contraindicated because of interactions with the experimental protocol, the following anesthetic regimens should be used since they provide effective surgical anesthesia and have a wide safety margin. A comprehensive list of anesthetic agents and dose rates is given in Table I.

Short-term Anesthesia

Very brief periods of general anesthesia (1–10 min) can be provided either by use of a volatile anesthetic, by intravenous injection of a short-

[15] C. J. Green, "Animal Anaesthesia." Laboratory Animals Ltd., London, 1979.
[16] T. Nicol, B. Vernon-Roberts, and D. C. Quantock, *Nature (London)* **208**, 1099 (1965).
[17] D. Tarin and A. Sturdee, *Lab. Anim.* **6**, 79 (1972).
[18] M. L. Norris and W. D. Turner, *Lab. Anim.* **17**, 324 (1983).

acting injectable anesthetic, or by the use of a reversible anesthetic combination. Provided that the apparatus is available, and the nature of the procedure does not prevent their use, then there can be little doubt that a volatile anesthetic represents the most simple and effective means of producing a short period of anesthesia in the mouse. If a volatile agent cannot be used, then intravenous administration of propofol or alphaxalone/alphadolone is recommended as the most suitable alternative.

Medium Duration Anesthesia

Moderate periods of general anesthesia (10–60 min) can be provided either by use of a volatile anesthetic, by intraperitoneal or intramuscular administration of injectable anesthetics, or by continuous intravenous infusion of a short-acting anesthetic. The most useful combinations for achieving medium duration surgical anesthesia are either fentanyl/fluanisone (Hypnorm) combined with midazolam or ketamine in combination with xylazine or medetomidine. The other anesthetics and anesthetic combinations described above are generally less satisfactory and have a narrower margin of safety. If the use of volatile anesthetics is practicable, then researchers are strongly recommended to consider their use since they allow easy alteration of depth of anesthesia and rapid recovery. When using any of these combinations, partial reversal of anesthesia with an appropriate antagonist (Table III) is strongly recommended.

Combined Injectable/Inhalational Regimens

Several of the disadvantages of producing anesthesia by the exclusive use of an inhalational agent or an injectable agent can be overcome by combining these techniques. For example, if an induction chamber is not available, then administration of a sedative or sedative/analgesic combination (Table I) can prevent any struggling or distress caused by administration of anesthetic by face mask. Administration of an injectable anesthetic, to induce anesthesia, followed by maintenance with a volatile agent also avoids problems of restraint during induction.

Some injectable anesthetic combinations may fail to provide sufficient analgesia for major surgery (e.g., laparotomy), and the addition of an inhalational agent at low concentration (e.g., 0.25–0.5% halothane) may be a safer and more convenient technique than trying to "top up" with additional injectable agent. Similarly, when using an inhalational agent as the major component of an anesthetic regimen, the concentration that is required to produce surgical anesthesia can be reduced by administering a potent analgesic (e.g., fentanyl). During prolonged procedures, this supplementation can be given intermittently during periods of major surgi-

TABLE III
ANESTHETIC ANTAGONISTS FOR USE IN MICE

Compound	Anesthetic regime	Dose rate	Comments
Atipamezole	For reversal of xylazine or medetomidine	1 mg/kg i.m., i.p., s.c., i.v.	Highly specific antagonist
Buprenorphine	Any regime using μ-opioids	0.05–1.0 mg/kg i.p. or s.c.	Slower onset than naloxone and nalbuphine, but longer duration of action
Doxapram	All anesthetics	5–10 mg/kg i.m., i.v., i.p.	General respiratory stimulant
Nalbuphine	Any regime using μ-opioids (e.g., fentanyl)	4–8 mg/kg i.p. or s.c.	Almost as rapid acting as naloxone; maintains postoperative analgesia
Naloxone	Any regime using μ-opioids (e.g., fentanyl)	0.05–0.1 mg/kg, i.v., i.m., i.p.	Reverses analgesia as well as respiratory depression
Yohimbine	For reversal of xylazine or medetomidine	2.1 mg/kg i.p.	Relatively nonspecific antagonist; not recommended

cal stimulation. The aim of this balanced anesthetic regimen is to minimize the interference to the animal's physiology caused by the drugs used, and so enable recovery to be as smooth and rapid as possible.

Patient Care

Preoperative Preparation

Irrespective of the choice of anesthetic, or the purpose of the research work, it is of fundamental importance to obtain mice that are in overt good health and free from subclinical infections. The animals should be housed for a 1- to 2-week acclimatization period, during which time food and water intake and growth rate can be assessed. A simple clinical examination should be carried out prior to anesthesia. It is unnecessary and undesirable to withhold food or water before anesthetizing mice.

Intraoperative Care

The provision of high standards of intraoperative care is essential for safe and effective anesthesia. Of special importance are the maintenance of body temperature, respiratory function, and the cardiovascular system, along with monitoring of the depth of anesthesia.

Maintenance of Body Temperature

Anesthesia depresses thermoregulatory mechanisms and causes a fall in body temperature. This can be a potentially lethal side effect of anesthesia in mice, since the high surface area relative to body mass results in rapid heat loss. Reversing hypothermia can be difficult, and it is preferable to take preventive measures to minimize losses of body heat. It is possible to insulate mice by wrapping them in "bubble packing" or aluminum foil, but this may hinder access to the surgical field. It is preferable to use a heating pad or heating blanket to maintain body temperature. Ideally, a thermostatically controlled blanket should be used; if this is not possible, then a temperature probe should be placed between the mouse and the blanket to ensure that overheating does not occur. Blanket temperatures should be in the range 37°–40°. If a heating pad is not available, then an infrared heat lamp or even a simple Anglepoise light can be used, but great care must be taken to prevent overheating of the animal. It is advisable to set up any heating lamps about 60 min prior to operating, so that the temperature will have stabilized at an appropriate level. It is important to continue to provide additional heating in the postanesthetic recovery period (see below). Hypothermia is probably the single most important cause of anesthetic mortality in mice, and attention to maintenance of body temperature is therefore of considerable importance.

Respiratory Function

The majority of anesthetics cause some depression of the respiratory system, causing both hypoxia (a reduction in arterial oxygen content) and hypercapnia (an increase in arterial carbon dioxide content). The effects of hypercapnia are made worse if the animal is also hypoxic; thus, providing oxygen during anesthesia can help reduce mortality. Even when using injectable anesthetics, one should provide oxygen via a face mask. The apparatus required is simple and inexpensive, consisting only of an oxygen cylinder and a combined regulator and flowmeter.

Even when oxygen is provided, respiratory obstruction can occur because of blockage of the airway with mucus or other material, or because of abnormal positioning of the head and neck. Respiratory obstruction can also arise because of compression of the chest or neck by the surgeon or by surgical instruments. Unfortunately, it is difficult to use electronic respiratory monitors in mice, since the majority of these instruments are too insensitive to register the small tidal volumes produced by these animals. A potentially useful technique for monitoring respiratory function is pulse oximetry, which provides an indication of the degree of oxygenation of blood. As smaller probes for commercially available instruments are developed, this may lead to a noninvasive method of monitoring mice.

Because of the lack of reliable respiratory monitors, monitoring of respiratory function will generally rely on clinical observations. Regular recording of the rate and patterns of respiration will alert the anesthetist to the development of respiratory depression. Mice invariably develop rapid respiration when handled prior to anesthesia, so estimation of normal resting rates can be difficult. Either a rate can be obtained immediately after the onset of anesthesia, or the resting rate can be estimated from published data (120–180 breaths/min). In albino mice, the muzzle and footpads should be inspected for evidence of cyanosis, but in pigmented mice such observations are difficult. Even in albino animals it is important to appreciate that obvious cyanosis (bluish discoloration caused by hypoxia) will not be noted until arterial oxygen content has been severely reduced.

If cyanosis is noted, or if the respiration falls to less than 50% of the estimated resting rate, then measures should be taken to assist respiratory function. First attempt to assess why respiratory depression or cyanosis is occuring. This may have arisen because of blockage of the oropharynx or upper respiratory tract with mucus, blood, or other material, or because of abnormal positioning of the head and neck. If so, the airway should be cleared using a suction device. A simple but effective suction apparatus can be constructed by connecting a 16-gauge catheter to a 20-ml syringe. The positioning of the head and neck should be checked, and any twisting or flexion corrected. The chest or neck may inadvertently be compressed by the surgeon or by surgical instruments. Very light pressure is all that is required to interfere seriously with chest movement. If oxygen is not already being administered, commencing oxygen therapy will be beneficial. If surgery has not commenced, consider giving an antagonist to reverse the anesthetic, or perhaps a general respiratory stimulant. If surgery has started, assist ventilation by manually squeezing the chest between your finger and thumb, at a rate of approximately 90 breaths/min, and try to complete the surgical procedure as rapidly as possible.

Cardiovascular Function

Cardiovascular failure can result from blood loss, hypothermia, or anesthetic overdose. Blood loss is a common cause of death during surgery in mice, simply because of the very small circulating volume. A mouse weighing 30 g has a blood volume of approximately 2 ml, and rapid loss of 0.5 ml of blood is usually sufficient to cause signs of cardiovascular failure ("shock"). It is therefore especially important to develop surgical techniques which minimize hemorrhage. During some surgical procedures, significant blood loss may be unavoidable, and in these circumstances blood should be transfused from a donor. This technique can also be valuable in maintaining circulatory volume during studies which require relatively large volumes of blood to be collected. It is preferable to collect blood in acid–citrate–dextrose (ACD) solution (three parts blood to one part ACD), as it can then be stored for several days at 4° until required.

Monitoring of the cardiovascular system can provide an early indication of impending cardiovascular failure, but such monitoring is rarely used during anesthesia of mice. If a long period of anesthesia is planned, or if unexpected mortality is encountered, then monitoring of heart rate should be considered. Monitoring of the electrocardiograph (ECG) is possible, but many of the monitors designed for use in humans are unable to detect the low-amplitude signals of mice. An additional problem is that many of these monitors are unable to register heart rates above 250 or 300 beats/min. One monitor capable of monitoring heart rate and the ECG in small rodents is the EC-60 available from Silogic Design Ltd. (Enterprise House, 181–189 Garth Road, Morden, Surrey, UK).

Postoperative Care

It is essential that continued care is provided to mice following completion of the surgical or other technique which required anesthesia. Hypothermia should be prevented by placing the mice in a suitable incubator, maintained at about 35°. As an alternative, a cage can be placed on a heating pad, or beneath a heat lamp, but great care should be taken to avoid overheating the animals. It is advisable to set up these postoperative facilities well in advance of completing surgery, so that temperatures will have stabilized and can be assessed. Mice should be placed on toweling or similar bedding (e.g., Vetbed) for recovery, not on sawdust, since this material can stick to the eyes and mouth of the animal.

Mice should be inspected frequently to ensure that respiratory depression does not develop. During surgery of large batches of animals this is easy to arrange, as those recovering from anesthesia can be reassessed each time a mouse is transferred to the recovery cage. If respiratory depression occurs, the mouse should be treated with doxapram and oxygen provided if necessary.

Abnormal quantities of body fluids may be lost during anesthesia, and in addition mice frequently reduce their food and water consumption following surgery. To minimize problems of fluid imbalance, it is advisable to administer saline by subcutaneous or intraperitoneal injection (0.5–1.0

TABLE IV
DOSE RATES FOR ANALGESICS IN MICE[a]

Drug	Dose
Buprenorphine	0.05 mg/kg s.c., 8–12 hourly
Butorphanol	1–5 mg/kg s.c., 4 hourly
Morphine	2–5 mg/kg s.c., 2–4 hourly
Nalbuphine	4–8 mg/kg i.m., 3 hourly
Pentazocine	10 mg/kg s.c., 3–4 hourly
Pethidine (meperidine)	10–20 mg/kg s.c., i.m., 2–3 hourly

[a] Dose rates are based on clinical experience, experimental analgesiometry [see P. A. Flecknell, *Lab. Anim.* **18**, 147 (1984) for review], and previously published data [W. V. Lumb and W. E. Jones, "Veterinary Anesthesia." Lea & Febiger, Philadelphia, Pennsylvania, 1984; C. E. Short, *in* "Principles and Practice of Veterinary Anesthesia" (C. E. Short, ed.), p. 28. Williams & Wilkins, Baltimore, Maryland, 1987].

ml per 30-g mouse). It is good practice to record body weight before and after surgery, so that the animal's recovery can be monitored in an objective fashion. It is also useful to record whether the mouse is passing urine or feces.

Provision of Postoperative Analgesia

Despite the small size of mice, there seems to be no reason not to presume that mice are capable of experiencing postoperative pain. The neurological mechanisms necessary for nociception are present in this species, and it is likely that there is sufficient cortical development for the animal to experience something analogous to human pain. Although this anthropomorphic approach is useful in that it ensures that some consideration is given to providing pain relief when it may be appropriate, it does not enable a rational choice of analgesic therapy to be instigated. To control pain effectively in mice, it is important to assess both the degree of pain and the animal's responses to analgesic therapy. It is only by assessing pain that the appropriate type of analgesic can be administered, at an appropriate dose, for an appropriate time.

Published data derived from analgesiometry in mice provide basic information concerning possible dose rates of analgesic drugs in mice.[19,20]

[19] P. A. Flecknell, *Lab. Anim.* **18**, 147 (1984).
[20] J. H. Liles and P. A. Flecknell, *Lab. Anim.* **26**, 241 (1992).

TABLE V

DOSE RATES OF NONSTEROIDAL
ANTIINFLAMMATORY DRUGS IN MICE[a]

Drug	Dose rate
Aspirin	120 mg/kg per os
Diclofenac	10.0 mg/kg per os
Flunixin	2.5–5.0 mg/kg s.c.
Ibuprofen	30 mg/kg per os
Indomethacin	1.0 mg/kg per os
Paracetamol	150 mg/kg per os
Phenylbutazone	30 mg/kg per os
Piroxicam	3.0 mg/kg per os
Suprofen	25 mg/kg per os

[a] Dosage based on efficacy in analgesiometry. See J. H. Liles and P. A. Flecknell, *Lab. Anim.* **26,** 241 (1992), for review.

There is, however, a lack of information concerning pain assessment in mice. Several approaches are possible. Simple clinical assessments can be made, based on the overall experience of the assessor. Animals believed to be abnormal because of postoperative pain can then be given an analgesic. A refinement of this approach is to develop a pain scoring system, as described by Morton and Griffiths.[21] Although this approach is helpful, it is clear from subsequent reports that specific criteria may need to be developed to enable this technique to be used in mice.[22] Objective measurements that may indicate the presence of pain have not been developed in mice, unlike rats,[23] but it is possible that similar measurements of food and water consumption and body weight may prove useful.

As a general guide, most mice appear to tolerate anesthesia and surgery very well in comparison to other species. If an animal appears less active, is reluctant to move, or becomes aggressive when handled, then postoperative pain may be present. Animals may fail to groom, may develop a ruffled fur coat, and may also adopt an abnormal posture. Failure to groom may also result in crusting of material around the nose, eyes, and urogenital area. All of the criteria discussed above should be considered in conjunction with the nature of the procedure that has been undertaken, as well as the previous behavior of the mouse.

[21] D. B. Morton and P. H. M. Griffiths, *Vet. Rec.* **116,** 431 (1985).
[22] A. C. Beynen, V. Baumans, A. P. M. G. Bertens, R. Havenaar, A. P. M. Hesp, and L. F. M. Van Zutphen, *Lab. Anim.* **21,** 35 (1986).
[23] P. A. Flecknell and J. H. Liles, *in* "Animal Pain" (C. E. Short and A. Van Poznak, eds.), p. 482. Churchill Livingstone, New York, 1992.

Section II

Germ Cells and Embryos

[3] Culture and Manipulation of Primordial Germ Cells

By JULIE E. COOKE, ISABELLE GODIN, CHARLES FFRENCH-CONSTANT, JANET HEASMAN, and CHRISTOPHER C. WYLIE

Introduction

Primordial germ cells (PGCs) are the founder cells of the germ line. Their descendants will form the functional gametes of the adult animal. Considering the necessary role played by PGCs in survival of the species, we know surprisingly little about the control mechanisms involved in their maintenance and proliferation.

In many organisms, including the mouse, the primordial germ cells are a migratory cell population. They arise extragonadally during early embryogenesis before moving toward, and colonizing, the developing gonad. The acquisition and subsequent loss of migratory activity by these cells during a short, defined period makes them an attractive model system in which to study the control of cell migration.

Owing to the inaccessibility of the mammalian embryo to experimental manipulation, PGCs are frequently studied in primary culture. The methods used for these studies are associated with many problems; for example, the germ cells are present only in small numbers in the early embryo, and they cannot yet be isolated as a pure population. Additionally, optimal survival and proliferation of PGCs in culture have been obtained only in the presence of a heterologous feeder layer, resulting in difficulties in distinguishing primary effects on PGCs from secondary effects mediated by accompanying somatic cells or feeder cells.

This chapter describes the isolation of PGCs from embryos of different ages, along with the techniques utilized in the culture of these cells. We hope that experiments utilizing the *in vitro* culture systems described below will help to elucidate the control mechanisms involved in the maintenance, proliferation, and guidance of PGCs.

Isolation of Primordial Germ Cells from Embryos of Different Ages

Mouse PGCs are first identifiable as a small cluster of alkaline phosphatase-positive cells at the posterior end of the primitive streak at about 7 days postcoitum[1] (dpc), where the day on which a vaginal plug is found

---

[1] M. Ginsburg, M. H. L. Snow, and A. McLaren, *Development (Cambridge, UK)* **110**, 521 (1990).

METHODS IN ENZYMOLOGY, VOL. 225

Copyright © 1993 by Academic Press, Inc.
All rights of reproduction in any form reserved.

is designated 0.5 dpc (copulation being presumed to take place at around midnight). The PGCs number 50–80 by the end of gastrulation. At 8–8.5 dpc the population of around 100 PGCs is found in the hindgut endoderm and at the base of the allantois. PGCs displaying a migratory phenotype are found in the hindgut epithelium at 9–9.5 dpc, by which time the population has increased to approximately 350. From here they migrate in a dorsal direction along the hindgut mesentery toward the developing genital ridges, which they begin to colonize at 10.5 dpc, when they number approximately 1000. Once within the gonad, the PGCs continue to divide, increasing their population to approximately 25,000 by 13.5 dpc.[2,3]

To enable PGCs to be studied *in vitro*, the region in which they are present at the age required must be dissected from the embryo and a cell suspension made from it before transfer to the appropriate culture conditions. All steps should be carried out under sterile conditions.

Collection of Embryos

1. Sacrifice females from timed pregnancies on the required day of embryonic development by cervical dislocation.

2. Remove the uterus as described by Damjanov and colleagues.[4]

3. Cut between the deciduoma to separate the embryos, and collect them in a sterile 9-cm petri dish containing filter-sterilized calcium- and magnesium-free phosphate-buffered saline (PBS; 16 g NaCl, 0.4 g KCl, 2.3 g NaH_2PO_4, 0.4 g KH_2PO_4, 2 liters distilled water, pH to 7.4). The embryos are now ready for dissection.

The technique for dissection of the PGC-containing region from the embryo varies according to the embryonic age. A dissecting microscope at low power with a separate fiber optic light source is generally used. Use fine forceps (Agar, T5034, Scientific Ltd., Essex, UK) to carry out the dissection.

Isolation of Cells from 8.5-Day Embryos

1. Using two pairs of fine forceps, grip the muscle wall of the uterus firmly at the point where it has been cut from the adjacent portion of the uterus and pull apart to expose the decidua. Discard the uterine wall tissue.

2. The position of the placenta is visible as a vascular streak about halfway up the decidual tissue. Using the forceps, cut across the decidua

---

[2] P. P. L. Tam and M. H. L. Snow, *J. Embryol. Exp. Morphol.* **64,** 133 (1981).

[3] J. M. Clark and E. M. Eddy, *Dev. Biol.* **45,** 136 (1975).

[4] I. Damjanov, A. Damjanov, and D. Solter, *in* "Teratocarcinomas and Embryonic Stem Cells: A Practical Approach" (E. J. Robertson, ed.), p. 5. IRL Press, Oxford, 1987.

at this point. It is important to make this cut accurately so that the 8.5 dpc embryo is not damaged before it is removed from the decidua.

3. The extraembryonic membranes should be visible in the upper half of the decidua, with the allantois protruding, and part of the yolk sac should be visible in the lower portion. Remove the embryo from the upper portion of the bissected decidua, taking great care not to damage the delicate embryonic tissue.

4. The embryo is folded within the membranes at this stage so that the anterior and posterior ends are closely apposed (Fig. 1A). Carefully cut the membranes so as to separate the two ends of the embryo and straighten it out. You should be able to distinguish the head process, somites, primitive streak, and allantois (Fig. 1B).

5. The PGCs reside at the posterior end of the primitive streak near the base of the allantois. Make a cut across the primitive streak so as to remove the posterior third of the embryo (Fig. 1C). Remove as much of the membrane from this dissected portion of the embryo as is possible without damaging the primitive streak or the allantois.

6. Collect the dissected PGC-containing regions.

Isolation of Cells from 10.5-Day Embryos

1. Remove the embryo (10.5 dpc) from the uterus by gently squeezing the uterine tissue. The muscle coat may have to be gently torn away from the decidua to enable access to the embryo.

2. Remove the extraembryonic membranes, taking extra care with the amnion, which is attached closely to the gut of the embryo. If caution is not exercised the gut may break and the mesentery could become disrupted, causing PGCs to be lost (Fig. 2A).

3. Make two cuts with the forceps, one just posterior to the anterior limb buds and the other just anterior to the posterior limb buds, dividing the embryo into three portions. Discard the head and tail ends.

4. Cut open the skin of the embryonic midregion to expose the internal organs. Open the skin out on either side, and, using the left-hand forceps with one forceps point on either side of the embryo, hold the embryo firmly down by its skin, with its ventral side uppermost.

5. The two genital ridges and the dorsal aorta which runs between them are closely apposed to the dorsal body wall (Fig. 2B,C). Also visible is the embryonic gut, which is attached to the dorsal aorta by the dorsal mesentery (Fig. 2B). Keeping as small an angle as possible between the right-hand forceps and the base of the petri dish, grip the genital ridges from the anterior end with one forceps point on either side and gently

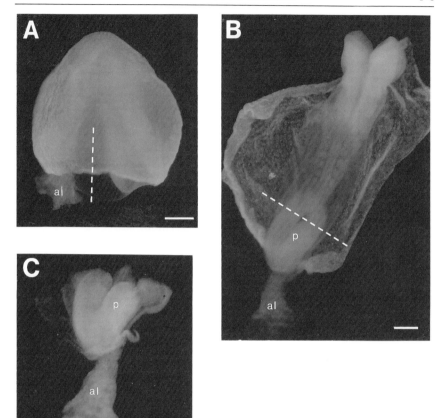

FIG. 1. Dissection of 8.5 dpc embryo. Dotted lines represent cuts made during the dissection (see text). (A) An 8.5 dpc embryo within the yolk sac. (B) Embryo indicating cut to be made when isolating the PGC-containing region. The neural fold, developing somites, notochord, and primitive streak can be seen in this late 8.5 dpc embryo. (C) Dissected PGC-containing region of 8.5 dpc embryo. al, Allantois; p, primitive streak. Bars: 250 μm.

pull upward, away from the dorsal body wall. With practice it is possible to remove the genital ridges, aorta, gut, and mesentery in one movement.

6. Snip the gut carefully away from its mesentery, and squeeze the blood out from the aorta to reduce the number of contaminating somatic cells. Collect the genital ridge/aorta/dorsal mesentery portions (PGC-containing regions, Fig. 2D).

Isolation of Cells from 12.5-Day Embryos

1. Gently squeeze the uterine tissue with the forceps to transfer the embryos (12.5 dpc) into the dish, then dissect the embryos free from their membranes (Fig. 3A).

2. Make two large cuts with the forceps, one just posterior to the anterior limb buds and the other just posterior to the posterior limb buds, to remove the anterior end and the tail which otherwise obscure the region to be dissected.

3. Cut open the skin to reveal the abdominal cavity of the embryo. By this stage of development the gut has elongated and is extensively looped, filling a large space within the abdomen. The liver and many other organs are also now more evident than they were at 10.5 dpc.

4. Hold the embryo on its back by pinning down the skin to the base of the dish with the points of the left-hand pair of forceps on either side of the body, then remove the viscera obscuring the genital ridges, which are more firmly attached to the dorsal body wall.

5. The genital ridges, which should now be visible, are closely associated with the developing kidney and mesonephros (Fig. 3B,C) and can be removed from the embryo along with this associated tissue. Again it helps to remove them from the anterior end first, keeping as acute an angle as possible between the forceps and the base of the dish.

6. If desired, the mesonephros can be removed from the genital ridge using fine hypodermic needles (Fig. 3D). Pricking the genital ridges with fine hypodermic needles can enable a high degree of purity of PGCs to be obtained. It disrupts the tissue of the genital ridge, resulting in the PGCs being released into the medium. The dislodged cells can be collected from the medium by pipette and spun down before resuspending in the medium required for use.

Formation of Single Cell Suspensions from Dissected Tissues

1. Collect all of the dissected PGC-containing regions together in one area of the dish. Pick them up in a siliconized Pasteur pipette and transfer to another dish containing fresh PBS. Swirl the fragments around and transfer through another wash of PBS before collecting together in a 15-ml centrifuge tube on ice.

2. Make the volume up to approximately 1 ml with fresh PBS. Add sufficient EDTA to give a 1 mM solution and let stand for 10 min (this step can be omitted for 8.5 dpc tissue fragments, which dissociate more easily than those from older embryos).

3. Break up the tissue fragments by pipetting up and down with a siliconized Pasteur pipette. Extra force is generated by holding the point

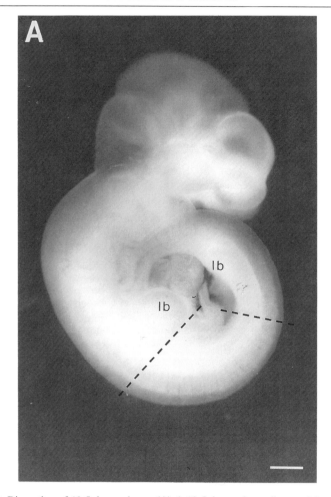

FIG. 2. Dissection of 10.5 dpc embryo. (A) A 10.5 dpc embryo dissected free of membranes. Dotted lines represent cuts made during the dissection (see text); lb, limb bud. (B) Dissected midregion from 10.5 dpc embryo, showing the dorsal body wall and somites, gut (g), and mesentery (m). (C) Same as (B) but viewed from a ventral aspect, and with the gut removed. The genital ridges (gr) can be seen lying either side of the aorta (a). (D) Dissected PGC-containing region of 10.5 dpc embryo. The genital ridges, aorta, and mesentery can be seen. Bars: (A)–(C), 500 μm; (D), 250 μm.

of the pipette very close to the base of the tube and expelling the contents rapidly, but care must be taken not to produce air bubbles. The suspension should be triturated in this manner about 50 times in order to obtain a single cell suspension.

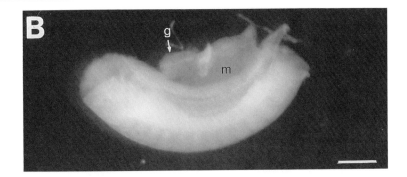

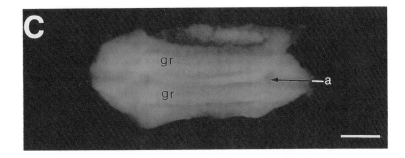

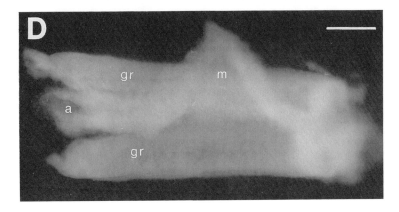

4. Dilute the cells to the required volume in medium which has been warmed to 37° before use. Different media can be used according to requirements. When serum is present in the culture conditions, we use Dulbecco's modified Eagle's medium (DMEM, Imperial Labs., Hamp-

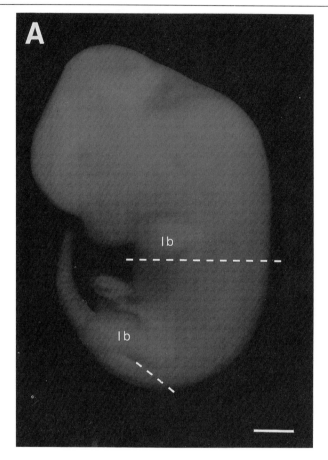

FIG. 3. Dissection of 12.5 dpc embryo. (A) A 12.5 dpc embryo dissected free of membranes. Dotted lines represent cuts to be made during the dissection (see text); lb, limb bud. (B) Genital ridges (gr) and surrounding tissues (a, aorta; k, kidney; mn, mesonephros) that have been removed from the embryo. (C) Urogenital ridges, consisting of genital ridges and mesonephric ridges. (D) Genital ridges which have been dissected free of mesonephric ridges with fine hypodermic needles. Bars: (A), 1 mm; (B)–(D), 500 μm.

shire, UK, 1-465-14) for resuspending the cells. However, when serum-free conditions are being observed, SF-1 medium (Northumbria Biologicals Ltd., Northumbria, UK, M642) should be used as it contains defined protein and lipids, not present in DMEM, which are required for optimal cell attachment and growth. We supplement all media used with 10 ml/liter penicillin/streptomycin solution (Sigma, St. Louis, MO, P0781). To this stock medium is added L-glutamine (Sigma, G6392) to give a concen-

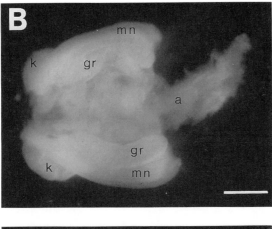

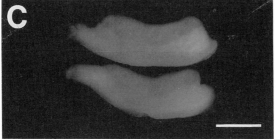

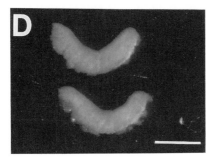

FIG. 3B–D.

tration of 4 mM and sodium pyruvate (Sigma, S8636) to give a 0.5 mM solution. To make serum-containing medium, we supplement DMEM with fetal calf serum (FCS, Northumbria Biologicals Ltd., S102) to give a concentration of 15%. All media are filter sterilized before use and can be stored for up to 2 weeks at 4°.

Identification of Primordial Germ Cells

To identify PGCs *in vivo* and to distinguish them in culture from accompanying somatic and feeder cells, specific endogenous markers are required. Researchers studying mammalian PGCs have long utilized the high surface activity of alkaline phosphatase as a convenient histological marker.[5] However, because fixation is a necessary step in the staining protocol, alkaline phosphatase can only be used as a retrospective marker, necessitating the use of other markers for the observation of live cells.

Staining of Cultures for Alkaline Phosphatase

1. Remove the medium from the cultures, taking care not to dislodge any cells, and fix with 4% (w/v) paraformaldehyde. Leave for 20 min.

2. Wash the cultures 3 times in Tris–maleate buffer (3.6 g Trizma base in 1 liter of distilled water, pH to 9.0 with 1 M maleic acid) for 10 min each wash.

3. Prepare the staining solution just prior to use:

Tris–maleate buffer	25 ml
10% $MgCl_2$	200 μl (0.5 mM)
Naphthol AS-MX phosphate (Sigma, N-5000)	10 mg (0.4 mg/ml)
Fast Red TR salt (Sigma, F2768)	25 mg (1 mg/ml)

4. Remove the Tris–maleate wash and add the staining solution. Allow the staining to develop for 15–20 min or until the PGCs are bright pink.

5. Remove the staining solution and stop the staining reaction by adding PBS, which brings the pH down to 7.4.

Paraformaldehyde-fixed frozen sections of mouse embryos can also be stained for alkaline phosphatase activity to visualize PGCs.

Extensive research has revealed many PGC surface antigens that can be recognized by specific antibodies. Several of these antibodies were raised against F9 teratocarcinoma and embryonal carcinoma (EC) cells, which have many features in common with PGCs,[6] and others were raised against early mouse embryos.[7,8] None of these markers are unique to germ cells; some are present on many cells of the early embryo but

[5] A. D. Chiquoine, *Anat. Rec.* **118**, 135 (1954).

[6] M. Evans, E. Robertson, A. Bradley, and M. H. Kaufman, *in* "Current Problems in Germ Cell Differentiation" (A. McLaren and C. C. Wylie, eds.), p. 139. Cambridge Univ. Press, Cambridge, 1983.

[7] E. M. Eddy and A. C. Hahnel, *in* "Current Problems in Germ Cell Differentiation" (A. McLaren and C. C. Wylie, eds.), p. 41. Cambridge Univ. Press, Cambridge, 1983.

[8] P. Stern, *in* "Current Problems in Germ Cell Differentiation" (A. McLaren and C. C. Wylie, eds.), p. 157. Cambridge Univ. Press, Cambridge, 1983.

TABLE I

Time Course of Expression of Useful Markers on Mouse Primordial Germ Cells during Embryogenesis[a,b]

Age (dpc)	Alkaline phosphatase	PG1[c]	Anti-Forssman[c]	SSEA-1[d]	SSEA-3[d]	TG-1[d]	M1 22/25[e]
7.5	−	nd	nd	nd	nd	−	nd
8.5	+	−	−	−	−	−	−
9.5	+	−	−	+	−	+	−
10.5	+	+	+	+	−	+	+
11.5	+	+	+	+	−	+	+
12.5	+	+	+	+	+	+	+
13.5	+	+	+	+	+	+	+
14.5	+	+	+	+	−	+	+
15.5 (F)	+ (weak)	−	−	−	−	−	−
15.5 (M)	+	+	−	−	−	−	−
17.5 (F)	−	−	−	nd	nd	nd	−
17.5 (M)	−	−	−	nd	nd	nd	−

[a] Sections containing PGCs were stained for alkaline phosphatase and double stained with one of six antibodies.

[b] +, Positive staining of PGCs; −, negative staining of PGCs; nd, not determined; dpc, days postcoitum.

[c] PG1 [J. Heath, Cell (Cambridge, Mass.) 15, 299 (1977)] and anti-Forssman [R. A. Karol, S. K. Kundu, and D. M. Marcus, Immunol. Commun. 10, 137 (1981)] are rabbit polyclonal antisera.

[d] SSEA-1 [D. Solter and D. P. Knowles, Proc. Natl. Acad. Sci. U.S.A. 75, 5565 (1978)], SSEA-3 [L. H. Shevinsky, B. B. Knowles, I. Damjanov, and D. Solter, Cell (Cambridge, Mass.) 30, 697 (1982)], and TG-1 [P. C. Beverley, D. Linch, and D. Delia, Nature (London) 287, 332 (1980)] are mouse monoclonal antibodies.

[e] M1 22/25 [K. R. Willison, R. A. Karol, A. Suzuki, S. K. Kundu, and D. M. Marcus, J. Immunol. 129, 603 (1982)] is a rat monoclonal antibody directed against the Forssman antigen.

are subsequently lost and reexpressed on migratory stage PGCs. Table I illustrates the timing of expression of some useful markers on mouse PGCs.

The TG-1 monoclonal antibody is used in our laboratory as a marker for PGCs. It was originally raised against human thymocyte glycoproteins[9] but also recognizes human neutrophils, murine teratocarcinoma cells, and migratory mouse PGCs. It stains no embryonic cells other than PGCs at the same sagittal level during PGC migration.[10] TG-1 recognizes the same epitope (or an overlapping one) as does anti-stage-specific embryonic

[9] P. C. L. Beverley, D. Linch, and D. Delia, Nature (London) 287, 332 (1980).

[10] C. C. Wylie, D. Stott, and P. J. Donovan, in "Developmental Biology: A Comprehensive Synthesis" (L. Browder, ed.), Vol. 2, p. 433. Plenum, New York, 1986.

antigen-1 (anti-SSEA-1). It is a specific sugar, 3-fucosyllactosamine.[11,12] We use indirect immunofluorescence techniques with TG-1 or anti-SSEA-1 and a rhodamine-conjugated second antibody to "live-label' and pick out PGCs.

Live Immunostaining of Cell Suspensions with TG-1 or Anti-SSEA-1

1. Prepare cell suspension (see above).

2. Spin the cells at 1000 rpm for 3 min, pour off the supernatant, and resuspend the cells in TG-1 supernatant or a 1/500 dilution of anti-SSEA-1 ascites fluid in PBS. Using a siliconized Pasteur pipette, ensure that the cells are not clumped together. Leave the tube on ice for 30 min.

3. Spin the cells at 1000 rpm for 3 min, pour off the supernatant, and resuspend in PBS.

4. Repeat Step 3, but this time resuspend the cells in tetra-methylrho-damine isothiocyanate-conjugated goat anti-mouse immunoglobulin (Ig) antibody (TRITC-GAM; Nordic Immunological Laboratories, Berkshire, UK), diluted to manufacturer's instructions. Using a siliconized Pasteur pipette, ensure that the cells are not clumped together. Leave the tube on ice for 30 min.

5. Spin the cells at 1000 rpm for 3 min, pour off the supernatant, and resuspend in PBS.

6. Repeat Step 5, resuspending in 1 ml of PBS. The labeled cells are now ready for observation and selection.

7. Place an aliquot of the suspension in a standing drop of PBS on a large coverslip, mounted using glycerin on the base of a rose chamber.

8. View the drop under an inverted fluorescence microscope using a ×16 lens and a rhodamine filter. PGCs, which express the TG-1 epitope, carry the rhodamine-conjugated second antibody and can be identified as fluorescent cells.

9. Pick out labeled cells using a siliconized glass Pasteur pipette which has been drawn out to form a fine point. Transfer the cells to warm medium and use as required.

Culture of Primordial Germ Cells

The choice of the *in vitro* system in which the PGCs are to be cultured depends on the questions to be addressed. Factors affecting the proliferation, survival, migration, and chemotaxis of PGCs should be considered.

---

[11] T. Feizi, *Nature (London)* **314**, 53 (1985).

[12] P. J. Donovan, D. Stott, I. Godin, J. Heasman, and C. C. Wylie, *J. Cell Sci. Suppl.* **8**, 359 (1987).

To achieve survival, proliferation, and migration of PGCs in culture, a heterologous feeder layer is required. Of various cell lines tested, the mouse embryonic fibroblast cell line STO,[13] previously used as a feeder layer for teratocarcinoma cells[14] and embryonal stem (ES) cells, was found to be the most satisfactory for supporting PGCs.[15] It has since been found that STO cells express two growth factors which play a role in PGC survival: leukemia inhibitory factor (LIF) and stem cell factor (SCF).[16,17]

Preparation of Feeder Cells

Growing Unirradiated Stocks of STO Cells

1. Plate the STO cells at a density of 10^4 cells/ml in DMEM plus 15% FCS into Falcon tissue culture flasks (Becton Dickenson, Oxford, UK). Use a volume of 10 ml per 250-ml flask or 30 ml per 750-ml flask. Passage the cells before they reach a confluent state as follows.

2. Wash 3 times in PBS or once in Hanks' balanced salt solution (Sigma, H8389).

3. Add 1.5 ml (250-ml flask) or 5 ml (750-ml flask) of trypsin/EDTA solution (Sigma, T5775) and shake the flasks horizontally to ensure a good covering by the trypsin/EDTA. Place the flasks in a 37° incubator for 2–3 min. Check the dissociation under an inverted microscope. If necessary, replace the flasks in the incubator for a few more minutes. In practice, this step should be carried out as quickly as possible.

4. Collect the cells in a 50-ml centrifuge tube to which 5–10 ml of FCS has been added, and wash the flask with DMEM plus 15% FCS, collecting all the medium in the 50-ml tube(s). The FCS inhibits the action of trypsin.

5. Spin the cell suspension for 3 min at 1000 rpm.

6. Discard the supernatant and resuspend the cells in 1 ml DMEM plus 15% FCS.

7. Count the cells in a hemocytometer and dilute to give 10^4 cells/ml.

8. Plate as before.

---

[13] STO cells are SIM mouse embryo fibroblasts, and they are available from the European Collection of Animal Cell Cultures (ECACC), PHLS Centre for Applied Microbiology and Research, Porton, Salisbury, Wiltshire SP4 0JG, UK.

[14] G. R. Martin and M. J. Evans, *Proc. Natl. Acad. Sci. U.S.A.* **72**, 1441 (1975).

[15] P. J. Donovan, D. Stott, L. A. Cairns, J. Heasman, and C. C. Wylie, *Cell (Cambridge, Mass.)* **44**, 831 (1986).

[16] Y. Matsui, D. Toksoz, S. Nishikawa, S. Nishikawa, D. Williams, K. Zsebo, and B. L. M. Hogan, *Nature (London)* **353**, 750 (1991).

[17] I. Godin, R. Deed, J. E. Cooke, K. Zsebo, T. M. Dexter, and C. C. Wylie, *Nature (London)* **352**, 807 (1991).

Irradiation. Before the STO cells are used as a feeder layer, they should be rendered incapable of proliferation by treatment with X rays, γ rays, or mitomycin C. We use X irradiation as the most convenient method. Large numbers of cells can be treated at once, and the irradiated stocks can be frozen down for storage. Our STO cells are subjected to a total dose of 50–55 Gy (>5000 rad), from an X-ray machine. Other methods for mitotic arrest of STO cells are described elsewhere.[18]

Freezing Irradiated Cells

1. Spin cells for 3 min at 1000 rpm.
2. Prepare the freezing medium: 10% dimethyl sulfoxide (DMSO) in DMEM plus 15% FCS.
3. Discard the supernatant, resuspend the cells in a small quantity of medium, and count the cells in a hemocytometer.
4. Resuspend the cells in a quantity of cell freezing medium sufficient to give a cell density of 2.5×10^6 cells/ml.
5. Place 1 ml of cell suspension in each cell freezing vial. Each vial will then provide 10 ml of cell suspension at the required density for plating onto multiwell plates. Store the vials at $-70°$ for a few days before transferring to a liquid nitrogen-containing cell freezer.

Preparation of STO Cell Feeder Layer

1. Warm 20–30 ml DMEM plus 15% FCS in a 37° water bath.
2. Thaw a vial of X-irradiated STO cells in a 37° water bath as quickly as possible.
3. Add the thawed cell suspension to a centrifuge tube containing 9 ml of warm DMEM plus 15% FCS. Spin at 1000 rpm for 3 min.
4. Discard the supernatant and resuspend the cells in 1 ml of DMEM plus 15% FCS. Dilute the suspension to 5 ml and count in a hemocytometer.
5. Dilute with DMEM plus 15% FCS to give a cell density of 2.5×10^5 cells/ml.
6. Plate the cells in Falcon 96-well tissue culture plates (Becton Dickenson) at 150 μl/well, in Falcon 24-well tissue culture plates at 1 ml/well, or in 8-well Labtek chamber slides (Nunclon, Denmark) at 300 μl/well.
7. Place the plates or slides in a 37°, humidified incubator in 5% CO_2 for 12–24 hr, by which time the STO cells should have flattened and spread sufficiently to form a confluent monolayer.

[18] E. J. Robertson, *in* "Teratocarcinomas and Embryonic Stem Cells: A Practical Approach" (E. J. Robertson, ed.), p. 76. IRL Press, Oxford, 1987.

TABLE II
REQUIREMENTS FOR PROLIFERATION ASSAY

Number of tests	Number of wells	Volume of STO cell suspension (ml)	Number of 8.5 dpc embryos	Number of 10.5 dpc embryos	Volume of PGC suspension (ml)
2	30	6.75	15	8	1.50
3	45	9.00	23	12	2.25
4	60	11.25	30	15	3.00
5	75	13.50	38	19	3.75
6	90	15.75	45	23	4.50
7	105	18.00	53	27	5.25
8	120	20.25	60	30	6.00

Plating Out STO Cells in Serum-Free Conditions

If required, the entire process can be carried out under serum-free conditions, although precautions must be taken to ensure that the STO cells stick properly to the wells and spread sufficiently to form a confluent monolayer.

1. Precoat wells with fibronectin at 40 μg/ml in PBS, and leave in a 37° humidified incubator for at least 2 hr or overnight. Aspirate off excess fibronectin and wash the coated wells twice with warm SF-1 medium just before use.

2. Use SF-1 serum-free medium instead of DMEM plus 15% FCS throughout.

Proliferation Assays

The following protocol is used to observe the effects of exogenous factors on the numbers of PGCs in culture over a period of a few days. The quantities referred to below and in Table II are those required for an experiment in which the PGCs are to be observed at three time points, and in which 5 wells of a 96-well plate are used for each treatment.

1. Plate out STO cells as described above in rows of 5 wells, the number of rows corresponding to the number of treatments to be tested, in three 96-well plates. Leave overnight at 37° in 5% CO_2.

2. Dissect embryos of desired age and prepare PGC suspensions as described above. For 8.5 dpc embryos, dissect out 0.5 embryonic PGC-containing regions per well; for 10.5 dpc embryos, dissect out 0.25 embryonic PGC-containing regions per well.

3. Dilute the PGC suspension in enough medium (DMEM or SF-1) to allow for 50 μl of suspension per well.

4. Plate out on previously prepared STO cell monolayers at 50 μl/well.

5. Feed each well with 100 μl of medium containing the required test substance, and replace the plates in the humidified incubator at 37° with 5% CO_2.

6. Remove one plate from the incubator at each of the required time points (we usually look at the cultures after 1, 3, and 5 days of incubation), and fix and stain for alkaline phosphatase as described above.

Incorporation of Bromodeoxyuridine

Proliferation kinetics can be studied using the thymidine analog bromo-deoxyuridine (BrdU), which becomes incorporated into S-phase nuclei. The cells which have incorporated the BrdU can be visualized immunocy-tochemically using a specific anti-BrdU monoclonal antibody.[19] We use the cell proliferation kit from Amersham International P.L.C., Bucking-hamshire, UK (RPN20), which contains most of the reagents required. The kit also contains notes on protocols, but we have found it necessary to include more blocking steps to reduce background staining. Cultures are set up as described above, on 8-well Permanox Labtek slides (Nunclon).

1. Prepare labeling medium by diluting BrdU reagent 1/1000 with the medium used to feed the cells. Warm to 37° before use, and replace existing culture medium with the labeling medium.

2. Incubate the cultures at 37°, 5% CO_2 for 1 hr.

3. Remove the labeling medium and wash cells briefly in PBS.

4. Fix and stain the cultures for alkaline phosphatase as described above.

5. Wash in PBS. At this stage the slides can be kept for several days at 4° to enable separate experiments to be immunostained together. Ensure that the slides do not dehydrate.

6. Incubate the cultures in 2% hydrogen peroxide in PBS for 1 hr.

7. Wash 3 times for 3 min (minimum) in PBS before starting the immu-nocytochemical staining.

8. Incubate the samples with blocking buffer (10% heat-inactivated sheep serum in PBS) for 1 hr.

9. Add sufficient anti-BrdU to cover the samples; let stand for 1 hr at room temperature.

10. Wash 3 times for 3 min (minimum) in PBS.

[19] G. Plickert and M. Kroiher, *Development* (*Cambridge, UK*) **103**, 791 (1988).

11. Cover the specimens with peroxidase-coupled sheep anti-mouse Ig at 1/100 dilution in blocking buffer; let stand for 30 min at room temperature.

12. During Step 10 thaw out one aliquot of 3,3'-diaminobenzidine tetrahydrochloride (DAB) and dilute into 50 ml of phosphate buffer (see cell proliferation kit protocols).

13. Wash 3 times for 3 min (minimum) in PBS; leave the slides in the last wash while preparing the staining reagents.

14. Add 5 drops substrate/intensifier (containing hydrogen peroxide, cobalt chloride, and nickel chloride; supplied with kit) to diluted DAB solution. Add this solution to the cultures for 5–10 min. If no definite staining appears within this time, keep the plates overnight at 4°.

15. Wash 3 times in distilled water (or more times if there is a high background).

16. Remove the plastic upper structure and silicone gasket from the Labtek slides and mount with a coverslip using aqueous mounting medium.

17. Record the proportion of alkaline phosphatase-positive cells which are also positive for BrdU incorporation; these will be pink cells with a brown/black nucleus.

Chemotropic Assays

The PGCs of all vertebrates arise as a population of cells which move toward their target tissues, the developing gonads, usually by a combination of morphogenetic movements and active migration. An assay system has been designed in our laboratory to test various factors for a potential role in PGC guidance. The classic chemotropic assays could not be used as PGCs require a feeder layer for movement and long-term survival in culture. Using the system described below, we have shown that mouse genital ridges, and medium conditioned by these tissues, influence the direction of movement of mouse PGCs *in vitro,* and that transforming growth factor β_1 (TGFβ_1) may be responsible for this effect.[20,21]

The assay system involves wells cut in agarose-coated petri dishes. PGCs are seeded onto a central well containing a STO cell feeder layer, and potential chemotropic agents are added to lateral wells (see Fig. 4). The distribution of PGCs in the central well is analyzed after 24 hr to see whether they have moved toward or away from the added agent, or indeed whether their distribution has altered at all.

[20] I. Godin, C. C. Wylie, and J. Heasman, *Development (Cambridge, UK)* **108,** 357 (1990).
[21] I. Godin and C. C. Wylie, *Development (Cambridge, UK)* **113,** 1451 (1991).

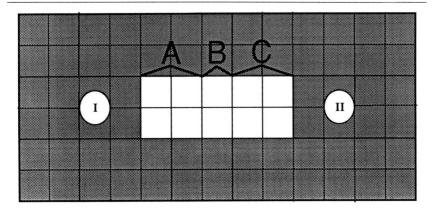

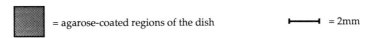

▮ = agarose-coated regions of the dish   ├────┤ = 2mm

FIG. 4. Layout of agarose-coated dishes for chemotropic assays. Wells cut into the agarose are represented as white circles (lateral wells) or as a white rectangle (central well). Test and control media are added to the lateral wells I and II. The central, rectangular well is covered by a monolayer of STO cells onto which is seeded the PGC suspension. For the purpose of analyzing the results of chemotropic assays, the central well is divided into three zones for counting, marked here as A, B, and C.

Setting up Chemotropic Assays

1. Prepare a 1.5% solution of agarose in distilled water and autoclave.

2. Prepare dishes: use 60-mm gridded petri dishes (Nunclon, 169558A). On the underside, outline a rectangle of 2 × 5 grid squares, in the center of the dish. If serum-free conditions are to be observed, coat the area of this rectangle with 100 μl of fibronectin at 40 μg/ml in PBS and leave in a 37° incubator overnight.

3. Melt the stock gel and mix with an equal quantity of 2× medium (DMEM + 15% FCS or SF-1 made from × 10 stock as required). Aspirate off the fibronectin and pour 11–12 ml of gel into each dish. Leave to set for 5–6 hr.

4. Using a sterile razor blade, cut out the central well, following the previously drawn outline. Remove the piece of gel with sterile forceps.

5. Using a 2 mm diameter sterile gel punch (Pharmacia, Piscataway, NJ, 19-3284-01), make two lateral wells, 2 mm in diameter, at either end of the central well, separated from the ends of the central well by 2 mm of gel (1 grid square). Remove the pieces of gel using suction.

6. Replace the dishes in the incubator so that the gel melts slightly around the central well to seal the cuts. Use suction to remove any liquid gel remaining in the central well before plating out the STO cells.

7. Plate X-irradiated STO cells in the required medium in the central well (as described above). A higher cell density (5×10^5 cells/ml) will be required to produce a confluent monolayer if serum-free medium is used. Replace the dishes in a 37°, humidified incubator with 5% CO_2 and let stand overnight.

8. Dissect embryos and make PGC suspensions (see above). For 8.5 dpc embryos, 2–3 PGC-containing fragments will be required for each dish or 0.5–1 for 10.5 dpc embryos.

9. Seed the PGC suspension as centrally as possible onto the central well.

10. Add the test substance to one of the lateral wells, control medium to the other. We have used both dissected organs[20] and soluble factors[21] as test substances.

11. If using soluble test substances, the medium in the lateral wells should be renewed at frequent intervals (e.g., 5 times in 24 hr) to ensure that a concentration gradient of test substance can be maintained. To avoid setting up fluid currents when changing the media, which may alter the position of PGCs, replace the media in both lateral wells simultaneously using a multichannel pipette.

12. After 24 hr, fix the dishes for 20 min with 4% paraformaldehyde without removing the gel. Then remove the gel with a spatula, fix for another 20 min, and stain for alkaline phosphatase as usual.

Counting and Analyzing Chemotropic Assays. To analyze the distribution of the PGCs in the central well, this well is divided into three zones for counting. End zones A and C each cover an area of 16 mm² (4 grid squares), and the central area, zone B, covers an area of 8 mm². Thus, zones A and C each represent 40% of the total area covered by the central well, and zone B represents 20% (Fig. 4). The number of PGCs in each zone is recorded and expressed as a percentage of the total number of PGCs present in the central well. After 24 hr during which the PGCs have migrated randomly around the well, 40% of the PGCs should be present in each of the end zones, and 20% in the central zone. This is exactly the result in control experiments (see Godin *et al.*[20]). If an agent which is chemotropic for PGCs is placed in one of the lateral wells (wells I or II in Fig. 4), then after 24 hr the proportion of PGCs in the end zone nearest to that lateral well (i.e., zone A or zone C) should be significantly higher than 40%. Again, this is seen when genital ridges, genital ridge-conditioned medium, or TGFβ₁ are added to one of the lateral wells.[20,21]

Emigration Assays

The aim of the emigration assay is to provide a method to address questions about the mechanisms regulating PGC migration. The assay was originally developed in order to explore the effects of soluble extracellular matrix components on migration, but it could equally well be used for the analysis of other soluble biological molecules such as growth factors. Another possible use is to explore the role of the feeder cell layer in PGC survival and migration by the use of blocking antibodies or by genetic manipulation of feeder cells so as to knock out selected molecules.

The system, involving explant cultures, arose from the observation that PGCs will migrate out of fragments of the developing gut of 9.5 dpc embryos which have been placed on feeder cell layers. Experiments using fluorescent dyes to label the explants prior to culture established that most, if not all, cell types other than the PGCs remain within the borders of the explant.[22] As a result, the migration of a relatively small number of PGCs can be assessed without cell purification. The age of the embryos used was found to be critical, as explants prepared from older or younger animals did not show PGC emigration. This presumably reflects the previous observations that 8.5 dpc PGCs have not yet started their migration while many 10.5 dpc PGCs are already postmigratory. The protocol followed is described below.

Preparation of Feeder Layers

1. Coat 13-mm diameter glass coverslips with poly(D-lysine) (50 μg/ml in sterile, distilled water) for 1 hr, wash with sterile, distilled water, and air dry for 6 hr. Coat the coverslips with laminin (Sigma) at 40 μg/ml in PBS for 1 hr and rinse twice before use. Place the coverslips into 24-well tissue culture plates.

2. Plate 2×10^5 STO cells in each well in SF-1 medium and leave overnight.

Preparation of Explants. At 9.5 dpc, migrating PGCs are in the hindgut, and the mesentery between the developing hindgut and the underlying aorta. The explants consist of a cylinder of tissue containing these structures.

1. Sacrifice females at 9.5 dpc by cervical dislocation.

2. Remove the reproductive tract from the female and dissect the embryos free of maternal tissue and extraembryonic membranes (see sec-

---

[22] C. ffrench-Constant, A. Hollingsworth, J. Heasman, and C. C. Wylie, *Development (Cambridge, UK)* **113,** 1365 (1991).

tion above on Isolation of Primordial Germ Cells from Embryos of Different Ages).

3. Expose the gut and aorta by peeling away the body wall on either side of the embryo. The anterior end of the developing gut and associated mesentery can be seen as a prominent tube of tissue standing proud of the dorsal body wall of the embryo.

4. Grasp the anterior end of the gut and mesentery with fine forceps and gently pull the gut and associated tissue free of the embryo in a rostral to caudal direction. It is usually possible to free the gut as far caudally as the remnant of the allantoic stalk.

5. Cut the excised tissue into three equal segments and transfer into SF-1.

6. Wash the fragments briefly in SF-1 before gently pipetting them onto the medium overlying the STO cells so that they sink down onto the center of the coverslip.

7. Leave the explants undisturbed for 1 hr before adding any test substances to the system.

Analyzing and Quantifying Gut Emigration Assays

1. The extent of PGC migration is assessed 1 or 2 days after plating. Fix the cultures and stain for alkaline phosphatase activity as described above.

2. Following staining, invert the coverslip onto a drop of glycerol-based mounting medium on a glass microscope slide, and seal the edges with nail varnish. The use of glass coverslips on microscope slides facilitates phase-contrast microscopy, enabling the edges of the explants to be seen. PGCs should be seen to have emigrated beyond the edges of the explants and to have moved individually over the STO cell monolayer.

3. Using a camera lucida attachment on the microscope, record the position of the PGCs and the explant edges for each individual explant. After appropriate calibration, the exact distances migrated (i.e., the distance from the PGCs to the explant edge) can be calculated. We routinely use 15 explants on each monolayer for calculating the mean distance migrated per cell within each experiment. As the extent of migration from each region of the developing hindgut may vary, it is important to include equal numbers of the three regions in one individual experiment.

A prerequisite for efficient analysis of the emigration assay is the demonstration that any manipulation of the system does not perturb the integrity of the monolayer, as we find that the PGCs do not migrate well over nonconfluent STO cell cultures. One must also ensure that cells from the explant (other than PGCs) do not migrate and carry the PGCs passively with them, creating the illusion of PGC migration. To exclude this possibility, we use carboxyfluorescein diacetate succinimidyl ester (CFSE) to label fluorescently all cells within the explant before culture (33 μM in PBS for 10 min). At the end of the assay period, a sheet of fluorescently labeled cells can be seen spreading out from the explant, and PGCs can be seen to have migrated beyond this labeled somatic cell sheet.

Problems and Future Goals

All of the above-described assay systems have problems in common when analyzing the results, namely, the presence of cell types other than PGCs within each system. We do not know the extent to which the STO cells, or the somatic cells which accompany the PGCs, contribute growth factors, extracellular matrix molecules, or other substances to these culture systems. As a result we cannot exclude the possibility that the effects of exogenous factors are enhanced, masked, or abolished by unidentified molecules already present within the system.

Solutions to such problems are presently being sought. Ultimately we would like to culture PGCs in a totally defined system, in the absence of feeder cells, and to be able to obtain pure populations of PGCs uncontaminated by somatic cells.

Acknowledgments

We are grateful to the Wellcome Trust for financial support.

[4] Isolation and Culture of Primordial Germ Cells

By MIA BUEHR and ANNE McLAREN

Aging, Staging, and Sexing Embryos

The location of primordial germ cells within the mammalian embryo is dependent on the age of the embryo, and at some stages a few hours can make a noticeable difference to their position. The phenotype and behavior of germ cells, and their developmental potential in culture, also vary according to their age. Accurate methods of aging and staging embryos are therefore central to any work involving primordial germ cells.

When reporting experiments with germ cells, care must always be taken to specify the age of the embryo as accurately as possible. However, there is at present no standard terminology for expressing embryo age,

Copyright © 1993 by Academic Press, Inc.
All rights of reproduction in any form reserved.

and different groups often use different systems. This can lead to confusion. For instance, an embryo "10 days old" is not equivalent to one "in the tenth day of development." Embryos defined as "day 10" can differ in age by 24 hr depending on whether the first day of pregnancy is termed "day 1" or "day 0." In this chapter we express the age of embryos in days postcoitum (dpc), assuming coitus to have taken place at 1 a.m. on the morning of finding the copulatory plug. Embryos removed between 9 a.m. and 5 p.m. are taken to be halfway through the day, so "10.5 dpc" indicates an age from 10 days 8 hr to 10 days 16 hr postcoitum.

The age of an embryo expressed in this way is often not enough to define its stage of development. Different strains of mice have very different rates of development, so embryos of the same age but of different strains may be at different developmental stages. Variation in the time of mating or ovulation can lead to considerable differences in stage between litters nominally of the same age, while embryos even within the same litter may differ markedly in developmental stage. Some means of staging embryos is needed to distinguish developmental phases that may last only a few hours. The method used will of course depend on the age and general developmental stage of the embryo. At 7.5 dpc four stages can be defined according to the criteria of Ginsburg et al.[1] In embryos 8 to 10 dpc, the number of somites provides the most accurate guide to developmental stage. In embryos older than 10 dpc, however, the number of somites is large, the body wall becomes opaque, and somite counts become increasingly tedious to perform and subject to error. From 10.5 to 15.5 dpc the appearance of the hind limb bud (and at later stages the hind foot) provides a convenient estimate of developmental stage. (See McLaren and Buehr,[2] Palmer and Burgoyne,[3] and Mackay and Smith[4] for detailed description of these stages.)

In some experiments it is necessary to know the sex of the embryo from which the germ cells are recovered. Gonads from embryos 12.5 dpc or older can be sexed under a dissecting microscope: male gonads contain testis cords, whereas ovaries appear uniform and undifferentiated. At 12.5 dpc, a small blood vessel can often be seen along the coelomic border of the testis, which is absent in the ovary. However, male and female gonads younger than 12.5 dpc cannot be distinguished morphologically, and other strategies must be used to sex them. Taketo-

[1] M. Ginsburg, M. H. L. Snow, and A. McLaren, *Development* (*Cambridge, UK*) **110,** 521 (1990).
[2] A. McLaren and M. Buehr, *Cell Differ. Dev.* **31,** 185 (1990).
[3] S. J. Palmer and P. S. Burgoyne, *Development* (*Cambridge, UK*) **113,** 709 (1991).
[4] S. Mackay and R. Smith, *J. Embryol. Exp. Morphol.* **97,** 189 (1986).

Hosotani *et al.*[5] have sexed embryos by dot hybridization with a Y chromosome-specific DNA probe. A quick and easy method for sexing embryos from 9.5 to 11.5 dpc is to examine amniotic cells for the presence of sex chromatin. A sex chromatin body indicates the presence of two X chromosomes[3,6] (see Protocol 1 below). Amnions from a dozen or so embryos can be processed and examined in about 1 hr, and during this time gonads can be kept in a suitable medium (e.g., Whittingham's M2) on the bench or in a refrigerator (5°) with no appreciable effect on their subsequent viability in culture. (The tissue is best stored as intact fragments rather than as dissociated cells.)

Procedures for Collecting Germ Cells

Primordial germ cells can first be detected by their relatively high alkaline phosphatase activity at 7 to 7.5 dpc, at the posterior end of the primitive streak.[1] At 8.5 dpc they are found in the stalk of the allantois and the margins of the open hindgut. At 9.5 dpc they are distributed along the length of the (now closed) hindgut, and by 10.5 dpc some are in the dorsal mesentery while many have already reached the coelomic angles and the genital ridges. At 11.5 dpc, and at all older stages, the germ cells are in the genital ridges, which now begin to differentiate into gonads. With time the germ cells proliferate, so that from a pool of about 150 cells at 8.5 days they increase to about 26,000 at 13.5 days (Fig. 1).[7]

Embryo Dissection

7.5 Days. The germ cells are located just posterior to the primitive streak, in the extraembryonic mesoderm at the base of the developing allantois. Remove the embryo from the deciduum, strip off Reichert's membrane, and remove the ectoplacental cone. The germ cells lie in the allantois and posterior part of the primitive streak.

8.5 Days. The germ cells lie in the allantoic stalk and around the margin of the open hindgut. Remove the embryo, trim off any large pieces of yolk sac, and cut open the amnion. If the embryo is bisected at the midpoint of the somites, all the germ cells will lie in the posterior half.

9.5 Days. Most of the germ cells lie within the wall of the hindgut. Remove the embryo from its membranes and trim off any fragments of

---

[5] T. Taketo-Hosotani, Y. Nishioka, C. M. Nagamine, I. Villalpando, and H. Merchant-Larios, *Development* (*Cambridge, UK*) **107,** 95 (1989).

[6] P. S. Burgoyne, P. P. L. Tam, and E. P. Evans, *J. Reprod. Fertil.* **68,** 387 (1983).

[7] P. P. L. Tam and M. H. L. Snow, *J. Embryol. Exp. Morphol.* **64,** 133 (1981).

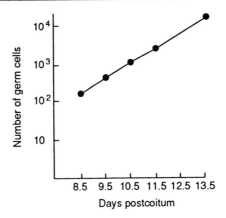

FIG. 1. Number of germ cells in embryos from 8.5 to 13.5 days postcoitum. Adapted from Tam and Snow.[7]

amnion and yolk sac. Cut the embryo across just posterior to the heart and discard the anterior portion. Working from the anterior end toward the tail, pull off the lateral body walls. The hindgut can now be seen as a low ridge projecting from the midline of the dorsal body wall, just under the neural tube, and can be cut away with a fine needle.

10.5 Days. Although most of the germ cells are in the urogenital ridges, some remain in the dorsal mesentery. The dissection procedure is similar to that for the 9.5 day embryo, but the gut is now seen to be suspended from the dorsal midline by the mesentery. Cut the urogenital ridges away from the body wall with a fine needle, trimming off any adhering fragments of extraneous tissue. If necessary, the urogenital ridges can be isolated from the mesentery, and, with experience, the developing gonadal rudiment can be separated from the mesonephros.

11.5 to 13.5 Days. All the germ cells are in the genital ridge or gonad. Cut the embryo across just behind the forelimb buds. Place the posterior half on its back, slit the abdominal wall if necessary, and pull out the gut and associated structures. The urogenital ridges can be seen attached to the back of the abdominal cavity (in older embryos, at the level of the developing kidney), and they can be cut out with needles or fine scissors or pulled out with forceps. Trim any adhering tissue and, if necessary, cut the genital ridge away from the mesonephros with a fine needle.

After 13.5 Days. In older embryos the gonads gradually shift to the locations they will occupy in the neonatal animal. The ovaries lie next to the kidneys, while the testes are found further posterior, on either side of the bladder.

Methods of Cell Disaggregation

Dissection of the appropriate part of an embryo will result in a block of tissue containing germ cells along with several kinds of somatic cells. If germ cells are to be isolated, they must first be released from this tissue by a disaggregation procedure. Common strategies of cell disaggregation can be summarized under three headings.

Mechanical Disruption of Gonads. Mechanical disruption is suitable only for embryos 10.5 dpc or older. Gonads are pierced with fine needles, slit open, squashed under siliconized coverslips, triturated, or otherwise mechanically disrupted so that the germ cells fall out. Cells can then be collected manually or by centrifugation. To improve efficiency, these procedures are usually carried out in a medium lacking calcium and magnesium. With these methods there is minimum interference with the surface properties of germ cells, and if carefully done the number of contaminating cells can be small. However, relatively few germ cells are released by simple mechanical disruption, and the method is suitable only when small numbers of cells are required.[8]

EDTA Treatment. If tissue containing germ cells is incubated in a calcium- and magnesium-free buffered saline containing the chelating agent EDTA, the bonds between germ cells and somatic cells are loosened, but the somatic cells (other than blood cells) remain attached to each other as an intact tissue. When EDTA treatment is followed by gentle mechanical disruption (such as pricking the gonad with a needle, or kneading it with forceps), a large sample of germ cells of very high purity can be obtained. If necessary, the entire tissue can be dissociated into a single cell suspension by trituration. With this method there is minimum interference with surface properties of the cells, and the procedure is efficient and quick. For this reason we recommend it in preference to any other, and a detailed protocol for germ cell collection by this method is given later in Protocol 3.

Enzyme treatment. In the mouse embryo there is minimal attachment between the germ cells and the stroma in which they lie, and aggressive enzyme treatment is normally not necessary to dissociate the cells. However, if germ cells are to be collected from perinatal gonads, some workers prefer to use enzyme treatment to dissociate all the cells (see Protocol 2 for some commonly used recipes). Total disaggregation of tissues containing germ cells will of course yield the maximum possible number of germ cells. The proportion of contaminating cells is also high, however, and the surface properties of the germ cells may be affected by enzyme treatment.

---

[8] M. De Felici and A. McLaren, *Exp. Cell Res.* **142,** 476 (1982).

Sample Collection

No disaggregation method so far devised will provide 100% uncontaminated germ cells, and some method of purifying the cell sample is necessary if a very pure sample of germ cells is required. Normally one must choose between a large sample with a large proportion of contaminating cells and a smaller but purer sample of germ cells.

If a large number of cells with a high proportion of somatic contaminants has been obtained (e.g., from disaggregation of whole gonads or urogenital ridges), some purification is possible by culturing the dissociated cells in plastic dishes overnight. The somatic cells adhere to the plastic, while the germ cells do not and so can be retrieved by subsequent centrifugation or collection by hand. This strategy can remove many somatic cells from the system, but unfortunately dead cells and red blood cells (which do not adhere to plastic) remain as contaminants.[8]

Some use has been made of Percoll gradients to separate germ cells from somatic contaminants (a detailed protocol is given by De Felici and McLaren[8]). However, the maximum purity of germ cell samples after Percoll separation was estimated to be between 80 and 90%, and samples of greater purity have been obtained with the collection procedure described in Protocol 3. Percoll is mildly toxic to germ cells, and cell viability in culture is reduced after Percoll separation. The Percoll separation system can be useful if large numbers of cells must be processed, if a high degree of purity is not required, and if the cells are not to be cultured after separation. However, we find that collecting cells by hand after EDTA treatment of the gonads is more effective in providing a sample of germ cells of moderate size (between several hundred and a thousand) with minimal contamination. The viability of germ cells collected in this way appears unaffected.

If a very pure sample of germ cells is required, there is at present no substitute for the collection of cells by hand from the mixture of cells resulting from the dissociation procedure. Depending on the proportion of germ cells in the mixture, an experienced worker can isolate from 100 to 1000 germ cells in about 15 min with a purity of 98% or more. If a greater number of cells is required, or if they must be collected more quickly, the purity of the sample will fall somewhat.

Identification of Germ Cells

At all stages of their normal development, germ cells mingle with somatic cells, and it can be difficult to distinguish the two cell types. At present, there are three ways by which germ cells are identified: by their morphological characteristics, with antibodies that recognize antigens specific to the germ cells, and by their expression of the enzyme alkaline phosphatase.

Morphological Identification

When tissue fragments containing germ cells are disaggregated, it is usually possible to distinguish the germ cells under the dissecting microscope. A magnification of 500× to 1000× is necessary if accurate identification is to be made. Freshly collected germ cells are round, smooth, and appear refractile (i.e., shiny). They are approximately 18 to 20 μm in diameter, somewhat larger than most somatic cells. They roll freely over the bottom of the dish, and though they themselves often aggregate in loose clumps, they do not at this stage adhere strongly to somatic cells. After a short time on the bench or in culture they may start to show a characteristic blebbing and the extrusion of short pseudopodia, but they still do not adhere to the bottom of the dish. At this stage it is easy to collect samples of germ cells with a finely drawn pipette. Inexperienced workers occasionally confuse germ cells with embryonic nucleated red blood cells, as both are round, nonadherent, and approximately the same size. However, blood cells have a distinctive reddish tinge, whereas germ cells are quite colorless.

After a few hours in culture, germ cells cultured on a monolayer adhere to the underlying cell layer and assume a fibroblast-like shape (Fig. 2). At this stage it is no longer possible to identify them with any degree of confidence unless they are visualized with antibodies or alkaline phosphatase staining. Germ cells cultured on surfaces to which they will not adhere (plastic, glass, or agar) can usually be identified as rounded, nonadherent cells even when cultured for some time. They do not, however, survive well in such conditions.

Identification with Antibodies

Germ cells express some surface antigens in a stage-specific sequence, as described by Donovan et al.,[9] and such antibodies as TG-1 and SSEA-1 have proved of value in the tagging of these antigens. The presence of germ cell-specific antigens can be followed over several days, and these antibodies provide the most reliable means yet known of identifying living germ cells in culture. Details of visualization of some antibodies are given by Donovan et al.[9] However, care must be taken to ensure that a particular

<hr>

[9] P. J. Donovan, D. Stott, L. A. Cairns, J. Heasman, and C. C. Wylie, Cell (Cambridge, Mass.) **44,** 831 (1986).

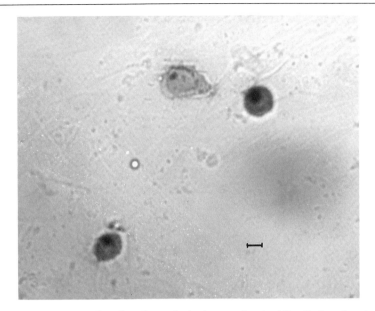

FIG. 2. Three germ cells cultured on a feeder layer and stained for alkaline phosphatase activity. Two are rounded, while the third has a fibroblast-like appearance. The dark spot of strong alkaline phosphatase activity is typical of germ cells at some stages of their development (see Ginsburg et al.[1]). Bar: 10 μm.

antibody reaction is truly specific to germ cells. Some polyclonal antibodies thought to react specifically with germ cells have subsequently been found to react with other cell types as well (see Ref. 10).

Identification with Alkaline Phosphatase Staining

For many years mouse primordial germ cells have been known to show strong alkaline phosphatase activity (Fig. 2).[11] Alkaline phosphatase activity can first be shown in primordial germ cells at 7 dpc, and the germ cells continue to express it until late in embryonic life. Alkaline phosphatase staining can be done on tissue sections, on whole mounts, on air-dried preparations, or on cultured cells, and the methods are described in Protocol 4. Although other cells in the embryo express alkaline phosphatase,[12] the appearance of stained primordial germ cells in tissue sections is very distinctive, and there are no other cells which could

---

[10] M. De Felici, J. Heasman, C. C. Wylie, and A. McLaren, *Cell Differ.* **18**, 119 (1986).

[11] D. A. Chiquoine, *Anat. Rec.* **118**, 135 (1954).

[12] H. Merchant-Larios, F. Mendlovic, and A. Alvarez-Buylla, *Differentiation* **29**, 145 (1985).

easily be confused with them. When isolated germ cells are cultured on a monolayer, however, the picture is not always so clear. Cells in a monolayer, especially of a primary culture, can sometimes unexpectedly express alkaline phosphatase, and though these cells can often be distinguished from primordial germ cells, care must always be taken when interpreting stained culture plates.

Media

Few studies have yet been made of the specific requirements of primordial germ cells in culture. A variety of standard tissue culture media, such as Eagle's minimal essential medium (EMEM), Dulbecco's modified Eagle's medium (DMEM), M199, or McCoy's medium have been used. No comparative study has been made of these media, but after many trials we find that the modification of DMEM described in Protocol 5 works consistently well in our hands.

Although a few groups use serum-free medium (e.g., Godin and Wylie[13]), most workers rely on heat-inactivated serum (usually 10%) for the provision of various growth factors and other substances necessary for successful germ cell culture. Horse, rat, and donor calf serum have been used, but most protocols for germ cell culture specify fetal calf serum. The requirements of rat and mouse fetal tissue may differ: when rat urogenital ridges are cultured, the inclusion of serum in the media has been reported to suppress the formation of testis cords.[14] However, serum seems to be necessary for the normal differentiation of testis cords in cultured mouse testes.[2,4,15]

Tissue culture media used for germ cell and gonad culture are normally bicarbonate buffered, include glutamine as an energy source, and contain penicillin and streptomycin as bacteriostatic agents. In addition, we find that the addition of mercaptoethanol at a concentration of 100 μM improves germ cell viability in culture. If fungal or yeast contamination is a problem, the antifungal agent Fungizone can be added, but as it is thought to have a detrimental effect on cultured cells it is best omitted.

Culture Conditions

Most germ cell culture work has been done under routine tissue culture conditions, that is, at 37°, with an atmosphere of 5% CO_2 in air. Occasion-

[13] I. Godin and C. C. Wylie, *Development (Cambridge, UK)* **113,** 1451 (1991).
[14] R. Agelopoulou, S. Magre, E. Patsavoudi, and A. Jost, *J. Embryol. Exp. Morphol.* **83,** 15 (1984).
[15] T. Taketo, C. D. Seen, and S. S. Koide, *Biol. Reprod.* **34,** 919 (1986).

have been made of other conditions. For instance, De Felici ren[16] reported improved survival or isolated 11 and 12 dpc germ n the cultures were maintained at 30°.

Handling Medium

A simple dissecting medium is needed for use on the bench, and this can be either phosphate or HEPES buffered, for instance, PBS (phosphate-buffered saline), PB1, or M2.[17] If the tissue is to be subjected to enzyme or EDTA treatment, and if collection can be done quickly, calcium- and magnesium-free PBS may be used, as it will not interfere with subsequent disaggregation procedures. If the tissues are expected to wait for some time on the bench, then a more complex medium such as M2 (which contains protein as well as an energy source) is preferable.

Germ Cell Culture

Culture of Isolated Germ Cells

Although germ cells can be maintained in isolation for several days, they do not survive as well as they do if cultured together with somatic cells (this is particularly true of younger germ cells[16]). Germ cells do not normally adhere to plastic, agar, or gelatin, and although the provision of fibronectin as a substrate increases germ cell adhesion to culture dishes,[18,19] neither fibronectin nor other extracellular matrix components appear to improve the long-term viability of germ cells. A low culture temperature may improve the survival of isolated germ cells from 11.5 and 12.5 dpc embryos,[16,20] but if germ cells are to be maintained in a viable condition for more than a day or two, they must be cocultured with somatic cells. Though conditioned medium may slightly improve the survival of isolated germ cells (Ref. 13; M. Buehr and A. McLaren, unpublished observations, 1991), contact between germ cell and feeder cell is necessary for optimal culture.[21]

---

[16] M. De Felici and A. McLaren, *Exp. Cell Res.* **144,** 417 (1983).
[17] D. G. Whittingham, *J. Reprod. Fertil. Suppl.* **14,** 7 (1971).
[18] A. Alvarez-Buylla and H. Merchant-Larios, *Exp. Cell Res.* **165,** 362 (1986).
[19] C. ffrench-Constant, A. Hollingsworth, J. Heasman, and C. C. Wylie, *Development (Cambridge, UK)* **113,** 1365 (1991).
[20] B. Wabik-Sliz and A. McLaren, *Exp. Cell Res.* **154,** 530 (1984).
[21] M. De Felici and S. Dolci, *Dev. Biol.* **147,** 281 (1991).

Culture of Germ Cells on Feeder Layers

Many different somatic cell types have been used as feeder layers for germ cells, including γ-irradiated STO cells,[9] lung fibroblasts (M. Buehr, A. Pearce-Kelly, and A. McLaren, personal communication), TM-4 cells,[21] and Sertoli and granulosa cells.[22] Feeder layers intended for germ cell culture are normally prepared and maintained according to standard methods (as described by Robertson[23]), although if a particular cell type is required as a feeder, specialized collection and culture methods may be necessary (see, e.g., De Felici and Siracusa[22]). When cultured on monolayers, germ cells can survive well, proliferate, and show many of the morphological and functional characteristics that they exhibit *in vivo*.[9,24]

Coculture of Germ Cells and Embryonic Gonad Cells

Germ cells cultured on monolayers derived from the cells of embryonic gonads survive well. Female germ cells in such cultures can enter meiotic prophase and (if maintained long enough *in vitro*) initiate a growth phase. A simple method of coculturing germ and somatic cells is to disaggregate the cells of the embryonic gonad and plate out the resulting cell suspension on a plastic dish. The somatic cells of the gonad adhere to the substrate, while the germ cells roll freely on the monolayer that eventually forms.[25,26] Alternatively, gonads can be cut into small fragments (0.5–1 mm) and cultured on plastic or other adherent surfaces.[2] Within a day or two, the somatic cells adhere and outgrow, releasing the germ cells to rest on the underlying cell layer.

Several authors have reported that female germ cells survive better in culture than do male cells, and certainly they can differentiate further (see, e.g., Refs. 2 and 16). Regardless of the culture system used, older germ cells will as a rule survive and differentiate better in culture than young ones, which may require special culture conditions. For instance, female germ cells from 12 and 13 dpc embryos cocultured with ovarian somatic cells in plastic dishes will enter meiosis in culture and later begin to grow, whereas germ cells in similar cultures from 10 and 11 dpc embryos will not.[26] However, if the younger cultures are maintained in dishes with gas-permeable bottoms (Petriperm dishes from Heraeus) some further germ cell differentiation will take place.[2]

[22] M. De Felici and G. Siracusa, *J. Embryol. Exp. Morphol.* **87,** 87 (1985).
[23] E. J. Robertson, *in* "Teratocarcinomas and Embryonic Stem Cells: A Practical Approach" (E. J. Robertson, ed.), p. 71. IRL Press, Oxford, 1987.
[24] I. Godin, C. C. Wylie, and J. Heasman, *Development (Cambridge, UK)* **108,** 357 (1990).
[25] R. Bachvarova, M. M. Baran, and A. Tejblum, *J. Exp. Zool.* **211,** 159 (1980).
[26] M. Buehr and A. McLaren, *Gamete Res.* **11,** 271 (1985).

Organ Culture of Embryonic Gonads

Although germ cells can be cultured satisfactorily when isolated and plated out over a feeder layer, maximum survival and differentiation of embryonic germ cells are obtained when the tissue or gonad containing the germ cells is maintained intact in an organ culture system. Under these conditions germ cells from both male and female embryos will readily undergo the first stages of germ cell differentiation, stages which isolated germ cells (in particular male germ cells) may never reach. The success of organ culture in supporting normal gonad and germ cell development depends largely on the age of the embryo from which the explant is obtained: older gonads generally differentiate better and to a more advanced stage than younger ones. Gonads from embryos 12.5 dpc and older present few problems.

In male gonads, the testis cords develop normally, and Leydig cells in the interstitial tissue secrete testosterone at the same age as they would *in vivo*. Whereas male germ cells maintained in isolated cultures or on feeder layers do not differentiate, germ cells in testis cultures progress to the T1 prospermatogonia stage. If cultures are maintained for a sufficient length of time, some later stages of germ cell differentiation (such as the appearance of T2, intermediate, and B spermatogonia) may also be seen in cultures derived from 12.5 or 13.5 dpc testes. No *in vitro* system yet devised for mammals has succeeded in supporting male meiosis beyond the early stages of meiotic prophase.[27]

Female germ cells in cultured ovaries from embryos 12.5 or 13.5 dpc will enter meiotic prophase, and some will later initiate a growth phase.[26] Follicle development in these cultures, however, is not normal unless exogenous follicle-stimulating hormone (FSH) and luteinizing hormone (LH) are added to the medium.[28] In the absence of these hormones, the oocytes grow surrounded only by a "primordial follicle" of undifferentiated cells.

Gonads younger than 12.5 dpc can be maintained in culture, but differentiation of both the somatic and germinal elements in them may not be entirely normal. Testes from 11.5 dpc embryos will differentiate normally only if cultured attached to a mesonephros[28a]: germ cells in cultures of isolated 11.5 dpc testes are not numerous, and they are not enclosed in normal testis cords. Isolated ovaries from 11.5 dpc embryos, though small, appear normal, and the germ cells within them enter meiosis as usual. Testes from 10.5 dpc embryos do not develop normal testis cords even

[27] A. Steinberger and E. Steinberger, *Exp. Cell Res.* **44,** 429 (1966).
[28] T. G. Baker and P. Neal, *Ann. Biol. Anim. Biochem. Biophys.* **13,** 137 (1973).
[28a] M. Buehr, S. Gu, and A. McLaren, *Development (Cambridge, UK)* **117,** 273 (1993).

when cultured with an attached mesonephros,[2] and the germ cells enter meiosis rather than differentiating normally as T1 prospermatogonia.

The maintenance of embryonic gonads in culture is generally straightforward. There are, however, two important points to bear in mind. First, if the gonad is to be cultured as a three-dimensional structure similar to a gonad developing *in vivo,* a substrate must be chosen to which the cells will not adhere. If embryonic gonads are cultured on tissue culture plastic dishes, for instance, they will adhere to the surface, and the cells will outgrow. The entire organ will then flatten onto the plastic and, in time, become a monolayer or a sheet of tissue a few cells thick. If the structure of the organ is to be maintained, it must be cultured on a substrate such as agar or a polycarbonate membrane to which the cells will not adhere.

Second, the cultured embryonic gonad appears to be extremely sensitive to poor gaseous exchange between the medium and the gas phase in the incubator. Gonads cultured on the bottom of dishes or wells with 1 or 2 mm of medium between them and the surface frequently develop large patches of necrotic cells, especially at the center of the organ. For optimal, healthy development, gonads should be cultured as close as possible to the gas–medium interface, or should even protrude slightly above it covered by only a thin film of fluid. To maintain the gonad in such a position and still retain a reasonably large reservoir of medium in the dish, the tissue is often supported on an agar block or metal grid[2] or is cultured on a polycarbonate filter floating on the surface of the medium.[29,30]

The embryonic gonad is extremely amenable to procedures involving dissociation and subsequent culture, and several workers have taken advantage of this to conduct studies in which gonadal cells are dissociated, combined with other cell types into heterotypic aggregates, and cultured.[30–32] The details of the protocols naturally vary with the question that is being addressed, and the cited papers give clear descriptions of the methods followed in each case. In general, however, cells are dissociated from appropriate tissues by one of the methods we have described, combined with other cells, and centrifuged. The pellet is then removed from the centrifuge tube and cultured. Often cells which have been dissociated and then collected in serum-containing medium are sticky and need no special treatment to ensure successful aggregation after centrifugation.

[29] T. Taketo and S. S. Koide, *Dev. Biol.* **84,** 61 (1981).

[30] D. Escalante-Alcalde and H. Merchant-Larios, *Exp. Cell Res.* **198,** 150 (1992).

[31] S. Dolci and M. De Felici, *Development (Cambridge, UK)* **109,** 37 (1990).

[32] N. Hashimoto, R. Kubokawa, K. Yamazaki, M. Noguchi, and Y. Kato, *J. Exp. Zool.* **253,** 61 (1990).

However, the addition of phytohemagglutinin to the medium can aid adhesion. A simple disaggregation–reaggregation method is given in Protocol 6.

Protocols

Protocol 1: Amnion Preparation for Sex Chromatin Analysis

This procedure is based on that of Palmer and Burgoyne.[3]

1. Prepare the fixative (3 : 1 (v/v), methanol–acetic acid), about 2 ml for each sample.
2. With a pair of fine forceps, pick up an amnion and place it on the inside of a centrifuge tube. Wash it down to the bottom of the tube with about 1 ml of fixative, and whirl briefly to mix. Repeat for each amnion.
3. Leave for at least 5 min (specimens can be left in fixative for several hours).
4. Centrifuge tubes at 10,000 rpm for 5 to 10 min.
5. Drain off fixative. Invert tubes and tap on a piece of paper towel to remove as much fixative as possible.
6. From a Pasteur pipette, drop 1–2 drops of 60% acetic acid directly on to the amnions, washing them down to the bottom of the tubes.
7. Leave the amnions for about 1 min or until they become wispy and opaque in appearance.
8. Add 1 ml of fresh fixative.
9. Centrifuge at 10,000 rpm for 5 to 10 min.
10. Pour off the supernatant, and resuspend the cells in the fluid that drains back down the sides of the tube. This should be no more than 1 or 2 drops.
11. Place a small drop on a clean microscope slide and allow to air-dry.
12. Place a drop of stain (1% aqueous toluidine blue) on the air-dried cells and cover with a coverslip.
13. The slides may be inspected immediately. A sex chromosome body at the periphery of a cell indicates the presence of an inactive X chromosome. One will be present in cells from normal females but absent in cells from normal males.

Protocol 2: Enzyme Methods for Gonad Disaggregation

Collagenase/DNase. The following method is adapted from Eppig.[33] Make up a collagenase/DNase solution (1 mg/ml collagenase–dispase and

[33] J. J. Eppig, *J. Exp. Zool.* **198**, 375 (1976).

0.1 mg/ml DNase) in a protein-free medium immediately before use. Rinse the gonads in protein-free medium, and then incubate in the above solution for 20 min at 37°. Disaggregate the tissue by gentle trituration and then return the cells to a medium containing serum to stop the action of the enzymes.

Embryonic tissue treated by this method disaggregates easily: the cells do not adhere to each other and readily form a single cell suspension. If the tissue is particularly resistant to disaggregation, however, longer incubation times may be necessary. More efficient disaggregation may also be obtained by making up the solution in calcium- and magnesium-free medium.

Lysozyme/Collagenase/Hyaluronidase. Mangia and Epstein[34] provide a method for tissue disaggregation using lysozyme, collagenase, and hyaluronidase. Solutions (0.5 mg/ml each) of lysozyme, collagenase, and hyaluronidase are made up immediately before use in calcium- and magnesium-free phosphate-buffered saline (PBS). Rinse the gonads in PBS, and then incubate them in the enzyme mixture for 10–20 min at 37°. Tease the tissue apart, allowing the germ cells to fall out, and collect them in a medium containing serum to stop the action of the enzymes. This method has been used primarily to isolate perinatal oocytes.

Pronase. The following procedure is adapted from Bachvarova *et al.*[25] Pronase (2 mg/ml) is made up in calcium- and magnesium-free PBS. Rinse the tissue in PBS and the incubate in the enzyme solution for 20 min at 37°. Return the tissue to PBS and disaggregate by trituration or by vigorous stirring on a magnetic stirrer. If the cells stick to one another, add 0.1 mg/ml DNase to the medium in which the disaggregation is done. The enzyme solution can be stored at −20° for several weeks.

Protocol 3: Procedure for Collecting Gonadal Germ Cells

This procedure is recommended for embryos 10.5 dpc or older.

1. Dissect out the urogenital ridges and cut the gonad or genital ridge away from the mesonephros or adjacent tissues.

2. Wash the gonads briefly in calcium- and magnesium-free PBS and incubate them in EDTA solution (see below) for 20 min at room temperature.

3. Wash the gonads in M2 medium (or a similar protein-containing medium).

---

[34] F. Mangia and C. J. Epstein, *Dev. Biol.* **45**, 211 (1975).

4. The gonads must now be gently disrupted. In a small volume of M2 medium, slit or prick them with a fine needle, or knead them gently with fine forceps. They should not be torn apart, as this can release large numbers of somatic cells.

5. Under a magnification of $500\times$ to $1000\times$, the germ cells can be seen spilling from the disrupted gonads. They can be picked up with a finely drawn Pasteur pipette with a tip diameter of about 25 μm. It is not necessary to siliconize the collecting pipettes.

6. If a sample of high purity is required, transfer the free cells to dishes containing a thick layer of agar (2% agar in PBS; 2 ml in a 35-mm dish) and collect a further sample. The agar facilitates the collecting of identified germ cells and protects the fine tip of the collecting pipette from damage.

7. Samples of cells can be passed through several changes of medium, each time collecting an increasingly pure sample of the germ cells as contaminating somatic cells are left behind.

EDTA Solution for Germ Cell Collection. The following solution keeps for several months at 5°. Sterilize before use.

EDTA, disodium salt	20 mg
NaCl	800 mg
KCl	20 mg
Na_2HPO_4	115 mg
KH_2PO_4	20 mg
Distilled water	To make 100 ml

Protocol 4: Staining for Alkaline Phosphatase Activity in Germ Cells

Alkaline phosphatase is a relatively stable enzyme, and it can retain its activity for long periods of time in fixed tissue. However, high or prolonged heat will affect activity, and every effort should be made to keep specimens as cold as possible through all stages of tissue processing. Fixation and dehydration should be done in a refrigerator, the time in molten wax should be as short as possible, and all specimens, embedded tissues, unstained slides, etc., should be stored cold.

Alkaline Phosphatase Staining of Tissue Sections

1. Fix small blocks of tissue (2–3 mm) in either (a) ice-cold 80% ethanol (if nuclear staining is not required) or (b) in freshly prepared (7:1, v/v) ice-cold 100% ethanol–acetic acid (if the slides are to be counterstained with a nuclear stain). Acetic acid will reduce the activity of alkaline phosphatase, so it should be omitted unless nuclear staining is necessary.

2. Fix tissue in ice-cold 100% ethanol–acetic acid (7 : 1) for 2 hr in the cold. Transfer to 70% ethanol for 2 hr, and then store in fresh 70% ethanol. (Tissues fixed in 80% ethanol may be stored in that solution.) Fixed tissue can be stored in 70 or 80% ethanol at 5° for several weeks without appreciable loss of alkaline phosphatase activity.

3. Dehydrate in two changes of absolute ethanol, 3 to 12 hr each.

4. Clear in two changes of chloroform, 1 hr each.

5. Embed after three changes of low melting point wax (54°), 45 min to 1 hr each.

6. Cut sections as required, from 5 to 10 μm thick. Ensure that the drying slides are not subjected to high temperatures. If slides must wait a long time before staining, place them briefly in a 60° oven. This will allow a thin film of wax to melt over the sections, helping to protect the enzyme from oxidative damage.

7. Hydrate slides as usual (e.g., two changes of 4 min each in xylene, absolute ethanol, 95%, and 70% ethanol, and distilled water).

8. Keep slides in distilled water while the stain (see below) is mixed.

9. Stain slides for 20–25 min in the dark, checking under a microscope to ensure that staining is satisfactory (cells with alkaline phosphatase activity will stain a dark reddish brown).

10. Wash in two changes of distilled water, 5 min each.

11. Counterstain if required. If the tissue has been fixed in 7 : 1 ethanol–acetic acid, a nuclear stain such as hematoxylin may be used. If a cytoplasmic counterstain is necessary, 0.25% (w/v) aqueous fast green provides a color contrast with the reddish brown of the enzyme stain.

12. Mount in an aqueous medium. (Commercially available aqueous mountants are preferable to glycerol, although this can be used in an emergency.) Aqueous mountants are essential, as alcohol will remove the reaction product.

Alkaline Phosphatase Staining of Air-Dried Germ Cells. The following procedure is from Brinster and Harstad.[35]

1. Collect germ cells, preferably in a medium containing no protein.

2. With a finely drawn Pasteur pipette, drop a sample of cells onto a clean slide. Allow to dry completely. (Slides may be stored in a refrigerator at this point.)

3. Place the slide in buffered acetone fix (see below) for 20–30 sec.

[35] R. L. Brinster and H. Harstad, *Exp. Cell Res.* **109**, 111 (1977).

4. Wash the slide in tap water for 10–20 sec and allow to dry.

5. Stain with the alkaline phosphatase stain (see below) for about 15 min, checking occasionally under a microscope. When the stain is satisfactory, rinse in distilled water and mount in an aqueous medium.

Buffered Acetone Fix for Air-Dried Germ Cells. Dissolve 8.3 g Tris and 4.6 g sodium citrate in 40 ml distilled water. While stirring, add 60 ml acetone. The fixative should be freshly prepared before use.

Alkaline Phosphatase Staining of Whole Mounts. The following procedure is adapted from Ginsburg *et al.*[1]

1. Collect young embryos intact (7 to 8.5 dpc), or dissect out appropriate pieces of older embryos (9.5 days or older).

2. Fix tissue in absolute ethanol for 12 to 24 hr.

3. Tissue samples may be cleared at this stage by passing them through two changes of xylene (or other clearing agent) of 6 to 12 hr each. This treatment renders the tissue more transparent after staining.

4. If specimens have been treated with xylene, rehydrate them through two changes of absolute ethanol and one of 70% ethanol (1 to 2 hr in each change of solution, depending on tissue size). Tissues that have remained in the absolute fix need only one change of 70% ethanol. Wash embryos in three changes of distilled water of 10 min each.

5. Transfer to freshly prepared stain (see below) for 15 to 20 min. The staining procedure should be monitored with a dissecting microscope and the embryos removed before background staining becomes too dark.

6. Wash the embryos with three changes of distilled water (5–10 min each) and mount in an aqueous medium.

Alkaline Phosphatase Staining of Cell Monolayers

1. Rinse the plates or wells 2 or 3 times in PBS to remove the culture medium.

2. Fix the plates for 2 to 18 hr in cold 80% ethanol. If nuclear counterstaining is required, fix for 2 hr in cold 7 : 1 absolute ethanol–acetic acid, followed by two changes of 70% alcohol, 2 to 18 hr each.

3. Hydrate the plates with two changes of distilled water, 10 min each.

4. Stain the plates with freshly prepared stain (see below), monitoring the procedure under a microscope.

5. When staining is satisfactory, rinse the plates in two changes of distilled water, 5 to 10 min each.

6. Any mounting or coverslipping of dishes must be done with an aqueous mounting medium.

Stain for Alkaline Phosphatase Activity

25 mg Fast Red TR (diazonium salt)
5 mg α-Naphthyl phosphate
44.6 ml Distilled water
0.3 ml of 10% $MgCl_2$ (aqueous solution)
5 ml of 4.5% Borax (aqueous solution)

Store the Fast Red TR and α-naphthyl phosphate at $-20°$. Mix the stain immediately before use; do not use if more than 30 min old. Mix the ingredients in the order indicated.

Protocol 5: Medium for Germ Cell Cultures

Composition

45 ml DMEM (Dulbecco's modified Eagle's medium, with 3.7 g/liter sodium bicarbonate)
5 ml Fetal calf serum, heat-inactivated
0.5 ml Glutamine stock
0.5 ml Penicillin–streptomycin stock
0.5 ml mercaptoethanol stock
(0.3 ml Fungizone, if necessary)

Stock Solutions

Glutamine: As supplied (100×). Dispense and store frozen.
Penicillin–streptomycin: 10,000 units/ml penicillin, 5 mg/ml streptomycin. Dispense and store frozen.
2-Mercaptoethanol: 7 μl/10 ml DMEM. Make up freshly before use.
Fungizone: As supplied (250 μg/ml). Dispense and store frozen. Protect from light.

This gives a medium with 10% (v/v) fetal calf serum, 584 mg/liter L-glutamine, 100 μM 2-mercaptoethanol, 100,000 units/liter penicillin, and 500 mg/liter streptomycin.

Protocol 6: Reaggregation of Gonadal Cells

1. Disaggregate gonads by any of the methods recommended. Collect the cells in a medium containing serum (and, if necessary, phytohemagglutinin: see recipe below) and add the cells with which the gonadal cells are to be reaggregated.
2. Place the cells in a suitable container for centrifugation. Large samples may be centrifuged in Eppendorf tubes, or in Gilson pipette tips that have been sealed at the end. If the sample is very small, it can be

introduced into a 10-μl microcapillary tube which has been flame-sealed at one end and cut down to a suitable length.

3. Centrifuge the sample at 10,000 rpm for 10 min at 18–20°.

4. Remove the pellet from the tube. This is most easily done by first submerging the entire tube in a dish of culture medium, and then introducing a fine pipette between the pellet and the inside of the tube. The pellet can then be dislodged with gentle currents of medium and washed out into the dish.

5. The pellet should be allowed to recover in an incubator for several hours without further handling. During this time the cells begin to attach to each other, and the pellet assumes some degree of structural integrity which enables it to be handled without breaking.

6. After the recovery period, the pellet can be cultured according to any protocol suitable for embryonic gonads.

Phytohemagglutinin for Cell Aggregations. Make up a stock solution of phytohemagglutinin in distilled water, as directed by the suppliers, and dilute this 1 : 100 in a suitable medium.

[5] Isolation and Culture of Oocytes

By JOHN J. EPPIG and EVELYN E. TELFER

Introduction

Oocyte growth occurs primarily in preantral follicles when the oocyte is arrested in prophase I of meiosis and is incompetent of resuming meiosis. About the time that a follicular antrum is formed, the oocyte nears completion of its growth phase and becomes competent of resuming meiosis. Therefore, antral follicles are the source of fully grown, immature, germinal vesicle (GV)-stage oocytes capable of undergoing spontaneous gonadotropin-independent maturation,[1] and preantral follicles are the source of growing, immature GV-stage oocytes incapable of spontaneous maturation

[1] Maturation is defined here as a process encompassing both the nuclear and the cytoplasmic events that occur in oocytes to prepare them for fertilization. The first morphological indication of maturation is the dissolution or breakdown of the nuclear envelope (germinal vesicle breakdown, GVB). Meiosis then proceeds to metaphase II where it is once again arrested, this time until sperm penetration initiates the completion of meiosis. Cytoplasmic events also occur that promote pronuclear formation and preparation for early embryogenesis. In this chapter, the terms oocyte or immature oocyte refer to GV-stage, primary oocytes, and the terms mature oocyte, ovum, or egg refer to secondary oocytes ready for fertilization.

METHODS IN ENZYMOLOGY, VOL. 225

Copyright © 1993 by Academic Press, Inc.
All rights of reproduction in any form reserved.

without further development. This chapter describes techniques for the collection and culture of both meiosis-incompetent and meiosis-competent oocytes to produce eggs able to undergo fertilization and development to live young.

Isolation and Maturation of Fully Grown, Meiotically Competent Germinal Vesicle-Stage Oocytes

Animals

Follicular development is stimulated in immature mice, around 24 days old, by an intraperitoneal injection of 5 IU of pregnant mare serum gonadotropin (PMSG) 48 hr before sacrificing them by cervical dislocation. Adult cycling mice can also be used, but the yield of oocytes is much smaller.

Isolation and Collection of Oocyte–Cumulus Cell Complexes

Ovaries are removed from animals and immersed in Waymouth MB752/1 medium (Table I) supplemented with 0.23 mM pyruvic acid, 50 mg/liter streptomycin sulfate, 75 mg/liter penicillin G (potassium salt), and 5% (v/v) fetal bovine serum (FBS). Initial studies of the developmental capacity of oocytes matured *in vitro* utilized minimum essential medium

TABLE I

PREPARATION OF WAYMOUTH MEDIUM MB752/1 FOR
OOCYTE GROWTH AND MATURATION[a]

Component	Amount
MB752/1 powder[b]	Contents of 1 liter kit
NaHCO$_3$	2240 mg
Pencillin G, potassium salt	75 mg
Streptomycin sulfate	50 mg
Pyruvic acid, sodium salt	25 mg
Fetal bovine serum	50 ml

[a] Dissolve MB752/1 powder kit in about 750 ml of water, then dissolve the NaHCO$_3$. Add the remaining components, except for the serum, and bring to 1 liter volume. Bubble 5-5-90 gas through the medium for 5 to 10 min, then add the serum and sterilize by filtration. Perfuse the storage bottle with 5-5-90 gas before refrigeration. Do not store longer than 2 weeks.

[b] GIBCO, 430 1400EB.

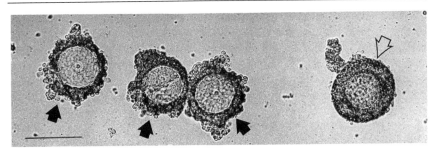

FIG. 1. Photomicrograph of oocyte–cumulus cell complexes (solid arrows) isolated from the antral follicles of a 24-day-old mouse. These can be confused with large preantral follicles (empty arrow) that are sometimes released from the ovary while puncturing the antral follicles with a needle. Note that the periphery of the preantral follicle is much smoother than that of the oocyte–cumulus cell complex. The cumulus cell-enclosed oocytes from antral follicles will undergo spontaneous maturation, but the growing oocyte in the preantral follicle will not. Bar: 100 μm.

(MEM),[2] but subsequent studies,[3] in which several media were tested, showed that oocytes matured in Waymouth MB752/1 medium had the highest percentage of subsequent development to the blastocyst stage. The large antral follicles are punctured using 25-gauge needles mounted on 1-ml syringe barrels. Wipe the needles with a Kimwipe saturated with 75% ethanol before use to remove lubricant. Do not use plastic tissue culture dishes, because they have been treated to encourage cell adhesion and adhesion makes manipulations difficult; Falcon petri dishes (No. 1008) are recommended. Collect the oocyte–cumulus cell complexes using glass pipettes drawn from 4 mm Pyrex glass tubing. Select only complexes having GV-stage oocytes completely enclosed by cumulus cells. Be careful to avoid selecting preantral follicles that are sometimes released from the ovary during antral follicle puncture (Fig. 1). Wash the complexes by serially transferring them through four dishes containing 2.5 ml of maturation medium.

Oocyte Maturation

Place the washed complexes in a petri dish containing 2.5 ml of maturation medium that is also supplemented with 1 μg/ml biological grade follicle-stimulating hormone (FSH). Place the dish in a modular incubator (Billups-Rothenberg) that can be flushed with a gas mixture of 5% O_2, 5% CO_2, and 90% N_2 (v/v; hereafter referred to as 5-5-90 gas). Incubate

[2] A. C. Schroeder and J. J. Eppig, *Dev. Biol.* **102**, 493 (1984).
[3] J. J. M. Van de Sandt, A. C. Schroeder, and J. J. Eppig, *Mol. Reprod. Dev.* **25**, 164 (1990).

at 37° for approximately 17 hr. Make sure the complexes are not clumped in the maturation dish, as aggregation has a deleterious effect on oocyte maturation. After maturation, wash the complexes in the fertilization medium and inseminate immediately. The medium used for the collection and maturation of oocytes is usually supplemented with 5% FBS to prevent hardening of the zona pellucida,[4-6] an event that would render the zona pellucida resistant to sperm penetration. If a serum-free medium is needed, medium supplemented with 1 mg/ml fetuin can be used.[7] Stock solutions of fetuin (10 mg/ml; Spiro preparation, GIBCO, Grand Island, NY) in maturation medium should be dialyzed 3 times against maturation medium (10 times the volume of the dialyzate) before use. Stock solutions of fetuin can be stored for 2 weeks at 4°.

Cumulus cell-enclosed GV-stage oocytes can be manipulated, by microinjection, for example, before maturation begins. Add 250 μM dibutyryl cyclic adenosine monophosphate or 100 μM 3-isobutyl-1-methylxanthine to the maturation medium to maintain meiotic arrest while manipulating the oocytes, then wash the complexes in maturation medium and proceed with oocyte maturation as described above.[8-10]

Expected Results

The embryonic developmental competence of mouse oocytes matured *in vitro* will vary according to the age of the donor mice,[11] the hormonal status of the mice donating the GV-stage oocytes,[12,13] FSH treatment of the maturing oocytes *in vitro*,[11] and the presence or absence of serum in the oocyte maturation medium.[7,14] The effect of the genotype of the oocyte donor has not been assessed. Nevertheless, when PMSG-primed (C57BL/6J × SJL/J)F$_1$ mice 24 to 26 days old provide the GV-stage oocytes for maturation in medium supplemented with FBS, the frequency of fertilization and cleavage to the 2-cell stage should be 75 to 95%. Moreover, 65 to

[4] M. DeFelici, A. Salustri, and G. Siracusa, *Gamete Res.* **12,** 227 (1985).

[5] J. G. Gianfortoni and B. J. Gulyas, *Gamete Res.* **11,** 59 (1985).

[6] S. M. Downs, A. C. Schroeder, and J. J. Eppig, *Gamete Res.* **15,** 115 (1986).

[7] A. C. Schroeder, R. M. Schultz, G. S. Kopf, F. R. Taylor, R. B. Becker, and J. J. Eppig, *Biol. Reprod.* **43,** 891 (1990).

[8] W. K. Cho, S. Stern, and J. D. Biggers, *J. Exp. Zool.* **187,** 383 (1974).

[9] C. Magnusson and T. Hillensjo, *J. Exp. Zool.* **201,** 138 (1977).

[10] S. M. Downs, A. C. Schroeder, and J. J. Eppig, *Gamete Res.* **15,** 305 (1986).

[11] J. J. Eppig and A. C. Schroeder, *Biol. Reprod.* **41,** 268 (1989).

[12] J. J. Eppig, A. C. Schroeder, and M. J. O'Brien, *J. Reprod. Fertil.* **95,** 119 (1992).

[13] A. C. Schroeder and J. J. Eppig, *Gamete Res.* **24,** 81 (1989).

[14] J. J. Eppig, K. Wigglesworth, and M. J. O'Brien, *Mol. Reprod. Dev.* **32,** 33 (1992).

85% of the 2-cell stage embryos should develop to the expanded blastocyst stage. Between 30 and 45% of the 2-cell stage embryos transferred to the oviducts of foster mothers should develop to live offspring.

Isolation, Growth, and Maturation of Growing, Meiotically Incompetent Germinal Vesicle-Stage Oocytes

Animals

Large numbers of oocyte–granulosa cell complexes can be isolated from the preantral follicles of 12-day-old mice. Use of the immature mice provides ovaries that are readily dissociated with collagenase and also relatively homogeneous populations of preantral follicles.

Isolation and Collection of Oocyte–Granulosa Cell Complexes of Preantral Follicles

Ovaries from 5 to 15 mice are immersed in 3.5 ml of culture medium containing 1 to 2 mg/ml crude collagenase (Worthington type CLS-1, Freehold, NJ, for example) and 0.02 mg/ml DNase (sterilized by filtration) and then are placed on a slide warming tray maintained at 35° and covered with a plexiglass box that is constantly flushed with 5-5-90 gas. After about 10 min, the ovaries are vigorously drawn in and out of an Eppendorf micropipette set at 1.0 ml and equipped with a sterile blue tip until the ovaries begin to dissociate. The ovaries are returned to the warming tray for 10-min intervals, at which times the dissociating ovaries are repeatedly drawn in and out of a micropipette set at 0.1 ml and equipped with a sterile yellow tip. This process is repeated until the ovaries are completely dissociated into individual oocyte–granulosa cell complexes. This is usually completed within 40 min. The complexes are washed by swirling them together in the center of the petri dish, drawing off the cloud of debris that overhangs the settling complexes, transferring the complexes to another petri dish containing 3.0 ml of medium, and repeating this process until the preparation is free of ovarian debris, usually four or five transfers.

This procedure results in a preparation such as that shown in Fig. 2. The complexes each consist of an oocyte, which is in mid-growth phase and incompetent of undergoing GV breakdown without further development, surrounded by one to three layers of granulosa cells. Most of the primitive theca layers and the basal lamina are removed by the collagenase digestion.

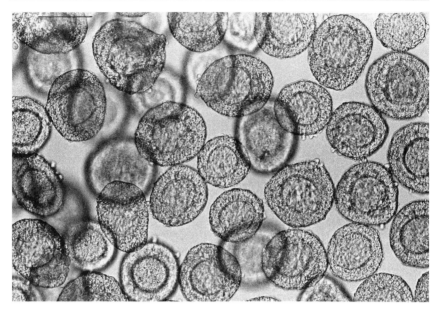

Fig. 2. Photomicrograph of preantral follicles produced by dissociation of the ovaries of 12-day-old mice using collagenase/DNase. Bar: 100 μm.

Oocyte Growth

The oocyte–granulosa cell complexes are distributed to the surface of either Transwell-COL membranes (Costar, Cambridge, MA; 3.0 μm pore size) or a rat tail collagen matrix prepared in the following way.[15,16]

Preparation of Rat Tail Collagen

1. Tails obtained from rats can be frozen until needed for preparation of collagen. They are sterilized overnight in 70% (v/v) ethanol immediately before use.

2. Dissect out tendons using two pairs of artery forceps. Starting from the tip of the tail clamp one end with artery forceps and clamp approximately 3 cm away from the other pair. After bending and pulling both pairs of forceps, the tendons will come loose from the skin. Cut the tendons loose and place them in a small beaker with 70% ethanol.

3. Wash tendons in sterile distilled water and blot dry with sterile filter paper.

[15] R. I. Ehrmann and G. O. Gey, *J. Natl. Cancer Inst.* **16,** 1375 (1956).
[16] M. Chambard, J. Gabrion, and J. Mauchamp, *J. Cell Biol.* **91,** 157 (1981).

4. Weigh tendons and place them in a large bottle with a sterile magnetic stir bar.

5. Add 100 ml of 1 : 1000 acetic acid–water per gram of tendons.

6. Stir tendons on a magnetic stirring plate in a cold room for 48 hr.

7. Transfer collagen to 50-ml centrifuge tubes and centrifuge at 4000 rpm at 4° for 60 min.

8. Dialyze the collagen solution against a 10-fold volume of 1 : 1000 acetic acid–water 3 times for 24 hr each time.

9. Transfer collagen to new 50-ml centrifuge tubes on ice.

10. Store collagen in a refrigerator or freezer.

Preparation of Collagen Gel Matrix for Culture

1. Keep all components on ice. Mix the following in order: 1.6 ml of rat tail collagen, 0.2 ml Dulbecco's phosphate-buffered saline (PBS), and 0.2 ml of 10× culture medium.

2. Add 70 to 100 μl of an ice-cold 0.5 M NaOH solution. Each lot of collagen is a little different, so the exact amount should be determined. The mixture should be homogeneous and pink in color.

3. Using a micropipette equipped with a sterile blue tip, add 0.250 ml of the collagen solution to the bottom of a well in a 24-well tissue culture cluster dish and distribute evenly. Keep cold during this process.

4. Set the culture dish on the slide warmer and allow the collagen to gel (>15 min).

5. Overlay the collagen gel matrix with about 1 ml of PBS and change every 5 min for 30 min. Add and remove the solutions very gently.

6. Overlay the collagen gel matrix with culture medium and incubate on the warming tray for 5 to 10 min.

7. Discard the medium and replace with 1 ml of fresh culture medium and distribute the oocyte–granulosa cell complexes.

Cultures are incubated at 37° in modular incubation chambers flushed with 5-5-90 gas and fed by exchanging approximately one-half of the medium with fresh medium every second day.

Maturation of in Vitro Grown Oocytes

After culturing the oocyte–granulosa cell complexes for 10 to 14 days, remove them from the surface of the Transwell-COL membrane or collagen gel matrix. This is done by sharply snapping the side of the dish with a fingernail to jolt the complexes free of the substratum. The complexes are collected with a drawn glass micropipette and matured as described above. After maturation, wash the complexes in the medium used for fertilization and inseminate immediately.

Expected Results

When oocyte–granulosa cell complexes are isolated from the preantral follicles of 12-day-old (C57BL/6J × SJL/J)F_1 mice and cultured for 10 days in FBS-supplemented medium, it should be anticipated that 60 to 80% of the oocytes will acquire competence to mature. Of the mature eggs, 50 to 60% will cleave to the 2-cell stage after insemination, and 55 to 70% of the 2-cell stage embryos will develop to the expanded blastocyst stage. Between 10 and 15% of the 2-cell stage embryos transferred to the oviducts of foster mothers should develop to live offspring.

The oocyte–granulosa cell complexes of preantral follicles can also be grown in serum-free medium supplemented with 1 mg/ml fetuin, 5 μg/ml insulin, 5 μg/ml transferrin, and 5 ng/ml selenium (ITS; Collaborative Research, Bedford, MA). Under these conditions, fewer oocytes competent of subsequent development to the blastocyst stage are produced.[14] These results suggest that serum-born growth factors are important for optimal oocyte development.

[6] Purification, Culture, and Fractionation of Spermatogenic Cells

By Anthony R. Bellvé

Introduction

The adult mammalian testis contains multiple populations of somatic and germinal cells that differ markedly in their stages of growth and differentiation. Germ cells, the major cell type of the testis, undergo a complex and fascinating process of differentiation.[1] The process, spermatogenesis, starts with the renewal of the rare spermatogonial stem cells (A_0/A_s), followed by the proliferative sequence of types A_{pr}, A_{al}, A_{1-4}, intermediate, and type B spermatogonia (Fig. 1). The latter form primary spermatocytes, which enter a prolonged meiotic prophase to effect pairing of homologous chromosomes, genetic recombination, and, with the two reduction divisions, random segregation of chromosomes to yield haploid spermatids. Then, during spermiogenesis, the spermatids undergo a remarkable sequence of differentiation that includes condensing the chromatin, reorganizing and shaping the nucleus, forming an acrosome replete with enzymes to aid fertilization, and assembling a tail with its microtubule

---

[1] A. R. Bellvé, *Oxford Rev. Reprod. Biol.* **1**, 159 (1979).

METHODS IN ENZYMOLOGY, VOL. 225
Copyright © 1993 by Academic Press, Inc.
All rights of reproduction in any form reserved.

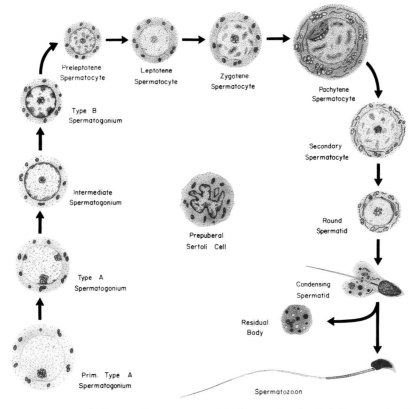

FIG. 1. Schematic diagram of spermatogenesis in the prepuberal and adult mouse, showing the relative volumes and characteristic features of the germ cells. The principal phases of spermatogenesis include the proliferation and renewal of spermatogonia (ascending axis), the meiotic events of primary spermatocytes (horizontal axis), and the morphological transformation during spermiogenesis (descending axis). The cell types shown can be isolated by sedimentation velocity at unit gravity, based on their different cross-sectional areas, and temporal appearance during development. Thus, Sertoli cells and primitive type A spermatogonia can be isolated from mice at day 6 after birth, type A and type B spermatogonia at day 8, preleptotene, leptotene, zygotene, and pachytene spermatocytes from day 17, and pachytene spermatocytes, spermatids, and residual bodies from adult animals. The different stages of spermatogonial stem cells and the diplotene spermatocyte are not shown. [Reproduced from A. R. Bellvé, J. C. Cavicchia, C. F. Millette, D. A. O'Brien, Y. M. Bhatnagar, and M. Dym, *Journal of Cell Biology* **74,** 68 (1977), by copyright permission of the Rockefeller University Press.]

doublets, structural fibers, and a gyre of mitochondria (see Chapter [7], this volume). The plasma membrane also undergoes major topographical reorganization to form specialized regional domains. Thus, the testis contains germ cells at all stages in a continuum of differentiation, along with a variety of supporting somatic cells.

The cellular complexity of the testis compromised early studies that tried to identify and characterize the unusual biochemical events of spermatogenesis. This limitation was offset, in part, by using testes at different periods of development, each chosen to contain cells at more advanced stages of spermatogenesis. This approach provided some insights into the biochemical processes, but the data were not definitive. It became necessary, therefore, to prepare homogeneous populations of germ cells at each stage of spermatogenesis, from stem cells to sperm, and in sufficient numbers to allow subcellular fractionation and biochemical studies. Only recently have there been advances in the short-term culture of germ cells, thereby enabling *in vitro* metabolic studies.

Theoretical and Historical Perspectives

Theory of Sedimentation Velocity at Unit Gravity

The sedimentation velocity (S) of a spherical cell descending through medium under the influence of gravity is defined by the following relationship:

$$S = \frac{2(\rho - \rho')}{9\eta} gr^2$$

where ρ and ρ' are the densities of the cell and the medium, respectively, η is the coefficient of viscosity, g is the acceleration due to gravity, and r is the radius of the cell. When the densities of the cell and medium are expressed in grams per cubic centimeter, η is in centipoise, and sedimentation velocity is in millimeters per hour. Given the viscosity of the medium at 5° and the density of the cell, it is possible to predict the sedimentation rate of the latter. Because the density of the medium is relatively low (~1.0 g/ml), the sedimentation velocity of a germ cell at unit gravity is primarily a function of its radius. This relationship has been verified experimentally and shown, for all practical purposes, to be independent of cell shape.[2] Given the marked differences that occur in diameter and therefore cross-sectional area of germ cells, sedimentation velocity is a

---

[2] R. G. Miller and R. A. Phillips, *J. Cell. Physiol.* **73**, 191 (1969).

logical procedure to apply for isolating discrete populations that represent different stages of spermatogenesis.

Separation of Germ Cells of Adult Testis

Germ cells of the adult mouse testis were first separated into quasi-homogeneous populations by taking advantage of differences in their sedimentation velocity at unit gravity.[3] The procedure was adapted from similar applications to isolate erythrocytes and spleen cells.[2] In its basic form, as the procedure evolved, testes or seminiferous tubules were dissociated into a monodisperse cell suspension by using mechanical or proteolytic procedures, in combination or sequentially. The resulting cells were layered over a linear gradient of bovine serum albumin (BSA) in medium, then allowed to sediment at unit gravity for a total period of 4 hr. The cells were collected, identified, and pooled into quasi-homogeneous populations.

In the procedure of Lam et al.,[3] adult mouse testes were dissociated by mincing the organ with a gang of 50 razor blades, each spaced 0.2 mm apart. The dispersed cells were separated into discrete populations by using sedimentation velocity on a linear gradient of 1–3% (w/v) BSA in phosphate-buffered saline (PBS), at unit gravity. Under these conditions, germ cells were found to sediment at rates linearly related to the respective cross-sectional areas. However, identification of cells in the isolated populations was based only on the relative volumes and kinetics of [^{3}H]thymidine incorporation. The cells were not identified by applying morphological criteria. In some cases identification may have been incorrect; for example, cells denoted as diplotene spermatocytes most likely consisted of advanced pachytene spermatocytes. Similar data were obtained on subjecting rat germ cells to sedimentation velocity, when allowing for differences in cell volumes and the kinetics of spermatogenesis in this species.[4]

The potential utility of the procedure became evident from studies on the kinetics of [^{3}H]arginine incorporation into protamines during the last steps of mouse spermiogenesis.[5] Unfortunately, however, the authors caused some confusion by misidentifying the testis-specific protein (cf. Kistler et al.[6,7]) and the protamines, P1 and P2 (cf. Bellvé et al.[8]). Finally,

[3] D. M. K. Lam, R. Furrer, and W. R. Bruce, Proc. Natl. Acad. Sci. U.S.A. 65, 192 (1970).

[4] V. L. W. Go, R. G. Vernon, and I. B. Fritz, Can. J. Biochem. 49, 753 (1971).

[5] D. M. K. Lam and W. R. Bruce, J. Cell. Physiol. 78, 13 (1971).

[6] W. S. Kistler, M. E. Geroch, and H. G. Williams-Ashman, J. Biol. Chem. 248, 4532 (1973).

[7] W. S. Kistler, C. Noyes, R. Hsu, and R. L. Heinrikson, J. Biol. Chem. 250, 1847 (1975).

[8] A. R. Bellvé, E. Anderson, and L. Hanley-Bowdoin, Dev. Biol. 47, 349 (1975).

sedimentation velocity was applied to assess the antineoplastic effects of cytosine arabinoside, cyclophosphamide, procarbazine, and vincristine on mouse spermatogenesis.[9] While useful, these early studies did not confirm the identity of the cells by applying cytochemistry or, better still, electron microscopy.

Two worthwhile improvements were soon introduced. First, mouse testes were dissociated by an incubation in 2.5% trypsin (w/v) and 20 μg DNase I/ml PBS, rather than by applying a mechanical method.[10,11] Tryptic dissociation, together with the hydrolysis of released DNA, resulted in less cell damage and consequently a greater yield of germ cells. Second, cytochemical evidence was reported to demonstrate purity of the different cell populations, thereby adding credence to the data. Third, estimates of cell viability were obtained based on the exclusion of trypan blue. Fourth, the incorporation of isotopic thymidine and amino acids, indicative of viable cells, were quantified as a function of cell differentiation.

From these data other limitations became apparent. The purity of the isolated cell populations was less than optimal, viability of the cells was low ($\leq$78%),[9] and multinucleated symplasts invariably contaminated most cell populations. Symplasts arose by the relaxation of intercellular cytoplasmic bridges and the subsequent confluence of germ cells from the same syncytial clone. The multinucleated cells had from 2 to over 30 nuclei and, because of their irregular volumes, contaminated all cell populations with sedimentation rates faster than the haploid cells (cf. Romrell et al.[12]).

The purity of mouse germ cells recovered by sedimentation velocity was improved substantially by the introduction of further modifications to the basic procedure.[12–14] These involved: (1) removing all cells from testicular blood vessels by perfusing the organ; (2) dissociating the decapsulated testes with a sequence of enzymes; (3) maintaining the germ cells in a defined culture medium; and (4) identifying the cells based on distinctive characteristics discernible by differential interference contrast light microscopy and by electron microscopy. A similar ultrastructural characterization of mouse germ cells isolated by sedimentation velocity or centrifugal elutriation was reported by Barcellona and Meistrich.[15]

[9] I. P. Lee and R. L. Dixon, *Toxicol. Appl. Pharmacol.* **23**, 20 (1972).

[10] M. L. Meistrich, *J. Cell. Physiol.* **80**, 299 (1972).

[11] M. L. Meistrich, W. R. Bruce, and Y. Clermont, *Exp. Cell Res.* **79**, 213 (1973).

[12] L. J. Romrell, A. R. Bellvé, and D. W. Fawcett, *Dev. Biol.* **49**, 119 (1976).

[13] A. R. Bellvé, J. C. Cavicchia, C. F. Millette, D. A. O'Brien, Y. M. Bhatnagar, and M. Dym, *J. Cell Biol.* **74**, 68 (1977).

[14] A. R. Bellvé, C. F. Millette, Y. M. Bhatnagar, and D. A. O'Brien, *J. Histochem. Cytochem.* **25**, 480 (1977).

[15] W. J. Barcellona and M. L. Meistrich, *J. Reprod. Fertil.* **50**, 61 (1977).

With these modifications, donor animals were perfused via the descending thoracic aorta with an enriched Krebs–Ringer bicarbonate medium (EKRB; defined below), to remove all blood cells. The desanguinated testes were decapsulated and incubated in 1.0 mg/ml collagenase in EKRB, for 15 min with gentle shaking. This step digested basement membranes and dispersed all interstitial cells without fragmenting the seminiferous tubules. The tubules were recovered and dissociated in 1.0 mg/ml trypsin, 1.0 μg/ml DNase in EKRB for another 15 min, to form a monodisperse population of germ cells. This gentler procedure resulted in less damage and substantially fewer symplasts. The viability of the germ cells remained high during the separation procedure, based on the consumption of O_2, exclusion of trypan blue, and maintenance of ultrastructural integrity.[12] As a result, the purity and viability of the isolated germ cells were improved substantially.

Isolation of Germ Cells from Prepuberal Mouse Testes

In the case of the adult seminiferous epithelium, the isolation of cells is limited to the most prevalent types present: pachytene spermatocytes, spermatids, and residual bodies. Germ cells at earlier stages of spermatogenesis are less abundant and therefore cannot be isolated, at least not by using sedimentation velocity at unit gravity. Early in development, however, the stem cells proliferate and differentiate gradually to yield the sequence of spermatogonia, spermatocytes, spermatids, and sperm (Table I).[13,14] Moreover, in the period immediately after birth, the stem cells rapidly decrease in volume, as the primitive type A spermatogonia (diameter 14–12 μm) pass through the series to form type B spermatogonia (diameter 8–10 μm). Thus, by using animals of different ages, it is possible to obtain testes in which the epithelium contains differentiating Sertoli cells and subsets of germ cells. More uniform growth rates of male pups, and hence better synchrony in testicular development, can be obtained by limiting litter size to 8 individuals, with the removal of an appropriate number of female pups at birth. This strategy was applied, successfully, to obtain Sertoli cells and primitive type A spermatogonia from mice at 6 days; type A spermatogonia and type B spermatogonia at 8 days; and preleptotene, leptotene/zygotene, and pachytene spermatocytes at 17 days after birth.[13] The germ cell populations isolated from prepuberal mice complement those from adult animals, thereby providing a continuum of cells at all stages of spermatogenesis.

Procedure for Separating Germ Cells by Sedimentation Velocity

The objectives for the separation of germ cells are (a) to eliminate selectively undesirable cells and cellular debris; (b) to produce a monodis-

TABLE I
TEMPORAL INCEPTION OF SPERMATOGENESIS DURING DEVELOPMENT OF MOUSE TESTIS[a]

Cell type	Days after birth								
	6	8	10	12	14	16	18	20	84
Primitive type A spermatogonia	16	—	—	—	—	—	—	—	—
Type A spermatogonia	—	17	7	7	6	9	3	4	1
Type B spermatogonia	—	10	11	8	6	8	7	6	3
Preleptotene spermatocytes	—	—	15	11	9	5	10	7	2
Leptotene spermatocytes	—	—	15	12	13	5	5	9	2
Zygotene spermatocytes	—	—	—	23	14	7	8	8	2
Pachytene spermatocytes	—	—	—	—	15	27	36	33	15
Secondary spermatocytes	—	—	—	—	—	—	1	1	1
Spermatids (steps 1–8)	—	—	—	—	—	—	1	4	31
Spermatids (steps 12–16)	—	—	—	—	—	—	—	—	40
Sertoli cells	84	73	52	39	37	39	29	28	3

[a] Cells were counted based on nuclei identified in 50 cross sections of seminiferous cords chosen at random from sections of testes of three mice at each of the designated ages. Data are expressed as a percentage of total cells in the seminiferous epithelium, excluding the infrequent necrotic cells. Modified from the data of Bellvé et al.,[13] by copyright permission of the Rockefeller University Press.

perse cell suspension; (c) to maximize the yield of germ cells; (d) to maintain morphological and biochemical integrity; and (e) to minimize the occurrence of multinucleated symplasts. The following procedure addresses these objectives (also see OBrien[16]).

Step 1: Preparation of Glassware. All glassware coming into contact with germ cells are washed thoroughly, including 15-ml test tubes, 125- and 250-ml flasks, 1200-ml fleakers, three-way valves, 20- and 50-ml syringe barrels used as cell loading chambers, and the sedimentation chambers, SP-120, SP-180, and SP-240 (Johns Scientific, Toronto, ON). Glass Pasteur pipettes can be used to transfer cells, but the tips must be fired to round off sharp edges. Disposable plastic pipettes with compressible bulbs can also be used.

All glassware and the stainless steel baffles are siliconized by placing them overnight in a large, sealed bell jar, along with a small glass dish containing 2 ml of Sigmacote (Sigma Chemical Co., St. Louis, MO). After siliconization, the glassware is washed thoroughly, then dried in a hood for 60 min. Glassware, other than Pasteur pipettes, can be used for 10–12

[16] D. A. O'Brien, *in* "Methods in Toxicology" (R. E. Chapin and J. Heindel, eds.), Vol. 3A, p. 246. Academic Press, San Diego, 1993.

separations, until cells begin to adhere to the surfaces. At this time the glassware needs to be acid washed and resiliconized. Glassware, plastic-ware, silicone tubing, stainless steel baffles, connectors, three-way valves, and cell loading and sedimentation chambers are sealed in autoclave bags, autoclaved at 250°–260° for 60 min, and stored until required.

Step 2: Preparation of Medium. An enriched Krebs–Ringer bicarbonate medium is prepared, consisting of 120.1 mM NaCl, 4.8 mM KCl, 25.2 mM NaHCO$_3$, 1.2 mM KH$_2$PO$_4$, 1.2 mM MgSO$_4$·7H$_2$O, 1.3 mM CaCl$_2$, supplemented with 11.1 mM glucose, 1 mM glutamine, 10 ml/liter of essential amino acids, 10 ml/liter of nonessential amino acids, 100 μg/ml streptomycin, and 100 U/ml penicillin (K$^+$ salt) (GIBCO BRL, Grand Island, NY). The medium is sterilized by filtration through a 0.30-μm filter into a sterile flask, and then aerated by bubbling 5% CO$_2$ in air through a sterile, cotton-tipped pipette for 5–10 min at 22°, until the pH is adjusted to 7.2 to 7.3. The volume of medium required to undertake all phases of the cell separation varies according to the number of testes and the volume of the sedimentation chamber being used (Table II). The medium can be made more efficiently by starting from concentrated stock solutions that withstand storage for several weeks.[16] Best results are achieved if glutamine and the essential and nonessential amino acids are added to the medium fresh from supplier stocks.

TABLE II

CONDITIONS USED FOR SEPARATION OF MOUSE SPERMATOGENIC CELLS BY SEDIMENTATION VELOCITY AT UNIT GRAVITY[a]

Parameter	Sedimentation chamber		
	SP-120	SP-180	SP-240
Chamber diameter (mm)	120	180	240
EKRB overlay (ml)	30	50	100
Cell suspension (ml)	10	25	50
2% BSA in EKRB (ml)	275	550	1100
4% BSA in EKRB (ml)	275	550	1100
Total cells loaded	2 × 10^8	5 × 10^8	10 × 10^8
Fraction volume (ml)	5	10	10
Number of fractions	115	118	235

[a] Cell separations are performed routinely under these conditions. However, it is recommended that only half the listed number of cells be applied to each chamber, until experience is gained both in preparing the germ cells and in applying the cell suspension to the BSA gradient. With experience it is possible to load 2- to 3-fold more cells.

Step 3: Preparation of Testes. Mice, 70 to 120 days of age, are anesthetized with CO_2 and then euthanized by cervical dislocation. Each animal is laid on its back, and the exposed ventral surface is wetted with 90–95% ethanol. A midline incision is made to open the thoracic and abdominal cavities. The animal is perfused by using a hypodermic syringe loaded with 10 ml of EKRB and equipped with an 18-gauge needle. The needle is inserted into the descending thoracic aorta, and the animal is perfused (1–2 ml) until the liver becomes blanched. Then, the apex of the heart is punctured to allow egress of the perfusate, and the remainder of the EKRB (~8 ml) is perfused slowly through the vascular system, until all blood tissue is removed from both testes. Prepuberal mice at postnatal days 6, 8, or 17 are sacrificed by cervical dislocation, and the testes are removed directly.

Testes are excised by using a pair of surgical scissors, while gently holding the attached epididymis or fat pad with fine forceps. After placing the testes in a drop of EKRB on a small sheet of aseptic Parafilm, the tip of the ventral pole is held with a pair of fine forceps, while the caudal pole is cut with a single 3–5 mm vertical incision. The contents of the testis are expressed through the caudal incision by applying a single gentle sweep with the backside of a scissor blade. The last step is performed gently to minimize the formation of symplasts, which can severely compromise the purity of the resulting germ cell populations.

Step 4: Collagenase Digestion. The testicular contents are placed in a sterile Erlenmeyer flask containing EKRB: 4–10 testes in 15 ml of medium in a 125-ml flask, 11–20 testes in 30 ml in a 150-ml flask, and 21–35 testes in 60 ml in a 250-ml flask. Dry collagenase (CLS-I; Worthington Corporation, Freehold, NJ) is added to the medium to a final concentration of 0.5 mg/ml, and the flask is closed with a vented silicone stopper. As opposed to purer preparations of the collagenase, CLS-I contains other proteases that facilitate tissue dissociation.[14] The flasks are shaken in a reciprocal shaking water bath for 15 min at 110–120 cycles/min at 32°, while being gassed slowly but continuously with humidified 5% CO_2 in air. During the last 3 min of incubation the dispersed seminiferous tubules are inspected. If necessary, the flask is swirled gently by hand, and/or the tissue is pipetted with a disposable plastic pipette, to disperse any remaining clumps of tubules. Prepuberal testes are processed similarly but in proportionately less EKRB and incubated for about 12 min only. Care must be taken to ensure that the fragile seminiferous cords of prepuberal testes are not fragmented excessively during the incubation. Once dispersed, the cords or tubules are allowed to settle to the bottom of the flask, and the spent medium containing interstitial cells and escaping germ cells is aspirated and discarded. The cords/tubules are resuspended in

EKRB, transferred to a siliconized test tube, and washed twice more in EKRB by sedimentation at unit gravity and decanting the supernatant.

Step 5: Tryptic Digestion. The cords/tubules are transferred back to a 125-ml or 250-ml flask, depending on the volume involved (Table II), and incubated with 0.25 mg/ml trypsin and 1 μg/ml DNase I of EKRB under the same conditions as described in Step 4. Again, after 12 to 15 min, the preparation is pipetted briefly and gently to help disperse the cells. The presence of clumped "strings" of cells indicates cellular aggregation due to DNA released from an excessive number of damaged cells. Aggregates of this nature are usually indicative of a defective medium or excessive enzymatic digestion. The problem may be corrected by adding more DNase I prior to the end of the usual incubation period. Aliquots from the suspension may be examined quickly by light microscopy to check the degree of tissue dissociation. When completed, the cells are sedimented by centrifugation at 500 g for 10 min at 4°, and the supernatant is aspirated.

The germ cells are resuspended in EKRB containing 0.5% (w/v) BSA (fraction V, Sigma), centrifuged at 500 g for 10 min, and resuspended in EKRB containing 0.5% (w/v) BSA, 1 μg/ml DNase I, and 100 DN-EP μg/ml soybean trypsin inhibitor (Sigma). The last reagent is needed to inhibit trypsin that is bound adventitiously to the cell surface. After another centrifugation, the cells are suspended in EKRB containing 0.5% (w/v) BSA and filtered through 80-μm mesh Nitex nylon filter cloth (Tetko, Inc., Lancaster, NY). The filtrate is brought to an appropriate volume in EKRB in 0.5% (w/v) BSA (Table II) for immediate application to the sedimentation chamber.

Step 6: Assembly of Sedimentation Apparatus. The cell sedimentation apparatus is assembled as follows (Fig. 2). Two sterile 1200-ml fleakers are used that were adapted to have one (fleaker 1) or two (fleaker 2) glass outlets at the base. Fleakers 1 and 2 are connected by 10 cm of thick-walled, silicone tubing ($\frac{1}{4}$ inch I.D.; $\frac{3}{8}$ inch E.D.). Clamp 1 is placed on the tubing to control the flow of medium between the two fleakers. Fleaker 2 is equipped with an 8-cm magnetic stirring bar, and it is connected at the second outlet via 10 cm of silicone tubing to the inlet of the three-way, T-bore, glass stopcock (valve 1). The barrel of a glass hypodermic syringe, 20 or 50 ml in volume depending on the size sample being loaded (Table II), is connected via fine silicone tubing to the vertical outlet of three-way valve 1. The syringe barrel acts as a cell loading chamber. The remaining horizontal outlet of three-way valve 1 is connected via 50 cm of silicone tubing to a three-way, glass/plastic, micrometering flow valve (valve 2) that is clamped to the base of the sedimentation chamber. Clamps are placed at points along the tubing for use in loading the EKRB through-

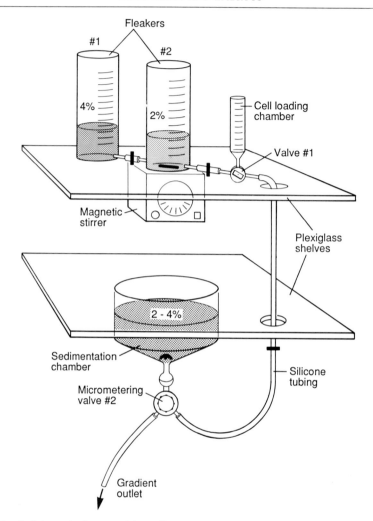

FIG. 2. Schematic diagram of the sedimentation system used to separate different populations of germ cells, at unit gravity. The apparatus is set by loading, in sequence: (a) 4% BSA in EKRB into fleaker 1 and connecting tubing; (b) 2% BSA in EKRB into fleaker 2 through to the base of the loading chamber; (c) EKRB (60 ml) throughout the silicone tubing and into the base of the sedimentation chamber; and (d) the cell suspension in 0.5% BSA in EKRB into the cell loading chamber. By switching valves 1 and 2, and releasing the appropriate clamps, the cells are loaded into the sedimentation chamber and underlayed by a gradient of 2–4% BSA in EKRB (see text for details). Flow rates into and out of the sedimentation chamber are regulated by adjusting micrometering valve 2. The gradient is unloaded via the outlet silicone tubing into a fraction collector. Two sedimentation systems are placed vertically in a 50-ft.[3] refrigerator, one in each side. For each half, an upper shelf is prepared with mounting brackets for holding the two fleakers, the loading chamber, and

out the system. The stainless steel baffle is placed carefully in the sedimentation chamber over the inlet at the base, to later dampen currents and spread incoming medium across the chamber. The chamber is leveled with the aid of three adjusting screws (see legend to Fig. 2).

Step 7: Preparing the Sedimentation Apparatus for Cell Loading. EKRB containing 4% (w/v) BSA is loaded into fleaker 1, to a predetermined volume (Table II). Clamp 1 is released partially to allow the 4% BSA in EKRB to pass through the outlet, into the silicone tubing, and to just enter the inlet at the base of flask 2. Clamp 1 is tightened. EKRB containing 2% (w/v) BSA is loaded into fleaker 2, and clamp 2 is released partially to allow the medium to flow through the outlet tubing and valve 1, then up into the base of the cell loading chamber. Valve 1 is closed, and the tubing at the outlet from fleaker 1 is clamped. EKRB without BSA is poured into the cell loading chamber, and valves 1 and 2 are opened to allow the medium to flow through the silicone tubing and to just enter into the sedimentation chamber. Valve 2 at the base of the sedimentation chamber is closed. Any air trapped in the silicone tubing is aspirated by piercing the wall with a sharp, 30-gauge hypodermic needle on a 10-ml syringe. Fine holes left in the tubing usually seal after withdrawing the needle. If not, any minor leaks can be closed by applying a thin film of silicone grease on the tubing in the area of the leak. Checks are made to ensure that sufficient medium is present in the syringe barrel to replace the aspirated air. The balance of the EKRB is loaded via the cell loading chamber through the silicone tubing and into the sedimentation chamber, to act as an overlay for the germ cells that are to follow. Valves 1 and 2 are closed when the level of EKRB drops to the top of the Luerlock of the cell loading chamber.

Step 8: Loading of Cells. Germ cells at a concentration of 2 to 5 $\times$ 10^6/ml in 0.5% BSA in EKRB are added into the cell loading chamber, in the volumes indicated (Table II). The concentration of cells can be increased as the operator gains experience. Valve 1 is opened in the direction of the sedimentation chamber, and the time is recorded as zero for the start of the cell separation. Loading of the germ cell suspension into the sedimentation chamber is set at a rate of 10 ml/min, by adjusting

---

the underslung magnetic stirrer. The lower shelf has a square cut out for inserting one of three plates, each with a circular hole suitable for holding the SP-120, SP-180, or SP-240 sedimentation chambers. Each plate has three adjustable leveling screws, one at the rear center and two in the front corners. The fraction collector is placed on the floor of the refrigerator, just below the second shelf.

micrometering valve 2. The flow rate is checked by timing the depletion of medium from the cell loading chamber with the aid of a stopwatch. The flow rate is adjusted carefully; if too slow, the cells will settle onto the base of the sedimentation chamber, and if too fast, the turbulence passing around the stainless steel baffle will disrupt those cells already layered beneath the EKRB. The common error of loading the germ cells retrograde into the fleakers can be avoided by marking the orientation of the stopcock on the external surface, relative to the inlet and outlet, and by keeping the clamps closed on the silicone tubing leading from the two fleakers.

Step 9: Sedimentation of Cells. Immediately after the cells have been applied, valve 1 below the cell loading chamber is switched to allow the gradient of 2–4% BSA to form under the suspension of germ cells in the sedimentation chamber, at a flow rate of 10 ml/min (Fig. 2). The magnetic stirrer is turned on to ensure mixing of the incoming 4% BSA with the 2% BSA in fleaker 2. After 5 min, the flow rate is increased to 40 ml/min, to give a total loading time of about 60 min. The germ cells, now layered at the top of the BSA gradient, sediment at velocities dictated by their respective cross-sectional areas. If the concentration of germ cells is too high, a "raining" effect occurs: the cells cluster into small droplets, like condensation forming on a cold surface. The "droplets" break away and sediment rapidly down through the gradient, giving the appearance of distant rain showers, a phenomenon that can be seen more readily by shining light across the gradient. The raining effect markedly decreases the level of purity achieved, and therefore must be avoided. The 2–4% BSA gradient, rather than 1–3% BSA, is more effective in stabilizing the top of the gradient, and thereby allows higher concentration of cells to be loaded (also see section on Possible Improvements for Separation of Germ Cells). In this regard, careful preparation of the cell suspension is beneficial, as DNA released from damaged cells is detrimental.

Step 10: Assessment of Cell Separation. After approximately 3 hr, the 2–4% BSA gradient is unloaded from the bottom of the sedimentation chamber, via valve 2 through 40 cm of silicone tubing, to a solenoid-activated, drop counter on a fraction collector (Fig. 2). The flow rate is adjusted to 40 ml/min, and the last fraction is collected 4 hr after first loading the cell suspension. Generally, fractions 1 to 20 do not contain cells. Fractions 21 to 100 are centrifuged at 500 g at 5° for 10 min, and the supernatants are aspirated down to 0.5 ml without disturbing the pellet. Cells in odd-numbered tubes are resuspended in the remaining volume. Aliquots are removed from three tubes, placed side-by-side on a microscope slide, and each sample is covered gently with a coverslip. The cells are observed by using differential interference contrast (DIC) light microscopy at a magnification of × 480 and identified on the basis of distinctive morphological features as well as differences in cell size measured with a calibrated reticule in an ocular lens of the microscope (Fig. 3).[12,13] When fractions containing the required cells are located, aliquots from even-numbered tubes are also examined to select all those needed for pooling. Fractions at the upper and lower limits of a peak of a particular cell type can be included or excluded depending on the need for maximal purity versus yield.

Step 11: Assessment of Cell Viability. The pooled populations of germ cells are reexamined to quantify final purity and cell viability, based on morphological features and vital staining. Dead cells stain blue on incubating with 0.4% (w/v) trypan blue in EKRB for 2–3 min. Vital staining can also be undertaken, perhaps more reliably, by using fluorescence microscopy with the fluorochromes fluorescein and ethidium bromide. Viable germ cells stain green on incorporating and hydrolyzing diacetyl-fluorescein to form the impermeant product fluorescein, whereas dead cells stain red on intercalating ethidium bromide into their DNA.

Step 12: Preparation and Counting of Cells. The pooled fractions are centrifuged at 500 g for 10 min at 5° to pellet the cells. The latter are suspended in a known volume of EKRB containing 0.5% BSA, then counted in a hemacytometer by using bright-field microscopy at a magnification of × 160. If necessary, the pooled fractions are washed in EKRB two or three times by repeated centrifugation to remove the 0.5% BSA. The cells then are available for physiological, biochemical, and molecular studies.

Morphological Characteristics of Mouse Germ Cells

Germ cells express characteristic features at different stages of spermatogenesis. The cells may be identified after staining with toluidine blue[17] or periodic acid–Schiff and hematoxylin.[18] It is also possible to identify germ cells by DIC light microscopy, based on differences in their size and morphology.[12-14] Correct identification of the cells by DIC was verified by two complementary methods. First, the distinctive features of germ cells at each stage of spermatogenesis were established by undertaking a detailed ultrastructural study of the developing[13] and adult[12] seminiferous epithelium. Second, germ cells in the isolated populations were characterized and identified by undertaking light and electron microscopy on alter-

[17] D. J. Wolgemuth, E. Gizang-Ginsberg, E. Engelmeyer, B. J. Gavin, and C. Ponzetto, *Gamete Res.* **12,** 1 (1985).

[18] M. L. Meistrich, *Methods Cell Biol.* **15,** 15 (1977).

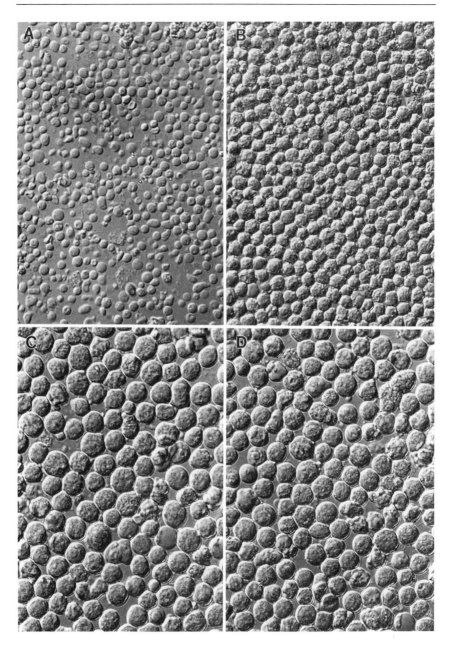

FIG. 3A–D. See legend on p. 100.

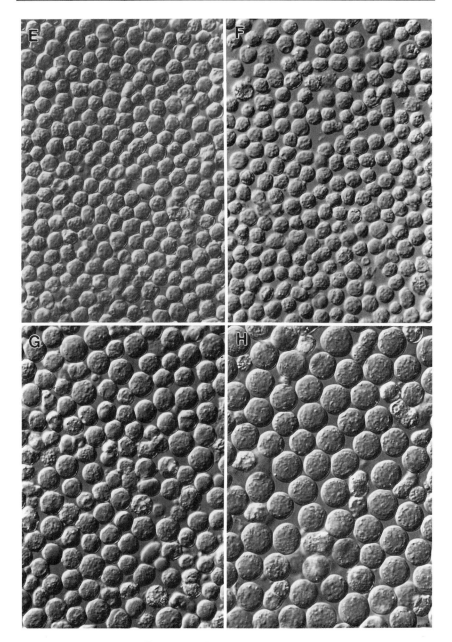

Fig. 3E–H. See legend on p. 100.

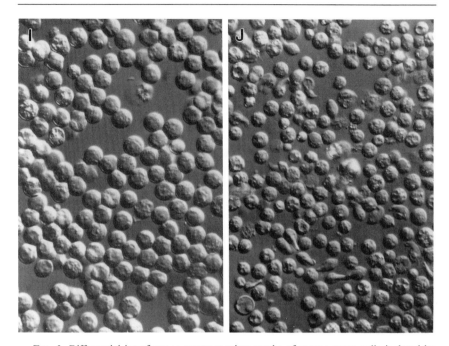

Fig. 3. Differential interference contrast micrographs of mouse germ cells isolated by sedimentation velocity at unit gravity. Testes of animals at the designated ages were dissociated by an incubation in collagenase followed by trypsin with DNase I, in the presence of EKRB. The dispersed germ cells were washed and layered onto a gradient of 2–4% BSA in EKRB, then allowed to sediment for 4 hr at 5°. After the gradient was unloaded, cells in the 10-ml fractions were identified, pooled, and suspended in approximately 0.5 ml of 0.5% BSA in EKRB (see Table III). Aliquots were taken from the different populations and mounted on microscope slides with glycerol. Photographs were taken with Panatomic X film (Eastman Kodak, Rochester, NY) at ASA 125, under DIC by using a Zeiss photomicroscope III. Photomicrographs include (A) cytoplasts; (B) prepuberal Sertoli cells; (C) primitive type A spermatogonia; (D) type A spermatogonia; (E) type B spermatogonia; (F) preleptotene spermatocytes; (G) leptotene/zygotene spermatocytes; (H) pachytene spermatocytes; (I) spermatids (steps 1–8); (J) residual bodies. Magnification: ×850. Modified from Romrell *et al.*,[12] and from Bellvé *et al.*[13] by copyright permission of Rockefeller University Press.

nate thick and thin sections. Based on these studies, the following ages and cellular features (see also Fig. 3) were found useful for isolating and identifying mouse germ cells by DIC (Table III).

Prepuberal Sertoli Cells. Day 6 testes, gradient fractions 66–76. The cells have a diameter of 7.5 to 8.2 μm, an irregular and granular appearance, dense chromatin, and nucleoli close to the nuclear membrane. The distinctive tripartite nucleolus is not evident at this stage of development. Number of cells recovered, 4.2×10^5/testis. Expected purity, over 99%.

TABLE III
COMPOSITION OF CELL POPULATIONS PURIFIED FROM DEVELOPING AND ADULT MOUSE TESTES[a]

Cell type	Cell populations (%)									
	PA	A	B	PL	L/Z	P	RS	CS	RB	S
Primitive type A spermatogonia	92	—	—	—	—	—	—	—	—	≤1
Type A spermatogonia	—	93	5	—	—	≤1	—	—	—	—
Type B spermatogonia	—	3	83	1	1	—	—	—	—	—
Preleptotene spermatocytes	—	—	—	93	7	—	—	—	—	—
Leptotene spermatocytes	—	—	—	3	35	—	—	—	—	—
Zygotene spermatocytes	—	—	—	—	23	≤1	—	—	—	—
Pachytene spermatocytes	—	—	—	—	10	91	1	—	—	—
Spermatids (steps 1–8)	—	—	—	—	—	—	94	5	3	—
Spermatids (steps 12–16)	—	—	—	—	—	—	2	67	6	—
Residual bodies	—	—	—	—	—	—	3	28	91	—
Sertoli cells	8	4	13	3	24	7	—	—	—	≤99

[a] Cell types were identified by applying a combination of light and electron microscopy and by using criteria established in the developmental study of mouse spermatogenesis (see Table I). Those cells (3–5%) remaining unidentified because of necrosis or an unfavorable section were not included in the data. The population of type B spermatogonia may contain intermediate spermatogonia (10- to 11-μm diameter) in S or G_2 phase that could not be identified with certainty. Data are presented as means (S.E. ≤ 8%) drawn from analyzing cell separations reported by Romrell et al.[12] and Bellvé et al.[13,14] and those undertaken by Bellvé through 1992. The cell populations include PA, primitive type A spermatogonia; A, type A spermatogonia; B, type B spermatogonia; PL, preleptotene spermatocytes; L/Z, leptotene/zygotene spermatocytes; P, pachytene spermatocytes; RS, spermatid (steps 1–8); CS, condensing spermatids; RB, residual bodies; S, prepuberal Sertoli cells. Modified from Romrell et al.,[12] and from Bellvé et al.[13,14] by copyright permission of the Rockefeller University Press.

Primitive Type A Spermatogonia. Day 6 testes, gradient fractions 36–50. The cells are spherical and have a diameter of 14 to 16 μm, a glassy to refractile appearance, and two or three nucleoli eccentrically placed in the nucleus. Number of cells recovered, 1.4×10^5/testis. Expected purity, over 90%.

Type A Spermatogonia. Day 8 testes, gradient fractions 40–45. The cells are spherical and have a diameter of 12 to 14 μm, a glassy to refractile appearance, and multiple nucleoli close to the nuclear membrane. Number of cells recovered, 1.6×10^5/testis. Expected purity, over 91%.

Type B Spermatogonia. Day 8 testes, gradient fractions 64–72. The cells are spherical and have a diameter of 8 to 10 μm, a glassy to refractile appearance, and crusts of heterochromatin adjacent to the nuclear membrane. This population may also contain intermediate spermatogonia. Number of cells recovered, 1.5×10^5/testis. Expected purity, over 76%.

Preleptotene Spermatocytes. Day 17 testes, gradient fractions 65–72. The cells are spherical and have a diameter of 7.6 to 8.2 μm, a slightly granular appearance, and fine strands of spiralizing chromatin. Number of cells recovered, 2.1 $\times$ 10^5/testis. Expected purity, over 94%.

Leptotene and Zygotene Spermatocytes. Day 17 testes, gradient fractions 48–56. The cells are spherical and have a diameter of 8.2 to 12 μm, a granular appearance, an increased nuclear to cytoplasmic ratio, and visible strands of thickened chromatin. Number of cells recovered, 7.3 $\times$ 10^5. Expected purity, over 78%, with Sertoli cells forming the major contaminant (Table III).

Pachytene Spermatocytes. Day 17 and adult testes, gradient fractions 25–40. The cells are spherical and have a diameter of 12 to 18 μm, a granular appearance, a thin rim of cytoplasm, a large nucleus with thick strands of bivalent chromosomes, and a prominent XY heterochromatic body adjacent to the nuclear envelope. Number of cells recovered, 7.5 $\times$ 10^5/prepuberal testis and 4.0 $\times$ 10^6/adult testis. Expected purity, over 90%.

Spermatids (Steps 1–8). Adult testes, gradient fractions 60–75. The cells are spherical and have a diameter of 8 to 10 μm, clear cytoplasm, smaller nucleus with homogeneous chromatin, and prominent central nucleolus. A small chromatoid body is located adjacent to the nucleus, and an acrosome is forming on the apical pole of the nucleus. Spermatid tails have been removed by the enzymatic dissociation, although some cells are capable of reassembling a tail in culture. Number of cells recovered, 12 $\times$ 10^6/testis; yield, 20–40 $\times$ 10^6/testis. Expected purity, over 94%.

Spermatids (Steps 12–16). Adult testes, gradient fractions 78–85. The cells are asymmetric and have a diameter of 7–8 μm. The nucleus is eccentrically located in the cell and is undergoing condensation. An acrosome is forming over the nucleus and protrudes from the body of the cell. Number of cells recovered, 25–50 $\times$ 10^6/testis. Expected purity, 54%. Residual bodies/cytoplasts form the principal contaminant.

Residual Bodies/Cytoplasts. Adult testes, gradient fractions 87–95. The small residual bodies or cytoplasts are spherical, but they have an irregular outline and a granular appearance. Residual bodies are anucleate, and the cytoplasm contains small, round vesicle-like structures. Number expected, 40 $\times$ 10^6. Expected purity, 89%.

Possible Improvements for Separation of Germ Cells

Separation of germ cells by sedimentation velocity has withstood the rigorous testing of 15 to 20 years of basic science research. In the formative years, many changes were introduced to improve the purity and integrity of the isolated cells. Further improvements are warranted. The volume of cell sample could be reduced to create a narrower zone to overlay the BSA gradient. This could improve resolution or could lead to equal resolution in a shorter sedimentation distance. Loading cells onto the top of a stable BSA gradient could circumvent problems caused by the turbulence that occurs during bottom loading, or by cells settling onto the chamber and later being swept indiscriminately into fractions while unloading the gradient. Precision top loading of samples to a depth of at least 0.3 mm has been achieved with a rotationally symmetric flow deflector.[19] To circumvent possible streaming, greater densities of cells might be loaded by changing the configuration of the BSA gradient to include an initial step of 1–3% (w/v) BSA in the first few millimeters, followed by 3–6% (w/v) BSA extending through the remaining volume.[2]

Limitations of Sedimentation Velocity at Unit Gravity

First, the protooncogenes *junB* and c-*jun* are activated during the isolation of germ cells, apparently in response to the loss of cell–cell contact.[20] Second, in the adult mouse, spermatogonia comprise less than 3% of all cells in the seminiferous epithelium, and therefore these cells cannot be isolated as a class by sedimentation velocity. Spermatogonial proliferation is also complex. Current models suggest that the sequence consists of types $A_0/A_s \rightarrow A_{pr} \rightarrow A_{al4} \rightarrow A_{al8} \rightarrow (A_{al16}) \rightarrow A_1 \rightarrow A_2 \rightarrow A_3 \rightarrow A_3 \rightarrow A_4 \rightarrow In \rightarrow B$ spermatogonia. Also, at the onset of spermatogenesis in the prepuberal mouse, the sequence involves more cell types than those currently separable (cf. Table III).[13] Separation of additional cell types by sedimentation velocity does not seem feasible. The need to isolate all stages of spermatogonia is stressed by the recent identification of growth and survival factors active on this cell type.[21–22] It might be possible, for instance, to develop antibodies to stage-specific, cell surface components and to apply these to sort spermatogonia by immunological methods.

Germ Cells from Other Species

Guinea pig germ cells have been isolated to high levels of purity from the adult testis, by applying the same refined procedures, to study biogene-

[19] A. T. Tulp, J. G. Collard, A. A. M. Hart, and J. A. Aten, *Anal. Biochem.* **105**, 246 (1980).
[20] A. A. Alcivar, L. E. Hake, M. P. Hardy, and N. B. Hecht, *J. Biol. Chem.* **265**, 20160 (1990).
[21] K. Manova, K. Knocka, P. Besmer, and R. F. Bachvarova, *Dev. Biol.* **110**, 1057 (1990).
[22] Y. S. Martinova, W. Zheng, and A. R. Bellvé, *J. Cell Biol.* **3**, 11a (1992).
[23] Deleted in proof.

sis of the acrosome.[24-26] In this case, Eagle's minimal essential medium, supplemented with 6 mM lactate, 1 mM sodium pyruvate, 2 mM glutamine, 100 μg/ml penicillin, 100 μg/ml streptomycin was used. Germ cells of this species sedimented at greater velocities than those of the mouse, owing to their larger cross-sectional areas. Thus, separation times were reduced to approximately 2 to 2.5 hr. The purities of the spermatocyte and spermatid populations[26] were lower than those achieved for the mouse (cf. Romrell et al.[12]).

Human spermatogenic cells have also been separated into discrete populations.[27] In this case, cell preparation proved to be more difficult, primarily because unidentified, enzyme-resistant, extratubular material prevented effective dispersal of germ cells from the seminiferous epithelium. The authors therefore had to resort to a combination of enzymatic and mechanical methods, with a resultant reduction in the purities of the isolated cell populations.

Rat germ cells also have been isolated to high levels of purity by applying sedimentation velocity at unit gravity, in the presence of PBS.[18] Like the mouse, in the rat greater purities can be achieved by first dissociating the testis with the sequence of collagenase and trypsin, then maintaining the cells in EKRB medium (A. R. Bellvé and P. Tomashefsky, unpublished, 1991). Rat germ cells are also larger than those of the mouse and therefore sediment at slightly greater velocities.

Other Cell Separation Procedures

Other separation methods have been reported with mixed success. Celsep (Wescor, Ogden, UT), a variation of the current sedimentation chamber, uses a comparable BSA gradient to achieve cell separation at unit gravity. It differs in that the chamber can be tilted for loading to decrease the cross-sectional surface area of the gradient. When the chamber is returned to a level position, the germ cells are dispersed over a greater surface area, which might allow more cells to be loaded onto the gradient. Tilting can cause undesirable turbulence, however, and later reduce the effective depth of the gradient. Both effects will compromise resolution of the different cell populations. In one report in which Celsep was used to isolate germ cells from the mouse testis, the average purities were lower: 87.6% for pachytene spermatocytes, 79.8% for early spermatids, and 71.8% for residual bodies.[17] Resolution may also have been

---

[24] C. E. Arboleda and G. L. Gerton, *Dev. Biol.* **125**, 217 (1988).

[25] O. O. Anakwe and G. L. Gerton, *Biol. Reprod.* **42**, 317 (1990).

[26] M. S. Joshi, O. O. Anakwe, and G. L. Gerton, *J. Androl.* **11**, 120 (1990).

[27] R. W. Shepard, C. F. Millette, and W. C. DeWolf, *Gamete Res.* **4**, 487 (1981).

compromised by the use of PBS, which is detrimental to the survival of germ cells (cf. Romrell et al.[12]).

Centrifugal elutriation has been used to separate populations of germ cells of adult rodent and ram testes. This procedure has the advantages of high capacity and speed for isolating early and late spermatids, cells that are recovered at the beginning of the elution sequence.[28,29] However, purification of the larger, fragile pachytene spermatocytes is suboptimal. Purity is generally compromised by contaminating spermatids eluting late in the sequence and by reduced viability, due to either the use of PBS or the shear forces that prevail in the elutriator chamber. Sedimentation velocity and elutriation have been used in sequence to obtain better levels of purity, but the time required to accomplish both methods, the reduced yield, and the associated reduction in cellular integrity offset the advantages.

Culture of Spermatogenic Cells

After being isolated by sedimentation velocity, germ cells can be cultured for periods up to 48 hr, sufficient time to allow anabolic labeling of cellular constituents. Short-term culture has been accomplished by using minimal Eagle's medium, supplemented with 15 mM HEPES, 1 mM sodium pyruvate, 6 mM sodium L-lactate, 100 U/ml pencillin, 100 μg/ml streptomycin, and 10% (v/v) fetal bovine serum (modified MEM).[30] Germ cells are cultured in plates or flasks at densities of 2 × 10^6/cm^2, under 5% CO_2 in air, but at 32° rather than the usual 37°, to avoid the adverse effects of the higher temperature on cellular metabolism. Under these conditions, germ cells remain viable (~85%) for periods of 15 to 20 hr, and they are able to replace surface components degraded by trypsin[31,32] and to reassemble tails.[33] Germ cells also remain viable in MEM lacking glucose and supplemented with 5 mM lactate, 0.4% (w/v) BSA, and nonessential amino acids.[34]

Germ cells cannot be cultured alone for longer intervals, but rather need to be cocultured as fragments or staged segments of tubules or cocultured with Sertoli cells. Viability of germ cells can be enhanced by adding spent medium from Sertoli cell cultures. The active components

[28] R. J. Grabske, S. Lake, B. L. Gledhill, and M. L. Meistrich, J. Cell. Physiol. 86, 177 (1975).

[29] M. Loir and M. Lanneau, Gamete Res. 6, 179 (1982).

[30] G. L. Gerton and C. F. Millette, Biol. Reprod. 35, 1025 (1986).

[31] C. F. Millette and C. T. Moulding, J. Cell Sci. 48, 367 (1981).

[32] C. F. Millette, D. A. O'Brien, and C. F. Millette, J. Cell Sci. 43, 279 (1980).

[33] G. L. Gerton and C. F. Millette, J. Cell Biol. 98, 619 (1984).

[34] J. A. Grootegoed, R. Jansen, and H. J. van der Molen, J. Reprod. Fertil. 77, 99 (1986).

of Sertoli cell spent media, presumably cytokines and/or growth factors, have not yet been identified.[35] However, it has been shown that the seminiferous growth factor (SGF)[36,37] is able to induce DNA synthesis in prospermatogonial stem cells in testicular segments of the day 3 mouse and rat testis.[22,23] In the rat, the percentage of stem cells undergoing DNA synthesis increases from 10 to 50%, a 5-fold induction, over the range of 6 to 12 pM bovine SGF. The effect is down-regulated by Mullerian-inhibiting substance, transforming growth factor β_1, and inhibin, but not by activin. These data demonstrate that growth factors are likely to have important roles in the regulation of mammalian spermatogenesis and therefore will be useful in the development of suitable culture systems.

Biochemical Fractionation of Germ Cells

Germ cells isolated by sedimentation velocity have been used for a variety of biological, biochemical, and molecular studies. Because the procedure is difficult, labor intensive, and expensive, applications were limited until the introduction of more sensitive biological, biochemical, and molecular techniques. Only basic methods peculiar to germ cells are presented here.

Purification of Germ Cell Plasma Membranes

Germ cells have been isolated from the adult mouse testis by sedimentation velocity, as described, and used successfully to purify plasma membrane fractions. Membrane proteins, including native, iodinated, and fucosylated constituents, have been characterized by two-dimensional electrophoresis. Unfortunately, procedures have not been developed for isolating plasma membranes from spermatogonia. The following protocol, taken from Millette *et al.,*[32] can be modified to include suitable protease inhibitors (see section on Isolation of Germ Cell Nuclei below).

Step 1. Purified mouse pachytene spermatocytes, spermatids, or residual bodies are washed in 0.5% BSA in EKRB, to yield 1.25×10^8 cells per tube, but not exceeding 1.8×10^8 cells. The cells are sedimented at 200 g for 5 min in a Beckman TJ-6R centrifuge or equivalent. The supernatant is aspirated, and all traces of the BSA–EKRB are removed from the walls of the tube by careful absorption with a paper towel.

[35] L. S. Kancheva, Y. S. Martinova, and V. D. Georgiev, *Mol. Cell. Endocrinol.* **69,** 121 (1990).

[36] L. A. Feig, A. R. Bellvé, N. Horbach Erickson, and M. Klagsbrun, *Proc. Natl. Acad. Sci. U.S.A.* **77,** 4774 (1980).

[37] W. Zheng, T. J. Butwell, L. Heckert, M. D. Griswold, and A. R. Bellvé, *Growth Factors* **3,** 73 (1990).

Step 2. The germ cells are swollen in hypotonic medium by the addition of a 1/10 dilution of 0.16 M NaCl, 3 mM MgCl$_2$, 5 mM KCl in 10 mM Tris-HCl (pH 7.4) (TBSS), at 4°. After addition of diluted TBSS, the cells are suspended by vigorous vortexing and incubated for exactly 5 min, at 4°.

Step 3. The cell suspensions are homogenized by three strokes in a small, glass Dounce homogenizer with a Teflon pestle (Thomas Scientific, Swedesboro, NJ, Size 0, Cat. No. 3431-E04), having a clearance of 0.05 to 0.1 mm. Immediately afterward, 0.08 ml of 10× TBSS is added to the sample to restore isotonicity and thereby protect the nuclei.

Step 4. The samples are placed in a small plastic tube and centrifuged at 1000 g for 30 sec in a microcentrifuge at 4°. Supernatants are carefully aspirated and transferred to a glass tube. The pellets are resuspended by adding 0.8 ml TBSS and centrifuged at 1000 g for another 10 sec. The second supernatant is pooled with the first supernatant to yield a total volume of 1.6 ml. The samples of enriched plasma membranes are held at 4° while preparing the sucrose gradients.

Step 5. Plasma membranes are purified by placing each sample in a discontinuous sucrose gradient in TBSS. An aliquot of 1.5 ml of plasma membrane suspension is mixed with an equal volume of 80% (w/v) sucrose in TBSS, to yield 3.0 ml of 40% sucrose in TBSS. Each preparation is layered over 3.0 ml of 45% (w/v) sucrose in TBSS in a cellulose nitrate tube. Then, 5 ml of 30% (w/v) sucrose in TBSS is layered carefully over the sample, followed by 1 ml of TBSS.

Step 6. The plasma membrane preparations are centrifuged in an SW 41 rotor at 125,240 g_{av} (32,000 rpm) for 90 min, by using a Beckman L8-M centrifuge or equivalent. The gradients are collected in 0.2-ml fractions. The latter are assayed as required for enzyme activity and by electron microscopy to identify the different membrane compartments.[32]

Step 7. Purified membranes are recovered in bands 1, 2, and 3 at the interfaces of the 0–30% sucrose, 30–40% sucrose, and 45–40% sucrose, respectively. Smooth plasma membranes vesicles (0.4–1.7 μm diameter) are recovered primarily in band 2 at the 30–40% sucrose interface. This conclusion is based on the distribution of ^{125}I-labeled lectins previously bound to the surfaces of intact cells, on the identification by electron microscopy of ribosome-free unit membranes, and on the detection of enzyme activities usually located at the cell surface. The anticipated yield of membrane protein is 280 μg protein from 10^8 germ cells per gradient.[32]

Isolation of Germ Cell Nuclei

Nuclei have been prepared from germ cells at all stages of spermatogenesis, from spermatogonia to spermatozoa. Procedures for preparing nuclei from mature spermatozoa are considered elsewhere in this volume.[37a]

[37a] A. R. Bellvé, W. Zheng, and Y. S. Martinova, this volume [7].

Step 1. Germ cells are isolated by sedimentation velocity, pooled, and washed in EKRB to remove residual BSA. The cells are suspended in 17 mM MgCl$_2$, 40 mM KCl, protease inhibitor cocktail (PIC) in 20 mM Tris-HCl (pH 7.7), for 1 min at 4°. PIC components are added to final concentrations of 5 μg/ml antipain, 0.5 μg/ml aprotinin, 2 mM benzamidine, 0.5 mM EDTA, 0.5 mM EGTA, 0.5 μg/ml leupeptin, 1 μg/ml pepstatin, and 1 mM phenylmethylsulfonyl chloride (PMSF; pH $\geq$ 7.0) or 10 μM p-aminophenylsulfonyl chloride (p-APMSF; pH $\leq$ 7.0). An optimal combination of inhibitors must be determined experimentally and will depend on the susceptibility of the particular protein(s) being studied. Crude nuclei are sedimented at 250 g for 5 min at 4°. Note that prior addition of 100 μg/ml soybean trypsin inhibitor (STI), immediately after dissociating the seminiferous tubules with trypsin, is critical for inhibiting activity of enzyme still adhering to the cell surface (see Step 5, p. 93).

Step 2. The crude nuclear pellet is suspended in 1% Triton X-100, 0.32 M sucrose, 10 mM MgCl$_2$, 40 mM KCl, PIC, 20 mM Tris-HCl (pH 7.7) (MKPT), for 10 min, at 4°. A 1-ml sample is layered over 1 ml of 1.1 M sucrose in MKPT, and centrifuged at 5000 g for 10 min at 4°. The pure nuclear pellet, as judged by electron microscopy,[38] is washed once in 0.32 M sucrose in MKPT, sedimented for 10 min at 135 g at 4°, and resuspended in the same buffer. The nuclei are used directly or stored at $-70°$ until required.

Preparation of Germ Cell Nuclear Matrices

Isolation of germ cell nuclear matrices involves the extraction of DNA, RNA, and protein.[38] Germ cell nuclei differ in structural stability at different stages of spermatogenesis, most likely owing to changes in components of the nuclear lamina. Consequently, three different procedures have evolved for preparing nuclear matrices from cells at specific stages of differentiation.[38,39] As presented here, procedure 1 is used on spermatogonia to preleptotene spermatocytes; procedure 2 is for preleptotene spermatocytes to spermatids (modified after procedure II of Ierardi *et al.*[38]); a different procedure again is used to prepare perinuclear matrices from mature sperm recovered from the epididymus and vas deferens.[37a,40]

Procedure 1. Nuclei from purified populations of type A spermatogonia, type B spermatogonia, or preleptotene spermatocytes are incubated in 50 μg/ml DNase I, 50 μg/ml RNase, 100 μg/ml STI, PIC, 10 mM MgCl$_2$ in 20 mM Tris-HCl (pH 7.4), for 15 min, at 22°. During the next 30 min, the total concentration of CaCl$_2$–MgCl$_2$ (3 : 2, mol/mol) is increased to 25, 50, 75, 100, and finally to 125 mM, by adding the appropriate volumes

[38] L. A. Ierardi, S. B. Moss, and A. R. Bellvé, *J. Cell Biol.* **96,** 1717 (1983).
[39] A. R. Bellvé, R. Chandrika, and A. H. Barth, *J. Cell Sci.* **96,** 745 (1990).
[40] A. R. Bellvé, R. Chandrika, Y. S. Martinova, and A. H. Barth, *Biol. Reprod.* **47,** 451 (1992).

from a stock solution, in 20 mM Tris-HCl (pH 7.4). The preparation is made 1.0 M in NaCl, incubated for another 15 min at 4°, and the residual nuclear structures are sedimented by centrifugation at 1000 g for 10 min.

The extracted nuclei in the pellet are suspended in 50 μg/ml DNase I, 50 μg/ml RNase, 125 mM CaCl$_2$–MgCl$_2$ (3 : 2, mol/mol), 100 μg/ml STI, PIC in 20 mM Tris-HCl (pH 7.4) and incubated for 30 min at 22°. The preparation is made 1.0 M in NaCl, incubated for 15 min at 4°, and pelleted at 1000 g for 5 min. Nuclear matrices are recovered after a final wash in 20 mM Tris-HCl (pH 7.4) at 4°.

Procedure 2. Preleptotene, peptotene-zygotene, and packytene spermatocytes, round spermatids (steps 1–8), and condensing spermatids (steps 12–16) are incubated in 50 μg/ml DNase I, 50 μg/ml RNase, 100 μg/ml STI, PIC, 10 mM MgCl$_2$ in 20 mM Tris-HCl (pH 7.4), for 60 min, at 22°. This preparation is sedimented by centrifugation at 500 g for 5 min at 4°.

The partially extracted nuclei in the pellet are resuspended in 1 M NaCl, 10 mM MgCl$_2$, 100 μg/ml STI, PIC, in 20 mM Tris-HCl (pH 7.4), and incubated for 30 min at 4°. After centrifugation at 800 g for 10 min at 4°, nuclei in the pellet are subjected again to another sequence of DNase–RNase digestion, NaCl extraction, followed by a final DNase–RNase digestion. Nuclear matrices are sedimented at 800 g for 15 min at 4°, and washed once in 20 mM Tris-HCl (pH 7.4) at 4°.

Characterization of Germ Cell Nuclear Matrices. Germ cell nuclear matrices have been characterized by light and electron microscopy, ethidium bromide staining, protein, DNA, and RNA analyses, and protein composition following sodium dodecyl sulfate–polyacrylamide gel electrophoresis (SDS–PAGE) and silver staining.[38] Extraction of germ cell nuclei to form matrices removes approximately 85% of the nuclear protein, 99% of the DNA, and 97% of the RNA. Nuclear matrix proteins range from M_r 8000 to 150,000 and undergo stage-specific changes as the germ cell differentiates from spermatogonium, spermatocyte, spermatid, to mature sperm (cf. Figs. 1 and 4). In addition, the germ cell matrix differs from the somatic counterpart, in lacking or having a reduced complement of somatic lamin proteins, lamins A, B, and C,[41a] and containing a number of novel germ cell-specific components (Fig. 4). The sperm nuclear matrix, in particular, differs substantially from those of cells at earlier stages of differentiation.[38–40]

Concluding Remarks

Utilization of germ cells, isolated by sedimentation velocity at unit gravity, has led to major advances in our understanding of the differentia-

---

[41a] S. B. Moss, B. L. Burnham, and A. R. Bellvé, *Mol. Reprod. Develop.* **34**, 164 (1993).

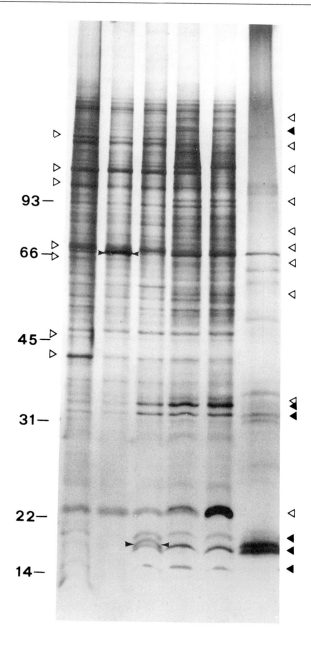

G　PL　LZ　P　R　S

tive processes of mammalian spermatogenesis. Pioneering electron microscopic studies had described the unusual morphological events occurring
in germ cells. These changes are now known to be attended by novel
components, either unique or in variant forms, that are not expressed in
somatic cells. Novel plasma membrane constituents have been identified
and characterized by immunological procedures[41] and two-dimensional
PAGE[42] and have been shown to undergo selective partitioning during
spermatogenesis.[41] Temporal changes also occur among the nuclear matrix
proteins during germ cell differentiation.[38] Synthesis of germ cell-specific
histones, TH2A (rat) and H2S (mouse), has been shown to occur uncoupled with DNA synthesis during meiotic prophase.[43–45] A family of novel,
immunologically related proteins, the thecins, have been localized to the
perinuclear theca of spermatids and maturing sperm and appear to undergo
specific endoproteolysis and relocalization on the surface of the nucleus
during spermiogenesis and epididymal maturation.[39,40]

   In the context of molecular biology, subtractive cDNA libraries have
been prepared from poly(A)$^+$ of isolated germ cells and used to define
patterns of differential gene expression during mouse spermatogenesis,[46]
including the three isozymes of lactate dehydrogenases, LDH A, B, and
C.[47] The relative abundance of HRPT, PGK-1, PGK-2, APRT, MTase,
and *Zfy* gene products have been compared by applying the quantitative

[41] C. F. Millette and A. R. Bellvé, *Dev. Biol.* **79**, 309 (1980).
[42] C. F. Millette and C. T. Moulding, *J. Cell Sci.* **48**, 367 (1980).
[43] W. A. Brock, P. K. Trostle, and M. L. Meistrich, *Proc. Natl. Acad. Sci. U.S.A.* **77**, 371 (1980).
[44] L. R. Bucci, W. A. Brock, and M. L. Meistrich, *Exp. Cell Res.* **140**, 111 (1982).
[45] Y. M. Bhatnagar , L. J. Romrell, and A. R. Bellvé, *Biol. Reprod.* **32**, 599 (1985).
[46] K. H. Thomas, T. M. Wilkie, P. Tomashefsky, A. R. Bellvé, and M. I. Simon, *Biol. Reprod.* **41**, 729 (1989).
[47] K. H. Thomas, J. Del Mazio, P. Eversole, A. R. Bellvé, S. L. Li, and M. I. Simon, *Development (Cambridge, UK)* **109**, 483 (1990).

FIG. 4. Electrophoretic analysis of nuclear matrix proteins prepared from germ cells at
different stages of mouse spermatogenesis. Discrete classes of germ cells were isolated from
testes at the designated postnatal ages, by sedimentation velocity at unit gravity (see text).
Nuclei were purified by sucrose gradient centrifugation, and then nuclear matrices were
prepared by salt extraction and DNase digestion, as described. The proteins were solubilized
in sample buffer, subjected to SDS–PAGE, and stained with silver. Solid and hollow arrowheads, respectively, signify proteins that decrease or increase in relative abundance as
spermatogenesis progresses. Samples included type A and type B spermatogonia (G), preleptotene spermatocytes (PL), leptotene/zygotene spermatocytes (LZ), pachytene spermatocytes (P), round spermatids steps 1–8 (R), and spermatozoa (S). (L. A. Ierardi, S. B. Moss,
and A. R. Bellvé, unpublished, 1985.)

polymerase chain reaction (PCR).[48] The temporal expression of several other gene products has been shown to occur transiently during meiotic prophase and/or spermiogenesis. These include germ cell-specific variants of the structural proteins α- and β-tubulin[49,50] and actin,[50,51] the heat-shock protein 70 gene family,[52,53] the homeobox-containing gene Hox 1.4,[54] the protooncogenes c-*myc*,[55,56] c-*abl*,[57,58] c-*mos*,[59-61] c-*raf*, c-*ras*, c-*jun*, and c-*fos*.[52] Several novel zinc-finger proteins, all potential transcription factors, also have been shown to be expressed transiently during mouse spermatogenesis, by using the isolated germ cells.[62-64] Last, the gene for the CREMτ (cyclic AMP-responsive element modulator) is expressed in premeiotic cells, with a splicing event leading to an antagonistic form of the regulatory protein.[65] In spermatocytes and spermatids, however, an alternate splicing leads to the abundant accumulation of the CREMτ activator. Thus, expression of CREMτ gene via splicing-dependent reversal provides an effective means of modulating germ cell differentiation.

Acknowledgments

The author gratefully extends appreciation to Drs. Y. Mohan Bhatnagar, Juan-Carlos Cavicchia, Martin D. Dym, Donald W. Fawcett, Lynn A. Ierardi, Clarke F. Millette, Deborah

[48] J. Singer-Sam, M. O. Robinson, A. R. Bellvé, M. I. Simon, and A. D. Riggs, *Nucleic Acids Res.* **18**, 1255 (1990).
[49] R. J. Distel *et al.* (1984).
[50] N. B. Hecht, K. C. Kleene, R. J. Distel, and L. M. Silver, *Exp. Cell Res.* **153**, 275 (1984).
[51] S. H. Waters, R. J. Distel, and N. B. Hecht, *Mol. Cell. Biol.* **5**, 1649 (1985).
[52] Z. F. Zakeri, D. J. Wolgemuth, and C. Hunt, *Mol. Cell. Biol.* **8**, 2925 (1988).
[53] Z. F. Zakeri, W. Welch, and D. J. Wolgemuth, *J. Cell Biol.* **111**, 1785 (1990).
[54] D. J. Wolgemuth, C. Viviano, E. Gizang-Ginsberg, M. Frohman, A. Joyner, and G. R. Martin, *Proc. Natl. Acad. Sci. U.S.A.* **84**, 5813 (1987).
[55] T. A. Stewart, A. R. Bellvé, and P. Leder, *Science* **226**, 707 (1984).
[56] H. Wolfes, K. Kogawa, C. F. Millette, and G. M. Cooper, *Science* **245**, 740 (1989).
[57] C. Ponzetto and D. J. Wolgemuth, *Mol. Cell. Biol.* **5**, 1791 (1985).
[58] C. Ponzetto, A. G. Wadewitz, A. M. Pedergast, O. N. Witte, and D. J. Wolgemuth, *Oncogene* **4**, 685 (1989).
[59] D. S. Goldman, A. A. Kiessling, C. F. Millette, and G. M. Cooper, *Proc. Natl. Acad. Sci. U.S.A.* **84**, 4509 (1987).
[60] G. L. Mutter and D. J. Wolgemuth, *Proc. Natl. Acad. Sci. U.S.A.* **84**, 5301 (1987).
[61] N. K. Herzog, B. Singh, J. Elder, I. Lipkin, R. J. Trauger, C. F. Millette, D. S. Goldman, H. Heiner, G. M. Cooper, and R. B. Arlinghaus, *Oncogene* **3**, 225 (1988).
[62] V. Cunliffe, P. Koopman, A. McLaren, and J. Trowsdale, *EMBO J.* **9**, 197 (1990).
[63] P. Denny and A. Ashworth, *Gene* **106**, 221 (1991).
[64] C. Hogg, M. Schalling, E. Grunder-Brundell, and B. Daneholt, *Mol. Reprod. Dev.* **30**, 173 (1991).
[65] N. S. Foulkes, B. Mellstrom, E. Benusiglio, and P. Sassone-Corsi, *Nature (London)* **355**, 80 (1992).

A. O'Brien, Stuart B. Moss, and Lynn J. Romrell, who made valuable contributions toward the development of techniques and research presented in this article. Many thanks also to Martin Seidensticker for his valuable comments on reviewing the manuscript. Preparation of this chapter was supported by National Institute of Child Health and Human Development Center Grant R01 HD 05077 and by Rockefeller Grant RF GA PS 9229.

[7] Recovery, Capacitation, Acrosome Reaction, and Fractionation of Sperm

By Anthony R. Bellvé, Wenxin Zheng, and Yordanka S. Martinova

Introduction

Sperm on leaving the testis pass through the caput, corpus, and cauda epididymis and into the vas deferens. While in transit the cells undergo maturation, acquiring the capacity of forward movement and becoming competent to fertilize an oocyte.[1,2] The biochemical events underlying these physiological processes are largely unknown. Components of the germ cell plasma membrane are modified during the early phase of sperm maturation. Surface glycoproteins undergo posttranslational changes and/ or become distributed to different domains on sperm of the mouse,[3] rat,[4,5] and guinea pig.[6] These changes are likely to involve endoproteolytic processing during early epididymal maturation.[6] In other instances, the sperm cell absorbs constituents from the surrounding epididymal secretions onto different domains of the cell surface.[7]

Proteins of the sperm nucleus and the enveloping perinuclear theca also undergo changes during epididymal maturation. Nuclear proteins of rat sperm, including the protamines, form intra- and intermolecular cystinyl bonds as the cells are transported through the epididymis.[8,9] The covalent bonds provide stability to the remodeled and condensed sperm

[1] A. R. Bellvé, *Oxford Rev. Reprod. Biol.* **1,** 159 (1979).

[2] A. R. Bellvé and D. A. O'Brien, *in* "Mechanisms of Mammalian Fertilization" (J. F. Hartman, ed.), p. 55. Academic Press, New York, 1983.

[3] K. L. Lakoski, C. P. Carron, C. L. Cabot, and P. M. Saling, *Biol. Reprod.* **38,** 221 (1988).

[4] C. R. Brown, K. I. von Glos, and R. Jones, *J. Cell Biol.* **96,** 256 (1983).

[5] D. M. Phillips, R. Jones, and R. Salgi, *Mol. Reprod. Dev.* **29,** 347 (1991).

[6] B. M. Phelps, D. E. Koppel, P. Primakoff, and D. G. Myles, *J. Cell Biol.* **111,** 1839 (1990).

[7] J. L. Daucheux and J. K. Voglmayr, *Biol. Reprod.* **29,** 1033 (1983).

[8] Y. Marushige and K. Marushige, *Biochim. Biophys. Acta* **340,** 498 (1974).

[9] Y. Marushige and K. Marushige, *J. Biol. Chem.* **250,** 39 (1975).

Copyright © 1993 by Academic Press, Inc.
All rights of reproduction in any form reserved.

chromatin. More intriguing transitions occur among other proteins associated with the nucleus. During mouse spermiogenesis, the thecins, a family of immunologically related proteins, are expressed on the anterior pole of the spermatid nucleus.[9a] At the onset of sperm maturation in the caput epididymis, the thecins of M_r 80,000, 77,000, and 75,000 undergo an apparent endoproteolytic processing to yield a 48,800 M_r protein(s). This truncated polypeptide assumes a bipolar distribution, with half of the protein molecules becoming restricted to an anteriodorsal margin and the others to a narrow posterioventral zone on the nucleus.[10] The processing and apparent relocalization events occur when the cell is transcriptionally and translationally inactive.

In this chapter, procedures are outlined for recovering mouse sperm from different regions of the excurrent ducts, namely, the caput, corpus, and cauda epididymides and the vas deferens. Different methods are described for purifying sperm heads, nuclei, and tails, for selectively solubilizing perinuclear theca proteins, and for preparing perinuclear matrices that retain the original conformation of the sperm nucleus. Operationally, for current purposes, the perinuclear theca refers to those cytoplasmic elements remaining on the nucleus after extraction of the cell with 1% (w/v) sodium dodecyl sulfate (SDS) (Fig. 1). Thereafter, the perinuclear theca can be selectively removed from the nucleus by a cationic detergent under reducing conditions. The perinuclear matrix is comprised of the theca and intranuclear transverse fibers that remain after the protamines and DNA have been displaced from the nucleus by an incubation with high salt and DNase I.[10]

Recovery of Spermatozoa from Excurrent Ducts

Sperm can be recovered from all regions of the mouse epididymides and vasa deferentia. The number of cells generally obtainable from the caput/corpus epididymis, cauda epididymidis, and vas deferens of 12- to 16-week-old mice is approximately 2×10^6, 25×10^6, and 5×10^6, respectively.[11] Less sperm can be recovered from 8- to 10-week-old animals, and perhaps 2- to 3-fold more from retired breeders. For physiological studies, care is needed to ensure the cells remain viable during recovery and preparation. For studies involving biochemical fractionation, strong denaturing and reducing conditions are required to solubilize structural components, owing to their stabilization by intra- and intermolecular disulfide bonds that are formed early in sperm maturation.[9] Thus, sperm recov-

[9a] A. R. Bellvé, R. Chandrika, and A. H. Barth, *J. Cell Sci.* **9b,** 745 (1990).
[10] A. R. Bellvé, R. Chandrika, Y. S. Martinova, and A. H. Barth, *Biol. Reprod.* **47,** 451 (1992).
[11] A. R. Bellvé, E. Anderson, and L. Hanley-Bowdoin, *Dev. Biol.* **47,** 349 (1975).

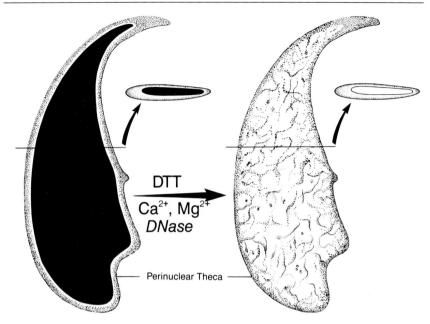

Fig. 1. Diagram of a mouse sperm nucleus and a perinuclear matrix. *Left*: Nucleus after separation from the sperm cell by treatment with 1% (w/v) SDS and S-EDTA (see text) and purification by centrifugation over a 1.6 M sucrose gradient. The nucleus consists of dense nucleoprotamine, and it is surrounded by elements of the perinuclear theca. *Right*: Perinuclear matrix after treatment of the nucleus with 25 mM DTT, 20 mM Tris-HCl (pH 8.3) for 20 min, followed by 50 μg/ml DNase I and $CaCl_2$–$MgCl_2$ (3 : 2, mol/mol), increasing stepwise in concentration from 125, 150, 175, 200, to 250 mM, at 10-min intervals. During this treatment protamines P1 and P2 are displaced from the nucleus, whereas the nonprotamines remain associated quantitatively with the perinuclear matrix. The resulting perinuclear matrices remain the same in size and configuration and are composed of elements of the cytoskeletal perinuclear theca and a network of karyoskeletal fibers. This structure consists of over 230 protein species, some of which are known to occur in distinct regional domains. (Modified after Bellvé *et al.*[10] Reproduced by copyright permission of *Biology of Reproduction* and Biologists, Ltd.)

ered from the various regions of the excurrent ducts differ in structural resilience. Cocktails of effective protease inhibitors are required to neutralize hydrolytic enzymes that are released on disrupting the acrosome,[10] particularly since the denaturing conditions used to dissociate the sperm also make proteins better substrates for proteases. It is with these caveats in mind that the following methods are presented.

Step 1. Mice are anesthetized by CO_2 inhalation and euthanized by cervical dislocation. The abdominal area is swabbed with cotton previously immersed in 70% ethanol. A transverse incision is made in the caudal abdominal skin, which then is drawn cranially. A similar incision is made in the underlying peritoneum. Both testes are brought forward from the scrotal sac. While holding the epididymal fat pad with a pair of forceps, light tension is applied to straighten out each vas deferens through to the inguinal canal. The epididymis and vas deferens are removed as a unit from the lateral wall of each testis by cutting the intervening tissue. When required for physiological studies, the excurrent ducts are held in enriched Krebs–Ringer bicarbonate medium (EKRB) (Table I) under paraffin oil, 5% CO_2 in air, at 37°. When required for biochemical fractionation, the ducts are held in phosphate-buffered saline (PBS) (pH 7.4) in a petri dish, at 4°.

Step 2. Sperm are expressed from the lumen of the vas deferens into PBS (pH 7.4, 4°) by gently sliding lightly compressed, smooth forceps along the length of the duct. Sperm of the cauda epididymis are recovered by cutting the duct into 1-mm segments and placing the segments into 5 to 10 ml of PBS (4°), depending on the number of epididymides. For studies on sperm maturation, caput and corpus epididymides are recovered separately and, if required, divided further into subregions. All cell and tissue preparations are vortexed briefly. Clumps of tissue are allowed to settle at unit gravity, and sperm in the supernatant are decanted. The tissue is resuspended in PBS, vortexed again, and allowed to settle. The second supernatant is decanted and pooled with the first supernatant. The sperm suspension is passed through a Nitex nylon monofilament mesh (80 μm; Tetko, Inc. Lancaster, NY) to remove residual clumps of somatic cells. Sperm are recovered by centrifugation at 500 g, for 10 min, at 4°.

Purification of Sperm Heads, Nuclei, and Tails

Sperm are resistant to most biochemical extraction techniques owing to the sclerotic nature of their structural elements. These include the following: in the sperm head, the polarized and highly condensed nucleus and the perinuclear theca; and in the tail, the basal plate, capitulum, mitochondrial shells, outer dense fibers, and fibrous sheath. Consequently, the cells can only be fractionated by using strong denaturing and reducing conditions[11,12] procedures that generally are not applied to somatic cells. The protocols presented below focus on isolating and fractionating mouse sperm heads or nuclei, and they are not always applicable to sperm of other species. Techniques for subfractionating tail structures are presented elsewhere.[13–15]

[12] A. R. Bellvé, D. J. McKay, B. S. Renaux, and D. H. Dixon, *Biochemistry* **27**, 2890 (1988).
[13] G. E. Olson and V. P. Winfrey, *J. Ultrastruct. Mol. Struct. Res.* **94**, 131 (1986).
[14] G. E. Olson and V. P. Winfrey, *J. Struct. Biol.* **103**, 13 (1990).

The following procedures are used to prepare (1) intact sperm heads and tails; (2) Triton-treated nuclei; (3) SDS-treated nuclei and tails; and (4) cetyltrimethylammonium bromide (CTAB)-treated nuclei. The procedures yield distinct products that are useful for different experimental purposes. The recommended protease inhibitor cocktail (PIC) consists of 1 mM phenylmethylsulfonyl fluoride (PMSF), 0.5 mM EGTA, 1.0 μg/ml aprotinin, 0.5 μg/ml leupeptin, 1.0 μg/ml pepstatin, and 5.0 μg/ml antipain.[10] Other inhibitors may be needed, depending on the type of protease activity that is encountered.

Isolation of Sperm Heads and Tails

Step 1. Mouse sperm are decapitated by an incubation with trypsin,[16] and the heads and tails are separated by isopycnic centrifugation on a gradient of metrizamide {[2-(3-acetamido-5-N-methylacetamido-2,4,6-triiodobenzamido)]-2-deoxy-D-glucose}.[17] Metrizamide forms solutions of high density and low viscosity, and is therefore useful for separating cells or their subcellular components. Sperm are collected in PBS and the cell concentration is adjusted to about 10^7/ml. Trypsin is added to a final concentration of 100 μg/ml, and the sperm are homogenized in a loose-fitting Dounce homogenizer for 1 to 2 min, at 22°, until the heads and tails are separated. A 10-fold excess of soybean trypsin inhibitor is added immediately, and the samples are placed on ice, to inhibit further tryptic activity. The heads and tails are washed 3 times in PIC, 50 mM Tris-HCl (pH 7.4), at 4°.

Step 2. The sperm preparation is subjected to centrifugation on a gradient of 40 to 60% metrizamide.[17] The gradient is prepared by diluting a stock solution of 90% metrizamide, with PIC, in 50 mM Tris-HCl (pH 7.4), to concentrations of 82.5, 75.0, 67.5, and 60.0%. The preparation of heads and tails is mixed with each of these solutions in a 1:2 ratio by volume, and 1 ml of each mixture is layered successively, beginning with the most dense solution, into 8.5 × 1.4-cm cellulose nitrate tubes. The samples are centrifuged in a Beckman L8-M centrifuge, equipped with an SW 41 Ti rotor, at 76,000 rpm, for 14 hr, at 4°.

Step 3. During centrifugation, the heads and tails respectively form narrow bands with buoyant densities corresponding to 62.5–64.0 and 45.5–47.0% metrizamide, based on measurements of the refractive index.[17] The side of each centrifuge tube is punctured with a syringe, and the heads and tails are recovered by aspiration. The heads are washed twice

[15] G. E. Olson and V. P. Winfrey, *Mol. Reprod. Dev.* **33**, 89 (1992).

[16] G. M. Edelman and C. F. Millette, *Proc. Natl. Acad. Sci. U.S.A.* **68**, 2436 (1971).

[17] F. M. Bradley, B. Meth, and A. R. Bellvé, *Biol. Reprod.* **24**, 691 (1981).

TABLE I
COMPOSITION OF MODIFIED EKRB MEDIUM FOR
THE RECOVERY CAPACITATION,[a] AND THE ACROSOME
REACTION OF MOUSE SPERM[b]

Component	Concentration (mM)	Amount (g/liter)
NaCl	120.00	7.012
KCl	2.00	0.149
$CaCl_2 \cdot 2H_2O$	1.70	0.250
$NaHCO_3$	25.00	2.100
$MgSO_4 \cdot 7H_2O$	1.20	0.296
$NaH_2PO_4 \cdot H_2O$	0.36	0.043
D-Glucose	5.60	1.009
Sodium pyruvate	1.10	0.121
Sucrose	18.50	6.333
TAPSO[c]	10.00	2.383
Pencillin G (sodium salt)	50.00 units	—
Streptomycin sulfate	50.00 units	—

[a] Composition of medium is from J. M. Neill and P. J. Olds-Clarke *Gamete Res.* **18,** 121 (1987) as modified by P. J. Olds-Clarke and R. Sego *Biol. Reprod.* **47,** 629 (1992). The pH of the medium is adjusted to 7.3 to 7.4 by aeration with humidified CO_2. Lactate has been omitted, because it has been shown to have an inhibitory effect on capacitation [J. M. Neill and P. J. Olds-Clarke, *Gamete Res.* **18,** 121 (1987)].
[b] Capacitation of sperm is induced by addition to the basic medium of 2% BSA (20 mg/ml BSA).
[c] TAPSO, 3-[N-Tris(hydroxymethyl)methylamino]-2-hydroxypropanesulfonic acid.

in 10 mM $MgCl_2$, 40 mM KCl, PIC, 50 mM Tris-HCl (MKPT), and treated for 15 min with 1% Triton X-100 in MKPT, at 4°, to remove soluble proteins.

Sperm heads and tails are recovered to purities of over 99 and 97%, respectively.[17] When examined by electron microscopy, the isolated heads from trypsin-treated sperm are seen to be intact, with a continuous plasma membrane and an attached basal plate. Purified tails may have discontinuities in the plasma membrane over the midpiece, but other structures of the tail are normal in morphology, except perhaps for some minor spatial rearrangement of the microtubules and the outer dense fibers. Few differences are evident by SDS–polyacrylamide gel electrophoresis (SDS–PAGE) and isoelectric focusing, when comparing proteins from tails derived by trypsin- versus SDS-induced cleavage, after the former have also been exposed to the detergent. This comparison indicates that trypsin does not penetrate the cell readily during the brief period of incubation. It is still possible that some tail constituents, particularly proteins of the plasma membrane, might be sensitive to the brief enzymatic digestion.

Isolation of Sperm Nuclei

Three procedures have been developed for isolating sperm nuclei. Procedure A, modified after Meistrich *et al.*,[18] involves sonication of sperm in the presence of Triton X-100. The procedure completely solubilizes the tail, but it is useful for isolating sperm nuclei at different stages of testicular and epididymal maturation. Procedure B, from Bellvé *et al.*,[9a-11] involves decapitating sperm with SDS, followed by separation of solubilized proteins, and insoluble tails and nuclei by sucrose density-gradient centrifugation. The latter procedure can only be applied to mature cells of the cauda epididymis and vas deferens, after all intra- and intermolecular disulfide bonds have formed among constituent proteins. Procedure C, modified from Balhorn *et al*,[19] utilizes CTAB and dithiothreitol (DTT) to solubilize all sperm structures other than the nucleus, which then can be purified by centrifugation.

Procedure A. Sperm are recovered in EKRB from different regions of the excurrent ducts, as described, and transferred to 1% Triton X-100 in MKPT (pH 7.7).[17] The samples are sonicated on ice, for three 10-sec bursts, each at 40 W. Sperm are examined by light microscopy to ensure the cells are decapitated and the tails fragmented. If necessary, the treatment is repeated.

The sperm preparation is layered over 1.1 *M* sucrose in MKPT in 50-ml polycarbonate centrifuge tubes and centrifuged at 5000 *g*, for 30 min. The supernatant of soluble proteins and residual tail fragments is decanted. The pellet of nuclei is recovered and washed twice in MKPT, by centrifugation. The purified nuclei are intact, with remnants of the inner acrosomal membrane, perinuclear theca, and nuclear envelope, but are devoid of tails and acrosomes.

Procedure B. Mouse sperm are decapitated by using 1% SDS, and the resulting nuclei and tails are separated by sucrose density-gradient centrifugation. SDS will also decapitate sperm of the guinea pig, rabbit, and human, but not those of the rat and bull.[11] In this protocol, sperm

[18] M. L. Meistrich, B. O. Reid, and W. J. Barcellona, *J. Cell Biol.* **64**, 211 (1975).
[19] R. Balhorn, G. L. Gledhill, and A. J. Wyrobek, *Biochemistry* **16**, 4075 (1977).
[20] M. Lalli and Y. Clermont, *J. Anat.* **160**, 419 (1981).

are washed 3 times in 75 mM NaCl, 24 mM EDTA, PIC, pH 6.0 (S-EDTA), by alternate suspension and centrifugation at 1500 g, for 15 min, at 4°. This lyses the few contaminating erythrocytes and epithelial cells. After the last wash, the cells are suspended in 1% SDS, S-EDTA, PIC, in 50 mM Tris-HCl (pH 7.2), for 5 min, at 22°, with gentle homogenization (10–20 strokes), to aid separation of heads and tails. Nuclei and tails are filtered through 80-μm mesh, Nitex filter cloth (Tetko). The nuclei are counted, and the number is adjusted to approximately 10^7/ml.

The preparation of nuclei and tails is layered over 8 ml of 1.1 M sucrose in 50 mM Tris-HCl, in 50-ml polycarbonate tubes. The lower concentration of sucrose is as effective for separating heads and tails[21] as is 1.6 M sucrose.[11] The samples are centrifuged in a Beckman J2-21 centrifuge equipped with a JS-18 rotor, or equivalent, at 3500 g, for 60 min, at 20°. At this rotor temperature the SDS remains in solution. On centrifugation, sperm tails form a band at the S-EDTA–sucrose interface, whereas nuclei pellet to the bottom of the tube.

The supernatant, tails at the interface, and the sucrose layer are aspirated separately. The centrifuge tubes are inverted to drain for 5 min. The walls are wiped carefully with a Kimwipe to remove any adhering tails. Sperm nuclei in the pellet are suspended in 0.25% SDS, PIC, 50 mM Tris-HCl (pH 7.2), and their number and purity are determined by counting an aliquot in a hemocytometer. The nuclei are washed once more in 0.25% SDS, 50 mM Tris-HCl (pH 7.2) and stored at $-70°$ until required. The low concentration of SDS prevents the nuclei from aggregating on centrifugation and freezing–thawing. Alternatively, the sperm nuclei are transferred to siliconized glass tubes, washed twice in 50 mM Tris-HCl (pH 7.2) to remove residual SDS, and used directly. The tails aspirated earlier are diluted out 10-fold in PIC, 50 mM Tris-HCl and recovered by centrifugation at 10,000 g for 30 min, at 4°. Both sperm nuclei and tails typically are recovered at purities exceeding 99% (Fig. 2[22]).

Procedure C. Sperm are recovered from the excurrent ducts as described. After being washed in PBS or 50 mM Tris-HCl (pH 7.2), the cells are incubated in 25 mM DTT, PIC, 50 mM Tris-HCl (pH 8.3), for 20 min, under N$_2$, at 22°. CTAB is added to 1% from a concentrated stock solution of 10% CTAB (1 : 10, v/v), PIC, 20 mM Tris-HCl (pH 8.3). The sperm are incubated for another 20 min, with occasional agitation to aid solubilization of structural elements. Nuclei, now depleted of the acrosome and perinuclear theca, are purified by sucrose density-gradient centrifugation,

[21] W. Zheng, M. A. Gawinowicz, A. H. Barth, and A. R. Bellvé, submitted for publication.
[22] D. A. O'Brien and A. R. Bellvé, *Dev. Biol.* **75**, 386 (1980).

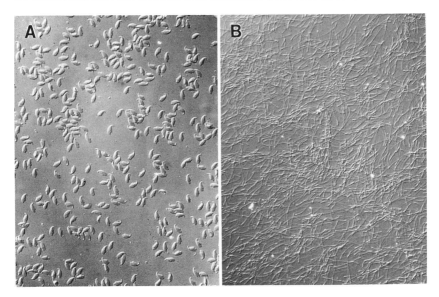

Fig. 2. Differential interference contrast light micrographs of mouse sperm nuclei (a) and tails (b) purified by sucrose density-gradient centrifugation. Sperm recovered from the cauda epididymis and vas deferens were washed in S-EDTA, decapitated in 1% SDS and S-EDTA with a brief homogenization, and purified by centrifugation over 1.6 M sucrose in S-EDTA. A gradient of 1.1 M sucrose was later found to be equally effective.[21] Sperm tails were retrieved from the S-EDTA–sucrose interface, and nuclei were recovered from the pellet. Magnification: ×300. [Reproduced by permission from D. A. O'Brien and A. R. Bellvé, *Dev. Biol.* **75**, 386 (1990).]

as described above. In this case, CTAB-soluble proteins in the supernatant include all nonnuclear proteins of the sperm cell.

Procedure A has been applied to study the ontogeny of thecins in nuclei isolated at different stages of spermatogenesis and epididymal maturation. This family of novel proteins appear during spermiogenesis on the apical pole of the transforming spermatid nucleus and gradually envelop the anterior two-thirds of the nucleus by step 16.[9a] When sperm leave the testis and enter the caput epididymis, the proteins assume a new distribution, becoming restricted to narrow margins on the anteriodorsal and the posterioventral poles of the nucleus (Fig. 3). Just preceding this morphological transition, thecin (M_r 80,000, 77,000, and 75,000) undergoes an apparent endoproteolytic cleavage to form a 48,000 M_r protein that stays on the nucleus for the remainder of sperm maturation (Fig. 4). Notably, the

[23] Deleted in proof.

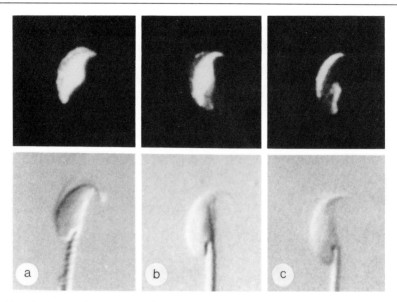

Fig. 3. Immunofluorescence micrographs of mouse sperm stained with a monoclonal antibody (MAb PNT-1) specific for thecin. Sperm were recovered from the testis (a), caput epididymis (b), and corpus epididymis (c), washed, attached to polylysine-coated slides, and fixed in 4% paraformaldehyde, 14 mM NaCl, 1 mM MgCl$_2$, 10 mM sodium phosphate (pH 7.4). The samples were blocked with 4% normal rabbit serum for 1 hr, washed, and reacted with MAb PNT-1 for 2 hr. The cells were washed thoroughly in PBS, then reacted with fluorescein-conjugated, anti-rat rabbit immunoglobulin G (FITC-IgG) and examined by indirect epifluorescence microscopy. Note that binding of MAb PNT-1 is restricted to the anterior pole of the nucleus of sperm from the testis, whereas the pattern seen in cauda sperm is bipolar, limited to the anteriodorsal and posterioventral margins of the organelle. Sperm of the caput epididymis show an intermediate pattern. Magnification: ×4000. [Reproduced by permission from A. R. Bellvé, R. Chandrika, and A. H. Barth, *J. Cell Sci.* **96,** 745 (1990).]

epitopes recognized by the monoclonal antibodies (MAbs), PNT-1 and PNT-2, appear to be conserved throughout the last stages of spermiogenesis and epididymal maturation, based on the comparable amount of immunoreactivity present in nuclear proteins of the different cell types. This conclusion is possible because both the gel and the immunoblot (Fig. 4) were loaded with protein from an equivalent number of haploid nuclei, to circumvent differences that exist in the protein content of the various cell types.

Procedure B, involving an extraction of mouse sperm with 1% SDS, removes the plasma membrane, acrosome, axoneme, and matrix and cristae of mitochondria.[9–11,23] The tail retains the basal plate, capitulum, outer dense fibers, mitochondrial shells, and fibrous sheath.[11,22] The residual sperm head consists of the nucleus and elements of the perinuclear

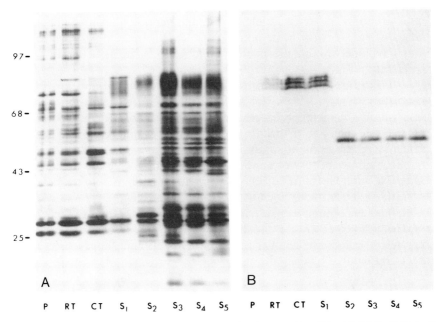

FIG. 4. Electrophoresis (A) and corresponding immunoblot (B) with MAb PNT-1 of nuclear proteins isolated from germ cells at different stages of meiotic prophase, spermiogenesis, and epididymal maturation. Pachytene spermatocytes (P), round spermatids (RT; steps 1–8), and condensing spermatids (CT; steps 12–16) were isolated by sedimentation velocity at unit gravity [L. J. Romrell, A. R. Bellvé, and D. W. Fawcett, *Dev. Biol.* **49**, 119 (1976); A. R. Bellvé, J. C. Cavicchia, C. F. Millette, D. A. O'Brien, Y. M. Bhatnager, and M. Dym, *J. Cell. Biol.* **74**, 68 (1977); A. R. Bellvé, C. F. Millette, Y. M. Bhatnagar, and D. A. O'Brien, *J. Histochem. Cytochem.* **25**, 480 (1977).] Sperm were recovered from the testis (S_1), the caput (S_2), corpus (S_3), cauda epididymis (S_4), and vas deferens (S_5) by dissecting the respective regions and removing the cells by gentle agitation. The sperm were suspended in 1% Triton X-100, 10 mM MgCl$_2$, 40 mM KCl, PIC, 0.32 M sucrose, 50 mM Tris-HCl (pH 7.7) (MKPT), for 10 min (see Procedure A in text). Nuclei were subjected to centrifugation over 1.1 M sucrose at 5000 g, for 20 min, at 4°, suspended in 0.32 M sucrose in MKPT, and centrifuged again. Washed nuclei were solubilized by addition of sample buffer, and the proteins were subjected to SDS–PAGE.[9a] The proteins were stained with silver or transferred electrophoretically to Immobilon membranes. The membranes were incubated in 10% nonfat dry milk to block nonspecific binding sites, then washed, and the proteins were reacted with MAb PNT-1, followed by horseradish peroxidase–IgG and the substrates, 4-chloro-1-naphthol and H$_2$O$_2$. Note that MAb PNT-1 recognizes three protein bands (M_r 80,000, 77,000, and 75,000) in spermatids and testicular sperm. By comparison, sperm from the epididymis and vas deferens express only a 48,000 M_r protein. [Reproduced by permission from A. R. Bellvé, R. Chandrika, and A. H. Barth, *J. Cell Sci.* **96**, 745 (1990).]

theca (Fig. 8a). Structures of the theca that are not present include the basal striations, posterior nuclear ring, and ventral spur of the posterior nuclear sheath.[11,22] The remaining elements of the theca still form a continuous envelope around the nucleus, except for a defined region at the implantation fossa.

In Procedure C, requiring DTT and CTAB, the perinuclear theca is completely removed from the nucleus, based on electron microscopic observations.[19] Nuclei prepared by this method are useful for two applications. First, removal of theca proteins facilitates purification of the two protamines,[19] and potentially other nuclear proteins. Second, removal of the theca allows access of antibodies to antigens within the nucleus that otherwise would be masked. This is the case for lamin B.[24] In the presence of the theca, an antibody specific for lamin B binds to a very limited region of the nucleus. Binding is restricted to a narrow band that extends from the apical pole, down along the ventral edge, and across the equatorial region to the dorsal edge of the nucleus. However, in the absence of the theca, antilamin B binds to the entire nuclear surface in an intense punctate pattern.[24] Similarly, several MAbs directed against intranuclear proteins bind to their respective epitopes only after the perinuclear theca has been removed.[10]

Recovery of Perinuclear Theca Proteins

The perinuclear theca is a highly structured, dense, "shell" enveloping the nucleus of mammalian sperm, except for a region at the implantation fossa, the site at which the tail normally is attached.[20,25] The theca is composed of two domains, the perforatorium on the anterior pole and the postacrosomal dense lamina enveloping the caudal pole. In rodent species, the perforatorium is a well-developed structure, covering the anterior pole and projecting forward of the nucleus with a slight curvature (Figs. 1 and 8a). The theca comprises at least 230 protein species, but little is known about the functions of the theca proteins, before or after fertilization.[10] Calicin, a prominent keratin-like protein of 60,000 M_r, has been localized to the postacosomal dense lamina over the posterior pole of the sperm nucleus in several species, and it is thought to serve a structural function in the theca.[25,26] Sperm thecin (M_r 48,800) has been localized to the perinuclear theca by biochemical fractionation[9a] and immunoelectron microscopy.[10]

[24] S. B. Moss, B. L. Burnham, and A. R. Bellvé, *Mol. Reprod. Dev.* **34,** 164 (1993).
[25] F. J. Longo, G. Krohne, and W. W. Franke, *J. Cell Biol.* **105,** 1105 (1987).
[26] J. Paranko, F. J. Longo, J. Potts, G. Krohne, and W. W. Franke, *Differentiation* **38,** 21 (1988).

Proteins of the perinuclear theca are recovered separately, by using a modification of the procedure originally reported by Balhorn et al.[19] Sperm recovered from the excurrent ducts are decapitated with 1% SDS, S-EDTA,[10–12] and the nuclei containing elements of the theca are purified by centrifugation. Thereafter, theca proteins are extracted from the nuclei with DTT and CTAB, and the nuclei and soluble proteins are recovered by centrifugation, as described below.

Step 1. Perinuclear theca proteins are extracted selectively from SDS-treated nuclei,[9a,10] by an incubation in DTT and CTAB. Sperm are treated with SDS, and the nuclei are purified by sucrose density-gradient centrifugation, as described. The purified nuclei, whether used fresh or from stocks frozen at $-70°$, are washed twice in 50 mM Tris-HCl (pH 7.2), to remove all traces of SDS. Otherwise, both the nucleus and the enveloping theca will decondense on adding DTT (see section on Isolation of Sperm Nuclei, Procedure B). The washed nuclei are suspended in 50 mM DTT, PIC, 50 mM Tris-HCl (pH 8.3) for 20 min, and then made 1% in CTAB by adding an appropriate volume from a stock solution of 10% CTAB (1 : 10, v/v), PIC, 50 mM Tris-HCl (pH 8.3). The samples are vortexed lightly to aid solubilization of the theca.

Step 2. The nuclei, now depleted of the perinuclear theca, are pelleted by centrifugation at 3200 g, for 30 min, at 4°. CTAB-soluble theca proteins in the supernatant are decanted and then precipitated in 75% ethanol, at $-20°$, for 4 hr. (Proteins precipitated in ethanol, as opposed to acetone, are solubilized more readily and resolved better by SDS–PAGE.) Proteins are pelleted by centrifugation at 5000 g for 30 min, at 4°, lyophilized, and stored at $-70°$. Alternatively, the proteins may be air dried and immediately dissolved in 6 M guanidine hydrochloride, 50 mM DTT, for alkylation with ethyleneimine[11,12] (also see below). The original centrifuge tubes containing sperm nuclei are inverted and allowed to drain for 5 min. Nuclei in the pellet are recovered, washed in 0.25% SDS, PIC, 20 mM Tris-HCl (pH 8.3) by centrifugation at 1500 g, and saved.

Preparation of Sperm Perinuclear Matrices

Sperm nuclei, with associated elements of the perinuclear theca, are used to prepare perinuclear matrices (Fig. 1). Nuclei are prepared by extraction of sperm with 1% SDS, S-EDTA and purified by sucrose density-gradient centrifugation.[10–12,21] After removing all residual SDS, the nuclei are digested with DNase I and extracted with salt, simultaneously, to remove DNA and protamines.[10] The structures formed consist of elements of the cytoskeletal perinuclear theca and intranuclear fibers and are composed of at least 230 protein species that account for over 4% of total nuclear protein. The various functions of these proteins in spermiogenesis, epididymal maturation, and/or events leading to fertilization and early embryogenesis are unknown. To aid in defining their etiology, properties, and functions, MAbs were prepared by immunizing female rats with mouse spermatocyte and spermatid nuclear matrices or with sperm perinuclear matrices.[9a,10] The use of matrices, which comprise a minor population of total nuclear protein, enabled the generation of MAbs against antigens of low abundance. These immunological probes have been applied to characterize antigenic matrix proteins peculiar to mouse spermatogenesis.[10,21]

Step 1. Sperm nuclei are washed to remove residual SDS and incubated with 25 mM DTT, PIC, 20 mM Tris-HCl (pH 8.3), for 20 min, at 22°, under N_2, to reduce disulfide bonds. On reduction, the nuclei are incubated for 10 min in 50 µg/ml DNase I (DN-EP; Sigma Chemical Co., St. Louis, MO), PIC, 20 mM Tris-HCl (pH 8.3) by addition from a 100-fold (w/v) stock solution of DNase I. Then, $CaCl_2$–$MgCl_2$ (3 : 2, mol/mol) is brought to 125, 150, 175, 200, and 250 mM, at 10-min intervals, by addition of appropriate volumes of 2.0 M $CaCl_2$–$MgCl_2$, PIC, 20 mM Tris-HCl (pH 8.3). DNase I activity requires 6.25 mM $MgCl_2$, and the enzyme remains active up to approximately 220 mM $CaCl_2$–$MgCl_2$ (3 : 2, mol/mol). During these incubations DNA is hydrolyzed as the protamines, P1 and P2, are displaced from the nucleus in a concentration-dependent pattern, with maximal displacement occurring at 225 mM $CaCl_2$–$MgCl_2$ (Fig. 5). The nonprotamine proteins remain associated, almost quantitatively, with the perinuclear matrix (Fig. 6).

Importantly, DNase I must be present at all stepwise increments in the concentration of divalent cations. If, as done in Step 2, $CaCl_2$–$MgCl_2$ (3 : 2, mol/mol) is added without DNase I through to 250 mM, P1 but not P2 is displaced from the nucleus (Fig. 6), which corresponds to only about 68% of total protamine (cf. Fig. 5). Moreover, some of the rapidly decondensing chromatin bursts through the surrounding perinuclear theca to form a halo around the nucleus.[10] In this case only remnants of the original perinuclear theca are recognizable.

Step 2. After the last incubation, the samples are subjected to centrifugation over 1.6 M sucrose, 250 mM $CaCl_2$–$MgCl_2$, 20 mM Tris-HCl (pH 7.2), at 100,000 g, for 12 hr, at 4°, either by using an SW 65 rotor in a Beckman L8-M ultracentrifuge, or a TLS 55 rotor in a Beckman LS-100 ultracentrifuge to the same relative centrifugal field. Proteins displaced from the nucleus are recovered by decanting the supernatant, precipitated

[27] L. A. Ierardi, S. B. Moss, and A. R. Bellvé, *J. Cell Biol.* **96**, 717 (1983).

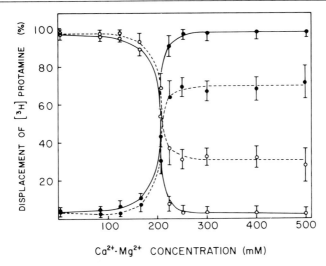

FIG. 5. Concentration-dependent displacement of ^{3}H-labeled protamines from mouse sperm nuclei by divalent cations and DNase I. In procedure 1 (solid lines), nuclei with an enveloping perinuclear theca were incubated in 25 mM DTT (pH 8.3) for 20 min, then in 50 μg/ml DNase I concurrently with stepwise increments in CaCl$_2$–MgCl$_2$ (3 : 2, mol/mol) to the concentrations specified. In procedure 2 (dashed lines), nuclei were incubated with 25 mM DTT (pH 8.3) and with divalent cations in stepwise increments through to the concentrations indicated, followed by 50 μg/ml DNase I. All samples were subjected to centrifugation over 1.6 M sucrose, at 100,000 g for 12 hr. In both procedures, aliquots were taken from the supernatants and solubilized pellets, to quantify displaced (●) and DNA/matrix-bound [^{3}H]protamine (○), respectively. Data points represent means ± S.E. [Reproduced by permission from A. R. Bellvé, R. Chandrika, Y. S. Martinova, and A. H. Barth, *Biol. Reprod.* **47**, 451 (1992).]

with 75% ethanol, for 4 hr, at $-20°$, and pelleted by centrifugation at 5000 g, for 30 min, at 4°. The centrifuge tubes with the pellet of perinuclear matrices are inverted, drained for 5 min, and the walls are wiped clean with a Kimwipe. Perinuclear matrices are recovered from the pellet, and washed in 25 mM Tris-HCl (pH 7.2) by centrifugation.

Characterization of Sperm Perinuclear Matrices

Sperm perinuclear matrices have been characterized by light, transmission, and scanning electron microscopy.[10] On removing DNA and protamines with DNase I and stepwise increments in the concentration of CaCl$_2$–MgCl$_2$ through to 250 mM, the sperm nucleus retains its original configuration, although it may swell 1.1- to 1.2-fold in size (Fig. 7). On reaching 250 mM CaCl$_2$–MgCl$_2$ (Fig. 7c), the nucleus is depleted of ethidium bromide-stainable DNA and consists of a peripheral shell of dense material that resists solubilization by SDS alone. Based on transmission

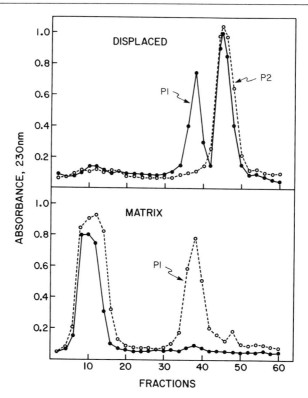

FIG. 6. Ion-exchange chromatography of mouse sperm nuclear proteins remaining bound to and displaced from DNA on exposure to divalent cations. Nuclei were incubated in 25 mM DTT (pH 8.3) for 30 min, then in $CaCl_2$–$MgCl_2$ (3:2, mol/mol) at stepwise increments in concentration from 125, 150, 175, 200, to 250 mM. This was done concurrently with (Procedure 1; ●) or followed by (Procedure 2; ○) digestion with 50 µg/ml DNase I. Nuclei/ perinuclear matrices were recovered by centrifugation over 1.6 M sucrose at 100,000 g for 12 hr. Displaced supernatant proteins and those associated with the pelleted perinuclear matrices were solubilized separately, in 6 M guanidine hydrochloride, 25 mM DTT, 1 mM PMSF, 0.5 M Tris-HCl (pH 8.6), and alkylated with ethyleneimine. Supernatant proteins in 18% guanidine hydrochloride, 1 mM PMSF, 0.1 M sodium phosphate buffer (pH 6.8) were applied to a column of Bio-Rex 70 (1 × 10 cm). After nonprotamine proteins were washed through the column, a gradient of 20 to 50% guanidine hydrochloride was applied to elute protamines P1 and P2. Note that, in procedure 1, both P1 and P2 are displaced when DNase I is present during treatment with divalent cations. In procedure 2, just P2 is displaced when DNase I is used only after the divalent cations. Both procedures result in the nonprotamine proteins remaining associated with the perinuclear matrices. [Reproduced by permission from A. R. Bellvé, R. Chandrika, Y. S. Martinova, and A. H. Barth, *Biol. Reprod.* **47**, 451 (1992).]

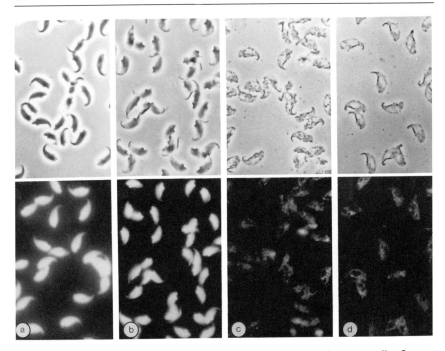

FIG. 7. Phase-contrast light micrographs of sperm nuclei (top) and corresponding fluorescence images (bottom) of DNA stained with ethidium bromide, before and after incubation with divalent cations and DNase I. Isolated nuclei were attached to polylysine-coated coverslips, incubated with $CaCl_2$–$MgCl_2$ (3 : 2, mol/mol) in stepwise increments from 125, 175, 200, 250, to 500 mM in 20 mM Tris-HCl (pH 8.3), at 10-min intervals, and stained with 1 μg/ml ethidium bromide. Fluorescence images shown in (c) and (d) required approximately 6-fold longer photographic exposures than for (a) and (b). Note that at 250 and 500 mM $CaCl_2$–$MgCl_2$ (c, d), the nuclei have increased 1.1- to 1.2-fold in size, are delineated by a prominent peripheral "shell," and have a translucent interior except for an area around the implantation fossa at the posterioventral pole. The transition in appearance between 125 and 250 mM $CaCl_2$–$MgCl_2$ (b, c) corresponds to the displacement of P1 and P2 (cf. Fig. 5) and the hydrolysis of DNA, as judged by the loss of ethidium bromide-stainable DNA. $CaCl_2$–$MgCl_2$ concentrations were as follows: (a) 6.25 mM, (b) 125 mM, (c) 250 mM, (d) 500 mM. Magnification: × 1000. [Reproduced by permission from A. R. Bellvé, R. Chandrika, Y. S. Martinova, and A. H. Barth, *Biol. Reprod.* **47**, 451 (1992).]

electron microscopy, the shell is composed of elements of the dense perinuclear theca including the cytoplasmic, subacrosomal perforatorium and postacrosomal dense lamina and internal tranverse fibers (Fig. 8b). The fibers are thought to be nuclear in origin. In scanning electron micrographs, the perinuclear matrix appears as a dense interwoven network of coarse fibers, with a denser domain occurring in the anterior pole demarcated by a line across the equatorial region.[10] The apical region of the perforatorium and the implantation fossa are evident on all matrices.

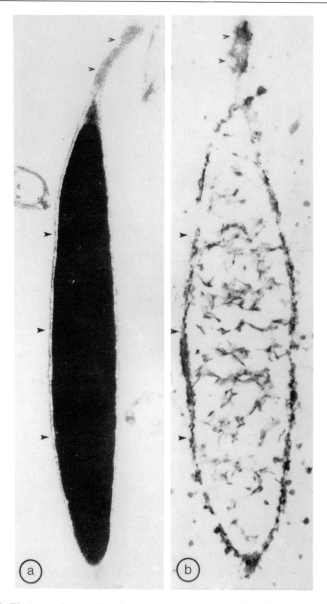

FIG. 8. Electron micrographs of a mouse sperm nucleus (a) and perinuclear matrix (b). Sperm were washed in S-EDTA and decapitated by brief homogenization in 1% SDS, S-EDTA. Nuclei were purified by sucrose density-gradient centrifugation and washed carefully to remove all residual SDS. (a) Nuclei were fixed in 1% glutaraldehyde, postfixed in 1% osmium tetroxide, double-stained with lead citrate, and sectioned for electron microscopy. (b) Sperm perinuclear matrices were prepared by embedding nuclei in 2% agarose

Perinuclear matrices are composed of multiple proteins that range in M_r from 70,000 to 8000, as judged by SDS–PAGE (Fig. 9). These can be resolved into at least 230 different protein species when subjected to isoelectric focusing and nonequilibrium pH gradient polyacrylamide gel electrophoresis and stained with silver.[10]

Fate of Thecin during Capacitation and Acrosome Reaction

Thecin remains distributed in an asymmetric bipolar pattern on the nucleus as sperm transit the length of the excurrent ducts. It is not known, however, if the protein undergoes further changes during capacitation, acrosome reaction, and fertilization. The molecular events underlying sperm capacitation in the female reproductive tract are not fully understood, although the process is prerequisite to a successful acrosome reaction.

The acrosome reaction occurs when sperm interact with ZP3, a glycoprotein of the zona pellucida of the oocyte,[28–30] an effect that is mimicked by the addition of the Ca^{2+}-specific ionophore A23187.[31] The acrosome reaction involves an exocytotic vesiculation and loss of the plasma and outer acrosomal membranes overlying the anterior region of the sperm head. Cells that have undergone the acrosome reaction can be identified and quantified by the lack of MAb M41 binding to proteins present in the plasma membrane that normally overlies the organelle. Viable sperm are required for these experiments, and therefore the cells must be processed gently in a suitable medium, to ensure metabolic and structural integrity. This includes the use of pyruvate, rather than lactate, as the latter has a detrimental effect on induction of capacitation and/or the acrosome reaction.[32] The following protocol is modified after that of Olds-Clarke

[28] P. M. Wasserman, *Science* **235**, 553 (1987).
[29] P. M. Wasserman, *Annu. Rev. Cell Biol.* **3**, 109 (1987).
[30] P. M. Wasserman, *Annu. Rev. Biochem.* **57**, 415 (1988).
[31] P. Primakoff, D. G. Myles, and A. R. Bellvé, *Dev. Biol.* **80**, 324 (1980).
[32] J. M. Neill and P. J. Olds-Clarke, *Gamete Res.* **18**, 121 (1987).

---

and incubating the sample with 50 μg/ml DNase I, 25 mM DTT (pH 8.3), for 20 min, to reduce disulfide bonds, followed by stepwise increments in $CaCl_2$–$MgCl_2$ (3 : 2, mol/mol) from 125, 150, 175, 200, to 250 mM. Note that the nucleus contains dense heterochromatin and is surrounded by residual elements of the cytoplasmic perinuclear theca, including the apical perforatorium. The perinuclear matrix consists of coarse intranuclear, karyoskeletal fibers and elements of the cytoplasmic perinuclear theca, including the subacrosomal perforatorium and postacrosomal dense lamina. Solid arrowheads mark the posterior region of the perinuclear theca, and hollow arrowheads, the apical perforatorium. Magnification: ×17,000. [Reproduced by permission from A. R. Bellvé, R. Chandrika, Y. S. Martinova, and A. H. Barth, *Biol. Reprod.* **47**, 451 (1992).]

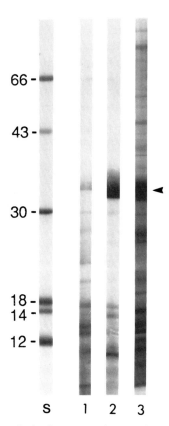

FIG. 9. Electrophoretic analysis of nonprotamine proteins associated with mouse sperm perinuclear matrices. Nuclei were homogenized briefly in 1% SDS, S-EDTA, and purified by centrifugation over a gradient of 1.6 M sucrose. The nuclei were washed extensively, then incubated with divalent cations alone, DNase I alone, or DNase I with $CaCl_2$–$MgCl_2$ (3 : 2, mol/mol) in stepwise increments from 125, 150, 175, 200, to 250 mM, as specified. Supernatants and perinuclear matrices were recovered by centrifugation over 1.6 M sucrose at 100,000 g, for 12 hr, at 4°. Arrowhead depicts exogenous DNase I. Lane 1, Supernatant from 125–250 $CaCl_2$–$MgCl_2$ alone; lane 2, supernatant from DNase I alone; lane 3, perinuclear matrices; lane S, size markers. [Reproduced by permission from A. R. Bellvé, R. Chandrika, Y. S. Martinova, and A. H. Barth, *Biol. Reprod.* **47,** 451 (1992).]

and colleagues,[32,33] and the acrosome reactions is scored by using MAb M41 in the procedure of Saling and Lakoski.[34]

Step 1. The procedure uses a modified Krebs–Ringer bicarbonate medium (EKRB), with a pH of 7.3 to 7.4 and an osmolality of 295 to 310

[33] P. J. Olds-Clarke and R. Sego, *Biol. Reprod.* **47,** 629 (1992).
[34] P. M. Saling and K. A. Lakoski, *Biol. Reprod.* **33,** 527 (1985).

mOsM, as defined in Table I. The medium may be frozen at $-70°$ and used for 1 month afterward. To induce capacitation, the medium is made 2% in bovine serum albumin (BSA, 20 mg/ml; fraction V, fatty acid-free; Pentex, ICN Immunobiologicals, Lisle, IL). Incubations are performed under paraffin oil and 5% CO_2 in air, at 37°.

Step 2. Sperm are taken from the caudae epididymides and vasa deferentia of individual, sexually active, adult mice, preferably 10 to 16 weeks old. The caudae epididymides from the same animal are placed in 150 μl of medium (Table I), under paraffin oil. The tissue is minced with a sharp pair of fine scissors, then incubated at 37° under CO_2 in air for 15 min, to allow egress of sperm. Sperm in the vas deferens are extruded into 100 μl of medium, by gently sliding a pair of lightly compressed forceps along the length of the duct. These cells are held separately under paraffin oil and 5% CO_2, at 37°, for 15 min.

Step 3. After a 10-min incubation, the samples are pipetted, briefly and gently, with a plastic or siliconized glass pipette, to aid release of sperm from the segments of cauda epididymis. After another 5 min (a total of 15 min), all tissue segments are removed with the aid of a fine plastic toothpick, and cauda sperm are pooled along with those from the vas deferens. Then, 200 μl of the sperm suspension is taken and diluted to 300 μl by adding an equal volume of fresh medium. The final concentration of sperm will be approximately 10^7 per 300 μl from each mouse. The sample is overlaid with 10–15 μl of paraffin oil and held under 5% CO_2 in air, at 37°.

Step 4. Two aliquots (10–20 μl) of a sperm suspension from an individual mouse are diluted 10- to 50-fold, placed on a prewarmed microscope slide, and quickly counted to determine percent motile sperm and total cell number. It is advantageous to precoat the hemocytometer slide with 2% BSA or 0.1% poly(vinyl alcohol) (PVA) (type II, M_r 10,000; Sigma), to prevent sperm from sticking to the surface. The slides are exposed to the 2% BSA in medium for 2 min, washed in deionized water, and air dried. Sperm samples of greater than 65% motile cells are pooled, mixed gently, and then used for experimental purposes. Samples with less than 64% motile sperm are discarded.

Step 5. Capacitation of sperm is induced by addition of 2% delipidated BSA. The experimental samples, along with the appropriate controls, are incubated at 37°, under 5% CO_2 in air, for 60 min, with occasional mixing. At the end of this period, a predefined number of samples are made 2 μM in A23187 to induce the acrosome reaction. [A23187 is kept as 100 μM stock solution in 100% dimethyl sulfoxide (DMSO) in a light-shielded container at $-20°$.] After another 30 min, the reaction is stopped by placing the cells on ice.

Step 6. Aliquots of sperm are taken from each sample and placed onto a super clean, poly(L-lysine)-coated microscope slide, and the cells are allowed to attach, at 37°, under 5% CO_2 in air. The slides are washed with filtered PBS, then lightly fixed with 2% paraformaldehyde in PBS (pH 7.4). Nonspecific sites are blocked by incubating the specimens in 2% normal rabbit (NRS) serum for 60 min. The specimens are washed 3 times in PBS and reacted with MAb M41 hybridoma (isotype IgG_{2a}) supernatant, diluted 1 : 4 (v/v), 2% NRS in PBS, for 2 hr, at 37°. After 3 washes in PBS, the sperm are reacted with a fluorescein-conjugated, anti-rat goat immunoglobulin G (IgG) for 45 min. The specimens are washed 3 times in PBS, mounted to prevent dehydration, and examined by epifluorescence and differential interference contrast microscopy. Sperm are scored for the presence or absence of an acrosome, based on the binding of MAb M41.

Step 7. The remainder of the sperm in each sample is processed for biochemical fractionation. The cells are washed, then extracted with 1% SDS and S-EDTA. The nuclei are purified by sucrose density-gradient centrifugation, washed extensively, and the perinuclear theca extracted with 1% CTAB, 25 mM DTT (see section on Recovery of Perinuclear Theca Proteins). Solubilized proteins are recovered, aminoethylated, subjected to SDS–PAGE, transferred to Immobilon membranes, and immunoblotted with MAb PNT-1 to detect thecin.

In this case, the Ca^{2+}-specific ionophore A23187 induced approximately 94% of capacitated mouse sperm to undergo the acrosome reaction, based on the loss of MAb M41 binding to the plasma membrane over the acrosome (Fig. 10A). By comparison, the incidence of spontaneous acrosome reactions in the presence and absence of BSA alone was approximately 12 and 26%, respectively. When perinuclear theca proteins were recovered from the same sperm samples, and submitted to SDS–PAGE and immunoblotting with thecin-specific MAb PNT-1, no changes, qualitative or quantitative, were detected in reactivity (Fig. 10B). Clearly, thecin remains unaffected by the extensive membrane vesiculation and release of acrosomal hydrolases from the overlying acrosome, events that are prerequisite to normal fertilization.

Conclusions

Procedures are given for recovering and fractionating mouse sperm for physiological and biochemical studies of nuclear function in epididymal maturation, capacitation, and the acrosome reaction. In a number of aspects these methods are peculiar to the sperm cell, with its many novel and unusual subcellular structures. Fractionation of the nucleus, along with the enveloping perinuclear theca, has enabled the identification and characterization of many new proteins of low abundance. Moreover, isola-

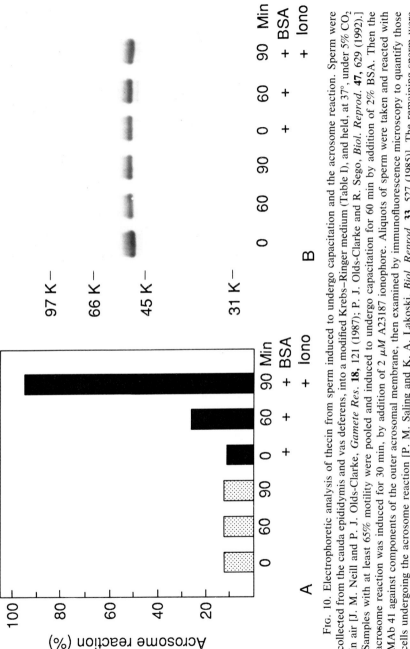

Fig. 10. Electrophoretic analysis of thecin from sperm induced to undergo capacitation and the acrosome reaction. Sperm were collected from the cauda epididymis and vas deferens, into a modified Krebs–Ringer medium (Table I), and held, at 37°, under 5% CO_2 in air [J. M. Neill and P. J. Olds-Clarke, *Gamete Res.* **18**, 121 (1987); P. J. Olds-Clarke and R. Sego, *Biol. Reprod.* **47**, 629 (1992).] Samples with at least 65% motility were pooled and induced to undergo capacitation for 60 min by addition of 2% BSA. Then the acrosome reaction was induced for 30 min, by addition of 2 μM A23187 ionophore. Aliquots of sperm were taken and reacted with MAb 41 against components of the outer acrosomal membrane, then examined by immunofluorescence microscopy to quantify those cells undergoing the acrosome reaction [P. M. Saling and K. A. Lakoski, *Biol. Reprod.* **33**, 527 (1985)]. The remaining sperm were used to prepare perinuclear theca proteins, which were subjected to SDS–PAGE and immunoblotting with MAb PNT-1. (A) Quantitation of acrosome-reacted sperm. (B) Corresponding SDS–PAGE/immunoblot.

tion of the perinuclear matrix has proved to be an invaluable first step toward preparing and obtaining an array of MAbs specific for many of these germ cell-specific proteins. Several of the MAbs were found to be directed against thecins, a family of novel, immunodominant proteins present in the perinuclear theca of haploid germ cells. The temporal expression and localization of thecins to the narrow layer between the nucleus and the developing acrosome is consistent with the proteins having a function in adhesion of the nuclear envelope with the inner acrosomal membrane. Alternatively, given the endoproteolytic processing and apparent relocalization of the proteins at the onset of epididymal maturation, it is plausible the truncated sperm thecin may serve a role in events after fertilization.

Acknowledgments

Appreciation is extended to the many collaborators and colleagues who have made valuable contributions toward the development of techniques and the research presented in this chapter. In particular, we wish to thank Drs. Everett Anderson, Donald W. Fawcett, Clarke F. Millette, Diana G. Myles, Deborah A. O'Brien, Stuart B. Moss, and Paul Primakoff. Preparation of this publication was supported by Grant P50 HD 05077, from the National Institute of Child Health and Human Development.

[8] Strategies and Methods for Evaluating Acrosome Reaction

By RICHARD A. CARDULLO and HARVEY M. FLORMAN

Introduction

The sperm of many animal species, including those of the mouse and all other mammals, contain within the apical region of the head a single secretory vesicle, or acrosome. Exocytosis of this granule (the acrosome reaction) is a terminal morphological alteration that must occur prior to penetration of the extracellular coat of the egg (zona pellucida) or fertilization.[1,2]

[1] H. M. Florman and D. F. Babcock, in "Elements of Mammalian Fertilization: Basic Concepts" (P. M. Wassarman, ed.), p. 105. CRC Press, Boca Raton, Florida, 1990.
[2] G. S. Kopf and G. L. Gerton, in "Elements of Mammalian Fertilization: Basic Concepts" (P. M. Wassarman, ed.), p. 153. CRC Press, Boca Raton, Florida, 1990.

Copyright © 1993 by Academic Press, Inc.
All rights of reproduction in any form reserved.

Recognition that the acrosome reaction occurs at the surface of the zona pellucida[3,4] has led to the isolation and subsequent molecular cloning of the glycoprotein agonist, ZP3.[5,6] ZP3 produces exocytosis through activation of sperm G proteins[7–9] with consequent elevation of sperm intracellular pH and Ca^{2+} (pH_i and Ca_i, respectively).[9] Additional or alternative agonists have been reported in mammals as well as in other species,[10–13] although the physiological relevance of these agents is not well understood. The mechanisms whereby these agonists promote exocytosis is subject to intensive investigation.

Assessment of acrosomal morphology and the progress of the acrosome reaction may be required for several reasons. First, exocytosis is an indirect indicator of the maturation of sperm populations. It is well known that mammalian sperm are maintained in an infertile state and acquire complete functional maturation (typically referred to as "capacitation") during subsequent incubations *in vivo,* within the female reproductive tract,[14,15] or *in vitro.*[16] In the mouse, this process is associated with an enhanced rate of spontaneous (agonist-independent) exocytosis that can, in turn, be used to monitor the progressive development of fertility. Thus, optimization of an *in vitro* fertilization protocol requires the ability to assess acrosomal status. Second, acrosomal exocytosis provides an accessible model for adhesion-dependent signal transduction and has recently been exploited with great success.[1,2]

Here, we consider briefly the morphology of the anterior sperm head and the various strategies available for the evaluation of sperm acrosomal exocytosis. When possible, the protocols chosen are specifically customized for the mouse system; cases where a given technique has only been established in other models are duely noted.

[3] R. B. L. Gwatkin, "Fertilization Mechanisms in Man and Mammals." Plenum, New York, 1976.
[4] P. M. Saling, J. Sowinski, and B. T. Storey, *J. Exp. Zool.* **209,** 229 (1979).
[5] P. M. Wassarman, *Development (Cambridge, UK)* **108,** 1 (1990).
[6] P. M. Wassarman and S. Mortillo, *Int. Rev. Cytol.* **130,** 85 (1991).
[7] Y. Endo, M. A. Lee, and G. S. Kopf, *Dev. Biol.* **119,** 210 (1987).
[8] Y. Endo, M. A. Lee, and G. S. Kopf, *Dev. Biol.* **129,** 12 (1988).
[9] H. M. Florman, R. M. Tombes, N. L. First, and D. F. Babcock, *Dev. Biol.* **135,** 133 (1989).
[10] P. Thomas and P. Meizel, *Biochem. J.* **264,** 539 (1989).
[11] P. F. Blackmore, S. J. Beebe, D. R. Danforth, and N. Alexander, *J. Biol. Chem.* **265,** 1376 (1990).
[12] S. Meizel and K. O. Turner, *Mol. Cell. Endocrinol.* **11,** R1 (1991).
[13] P. F. Blackmore, J. Neulen, F. Lattanzio, and S. J. Beebe, *J. Biol. Chem.* **266,** 18655 (1991).
[14] M. C. Chang, *Nature (London)* **168,** 697 (1951).
[15] C. R. Austin, *Aust. J. Sci. Res.* **4,** 581 (1951).
[16] B. J. Rogers, *Gamete Res.* **1,** 165 (1978).

Morphology of Acrosomal Exocytosis

The acrosome lies in the apical region of the sperm head, overlying the nucleus and immediately subjacent to the plasma membrane. The vesicular space, which appears as a fibrillar network on ultrastructural examination, contains a variety of constituents: of these, several have known enzymatic activities and many appear to be sperm-specific.[2,17] Acrosome reactions consist of (1) an elevation of sperm intracellular ionic mediators of exocytosis that include Ca_i and pH_i and that probably occur in response to a stimulatory agonist; (2) an initial stage, characterized by multiple membrane fusion–fission events involving the outer acrosomal membrane and the overlying plasmalemma and resulting in the formation of hybrid vesicles from the two membrane domains and in the dehiscence of acrosomal contents; and (3) a latter stage, wherein the hybrid vesicles are sloughed from the sperm surface and in which more extensive release of acrosomal contents is realized. A more complete consideration of the physical and functional aspects of the acrosome reaction is available elsewhere.[2,17,18]

Assay Strategies

Given this overview, there are several available strategies for the determination of sperm exocytosis. Assays currently in use detect Ca_i elevations that precede the agonist-induced acrosome reaction; alterations in the plasma membrane domain prior to membrane fusion; loss of functional groups (epitopes, lectin binding sites, enzymes) from the plasma membrane, outer acrosomal membrane, or soluble acrosomal contents; or exposure of functional groups on or associated with the inner acrosomal membrane. Selected assays are described in detail in this chapter.

One determining principle in selection of an assay protocol must obviously be experimental need. For example, when assessing the acquisition of the fertile state by sperm populations, as may be required to optimize *in vitro* fertilization methods, then a population-based assay is most appropriate. Alternatively, sperm signal transduction mechanisms may be masked by extensive heterogeneity, with regard both to physiological state and to agonist response,[9] and may require analysis at the single cell level. Additional considerations include determinations in living and in fixed cells.

[17] R. Yanagimachi, *in* "The Physiology of Reproduction" (E. Knobil and J. Neill, eds.), p. 135. Raven, New York, 1988.
[18] S. Meizel, *Am. J. Anat.* **174,** 285 (1985).

Specific Assays

Bright-Field Assays

By far the most convenient and rapid assay for acrosome reactions is to use a conventional light microscope [equipped with bright-field, phase-contrast, or differential interference contrast (DIC) optics] and to look for the presence or absence of an acrosome. This is easily achieved with species possessing large acrosomes (e.g., guinea pig) but is not reliable for most other species. Some investigators have had success scoring rodent sperm, most notably those of the mouse and hamster, using either phase-contrast or DIC microscopy (Fig. 1A,B).[19-21] The following protocol utilizes DIC and is the most sensitive of the conventional light microscopy assays for use with mouse sperm.

Materials

Poly(L-lysine), 70–100 kDa (Sigma, St. Louis, MO): 1 mg/ml in distilled water, made freshly

Method

1. Coat coverslips with poly(L-lysine) solution (10 min, room temperature). Rinse several times in distilled water and store wet until used.

2. Adjust the sperm concentration to approximately 10^6/ml in culture medium.

3. Add aliquots of sperm to coverslip, incubate 5 min, and gently wash in culture medium to remove loosely adherent sperm.

4. View sperm under DIC with at least $1000\times$ total magnification, using a $\times 100$ Plan Neofluar or Plan Apochromat objective at high numerical aperture (e.g., 1.3 N.A.). A short working distance, high numerical aperture oiled condenser should also be used. If the optics are adjusted correctly, the "barber pole stripes" on the midpiece should be resolvable.

5. Adjust the polarizers so that on acrosome-intact sperm the acrosomal crescent appears white and the remainder of the head appears gray. Under these conditions acrosome-reacted sperm will display no white acrosomal crescent.

Comments. (1) Determining acrosome reactions using bright-field microscopy is difficult, and, unless the highest quality optics are available (e.g., $\times 100$, 1.3 N.A. Plan Neofluar objectives and 0.75–1.4 N.A. condensers), scoring acrosome reactions in rodent sperm is unreliable. (2) If

[19] S. Meizel, *Biol. Rev.* **59**, 125 (1984).
[20] J. D. Bleil and P. M. Wassarman, *Dev. Biol.* **102**, 1363 (1986).
[21] N. L. Cross and S. Meizel, *Biol. Reprod.* **41**, 635 (1989).

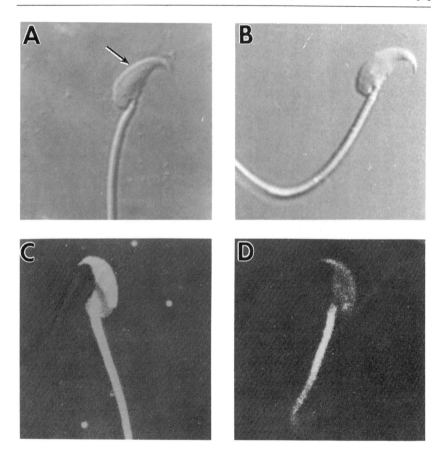

Fig. 1. Microscopic methods for assaying acrosome reactions. (A, B) Nomarski differential interference contrast microscopy of (A) acrosome-intact and (B) acrosome-reacted sperm. The intact acrosome is barely resolvable as a faint white stripe overlying the anterior head (arrow). In acrosome-reacted sperm, no such stripe is detected, and there is a subtle decrease in the curvature of the sperm head. (C, D) Fluorescent detection using chlortetracycline. Capacitated and acrosome-intact sperm (C) display fluorescence on the anterior head and a dark band over the postacrosomal region. Acrosome-reacted sperm (D) reveal a weak fluorescence over the entire head. (CTC photomicrographs donated by Dr. Bayard Storey, University of Pennsylvania.)

sperm are stuck on poly(L-lysine)-coated coverslips, no adjustments should be made with the focusing knob while scoring. Even minor adjustments of the focusing knob can lead to a change in the interference patterns and result in erroneous scoring. (3) All assays should first be calibrated using positive (10 μM A23187 ionophore in a Ca^{2+}-containing medium for

30 min) and negative (no additions in a Ca^{2+}-free medium) controls. Slides should be scored blind, and all sperm in a field must be counted. Again, it is imperative that the focus not be adjusted within a field, as this can lead to false positives. (4) If possible, the assay should be calibrated using either transmission electron microscopy or an acrosin release assay (see below) to check the reliability of the scorer.

Assays That Measure Acrosin Release

The end point of the acrosome reaction is accompanied by a release of acrosin, a trypsinlike protease that is enclosed within the acrosomal vesicle. In general, any enzymatic assay that can be used to detect trypsin can be used to quantify acrosin release as well. These assays employ either chromogenic[22,23] or fluorogenic[24] groups that are complexed to a number of arginyl or lysyl residues and that display altered spectral characteristics after hydrolysis. Sperm concentrations as low as 10^5/ml (in 1-ml volumes) can be assayed. The two assays described below have been used for sperm from several species, including human, mouse, and ram.[25]

Spectrophotometric Assays of Acrosin Release

Materials

N^α-Benzoyl-L-arginine ethyl ester hydrochloride (BAEE; ICN, Costa Mesa, CA): Make up a 50 mM stock solution in dimethyl sulfoxide (DMSO) and then dilute to 5 mM in culture medium
A23187 (Sigma): 10 mM stock in DMSO, freshly made

Method

1. Adjust sperm concentration to 5×10^5/ml in culture medium and add either agonist, positive control (10 μM A23187 final concentration), or negative control (equal volume of culture medium).

2. At intervals (no greater than 30 min), pellet sperm at 500 g at 4° and collect the supernatant. Store on ice.

3. Adjust each tube to 3 ml and place in a 1 cm path length quartz cuvette. The reference cuvette should contain 3 ml of culture medium and 50 μl of the BAEE stock solution.

[22] G. W. Schwert and Y. Takenaka, *Biochim. Biophys. Acta* **16**, 570 (1955).
[23] E. L. Smith and M. J. Parker, *J. Biol. Chem.* **233**, 1387 (1958).
[24] Y. Kanaoka, T. Takahashi, H. Nakayama, K. Takada, T. Kimura, and S. Sakakibara, *Chem. Pharm. Bull.* **25**, 3126 (1977).
[25] C. A. McKinnon, F. E. Weaver, J. A. Yoder, G. Fairbanks, and D. E. Wolf, *Mol. Reprod. Dev.* **29**, 200 (1991).

4. Record A_{253} (absorbance at 253 nm) for sufficient time to establish rates of nonenzymatic hydrolysis. Add 50 μl BAEE stock solution and monitor A_{253}. If the BAEE stock solution is adjusted correctly, the change in A_{253} with time should be linear: if the rate is too high, dilute the stock solution so that linearity is achieved. In general, the BAEE stock solution should be adjusted so that the A23187-treated sperm exhibit a change in absorbance of approximately 1 OD unit in 5 min.

5. Determine the slope of each line. Report values as normalized activities, that is, (slope$_{agonist}$ − slope$_{medium}$)/(slope$_{ionophore}$ − slope$_{medium}$), where the terms indicate the measured slopes in the presence of agonist, culture medium (negative control), and A23187 (positive control), respectively. In this way, all activities should range between 0 and 1.

Comments. (1) Because acrosin, like most proteases, will autodigest, it is imperative that all solutions be kept at 4°. (2) Determine appropriate BAEE concentrations first using trypsin (type I, Sigma). Trypsin concentrations of approximately 0.01 mg/ml can be used to determine the optimal [BAEE]. (3) There have been reports of using this type of assay without first pelleting the sperm (i.e., with sperm in the cuvette). Although changes in optical density can be seen, the changes are often attributable to light scattering, which can be significant at 253 nm. In cases where it is essential to use sperm suspensions, a fluorometric assay provides a reasonable alternative (see below). (4) A number of different chromogenic substrates are available besides BAEE. One performing particularly well is N^{α}-benzoyl-DL-arginine-*p*-nitroanilide hydrochloride (ICN).

Fluorometric Assays of Acrosin Release

Materials

Benzoyl-L-arginine-4-methylcoumaryl-7-amide hydrochloride (Bz-Arg-MCA; ICN): Make 50 mM stock solutions in DMSO and dilute to 5 mM in culture medium. Make fresh
A23187 (Sigma): 10 mM stocks in DMSO, freshly made

Method

1. Adjust slit widths to less than 10 nm on excitation and emission monochromators to reduce stray light artifacts. In our experience, excitation and emission slit settings of 10 and 5 nm, respectively, are ideal.

2. Set excitation and emission monochromators at 350 and 450 nm, respectively. Following hydrolysis, coumarin emission is blue-violet in color.

3. Adjust the sperm concentration to approximately 10^5/ml in culture medium and add agonist. Negative and positive controls receive equivalent volumes containing culture medium or A23187, respectively.

4. Sperm are sedimented (500 g, 4°) at selected intervals (0–30 min), and supernatants are stored on ice.

5. Adjust each tube to 3 ml and place in a 1 cm path length quartz cuvette. Reference cuvettes receive 3 ml of culture medium and 50 μl of Bz-Arg-MCA stock solution.

6. Record the fluorescence emission intensity as a function of time for each sample. Intensity changes should be linear with time if the Bz-Arg-MCA is in the correct concentration range. Rates of intensity change that are too high may be modulated by diluting stock Bz-Arg-MCA until linearity is achieved.

7. Measure the slope of each line. Report values as normalized activities, that is, $(slope_{agonist} - slope_{medium})/(slope_{ionophore} - slope_{medium})$, where the terms indicate the determined slopes in the presence of agonist, culture medium (negative control), and Ca^{2+}/H^+ ionophore (positive control), respectively. In this way, all activities should range between 0 and 1.

Comments. (1) In general, the fluorometric assay is more sensitive and can be employed with lower concentrations of enzyme (i.e., sperm) and substrate than the spectrophotometric version. (2) As with the spectrophotometric assay, optimal substrate concentrations should first be determined with trypsin. In this assay, 1 μg/ml trypsin (type I, Sigma) should be used for the standard. (3) It is not necessary to remove sperm by sedimentation prior to assays, in contrast to the situation described for the spectrophotometric assay (see above). However, several precautions must be met if the assay is to be carried out in sperm suspensions. First, assays should be performed at temperatures used for sperm cultures. In the case of mouse sperm, this means heating the cuvette holder to 37°. Second, dead cells in sperm suspensions continually settle, and so the cuvette must be stirred continually. Stirring rates must be carefully controlled, as excessive stirring can produce cell breakage and death.[26,27] (As such damage releases acrosin, these assays can be used to evaluate the extent of cell death.) Third, the slit widths must be as small as possible (2–5 nm); although this minimizes stray light, it also significantly decreases the amount of signal. Given these considerations, it is often better to use the cell-free assay described above than to use cell suspensions.

[26] R. A. Cardullo and R. A. Cone, *Biol. Reprod.* **34,** 820 (1986).
[27] J. M. Baltz, P. O. Williams, and R. A. Cone, *Biol. Reprod.* **43,** 485 (1990).

Assays Employing Fluorescent Antibodies

Monoclonal and polyclonal antibodies have been used to assay acrosome reactions.[28–32] Useful antibodies fall into two classes: (1) those that recognize antigens prior to the acrosome reaction, with loss of antigen following exocytosis (e.g., HS21[28,29]); and (2) those that change their distribution following exocytosis (e.g., guinea pig sperm PH-20,[30] ram sperm ESA-152,[25] and mouse sperm M-42[32] and β-1,4-galactosyltransferase[31]). These antibodies are, at present, not available commercially and must be obtained from individual investigators. The assay below is a generic protocol for indirect immunofluorescent labeling of the sperm surface, using a fluorescent secondary antibody directed against an unlabeled primary antibody. The assay is designed for small volumes and is performed directly on poly(L-lysine)-coated coverslips. Larger volumes (>100 μl) are required to perform the assay in solution.

Materials

Poly(L-lysine), 70–100 kDa (Sigma): 1 mg/ml in distilled water, made freshly

2% Formaldehyde (reagent grade, Sigma), in culture medium

Primary antibody solution (e.g., monoclonal or polyclonal IgG): Adjust antibody to 50 μg/ml in culture medium containing 10 mM NaN$_3$. Solubilized antibody is stable for up to 6 months at 4°. Antibody integrity should be checked by polyacrylamide gel electrophoresis, and specificity should be confirmed. Sediment (3000 g; 30 min; 4°) and discard large particulates prior to use

Fluorescently conjugated secondary antibody solution [e.g., fluorescein isothiocyanate (FITC)-labeled goat anti-rabbit IgG; ICN]: Adjust antibody to 500 μg/ml in culture medium containing 10 mM NaN$_3$. Remove unconjugated fluorophore using a 10–15 ml column of BioGel P-6 (Bio-Rad, Richmond, CA) and collect the void volume. Dilute solution to approximately 200 μg/ml in culture medium containing 10 mM NaN$_3$. Make fresh

Method

1. Adjust the sperm concentration to 5 × 10^6/ml and sediment cells (500 g). Gently resuspend cells in 2% formaldehyde solution for 10 min and then sediment fixed cells.

[28] H. M. Florman, K. B. Bechtol, and P. M. Wassarman, *Dev. Biol.* **106**, 243 (1984).
[29] D. P. Wolf, J. Boldt, W. Byrd, and K. B. Bechtol, *Biol. Reprod.* **32**, 1157 (1985).
[30] B. M. Phelps and D. G. Myles, *Dev. Biol.* **123**, 63 (1987).
[31] L. C. Lopez and B. D. Shur, *J. Cell Biol.* **105**, 1663 (1987).
[32] K. A. Lakoski, C. P. Carron, C. L. Cabot, and P. M. Saling, *Biol. Reprod.* **38**, 221 (1988).

2. Cells are resuspended in culture medium containing 50 mM glycine and 10 mM NaN$_3$, incubated 30 min, washed by sedimentation, and resuspended at 10^6 cells/ml in culture medium.

3. Affix cells to poly(L-lysine)-coated coverslips and allow 10 min for settling. Remove loosely adherent sperm by dipping coverslips in culture medium 3 times.

4. Pipette 25 μl of primary antibody solution onto the coverslip, cover with a petri dish, and incubate for 1.5–2 hr at 4°.

5. Gently siphon off primary antibody solution and dip coverslips in culture medium 3 times.

6. Pipette 25 μl of secondary antibody solution onto the coverslip, cover with a petri dish, and incubate for 0.5–0.75 hr at 4°.

7. Gently siphon off secondary antibody solution, dip coverslips in culture medium 3 times, wick remaining solution from coverslips with tissue paper, and mount coverslips onto microscope slides using Gelvatol (Monsanto).

8. Find sperm using bright-field optics and then switch to epifluorescence optics. Score all cells in a field according to acrosomal status.

Comments. (1) Intact immunoglobulins G (IgGs) are notorious for cross-linking antigens on the cell surface, leading to patching and capping events. In some cases cross-linking receptors will actually lead to activation of signaling cascades.[33] Indeed, it has been proposed that cross-linking of ZP3 glycopeptide binding sites on the sperm surface is sufficient to initiate the acrosome reaction.[34] Cross-linking of other sperm surface constituents by bivalent IgG can lead to lateral redistribution[25] and to exocytosis.[25,35,36] Because of these potential artifacts, Fab fragments should be used instead of bivalent IgGs for both primary and secondary antibodies whenever possible. If intact antibodies must be used, all labeling should occur at 4° and in the presence of fixative. (2) If 2% formaldehyde is inappropriate for fixation, then 3% glutaraldehyde or 70% ethanol can be used instead. Beware that even electron microscopy (EM)-grade glutaraldehydes are somewhat autofluorescent, especially when used with fluorescein filters. (3) Many antigens are present on the cell surface at low copy number, making observation of immunofluorescence patterns difficult. This problem was resolved with the advent of low light level video cameras (e.g., SITs or cooled CCDs), which capture images in seconds where photographic film may require minutes. (4) A necessary

---

[33] D. L. Cadena and G. N. Gill, *FASEB J.* **6**, 2332 (1992).
[34] L. Leyton and P. M. Saling, *J. Cell Biol.* **108**, 2163 (1989).
[35] D. Aarons, H. Boettger-Tong, and G. R. Poirier, *Mol. Reprod. Dev.* **30**, 258 (1991).
[36] M. B. Macek, L. C. Lopez, and B. D. Shur, *Dev. Biol.* **147**, 440 (1991).

Method

1. Centrifuge all dyes in a microcentrifuge prior to use to remove particulates.

2. Adjust the sperm concentration to 5×10^5/ml in culture medium. Mix 200-μl droplets of the sperm suspension 1 : 1 with the 1% trypan blue solution for 10 min under oil at 37°.

3. Transfer sperm to microcentrifuge tubes and dilute to 1.5 ml with culture medium. Concentrate sperm by sedimentation (500 g) and remove residual trypan blue in a second wash with BSA-free culture medium.

4. Fix sperm for 30 min in 3% glutaraldehyde in 0.1 M cacodylate buffer (pH 7.4).

5. After fixation, wash sperm 2 times in deionized water and air-dry onto microscope slides.

6. Gently stain slides in 0.8% Bismarck brown for 5 min and wash with distilled water.

7. Stain slides in 0.8% rose Bengal for 15 min and then wash in distilled water, dehydrate in ethanol, and clear in xylene. Mount with a coverslip and observe using bright-field optics.

8. Score sperm as follows: (a) in dead sperm with intact acrosomes the head appears pale blue or gray with a dark pink apical acrosomal ridge; (b) in degenerative, acrosome-reacted sperm the head is gray with no pink staining in the acrosomal region; (c) in "live" unreacted sperm the postacrosomal region stains brown, the acrosome stains pink, and the apical region of the acrosome is a dark pink; and (d) in "live" acrosome-reacted sperm the postacrosomal region stains pale brown, the acrosome is white or pale pink, and the apical ridge is not stained.

Comments. (1) All cells in a field must be counted in order to avoid bias. (2) Care should be exercised during trypan blue staining, as cells are easily damaged and the resultant dye accumulation can mislead investigators concerning the viability of the sperm preparation. (3) Moribund sperm slough plasma membranes, including the acrosomal membranes. These so-called false acrosome reactions contribute to background levels of exocytosis and can be a particular problem during assays of fixed sperm. In this regard, other differential staining protocols have been described and validated by ultrastructural analysis.[37] The significant advantage of the triple stain lies in the incorporation of a viability assay (trypan blue), thereby permitting discrimination between cells that have lost their acrosomes while still alive and those that have simply died.

Chlortetracycline Fluorescence Assay

Chlortetracycline (CTC) is a fluorescent, chelate probe of Ca^{2+} that was introduced as an antibiotic (Aureomycin) and later applied to examine mitochondrial Ca^{2+} uptake. Subsequently, it was used to determine acrosomal morphology in single mouse sperm, first as a vital stain[41] and then in fixed cells.[42] Functionally mature mouse sperm (i.e., those able to fertilize eggs and undergo zona pellucida-induced acrosomal exocytosis) display a characteristic CTC fluorescent pattern, consisting of intense emission from the acrosomal crescent in the anterior head and from the midpiece (Fig. 1C,D). Anterior head fluorescence is lost following exocytosis (the AR or acrosome-reacted pattern). Apparent intermediate stages can be detected using the fixed-cell version of the assay, owing to the higher [CTC] employed.

Materials

Chlortetracycline (Sigma): Store desiccated at $-20°$
L-Cysteine (Sigma): Store desiccated at room temperature
Glutaraldehyde (12.5%, EM Grade; Polysciences): Make fresh

Method

1. Prepare solution I (500 μM CTC, 5 mM L-cysteine, 1 M Na–HEPES, pH 7.5) immediately before use. Store at 4° in a light-shielded container.

2. To a warmed microscope slide, add 10 μl sperm solution and 10 μl solution I. Mix gently and immediately add 1 μl of 12.5% glutaraldehyde.

3. CTC fluorescence (excitation and emission maxima of 395 and 510 nm, respectively) is stable for at least 1 day if slides are stored in a light-shielded container at 4°. To assess acrosomal morphology, first focus on a field under bright-field optics and then switch to fluorescence. Do not adjust the focus while scoring a field, as this can alter assessments. Score cells as follows: (a) acrosome-intact sperm have a dark posterior head and a brightly fluorescent acrosomal crescent; (b) sperm in an early intermediate stage of the acrosome reaction show a banded pattern, with a dark posterior head, a band of fluorescence (banded, or B pattern) in the region of the equatorial zone, and a dark apical head; (c) sperm in a later intermediate stage have dark apical and posterior head regions and with punctate fluorescence (stippled, or S pattern); and (d) sperm that have completed the acrosome reaction have an entirely dark head (AR pattern).

Comments. (1) The protonated, uncomplexed form of CTC rapidly permeates cellular membranes, whereas the deprotonated anion and Ca^{2+}-complexed forms are relatively impermeable. The latter form binds tightly to membranes, with marked fluorescence enhancement. As a result, CTC reports from intracellular as well as surface membranes. In addition, CTC is a relatively nonselective indicator and also shows fluorescence enhance-

[41] P. M. Saling and B. T. Storey, *J. Cell Biol.* **83**, 544 (1979);
[42] C. R. Ward and B. T. Storey, *Dev. Biol.* **104**, 287 (1984).

ment on binding other metal cations (e.g., Mg^{2+}). These forms do have unique spectral characteristics, although discrimination typically requires specialized excitation conditions (i.e., customized filters or a scanning monochromator).[43] Finally, the dye may also report alterations in membrane potential.[44] Users are therefore cautioned that this probe is useful for assaying acrosomal exocytosis, but that the ionic and biological basis for altered emission patterns (and specifically for intermediate forms) is not understood.

(2) It is often possible to combine filters and dichroic mirrors from other applications to the detection of CTC. The choice of an excitation filter depends in part on the light source. With xenon arc lamps, appropriate filters should transmit light at 395 nm (CTC excitation maximum). Nevertheless, the excitation spectrum has a broad peak, and the 366 nm line of mercury sources can be used. Exciter filters designed for Fura-2 or for Hoechst 33342 typically can be used with CTC. The emission peak for CTC is very close to that of fluorescein, thereby permitting the use of dichroic mirrors and barrier filters designed for FITC.

(3) CTC fluorescence photobleaches rapidly, and precautions include using fresh (prepared within 10 min) solutions that are kept on ice and shielded from light. (4) Glutaraldehyde is notoriously fluorescent, and only EM grades should be used. In all experiments using glutaraldehyde, control slides containing sperm with no CTC should be scored separately. (5) To avoid bias it is important to account for all cells in each field. (6) The CTC assay was developed for mouse sperm and has also been applied to human[45] and equine sperm.[46] In contrast, the fluorescence patterns observed in bovine and ovine sperm do not correlate with acrosomal morphology (R. A. Cardullo and H. M. Florman, unpublished results, 1993).

Assays Employing Fluorescent Indicator, Fura-2, for Measuring Agonist-Promoted Elevations of Internal Calcium

Elevation of Ca_i is a characteristic of the acrosome reaction induced in sperm of mammals by agonists from the zona pellucida[9,47,48] and from

[43] R. Y. Tsien, in "Fluorescence Microscopy of Living Cells in Culture, Part B: Quantitative Fluorescence Microscopy-Imaging and Spectroscopy" (D. L. Taylor and Y.-l. Wang, eds.), p. 127. Academic Press, New York, 1989.

[44] S. Tang and T. Beeler, Cell Calcium 11, 425 (1991).

[45] M. A. Lee, G. S. Trucco, K. B. Bechtol, N. Wummer, G. S. Kopf, L. Blasco, and B. T. Storey, Fertil. Steril. 48, 649 (1987).

[46] D. D. Varner, C. R. Ward, B. T. Storey, and R. M. Kenney, Am. J. Vet. Res. 48, 1383 (1987).

[47] B. T. Storey, C. L. Hourani, and J. B. Kim, Mol. Reprod. Dev. 32, 41 (1992).

[48] H. M. Florman, M. E. Corron, T. D.-H. Kim, and D. Babcock, Dev. Biol. 152, (1992).

follicular fluids[10,13,49,50] as well as that in sea urchin sperm promoted by the fucosyl sulfate glycoconjugate agonist of egg jelly.[51] Moreover, analysis of single sperm responding to zona pellucida agonist shows that Ca_i elevations are restricted to those cells that subsequently undergo the acrosome reaction,[9] and they thus provide an effective exocytosis assay. This assay may be applied to sperm populations, as well as to individual sperm cells, in order to determine the exocytotic response to agonist. This approach permits analysis of Ca_i-coupled signal transduction in living sperm, and it provides a relative measure of exocytosis. The following protocol has been successfully applied to bovine sperm[9] and, in modified form, to the mouse.[47]

Materials

Fura-2 acetoxymethyl ester (F2/AM; Molecular Probes, Eugene, OR): Resuspend at 1 mg/ml in dry (4 Å molecular sieves), spectroscopic grade DMSO (1 mM final concentration). The resulting solution will either be clear or have a slightly yellowish tint. Effective solubilization of dye can be confirmed by the development of an orange color following alkaline (1 N NaOH) deesterification. Solutions of F2/AM may be stored in inert environments at $-20°$ in a light-shielded container

Poly(L-lysine), 70–100 kDa (Sigma): 1 mg/ml in distilled water, made fresh

$MnCl_2$ (Sigma): 10 mM stock solutions are stored at room temperature

$CaCl_2$ (Sigma): 1 M stock solutions are stored at $-20°$

Diethylenetriaminepentaacetic acid (DEPTA; Sigma): 50 mM stock solution in 100 mM HEPES (pH 7.4), stored at room temperature

Ethylene glycol bis(β-aminoethyl ether)-N,N,N',N'-tetraacetic acid (EGTA; Sigma): Resuspend 15.12 g EGTA in 100 ml of 2 M TAPS [N-tris(hydroxymethyl)methyl-3-aminopropanesulfonic acid; 48.66 g/100 ml water, pH 8.7; Sigma]. Store at room temperature

Method

1. Coat coverslips with poly(L-lysine) solution (10 min, room temperature). Rinse several times in distilled water and store wet until used.

2. Incubate functionally mature ("capacitated") sperm (1–5 × 10⁷ sperm/ml) with 1 μM F2/AM for 30 min at physiological temperature.

[49] P. Thomas and S. Meizel, *Gamete Res.* **20,** 397 (1988).
[50] L. E. Franklin, C. Barros, and E. N. Fussell, *Biol. Reprod.* **3,** 180 (1970).
[51] R. W. Schackmann, *in* "The Cell Biology of Fertilization" (H. Schatten and G. Schatten, eds.), p. 3. Academic Press, New York, 1989.

Collect sperm by sedimentation (300 g, 5 min) and incubate an additional 30–60 min to permit more complete deesterification.

3. Add aliquots of sperm to the coverslip, incubate 5 min, and gently wash in culture medium to remove loosely adherent sperm.

4. Place the coverslip in a fluorescence microscope and collect the fluorescence emission (510 nm) from "resting" (i.e., not stimulated by agonist) cells. For an "excitation ratioing" dye, such as Fura-2, this entails alternate excitation at wavelengths that excite the Ca^{2+}-bound and Ca^{2+}-free forms of the dye, respectively. Add agonist and monitor alterations in emission.

5. Evaluate the contribution of extracellular dye by recording the fluorescence intensity during sequential addition of 10 μM $MnCl_2$ (1 μl/ml) and 100 μM DEPTA/HEPES (2 μl/ml). Mn^{2+} quenches F2 fluorescence[52] and does not enter sperm rapidly.[9] DEPTA is a poorly permeable, Mn^{2+}-selective chelator. Therefore, Mn^{2+}-quenchable, DEPTA-rescued fluorescence originates from extracellular dye.

6. Fluorescence is calibrated by addition of 10 μM ionomycin, a Ca^{2+}/H^+ ionophore, to sperm in medium containing saturating $[Ca^{2+}]$. For example, typical mammalian sperm culture media contain 1–3 μmol/ml Ca^{2+}, and this can be chelated by addition of 15–20 μl/ml EGTA–TAPS (6–8 μmol/ml). By definition, maximum and minimum fluorescence (F_{max} and F_{min}) are obtained after ionophore addition and chelation, respectively.

7. Repeat Steps 4–6 with cells that were *not* loaded with Fura-2 in order to determine levels of cellular autofluorescence.

8. Ca_i is determined as follows:

$$Ca_i/K_D = (R_{max} - R)/(R - R_{min})B \tag{1}$$

where R, R_{max}, and R_{min} are the observed fluorescence ratios (F_{340}/F_{380}) in experimental samples, in medium containing saturating Ca^{2+}, and in medium containing minimal Ca^{2+}; B is equal to the ratio of Ca^{2+}-bound to Ca^{2+}-free fluorescence emission following excitation at the Ca^{2+}-insensitive wavelength (i.e., a scaling factor); and K_D is the dissociation constant for Fura-2–Ca^{2+} complexes in the intracellular environment.[52–54]

Comments. (1) Conditions for loading (time, [F2/AM], [sperm]), dye efflux, and permeabilization should be determined when applying this protocol to each new species. In particular, note that rodent sperm do not generally tolerate the sedimentation conditions applied in other species (e.g., bovine,[9,48] ovine[55]), nor do they retain deesterified dyes so well

[52] G. Grynkiewicz, M. Poenie, and R. Y. Tsien, *J. Biol. Chem.* **260**, 3440 (1985).
[53] R. Y. Tsien, T. J. Rink, and M. Poenie, *Cell Calcium* **6**, 145 (1985).
[54] P. H. Cubbold and T. J. Rink, *Biochem. J.* **248**, 313 (1987).
[55] D. F. Babcock and D. R. Pfeiffer, *J. Biol. Chem.* **262**, 15041 (1987).

(R. A. Cardullo and H. M. Florman, unpublished results). (2) Fluorescence emission may be collected either with an appropriate camera (e.g., SIT), with subsequent analysis using digital image processing algorithms, or with a photomultiplier tube. (3) Absolute calibration of intracellular dye is difficult, owing to uncertainties regarding the physical nature of the internal milieu (viscosity, ionic composition,[52–54] protein interactions[56]) and how this may influence dye–Ca^{2+} affinity and fluorescence characteristics. It is convenient to express results as the dimensionless quotient, Ca_i/K_D, as all values on the right-hand side of Eq. (1) are determined experimentally. Calibration problems have been discussed in great detail elsewhere.[52–54] (4) Spectral properties of intracellular dyes should be determined directly. (5) New and improved dyes are being developed constantly. Dyes such as Calcium Green, Calcium Orange, and Calcium Crimson (Molecular Probes) show enhanced fluorescence emission intensity when chelated with Ca^{2+}. In addition, these dyes are excited with visible light, do not require specialized excitation filter arrangements, and have a higher dynamic range than Fura-2.

[56] M. Konishi, A. Olson, S. Hollingworth, and S. M. Baylor, *Biophys. J.* **54,** 1089 (1988).

[9] Culture of Preimplantation Embryos

By JOEL A. LAWITTS and JOHN D. BIGGERS

Introduction

Attempts to culture preimplantation mammalian embryos began in 1912 when Brachet described the behavior of cultured rabbit blastocysts.[1] Partial success in the culture of the preimplantation stages of the mouse was not achieved until 1949 when Hammond reported that 8-cell mouse embryos would differentiate into blastocysts in a complex medium of biological fluids.[2] This result stimulated the work of Whitten[3] who showed that outbred 8-cell mouse embryos can grow into blastocysts in a simple, chemically defined medium, namely, a Krebs–Ringer bicarbonate[4] supplemented with carbon sources and bovine plasma albumin. McLaren and Biggers[5] then demonstrated that blastocysts produced by Whitten's tech-

[1] A. Brachet, *C.R. Hebd. Seances Acad. Sci.* **155,** 1191 (1912).
[2] J. Hammond, Jr., *Nature (London)* **163,** 28 (1949).
[3] W. K. Whitten, *Nature (London)* **177,** 96 (1956).
[4] H. A. Krebs and K. Henseleit, *Z. Phys. Chem.* **210,** 33 (1932).
[5] A. McLaren and J. D. Biggers, *Nature (London)* **182,** 877 (1958).

METHODS IN ENZYMOLOGY, VOL. 225

Copyright © 1993 by Academic Press, Inc.
All rights of reproduction in any form reserved.

nique could develop into normal young after being transferred to the uterus of a surrogate mother.

Considerable progress in the culture of mouse preimplantation embryos has been made in the years that have elapsed since the pioneer studies were done. The progress has largely involved the improvement of media. From an historical point of view three periods can be recognized. The first period up to 1971 has been described in detail elsewhere.[6,7] During this time the basic techniques, many of which are still used today, were developed. Often these techniques were dual, such as the test-tube method versus the microdroplet method and 5% oxygen versus 21% oxygen in the gas phase. Two popular media adopted or developed during this period were medium BMOC[8] and Whitten's medium.[9] The 2-cell block phenomenon was also recognized during this period.[10] Until 1968 it had been impossible to culture mouse zygotes (1-cell mouse embryos) beyond the 2-cell stage. Then Whitten and Biggers[11] showed that the zygotes of F_1 hybrids between two inbred lines would develop into blastocysts, whereas the zygotes of inbred lines and outbred animals would not. The second period, from 1971 to 1989, was a time of empirical improvement. After 1978 this work, which was often anecdotal in nature, was driven by laboratories associated with human *in vitro* fertilization clinics. During this time, however, several media became popular with scientists involved in basic research; among these were BWW[6] and M16.[12] The third period began in 1989. This period, which is still ongoing, has been characterized by a more thorough understanding of the physiological principles involved in the improvement of embryo culture techniques. Media are now available that allow the development of the zygote to proceed without interruption through the 2-cell stage to the blastocyst of outbred and inbred strains of mouse. Among these media are CZB[13,14] and SOM,[15] which contain

[6] J. D. Biggers, W. K. Whitten, and D. G. Whittingham, *in* "Methods in Mammalian Embryology" (J. C. Daniels, ed.), p. 86. Freeman, San Francisco, 1971.

[7] J. D. Biggers, *in* "The Mammalian Preimplantation Embryo" (B. D. Bavister, ed.), p. 1. Plenum, New York, 1987.

[8] R. L. Brinster, *J. Reprod. Fertil.* **10,** 227 (1965).

[9] W. K. Whitten, *Adv. Biosci.* **6,** 129 (1971).

[10] J. D. Biggers, *in* "Implantation in Mammals" (L. Gianaroli, A. Campana, and O. Trounson, eds.), Serono/Raven, Rome, in press, 1993.

[11] W. K. Whitten and J. D. Biggers, *J. Reprod. Fertil.* **17,** 399 (1968).

[12] D. G. Whittingham, *J. Reprod. Fertil.* **14** (Suppl.), 7 (1971).

[13] C. L. Chatot, C. A. Ziomek, B. D. Bavister, J. L. Lewis, and I. Torres, *J. Reprod. Fertil.* **86,** 679 (1989).

[14] C. L. Chatot, J. L. Lewis, I. Torres, and C. A. Ziomek, *Biol. Reprod.* **42,** 432 (1990).

[15] J. A. Lawitts and J. D. Biggers, *Mol. Reprod. Dev.* **31,** 189 (1992).

glutamine, and EDTA, as a chelator and for the scavenging of free radicals.[16,17]

An investigator has some liberty in choosing a particular technique for the culture of the preimplantation mouse embryo. This flexibility stems from the ability of the embryos to adapt to the artificial environments that are inevitably imposed on them when placed in culture.[18] The most important concern in each particular investigation is that the chosen technique does not result in artifacts that could bias the results (see below). Gradual improvements will be made by pooling the cumulative experience obtained in each application. The techniques described in this chapter are those that have proved useful in our laboratory.

Superovulation

Superovulation is used for producing larger batches of synchronous embryos than is possible with natural mating. Animals should be housed in a room capable of controlling photoperiod, and should be maintained on a 14 hr light and 10 hr dark cycle (lights on at 0500 hr). The female mice are superovulated by an intraperitoneal (i.p.) injection of 5 IU pregnant mare serum gonadotropin (PMSG), followed by an i.p. injection of 5 IU human chorionic gonadotropin (hCG) 46–48 hr later. Immediately after the hCG injection, females are placed in cages with fertile males and examined the following morning for the presence of a vaginal plug. These mice are assumed to be pregnant, and the time is denoted day 1. Mice have been successfully superovulated in our laboratory with injections of PMSG given from 1100 to 1600 hr. Thus, the time of injection can be integrated with the time of embryo retrieval to maximize development *in vitro*. In our experience the older the embryos are within a given cell stage when they are removed from the donor, the better they will grow *in vitro*. Therefore, embryos removed 24 hr post-hCG may not survive as well as embryos removed at 28 hr.

Culturing Zygotes and Embryos

General Considerations

Successful embryo culture requires skills in mouse handling, microdissection, and embryo handling. Minimizing the time between sacrificing the embryo donors and putting the embryos into culture will result in less

[16] M. Nasr-Esfahani, M. H. Johnson, and J. R. Aitkin, *Hum. Reprod.* **5,** 997 (1990).
[17] Y. Goto, Y. Noda, K. Narimoto, Y. Umaoka, and T. Mori, *Free Radical Biol. Med.* **13,** 47 (1992).
[18] J. D. Biggers, *in* "Preimplantation Genetics" (Y. Verlinsky and A. Kuliev, eds.), p. 25. Plenum, New York, 1991.

TABLE I
COMPOSITION AND OSMOLARITY
OF MEDIUM kSOM[a]

Component	Concentration (g/liter)
NaCl	5.55
KCl	0.186
KH_2PO_4	0.0476
$MgSO_4$	0.0493
Lactate	1.87
Pyruvate	0.022
Glucose	0.036
Glutamine	0.146
BSA[b]	1.0
EDTA (tetrasodium salt)	0.0038
$NaHCO_3$	2.10
$CaCl_2$ (dihydrous)	0.251

[a] Osmolarity of kSOM medium is 256.
[b] BSA concentration is in mg/ml. Penicillin and streptomycin are also added at concentrations of 100 units/ml and 50 μg/ml, respectively.

variation between replicates in experiments and overall increased embryo survival. It is sound practice to monitor the time interval between sacrificing the donors and getting the embryos into culture, and to work toward decreasing this interval. However, it is just as important to make sure that embryos have been washed properly to remove extraneous cells and other debris, as well as any media except for the final culture medium. These aspects of quality control should not be compromised in order to save time.

The choice of a medium for embryo culture depends on the period of embryo development. For preimplantation growth a relatively simple medium, with basic salts and energy sources, will usually be sufficient. Medium kSOM[15] (Table I) was developed from a procedure called simplex optimization, which has the benefit of being able to optimize several components simultaneously.[19] The resulting medium (SOM) has widely different concentrations of components compared to other media presently being used, such as medium M16[12] or CZB.[13] Medium SOM contains comparatively low concentrations of NaCl, KCl, KH_2PO_4, lactate, and glucose (Table I). It has been used to grow zygotes into blastocysts from CF1, CD1, FVB, NOD, and C57BL/6 strain mice, and it will also support *in vivo* survival to term if the embryos are transferred after culture.[20] If

[19] J. A. Lawitts and J. D. Biggers, *J. Reprod. Fertil.* **91,** 543 (1991).
[20] J. A. Lawitts, unpublished observations (1993).

the goal of the research is to grow postimplantation embryos *in vitro*, more complex media are used after the blastocyst stage, often containing some type of serum.[21-23] The nutrient requirements for growth of postimplantation embryos are largely unknown.

Preparation of Culture Media

When experiments are designed to compare the effects of several media simultaneously, it may be beneficial to prepare each medium from individual stock solutions of each component. In our laboratory, as many as 27 different media have been made in this way within a period of 2 hr. Most components for a simple medium such as kSOM are stored as 0.1 M solutions, with the following exceptions. NaCl and $NaHCO_3$ are stored at 1.0 M, whereas bovine serum albumin (BSA; fraction V) is stored at a concentration of 100 mg/ml, $CaCl_2$ at 0.171 M, EDTA at 1.0 mM. Penicillin G and streptomycin sulfate are stored as one solution at concentrations of 10,000 units/ml and 5.0 mg/ml, respectively. All components are stored at 4°, in 50-ml conical centrifuge tubes except for glutamine and the antibiotic mixture, which are stored at $-15°$ in 5-ml polypropylene tubes. Fresh solutions of $NaHCO_3$, pyruvate, lactate, and glucose are prepared weekly. BSA is prepared every time media are made. Water for making up culture media components is first deionized by a filter system which produces 18 MΩ water, which is then glass-distilled.

Culture medium is prepared in 10-ml volumes. The concentrations of the components (in g/liter) in medium kSOM are shown in Table II. All components, except for $CaCl_2$, are added to polystyrene culture tubes, distilled water is added to bring the volume of each tube to 9.0–9.5 ml, and then $CaCl_2$ is added and the tube quickly capped and inverted twice. Finally, distilled water is added to bring the final volume of each medium to 10 ml. Each tube is inverted twice to mix the components; then the contents are drawn into a disposable syringe, and the medium is pushed through a 0.2-μm filter into a second tube. Although culture media can be routinely used for 1 week, in critical experiments comparing different media simultaneously, media are usually made fresh for each trial of the experiment.

Embryo Collection

Collecting embryos requires the use of a medium whose pH is stable when exposed to air. Therefore, most culture media, which are bicarbon-

[21] Y. Hsu, J. Baskar, L. Stevens, and J. Rash, *J. Embryol. Exp. Morphol.* **31**, 235 (1974).
[22] Y. Hsu, *Dev. Biol.* **68**, 453 (1979).
[23] E. S. Hunter, W. Balkan, and T. W. Sadler, *J. Exp. Zool.* **245**, 264 (1988).

TABLE II
COMPOSITION OF MEDIA kSOM, CZB, AND M16[a]

Component	Concentration in medium (mM)		
	kSOM	CZB	M16
NaCl	95	82	94.7
KCl	2.5	4.86	4.78
KH$_2$PO$_4$	0.35	1.17	1.19
MgSO$_4$	0.2	1.18	1.19
Lactate	10.0	30.1	23.3
Pyruvate	0.20	0.26	0.33
Glucose	0.20	0.0	5.56
BSA	1.0	5.0	4.0
NaHCO$_3$	25	25	25
CaCl$_2$	1.71	1.71	1.71
Glutamine	1.0	1.0	0.0
EDTA	0.01	0.1	0.0

[a] Components are in millimolar, except for BSA, which is in mg/ml. Penicillin and streptomycin are also added at concentrations of 100 units/ml and 50 μg/ml, respectively.

ate buffered, are insufficient for maintaining embryos in air for even short periods of time. One commonly used flushing medium is Dulbecco's phosphate-buffered saline (DPBS; Table III). A flushing–holding medium (FHM; Table III) has been developed that is a modification of medium kSOM, with decreased NaHCO$_3$ and additional HEPES buffer. Zygotes are released from the oviducts by inserting a 30-gauge needle in the ostium of the oviduct and injecting approximately 0.1 ml of flushing medium. Alternatively, the swollen portion of the ampulla can be found and pierced with a needle to release the zygotes. Pooled zygotes from several donors are washed through one drop of flushing medium containing 0.65 mg/ml hyaluronidase to remove any cumulus cells, followed by two drops of DPBS. If the experiments involve relatively small numbers of embryos, it is possible to retrieve zygotes from the oviduct using a flushing medium containing hyaluronidase. This eliminates one washing step and therefore decreases the time needed to get the embryos into culture media and then into the incubator. Although it may seem desirable to do final washes in culture media before transferring embryos to culture vessels, this is discouraged because of the rapid pH change which culture media undergo when exposed to air.

TABLE III
COMPOSITION OF MEDIA FHM AND DPBS

Component	FHM		DPBS	
	mM	g/liter	mM	g/liter
NaCl	95.0	5.55	136	8.0
KCl	2.5	0.186	2.68	0.2
KH$_2$PO$_4$	0.35	0.0476	1.47	0.2
Na$_2$HPO$_4$	0.0	0.0	8.1	1.15
MgSO$_4$	0.20	0.0493	0.0	0.0
MgCl$_2 \cdot$6H$_2$O	0.0	0.0	0.49	0.1
Lactate	10.0	1.87	0.0	0.0
Pyruvate	0.2	0.022	0.0	0.0
Glucose	0.2	0.036	0.0	0.0
Glutamine	1.0	0.146	0.0	0.0
BSA[a]	1.0	1.0	3.0	3.0
EDTA (tetrasodium salt)	0.01	0.0038	0.0	0.0
NaHCO$_3$	4.0	0.336	0.0	0.0
HEPES	20.0	4.76	0.0	0.0
CaCl$_2$ (dihydrous)	1.71	0.251	0.9	0.133

[a] BSA concentration is in mg/ml. Penicillin and streptomycin are also added at concentrations of 100 units/ml and 50 μg/ml, respectively.

Embryo Handling

Embryo handling involves collecting embryos after flushing, washing the embryos through various wash drops, and then transferring them into culture medium. Developing good embryo handling skills will not only improve embryo viability by decreasing the time it takes to transfer embryos into culture, it will also minimize the loss of embryos, an important consideration when comparative experiments are being done which require that adequate and preferably equal numbers of embryos be allotted to each medium.

Embryos can be pipetted by mouth using a Pasteur pipette pulled in a flame to a narrow bore and then attached to tubing and a mouthpiece as shown in Fig. 1. Capillary tubing can also be used to make pipettes; however, Pasteur pipettes are preferable because they have thicker glass, which can be held more securely like a pencil, and there is less chance of breaking the pipette at the point where it is held in the hand. The pipette is rolled in a small flame, approximately 1 cm high, until the glass gets noticeably soft. The pipette is then removed from the flame and pulled. This results in a long, narrow shaft that changes diameter gradually over

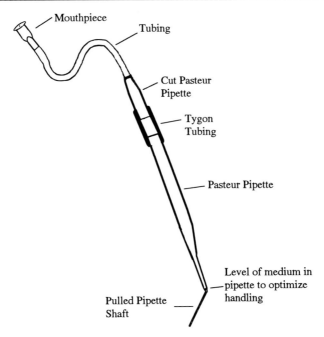

FIG. 1. Mouth-pipetting apparatus consisting of a mouthpiece, tubing, and a Pasteur pipette pulled in a flame to produce a long, narrow shaft that changes diameter gradually.

the pulled length of the pipette. If the pipette remains in the flame as it is pulled, a short shaft with a rapidly changing diameter will result. The pipette is broken by pulling the shaft straight out while pinching slightly. This will often generate a tip which has broken evenly. If the pipette is broken by bending along the shaft, a jagged tip is often produced, which will make it difficult to pick up embryos. A long, narrow shaft approximately 2–3 cm long with a gradually changing diameter and an internal diameter of 200 to 250 μm is optimal for controlling the embryos and keeping them in the tip of the pipette, reducing the chance that embryos will be lost in the pipette, and also lessening the amount of medium being transferred from drop to drop.

Prior to picking up embryos, clean medium is pulled into the pipette up to the first location where the diameter changes significantly (Fig. 1). This procedure reduces the force of capillary action, which rapidly draws the embryos into the pipette and increases the chance of losing embryos. As a result of filling the pipette prior to picking up the embryos, it is possible to keep the embryos near the tip of the pipette and reduce loss as well as decrease the amount of medium transferred between drops. If large numbers of embryos are pulled into the pipette simultaneously, it is still possible, even with the best technique, that embryos may stick to the side of the pipette. Those embryos can often be dislodged by rapidly pulling and then pushing medium into and out of the pipette, being careful not to push all the medium out of the pipette, causing bubbles. During early training, it is inevitable that bubbles are blown into either a wash drop or a culture drop, making it difficult to observe the embryos. Bubbles can be removed by focusing the microscope on top of the bubble and then placing the pipette directly on top of the bubble while applying suction to draw the bubble into the pipette. If removing the bubbles from a culture drop results in a loss of a portion of the medium, the embryos should then be transferred into another drop.

Whenever the pipette contains embryos and is not in a drop of medium, the hole in the mouthpiece should be covered with the tongue. This precaution prevents drawing air into the pipette, which will cause the embryos and medium to move into the area of the pipette where the diameter is large. It is often impossible to retrieve embryos after that happens.

Embryo Culture System

The culture system consists of a tissue culture incubator at 37° equilibrated with 5% CO_2 in air and individual culture chambers within the incubator that can each be equilibrated with any desired atmosphere, produced by a gas mixer (Fig. 2). A line from the gas mixer enters the incubator and connects to a gas washing bottle, which in turn is connected to the culture chamber. The gas washing bottle is filled with distilled, deionized water that warms the gases as they pass through the bottle, thereby decreasing fluctuations in temperature within the chamber as it is being equilibrated. Use of individual culture chambers facilitates experiments where atmosphere is a variable. Also, in situations where the oven is opened several times throughout the day, individual chambers provide a more consistent temperature, provided they are not removed from the incubator. A glass vacuum desiccator with a ground glass joint makes an excellent culture chamber. The tube from the gas washing bottle is attached to the sidecock of the desiccator, and the chamber is gassed with the lid slightly open. The chamber is gas equilibrated for several minutes and then sealed, maintaining a constant atmosphere and eliminating possible pH fluctuations in the culture medium. This system is most useful for cultures with several hours between observations. Otherwise, its advantages are somewhat diminished.

Zygotes are cultured in 60-mm tissue culture dishes in 50-μl drops of medium under 5 ml of silicone oil (dimethylpolysiloxane, 50 centistokes),

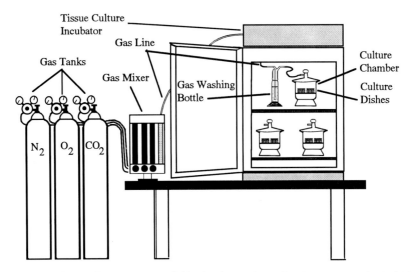

FIG. 2. Embryo culture system. Individual embryo culture chambers are contained within a tissue culture incubator. Individual gases are combined using a gas mixer and enter the culture chambers after passing through a gas washing bottle filled with deionized, distilled water. This system helps to prevent fluctuations in temperature and atmosphere when the tissue culture incubator is opened constantly.

or 35-mm culture dishes with 2.5 ml oil. Oil is prepared by filtering 20–30 ml of deionized, distilled water into a 1-liter bottle of oil. The bottle is swirled vigorously until the water is dispersed throughout the oil, and then the water is allowed to settle out for approximately 2 weeks. Dishes are made up the day prior to the start of each trial and are equilibrated overnight in an atmosphere of 6% CO_2, 5% O_2, and 89% N_2. Each culture dish contains four drops. Zygotes are transferred from flushing medium into one of the drops of culture medium. Any remaining DPBS in the transfer pipette is expelled, and the pipette is refilled from another drop of culture medium. The zygotes are then transferred into the second drop, the medium is expelled from the pipette, and new medium is drawn up from a third drop. The zygotes are then transferred into the third drop. The fourth drop is a spare. In experiments testing different media, this procedure allows for three washes through the appropriate culture medium, in a gas equilibrated state, reducing the chance of pH fluctuation and thereby increasing embryo viability. The culture dishes are returned to the culture chamber at 37°, which is gas equilibrated and then sealed.

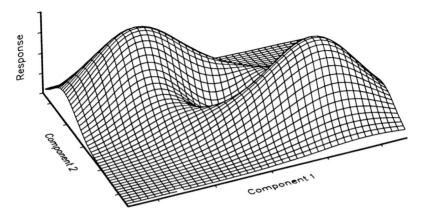

Response

Component 2

Component 1

FIG. 3. Theoretical concentration–response surface demonstrating that there may be different optima depending on the concentration of two components.

Recording Developmental Progress

An experimental end point for culturing preimplantation embryos is often development to the blastocyst stage. However, when developing culture media or when testing embryos from a strain of mouse new to the laboratory, it may be beneficial to score embryo development at 12- or 24-hr intervals. Removing embryos from the incubator several times a day will impede development. When recording developmental progress, minimize the time embryos remain outside the culture chamber. Take out only the number of dishes that can be scored in 5 min or less (3 or 4 dishes). Try to limit the quantity of embryos in each drop to a number that can be scored without much difficulty (10–12). If larger numbers are necessary, use the fourth drop in the dish and divide the embryos into two drops. When the culture dishes are outside of the culture chamber and not on the microscope stage being scored, they can be maintained at 37° in a small bench-top incubator to decrease temperature fluctuations.

Beware of Artifacts

The transfer of preimplantation embryos from a maternal environment to a culture medium imposes stress. Embryos *in vitro* are required to develop in an environment which is artificial. This environment is a far cry from the highly integrated maternal environment in which change is normal and physiologically controlled. If the stress *in vitro* is severe, the embryos will soon succumb. If the stress is mild, the embryos may adapt[18] and proceed through development at a rate less than normal. The fact that several media of different composition support development implies that the embryos can adapt to the artificial environments. The embryos may also arrest at particularly sensitive times, from which they can be rescued. The 2-cell block is a particular example of this phenomenon.[11]

Unifying Mathematical Model for Media Development

Consider a medium prepared from n components. By varying the concentrations of each component, an infinity of media can be prepared. Conceptually the responses to each of these media can be represented by a concentration–response surface in $(n + 1)$ dimensions.[6,24] Figure 3 shows a hypothetical surface for the trivial case where $n = 2$. A medium that gives a maximum response is clearly one of interest. A practical experimental technique that can locate a maximum response is sequential simplex optimization.[25,26] This technique was used by Lawitts and Biggers[19] to identify media SOM and kSOM. The application of this experimental strategy, together with new knowledge regarding the physiology of the preimplantation embryo which will identify important components, is likely to result in better media for use in the future.

Acknowledgment

Supported by the National Cooperative Program on Non-Human *in Vitro* Fertilization and Preimplantation Development, National Institutes of Health Grant Number HD21988.

[24] J. D. Biggers, L. M. Rinaldini, and M. Webb, *Symp. Soc. Exp. Biol.* **11,** 264 (1957).
[25] W. Spendley, G. R. Hext, and F. R. Himsworth, *Technometrics* **4,** 441 (1962).
[26] F. H. Walters, L. R. Parker, S. L. Morgan, and S. N. Deming, "Sequential Simplex Optimization." CRC Press, Boca Raton, Florida, 1991.

[10] Isolation and Culture of Whole Postimplantation Embryos and Germ Layer Derivatives

By KARIN STURM and PATRICK P. L. TAM

Introduction

The successful culture of postimplantation mouse embryos and germ layer tissues has allowed unprecedented control over the conditions affecting embryonic development and histogenesis. Advances in whole-embryo culture techniques have enabled mouse and rat embryos to develop outside the confine of the uterus for 24–72 hr during which gastrulation and early organogenesis are achieved in the same rate as *in vivo* between 6.5 and

Copyright © 1993 by Academic Press, Inc.
All rights of reproduction in any form reserved.

10.5 days postcoitum (dpc). Early pregastrulation (younger than 6.5 dpc) and late organogenesis stage mouse embryos (beyond 10.5 dpc) can also be cultured, though less successfully and for a much shorter period of time (24–36 hr). Nevertheless, such achievement in embryo culture has already rendered rodent embryos, particularly those of the mouse, amenable to direct experimental manipulation that heretofore could only be performed on amphibian and avian embryos.

Although whole-embryo culture is invaluable for studying morphogenesis and histogenesis in an anatomically intact environment, *in vitro* culture of isolated germ layer derivatives offers an additional avenue for investigating the developmental fate of specific cell populations when a much more extended period than that provided by whole-embryo culture is required to obtain full tissue differentiation. In addition, it is invaluable for the analysis of the differentiation pattern of specific cell populations under a defined set of experimental conditions.

In this chapter, procedures for the isolation and the culture of whole embryos and germ layer derivatives are described. Specific emphasis is given to the gastrulation and early organogenesis stages (7.5 to 9.5 dpc); embryos at these developmental stages can be grown most successfully with the existing *in vitro* techniques, and they are the most commonly used for manipulative experiments.

Isolating Embryos for Whole-Embryo Cultures

The viability of embryos in culture depends critically on the ability to exchange nutrients and metabolic waste with the culture medium by diffusion through the investing membranes to sustain growth. Embryos at gastrulation and organogenesis stages rely almost entirely on the nutritive functions of the yolk sacs rather than the chorioallantoic placenta, which is not fully functional until the fetal stages. However, the parietal yolk sac, which is made up of trophoblasts, Reichert's membrane, and the parietal endoderm, seldom expands properly *in vitro* and thus impedes embryonic growth. The strategy of the explantation procedure therefore involves (1) extracting the decidua from the uterine horns, (2) dissecting the investing decidual mass away from the conceptus, (3) removing the parietal yolk sac and the trophoblasts which interfere with embryonic growth, and (4) isolating the embryo while keeping the visceral yolk sac, the amnion, and the ectoplacental cone intact.

Timing of Embryonic Ages

Breeding colonies of mice are maintained on a normal daily cycle of 12 hr light and 12 hr darkness (the switch to the light period is set at 6–7 a.m. and the dark period at 6–7 p.m.). Animals normally mate in the middle of the dark period, and this is ascertained by the presence of copulation plugs in the vagina the morning after. Mice are 0.5 days pregnant on the day plugs are detected, and embryos at the appropriate stages of development are obtained by dissecting the uterus of the pregnant animals at 6.5 to 9.5 dpc. Alternatively, mouse colonies can be maintained on a reversed light cycle so that the animals are in the dark phase of the cycle during our normal daytime. This 12 hr shift in the light cycle results in a corresponding shift in the time of the mating. If plugs are checked at the routine time as for animals under normal light cycle, the day of plug detection is already 1.0 day pc. Therefore, when the pregnant animals maintained under reversed light cycles are dissected during normal working hours, embryos at 6.0 to 9.0 dpc are obtained. This is particularly useful for experiments which require the studying of a narrow window in a rapid succession of morphogenetic events such as those of neurulation, axial rotation, and somite formation.

Dissecting Decidua

Pregnant mice are sacrificed by cervical dislocation and placed in a supine position. The fur over the abdomen is cleansed with 70% (v/v) ethanol, which also helps to keep any loose hairs from flying around and sticking to the internal viscera. A V-shaped incision is made through the skin and the underlying muscles, with the apex at the pubic symphysis and extending laterally and rostrally to the costal margin. After reflecting the skin and muscle flaps over the thorax, the intestines are displaced aside to expose the uterine horns. The cervical segment of the uteri is grasped with a pair of blunt forceps and is severed from the vagina with scissors. The two uterine horns, which are still connected by the cervical stump, is then lifted clear of the peritoneal cavity, and the mesometrium together with any adhering fat is trimmed away along the length of the horns (Fig. 1a). Finally, the uteri are explanted by cutting their connection with the oviducts. The uteri are rinsed in 2–3 changes of explantation medium (PB1 or M2 medium, prewarmed to 25°–30°) and finally placed in fresh medium in disposable 60-mm petri dishes (Corning, New York, NY). The composition of these media is given in Table I. It is important to check regularly the quality of every new batch of media, which ordinarily remains good for at least 2 weeks if kept at 4°.

The uterine horns are opened by splitting the uterine wall along the antimesometrial side (Fig. 1b). This is best done by teasing the muscular tissue along the long axis of the horns using two pairs of watchmaker's forceps under a dissecting microscope. Care should be taken to avoid any excessive compression of the decidua during the dissection, which may

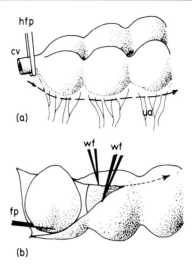

FIG. 1. Explantation of the decidua. (a) Longitudinal cut at the root of the mesometrium (dashed line, arrows show the direction of the cut; ua, uterine arterial network) while holding the cervical segment (cv) with the holding forceps (hfp). (b) Splitting of the uterine wall along the antimesometrial side (dashed line and arrow) with two pairs of watchmaker's forceps (wf) and dislodging of the deciduum from the uterine wall by pinching the tissue with forceps (fp).

deform the embryos and compromise their developmental potential. After the uterus is torn open, the uterine walls are pushed apart to expose the decidua, which are then freed by carefully sliding the points of a pair of forceps underneath the decidua and pinching them off the uterine wall (Fig. 1b). The decidua are transferred gently with fine forceps or a wide-bore Pasteur pipette to fresh explantation medium.

Dissection of Primitive Streak Stage Embryos

Embryos at the primitive streak stage (6.5–7.5 dpc) are found as cylindrical structures in the antimesometrial half of the pear-shaped decidua (Fig. 2a). Dissection of the decidual tissues therefore usually starts with the broader end in order to avoid damaging the embryo. The deciduum is held lightly by a pair of watchmaker's forceps (holding forceps), and a cut is made at one-third the height of the deciduum in the midplane (Fig. 2a) by pressing together the points of another pair of watchmaker's forceps (dissecting forceps). The deciduum is then split into halves by grasping and pulling apart the flaps produced by the vertical cut with forceps. The egg cylinder, with the reddish ectoplacental cone and a fluffy trophoblastic covering, is often left embedded in one-half of the deciduum. The half-

TABLE I

COMPOSITION OF PB1 AND M2 MEDIA FOR EXPLANTING MOUSE EMBRYOS

Component	PB1 medium		M2 medium	
	mM	g/liter	mM	g/liter
NaCl	136.9	8.0	94.7	5.53
KCl	2.68	0.2	4.78	0.36
$MgSO_4 \cdot 7H_2O$	—	—	1.19	0.16
$MgCl_2 \cdot 6H_2O$	0.49	0.1	—	—
$Na_2HPO_4 \cdot 12H_2O$	8.04	2.88	—	—
KH_2PO_4	1.47	0.2	1.19	0.16
$CaCl_2 \cdot 2H_2O$	0.90	0.13	1.71	0.25
Glucose	5.56	1.0	5.56	1.0
Sodium pyruvate	0.33	0.036	0.33	0.036
Sodium lactate (60% syrup)	4.35	—	—	23.28
$NaHCO_3$	—	—	4.15	0.35
HEPES	—	—	20.85	4.97
BSA[a]		4.0		4.0
Penicillin		0.06		0.06
Streptomycin sulfate		0.05		0.05
Phenol red		0.01		0.01

Make up to 1 liter with twice distilled water and sterilize by filtering through a 0.22-μm
   membrane filter (Millipore); final pH should be 7.3–7.4 and osmolarity at 286–292
   mOsmol/liter for PB1 medium and 284–286 mOsmol/liter for M2 medium

[a] BSA (bovine serum albumin, Sigma or Miles Laboratory) is purified by removing the
   free fatty acid and other contaminants through extensive dialysis with distilled water
   followed by lyophilization. Fetal calf serum or heat-inactivated rat serum may be added
   at 10% (v/v) in place of BSA.

deciduum is pinned down with the holding forceps, and the egg cylinder
is scooped out by teasing the tissues surrounding the egg cylinder with
the closed points of the dissecting forceps. The entire explantation procedure can be expedited by cutting the decidua into halves with a pair of
Wecker's iridectomy scissors. The plane of cutting is positioned slightly
off the midplane for 6.5-day egg cylinders (Fig. 2a) and is slanted more
to one side for the larger 7.5-day egg cylinders in order to avoid cutting
the embryo. The overlying tissue layers are removed, and the egg cylinder
is then taken out of the decidua as described before.

The next step is to remove Reichert's membrane and the associated
trophoblasts and parietal endoderm. Reichert's membrane is grasped with
the dissecting forceps near the junction between the embryonic and extraembryonic portions of the egg cylinder, where usually there is enough
clearance of Reichert's membrane from the visceral endoderm to allow

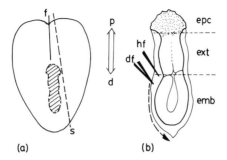

FIG. 2. Dissection of primitive streak stage embryos. (a) Positions of the egg cylinder (shaded), the vertical cut (f) with forceps, and the slicing of deciduum (s) with a pair of iridectomy scissors. (b) Tearing of Reichert's membrane with the dissecting forceps (df) in the distal direction with the membrane pinned down by holding forceps (hf) next to the junction between the extraembryonic (ext) and embryonic (emb) portions of the egg cylinder. Reichert's membrane is finally trimmed back to the edge of the ectoplacental cone (epc). The double arrow shows the proximal (p)–distal (d) axis of the egg cylinder.

the forceps to pierce through without damaging the embryo. Next, the egg cylinder is fixed by pinning Reichert's membrane down on the petri dish with the holding forceps at a position immediately proximal to that of the dissecting forceps. The membrane is then torn by gently bringing the dissecting forceps, which is still grasping the membrane, toward and then around the distal tip of the egg cylinder (Fig. 2b). Once it is reflected over the tip of the cylinder, Reichert's membrane will retract on its own accord onto the extraembryonic portion of the egg cylinder. The loosened Reichert's membrane is then trimmed away from the edge of the ectoplacental cone to prevent the embryo from sticking to the culture vessel or to another embryo during culture.

Embryos may be left in the explantation medium for 30–45 min at room temperature without any significant effect on developmental potential. However, it is advisable to transfer dissected embryos using a sterile Pasteur pipette to preequilibrated culture medium (see later sections) immediately after a batch of 10–15 have been harvested. To avoid diluting the culture medium, only a minimum amount of explantation medium should be used in transferring embryos.

Dissection of Early Organogenesis Stage Embryos

Embryos at the early organogenesis stage (8.5–9.5 dpc) are at least 3–4 times the size of the egg cylinder and have an expanded visceral yolk

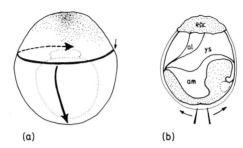

(a)   (b)

FIG. 3. Dissection of the 8.5-day embryo. (a) Position of the conceptus (dotted lines) and the cuts around and down the deciduum, starting from the point indicated by the small arrow. (b) Tearing of Reichert's membrane by two pairs of watchmaker's forceps by first splitting the membrane at the embryonic pole of the conceptus. epc, ectoplacental cone; al, allantois; ys, visceral yolk sac cavity; am, amniotic cavity.

sac and amniotic cavity (Fig. 3a). Exceptional care must be taken during dissection not to compress the fluid-filled sacs or to damage the lining epithelia, as this will inevitably result in the leakage of fluid and the collapse of the yolk sac. Dissection of the decidual mass should start at the site overlying the junction of the ectoplacental cone and the yolk sac (Fig. 3a) which is demarcated by the change in opacity of the decidua from a dull red mass to a translucent area over the fluid-filled conceptus. While steadying the decidua with the holding forceps, the decidual tissues are slit with the points of the dissecting forceps first around the equator and then along the meridian of the decidua (Fig. 3a). Finally, the conceptus is explanted by peeling off the adhering decidual tissues in small pieces from the surface of the parietal yolk sac and the ectoplacental cone. A small hole is made in Reichert's membrane over the embryo proper, and the membrane is torn apart as shown in Fig. 3b to expose the visceral yolk sac. The freed Reichert's membrane is trimmed back to the edge of the ectoplacental cone.

Methods of Whole-Embryo Culture

Survey of Culture Protocols

Mouse embryos are usually cultured in chemically defined media supplemented with rat, mouse, fetal calf, human fetal cord, or adult human sera. The media are maintained at the normal osmolarity of mammalian body fluids (280–290 mOsmol/liter) and at physiological pH (7.3–7.4) and temperature (37°). Buffering of the medium is provided by the chemical reaction between $NaHCO_3$ and the weak acid produced by the dissolved

CO_2 (5% by volume in the gas phase). The level of O_2 in the gas phase is changed in accordance to the metabolic requirement of embryos at different ages. Gas mixtures are fed to the culture either intermittently or continuously. There is a further choice between the static culture system, where the medium is placed in a petri dish or in wells of chamber slides, and the flowing system, where the medium is kept moving in rotating bottles or tubes.

Table II summarizes the different protocols used and the results obtained by several laboratories involved with culturing postimplantation mouse embryos.[1-10] The general consensus is that mouse embryos develop best when they are explanted at the late primitive streak (7.5 dpc) to early somite stages (8.5 dpc) and cultured in chemically defined media containing either 50% rat serum[1,6] or a combination of 50% rat serum and 25% mouse serum.[5,7] In over 90% of the embryos cultured from these stages, tissue growth and morphological development during the first 36 hr *in vitro* are remarkably similar to those *in vivo*. However, embryos begin to display obvious developmental retardation and morphological anomalies after 48 hr in culture. This hurdle can be partially overcome by adopting a continuous gassing system, changing the culture medium at 24-hr intervals, culturing fewer advanced embryos per volume of media, and raising the content of serum to as high as 75%.

In the best experiment, 7.5-day egg cylinders have been cultured for 72 hr, using a combination of roller and rotating drum methods with daily changes to fresh medium, and they developed to the hind limb bud stage (29–31 somites), which is equivalent to 10.5 dpc (Fig. 4). Cultures of younger embryos (6.5 dpc) give less satisfactory results, and only 55–82% of them will develop to the early somite stage after 48 hr *in vitro* (Table II). Much more consistent results are obtained when the 6.5-day embryos are grown in medium supplemented with freshly prepared (stored for less than 4 weeks) human cord serum. The static culture method seems adequate for the early egg cylinder, and the use of roller bottles becomes critical only when the culture is extended beyond the primitive streak

[1] P. P. L. Tam and M. H. L. Snow, *J. Embryol. Exp. Morphol.* **59,** 131 (1980).
[2] K. A. Lawson and R. A. Pedersen, *Development (Cambridge, UK)* **101,** 627 (1987).
[3] K. A. Lawson, J. J. Maneses, and R. A. Pedersen, *Development (Cambridge, UK)* **113,** 891 (1991).
[4] P. P. L. Tam and R. S. P. Beddington, *Ciba Found. Symp.* **165,** 27 (1992).
[5] P. P. L. Tam, *Development (Cambridge, UK)* **107,** 55 (1989).
[6] P. P. L. Tam and R. S. P. Beddington, *Development (Cambridge, UK)* **99,** 109 (1987).
[7] E. S. Hunter III, W. Balkan, and T. W. Sadler, *J. Exp. Zool.* **245,** 264 (1988).
[8] T. W. Sadler and D. A. T. New, *J. Embryol. Exp. Morphol.* **66,** 109 (1981).
[9] T. W. Sadler, *J. Embryol. Exp. Morphol.* **49,** 17 (1979).
[10] P. P. L. Tam, W. Y. Chan, K. R. Mao, and T. Chiu, *Fertil. Steril.* **48,** 834 (1987).

TABLE II

In Vitro Culture of Postimplantation Mouse Embryos Using Different Protocols

| Culture methods | | | Embryonic development (no. of embryos studied) | | | | Ref. |
| | | | 22–24 hr | | 44–48 hr | | |
Medium[a]	Type[b]	O$_2$ level (%)	Normal %	Stage[c] somite no.	Normal %	Somite no.	
6.5-day embryos							
DR50	Dish	20	88 (64)	Late PS	55 (64)	6.2 ± 0.6 (35)	1
DR50	Dish	20	92 (160)	Late PS/HF	82 (162)	Mean = 5.67	2, 3
DRH	Well	20	86 (129)	Late PS	69 (51)	Early somite	4
7.5-day embryos							
DR50	Dish	20	84 (26)	6.5 ± 0.4 (84)	ND[d]	ND	1, 5
DR50	Roller-i	5	79 (69)	5.0 ± 0.4 (50)	ND	ND	6
DRM	Roller-i	5 → 20	90 (177)	5.6 ± 0.4 (160)	87 (139)	15.7 ± 0.6 (121)	5
TRM → TR75	Roller-i	5 → 95	ND	ND	95 (21)	23.7 ± 0.6 (20)[e]	7
8.5-day embryos							
RS	Roller-i	5 → 20		17.8 ± 0.9 (17)	79 (27)	28.2 ± 0.9 (13)	8, 9
DR50	Roller-i	5 → 20	86 (99)	18.1 ± 0.4 (99)	ND	ND	6
DHu50	Roller-i	5 → 20	91	17.1 ± 0.2 (295)	ND	ND	10
DR75	Drum-c	5 → 20	ND	ND	66 (63)	31.1 ± 0.4 (41)	unpublished[f]
9.5-day embryos							
DR75	Drum-c	20 → 40	74 (65)	31.4 ± 1.9 (48)	ND	ND	unpublished[f]

[a] Medium: D, Dulbecco's modified Eagle's medium; R, rat serum; M, mouse serum; H, human cord serum; Hu, adult human serum; T, Tyrode's saline. DR50, D plus 50% R; DRH, D plus 50% R and 25% H; DRM, D plus 50% R and 25% M; RS, 100% R; DHu50, D plus 50% Hu; TRM, T plus 50% R and 25% M; TR75, T plus 75% R.

[b] Types: Dish, static culture in dishes; well, 4-well chamber slides; roller-i, rotating bottles with intermittent gassing; drum-c, bottles on rotating drum with continuous gassing.

[c] PS, Primitive streak; HF, head fold stage.

[d] ND, not determined.

[e] Cultures were extended to 72 hr: 97% (31) of embryos developed normally and had made 32.1 pairs (±0.6, n = 30) of somites.

[f] K. Strum and P. L. Tam, unpublished results.

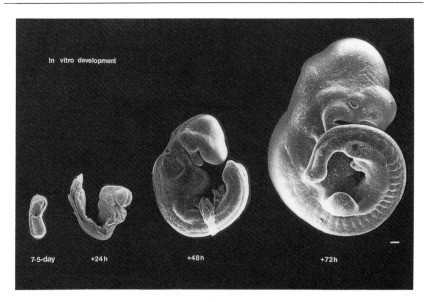

In vitro development

7·5-day   +24 h   +48h   +72h

FIG. 4. Scanning electron micrographs of the best embryos cultured in one experiment, showing the extent of growth and morphogenesis that may be achieved *in vitro* (bar: 100 μm). The 7.5-day embryos were cultured in rotating bottles containing DR50 medium (which was changed after 24 hr of culture) and gassed intermittently with 5% O_2, 5% CO_2, and 90% N_2 for the first 38 hr, and then cultured in fresh DR50 medium in a gas phase with 20% O_2, 5% CO_2, and 75% N_2 until 48 hr. Embryos were then cultured in rotating bottles gassed continuously with 20–40% O_2 until 72 hr *in vitro*.

stage. For embryos older than the forelimb bud stage, the method of flowing medium and continuous gassing produces the best results. Mouse embryos at 11.5 days have been cultured with Waymouth's medium containing 50% fetal calf serum. Satisfactory embryonic development was obtained only for the first 6 hr, with embryos still within intact yolk sacs. Results of *in vitro* development for 24 hr were much more variable, even after embryos were exposed directly to the medium by opening up the yolk sac and amnion.[11] Further improvement of the culture of 11.5-day embryo was achieved in roller bottles containing 25% rat serum in Waymouth's medium (which contains a low Ca^{2+} content, 0.822 mM) and gassed with 95% O_2.[12]

[11] L. Martinez de Villarreal and D. L. Fisher, *J. Exp. Zool.* **223**, 189 (1982).
[12] K. Nakashima and Y. Fujiki, *J. Exp. Zool.* **241**, 191 (1987).

TABLE III
CULTURE MEDIA FOR POSTIMPLANTATION MOUSE EMBRYOS

Dulbecco's modified Eagle's medium (DMEM)
  To make 1 liter from dry powder:
    Dissolve contents of packet in about 800 ml water
    Add to this solution 0.06 g penicillin G, 0.05 g streptomycin
      sulfate, 3.7 g NaHCO$_3$
    Bring volume to 1 liter and check that osmolarity is 320–340
      mOsmol/liter and pH after equilibration with 5% CO$_2$ is
      7.2–7.4
    Sterilize medium by passing through 0.22-μm membrane fil-
      ters and store at 4°
Serum
    Thaw just enough aliquots of serum and mix thoroughly to
      dissolve any crystallized protein
    Heat-inactivate serum with cap of tube partly loosened (to
      allow ether vapor to escape) in water bath at 56° for 35 min
    Add appropriate amount of serum to final culture medium,
      no further filter sterilization is necessary
Culture media[a]
  For 10 ml add x ml of:

	DMEM	RS	MS	HCS
DH	2.5	—	—	7.5
DRH	2.5	5.0	—	2.5
DRM	2.5	5.0	2.5	—
DR50	5.0	5.0	—	—
DR75	2.5	7.5	—	—

[a] For medium types, see Table II and text.

Medium Preparation

Dulbecco's modified Eagle's medium (DMEM, Flow Laboratories, McLean, VA) is prepared from prepackaged powder formulas as described in Table III. Each batch of medium should be tested for sterility (by incubating a fresh aliquot at 37° for 3–5 days to reveal the growth of any contaminants), pH values, and osmolarity. DMEM has a shelf life of 4 weeks at 4°.

Five different media are used in our laboratory for culturing postimplantation mouse embryos (Table III). A mixture of DMEM and rat serum (RS) is regularly used for culturing 8.5-day and 9.5-day embryos (DR50 medium is DMEM:RS, 1:1, v/v; DR75 medium is DMEM:RS, 1:3, v/v). Primitive streak stage embryos clearly benefit from a higher serum content (up to 75%) and also the inclusion of either mouse serum (MS) or human cord serum (HCS) in the medium. A 1:2:1 mixture of DMEM:RS:MS (i.e., DRM medium) or DMEM:RS:HCS (i.e., DRH medium) consistently supports better development of the egg cylinders than media supplemented with rat serum only. A 1 : 3 mixture of DMEM and human cord serum (DH medium) is also used when 6.5-day pre-primitive streak stage embryos are cultured.

Serum is prepared by the centrifugation of blood immediately after collection to separate the cells from the plasma. A fibrin clot is formed during centrifugation, and the serum is obtained by squeezing the clot, followed by a second round of centrifugation.

For human cord serum, fresh cord blood is collected in the maternity ward from placentas of cesarean deliveries. The blood is drawn with syringes from the umbilical vessels and transferred to sterile plastic culture tubes (15-ml tubes, Sterilin) and kept on ice until further preparation (see below).

Mouse embryos develop equally well in culture media supplemented with either rat or mouse serum, or even in whole serum.[13] The embryotrophic activity of the animal sera is not significantly affected by the sex of the animals or, in females, by the stage of the estrous cycle or the age of gestation.[14] During the exsanguination of the animal, we prefer using diethyl ether or halothane over other chemical agents for anesthesia because the dosage of anesthetics can be easily regulated during the operation and residual ether in the serum may be readily removed by vaporization during heat inactivation.[14a]

The etherized animal is placed in a supine position and maintained under deep anesthesia by bringing a piece of ether-soaked cottonwool in a beaker over its snout. The abdominal wall is sprayed with 70% ethanol and then cut open with a large vertical incision. The viscera are displaced to one side, and the fasciae and fat covering the vena cava and the dorsal aorta are dissected with fine forceps to expose the blood vessels in the dorsal body wall. The right common iliac artery is held firmly with a pair of curved blunt forceps, and the abdominal aorta is punctured with needles (1.00 × 38 mm for rats and 0.50 × 16 mm for mice) at its bifurcation into common iliac arteries. Blood is drawn gently (to prevent hemolysis which diminishes the embryotrophic activity of the serum) into a syringe (20-ml syringe for rats of 300–450 g, 2-ml syringe for mice of 35–50 g). It is important to keep the bevel of the needle down during puncture and to leave the needle in the vessel or, when the puncture is not successful, in the perivascular space, otherwise blood will spurt out all over with amazing

---

[13] G. Van Meele-Fabry, J. J. Picard, P. Attenon, P. Berthet, F. Dehaise, M. J. P. Govers, P. W. J. Peters, A. H. Piersma, B. P. Schmid, J. Stadler, A. Verhoef, and C. Verseil, *Reprod. Toxicol.* **5,** 417 (1991).

[14] M. K. Sanyal and F. Naftolin, *J. Exp. Zool.* **228,** 235 (1983).

[14a] Other procedures can also be recommended, such as using a 100% CO_2 environment for anesthetizing. See also [2] in this volume.

force. The animal usually stops breathing after it has lost about two-thirds of its blood. However, it is advisable to routinely sacrifice the animal by cervical dislocation after blood is collected.

Usually 10–13 ml blood can be collected from each adult rat. After removing the needle from the syringe, the blood is transferred gently in 10-ml aliquots to sterile plastic culture tubes and spun for 15 min at 2000 *g* (about 3600 rpm on an MSE benchtop centrifuge using a swing-out head). In contrast to the rat, each mouse gives only 1.0–1.2 ml of blood, and this volume is transferred to sterile 1.5-ml Eppendorf micro test tubes and spun at the top speed of the Eppendorf centrifuge for 15 min. The fibrin clot formed after centrifugation on top of the column of blood cells is held by a pair of sterile long-pointed forceps and squeezed by winding it around the forceps. The fibrin clot is then discarded, and the tubes are respun to obtain the so-called immediately centrifuged serum for embryo culture. The serum is aspirated with sterile Pasteur pipettes and stored frozen at $-20°$ until use. Before adding to DMEM, serum is thawed and heat-inactivated for 30–35 min at $56°$ to remove complement activities (Table III). The final medium is equilibrated overnight at $37°$ in 5% CO_2 in air.

Culture Protocols

Pre- and early primitive streak stage mouse embryos (6.5 dpc) are cultured in static medium either in 35-mm culture dishes (Corning) or in the wells of 4-well chamber slides (Nunc, Naperville, IL) until they develop to the late primitive streak stage. About 8–10 embryos are cultured in 1.5 ml of medium in a dish, and about 4–5 are kept in 0.75 ml of medium in each well of the chamber slide. The culture dishes and slides are placed in an ordinary incubator gassed with 5% CO_2 in air, or alternatively inside an air-tight anaerobic jar filled with a gas mixture of 5% CO_2, 5% O_2, and 90% N_2 (Table IV). No significant differences in embryonic development are found when embryos are cultured in different gas phases.

Late primitive streak stage (7.5 dpc) to early somite stage embryos (8.5 dpc) are cultured either in disposable culture tubes (Fig. 5A) or in glass bottles (Fig. 5B) rotating at 30 rpm on a roller apparatus in a BTC Engineering (Cambridge, UK) embryo culture incubator. Each 50-ml glass bottle (Wheaton, Millville, NJ) contains 4 ml of medium, which is sufficient for culturing eight 7.5-day or four 8.5-day embryos for 24 hr.[14b] The Universal tube (Sterilin) has a 30-ml capacity and takes 3 ml of medium, enough

---

[14b] The glass bottles are cleaned by soaking briefly (30–45 min) in 7X detergent (Flow Laboratories), rinsed with 5 changes of distilled water, sonicated in an ultrasonic water bath for 10 min, rinsed again with 5 changes of distilled water, and autoclaved.

TABLE IV
EXPERIMENTAL PROTOCOLS FOR CULTURING POSTIMPLANTATION MOUSE EMBRYOS

Time in culture (hr)				Culture		Gas phase	
6.5 days	7.5 days	8.5 days	9.5 days	Method[a]	Medium[b]	Oxygen (%)	Delivery
0–24				Static	DRH/DH	5	Continuous
24–48	0–24			Roller	DRM/DR50	5	Intermittent
	24–42	0–18		Roller	DRM/DR50	5	Intermittent
	42–48	18–24		Roller	DRM/DR50	20	Intermittent
	48–66	24–42	0–18	Drum	DR75	20	Continuous
		42–46	18–22	Drum	DR75	40	Continuous
			>22	Drum	DR75	95	Continuous

[a] Static, Medium in dishes or in wells of chamber slides; roller, medium in rotating bottles or tubes on a roller apparatus; drum, medium in bottles on a rotating drum with continuous gassing.
[b] For medium types, see Table III.

for culturing six 7.5-day or three 8.5-day embryos. The gas requirements of embryos at these stages are given in Table IV. After the embryos have been transferred into the bottles or tubes, gassing is done by gently blowing a stream of the appropriate mixture through a cotton-plugged Pasteur pipette (or a disposable Eppendorf pipette tip) over the surface of the medium for about 2–3 min. The bottles are then sealed tightly with sterile silicone rubber stoppers. For the Universal tubes, a smear of petroleum jelly or silicone grease is put on the brim of the tube, and the cap is screwed tightly after gassing. The gas phase should be replenished or replaced intermittently according to the schedule described in Table IV. The bottles and tubes are then placed on the roller apparatus, which is slightly tilted to ensure a good depth of medium is maintained at the bottom of the vessel for the growing embryos.

Forelimb bud stage embryos (9.5 dpc) can be cultured in rotating bottles with intermittent gassing as for younger embryos, except that at least 2 ml medium is used for each embryo and a higher O_2 level in the gas mixture (40–90%) is required (Table IV). However, a more sophisticated system of rotating bottles with continuous gassing gives much better results. The culture bottles are attached by hollow silicone rubber stoppers to the apertures on the faces of a rotating drum (Fig. 5C). The gas mixture is then fed into the bottles via the inlet on the rotator after first passing through a glass fiber filter and being humidified by bubbling through distilled water. In contrast to the case for sealed bottles, the gas phase in this system is continuously replenished and can be changed easily by connecting to different gas cylinders without interrupting the culture process. The rotating drum apparatus (available from BTC Engineering or

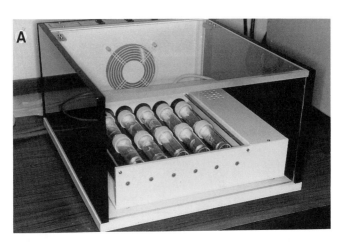

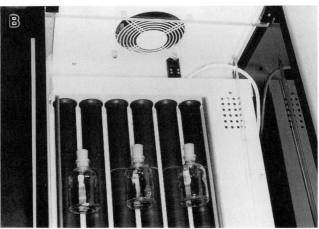

Ikemoto Scientific Technology, Tokyo, Japan) can be fitted into the same chamber as for the roller apparatus or in an ordinary dry incubator.

The timing of gas changes as given in Table IV should be adhered to as much as possible, but considerable variations in the rate of development of embryos in the same experiment may often require some adjustments to the schedule of gas changes. For extended cultures, it is important not to overcrowd the culture with embryos, and to replenish with fresh medium or change to a different medium at 24 hr intervals. Depletion of the nutrients (e.g., glucose, pyruvate, and L-glutamine) in the culture medium with time, uncontrolled pH fluctuations (owing to the accumulation of waste products or inadequate control of the CO_2 level), and inappropriate levels of oxygen are the major causes of malformations and growth retardation.[15]

Analysis of Results

Assessment of embryonic growth[16] is normally done by determination of the total protein and DNA contents of embryos and by the measurement of embryonic sizes. Several parameters such as crown–rump length, head dimensions, and axial length can be used as indicators of embryonic sizes. Morphological development of the embryos is studied by comparing the anatomy with that of embryos which have developed for an equivalent period of time *in vivo,* and the results can be summarized using a semiquantitative scoring system originally devised for teratological studies. Furthermore, routine histology is carried out for a detailed examination of the extent of histogenesis of specific tissues and organs of the embryo.[17]

In Vitro Culture of Germ Layer Derivatives

Various culture methods are available which mimic closely the *in vivo* environment, allowing the study of normal differentiation under defined

[15] M. K. Sanyal, *J. Embryol. Exp. Morphol.* **58,** 1 (1980).
[16] N. A. Brown, *in* "Postimplantation Mammalian Embryos" (A. J. Copp and D. L. Cockroft, eds.), p. 93. IRL–Oxford Univ. Press, Oxford, 1990.
[17] M. H. Kaufman, *in* "Postimplantation Mammalian Embryos" (A. J. Copp and D. L. Cockroft, eds.), p. 81. IRL–Oxford Univ. Press, Oxford, 1990.

FIG. 5. BTC embryo culture chamber and roller apparatus with (A) Universal culture tubes and (B) glass bottles sealed with silicone rubber stoppers. In both cases, a maximum of ten cultures can be accommodated on the roller apparatus, which rotates at 30 rpm. (C) Bottles attached to a rotating drum with continuous feeding of gas (inlet, open arrowhead; outlet, solid arrowhead). The drum takes a total of twenty 5-ml bottles as shown or ten larger, 10-ml bottles. The adaptor ring (r) on the left end of the shaft can hold ten culture tubes.

conditions. Alternatively, culture experiments can be designed in such a way that only the effects of specific factors which may influence differentiation can be studied. The most important factors that influence the outcome of both types of culture experiments are (1) the method of isolation of the specific germ layer derivatives and (2) the conditions under which the tissue fragments are allowed to differentiate.

Isolation of Germ Layer Tissues

Embryonic tissues can be separated into germ layers by microsurgical dissection, enzymatic digestion, or by a combination of both methods. Mechanical dissection is performed using metal and glass needles to cut embryonic fragments. Tissue fragments can be passed through a fine metal mesh or through a syringe equipped with an 18- to 23-gauge needle to obtain small fragments and single cells.

To aid the separation of tissues while maintaining each germ layer intact, gentle digestion of the basement membrane with proteolytic enzymes is combined with mechanical dissection of embryonic tissue fragments. Trypsin treatment alone is not sufficient for digesting the basement membrane and separating the germ layers. The tissue pieces become very sticky owing to cell lysis and the mucilaginous products of partially digested matrix materials. This problem can be overcome by using a combination of trypsin with pancreatin, collagenase, dispase, or hyaluronidase. In addition, tissue fragments can be incubated in calcium- and magnesium-free saline or in Tyrode's solution prior to digestion. By varying the temperature and duration of digestion, as well as the concentration and combination of enzymes, embryonic tissues can be isolated from egg cylinder to late organogenesis stage embryos. It is important to monitor the digestion regularly, since overdigestion will lead to complete disaggregation of cell layers and can also cause cell death. Digestion is terminated by transferring the tissue fragments to serum-containing medium. To minimize damage to the isolated tissues, the least amount of digestion required for tissue separation should be used. However, if a single cell suspension of a germ layer is desired, the isolated cell sheet can be dissociated further in a more controlled fashion by trypsin/EDTA treatment.

For transplantation of fragments to other embryos[18] or to ectopic sites, where it is important to have a pure population of cells or small pieces of tissue, and where even slight stickiness of the tissue may hamper transplantation, mechanical dissection is the method of choice. In experiments in which it is critical that the tissue fragment is intact and cell loss

---

[18] P. P. L. Tam and S. S. Tam, *Development (Cambridge, UK)* **115,** 703 (1992).

is minimal, combined mechanical and enzymatic separation would be the most appropriate.

Dissecting Instruments

In addition to the fine watchmaker's forceps used in isolation of whole embryos, metal or glass needles are required for the dissection of embryonic tissue fragments. Metal needles are fabricated from orthodontic alloy wire (Rocky Mountain Orthodontics, Series 380 or 381, Denver, CO). The wire is first fixed to a handle either by inserting it into the molten tip of a Pasteur pipette or by attaching it to a wooden holder with glue. The wire is then sharpened electrolytically in an orthophosphoric acid bath, then dipped into a 1 M sodium carbonate solution to neutralize the acid, rinsed well with distilled water, and sterilized with 70% ethanol. The sharpness of the blade is judged under a dissecting microscope. The tips should be fine but not hair thin, and the shaft should be flat with a sharp cutting edge. The needles should be cleaned and resharpened for every experiment.

Glass needles with a tip of 0.5–1 cm are made from thick-walled capillaries (Leitz, Heidelberg, Germany) that have been fused in the middle over a microflame to produce a section of solid glass, using a vertical micropipette puller (Kopf, Tujunga, CA). It helps, but is not necessary, if a bend is introduced in the shaft just beyond the pulled tip of the needle using a microforge (Narishige, Tokyo, Japan), such that the sharp edge of the needle can be moved parallel to the bottom of the dish during dissection, while the body is held at an angle. To prevent tissue fragments from sticking to the needles during dissection, the pulled glass needles are siliconized by dipping them into Repelcote (BDH, Poole, UK) followed by several rinses in distilled water and drying.

Isolation of Tissues from Gastrulating Mouse Embryos

Gastrulating embryos (7.5 dpc) are recovered from the uterus and are freed of Reichert's membrane as previously described.[19,20] The method for isolation of the germ layers (endoderm, ectoderm, and mesoderm) is shown schematically in Fig. 6. Embryonic and extraembryonic regions are separated by transecting the conceptus with two glass needles just proximal to the amnion (or amniotic folds in younger embryos). The embryonic portion is split longitudinally along the primitive streak (ps),

---

[19] M. H. L. Snow, *in* "Methods in Mammalian Reproduction" (J. C. Daniel, Jr., ed.), p. 167. Academic Press, New York, 1978.

[20] R. S. P. Beddington, *in* "Mammalian Development" (M. Monk, ed.), p. 43. IRL–Oxford Univ. Press, Oxford, 1987.

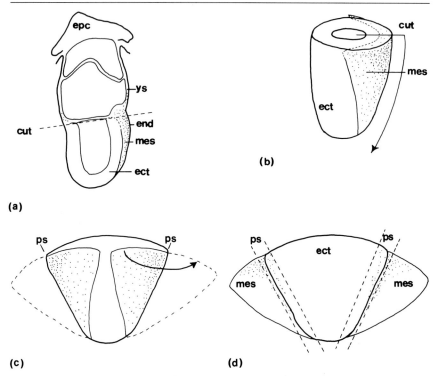

FIG. 6. Dissection of germ layers of gastrulating mouse embryos. (a) Separation of embryonic and extraembryonic regions by transecting the egg cylinder proximal to the amnion with two glass needles. epc, Ectoplacental cone; ys, yolk sac; end, endoderm; mes, mesoderm; ect, ectoderm. (b, c) The primitive streak (ps) is split longitudinally to flatten the egg cylinder, and the mesodermal wings are lifted from the ectoderm. (d) The ectoderm and mesoderm are separated by cutting with glass needles lateral to the primitive streak (dashed lines).

the site of mesoderm ingression in the embryonic ectoderm (Fig. 6b). The flattened egg cylinder is then incubated in a solution of 0.5% (w/v) trypsin, 2.5% (w/v) pancreatin in calcium- and magnesium-free PBS (pH 7.6) at 4° for 10–20 min. To stop the enzymatic activity, the fragment is transferred to cold PB1 containing 10% rat or fetal calf serum. The endoderm, which usually detaches easily from the egg cylinder, tends to roll up around other pieces of tissue. Prior to culture or enzymatic analysis, the endoderm has to be unrolled to remove nonendodermal cells. On the remaining tissue fragment, the mesodermal wings are lifted from the ectoderm and can be trimmed cleanly from the embryonic ectoderm by cutting alongside the primitive streak with glass needles (Figs. 6c,d). The primitive streak remnants are then cut away to obtain clean fragments of primitive streak and an intact layer of ectoderm.

The yolk sac, which makes up the extraembryonic portion of the conceptus, can be separated into visceral endoderm and mesoderm using the same trypsin–pancreatin solution as above, but prolonging the incubation at 4° for 1–3 hr. After stopping the digestion by transferring the fragments to serum-containing medium, the two layers can be torn apart using fine watchmaker's forceps. Alternatively, the digestion time in the enzyme solution can be reduced to 20–30 min, followed by incubation in DMEM containing 10% fetal calf serum at 37° for 1–2 hr.[21] The two tissue layers usually will come apart on their own accord and can be separated easily with fine forceps.

Isolation of Mesodermal and Ectodermal Tissues from Organogenesis Stage Embryos

The isolation of the somitic tissues and the caudal neural tube of 9.5-day embryos using a combination of mechanical dissection and enzymatic digestion is described as a standard protocol, which applies equally well to 9.5- to 13.5-day embryos. It is a good practice that only embryos from one litter are processed at a time to ensure that freshly dissected tissues can be placed into culture as soon as possible for optimal viability.

After retrieving the embryos from the decidua (Fig. 2), the yolk sac is removed by first opening it with fine watchmaker's forceps at the base of the placenta and then pulling it around the embryo. The amnion is removed similarly, first by introducing a tear in the membrane just next to the face of the embryo, where the amnion stays free of the embryo. To avoid damaging the posterior part of the embryo when removing the membranes, the yolk sac stalk is held close to its attachment to embryo with a pair of forceps. This will prevent the tearing of membrane into the body of the embryos when they are dissected away with the second pair of forceps.

For each experiment, the developmental stage of the isolated embryos should be carefully determined[16,17] to ascertain uniformity of experimental materials. Individual litters must be assessed properly for developmental stages before any pooling of embryos can be done for different experiments. Two useful criteria for staging embryos are the number of somites and brain morphology.

A posterior fragment of 9.5-day embryos (Fig. 7a) is obtained by transecting the midtrunk of the embryo with fine watchmaker's forceps. The coelomic wall and the hindgut are then trimmed away using alloy needles to produce a smaller fragment (Fig. 7b). Three cuts are made: cut (i)

[21] B. L. M. Hogan, *Dev. Biol.* **76**, 275 (1980).

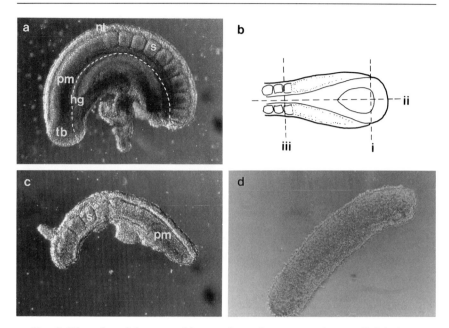

FIG. 7. Dissection of the presomitic mesoderm of organogenesis stage (9.5 dpc) mouse embryos. (a) Posterior region of the embryos containing the somites (s) and presomitic mesoderm (pm); nt, neural tube; tb, tail bud/primitive streak; hg, hindgut. In the next step the hindgut and coelom are trimmed away (dashed line). (b) Fragments containing the presomitic mesoderm are obtained by cuts at position (i) to remove the primitive streak, position (ii) to bisect the fragment into halves, and position (iii) to remove the existing somites. (c) Posterior fragment after cuts (i) and (ii). (d) Presomitic mesoderm is isolated from the fragment in (c) by enzymatic treatment followed by the removal of the neural tube, the dorsal aorta, and the surface ectoderm with glass needles.

removes the primitive streak; cut (ii) bisects the fragment along the ventral groove of the neural tube by gently pressing the needle through the tissue layers, followed by cut (iii) made immediately posterior to the last segmented somite (Fig. 7b). During the dissection, the fragment is held with the neural tube facing upward against the bottom of the dish using fine watchmaker's forceps. The resulting fragments (Fig. 7c) are then transferred in a minimal amount of medium using a pulled mouth pipette to a dish of 0.5% trypsin, 0.25% pancreatin, 0.2% glucose, and 0.1% polyvinylpyrrolidone (PVP) in calcium- and magnesium-free PBS at 4°. Only a few fragments are transferred at a time since the digestion is rapid and overdigestion produces very sticky tissues. As soon as the neural tube starts to separate from the piece, the fragments are removed from the enzyme solution one by one and are placed into a large volume of cold PB1 supplemented with 10% rat serum.

After all fragments have been digested, the germ layer derivatives are further dissected using glass needles. By pinning down the fragment with one needle, the neural tube is peeled off from the fragment as a whole piece with the other needle. To isolate somites and presomitic mesoderm, the cutting needle is run along the ventral side of the somitic tissues, and cuts are made above the dorsal aorta. Because the surface ectoderm which still covers the fragment is elastic and quite resilient to cutting, it can be pinned down in order to hold the fragment during dissection. Individual somites can then be separated from the surface ectoderm by drawing the tip of the needle between somite and ectoderm. Separation of the presomitic mesoderm is done in the same way, but care must be taken that the fragment does not break, as cells in this tissue are loosely packed and do not form an epithelium. This may prove to be difficult for 9.5-day embryos, in which the presomitic mesoderm is longest, so extra care is required during the dissection.

Assessing Quality of Isolated Fragments

To assess the cleanliness of the dissection, as well as the integrity of the isolated fragment, histological analysis is performed. Representative fragments are analyzed for cell number, and the dimensions of all fragments is recorded. Video recording is an economical way to get this information, since a variety of measurements can be performed on the stored images, which may be taken continually during the experiment.

The presence of tissue-specific molecular markers can be used to determine whether the method of dissection employed yields fragments of the desired purity.[22] The presence of tissue-specific mRNAs can be detected by *in situ* hybridization or reverse transcriptase–polymerase chain reaction (RT–PCR) analysis,[23] and tissue-specific antigens can be analyzed by immunofluorescence. Tissue-specific markers include *Brachyury* for mesoderm, specific combinations of *Hox* genes for somites at given axial positions, myosin heavy chain, *MyoD, myogenin* for the muscle lineage, α-fetoprotein and cytokeratin markers for endoderm, and neurofilament markers for neural tissues. Molecular analyses such as these tend to be destructive, however, and therefore are not appropriate for assaying individual fragments before culture.

[22] R. S. P. Beddington and K. A. Lawson, *in* "Postimplantation Mammalian Embryos" (A. J. Copp and D. L. Cockroft, eds.), p. 267. IRL–Oxford Univ. Press, Oxford, 1990.
[23] D. A. Rappolee, D. Mark, M. J. Banda, and Z. Werb, *Science* **241,** 708 (1988).

Culturing Germ Layer Derivatives

Media

Embryos are dissected in PB1 medium (Table I) containing 10% rat serum or 10% fetal calf serum. For separation of the germ layers, embryo fragments are incubated in cold dissection medium after enzymatic treatment. Incubation in serum-free medium should be avoided.

A variety of defined media have been used for culture of embryo fragments, including modified Eagle's media (MEM, DMEM), Ham's F12, RPMI 1640, and BME. The media usually contain antibiotics (penicillin, streptomycin, achromycin) and are supplemented with different sera (fetal calf, horse, rat, mouse, or combinations thereof) at concentrations ranging from 2.5 to 50%. As a source of a variety of embryonic growth factors, 3–5% chick embryo or rat embryo extracts or conditioned medium from embryonic cell lines or other embryo fragments can be added to the culture medium.[24-26] Media have also been supplemented with general growth-stimulating agents like glucose, nonessential amino acids, glutamine, and taurine. Addition of specific growth factors or morphogens promotes proliferation or differentiation of some tissues in culture. These include epidermal growth factor (EGF, 10 ng/ml) and insulin (2–5 μg/ml) that are added to cultures of intestinal epithelium,[27] transferrin (4–10 mM) added to neural cultures and to cocultures of neural and mesenchymal tissues,[28,29] and retinoic acid (2 μM) added to spinal cord cultures.[30] Optimal culture conditions and supplementation for specific tissues have to be determined empirically depending on the experimental question. For additional specific media, see also Tam.[31]

Preparation of Culture Substrates

Tissue fragments can be cultured in suspension roller cultures, in semisolid medium, such as Matrigel or thick collagen gel, or on thin matrix substrates on solid supports.[31,32] Glass coverslips to be used as matrix-

[24] E. Vivarelli and G. Cossu, *Dev. Biol.* **117**, 319 (1986).

[25] J. R. Sparrow, D. Hicks, and C. J. Barnstable, *Dev. Brain Res.* **51**, 69 (1990).

[26] F. Alliot, E. Lecain, B. Grima, and B. Pessac, *Proc. Natl. Acad. Sci. U.S.A.* **88**, 1541 (1991).

[27] G. S. Evans, N. Flint, A. S. Somers, B. Eyden, and C. S. Potten, *J. Cell Sci.* **101**, 219 (1992).

[28] E. Buse and B. Krisch, *Anat. Embryol.* **175**, 331 (1987).

[29] G. Klein, M. Langegger, C. Goridis, and P. Ekblom, *Development (Cambridge, UK)* **102**, 749 (1988).

[30] L. Wuarin and N. Sidell, *Dev. Biol.* **144**, 429 (1991).

[31] P. P. L. Tam, *in* "Postimplantation Mammalian Embryos" (A. J. Copp and D. L. Cockroft, eds.), p. 317. IRL–Oxford Univ. Press, Oxford, 1990.

[32] P. P. L. Tam, *J. Embryol. Exp. Morphol.* **92**, 269 (1986).

coated supports have to be cleaned by soaking in chromic acid first. For this, coverslips are placed individually in a slide rack and are immersed in chromic acid for several hours. They are then washed in running water for 5 hr and dried by baking at 80° for 2 hr. Chamber slides (Nunc) are convenient for culturing small numbers of tissue samples in one experiment in different medium preparations or substrates. Both glass coverslips and glass chamber slides are suitable for experiments that involve detection of cell-specific antigens with fluorescent antibodies. Alternatively, Thermanox plastic coverslips (Miles Laboratory, Naperville, IL) can be used as solid support. The plastic slips can be peeled off from the block after the specimen has been embedded in methacrylate resin for histology.

Extracellular matrix components (obtained from Collaborative Research, Bedford, MA; Boehringer-Mannheim, Indianapolis, IN; Sigma, St. Louis, MO; or Calbiochem, San Diego, CA) are resuspended in water or buffered saline and stored in aliquots. For thin substrate coating, the diluted stock solution is pipetted onto slide or coverslip and is allowed to dry down completely in a sterile hood overnight, or, alternatively, the coverslip is incubated in the solution at 37° for 3–4 hr, then the residual solution removed and dried. Details about the preparation of substrates and the coating of slides for a collection of matrixes are given in Table V. Before use, the coated slides or coverslips are rinsed twice with PBS, then incubated with culture medium for 1 hr. The medium is replaced with fresh equilibrated medium prior to culture of the explants. Some examples of the growth of explanted presomitic mesoderm of 9.5-day mouse embryos on different substrates are shown in Fig. 8.

Experiments Using the Described Dissection and Culture Methods

A number of experiments using isolated tissue fragments and germ layer derivatives, which have previously only been possible with amphibian and avian embryos, can now be done using mammalian material. By analyzing mesodermal cells from normal and T/T mutant embryos under different culture conditions, the developmental defects of T/T embryos were found to be due to abnormal migration of mesoderm cells leaving the primitive streak.[33] In *in vitro* differentiation experiments of somites and limb buds, the expression of putative regulator molecules *MyoD* and *myogenin* in relation to genes which are expressed later during myogenesis could be investigated.[34] From a culture of neural tubes from 8.5 dpc embryos, neural crest cells were isolated that could participate in normal

---

[33] K. Hashimoto, H. Fujimoto, and N. Nakatsuji, *Development (Cambridge, UK)* **100,** 587 (1987).

[34] M. G. Cusella-DeAngelis, G. Lyons, C. Sonnino, L. DeAngelis, E. Vivarelli, K. Farmer, W. E. Wright, M. Molinaro, M. Bouchè, M. Buckingham, and G. Cossu, *J. Cell Biol.* **116,** 1243 (1992).

TABLE V
PREPARATION OF EXTRACELLULAR MATRIX-COATED SLIDES

Substrate	Diluent for stock	Concentration (μg/ml)	Storage of diluted solution	Amount required for coating[a]	Coating procedure	Storage of coated slides
Fibronectin	H$_2$O	100	−20°, 2 weeks	1–2 μg/cm^2	2 hr at 37°, aspirate, then dry	Dry, 4°, 1 month
Laminin	PBS	10–20	−70°, 9 months	1–2 μg/cm^2	2 hr at 37°, aspirate, then dry	Dry, 4°, 1 month
Collagen I (thin coating)	0.02 N acetic acid	50	4°, 9 months	5–10 μg/cm^2	2 hr at room temperature, aspirate, rinse well with PBS, then dry	Dry, 4°, 1 week
Collagen I (thick gel)	0.02 N acetic acid	50	4°, 9 months	10 μg/cm^2	Gel 3 min in ammonia vapor, rinse well in H$_2$O[b]	In DMEM or PBS, 10% FCS, 4°, 1 week
Collagen IV (thin coating)	0.05 N HCl	50	−70°, 9 months; 4°, 1 week	1–10 μg/cm^2	1 hr at room temperature, aspirate, rinse well with PBS, then dry	Dry, 4°
Gelatin	H$_2$O	100	4°, 1 month	1–10 μg/cm^2	Autoclave, add to plate, dry in flow hood	Dry, room temperature, 2 months
Matrigel (thick gel)	Undiluted	—	−20°, 9 months	50–200 μl/cm^2	Thaw at 4°, gel 30 min at 37°, equilibrate with medium	Use immediately
Matrigel (thin coating)	PBS	1:50	4°, 1 week	1–4 μl stock/cm^2	2 hr at room temperature, aspirate, then dry	Dry, 4°

[a] Concentrations have to be optimized for each cell type cultured.

[b] Collagen can also be cross-linked with riboflavin and UV light [E. B. Masurovsky and E. R. Peterson, *Exp. Cell Res.* **76**, 447 (1973)].

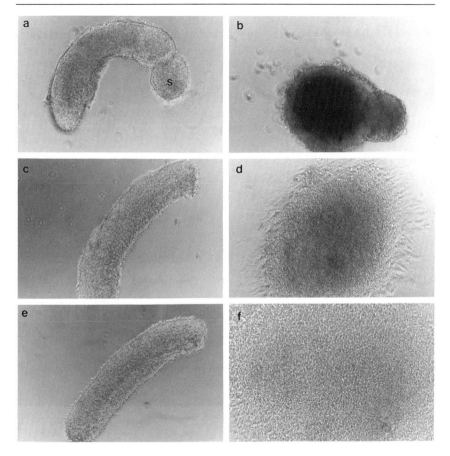

FIG. 8. Growth of isolated pieces of presomitic mesoderm on different substrates. Wells of 8-chamber slides were either uncoated (a, b), coated with collagen (c, d), or coated with fibronectin (e, f). Fragments are shown immediately after explantation (a, c, e) and after 16 hr of culture in DR50 medium (b, d, f). The fragment in (a) still has the last formed somite (s) attached.

development after injection into postimplantation embryos.[35] By using transgenic cells or by marking cells with retroviruses,[36] cell surface markers such as wheat germ agglutinin conjugated to gold particles, or fluorescent dyes like DiI or DiO (Molecular Probes, Eugene, OR), it is possible to study cell migration and cell–cell interaction in cocultures of two cell

[35] D. Huszar, A. Sharpe, and R. Jaenisch, *Development (Cambridge, UK)* **112,** 131 (1991).
[36] E. Y. Snyder, D. L. Deitcher, C. Walsh, S. Arnold-Aldea, E. A. Hartweig, and C. L. Cepko, *Cell (Cambridge, Mass.)* **68,** 33 (1992).

populations from different germ layer derivatives *in vitro*. Experiments of this type will aid greatly in the understanding of the inductive mechanisms necessary for normal organ development in mammals.

Acknowledgments

Part of the work reported here is supported by grants of the National Health and Medical Research Council of Australia. We thank Professor Rowe for comments on the manuscript.

[11] Histogenetic Potency of Embryonic Tissues in Ectopic Sites

By Patrick P. L. Tam

Introduction

Direct experimental manipulation and continuous monitoring of embryonic development have been hampered by the inaccessibility of the mouse embryo once it has implanted in the uterus. For example, *in situ* cell marking and tissue transplantation experiments that can be performed readily on developing amphibian and avian embryos are technically difficult for the mouse, especially during the immediate postimplantation stages of development. However, advances in embryo culture technology have made it feasible to maintain embryos explanted from the uterus *in vitro* for a considerable length of time, during which gastrulation and early organogenesis can be accomplished. Embryos in culture are now amenable to a variety of experimental manipulations and most importantly allow direct and uninterrupted observation of details of the process of embryogenesis. Using this approach, enlightening information on the morphogenesis, tissue differentiation, and developmental fate of specific cell populations in mouse and rat embryos have been obtained.

Even with the best culture system, however, mouse embryos can only be grown to a certain stage of organogenesis, beyond which development *in vitro* becomes abnormal. Because of this limitation, it is therefore impossible to obtain sufficient histological differentiation for some embryonic tissues so that they produce the definitive phenotypes or express tissue-specific gene products. The difficulty posed by the lack of terminal differentiation can be overcome in some cases, such as for myogenic and chondrogenic cells, where early lineage-specific markers and reporter transgenes can be used for a positive identification of the tissue type prior

Copyright © 1993 by Academic Press, Inc.
All rights of reproduction in any form reserved.

to histogenesis. More often than not, however, tissue type identity can only be established after a prolonged period of development, longer than that permitted by whole-embryo culture.

Two alternative experimental approaches are available to study the long-term histogenesis of embryonic tissues, namely, growing the tissues either in tissue/organ culture or in ectopic sites.[1] As far as histological differentiation is concerned, embryonic tissues thrive much better in ectopic sites than in cultures. This is presumably due to the provision of richer and less limiting source of trophic factors by the animal host than by the culture media. Although the environment in the ectopic sites is far from being neutral in its influence on cell differentiation, tissues grown in these sites are more likely to exhibit their maximal histogenetic potency and to undergo organotypic differentiation.

Examples of the variety of differentiated tissues are shown in Fig. 1. Fragments of the primitive streak of early somite stage mouse embryos grow into teratomas that consist of derivatives of all three definitive germ layers. Nodules of tissues in the tumor are organized into structures resembling the skin and the intestine. Some tissues, such as muscles and bone, also seem to exhibit normal sequences of differentiation, for example, fusion of the myoblasts to striated myotubes and cartilage formation followed by bone deposition. The ectopic transplantation approach is therefore ideal for studying the histogenetic potency of a defined cell population, tissue, or organ primordia.

In most cases, the precise tissue composition of the original transplant is not known. Therefore, the production of a diverse range of tissue types in the tumor does not necessarily reflect the pluripotential nature of the progenitor tissue, but may instead reveal the heterogeneity of the initial population. The direction of differentiation of tissues in the developing transplant is also influenced by the nature of cell–cell interactions during the initial phase of tissue disaggregation and reorganization, which are common features of embryo-derived teratomas of rats and mice.[2,3] Histogenetic potencies of cells from mutant embryos and experimental embryos (e.g., teratogen-treated or surgically manipulated) have also been tested by ectopic transplantation. Furthermore, because of the possibility of recombining tissues from different lineages and from different developmental stages into a single transplant, the developmental potential of tissues subjected to atypical or asynchronous interactions can also be

[1] P. P. L. Tam, in "Postimplantation Mammalian Embryos" (A. J. Copp and D. L. Cockroft, eds.), p. 317. IRL–Oxford Univ. Press, Oxford, 1991.
[2] P. P. L. Tam, J. Embryol. Exp. Morphol. 82, 253 (1984).
[3] A. Svajger, B. Levak-Svajger, and N. Skreb, J. Embryol. Exp. Morphol. 94, 1 (1986).

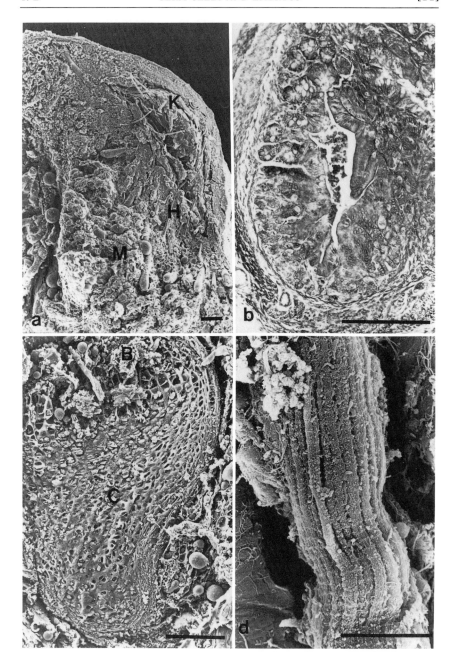

TABLE I

FORMATION OF TUMORS BY TRANSPLANTING GERM LAYERS AND TISSUE FRAGMENTS OF
7.5-DAY C57BL MOUSE EMBRYOS TO KIDNEYS AND TESTES

Germ layers/fragments	Number transplanted	Number of tumors recovered	Transplants developed (%)
Whole embryos	89	71	80[a]
Half-embryos	84	64	76[a]
Ectoderm	61	50	82[a]
Mesoderm	16	4	25[b]
Endoderm	38	2	5[c]
Ectoderm + mesoderm	8	7	86[a]
Endoderm + mesoderm	28	16	57[d]
Endoderm + ectoderm	8	5	63[a]
Ectoderm fragment	29	18	62[a]

[a] Extensive differentiation of tissues belonging to all three germ layers.
[b] Adipose tissue and connective tissue only.
[c] Very limited differentiation.
[d] Gut, muscles, and connective tissues.

assessed by ectopic transfer. An example of such a study is given in Table I, where the differentiative capacity of the germ layers of gastrulating mouse embryos were tested by growing the various germ layer recombinants in ectopic sites. Extensive tissue differentiation occurs when the graft consists of ectoderm alone and ectoderm recombined with one other germ layer. Mesoderm and endoderm alone or in combination produce only a limited repertoire of tissues.

This chapter describes procedures for transplanting embryonic tissues to ectopic sites and types of analyses that are applicable to tumors produced by the grafts. Methods of isolating tissue fragments from postimplantation mouse embryos are only briefly described here. For more details, the reader should refer to chapter [10] in this volume.

General Strategy

The purpose of transplanting embryo or tissue fragments to an ectopic site is to enable tissue differentiation and, in certain cases, morphogenesis

FIG. 1. Tissues found in teratomas derived from transplants of the primitive streak and tail bud of early organogenesis stage mouse embryos to kidneys: (a) epidermal structures including keratinized epithelium (K), hair follicles (H), and dermal mesenchyme (M); (b) gutlike structure with lining columnar epithelium, intestinal crypts of the submucosa, and circular muscle layers; (c) differentiating hyaline cartilage (C) and spongy bone (B); (d) bundles of striated muscle fibers. Bar: 100 μm.

to occur in an *in vivo* environment. Mouse embryos are isolated from the uterus of pregnant animals at specific times during gestation. Whole early postimplantation embryos can be used for transplantation. Because of the space constraints of the ectopic sites, however, only dissected fragments of the much bulkier organogenesis stage embryos can be transplanted.

Cells (e.g., myoblast, chondrocyte, neural crest cells, primordial germ cells), tissues (e.g., epithelia, mesenchyme), and organ primordia (e.g., branchial arch, neural tube, tail bud, limb bud) to be tested are isolated from defined regions of the embryo by mechanical dissociation or by a combination of enzymatic digestion and microdissection.[1] The tissue is trimmed into small fragments or disaggregated into clumps and then transplanted surgically to one of the three common ectopic sites in adult mice, namely, the subcapsular space of the kidney, the interstitium of the testis, and the anterior chamber of the eye. The grafts are harvested as teratomas at different times after transplantation, then processed for histological or biochemical studies. Tumors can also be retransplanted for an assessment of the neoplastic potency.

Choice of Mouse Strains for Transplantation Experiments

To avoid complications due to host-versus-graft reaction, syngeneic animals are always preferred as recipients, especially for long-term experiments. Nevertheless, it is possible to graft embryos or tissue fragments to allogeneic or xenogeneic recipients for studies of less than 2 weeks without eliciting any serious rejection response. The sex and the age (normally 6–8 weeks) of the recipient do not seem to affect the differentiation of the graft, although the precise effect of the sex of the host, which may be mediated hormonally, on the differentiation of the gonadal primordia is as yet unknown.

Embryos and tissue fragments of nearly all strains of mice will grow after transplantation to ectopic sites of syngeneic hosts and will produce teratomas containing a great variety of tissues and organ-like structures. Special strains of mice will have to be used for certain types of experiments. For example, those studies addressing the differentiation of retinal pigment cells and neural crest-derived melanocytes require the use of mice with pigmented coats (e.g., C3H, C57BL, DBA) so that the pigmented population can be traced.[4] Mouse mutants with known deficiencies in certain tissue types (e.g., *Brachyury* and t-complex mutants with poor mesoderm differentiation) are used for studying the cause of the mutant phenotypes with respect to the histogenetic potential of embryonic cells.

[4] W. Y. Chan and P. P. L. Tam, *J. Embryol. Exp. Morphol.* **96,** 183 (1986).

TABLE II
TUMOR FORMATION OF EMBRYO FRAGMENTS IN DIFFERENT ECTOPIC SITES[a]

Germ layers/fragments	Number transplanted	Number of recovered	Transplants developed (%)
Kidney capsule			
Isolated germ layers	82	50	61
PS[b] stage embryos	62	58	93
PS stage half-embryos	50	25	50
Neural plate[c]	520	272	52
Caudal fragments[d]	158	128	81
Testis			
Isolated germ layers	55	23	42
PS stage embryos	42	38	90
Intraocular chamber[e]			
Isolated germ layers	84	37	44
7.5-Day ectoderm fragments	226	116	51

[a] P. P. L. Tam, unpublished results (unless noted otherwise).
[b] PS, Primitive streak.
[c] From Ref. 4.
[d] From Ref. 2.
[e] From Ref. 7.

Transgenic mice that show lineage- or tissue-specific expression of an integrated gene (such as the *lacZ*, chloramphenicol acetyltransferase gene, or diphtheria toxin gene) are potentially useful in tracing the process of lineage specification and differentiative pathways of progenitor populations in teratomas.

Teratomas derived from certain strains [e.g., C3H, 129J/Sv, A/He, CBA/H-T6, AKR, and (129/Sv × A/He)F_1 and (C57BL × CBA)F_1 hybrids] may give rise to transplantable tumors that contain embryonal carcinoma cells.[5,6] These "permissive" mouse strains should be used for experiments if a test of neoplastic activity is required.

Choice of Ectopic Sites

Whole embryos and tissue fragments generally grow with comparable frequencies in different ectopic sites. Table II summarizes the frequency with which tissues of gastrulating embryos formed teratomas in three ectopic sites. For transplants such as the germ layers, about 42–61% of the

[5] D. Solter, I. Damjanov, and H. Koprowski, *in* "The Early Development of Mammals" (M. Balls and A. E. Wild, eds.), p. 243. Cambridge Univ. Press, London, 1975.
[6] D. Solter, N. Adams, I. Damjanov, and H. Koprowski, *in* 'Teratomas and Differentiation" (M. I. Sherman and D. Solter, eds.), p. 139. Academic Press, London, 1975.
[7] W. Y. Chan, *J. Anat.* **175,** 41 (1991).

grafts formed teratomas, but over 90% of grafts involving whole embryos developed in either the kidney or the testis. Therefore, as far as embryo-derived teratomas are concerned, there is no significant difference in the growth capacity of similar grafts made to different ectopic sites. An exception is found, however, with the transplant of genital ridges.[6] The incidence of teratomas was much higher in grafts made to scrotal sites (testis, epididymis, epididymal fat pads) than in those grafted to nonscrotal sites (spleen, liver, kidney, ovarian fat pad, etc.). The genital ridges also form teratomas more often in scrotal testes than in cryptorchid testes. In view of this possible bias in the development of the transplant, it is important during the preliminary runs of the study to test the differentiative capacity of the embryo/tissue in several ectopic sites before any particular one is chosen for further experimentation.

The choice of ectopic sites is also influenced by considerations such as the size of the graft and whether a continuous monitor of growth is required. If extensive growth of the fragment is required for the initiation of tissue differentiation and morphogenesis, then the kidney or testis should be chosen for transplantation of the fragment. The less spacious anterior chamber of the eye normally can accommodate only relatively small transplants, but it permits frequent microscopic observations of the developing graft and, therefore, a continuous monitoring of tissue growth not possible with the two intraabdominal sites.

Preoperative Preparation of Host Animals

Adequate anesthesia and proper surgical procedures are absolutely crucial to the success of the experiments. The primary aim is to ensure good access to the ectopic sites without inducing excessive stress and surgical trauma to the animal.

Deep anesthesia is essential to maintain the animal in a compliant state during surgery. The state of anesthesia can be assessed by the absence of the corneal reflex and the abolition of pain reflex when the tail tip is grasped by a pair of blunt forceps. Nembutal and Avertin are the two most commonly used anesthetics. It is important to test any freshly prepared anesthetics on three or four trial animals to establish the potency and toxicity of the preparation (especially for Avertin, which may decompose to toxic products) before use in the actual experiment. The anesthetics, which should be kept at 4° and protected from light, must be brought to body temperature and shaken well before use.

Pentobarbital Sodium (Nembutal). Nembutal is prepared either by diluting a stock solution (60 mg/ml, Abbott, North Chicago, IL) with sterile saline (0.9% sodium chloride) or dissolving the powder (Sigma, St.

Louis, MO) in sterile saline. The concentration of the anesthetic in the working solution should be in the range of 6.0–7.5 mg/ml. It is injected intraperitoneally at a dose of 600–750 μg/10 g body weight. A mouse weighing 30–35 g therefore receives 0.3–0.35 ml of the Nembutal solution.

Tribromoethanol (Avertin). Avertin is prepared by dissolving 2.5 g of 2,2,2-tribromoethanol (Aldrich, Milwaukee, WI) in 5 ml of 2-methyl-2-butanol (tertiary amyl alcohol, Merck, Darmstadt, Germany) on a heated stirrer. Two hundred milliliters of distilled water is then added, and the mixture is kept stirred until the amyl alcohol is completely dispersed. Approximately 0.2 ml/10 g of the anesthetics is given intraperitoneally, which amounts to 0.6–0.7 ml for a 30–35 g mouse.

Intraperitoneal injection is usually made without restraint of the animal. The mouse is lifted by its tail while allowing it to hold on to the top of a cage with its front legs. Injection is then made to the lower abdominal region of the mouse. Care must be taken to keep the needle from going too deeply into the abdomen, which may result in severe shock to the animal when the internal viscera or major blood vessels are accidentally impaled.

Prior to surgery, the skin over the incision area is swabbed with sterile gauze soaked with 70% ethanol. This will serve to cleanse and wet the fur and prevent it from flying around and sticking to instruments and tissues. Alternatively, the skin over the incision area can be shaved to remove the fur.

For ease of handling the animal during surgery, the mouse is put onto a piece of Whatman (Clifton, NJ) filter paper or on the lid of a 90-mm petri dish so that it could be moved around without changing its posture. The animal on the filter paper or dish lid is then placed on the stage of a dissecting microscope. Transplantation to kidney and testis is done usually at a magnification of 4–10×, but transfer to the anterior ocular chamber is best performed at 40×. Shadowless incident illumination is provided by a cold light source (Volpi fiber optics) through a ring light guide attached to the objectives of the microscope.

Transfer to Ectopic Sites

Transfer Pipette

Transplantation of embryos/tissue fragments is done by transferring the grafts in a tiny volume of culture medium using a mouth-controlled micropipette. The micropipette is made as follows. A microburner is constructed by fitting a Pasteur pipette on a Bunsen burner or by attaching a 19-gauge needle to a 2.5-ml syringe connected to the end of a piece of rubber tubing from the LP gas tap. The shaft of a Pasteur pipette is heated using the microburner, and, once molten, the shaft of the Pasteur pipette is pulled into a capillary with an external diameter of 200–400 μm. The capillary is then broken by snapping with fine watchmaker's forceps at a point about 3–4 cm from the shoulder of the Pasteur pipette. The tip of the pipette is then heat polished by a microforge (Beauduoin, Paris, France) to prevent it from inflicting any damage to the host tissue during transplantation.

The micropipette is connected to a mouth-operated aspiration device made up of a mouthpiece, a good length (at least 80 cm) of flexible rubber tubing, and a cotton-plugged pipette adaptor. A more sophisticated braking pipette system can also be used.[8] The micropipette is filled with the transfer medium by dipping the tip into the medium and allowing the pipette to be filled up with the solution by capillary action. A 2–3 mm column of air is then drawn into the pipette to create an air-brake. The embryo or tissue fragment is then picked up by capillary action in a small volume of fluid and is positioned at about 3–4 mm from the tip of the micropipette. The loaded micropipette should be placed within easy reach by pressing it into a piece of Bostik blu-tack stuck onto the bench surface or microscope stage.

Transplantation to Kidney

A 1-cm incision is made through the skin (Fig. 2A) on the lateral flank of the trunk about 1 cm from the spine and immediately caudal to the last rib. The kidney can be located by searching for a dull red, solid mass underneath the muscle layers as the incision is moved over the lateral lumbar regions. An incision of 0.5–0.7 cm is then made through the muscle layers to expose the kidney. The fat pad adhering to the cranial pole of the kidney is grasped and pulled with a pair of blunt forceps to bring the kidney out of the peritoneal cavity. A slight pressure applied by pressing the flank of the body also facilitates the manipulation of the kidney through the incision.

It is important to make the incision in the muscle wall just larger than the smaller diameter of the kidney so that the kidney can be eased out of the abdominal cavity along its long axis through the incision. The small incision also helps to immobilize the kidney: by turning the exposed kidney 90°, it can then be lodged against the incision (Fig. 2A) without using any additional instruments to stop it from sliding back into the abdominal cavity. If fixation of the kidney becomes a problem, especially when too

[8] I. Damjanov, A. Damjanov, and D. Solter, in "Teratocarcinomas and Embryonic Stem Cells" (E. J. Robertson, ed.), p. 1. IRL Press, Oxford, 1987.

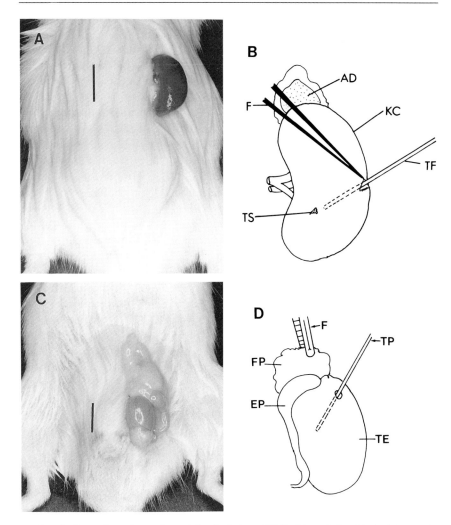

FIG. 2. Transplantation to intraabdominal sites. (A) Position of the skin incision at the lateral flank and the exposed kidney; (B) arrangement of instruments during transplantation: (F) holding forceps, (TF) transfer pipette, (TS) tissue fragment, (KC) kidney capsule, (AD) adrenal gland; (C) position of the skin incision in the inguinal region and the exposed testis and associated structures; (D) arrangement of instruments during transplantation: (F) holding forceps, (TP) transfer pipette, (FP) fat pad, (EP) epididymis, (TE) testis capsule. (Figure 2A reproduced from Tam[1] by permission of Oxford University Press.)

large an incision has been made, the kidney can be immobilized with a Desmarres chalazion forceps.[8]

The exposed kidney is allowed to stand for about 1 min to let the capsule dry out so that it can be grasped more easily by the forceps. The kidney capsule is first picked up with a pair of fine watchmaker's (holding) forceps, and a hole is made by puncturing the capsule about 2–3 mm away from the holding forceps with the closed point of another pair of forceps. The points of the puncturing forceps are then allowed to spring apart to produce a tear in the kidney capsule. The micropipette containing the tissue is then brought close to the tear in the capsule. The holding forceps is lifted slightly to create a tent in the capsule (Fig. 2B). The micropipette is then inserted through the opening and pushed for about 3–6 mm into the space between the capsule and the kidney surface. The tissue is expelled slowly by blowing into the mouthpiece as the pipette is carefully withdrawn. A positive presure is maintained until the pipette is completely out of the capsule, in order to prevent any reflux of fluid that may carry the tissue back into the pipette. The holding forceps is then released to let the capsule snap back and trap the graft between the capsule and the kidney surface. The micropipette should then be checked to ensure that the tissue has actually been transplanted.

The kidney is then inspected under the microscope to ascertain the location of the graft. (It is possible to transplant more than one tissue fragment to each kidney, especially when the grafts are small and the period of growth is no more than 2 weeks. Up to three different grafts have been made in some cases. Teratomas derived from each graft can be distinguished easily, provided that the original grafts are positioned far apart from one another in the kidney.) The kidney is dislodged by bringing the caudal pole to the wound opening and allowing it to slide back into the peritoneal cavity on its own accord. The muscle layers are closed with absorbable sutures (coated vicryl J-303, Ethicon) and the skin with silk sutures (Ethilon 668, Ethicon; both sutures obtained from Johnson & Johnson, Sydney, Australia) or Michel wound clips (Aesculap). A similar transplantation can be done on the other kidney if necessary.

Transplantation to Testis

The anesthetized animal is placed in a supine position, and a 1-cm paramedian incision (1–1.5 cm from the midline) is made through the skin and the muscle layers in the inguinal area (Fig. 2C). The vas deferens, which is a thick muscular tube with accompanying arteries (deferential and testicular arteries), is located by searching the abdominal contents with blunt forceps. The testis is then retracted from the scrotal sac into the abdominal cavity by gently pulling at the vas deferens. The fat pads over the caput epididymis are grasped by blunt forceps, and the testis is taken out of the abdomen and placed in its normal anatomical position on the skin (Fig. 2C). Precaution must be taken not to sever the blood supply to the testis during the handling of the vas deferens and the arteries.

While holding the testis with blunt forceps, a puncture is made in the tunica albuginea with a pair of watchmaker's forceps or a 25-gauge hypodermic needle. The testis is then held by its fat pad, and the transfer pipette is inserted through the hole into the testicular interstitium (Fig. 2D). The tip of the micropipette is kept close to the underside of the tunica and is inserted for about 5 mm into the testis. The tissue is expelled from the micropipette as it is slowly withdrawn. Both the micropipette and the testis should then be checked to ensure that the transplantation has been properly performed. The testis and the associated structures are then returned to the abdominal cavity.

After the animal has recovered and begins to walk around, the testis will normally return to the scrotal sac, but sometimes it is necesary to fix the testis into the abdominal cavity by suturing the fat pad onto the muscle wall. This is because teratomas may grow to a massive size, especially in long-term experiments, and a swelling testis trapped in the scrotal sac may cause serious discomfort to the animal. The muscle wall and the skin are then closed with sutures as for transplantation to the kidney. Michel wound clips should not be used when closing the skin, however, as they may cause tightening of the skin and interfere with genital functions. Transplantation can also be done to the other testis if required.

Transfer to Anterior Ocular Chamber

An important step in the preoperative preparation is to treat the eye with one drop of 1% (w/v) atropine solution before anesthetizing the animal. The atropine dilates the pupil and prevents the herniation of the iris through the incision in the cornea. The anesthetized mouse is placed on its side under a stereomicroscope at a magnification of 40×. The cornea is exposed by retracting the eyelids with fingers, and a horizontal incision is made by stabbing the ventral area of the cornea (Fig. 3a) with the point of a No. 11 surgical blade. As little pressure as possible should be applied to retract the eyelids; otherwise, this may result in the leakage of aqueous humor and herniation of the iris when the incision is made.

Transplantation should be done quickly before the cornea dries up and turns sticky. The loaded micropipette is inserted through the incision into the anterior ocular chamber, and the transplant is deposited near the iridocorneal angle in the caudal or dorsal aspect of the chamber (Fig. 3b). The transplant can also be repositioned later by massaging the area of the

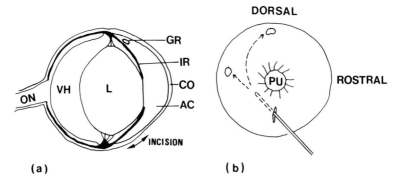

FIG. 3. Transplantation to the anterior ocular chamber. (a) Cross section of the eyeball showing the position of the incision; (b) transfer of tissue to the iridocorneal angle of the chamber. ON, Optic nerve; VH, vitreous humor; L, lens; AC, anterior chamber; CO, cornea; IR, iris; GR, graft; PU, pupil.

cornea overlying the transplant with the polished end of another micropipette. The transplant should be in direct contact with the iris to ensure good vascularization, which is crucial to promote maximal tissue differentiation. Transplantation should only be performed to one eye, and not to both eyes of the animal, to prevent any loss of vision which may interfere with feeding and locomotion.

Postoperative Care

Hypothermia always sets in after anesthesia. The animal should be kept warm during the operation with the stage lamp of the dissecting microscope and, after the operation, by leaving it to recover under a table lamp or on a warm plate until it fully regains consciousness. The sutured wound should be cleansed with a swab containing 70% 2-propanol (Briemar, Nominee, Victoria, Australia). Operated animals should be kept on autoclaved beddings and should be checked regularly for wound infection and general well-being. For long-term experiments the size of the tumor in the kidney or testis of animals should be checked regularly by palpation. Animals should be sacrificed if there are obvious signs of discomfort and physical handicap due to tumor growth.

The growth of the graft in the anterior ocular chamber should be examined daily under a dissecting microscope during the course of the experiment for detection of postoperative complications such as scarring or vascular invasion of the cornea. Mice with grafts grown to cover more than 60% of the area under the cornea must be sacrificed immediately to

TABLE III
ANALYSIS OF TUMORS DERIVED FROM TRANSPLANTS OF EMBRYOS OR TISSUE
FRAGMENTS TO ECTOPIC SITES

Parameter	Methods
Tissue composition	Histology of tumors using a variety of general staining methods (e.g., hematoxylin and eosin, Masson's trichrome, toluidine blue), special histochemical staining of tissues or matrix components (e.g., neural tissues by silver stains, connective tissue by van Gieson's staining, secretory cells by alcian blue and periodic acid–Schiff's reagent), and immunolocalization of specific molecules on tissue sections
	Transmission and scanning electron microscopy for ultrastructural characterization of cell and tissue types and application of cytochemical methods to detect cellular and extracellular contents
Neoplastic potency	Derivation of cell lines in culture
	Testing for retransplantability of tumor cells
Gene expression	Molecular analysis of specific mRNAs
	Tissue-specific expression by *in situ* RNA hybridization
Lineage analysis	Analysis of clones derived from cells marked by lineage-specific transgenes or by retroviral introduction of reporter genes

prevent the development of intraocular hemorrhage or elevated chamber pressure due to blockage of drainage channels. Intraocular grafts usually reach their maximum size by about 14–21 days.[7]

Analysis of Results

Fully differentiated teratomas are usually obtained 30–45 days after transplantation. Transplants may, however, be harvested at earlier times if it is necessary to study the intermediate stages of tissue differentiation or morphogenesis. A multitude of tissue types are usually found in the teratoma.[1] Teratomas are usually examined (see Table III) either microscopically following routine histological preparation and staining or ultrastructurally using transmission and scanning electron microscopy. Special tissue types can be studied in the histological or electron microscopical preparations by using a variety of histochemical, enzymatic, and immunological procedures. The tissue composition of the teratoma is usually analyzed by noting the presence or absence of certain tissue types. Often this is sufficient to reveal qualitative differences between teratomas displaying different propensities to differentiate along various pathways. Some morphometric quantitation of tissue types can be done to compare the histogenetic capacity of different grafts. The results, however, are often difficult to interpret because of variations in the growth rate and in the relative abundance of different types of tissue, even in teratomas derived from similar transplants.

A largely unexplored approach to analyze the teratomas is to examine the pattern of expression of tissue- or lineage-specific genes in the differentiating tissues and thus provide models to study lineage specification and differentiative pathways of specific groups of progenitor cells in the embryo. This could be done either by *in situ* hybridization of gene transcripts or by molecular analyses of mRNAs performed in conjunction with morphological studies of the teratoma. Nevertheless, it remains to be established whether tissues in ectopic sites differentiate in a manner similar to their normal counterparts before an extrapolation of the ectopic results to *in vivo* differentiation is warranted.[9] If individual progenitor cells could be specifically labeled with genetic markers, such as the *lacZ* gene introduced through retroviral vectors or by microinjection, then it would also be possible to perform clonal analyses and fate mapping of specific cell populations by following their differentiation in the teratoma.

Teratomas derived from early postimplantation embryos or genital ridges often contain nests of undifferentiated embryonal carcinoma cells, which are suggestive of malignancy. A direct test of the neoplastic potential is to examine the retransplantability of the teratoma.[9] Briefly, tissue fragments are obtained from the teratoma by mincing the teratoma in phosphate-buffered saline (pH 7.2). The small fragments are further dissociated in the same buffer by homogenizing the tissue gently in a 1-ml siliconized all-glass homogenizer or by forcing the tissue 2–3 times through a 21-gauge hypodermic needle. The fragments are then pelleted by centrifugation through 1 ml fetal calf serum at 400 g for 10 min at 4° to remove debris and dead cells. The pellets are transferred to another mouse either by subcutaneous injection to the lateral flank to produce a solid tumor or by intraperitoneal injection to produce ascites. Tumors obtained from this retransplantation can be further propagated in animals or used as sources of stem cells to establish teratocarcinoma cell lines.[10]

Acknowledgments

The author wishes to thank Dr. W. Y. Chan for advice on the transplantation to anterior ocular chamber and Professor Peter Rowe for reading the manuscript.

[9] R. S. P. Beddington, *J. Embryol. Exp. Morphol.* **75**, 189 (1983).
[10] M. A. Rudnicki and M. W. McBurney, *in* "Teratocarcinomas and Embryonic Stem Cells" (E. J. Robertson, ed.), p. 19. IRL Press, Oxford, 1987.

Section III

Fertilization

[12] Micromanipulation of Gametes and Embryos

By Jon W. Gordon

This chapter serves as an instructional guide for mouse gamete and embryo micromanipulation. The written component describes each manipulation procedure as well as relevant biological and technical principles that impact on its successful practice. The pictorial component supplements the written portion, in many cases providing a stepwise depiction of the procedure under discussion. All equipment, supplies, reagents, and media required for every commonly used manipulation procedure are also listed in tabular form in order to facilitate ordering and rapid assembly of relevant materials. The Appendix at the end of this chapter includes Tables A.I–A.XIII which list additional information regarding formulations, such as formulas for media. Mouse *in vitro* fertilization (IVF) media and the timing of hormone administration and embryo replacement are given in recognition of the possibility that the reader may be introducing such techniques into the laboratory for the first time, and because many of the micromanipulation methodologies are used to investigate the physiology of fertilization. As such, it is often important to perform *in vitro* fertilization in conjunction with manipulation.

This chapter does not discuss production of transgenic mice. This technique is addressed in [45] in this volume. Embryo transfer techniques are also described in greater detail in [45].

Introduction

General Background

The ability to manipulate mammalian gametes and embryos has revolutionized the study of mammalian fertilization and early development. Not only has this technology made feasible novel and powerful studies of this crucial phase of the life cycle, it has fostered the development of important new clinical approaches to reproductive disorders in humans, and has led to improved procedures for reproductive management of experimental animals and livestock. Thus, micromanipulation has assumed a central role in the basic and applied science of mammalian reproductive physiology and developmental genetics.

Of course micromanipulation is only a tool, and its power is dependent on concomitant advances in other areas, most notably, recombinant DNA

METHODS IN ENZYMOLOGY, VOL. 225
Copyright © 1993 by Academic Press, Inc.
All rights of reproduction in any form reserved.

technology and embryo culture. Moreover, its successful use, for either basic experiments or practical procedures, depends on an appreciation of the fundamentals of reproductive biology. Thus, it behooves the practitioner of micromanipulation to acquire a basic knowledge of fertilization and early development in the species which is to be the object of study. In general, most basic investigations involving micromanipulation utilize mice and, to a lesser extent, rabbits, rats, hamsters, and livestock. Clinical micromanipulation obviously involves humans. It is beyond the scope of this chapter to address all of these specialized adaptations of micromanipulation. Accordingly, only mice are discussed here, but some of the procedures are described with the understanding that mice are used as models for developing techniques for human IVF. Many of the basic techniques used in mice can readily be adapted to other species, including humans. Before discussing the various protocols in detail, some general biological features of fertilization and early development in mice are discussed.

Biology of Fertilization and Early Development

In vivo, fertilization results from a carefully orchestrated series of functions on the part of both the sperm and egg, and it is critically dependent on timing. Therefore, it is not surprising that fertilization *in vitro* must also be properly timed. Oocytes must be at a suitable stage of maturation, but they cannot be excessively aged at the time of insemination. Similarly, the sperm must be preincubated so as to achieve a state of capacitation prior to fertilization, but they do not survive indefinitely after this time. While it is true that under most circumstances the human spermatozoon will survive for more than 24 hr *in vitro,* cases of male factor-related infertility, for which micromanipulation-assisted fertilization (MAF) may be employed as a clinical remedy, are often associated with a reduced half-life of sperm. Thus, it is important, for all species, to time micromanipulation of gametes in such a way as to retain viability and the full repertoire of biological functions. In mice, sperm are generally difficult to work with: they survive for only a few hours in culture, and they are very difficult to freeze.

In general, the oocyte is far more delicate than the embryo. Therefore, when micromanipulation is to be performed on oocytes, greater care must be taken to maintain a stable and supportive environment. In addition, if IVF is to be performed in the mouse either via or in conjunction with micromanipulation, the oocyte must be inseminated soon after ovulation in order to retain viability and a high rate of fertilization. The spermatozoon is also a delicate cell. In our experience, sperm are very susceptible to conditions of low pH, such that below pH 7.2 viability is rapidly lost. In addition, because incubation is required to induce capacitation, but the survival time of sperm is limited, the timing of release of sperm from the male tract as well as insemination must be carefully planned (Table IV). Once fertilization has taken place, the zygote can tolerate surprisingly severe changes of pH and/or temperature without losing developmental potential.

The MAF technique is almost invariably performed using oocytes retrieved after superovulation of the female. Therefore, many eggs may be abnormal, and it is important to select oocytes with an identifiable polar body and a uniform cytoplasm that does not contain inclusion bodies. It should be emphasized, however, that no morphological abnormality of oocyte or spermatozoon has yet been correlated with birth defects in the human.

Another important point is that MAF, either by zona drilling (ZD) or by subzonal sperm insertion (SZI), entails exposure of the oocyte surface to sperm which have not traversed the zona pellucida (ZP). Normally the process of fertilization is characterized by binding of acrosome-intact spermatozoa to the ZP, induction of the acrosome reaction (AR) by the zona protein ZP3, and penetration of the zona.[1] Thus, only acrosome-reacted sperm encounter the egg surface. When MAF is performed the ZP is bypassed, and, thus, acrosome-intact sperm can reach the egg. It is presumed that such sperm cannot fertilize, and therefore, that MAF will only work if spontaneously acrosome-reacted sperm are present. When mouse sperm are capacitated by incubation in medium, spontaneous acrosome reactions rarely occur in more than 30% of the sperm. Thus it is important in SZI methods to assure that sufficient numbers of sperm are microinjected in order to guarantee that one which has undergone the AR is made available for fertilization.

The *in vitro* development of preimplantation embryos is almost always slower than development *in vivo*. The simple act of removing cleaving embryos from the mouse slows development by as much as 24 hr. These delays, which may be further exacerbated by prolonged micromanipulation procedures that are often carried out at room temperature, must be kept in mind when planning the implantation of embryos into the female. In general, pseudopregnant female recipients are mated one or more days later than the "donor" females from which the oocytes or embryos are obtained. Table IV gives the timing of superovulation, mating, and embryo transfer for all procedures described in this chapter.

[1] J. D. Bleil and P. M. Wassarman, *Dev. Biol.* **95,** 317 (1983).

Up until at least the 4-cell stage, the developmental totipotency of blastomeres can be demonstrated,[2] and, even after this time, totipotency is clearly retained for cells destined to become the embryo proper. Therefore, manipulations that entail loss of removal of one or more cells are generally not harmful to the embryo. However, the process of blastulation is dependent on the age of the embryo rather than the number of cells. As a result, removal of excessive numbers of cells can lead to development of "trophoblastic vesicles"[2] that lack an inner cell mass (ICM) and that cannot develop into live young.

Embryo cleavage is very rapid; thus, cells that appear to be distinct within the zona may actually be still in the process of completing the previous cell division. As a result, apparently distinct blastomeres may actually retain cytoplasmic continuity with their neighbors, and forcible separation of blastomeres under such circumstances can lead to cell death. In addition, when manipulation is performed for the purpose of retrieving cells for genetic analysis, care must be taken to ensure that the removed material contains at least one complete diploid genome. The problems of incomplete cleavage and fragmentation can best be addressed by removing blastomeres with a visible nucleus. This reduces the chances of retrieving a cytoplasmic fragment, and also increases the likelihood that the previous cleavage is complete.

Compaction of the morula depends on a favorable spatial relationship between the blastomeres. This spatial arrangement is usually provided by the ZP. If micromanipulation is performed prior to compaction of the morula, it is important that a relatively intact ZP or suitable substitute for it be retained.

Equipment

Regardless of the species or the micromanipulation procedure, the same basic equipment is needed. Table I contains a list of equipment for use in both MAF and embryo manipulation. Before proceeding, some comments on the equipment list may be helpful. With regard to the microscope fitted with interference optics, it is important to be aware that use of Nomarski optics requires glass slides, whereas Hoffman interference optics function through disposable plastics. Hoffmann optics are more convenient, because use of plastics eliminates the need for the repeated sterilization of glassware and careful cleaning of materials required for techniques involving enzymatic amplification of DNA from single cells.

[2] A. K. Tarkowski and J. Wroblewska, *J. Embryol. Exp. Morphol.* **18,** 155 (1967).

TABLE I

 EQUIPMENT LIST FOR MICROMANIPULATION

Microscope, preferably inverted, equipped with
    interference optics (Nomarski, Hoffmann, or phase
    contrast)
Two micromanipulators
Marble table for manipulation[a]
Two syringe micrometers for applying suction or
    positive pressure through micromanipulator system
Video camera for teaching and recording[a]
Microforge
Pipette puller
Needle grinder
Dissecting microscope equipped with transmitted light
    and overhead illumination
Microburner or alcohol lamp
Bench-top microcentrifuge

---

[a] Optional.

There are many different designs of micromanipulators, some power operated. Unless it is absolutely necessary to record the precise volume of solution delivered during a procedure, power equipment is not needed, and in fact has disadvantages, including expense, space consumption, the requirement for maintenance, and the need, in the case of power microinjectors, for a gas supply. All these additional items add a level of complexity to the system for which increased convenience and/or control do not compensate. Manually operated equipment offers one of the most sensitive instruments available: the human hand. This device is obtained at no additional cost, occupies no additional space, and is self-maintaining. Therefore, the value of this apparatus should not be underestimated.

With regard to the syringe micrometers, most micromanipulation procedures can be carried out with only a single micrometer, usually affixed to the holding pipette. The system controlling the microneedle either can be connected to a syringe or can be operated by mouth. For some complex procedures, however, two micrometers may prove useful.

There are two pieces of equipment listed as optional (Table I). The first is a marble table or air table. A variety of tables are available for providing a stable base for the micromanipulator. These can be costly, are often very heavy, and may require for their installation some renovation of the laboratory (e.g., removal of preexisting bench). In my own experience they are not necessary. If the purchase of a table is contemplated, first test the system on an ordinary table or laboratory bench. If vibration appears to be a problem, then a special table can be purchased.

TABLE II
REAGENTS FOR MICROMANIPULATION[a]

Concentrated HCl
Hydrofluoric acid (HF), 48% (Mallinckrodt, HF 2640)
Fluorinert (3M, FC-77)
Mineral oil (e.g., Mallinckrodt, 6358)
Sucrose (Sigma, S 1888)
NaOH, 5 N
Pregnant mare serum gonadotropin (PMSG) for mouse
  superovulation
Human chorionic gonadotropin (hCG) for mouse
  superovulation
Sodium pentobarbital stock solution (50–60 mg/ml)
Ethanol, 100%
Propylene glycol
Methyl cellulose to give viscosity of 400 centipoise in a
  2% (w/v) solution (Sigma, M 0262)
Colcemid (demecolcine) Sigma, D 7385)
Cytochalasin D (Sigma, C 8273)
Hyaluronidase (Sigma, H 3606)
Bovine serum albumin (BSA, fraction V)

[a] List does not include regents contained in basic media
(see Appendix).

The video recording equipment is obviously not needed for micromanipulation and is accordingly listed as optional. However, such equipment is extremely valuable for teaching, and is therefore highly recommended.

Reagents

Reagents required for production of basic culture media are listed in the tables in the appendix. These tables also describe how each individual medium is formulated. Other reagents that are added to produce specialized media are listed in Table II. Suppliers and stock numbers are given only as examples and for convenience. They are not intended to indicate that other sources of materials would not be satisfactory.

Equilibration of oil is carried out by adding protein-free medium to the oil at 20% (v/v), shaking the flask, gassing the solution for several minutes by bubbling 5% CO_2 through the solution, and storing the oil in the incubator with the cap of the flask loose. When the medium has settled to the bottom of the flask, it should have a color consistent with complete gas equilibration. After the oil is used it need not be reshaken, but it should be gassed again before returning it to the incubator. After the flask has been gassed 3 times it is discarded. It also should be emphasized that

TABLE III
Supply List for Micromanipulation

Rubber tubing equipped with mouthpiece for pipetting small volumes

Filters (0.22 μm) to fit rubber mouthpiece-operated tubing (e.g., USA Scientific, No. F192)

Sterile conical centrifuge tubes, 50 and 15 ml

Microcentrifuge tubes, 1.5 ml

Sterile plastic tissue culture dishes of suitable size for micromanipulation and other procedures

Disposable filters for sterile filtration of volumes from 100 to 1000 ml

Watchmaker's forceps, 2 pair

Small iris scissors, 1 pair

Small hemostat

4-O silk thread for closing wounds after mouse surgery

Small needles with cutting edge for closing mouse wounds (e.g., American Hospital Supply, Anchor No. 1834-8)

Microcapillary tubing for producing microneedles used for sperm injection, and for holding pipettes (e.g., Mertex, 100 × 0.5 mm ID, No. MX-999, Mercer Glass Works, NY)

Self-filling microcapillary tubing for zona drilling or DNA microinjection (e.g., 12111 Omega dot tubing, Glass Co. of America, Millville, NJ)

Spinal needle for filling barrel of microcapillary tubing used for zona drilling or DNA injection (e.g., Popper & Sons, Quinke Babcock 26-gauge spinal needle)

Final polyethylene tubing for drawing HF into tips of needles used for sperm injection (e.g., Intramedic, PE60, PE90, PE160, PE190)

Plastic tubing adaptor for PE tubing (e.g., Clay Adams, Parsippany, NJ, A-1026)

    Size B fits PE60 and PE90

    Size C fits PE160 and PE190

Glass depression slides for harvesting mouse oocytes or embryos, or, if Nomarski optics are used for micromanipulation, for use on this microscope [e.g., Fisher, Pittsburgh, PA, No. 12-560A (thin) or 12-565A (thick)]

Anesthesia for mice (e.g., sodium pentobarbital, 6 mg/ml; see Table A.XIII for formulation)

Sealing wax

Automatic pipettors capable of delivering 0–200 and 0–20 μl

Pasteur pipettes, 9 inch (Kimble brand preferred, Baxter, Edison, NJ)

Diamond pen for scoring and breaking glass

---

different laboratories employ different manipulation approaches. As such, the reagents and media suggested are exemplary of the methods used in our laboratory.

Supplies

A supply list for micromanipulation is given in Table III. As with the reagents, brands and stock numbers are given for convenience only.

However, it is easier to score and break finely drawn 9-inch Pasteur pipettes of the Kimble brand than of other varieties.

Toolmaking

All micromanipulation procedures are most successful when tools that allow minimal distortion of the oocyte or embryo are used. To limit mechanical stress, two general principles apply. First, the aperture of the holding pipette should be of sufficient size to immobilize the oocyte/embryo, but not so large as to allow aspiration into the shaft of the holding pipette. An aperture approximately 20% the diameter of the oocyte/embryo is suitable. The aperture can be adjusted to compensate for variations in the shaft size; that is, a large shaft which can exert strong suction can be rendered less dangerous by narrowing the aperture, and a narrow shaft for which insufficient suction might be expected can be corrected by leaving a relatively wide aperture. The principle underlying these modifications are embodied in Poiseuille's law:

$$Q = \pi \Delta P r^4 / 8 \eta l \tag{1}$$

where Q is the flow, ΔP is the pressure gradient, r is the radius of the conduit for fluid flow, η is the viscosity of the solution, and l is the length of the conduit. Examination of Eq. (1) shows that small changes in the radius dramatically change the flow rate: reducing the radius by one-half increases the resistance to flow by a factor of 16. Thus, the flow characteristics of the holding pipette can be substantially modified by changing the radius of the shaft. These features are demonstrated in Fig. 1. Note that r_1, the radius of the aperture, is the greatest determinant of flow because it is narrowest, and also note that flow can accordingly be most profoundly regulated by changing r_1. In Fig. 1, ΔP is shown as $P_1 - P_2$.

Another important principle is that, under ideal conditions, the shaft of the holding pipette makes contact with the floor of the culture dish (Fig. 1). This is because the microneedle almost always approaches the oocyte or embryo from an angle (e.g., Figs. 1 and 2). Thus, when free space exists under the holding pipette, it is possible to push the oocyte or embryo under the pipette, thereby destroying it. Because it is best to keep the shaft of the holding pipette in contact with the dish during manipulation, it is convenient for the outside diameter of the holding pipette to be approximately the same as the outside diameter of sphere described by the zona pellucida (Fig. 1). Some room for error exists here, but when the holding pipette is smaller in diameter than the outside diameter of the oocyte with its zona, it is important to make sure that the oocyte makes contact with the dish prior to beginning micromanipulation.

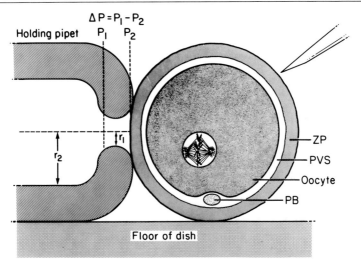

FIG. 1. Diagram of a metaphase II oocyte in the holding pipette. Note that the pressure gradient across the holding pipette aperture is $P_1 - P_2$ and that the radius of the aperture, r_1, is the most important determinant of flow [see Poiseuille's law, Eq. (1)]. Also note the size relationship of the holding pipette and oocyte, with equal outside diameters of the holding pipette and ZP. Although the microneedle appears to be parallel to the floor of the culture dish when examined under the microscope, it is actually approaching the egg at an angle. For this reason, use of the culture dish floor to support the egg is helpful. PB, Polar body; PVS, perivitelline space; ZP, zona pellucida.

It is also quite useful, though not necessary, if the shaft of the holding pipette is long (Fig. 2). This allows the holding pipette to reach the micro-drop when overhead objectives are used. In addition, the pipette is flexible and can "slide" across the floor of the dish when pressed against it. Finally, the thin shaft minimizes optical distortion by remaining submerged in the microdrop and thereby leaving the interface between medium and overlying oil, within the microscope field, undisturbed.

With these issues in mind, we now proceed to discuss the various micromanipulation procedures.

Zona Drilling

General Considerations

Originally developed as a method for assisting fertilization,[3,4] zona drilling (ZD) is a highly versatile procedure with a number of applications.

[3] J. W. Gordon and B. E. Talansky, *J. Exp. Zool.* **239,** 347 (1986).
[4] J.W. Gordon, *Ann. N.Y. Acad. Sci.* **541,** 601 (1987).

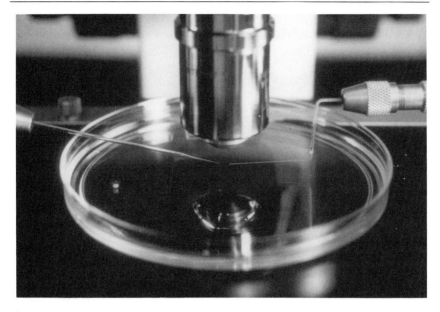

FIG. 2. Photograph of microtools poised over the micromanipulation dish. In this case, with the overhead objective, the long thin shaft of the holding pipette (right) allows the experimenter to enter the microdrop without the objective presenting an obstruction. For this and several other reasons (see text), such holding pipettes are advantageous. Also note the angle of the microneedle (left).

The term "zona drilling" usually refers to the localized dissolution of the zona pellucida (ZP) with a solvent, usually acidified Tyrode's solution (ATS). However, the principle embodied by zona drilling, namely, that opening of the ZP can modify the processes of fertilization without compromising embryonic development, can be extended to several other intricate manipulations of the egg and embryo. This versatility is reflected in several subsequently developed procedures wherein the ZP is opened mechanically, enzymatically, or with a laser.

For typical ZD, ATS is needed. Table A.I (in the Appendix) shows a formula for this solution that is compatible with both mouse and human embryos. We generally prepare the solution without acidification and acidify immediately before use. Acidification is accomplished by adding concentrated HCl to about 0.66% (v/v) (e.g., 900 μl of Tyrode's solution, 6 μl of HCl). After the acid is added, the solution is filtered through a 0.22-μm filter. The microneedle is then loaded with the ATS in preparation for drilling. We generally fill the tip of the microneedle from the back by lowering the base of self-filling tubing (e.g., Omega dot, Table III), then

TABLE IV
SCHEDULE FOR MOUSE *in Vitro* FERTILIZATION

Step	Schedule
Light cycle for animal room	Lights off, 10 PM
	Lights on, 8 AM
PMSG injection	Day −3; 1600 hr
hCG injection	Day −1: 1900–2100 hr
Sacrifice and manipulation	Day 0
	Sacrifice of female: 12.5 hr after hCG, or 0730–1030 hr
	Sacrifice of male and recovery of sperm: 30–60 min prior to insemination
	Oocyte insemination: 0800–1100 hr
	Examination for pronuclei: 1500–1800 hr
	Embryo transfer: anytime between appearance of pronuclei to blastocyst stage
Mating of foster recipients (if embryo transfer is planned)	Day 0: 1300–1600 hr
Transfer of embryos	Days 0–4
Examination for birth of newborns	Days 18–21

use a spinal needle and syringe to fill a substantial portion of the barrel of the needle. This allows for manipulation of numerous eggs without refilling of the microneedle.

Assisted Fertilization

Females are superovulated according to the schedule outlined in Table IV. Oocytes are recovered and treated with hyaluronidase to remove the cumulus cells. Hyaluronidase can be dissolved either in mouse fertilization medium (MFM, see Tables A.II–A.V, A.VII, and A.VIII for formulation of concentrated stocks and final medium formulation)[5] or in M16 medium (Tables A.II–A.V, A.VII and A.VIII).[3] Preparation of hyaluronidase is described in Table A.IX. After the eggs are washed, those with first polar bodies are loaded into a rectangular microdrop, under oil, composed of M2 medium (Tables A.II–A.VIII).[3] Eggs are then grasped in the holding pipette, the microneedle is pressed tangentially against the zona, and the acid solution is expelled until a suitably sized hole in the zona is produced (Fig. 3). The drilled oocyte is then moved to the opposite end of the microdrop, and the procedure is repeated until all eggs are drilled. After drilling, eggs can be fertilized *in vitro* according to standard procedures.[4]

[5] P. Quinn, C. Barros, and D. H. Whittingham, *J. Reprod. Fertil.* **66,** 161 (1982).

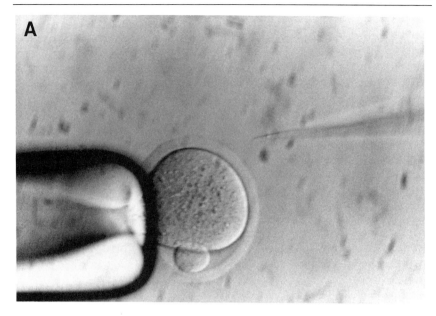

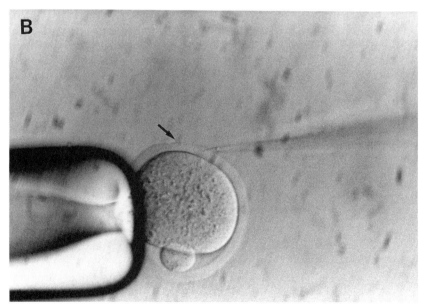

FIG. 3. Zona drilling. (A) Oocyte held in place prior to zona drilling. The first polar body is at 6 o'clock. (B) Zona drilling in progress. Note that the zona is thinned by Tyrode's solution (arrow). (C) Zona drilling complete. Immediately after the procedure, the oocyte may bulge through the opening in the zona. The gap in the zona is indicated by the two arrows.

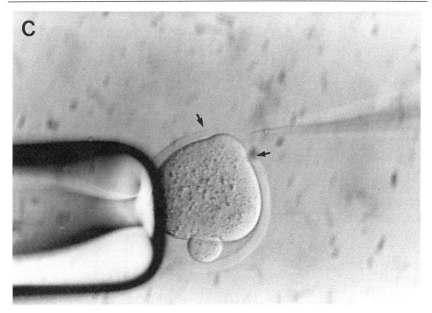

FIG. 3C.

When ZD is followed by IVF, we generally sacrifice the male and begin sperm capacitation 30–60 min prior to insemination. Thus, sperm are released from the male tract during the ZD procedure.

Assisted Hatching

Assisted hatching was first conceived by J. Cohen as a method for promoting implantation of human embryos,[6] and it was based on the findings[7,8] that ZD could lead to precocious hatching of normal mouse embryos. The procedure is not widely used clinically as yet, because it is uncertain which embryos might actually be assisted in implanting after the procedure. Nonetheless, assisted hatching has received some attention as a potential clinical tool.

Assisted hatching is best performed in the same way as zona drilling of unfertilized oocytes. The embryo is placed in the micromanipulator, and ATS is expelled from the microneedle until a readily vislble hole is created. This hole should not be too small, as small holes might actually

[6] J. Cohen, *J. In Vitro Fert. Embryo Transf.* **8,** 179 (1991).
[7] B. E. Talansky and J. W. Gordon, *Gamete Res.* **21,** 277 (1988).
[8] Y. Odawara and A. Lopata, *Fertil. Steril.* **51,** 699 (1989).

impede normal hatching.[6,7] If there are cytoplasmic fragments under the zona, these can be removed after drilling by drawing them into the micro-needle. It is unclear whether removal of such material is efficacious. After assisted hatching, the embryo is washed and returned to culture for observation or until the time of transfer.

Polar Body Biopsy

Polar body biopsy can be readily accomplished in the mouse with the assistance of zona drilling.[9] For this procedure, the oocytes are treated in the same way as for ZD. After washing the eggs free of hyaluronidase and cumulus cells, the eggs can be loaded either into M2 medium or into M2 supplemented with 100 mM sucrose (0.34 g/10 ml).[9] The polar body is then either at 12 or 6 o'clock, and the zona is probed with the microneedle until it is determined that contact with the zona immediately outside the position of the polar body is made. Acid is then expelled until a hole is produced in the zona of sufficient size to allow extrusion of the polar body. If this procedure is done correctly, the polar body will then spontaneously emerge from the ZP. If the polar body does not extrude spontaneously, it can be pushed out by applying pressure with the microneedle at a point distant from the drilling site.[9]

If polar body biopsy is performed for the purpose of genetic analysis, two important aspects of the procedure must be kept in mind. First, each oocyte must be manipulated in a separate microdrop to track every oocyte and its corresponding polar body. Second, it is absolutely essential that residual cumulus cells are not allowed to accompany the polar body to the polymerase chain reaction (PCR) tube. Each of these cells contains an entire diploid genome, and the DNA would thus contaminate the PCR reaction. To avoid this, we observe during the procedure for cumulus cell detachment, and, when this occurs, we immediately aspirate the detached cells into the holding pipette.

Embryo Biopsy

Because the human embryo tolerates ATS very well,[10] it is possible to perform the following embryo biopsy procedure on either mouse or human oocytes. Although many laboratories do not place embryos in special media for biopsy,[10] we feel that such media are very helpful. We employ biopsy medium (BXM), which lacks calcium or magnesium and

[9] J. W. Gordon and I. Gang, *Biol. Reprod.* **42,** 869 (1989).

[10] A. H. Handyside, E. H. Kontogianni, K. Hardy, and R. M. L. Winston, *Nature (London)* **344,** 768 (1990).

which contains both EDTA and 100 mM sucrose (Table A.X).[9] After 30–45 min of incubation in BXM, the embryos are placed on the micromanipulator in the same medium. The holding pipette and microneedle can then be used to rotate the embryo such that a nucleated cell is positioned optimally for biopsy. ZD is then performed at the site on the zona to which that cell abuts. The microneedle can then be used to push against the zona, as with polar body biopsy, so as to extrude the blastomere from the drilled hole (Fig. 4). After the 4-cell stage of development, this procedure can be used to remove more than one cell.[9]

It is also possible to biopsy cells from the trophoblast layer of the expanding blastocyst. When ZD is performed over the trophoblast region, several cells will be spontaneously extruded through the hole in the zona. These cells can then be teased away. Manipulation of the blastocyst often leads to contraction of the blastocoel. However, blastocysts usually reexpand rapidly, and contraction does not indicate embryo demise.

As with polar body biopsy, it is essential to avoid contamination by other genetic material if the PCR is to be performed. An important potential source of contamination is sperm attached to the ZP. A single spermatozoon will contain an entire haploid genome from the male parent, and it will have a 50% chance of carrying a Y chromosome. Thus, great care should be taken to avoid accidental inclusion of sperm in the PCR. In addition, it is always advisable to remove a nucleated cell for PCR in order to assure that one is not retrieving a cytoplasmic fragment that lacks DNA.

Partial Zona Dissection

Partial zona dissection (PZD) is a method of zona drilling applied to human eggs in order to avoid the toxic effects of ATS.[11] This procedure is therefore used almost exclusively for assisting fertilization of, or removing of polar bodies from, human oocytes. However, because mice serve as an important experimental test system for human MAF, this procedure is described and shown here for mouse oocytes. Although it is not necessary to add sucrose to the medium for PZD, we find that 100 mM sucrose (0.34 g/10 ml) greatly facilitates the procedure and protects the oocyte from excessive distortion as the microneedle is passed through the perivitelline space (PVS).

The PZD procedure is shown in Fig. 5. For PZD, the oocyte is grasped such that a region of the zona free of cumulus cells is positioned at either 6 or 12 o'clock. The microneedle is then used to probe the zona such that

[11] H. E. Malter and J. Cohen, *Fertil. Steril.* **51**, 139 (1989).

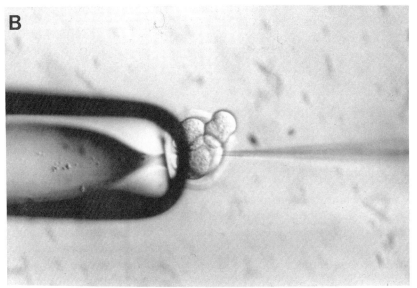

FIG. 4. Blastomere biopsy at the 4-cell stage. (A) After preincubation in biopsy medium (BXM, Table A.X), the embryo is subjected to zona drilling. The point of drilling is shown by the arrow. (B) By pushing on the zona at a point distant from the drilling site, a blastomere is extruded through the drilled hole. (C) Biopsy complete, with a single blastomere removed from the embryo.[8]

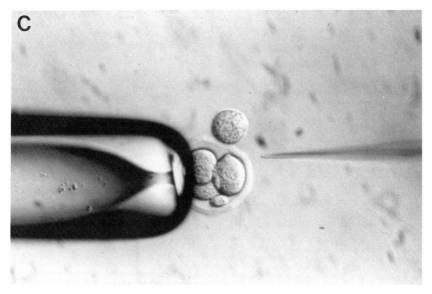

FIG. 4C.

the needle is at the appropriate distance from the floor of the culture dish. The microneedle is then inserted into the PVS at an angle tangential to the surface of the spherical oocyte. Using the joystick and x- and y-axis stage micrometers, the microneedle is thrust forward and maneuvered such that it makes a second hole in the zona and the tip leaves the PVS. This process isolates a region of the ZP between the microneedle and the oocyte (Fig. 5A). At this point the oocyte is released from the holding pipette and remains attached to the microneedle (Fig. 5B). The oocyte is now ready for initiation of the dissection procedure.

Before beginning to dissect the zona, it is very important that the holding pipette be adjusted such that it is making contact with the floor of the culture dish. Otherwise, during dissection, the oocyte can become trapped under the microneedle and destroyed.

With the holding pipette remaining motionless, the joystick controlling the microneedle is then used to trap the isolated region of ZP against the outside wall of the holding pipette (Fig. 5B). The joystick is then either moved back and forth (along the x axis) or raised up and down relative to its distance from the floor of the dish to achieve maximum contact between the glass wall of the holding pipette and the microneedle. This rubbing action is repeated until the oocyte detaches from the microneedle and drifts down to the floor of the dish. At this point the hole in the ZP

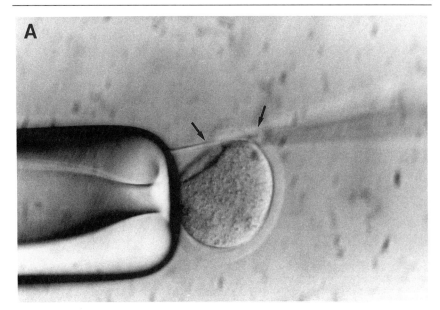

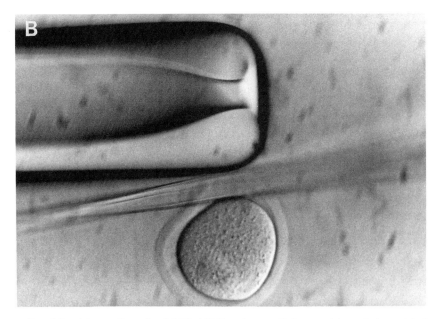

FIG. 5. Partial zona dissection (PZD). (A) The microneedle is passed through the perivitelline space, thus isolating a portion of the zona from the egg. The entry and exit points of the microneedle are shown by arrows. (B) The isolated region of the zona is pressed against the holding pipette, and the needle is moved up and down in the microdrop, thus opening the zona by abrasion. (C) PZD complete, with the hole in the zona clearly visible at 1 o'clock.

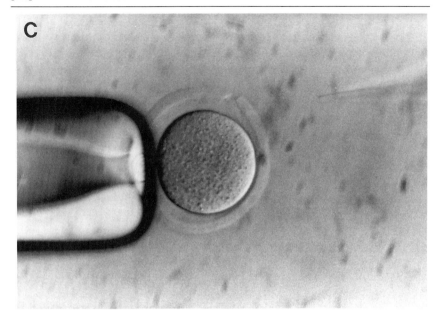

FIG. 5C.

has been created (Fig. 5C), even though it frequently cannot be seen. It is not necessary to repeat the procedure or spend excessive amounts of time trying to visualize the hole. Data from our laboratory and others show that visibility of the hole after PZD does not correlate with increased fertilization rate. After PZD or any other micromanipulation procedure in which sucrose is used, the oocyte is washed immediately and extensively in MFM. We typically transfer the egg to a 50-μl microdrop under oil and move the egg to a fresh drop 3 times with a minimal transfer volume before transferring it to the insemination dish.

Subzonal Sperm Insertion

General Considerations

Subzonal sperm insertion (SZI) may seem a very demanding procedure, but, if performed properly with adequate microtools, it is actually simple and safe. Early work with the mouse showed that contact between the microneedle and the sperm can lead to a loss of motility, and that such loss irrevocably destroys the ability of the sperm to fuse with the egg.[12] Thus, it is essential that the microneedle be of sufficient size to allow free passage of the sperm. Balanced against this imperative is the fact that increased size of the microneedle is associated with more severe trauma to the egg during the SZI procedure. Thus, it is necessary to minimize the size of the microneedle in order to avoid damage to the oocyte. When producing microtools for SZI and performing sperm insertion, these opposing factors must be correctly balanced to achieve success.

Fortunately, the SZI technique in the human, which is employed much more commonly than in the mouse, is relatively easily done. The oocyte is large, the spermatozoon small, and the sperm less sensitive to the harmful effects of contact with the glass microneedle. SZI in the mouse can be facilitated by use of agents such as methyl cellulose (Table A.XI)[13] or polyvinylpyrrolidone (PVP) to coat the surface of the sperm and prevent loss of motility.

One of the most difficult aspects of SZI is the necessity that a large-bore pipette be inserted under the zona without damaging the oocyte. Unless the entire microneedle system is filled with fluid, capillary action tends to draw material into the large microneedle in the absence of positive pressure. The result of this spontaneous filling is that, if time is required to safely position the microneedle under the zona, a significant volume of fluid that is devoid of spermatozoa fills the tip. Consequently, the accumulated fluid must be expelled before sperm appear in the PVS, and the excessive flow in the PVS exerts prolonged pressure on the egg and also lengthens the time required for SZI. Another by-product of this capillary action is that when the PVS in entered the oocyte can be drawn to the microneedle and destroyed. Problems associated with passive filling of the microneedle contribute more than any others to oocyte damage during SZI.

The best way to control flow of material through the microneedle is by mouth. Accordingly, the instrument collar should be connected to rubber tubing fitted with a 0.22 μm filter and a mouthpiece at the opposite end. Maintenance of constant, light positive pressure on the microneedle that is sufficient to overcome capillary action can be accomplished with a syringe micrometer, but it is far more easily done by mouth.

Microneedle Production

Production of a suitable microneedle is essential for successful SZI. In the mouse and the human, many steps are traditionally required to

[12] P. E. Barg, M. Z. Wahrman, B. E. Talansky, and J. W. Gordon, *J. Exp. Zool.* **237,** 365 (1986).
[13] J. R. Mann, *Biol. Reprod.* **38,** 1077 (1988).

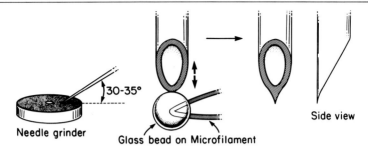

FIG. 6. Diagram for production of SZI microneedles. The needle is first ground to the appropriate size, with a bevel angle of 30°–35°. Next the needle is placed on the microforge and lowered onto the heated glass bead, which is on the microfilament (center). The double arrow indicates the motion by which the microneedle tip is touched to the heated filament. The needle can then be used directly or etched with hydrofluoric acid (see text).

fashion the microneedle. These processes are described below and illustrated in Fig. 6. Our laboratory, however, has devised an alternative method for rapid, reproducible, and simple construction of microneedles for human SZI. This is shown in Fig. 7 and is also described.

To produce a microneedle for mouse SZI, the needle is first pulled with the pipette puller, then transferred to the needle grinder. The grinder is set at an angle of 30°–35°, and the needle is then ground to introduce a bevel of appropriate size (an opening ~2–3 times the diameter of the sperm head). The microneedle is then placed on the microforge such that the tip points toward the floor and is situated above the filament. For all procedures with the microforge, the filament should have a ball of glass fused to it.

The microforge is then activated, and sufficient heat is applied such that the glass ball just begins to glow red. As the microneedle tip is lowered to the glass bead, the blower of the microforge is used to prevent rising heat from fire-polishing the edges of the microneedle. The microneedle is then lowered, and the tip is briefly brought into contact with the glass. As the needle is withdrawn, a fine tip is pulled (Fig. 6). The needle is then ready for use.

Rapid Procedure for Microneedle Production. We have used the following rapid method for making microneedles, which has proved to be very safe, with fewer than 0.5% of oocytes destroyed during SZI. In addition, our human IVF program has established several pregnancies after use of this needle. The advantages of this method is that no microforge or needle grinder is required, and a functional needle can be produced in about 60 sec (Fig. 7).

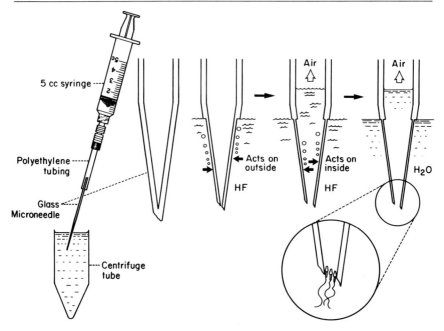

FIG. 7. Fast method for SZI microneedle production. The needle is pulled and affixed to plastic tubing which in turn is connected to a syringe (left). The tip is lowered into hydrofluoric acid (center) which thins the wall from the outside. Suction is then applied, and after the tip is opened by the action of HF, the acid can be seen entering the needle (center right). This confirms that the needle is open, and, at the same time, the HF thins the inside wall of the glass. After this procedure, the microneedle is rinsed by suction in distilled water. The final size of the opening should accommodate about three spermatozoa (lower right).

For this method, the needle is pulled on the pipette puller in the usual fashion. However, the back end of the needle is fire-polished by brief exposure to the microburner, then placed in the polyethylene tubing with the attached syringe. The needle tip is then placed in the hydrofluoric acid (HF) for a few seconds to thin the outside wall of the glass, after which suction is applied with the syringe (Fig. 7, left). When HF enters the tip, it thins the inside wall (Fig. 7, center). As soon as HF is seen entering the needle, the tip is transferred to distilled water, which is drawn into the tip to wash out the HF (Fig. 7, right). When the tip is in the water, suction can be applied until the water is drawn completely out the back end of the needle into the polyethylene tubing. With a minimum of experience, the size of the microneedle tip can be estimated by the ease with which the water flows through the microneedle barrel. When the washing is complete (7–8 washes in distilled water), the needle is ready for immediate use. Although these needles rarely have a regular tip, the glass is so thin, having been attenuated both inside and out with the HF, that if the aperture is of the correct size (opening large enough to accommodate about 3 sperm; Fig. 7, lower right), the tip is very sharp and can be readily inserted through the zona.

Preparation of Sperm and Subzonal Sperm Insertion

A demonstration of SZI using mouse oocytes and human sperm is shown in Fig. 8. For mouse SZI, sperm are released from the caudae epididymides and vasa deferentia into mouse capacitation medium (MCM, see Tables A.II–A.V, A.VII, and A.VIII)[12–15] to capacitate for at least 30 min. Motile sperm are placed in a microdrop of M2 medium supplemented with methyl cellulose (MSZIM) as described by Mann[13] (Table A.XI). Sperm move quite slowly in the methylcellulose, and those which moved best should be selected for aspiration into the microneedle. The needle is then moved to the microdrop containing the oocytes, which have been selected for the presence of a first polar body (Fig. 8A).

The zona is pierced and the PVS entered. The zona should not be penetrated at the so-called equator of the oocyte (Fig. 8B), as excessive distortion prior to piercing the zona can destroy the oocyte. As the PVS is entered, positive pressure is applied through the microneedle system such that suction of the oolemma against the needle and subsequent oocyte destruction are avoided (Fig. 8B). Once access is gained into the PVS, positive flow can be appreciated by the apparent swelling of the PVS. Positive pressure is then maintained until the desired number of motile sperm are expelled (Fig. 8C). The eggs are then placed in M16 medium or mouse fertilization medium (MFM, Table A.VIII) and observed 6–8 hr later for appearance of pronuclei.

Pronucleus Removal

Removal of pronuclei from fertilized eggs is can be accomplished either by enclosing the pronucleus within a vesicle composed of the vitelline membrane[16] or by entering the cytoplasm and drawing the pronucleus directly into the microneedle.[17,18] In either case, a sharp microneedle with a fairly large aperture is needed (e.g., as in Fig. 7). In addition, it is

[14] B. E. Talansky, P. E. Barg, and J. W. Gordon, *J. Reprod. Fertil.* **79,** 447 (1986).

[15] V. M. Thadani, *J. Exp. Zool.* **219,** 277 (1982).

[16] J. McGrath and D. Solter, *Cell (Cambridge, Mass.)* **37,** 179 (1983).

[17] C. Anderegg and C. L. Markert, *Proc. Natl. Acad. Sci. U.S.A.* **83,** 548 (1986).

[18] J. W. Gordon, L. Grunfeld, G. J. Garrisi, D. Navot, and N. Laufer, *Fertil. Steril.* **52,** 367 (1989).

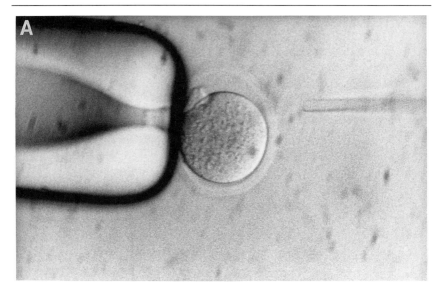

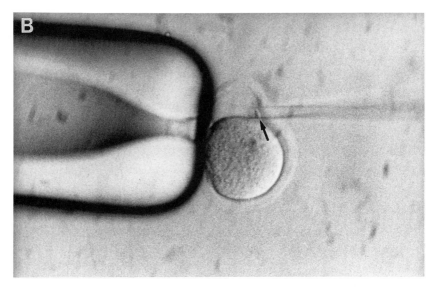

Fig. 8. Subzonal sperm insertion. In this pictorial demonstration, a mouse oocyte is injected with a human sperm, which has been immobilized for the purpose of photography. (A) Oocyte prepared for SZI, with the microneedle, loaded with sperm, in position. (B) Microneedle in the perivitelline space, with sperm (arrow) being expelled by mouth. (C) SZI complete, with a spermatozoon visible in the PVS (arrow).

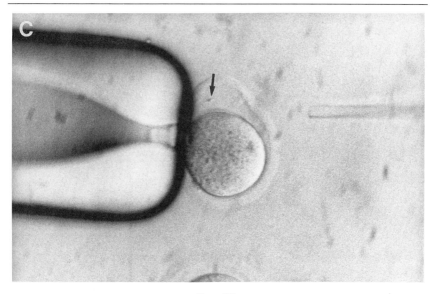

FIG. 8C.

helpful to add cytochalasin and Colcemid as cytoskeletal relaxing agents to produce pronucleus extraction medium (PNM, Table A.XII) prior to manipulation.[16,17] This treatment reduces mortality to the zygote. Pronucleus removal from a human zygote is illustrated in Fig. 9.

To fashion the microneedle, a procedure similar to that for SZI is used, except that the aperture should be 50–75% the diameter of the pronucleus (Fig. 9A). After the PVS is entered, the pronucleus is then drawn into the microneedle by applying suction to the overlying membrane of the zygote, or by entering the zygote and aspirating the pronucleus directly (Fig. 9B). After the pronucleus is in the shaft of the microneedle, suction is relaxed, and the needle is slowly withdrawn. Slow removal of the microneedle allows the large gap in the zygote membrane to reseal (Fig. 9C).

Sperm Injection

Direct injection of spermatozoa into the ooplasm rarely leads to successful fertilization in rodents,[19] but it is far more likely to succeed in the

[19] V. M. Thadani, *J. Exp. Zool.* **212,** 435 (1980).

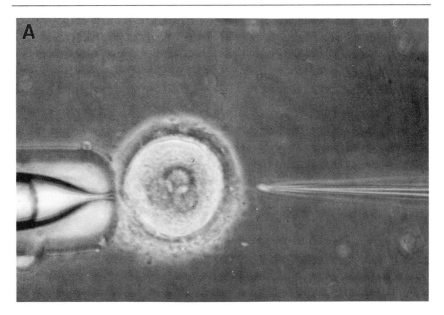

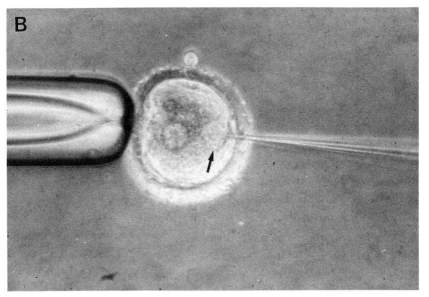

FIG. 9. Pronucleus removal. (A) Tripronuclear human zygote affixed to the holding pipette in preparation for extraction of a pronucleus. (B) Microneedle dragging a pronucleus out of the cell (arrow). (C) Pronuclear extraction complete, with one pronucleus on the end of the microneedle and two others inside the zygote.

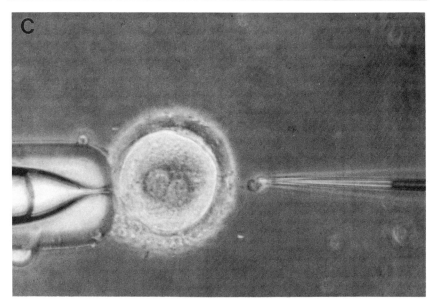

FIG. 9C.

human[20] and has been reported to work in several other species. Methods of sperm preparation vary, but in all cases it is advisable to assure that the acrosome has been lost from the sperm destined for cytoplasmic insertion. This can be accomplished by sonication,[18] by freeze–thawing, or, as reported in a recent publication involving the human, by administration of an electrical charge after prolonged incubation.[19] Often sperm are simply incubated in insemination medium and used directly. Sonication usually results in loss of the tail in addition to the acrosome. An important point is that addition of volume to the unfertilized egg is very hazardous.

   We perform cytoplasmic injection with a microneedle that has an aperture slightly less than the diameter of the sperm head. We first enter the PVS and expel medium so as to create a channel in the zona. Then a sperm is affixed to the microneedle by suction, the channel is reentered, and, with continued application of suction, the microneedle with the protruding sperm is pushed into the oolemma. The suction introduces a break in the vitellus. When a small amount of cytoplasm is seen in the microneedle, a slight positive pressure is applied until the cytoplasm can be

---

[20] G. Palermo, H. Joris, P. Devroey, and A. C. Van Steirteghem, *Lancet* **340,** 17 (1992).

seen slowly flowing out of the microneedle. At this point the microneedle is slowly withdrawn, leaving the sperm in the cytoplasm. This approach minimizes the addition of volume to the oocyte.

Summary

The procedures described in this chapter are applicable to problems of oogenesis, fertilization, and early development. Many important procedures have not been described (e.g., pronuclear exchange). However, nearly every experimental manipulation thus far published for the mouse either directly involves the techniques included here or is closely related to them. Therefore, if all of the skills outlined in this chapter are successfully mastered, it should be possible to adapt that knowledge to most or all known experimental protocols. Successful establishment of micromanipulation is certain to have a significant and lasting positive impact on nearly any research program in mammalian development.

Appendix

This appendix consists of a number of tables detailing formulations for various stock solutions and media discussed in this chapter.

TABLE A.I
Acid Tyrode's Solution (ATS)[a]

Compound	Amount (g)
NaCl	0.800
KCl	0.020
$CaCl_2 \cdot 2H_2O$	0.020
$MgCl_2 \cdot 6H_2O$	0.010
$Na_2HPO_4 \cdot H_2O$	0.005
Glucose	0.100
Polyvinylpyrrolidone	0.400

[a] Add distilled water to 100 ml. Before use, add approximately 0.66% concentrated HCl. Filter through a 0.22-μm filter immediately before use.

TABLE A.II
M2 AND M16 CONCENTRATED STOCKS: STOCK A[a]

Compound	Amount (g)
NaCl	5.534
KCl	0.356
KH_2PO_4	0.162
$MgSO_4 \cdot 7H_2O$	0.293
Sodium lactate, 60% syrup	4.439 (g or 2.610 ml)
Glucose	1.000
Penicillin	0.060
Streptomycin	0.050

[a] Stock A (100 ml), $10\times$ solution. Storage time: 3 months.

TABLE A.III
M2 AND M16 CONCENTRATED STOCKS: STOCK B[a]

Compound	Amount
$NaHCO_3$	2.101 g
Phenol red, 0.5% solution	0.010 ml

[a] Stock B (100 ml), $10\times$ solution. Storage time: 2 weeks.

TABLE A.IV
M2 AND M16 CONCENTRATED STOCKS: STOCK C[a]

Compound	Amount
Sodium pyruvate	0.180 g

[a] Stock C (50 ml), $100\times$ solution. Storage time: 2 weeks.

TABLE A.V
M2 AND M16 CONCENTRATED STOCKS: STOCK D[a]

Compound	Amount
$CaCl_2 \cdot 2H_2O$	0.252 g

[a] Stock D (10 ml), $100\times$ solution. Storage time: 3 months.

TABLE A.VI
M2 AND M16 CONCENTRATED STOCKS: STOCK E[a]

Compound	Amount
HEPES solid	5.958 g
Phenol red, 0.5% solution	0.010 ml

[a] Stock E (100 ml), 10× solution. Storage time: 3 months.

TABLE A.VII
M2 AND M16 CONCENTRATED STOCKS: MAKING UP STOCKS

Stock	Procedure
Stocks A–D	Weigh solids into a clean flask or appropriate sized disposable conical centrifuge tube and add appropriate quantity of distilled water (18 mΩ). If sodium lactate syrup is weighed (stock A), use a small weighing boat and decant contents into flask, then rinse boat 2–3 times with distilled water and decant rinse into flask
Stock E	Weigh out HEPES and add phenol red. Add 50 ml distilled water, then adjust pH to 7.4 with 5 N NaOH. Add distilled water to 100 ml

TABLE A.VIII
M16, MOUSE FERTILIZATION MEDIUM (MFM),
MOUSE CAPACITATION MEDIUM (MCM), AND
M2 MEDIUM FROM CONCENTRATED STOCKS

M16 (500 ml)

Stock	Amount
A	50 ml
B	50 ml
C	5 ml
D	5 ml
Water	390 ml
BSA (fraction V)	2 g

Rinse all pipettes, etc., thoroughly into final flask, then add water to 500 ml. Add 2 g BSA, dissolve without excessive frothing, and sterile filter into 50-ml sterile disposable conical centrifuge tubes

MFM

Mouse fertilization medium is identical to M16 except for the following changes in stock A:

NaCl	5.97 g
Sodium lactate syrup	None

TABLE A.VIII (*continued*)

MCM

For capacitation of sperm, sodium lactate
 syrup may be added to MFM (4–6 μl/2 ml)
 immediately prior to addition of sperm

M2 (100 ml)

Stock	Amount
A	10.0 ml
B	1.6 ml
C	1.0 ml
D	1.0 ml
E	8.4 ml
Water	78.0 ml

Make up as for M16, but recheck pH (final pH
 should be 7.3–7.4) prior to sterile filtration

TABLE A.IX
MEDIUM SUPPLEMENTED WITH HYALURONIDASE
FOR CUMULUS CELL REMOVAL

Add hyaluronidase to a final concentration of 1–2
 mg/ml to M2 or M16
Filter sterilize before use

TABLE A.X
CHELATING MEDIUM FOR BIOPSY (BXM)[a]

Compound	Amount
10× stock (see below)	10.0 ml
EDTA, 200 mM, pH 7.5	1.0 ml
Sodium lactate, 60% syrup	426.0 μl
Stock C (Table A.IV)	1.0 ml
BSA	0.4 g
Sucrose	3.42 g
For 10× stock	
NaCl	6.2 g
KCl	0.2 g
KH_2PO_4	0.2 g
Na_2HPO_4	1.15 g
Glucose	1.0 g

[a] Recipe makes 100 ml BXM.

TABLE A.XI
M2 with Methyl Cellulose for Mouse
Subzonal Sperm Insertion[a]

1. Add methyl cellulose to distilled water at 2.2%
   (w/v) and boil while stirring
2. Place suspension on ice for 30 min with continued stirring. The solution will become almost
   clear, but will remain slightly cloudy
3. Centrifuge solution at 20,000 g for 15 min at 4°
4. Add 7.8 ml of this final solution to 2.2 ml of
   $10\times$ filter-sterilized M2 containing $10\times$ BSA (40
   mg/ml)

[a] The composition of this medium (MSZIM) is
from Ref. 8.

TABLE A.XII
Cytoskeletal Relaxant Medium for Pronuclear Extraction (PNM)

Add demecolcine to final concentration of 0.1 μg/ml and cytochalasin D to 1.0 μg/ml (final) in medium used for manipulation
microdrop [M2 for mice, phosphate-buffered saline (PBS) for
humans]. Solids can be added directly, or $100\times$ solution (10
μg/ml demecolcine, 100 μg/ml cytochalasin D) can be made
in distilled water. This solution, which should be stored at
$-20°$, can be added to microdrop medium immediately before
use at 1% (v/v).

TABLE A.XIII
Preparation of Pentobarbital Anesthesia
for Mouse Surgery[a]

Compound	Amount (ml)
100% Ethanol	10
Propylene glycol	20
Water	70

[a] To make 100 ml, mix the above components,
then add sodium pentobarbital stock solution
(see Table II in text) to a final concentration of
6 mg/ml. Administer to mice by intraperitoneal
injection at 60 mg/kg, or 0.1 ml/10 g body
weight.

Acknowledgments

Work toward preparation of this chapter was supported by National Institutes of Health
Grants HD25136, HD20484, and CA42103.

[13] *In Vitro* Capacitation and Fertilization

By Lynn R. Fraser

Introduction

Successful fertilization *in vitro* has a minimum of two basic requirements: (1) capacitated sperm and (2) mature, unaged oocytes. Depending on the aims of the experiment, further requirements may exist. For example, if development beyond the 2-cell stage is required, the genetic background of the oocyte donor is important. If care is taken to ensure that fairly rigorous methods of gamete preparation and coincubation are followed, an *in vitro* system can be used for a wide range of investigations focusing on gamete competence (e.g., requirements for either sperm capacitation or oocyte maturation). Furthermore, when fully capacitated sperm are incubated with mature oocytes, sperm penetration is rapid and synchronous among the whole population of oocytes. Such synchrony in the initiation of development provides a distinct advantage over fertilization *in vivo,* where asynchrony predominates; the latter is due, in part, to the spread in time over which ovulation occurs. It is also possible, within limits, to choose the time of fertilization and hence initiation of development.

Culture Media and Methods

Medium Composition

Most media that have been used successfully for *in vitro* fertilization are fairly simple solutions which can be prepared in the laboratory. The two most common are a modified Tyrode's medium and a Krebs–Ringer medium; both of these are buffered with sodium bicarbonate. The composition of the former is given in Table I. The original modified Tyrode's medium contains glucose, pyruvate, and lactate, as detailed in Table I. For investigations focusing on sperm capacitation, where culture of oocytes is very restricted, pyruvate and lactate can be omitted; the correct osmolality is retained by increasing the NaCl concentration as noted. Mouse sperm cannot fertilize oocytes in the presence of either pyruvate or lactate alone, but they can do so with only glucose present.[1]

---

[1] L. R. Fraser and P. J. Quinn, *J. Reprod. Fertil.* **61,** 25 (1981).

Copyright © 1993 by Academic Press, Inc.
All rights of reproduction in any form reserved.

TABLE I

MODIFIED TYRODE'S MEDIUM FOR *in Vitro* FERTILIZATION AND CULTURE

Component	Complete (mM)	Pyruvate- and lactate-free (mM)
NaCl	99.23	124.73
NaHCO$_3$	25.00	25.00
Sodium DL-lactate	25.00	—
KCl	2.68	2.68
NaH$_2$PO$_4$ · 2H$_2$O	0.36	0.36
MgCl$_2$ · 6H$_2$O	0.49	0.49
Sodium pyruvate	0.50	—
CaCl$_2$ · 2H$_2$O	1.80	1.80
Glucose	5.56	5.56
Penicillin, sodium salt	100 units/ml	100 units/ml
Phenol red (0.5% in 154 mM NaHCO$_3$)	0.1 ml/100 ml	0.1 ml/100 ml

The salts used should be of high purity (e.g., Analar in the UK). The water used for media needs to be pure, but we have had little difficulty with water from a variety of sources, for example, deionized water supplied for making up tissue culture media from the commercially available dry salts or sterile water supplied to hospitals for intravenous administration (e.g., from Boots, Nottingham, UK). Currently we use water obtained from a Milli-Q system (Millipore, Bedford, MA) in which water is initially purified by reverse osmosis, then passed through several resin beds to remove various ions, metals, etc. This has proved to be excellent for our on-going sperm-oriented studies. Extended culture of embryos produced by *in vitro* fertilization might prove to be more demanding.

Routinely, these media are supplemented with serum albumin, usually bovine (BSA). This is an important constituent for fertilization per se because mouse sperm require it in order to undergo the acrosome reaction.[2] There is a considerable mystique attached to BSA, although this relates more to embryo culture than to fertilization. We have used crystallized BSA (purer than fraction V) and never experienced obvious variation in batches from the same supplier or indeed between suppliers. We have used crystallized, lyophilized BSA from Sigma (St. Louis, MO; No. A4378) for several years with no obvious problems. Within a wide range, the concentration of BSA is probably not critical: similar proportions of sperm underwent spontaneous acrosome reactions when incubated for 2 hr in 0.1, 0.25, 1, 4, or 30 mg/ml BSA. Although earlier studies from our laboratory used 32 mg BSA/ml for fertilization, we have been using 4 mg/

[2] L. R. Fraser, *J. Reprod. Fertil.* **74,** 185 (1985).

ml with consistent success for several years. There is no evidence that very high concentrations of BSA accelerate capacitation, but in the absence of BSA capacitation is delayed.

Because of the bicarbonate buffering in the medium, a CO_2-containing gas mixture is needed to maintain a steady pH. With 25 mM NaHCO$_3$, 5% CO_2 in the gas phase is needed. Either 5% CO_2 in air or 5% CO_2–5% O_2–90% N_2 will serve for fertilization; in some instances, the latter mixture may prove superior for embryo culture. Do not substitute bicarbonate with another buffer: it is required for successful fertilization. It is best to warm the gas mixture before use. Using tubing, connect the gas tank to a gas washing bottle kept in a 37° water bath. The gas goes into the distilled water-containing bottle, which effectively "washes" and warms the gas, and then out through the air space into a second piece of tubing. This can be connected to adaptors to permit gassing of medium, gametes, etc.

Medium Preparation

It is better to make up fresh medium at relatively short intervals than to store it refrigerated for long periods of time. We routinely make up a volume of medium adequate for 1–2 weeks of use; after 2 weeks, any remaining is then discarded. Add dry salts to the water (the reverse may cause certain salts to form nearly insoluble clumps) in a sterile beaker (sterilized by dry heat, if glass); mix by swirling and pouring between the beaker and measuring cylinder used for water. Filter medium (in our case, using a sterile disposable 0.45-μm Millex (Millipore, Molsheim, France) filter unit attached to a large syringe) into a sterile bottle, gas for 5 min with the 5% CO_2 mixture, using a sterile Pasteur pipette, to equilibrate it and then store, capped, in the refrigerator.

For individual experiments, place the desired amount of medium in a sterile plastic container and add BSA, at 4 mg/ml. After capping, put the container into a 37° water bath and leave, undisturbed, for the BSA to go into solution. Do not shake to mix: the high protein content will cause frothing. The BSA lowers the pH (the phenol red color changes from pink to straw); add approximately 1–2 drops of 0.2 M NaOH per 5 ml of medium, then gas for approximately 15 sec to achieve an orange-pink color (pH ~7.6). Leave in the water bath and always regas before use. To gas the medium, we use a 19-gauge syringe needle, attached to a fixed syringe, which is in turn connected to the gas washing bottle tubing. Allow the gas to play over the surface: do not put the needle into the medium as this will cause frothing and denaturation of the BSA. Replace the cap and keep in the water bath.

Culture Methods

All incubations can be carried out in 35-mm sterile plastic culture dishes in which the medium is overlaid with sterile (autoclaved) liquid paraffin. We have used paraffin from Boots with consistent success. Some liquid paraffins may contain toxic materials, so if culture difficulties are encountered, try changing suppliers. Equilibrate the liquid paraffin with BSA-free medium and 5% CO_2 by adding 2 ml medium and then approximately 18 ml paraffin to a sterile plastic container, followed by bubbling through the gas mixture, using a sterile Pasteur pipette, for 4 min. Centrifuge the mixture at approximately 750 g_{max} for 10 min at room temperature to spin down the medium, and use the top layer of paraffin for culture. The paraffin overlay helps to stabilize the medium, minimizing evaporation and pH changes. For all incubations we use stainless steel anaerobic culture jars fitted with a removable shelving unit ("homemade") which holds the dishes. We gas a jar for 3–5 min at a flow rate of 3 liters/min, seal it, and place it in the incubator. Any other unit which functions in the same way, that is, can be gassed and sealed, is equally suitable. Alternatively, a CO_2 incubator can be used, if available.

Oocytes

Donor Genotype

If culture of embryos beyond the 2-cell stage is required, then the genotype of the oocyte donors needs to be considered. Embryos derived from oocytes of most inbred, outbred, and random bred mouse strains do not develop well beyond the 2-cell stage, many exhibiting the "2-cell block." This problem can be overcome by using F_1 hybrid females as oocyte donors; those with C57BL as one of the parental strains develop particularly well *in vitro*, even when fertilized *in vitro*. Using oocytes from (C57BL/10 × CBA)F_1 females, we were able to achieve very good fertilization (usually 80–100% of oocytes fertilized) and subsequent development *in vitro* (with usually >80% of 2-cell embryos reaching the blastocyst stage). It is possible to obtain improved development of embryos from a "blocking" strain by altering the composition of the culture medium,[3] although the proportion reaching the blastocyst stage is reportedly lower than that obtained using a "nonblocking" genotype.

[3] C. L. Chatot, C. A. Ziomek, B. D. Bavister, J. L. Lewis, and I. Torres, *J. Reprod. Fertil.* **86**, 679 (1989).

Hormonal Stimulation

Unfertilized oocytes are obtained from females treated with exogenous hormones, that is, "superovulated"; this both increases the yield and allows specific timing of collection to avoid aging *in vivo*. Animals should be maintained under a constant light–dark regime (usually 12–12 hr or 14–10 hr). Ovarian stimulation is achieved by injecting an aqueous solution of pregnant mare serum gonadotropin (PMSG) intraperitoneally into females, preferably those at metestrus or diestrus of their cycle. In many strains of mice, it is easier to fit in with the endogenous hormone cycle than to try to override it with the exogenously administered hormones. Animals can be selected quite easily by examing the vaginal region: when the area surrounding the vaginal opening is pale to slightly purple in color, dry, and unswollen, the females should respond to the PMSG. If the area is either moist and noticeably swollen with a bright pink color (proestrus) or somewhat paler and more corrugated, owing to recent loss of the edema (estrus), females will respond less well. Ovulation is induced by injecting an aqueous solution of human chorionic gonadotropin (hCG), again intraperitoneally.

The amount injected and timing of administration are important. It is best to check the response of a strain to PMSG/hCG. For PMSG, 5 or 7.5 IU/mouse are most commonly used. However, each strain will respond differently, and it is advisable to do a small series of injections to determine the dose response. We routinely use 7.5 IU because that is optimal for the outbred TO strain we use, but other strains may respond optimally to 5 IU. Some strains do not respond well to any dose; in such cases, a low dose of 1–2.5 IU/PMSG per mouse may at least serve to synchronize the timing of follicular development. The hCG, most commonly used at 5 IU/mouse whatever the amount of PMSG, should be given 48–54 hr after the PMSG. A longer or shorter interval may interfere with the response. If sperm suspensions are to be sampled at 2 or more time points, hCG can be administered asynchronously to females to ensure the availability of unaged oocytes. Ovulated, unfertilized oocytes should be collected and used shortly after ovulation (typically 12–14 hr after hCG); again, strain differences exist. Oocytes collected more than 16 hr after hCG will have begun to age. In some strains, aged oocytes are prone to spontaneous activation on release into culture medium; in others, they simply become resistant to fertilization.

Hormones are usually obtained as dry powders. Reconstitute in sterile water to the appropriate concentration. We make up PMSG and hCG so that the desired amount, per mouse, is present in 0.1 ml, a convenient volume to administer. The hormone solutions are transferred to small plastic vials, with enough per vial to inject a reasonable number of mice, and then frozen. For use, hormones can be thawed as needed and injected. The frozen hormone solutions will maintain potency for at least 1 year, in our experience.

Collection

To collect the oocytes, remove the oviducts from hormonally stimulated females. If the mouse has responded, the distended ampulla containing the mass of oocytes (the "cumulus clot") can be seen with the eye as a blister on the oviduct. Release oocytes by slitting the ampulla, which is easily achieved using a dissecting microscope and 25-gauge needles on 1-ml syringes. Oocytes can be released directly into sperm suspensions. An alternative, and probably more desirable, approach is to release all cumulus clots into approximately 4 ml medium covered with liquid paraffin. This permits a brief washing of the clots and more equal distribution of the oocytes among sperm suspensions receiving different treatments. Keep the culture dish containing the cumulus clots on a warming tray (~37°) while sperm suspensions are being readied for oocyte introduction.

In some experiments, it might be desirable to remove cumulus cells or zonae pellucidae. The following manipulations are most easily carried out using Pyrex watch glasses (preferably flat-bottomed) sterilized by dry heat. The curved sides help to keep the oocytes near the center of the dish: gentle rotation to swirl the medium will bring most oocytes to the center for easy inspection and selection for transfer to dishes with fresh medium as detailed below. Use 1 ml of appropriate medium and cover with paraffin in all instances. Commercially available hyaluronidase (e.g., bovine testicular hyaluronidase, Sigma) can be used to remove cumulus cells; about 300 units hyaluronidase/ml medium are commonly used. Transfer cumulus clots into this, let stand until cells have dispersed, then transfer through two washes of fresh, hyaluronidase-free medium. To remove zonae, first remove cumulus cells. Then transfer oocytes, in batches, to medium containing 0.04% pronase. When the zonae begin to distend and buckle, remove in a minimal volume of pronase solution to fresh medium; swirl to disperse residual pronase. Dissolution of zonae is asynchronous, so several repeats of this procedure may be needed for each batch of oocytes. Tease away the remnants of zona material, using a 25-gauge needle, and transfer immediately to fresh medium. All of the solutions contain BSA at 4 mg/ml.

Prepare a stock solution of 2% pronase in phosphate-buffered saline (PBS), pH 7.4; dialyze overnight against PBS, filter through a 0.45-μm filter to sterilize, and store in the refrigerator. Just before use, dilute the stock solution to the above concentration with BSA-containing medium.

Alternatively, use acidified Tyrode's medium to remove zonae. The composition of this (in g/100 ml) is as follows: NaCl, 0.80; KCl, 0.02; $CaCl_2 \cdot 2H_2O$, 0.024; $MgCl_2 \cdot 6H_2O$, 0.01; glucose, 0.10; polyvinylpyrrolidone or polyvinyl alcohol, 0.40. Adjust the pH to 2.1–2.5 using 0.2 M HCl, divide into small aliquots, and freeze. To use, thaw and warm, then place in a watch glass and cover with liquid paraffin. Add oocytes as above and, as soon as zonae begin to dissolve, transfer to fresh, warm, BSA-containing medium. Wash one more time before use.

Sperm Suspensions

Preparation

Remove caudal epididymides from the required number of males. We routinely use at least two males and a set volume of medium, 0.5 ml, per epididymis in preparing the sperm suspensions. Thus, with two males use 2 ml medium. Squeeze out the epididymal contents using two pairs of watchmaker's forceps and discard the remaining tissue. If different media are to be used in order to compare effects on sperm function, it is desirable to make a single initial suspension in control medium, divide this into subsamples, and then add concentrated stock solutions of compounds to be tested. This approach avoids differences due to biological variation among males. If this is not possible, use some sperm from each male, for example, 1 epididymis from each of two males into two different media.

Capacitation

Preincubate sperm suspensions to allow sperm to undergo capacitation, that is, acquire fertilizing ability. Using a complete culture medium, capacitation should occur *in vitro,* although the amount of time required may vary among strains of mice just as it varies among species of mammals. This can be determined by preincubating sperm suspensions for varying amounts of time (e.g., 30, 60, 90, 120, 150 min) and then adding freshly ovulated, unfertilized oocytes. The earliest point at which *rapid* fertilization of the large majority of oocytes by preincubated sperm can be obtained within a limited period of coincubation is the optimal capacitation time. For the outbred TO strain, this time is 120 min: preincubation of sperm for 120 min and then mixing with unfertilized oocytes usually lead to fertilization of over 80% of oocytes, with almost all sperm heads being fully decondensed within 75 min of gamete mixing. This high degree of synchronous penetration may be of considerable value in certain experimental protocols, especially where precise timing of initiation of embryonic development is important.

When preparing sperm suspensions, it is important to handle them carefully and minimize any manipulations unless specific effects are desired. This is particularly true for "washing" of sperm. Centrifuging recently prepared mouse sperm suspensions and resuspending the cells in fresh medium to wash them will change them functionally. A major feature of capacitation is the loss of decapacitation factor molecules (DFs) from the sperm surface, which in turn promotes sperm fertilizing ability. Washing mouse sperm in a typical culture medium used for *in vitro* fertilization promotes loss of the DFs and significantly increases the fertilizing ability of the washed suspensions, compared with unwashed counterparts.[4]

We routinely capacitate sperm by incubating them in a relatively concentrated suspension (equivalent of 1 male/ml, $\sim$1–4 $\times$ 10^7 cells/ml); this is then diluted for fertilization. Using this method, capacitation and subsequent fertilization can be achieved consistently with sperm from a range of mouse strains; at least some strains function less well if preincubated at the lower concentration from the beginning.[5]

Coculture for Fertilization

Remove preincubated sperm suspensions from the incubator. Whenever gametes are not in the incubator, keep the culture dishes on a warming tray ($\sim$37°) to prevent temperature shock. Gently mix preincubated sperm suspensions and then dilute approximately 10-fold to a concentration of about 1–2 $\times$ 10^6 cells/ml, although greater dilution and hence a lower concentration can be used if desired. Do not use pipettes with very narrow apertures; excessive shearing can damage the cells. Transfer the diluted suspension to a fresh culture dish and cover with equilibrated liquid paraffin. Transfer freshly ovulated oocytes into the suspension and, after gassing, incubate at 37° for the desired length of time. For transferring oocytes, we use fine glass pipettes, prepared from Pyrex glass tubing by heating in a flame and then pulling to draw out to fine tapers. These pipettes are inserted into tubing attached to a micrometer dial for easy manipulation; in other laboratories, tubing attached to a plastic mouthpiece is used.

[4] L. R. Fraser, *J. Reprod. Fertil.* **72**, 373 (1984).
[5] L. R. Fraser, *J. Reprod. Fertil.* **49**, 83 (1977).

Assessment

General

Assessment of fertilization depends on the nature of the experimental questions being asked. If analysis of sperm function is desired, then gametes should be coincubated for a short time (<2 hr) so that the ability of sperm to fertilize oocytes at specific times can be measured. A longer coincubation will blur the distinctions between sperm fully capacitated at the time of mixing with oocytes and those reaching that state later in the incubation. With the TO strain of mice, a 2-hr preincubation of sperm will lead to maximal, rapid fertilization within 75 min: the majority of fertilized oocytes will have a fully decondensed fertilizing sperm head visible in the cytoplasm when fixed and stained. Sperm that are slower at achieving fertilization will be at earlier stages of decondensation. In contrast, longer coincubation times for gametes will produce pronuclear stage zygotes. It is much more difficult to analyze development precisely at this point since the pronuclei are present for over 12 hr, whereas sperm head decondensation requires only about 30 min. Thus, the latter offers a more rigorous assessment method for analyzing sperm function.

If the experiment involves further culture, it is best to transfer oocytes from the sperm suspension after a few hours (~3–4 hr) to fresh medium. If paraffin is put in the dish with subsequent addition of the medium, the medium should form and hold a droplet shape, which makes visual assessment easier. At transfer, it should be possible to observe the presence of two pronuclei in fertilized zygotes, in addition to the easily distinguished second polar body (the first polar body usually begins to disintegrate about the time of ovulation and hence is less prominent under the dissecting microscope). The presence of a single, fairly large pronucleus is usually indicative of spontaneous oocyte activation (common in aged oocytes).

To assess embryonic development, continue incubation at 37° and carry out visual assessment at fixed intervals. Zygotes fertilized on day 1 should be at the 2-cell stage early on day 2, etc. The liquid paraffin over the droplet of medium containing the embryos usually helps to keep the cultures free from contamination by microorganisms.

Fixing and Staining

To visualize the fertilizing sperm at early stages of gamete interaction, we routinely remove oocytes from the sperm suspensions approximately 65 min after mixing and transfer to small droplets of fresh medium covered with paraffin. If cumulus cells are still present, use droplets of hyaluronidase-containing medium to remove them *in situ*. All treatment groups are fixed at 75 min with phosphate-buffered formalin (4% formaldehyde, v/v), which is added with a 10-ml syringe and 21-gauge needle. This approach allows essentially simultaneous fixation of all groups and thus permits accurate temporal comparisons of different experimental conditions.

Leave oocytes/zygotes at least 30 min in the fixative so that zonae will harden somewhat; immediate processing will cause cells to flatten excessively, making analysis more difficult. If cells cannot be stained and assessed on the day of fixation, transfer to dilute formalin (1/6 to 1/7 dilution with PBS) after about 2 hr in initial fixative. Otherwise zonae will harden excessively and make visualization of oocytes per se difficult if many sperm are attached to the zona surface. The acetic acid in the stain dissolves mildly fixed zonae, leaving only the denuded oocyte/zygote.

We stain with 0.75% (w/v) orcein in 45% (v/v) acetic acid; this is mixed well, then filtered at the time of preparation. A small volume is again filtered immediately before use. Slides should be cleaned thoroughly (e.g., in 70% ethanol). Using a fine pipette, transfer one to several oocytes in a minimal volume of fixative to a small drop of aceto-orcein on a slide; add another drop of stain, followed by a coverslip. Remove any excess stain with a tissue at the edges of the coverslip. The oocytes on the slide can be seen under the dissecting microscope: holding the slide, turn it over and draw a ring around the oocytes with an indelible marker. This simplifies searching for the oocytes, which may have become scattered during the staining procedure. Examine within a few minutes using a ×40 objective on a research microscope. Both bright-field and phase-contrast illumination can be informative, depending on the stage of sperm head decondensation.

Because sperm–oocyte fusion triggers oocyte activation, that is, resumption of the second meiotic division, both oocyte chromosomes and sperm head decondensation can be assessed in fertilized cells.[6] Furthermore, because the changes in sperm chromatin and oocyte chromosomes proceed in tandem, certain associations of staging are observed in most cases. First, an unfertilized oocyte has no sign of a fusing sperm head, and the chromosomes are clearly at metaphase of the second meiotic division, in a zipperlike or radiating spokelike configuration, depending on the orientation of the spindle in that particular preparation. Once activation has been triggered, the oocyte chromosomes begin to separate and can be seen in early, mid, and late anaphase (with increasing amounts of space between the two sets of chromosomes as they move to opposite ends of the spindle) and then telophase/second polar body (chromosomes

[6] L. R. Fraser, *J. Reprod. Fertil.* **69**, 419 (1983).

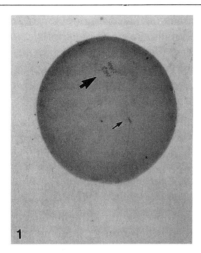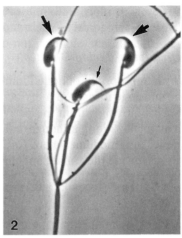

Fig. 1. Recently fertilized zygote as seen after fixation and staining with aceto-orcein. The oocyte chromosomes (large arrow) are at mid anaphase of the second meiotic division, while the sperm head (small arrow) has just begun to decondense. Only sperm actually fusing with oocytes will take up stain and hence become visible with bright-field illumination.

Fig. 2. Three mouse sperm as seen after fixation and visualization under phase-contrast illumination ($\times 100$ objective, oil). The two outer cells both have an acrosomal cap (large arrows) on the convex side, while the middle one does not. The head of the latter is narrower, and the perforatorium can be seen at the distal point of the head (small arrow).

form rosettelike clusters at the ends of the spindle and may be seen being extruded from the surface, again depending on orientation).

Sperm head decondensation begins in the posterior region and travels toward the anterior tip. The regions of membrane fusion are clearly stained (only fusing sperm take up stain), but the decondensing chromatin stains less intensely since its relative concentration is less. Initial stages of decondensation can be seen in association with anaphase-stage oocyte chromosomes, particularly at early and mid anaphase. A zygote at mid anaphase with a sperm at an early stage of fusion and decondensation is shown in Fig. 1. Midway stages of decondensation are mostly associated with late anaphase, while completely decondensed sperm heads are usually associated with telophase stage chromosomes. Oocyte chromosomes are most readily observed with bright-field optics, as are early and mid stages of decondensation. Fully decondensed heads can be seen this way, but if staining is pale, phase contrast is helpful. The neck region and the perforatorium, the V-shaped anterior tip of the head which is formed by the inner acrosomal membrane, both show up as dark structures with this

the original description of the method,[9] slides were prepared one at a time, mixing the dye and live cells on a slide, then fixing and adding a coverslip. We have modified this to allow staining and fixing in suspension, with slide preparation occurring at a later, convenient time; this approach permits accurately timed samples to be taken. The buffer, consisting of 130 mM NaCl and 20 mM Tris-HCl, can be kept refrigerated for 2 weeks or so; 5 mM cysteine is added at the time of making up the CTC staining solution. CTC solution, containing 750 μM CTC (Sigma) in buffer (final pH 7.8), is prepared on the day of use, kept wrapped in foil to exclude light, and maintained on ice. To avoid cold shock to sperm, warm up an aliquot to room temperature before use. To stain the sperm in suspension, place 45 μl sperm in a small (0.5 ml) microcentrifuge tube wrapped in foil, then add 45 μl CTC solution. After mixing well, add 8 μl of 12.5% (w/v) paraformaldehyde in 0.5 M Tris-HCl buffer (final pH 7.4) with mixing. Prepare slides by placing 10 μl of this suspension on a clean slide. Add a drop of 0.22 M 1,4-diazabicyclo[2.2.2]octane (Sigma) in glycerol : PBS (9 : 1) and mix in carefully to retard fading of fluorescence. Add a coverslip and gently compress slide plus coverslip between tissues to remove excess fluid. This helps to maximize the number of sperm cells lying flat on the slide. Seal the edges of the coverslip with nail varnish and store in the dark in the cold. It is best to assess the slides on the same day or the next, but slides are still scorable for up to 4–5 days, if maintained carefully. Observe under violet light. We use an Olympus (Tokyo, Japan) BHS microscope where the mercury excitation beam is passed through a 405 nm bandpass filter and CTC fluorescence emission is observed through a DM 455 dichroic mirror.

Three basic staining patterns on the sperm head can be distinguished: F, with uniform fluorescence, which is characteristic of uncapacitated, acrosome-intact cells; B, with a fluorescence-free band in the postacrosomal region, which is characteristic of capacitated, acrosome-intact cells; and AR, with dull or absent head fluorescence, which is characteristic of acrosome-reacted cells. If the fluorescence microscope has phase-contrast optics, it is possible to verify the presence or absence of the acrosomal cap as detailed above. In all sperm, the midpiece region of the tail exhibits bright fluorescence, irrespective of the pattern on the head.

Troubleshooting

Failure to Obtain Superovulation

If you have not done so, try a dose–response check of superovulation as detailed earlier. You may be using too little or too much PMSG; too

[9] C. R. Ward and B. T. Storey, *Dev. Biol.* **104**, 287 (1984).

much can inhibit follicular response. Also ensure that the interval between PMSG and hCG is approximately 48–54 hr, and that you are not trying to recover oocytes too early after the hCG injection, before completion of ovulation.

Failure to Obtain Fertilization

There could be many possible reasons for failure to obtain fertilization. It is always helpful, if possible, to use strains which are highly fertile *in vivo*. Some inbred strains have clearly been selected for traits other than good reproductive performance and consequently do poorly *in vitro*. The BALB/c strain, for example, falls into this category. The females do not respond well to exogenous hormones, and the sperm are usually very poor, with many morphological abnormalities.

Always check sperm motility before coincubation with oocytes: a suspension with almost no motile cells is unlikely to be successful. If you are having problems, also check motility as soon as the suspension has been prepared. It is not normal to have a marked decline in proportion of motile cells between the initial and final stages of sperm preincubation. If you observe such a change, there could be problems with the medium.

Technical problems might involve the water used to prepare the medium: try another source. Liquid paraffin is sometimes toxic, although this is more often picked up during culture than at fertilization. Again, another source should be tried. Also, check that the recipe used to prepare the medium is correct. Most media have a calculated osmolality of approximately 300–315 mOsm and an osmolality of approximately 285–295 mOsm when measured with an osmometer.

In our experience, once a system for *in vitro* capacitation and fertilization has been established successfully, it is usually trouble-free. Individual experiment failure can usually be attributed to a poor sperm suspension; this makes checking the suspension for motility essential.

[14] *In Vitro* Fertilization

By JEFFREY D. BLEIL

Introduction

In vitro fertilization (IVF) of mouse eggs has been used by investigators studying morphological and molecular details of sperm–egg interaction. Interest in mouse IVF has increased in laboratories using the mouse as

METHODS IN ENZYMOLOGY, VOL. 225

Copyright © 1993 by Academic Press, Inc.
All rights of reproduction in any form reserved.

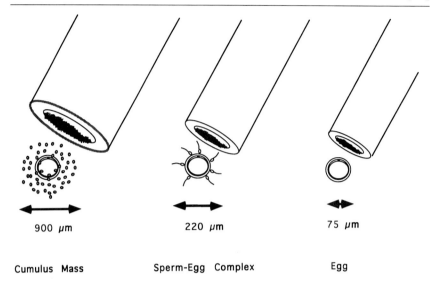

FIG. 1. Comparison of micropipette bore sizes required for gamete manipulation.

a model system for developing methods in human IVF programs and for the purpose of propagating mice which have been rendered infertile by mutations (e.g., infertile transgenic mice). This chapter is designed to provide investigators with a simplified, standard method for successful mouse IVF. Media, methods, and mouse strains described here have been chosen on the basis of success in our laboratory, ease of setup, and minimal cost.

Materials

Glassware and Water

Glassware should be thoroughly washed, siliconized, rinsed, and sterilized at 200° for at least 30 min. Siliconized, sterilized Pasteur pipettes (long tipped) for transfer of cumulus masses are simply rounded at the tip with a flame and used with a pipette bulb. Micropipettes for transfer of gametes and embryos are formed by heating a Pasteur pipette over a sharp flame at a position 3 cm from the tip. When the glass is soft enough to bend easily, the pipette is removed from the flame, and both ends of the pipette are quickly pulled until the tube is necked to the proper diameter. After cooling, the pipette is pulled firmly, in order to create a flat break in the necked portion of the tube. The inner bore diameter of micropipettes will vary, depending on the cells to be transferred (see Fig.

1). Micropipettes are used by attaching thin latex tubing to the blunt end (minimum 25 cm length), attaching a mouthpiece to the free end of the tubing, and using mouth pressure to control the flow of solution. Before picking up gametes, a column of medium (between 10 and 500 μl, depending on the application) should be pulled into the pipette.

Culture grade water should be used for glassware rinsing and solutions. Water prepared by Milli-Q (Millipore, Bedford, MA) or Nano-pure (Barnstead, Dubuque, IA) systems is ideal, although commercially available, bottled HPLC-grade water is equally suitable.

In Vitro Fertilization Media

Either of two culture media will support successful IVF. M199M[1] (a modification of GIBCO, Grand Island, NY, Medium 199, No. 3201150AG), a bicarbonate-buffered medium, is used in a 5% CO_2 (in air) tissue culture incubator (37°, $\geq$90% humidity). Alternatively, laboratories not equipped with such an incubator can use HH medium[2] (a modification of Ham's medium, GIBCO, No. 3301955AG), a HEPES-buffered medium which is simply used in atmospheric CO_2 (37°, $\geq$90% humidity). HH medium can be used in a humidified 37° incubator or on a covered, humidified warm plate. Modifications of these media are presented in Table I.

All tissue culture is performed in Falcon tissue culture dishes (60 or 100 mm diameter) under a 1 cm layer of light paraffin oil in order to retain temperature and gas pressure during transfer and observation. Round (30–100 μl) or flattened (0.5–1.0 ml) drops of medium are placed on the bottom of the dish and covered with warmed, equilibrated paraffin oil and incubated at least 30 min at 37° prior to addition of gametes.

It should be noted that care must be taken to minimize the time dishes are out of the incubator. Both eggs and sperm are subject to temperature shock and are deleteriously affected by exposure to bright light. In the case of M199M, prolonged periods ($\geq$5 min) outside the incubator can be lethal to gametes, owing to a drop in CO_2 pressure and consequent rise in pH.

Paraffin Oil

Light paraffin oil (mineral oil, without additives) is prepared at least 1 week beforehand by mixing, in tissue culture bottles, with 1/10 volume of 150 mM NaCl and steam sterilizing. Sterilized oil is allowed to equili-

---

[1] J. D. Bleil and P. M. Wassarman, *Cell (Cambridge, Mass.)* **20**, 873 (1980).
[2] B. R. Behr, C. J. Stratton, W. D. Foote, V. Knutzen, and G. Sher, *J. In Vitro Fertil. Embryo Transf.* **7**, 9 (1989).

TABLE I

In Vitro FERTILIZATION MEDIA

Components	M199M	HH
Commercial medium	Medium 199 with Earle's salts with bicarbonate with glutamine[a]	Ham's F10 medium ($10\times$) without bicarbonate with glutamine[b]
Modifications		
Pyruvic acid	30 μg/ml	—
Bovine serum albumin (BSA; Sigma, B2518)	4 mg/ml	4 mg/ml
Calcium lactate	—	245 μg/ml
NaHCO$_3$	—	500 μg/ml
HEPES	—	2.4 mg/ml
Streptomycin sulfate	75 μg/ml	75 μg/ml
Penicillin G	75 μg/ml	75 μg/ml

[a] Antibiotics are optional. Modified medium is made freshly, filter sterilized (0.2-μm filter), and stored overnight in the incubator to equilibrate CO_2.

[b] Antibiotics are optional. Modified medium is made freshly, brought to 280 mOs*m*, filter sterilized (0.2-μm filter), and incubated at least 2 hr to bring temperature to 37°.

brate in the tissue culture incubator for at least 6 days. During this time, turbidity in the upper paraffin layer should be lost, and gas pressure will equilibrate. Equilibrated paraffin oil can be stored in the incubator for at least 2 months.

Mice

Outbred mice, CD-1 (Charles River Labs, Wilmington, MA), or HSD : ICR (Harlan-Sprague Dawley, Indianapolis, IN), are recommended for IVF because of availability and cost. However, other strains will produce more gametes and/or evidence higher efficiency of fertilization. In any case, procedures described here should be applicable to virtually any mouse strain. Female mice (30–100 days old, 5–10 per cage) and male mice (8–40 weeks old, caged individually) are kept on a 12 hr light/12 hr dark schedule at 22° for at least 2 weeks prior to IVF procedures, since newly caged animals tend to produce low numbers of gametes. It should be noted that a considerable drop in numbers of superovulated eggs in CD-1 and HSD : ICR mice have been observed during the months of June–August (to as low as an average of 3 eggs per female), regardless of geographical location and care in regulation of feed, light, and temperature.

Procedures

Gamete Isolation and Treatment

Egg Isolation. Female mice are superovulated by serial injections of sterile saline containing pregnant mare serum gonadotropin (PMSG, Calbiochem, La Jolla, CA, 2500 IU/mg), followed by human chorionic gonadotropin (hCG, Sigma, St. Louis, MO, CG-2). Approximately 5 IU in 100 μl is delivered into the peritoneal cavity using a 30-gauge needle. PMSG is injected at 3 PM on day 1. HCG is injected at 5 PM on day 3.

Ovulated eggs are isolated at 6–7 AM on day 4. A superovulated female mouse is sacrificed by cervical dislocation, the peritoneal cavity exposed, and the entire oviduct removed to a 100-μl drop of medium under oil. Ovulated eggs in the oviduct are embedded within a loosely cross-linked mass of cumulus cells. These cumulus masses create a "bulge" in the oviduct, which is easily observed under a dissecting microscope. The oviduct is torn at this position, using a 25- to 30-gauge needle, while the oviduct is held rigid with watchmaker's forceps. The cumulus masses, which squeeze out through the tear in the oviduct, are removed with a Pasteur pipette to another 100-μl drop of medium and sperm added directly, or cumulus cells are removed with hyaluronidase (see below) before exposure to sperm.

Cumulus masses can also be isolated from anesthetized females by surgically exposing the oviduct, located at the animal's lower back, near the kidney. The oviduct is torn at the "bulge" with a needle, and released cumulus masses are transferred to medium with watchmaker's forceps.

Under the conditions described above, eggs are ovulated between 2 and 4 AM. Cumulus masses isolated at 7 AM contain eggs which are fertile before 8:30 AM. After this time, the eggs become progressively infertile. Therefore, it is of importance to mix sperm and eggs as quickly as possible after egg isolation.

Sperm Isolation. A male mouse is sacrificed by cervical dislocation, the peritoneal cavity exposed, and both cauda epididymides removed to a 250-μl drop of medium under oil. The cauda is held in place with watchmaker's forceps and cross-sectioned at least 3 times using a pair of microscissors or a sharp-pointed scalpel. Sperm are allowed to swim out for 15 min at 37°, and the entire drop, less cauda, is gently transferred under a 1.5-ml volume of medium in a 12-ml culture tube (Falcon, round-bottomed). Sperm are allowed to swim up for at least 45 min. During this time, motile sperm become "capacitated."[3] At the end of this period,

[3] H. M. Florman and D. F. Babcock, *in* "Elements of Mammalian Fertilization" (P. M. Wassarman, ed.), Vol. 1, p. 105. CRC Press, Boca Raton, Florida, 1991.

sperm in the upper one-quarter of the medium should be fertile, 70–100% motile, and at a concentration of at least 5×10^5/ml, as determined with a hemocytometer. Sperm from the upper one-quarter of the medium are used immediately for insemination.

During the course of isolation, sperm undergo an easily observed change in motility, which is associated with capacitation and gaining of the ability to fertilize eggs. Immediately after release from the cauda, sperm move in a straight-line direction, with little lateral movement of the tail. Capacitated sperm, which have swum up in the culture tube, move with a whiplike motion and have little forward progression. Sperm that attain this motility after the swim-up period are largely fertile and a good indicator that the medium will support IVF.

Sperm can also be isolated from anesthetized males by surgically exposing the cauda epididymis, puncturing it with a 30-gauge needle, and gently drawing a small volume of released sperm from the epididymal surface. In this case, no swim-up is performed, but sperm are allowed to incubate 1 hr in 30–100 μl medium before mixing with eggs. Alternatively, if surgery on a particularly valuable male is undesirable, small numbers of sperm can be isolated from the uterine horns of a mated female by "flushing" the uterine lumen. (In this case, females are mated with individually caged males, starting at 8 PM, and uteri are dissected between 4 and 6 AM.) This is accomplished by dissecting out oviducts along with uterine horns (cut above the cervix), inserting a blunt 30-gauge needle into the infundibulum (end of the oviduct), and injecting about 0.3 ml of medium into the oviduct. Although sperm which flow out of the uteri will be mixed with clots of semen, at low concentration, and will evidence a low percentage of motile cells, sufficient numbers of capacitated, fertile sperm should be available for successful IVF. After concentrating (see below), these sperm can be added directly to the eggs.

In the case of extremely low sperm concentrations ($<10^5$/ml), sperm can be concentrated prior to insemination. This can be accomplished by banding the cells onto a cushion of Percoll. An 85% Percoll solution is made up with 8.5 ml Percoll (Pharmacia, Piscataway, NJ), 1 ml Ham's F10 ($10\times$ concentrate), 0.1 ml of 24.5 mg/ml calcium lactate, 0.1 ml of 50 mg/ml NaHCO$_3$, and 0.3 ml of 80 mg/ml HEPES. The 30% Percoll is made up with 3.5 ml of 85% Percoll and 6.5 ml HH medium. Approximately 0.5 ml of 30% Percoll is transferred to a conical tissue culture tube (Falcon), and 0.1 ml of 85% Percoll is gently layered underneath. After incubating at least 15 min to bring the temperature to 37°, 1–5 ml medium containing diluted sperm is layered atop the 30% Percoll. Sperm are banded at the 30–85% Percoll interface by spinning in a tabletop centrifuge (preferably in a 37° room) at approximately 3000 rpm for 5–10 min. In as small a volume as possible (e.g., 30 μl), the sperm band is gently removed from the interface and transferred to a 100-μl drop of medium under oil, and these concentrated sperm are used to inseminate eggs. This procedure can be scaled down by using 0.5-ml conical microcentrifuge tubes and spinning in a tabletop centrifuge fitted with adaptors.

Removal of Cumulus Cells. The cumulus apparently serves as a medium to "filter" fertile sperm and also to partially immobilize sperm as they interact with the surface of the zona pellucida.[4] For these and, perhaps, other reasons, we have found that cumulus-intact eggs are more efficiently fertilized *in vitro* than cumulus-free eggs. For this reason, it is recommended that sperm be mixed directly with cumulus masses. However, certain conditions (e.g., zona drilling, see below) require that insemination of cumulus-free eggs be performed. In this case, cumulus masses are placed into medium (5 μl per egg, using a volume up to 1 ml) containing hyaluronidase (500 μg/ml, Sigma type VI-S, 3000–15,000 IU/mg) and soybean trypsin inhibitor (10 μg/ml, Sigma, T9003). At 37°, all cumulus cells disperse within 2–8 min, and eggs are immediately washed through 3 drops (100 μl each) of fresh medium before insemination.

Zona Drilling. In some cases of male infertility and/or low sperm concentration, and assuming at least some caudal sperm can be isolated, it may be necessary to assist sperm fusion with the egg. A procedure known as zona drilling, which involves creating a small hole in the zona pellucida, has been shown to effectively "rescue" sperm evidencing low motility, abnormal morphology, or oligospermy.[5] During the course of interaction of normal sperm with the egg, sperm first recognize and bind to the zona pellucida in a species-specific manner, subsequently undergoing the acrosome reaction at the zona surface. Finally, under the forward thrust of the flagellum, and catalyzed by proteolytic digestion of the zona in the region of the sperm head, the sperm bores through the zona, gaining entrance to the egg plasma membrane. Thus, acrosome-reacted sperm defective in any of these normal functions can gain direct access to the plasma membrane of zona-drilled eggs and, in many cases, successfully fertilize the egg.

Zona drilling requires micromanipulators (a holding pipette and an injection pipette) and a high-power inverted microscope. Briefly, cumulus-free eggs are positioned for micromanipulation, and a microneedle containing acid Tyrode's solution[6] is pressed tangentially against the zona pellucida. A small volume of the acid dissolves a hole in the zona. Eggs treated

---

[4] J. D. Bleil, J. M. Greve, and P. M. Wassarman, *Dev. Biol.* **128,** 376 (1988).
[5] J. W. Gordon, *In Vitro Fert. Embryo Transfer* **7,** 223 (1990).
[6] J. W. Gordon and B. E. Talansky, *J. Exp. Zool.* **239,** 347 (1986).

in this manner are washed through two drops of medium and inseminated as described below.

Sperm–Egg Incubation

Immediately after isolation, cumulus-intact or cumulus-free eggs (as well as zona-drilled eggs) are placed into 25-μl drops of medium (10–20 eggs per drop) under oil. An equal volume of motile, capacitated sperm is added, and the mixture incubated for 3–6 hr at 37°. The final sperm concentration should be 10^4 to 2×10^5/ml.

After 3–6 hr, fertilized eggs will contain two pronuclei. Observation of two pronuclei within the egg at this time is arguably the best criterion of successful fertilization. Pronuclei can be seen in the original incubation drop by adding 200 μl fresh medium (to expand the drop circumference, dilute sperm and cumulus cells, and replenish nutrients), drawing off sufficient medium to "flatten" the drop, and then observing eggs with a $\times 40$–60 objective on an inverted microscope set up for phase-contrast or Hoffman interference illumination. Only fertilized eggs will evidence two pronuclei, and polyspermic eggs will have more than two. During the course of sperm–egg incubation, cumulus masses will generally disperse within 4 hr, owing to sperm-associated hyaluronidase. If masses persist after 4 hr (making it difficult to visualize pronuclei), they can be dispersed by hyaluronidase as described above. It is reasonable to expect 50–80% of cumulus-intact eggs and 30–60% of cumulus-free eggs to be fertilized under the conditions described here. It is recommended that eggs containing two pronuclei be immediately transferred to pseudopregnant foster females, as described below.

During an IVF experiment, small numbers of gametes can be removed for observation in order to determine whether interaction between sperm and egg is occurring as expected. Interaction between sperm and cumulus-enclosed eggs can be observed with a high-power inverted microscope. Glass beads (200 μm diameter) are loosely dispersed on the bottom of a tissue culture dish, and a 5-μl drop of medium containing sperm and one cumulus-enclosed egg is transferred to the dish. After a cover glass is placed atop the medium drop, the air space (created by the beads) between dish and cover glass is filled with equilibrated paraffin oil. The glass beads hold the cover glass a sufficient distance to displace cumulus cells to the side of the egg without crushing the egg, thereby allowing a fairly unobstructed view of the egg. Sperm are observed using phase-contrast or Hoffman interference. The same procedure, using siliconized microscope slides (suspended upside down) instead of tissue culture dishes, allows the use of Nomarski differential interference contrast (DIC) and, consequently, a high-resolution image devoid of the "background noise" created by cumulus cells out of the plane of focus. Within 20–30 min following insemination, sperm are observed traversing the cumulus mass. Approximately 20 min following insemination, at least one sperm should have contacted the zona pellucida surface. This contact is an excellent indicator that IVF is proceeding normally. Subsequent events (penetration and fusion) are difficult to observe without additional equipment.

Interaction between sperm and cumulus-free eggs can be observed in the insemination drop itself or in a parallel drop in another tissue culture dish (under oil) with the dissecting microscope. Observations of sperm–egg interaction in the dissecting microscope are best performed using dark-field microscopy, where egg and sperm are visible but the zona pellucida is invisible. Within seconds after insemination, sperm contact the surface of the zona pellucida, and the number of sperm associated with the zona reaches a maximum at about 5–10 min after insemination. By 10 min, the "ball" of sperm associated with the egg will begin to drive the egg across the dish surface. Movement of the sperm–egg balls across the dish is an excellent indicator that IVF is proceeding normally. Subsequent events are extremely difficult to observe without additional equipment.

Development of Embryos beyond Pronuclear Stage

In cases when it is of interest to observe development past the pronuclear stage, 1-cell embryos must be transferred to a medium that will support early embryonic development. A number of mouse strains are blocked from progressing past the 2-cell stage in media which support IVF.[7] Consequently, a medium which supports development of 1-cell embryos through the 2-cell stage has been developed.[8] This medium, CZB, which requires a CO_2 incubator and is defined in Table II, is used for early embryo culture as described below.

Pronuclear (1-cell) stage embryos, free of cumulus cells, are washed through 3 drops of CZB medium, incubated in 50-μl drops of CZB medium (10 embryos per drop) for 24–27 hr, and scored for development of 2-cell embryos. Since apparent 2-cell embryos can develop during this time owing to parthenogenesis or fragmentation of unfertilized eggs, scoring of the 2-cell stage must be interpreted cautiously. During all embryo culture, approximately 40 μl medium is removed by micropipette and replaced with fresh medium every 12 hr.

[7] M. J. Goddard and H. P. M. Pratt, *J. Embryol. Exp. Morphol.* **73**, 111 (1983).
[8] C. L. Chatot, C. A. Ziomek, B. D. Bavister, J. L. Lewis, and I. Torres, *J. Reprod. Fertil.* **86**, 679 (1989).

TABLE II
CZB MEDIUM

Component	Concentration (mM)
NaCl	81.62
KCl	4.83
KH_2PO_4	1.18
$MgSO_4 \cdot 7H_2O$	1.18
$NaHCO_3$	25.12
$CaCl_2 \cdot 2H_2O$	1.70
Glucose	0
Sodium lactate	31.30
Sodium pyruvate	0.27
EDTA (disodium salt)	0.11
Glutamine	1.00
BSA (mg/ml)	5.00
Penicillin G, sodium salt (U/ml)	100.0
Streptomycin (mg/ml)	0.70

Approximately 48 hr after the 1-cell stage (when most embryos have developed to the 4-cell stage), embryos are transferred to CZB medium supplemented with 5.56 mM glucose and incubated in this supplemented medium thereafter. Addition of glucose is required at this time, presumably owing to a change in metabolic requirements through the early morula stage.[8] Up to and including development to the blastocyst stage, embryos can be transferred to pseudopregnant foster females, as described below.

Embryo Transfer to Pseudopregnant Females

To develop embryos to term, *in vitro* fertilized embryos are transferred to the oviduct or uterine horn of a pseudopregnant foster female. In most cases, it is preferable to transfer *in vitro* fertilized embryos to a foster female as soon as possible after fertilization (i.e., at the pronuclear stage, when success can first be determined). Transfer at this time minimizes the effect of medium and culture manipulation on embryos. In addition, foster females can be "timed," so that the timing of their reproductive system corresponds almost precisely with the "expected" stage of embryonic development (i.e., the pronuclear stage embryo).

Pronuclear stage embryos are directly transferred to the oviduct of a pseudopregnant female. A female which has successfully mated (e.g., has an observable semen plug at the vaginal opening) with a vasectomized male on the evening prior to IVF is anesthetized, and the oviduct is surgically exposed. Under the dissecting microscope, a small puncture is made in the inflated bursa with a sharp needle. The oviduct is held in place with watchmaker's forceps while a micropipette containing 5–10 embryos in 0.5–1.0 μl medium is pushed through the tear in the bursa and inserted into the infundibulum. Embryos are gently transferred into the oviduct, along with a small volume (~0.5 μl) of air (to discourage discharge of embryos from the oviduct).

In some cases, it may be of interest to transfer embryos which have developed past the pronuclear stage. Transfer of embryos up to and including the morula stage is performed as described above. Blastocysts, on the other hand, are transferred into the uterine horns of a pseudopregnant female. The pseudopregnant female is mated on the evening following IVF, anesthetized 3 days following IVF (e.g., when blastocysts have developed in culture), and the ovary, oviduct, and uterine horn surgically exposed. While holding the fat pad overlying the ovary, blastocysts, in the bottom quarter of a micropipette containing about 5 μl medium, are transferred through a small hole in the uterine horn (made with a 30-gauge needle). For a more detailed description of embryo transfer, see Hogan *et al.*[9]

[9] B. Hogan, F. Costantini, and E. Lacy, "Manipulating the Mouse Embryo," p. 135. Cold Spring Harbor Laboratory, Cold Spring Harbor, New York, 1986.

Section IV

Gene Expression: Messenger RNA

[15] Radioisotope Labeling of UTP and ATP Pools and Quantitative Analysis of RNA Synthesis

By KERRY B. CLEGG and LAJOS PIKÓ

Introduction

Radioisotope labeling is a highly sensitive procedure to obtain information on the synthesis of various classes of RNA during early embryogenesis. A problem of particular interest is the transcriptional activation of the zygote genome following fertilization and the transition from the use of maternally derived RNA components to those derived from the zygotic genome. Early mouse embryos lend themselves readily to radioisotope labeling experiments because they can be cultured *in vitro* in a simple, chemically defined medium, and thus labeled precursors can be introduced under controlled conditions. However, to be able to design such experiments efficiently and interpret the RNA synthetic data in a quantitative fashion, one has to know the behavior of the precursor ribonucleoside triphosphate (NTP) pools and characterize the labeled products in some detail. The parameters that one needs to consider include the size of the endogenous NTP pool; rate of uptake and conversion of the labeled nucleoside to a phosphorylated form that can be incorporated into RNA; equilibration, specific activity, and turnover of the intracellular precursor pool; distribution of the labeled precursor in the RNA product [e.g., incorporation of labeled adenosine into the poly(A) tail of mRNA versus internal locations in the molecule]; and possible nonspecific incorporation of label into molecules other than RNA.

Here, we briefly summarize the information we have obtained on the behavior of the UTP and ATP pools of early mouse embryos and the use of labeled precursors of these pools in the analysis of the overall rates and pattern of RNA synthesis. We hope that a description of the principles and some of the techniques involved in this work will facilitate the planning of radioisotope labeling experiments not only in mice but also in early embryos of other mammalian species.

Materials and Methods

Reagents

Radioactive nucleosides and nucleotides (Du Pont–New England Nuclear, Boston, MA, or Amersham–Searle, Arlington Heights, IL)

Copyright © 1993 by Academic Press, Inc.
All rights of reproduction in any form reserved.

Nonradioactive nucleosides and nucleotides (Sigma, St. Louis, MO)

Basic salt solution (BSS): 120 mM NaCl, 4.6 mM KCl, 1.6 mM CaCl$_2$, 1 mM KH$_2$PO$_2$, 1 mM MgSO$_4$, 25 mM NaHCO$_3$, pH 7.6

Tris-buffered saline (TBS): 145 mM NaCl, 10 mM Tris-HCl, pH 7.6

Bovine serum albumin (BSA), crystalline (ICN, Costa Mesa, CA)

Perchloric acid (PCA), 0.5 N

Standardized KOH solution: 0.83 N KOH in 40 mM Tris-HCl, pH 8.3; solution is tested by titration against PCA to ensure perchlorate precipitation and final pH of 8.0–8.3

Dowex 1-X8 resin (200–400 mesh, chloride form) (Bio-Rad Laboratories, Richmond, CA)

Polyethyleneimine (PEI) cellulose (20 × 20 cm plastic-backed thin-layer plates, UV indicator) (EM Science, Gibbstown, NJ, or Baker, Phillipsburg, NJ)

Poly(dA-dT), double-stranded alternating copolymer (Sigma)

RNA polymerase (*Escherichia coli*) (United States Biochemical, Cleveland, OH, or Promega, Madison, WI)

RNA polymerase diluent: 10 mM Tris-HCl, pH 8.0, 1 mM MgCl$_2$, 1 mM 2-mercaptoethanol, 50 μM EDTA, 0.1 mg/ml BSA

10× RNA polymerase assay buffer: 400 mM Tris-HCl, pH 8.0, 100 mM MgCl$_2$, 100 mM 2-mercaptoethanol

RNA extraction buffer: 100 mM NaCl, 50 mM Tris-HCl, pH 7.5, 5 mM EDTA, 0.5% sodium dodecyl sulfate (SDS), 0.5 mg/ml proteinase K

LiCl, reagent grade (Baker)

Cetyltrimethylammonium bromide (hexadecyltrimethylammonium bromide; CTAB) reagent grade (Baker)

Poly(U)-Sepharose 4B (Pharmacia, Piscataway, NJ)

Poly(U) binding buffer: 500 mM NaCl, 10 mM Tris-HCl, pH 7.5, 1 mM EDTA, 0.05% SDS

Poly(U) elution buffer: 90% deionized formamide/10% (10 mM Tris-HCl, pH 7.5, 1 mM EDTA, 0.05% SDS) (v/v)

RNase A and RNase T1 (Boehringer, Indianapolis, IN)

RNase A and Ti digestion buffer: 0.5 M NaCl, 50 mM Tris-HCl, 5 mM EDTA, pH 7.5

Polysome isolation buffer: 150 mM NaCl, 50 mM Tris-HCl, pH 7.5, 20 mM MgCl$_2$, 10 mM EGTA, 20 μg/ml emetine hydrochloride, 20 μg/ml cytochalasin B, 0.5% Triton X-100, 100 U/ml heparin, 50 U/ml RNasin, 2 mM 2-mercaptoethanol

Production of Embryos

In our experience, the following procedures yield reproducible batches of embryos with respect to stage of development. Female mice 3–4 weeks of age (our usual supplier is Charles River Breeding Laboratories, Inc., Wilmington, MA) are maintained on a 14 hr light/10 hr dark period with the midpoint of the dark period at midnight. Breeding age males are individually caged and maintained on the same light schedule. At 5–6 weeks of age, females are superovulated by intraperitoneal injection of 5 IU of pregnant mare serum gonadotropin followed 44–48 hr later (10 AM–12 noon) by 5 IU of human chorionic gonadotropin (hCG) and mated overnight two females per male. Successfully mated females are identified by the presence of vaginal plugs on the following morning. The embryos are flushed from the oviducts with a modified pyruvate–lactate medium (PLM)[1] containing 50–100 μg/ml gentamicin sulfate, 0.1% BSA, and 25 mM HEPES, pH 7.4, at various days of development up to the 8-cell stage and are used for experiments either immediately or after culture for 1–2 days in PLM at 37°, 5% (v/v) CO_2 in air, from the 8-cell stage to the blastocyst.

The following developmental stages are obtained at the indicated times after hCG injection: 22–24 hr = day 0 (when the vaginal plug is found), 1-cell stage; 44–46 hr = day 1, 2-cell stage; 70–72 hr = day 2, 8-cell stage; day 2 + 24 hr culture = day 3, early cavitating blastocyst (average of 32 cells); day 2 + 48 hr culture = day 4, expanded blastocyst (about 60 cells). The embryos should be carefully monitored for stage of development and stage of the cell cycle because uptake of precursors and ribonucleotide pool sizes can vary considerably during early development, for example, with respect to the UTP pool.[2] Cell numbers in samples of day 3 and day 4 embryos can be determined by gently squashing 1% acetocarmine-stained embryos under a coverslip and counting the nuclei. If oocytes or newly ovulated eggs are used, it is important to remove all adhering follicle cells by repeated washes with PLM containing about 800 U/ml bovine testicular hyaluronidase; the removal may be facilitated by an additional wash with 0.25% subtilisin (bacterial protease type VII, Sigma) in basic salt solution.[2]

Radioactive Labeling of Embryos

If desired, mating schedules can be arranged so that multiple developmental stages are available on any given day and several labeling experiments can be carried out simultaneously. The staged embryos (generally 20–200 embryos per group) are set up in a 45-μl drop of PLM (40-μl drop plus embryos transferred in 5 μl) in a plastic culture dish under mineral oil saturated with basic salt solution (BSS). The nucleotide pools are

[1] R. L. Brinster, *J. Reprod. Fertil.* **10,** 227 (1965).
[2] K. B. Clegg and L. Pikó, *Dev. Biol.* **58,** 76 (1977).

labeled by culturing the embryos for 2 hr in the presence of 5–50 μM [5,6-^{3}H]uridine or [2,8-^{3}H]adenosine (usually the highest specific activity available, 40–50 Ci/mmol). The isotopes are lyophilized to dryness or dried in a SpeedVac centrifuge (Savant Instruments, Farmingdale, NY), then reconstituted at 10 times the final desired concentration in sterile BSS or prewarmed PLM; 5 μl of this solution is added to the 45-μl droplet. At the end of the labeling period, the exogenous label is removed from the medium by about six exchange washes with 100–200 μl of TBS (rate of dilution about 1 : 10 per wash); if the embryos are used for the extraction of nucleotides, the wash fluid also contains 5 mg/ml of BSA to serve as a precipitation carrier (TBS-BSA). We routinely check an aliquot of the last wash fluid by scintillation counting to ensure that it is free of radioactivity above background levels.

The embryos are handled and the washes performed with a finely drawn Pasteur pipette operated by a mouth-held rubber tube.[3] Microtransfer pipettes are prepared by pulling 25-μl disposable micropipettes (VWR Scientific, Philadelphia, PA) over a flame to give a stem of 10–20 mm with an opening at the tip of 100–150 μm (checked under a dissecting microscope equipped with an ocular micrometer). The pipettes are calibrated for the desired volume (e.g., 5 μl) by drawing a measured droplet of water off a surface of Parafilm and marking the meniscus with a fine black ink mark. The calibrated pipettes are rinsed with several changes of acetone to dry them and then stored on a piece of double sticky tape or modeling clay (e.g., Permoplast, Nasco, Modesto, CA) attached to a microscope slide in a tightly closed box until use.

Extraction of Soluble Nucleotides

A flow chart showing the overall plan for the analysis of labeled ribonucleotides (with respect to uridine nucleotides) in embryo extracts is given in Fig. 1. The procedural details are described below (see also Ref. 2).

Labeled embryos are transferred[4] in a measured volume of TBS-BSA (see above) to 40 μl of 0.5 N PCA in a microcentrifuge tube, and the nucleotides are extracted by freezing and thawing 5 times in dry ice–ethanol. The PCA-precipitable material is collected by centrifugation at 12,000 g, 0°, for 10 min. The supernatant is carefully removed to a second microcentrifuge tube, and 40 μl of the extract is neutralized to approximately pH 8 by adding 20 μl of a standardized KOH solution. The insoluble

---

[3] J. W. Gordon and F. H. Ruddle, this series, Vol. 101, p. 411.
[4] To transfer embryos in as small a volume as possible, the stream from the washing fluid should be used to gently direct the embryos into a pile in the center of the drop. Additionally, the pipette can be used to push the embryos together. Several hundred embryos can then be aspirated into the pipette in as little as 5 μl.

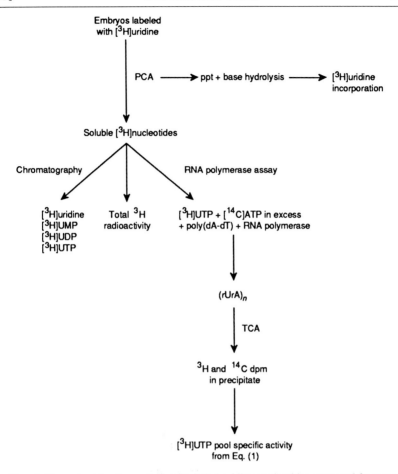

Fig. 1. Flow chart for the analysis of labeled uridine nucleotides extracted from mouse embryos cultured *in vitro* in the presence of [³H]uridine. The fate of the ³H label is followed into nucleotide derivatives and into RNA (i.e., acid-precipitable, alkali-sensitive radioactivity). The [³H]UTP pool specific activity can be derived from an RNA polymerase assay and used for the calculation of the UTP pool size (if the amount of [³H]UTP is known) and for the quantitation of RNA synthesis. For methodology, see text.

potassium perchlorate salt thus formed is precipitated by chilling to 0° on ice for 5–10 min and is removed by centrifugation as above. Approximately 40–50 μl of the 60-μl extract can be recovered. A 5-μl sample of the extract is counted directly to determine the total soluble radioactivity, and the remainder is used for chromatography and for the RNA polymerase assay (see below).

If desired, the PCA-precipitable material can be further assayed for label incorporated into RNA (i.e., alkali-sensitive material) as follows. The precipitate is washed at least 5 times (by repeated vortexing and recentrifuging) with ice-cold 5% (w/v) trichloroacetic acid (TCA) to remove residual soluble counts and hydrolyzed with 100 μl of 0.5 N NaOH for 1 hr at 37°. The hydrolyzate is reprecipitated with 50 μl 30% TCA, recentrifuged, and the radioactivity in the supernatant determined by scintillation counting.

Chromatography of Uridine and Adenosine Nucleotides

The uptake of labeled uridine and adenosine by the mouse embryo is followed by endogenous phosphorylation, and only a portion exists as UTP or ATP available for nucleic acid synthesis. Therefore, it is important to know the distribution of the total label among the various nucleotide intermediates. For example, after a 2-hr culture in the presence of [³H]uridine, about 30% of the acid-soluble label was found as UTP in 1-cell mouse embryos versus about 70% in embryos from the 2-cell stage through the early blastocyst stage.[2] In contrast, in similar labeling experiments with [³H]adenosine, mouse embryos contained 85–90% of the acid-soluble label in ATP at all stages from 1-cell through blastocyst.[2] Nucleotide forms can be separated by various liquid- or solid-phase chromatographic methods, but we have found Dowex 1 ion-exchange chromatography for uridine and PEI cellulose thin-layer chromatography for adenosine to be fast, simple, and reliable techniques. PEI chromatography can easily accommodate multiple samples, thus ensuring uniformity, and the Dowex method described below utilizes a step elution that is rapid and reproducible.

The distribution of label among the nucleotides of uridine is examined by chromatography of 20 μl of the PCA extract on a 0.7 × 2.6 cm Bio-Rad Econo column of Dowex 1-X8 ion-exchange resin.[5] The sample is mixed with 20 μmol each of uridine, UMP, UDP, and UTP (in 50 mM NH₄Cl) as carrier, and the mixture (about 100 μl) is applied to the Dowex column. The column is washed with 5 ml of 50 mM NH₄Cl to remove the free uridine, and the nucleotides are eluted sequentially by the stepwise addition of 5 ml each of 0.1, 0.175, and 1.0 M NH₄Cl. Fractions of 1 ml are collected and counted in a suitable scintillation cocktail such as Biofluor (Du Pont–NEN). It is important to determine the single channel counting efficiency for ³H in order to convert the total counts obtained into picomoles nucleoside incorporated. Additionally, any difference in

[5] R. H. Lindsay, M.-Y. Wong, C. J. Romine, and J. B. Hill, *Anal. Biochem.* **24,** 506 (1968).

quench of the samples should be adjusted for in all determinations if not automatically compensated for by the counter. However, we have seen little or no difference in the quench of the various samples assayed by this technique. Under the above conditions, the additional uridine metabolites uracil and UDP-glucose[6] would be collected together with uridine and UMP, respectively. Conversion of uridine to cytidine appears to be insignificant in the mouse embryo.[6]

Thin-layer chromatography (TLC) of the adenosine nucleotides is carried out by spotting 20 μl of the PCA extract mixed with 10 μl aqueous standard containing 20 μmol each of adenosine, AMP, ADP, and ATP onto PEI cellulose plates in small 1–2 μl portions (under a hot air stream from a hair dryer) at the origin 1.5 cm from the edge. The volume spotted should be such that the diameter of the spot is approximately 3 mm. The plates are prepared prior to use by developing in 100% methanol, air drying, and developing in deionized water to remove potentially interfering impurities to the top of the plate. The plates are then scored to the plastic base at 1.5-cm intervals to provide lanes for up to 12 samples. After spotting, the plates are developed for 1 hr in distilled water, dried, and rechromatographed for 1.5 hr in 1.2 M LiCl.[7] All chromatography is carried out at room temperature in air-tight TLC tanks equilibrated with about a 1 cm depth of the developing solution.

Following chromatography, the plates are again dried, and the individual lanes are separated by cutting with a pair of scissors. The nucleotide spots are visualized by illumination from above with short-wave UV irradiation (e.g., from a hand-held black light lamp, UV Products, Inc., San Gabriel, CA), and the lanes are cut into 1 cm pieces and placed in scintillation vials. The sample is covered with 0.3 ml of a tissue and gel solubilizer (e.g., Solvable, NEN–Du Pont) for about 10 min to solubilize the nucleotides. A compatible scintillation fluid is then added, and the vials are counted. The percentage of label found in UTP and ATP is determined and can be used in the calculation of the pool sizes for these nucleotides as described below.

RNA Polymerase Assay of Nucleotide Pool Specific Activity

The specific activities of the UTP and ATP pools can be determined with an *E. coli* RNA polymerase assay using a dual-labeling technique with an alternating poly(dA–dT) template.[2,8,9] We have found this assay

[6] H. R. Woodland and C. F. Graham, *Nature (London)* **221**, 327 (1969).

[7] K. Randerath and E. Randerath, this series, Vol. 12A, p. 323.

[8] M. Sasvári-Székely, M. Vitéz, M. Staub, and F. Antoni, *Biochim. Biophys. Acta* **395**, 221 (1975).

[9] R. E. Maxson and R. S. Wu, *Eur. J. Biochem.* **62**, 551 (1976).

to be specific and highly sensitive, capable of detecting picomole quantities of nucleotide.[2] Since both the driving nucleotide (nucleotide in excess) and the nucleotide in the experimental sample are labeled, the measurements are unaffected by the rate of synthesis, the length of incubation, or the concentration of the nucleotides in the enzyme assay.

Commercial RNA polymerase is diluted to 120 U/ml, immediately before use, with RNA polymerase diluent. Alternating poly(dA–dT) is dissolved at 2 μmol/ml in 50 mM NaCl, 10 mM Tris-HCl, pH 8.0, and is stored at $-20°$ until use. The ^{14}C-labeled NTP is added in excess, generally undiluted, in order to maintain a sufficiently high molar concentration (at least 20 times the concentration of the endogenous NTP from the embryo extract), which minimizes potential dilution effects from embryonically derived NTPs. The assay mixture contains, in a final volume of 50 μl, the following components added in order with gentle vortexing: 25 μl of neutralized PCA extract from embryos that have been cultured *in vitro* in the presence of ^{3}H-labeled uridine or adenosine, 5 μl of poly(dA-dT), 10 μl (about 1 nmol) of [14C]ATP or [14C]UTP, 5 μl of 10× assay buffer, and 5 μl of diluted RNA polymerase. Incubations are carried out at 37° for 30 min and are terminated by the addition of 0.5 ml of 0.1 M $Na_4P_2O_7$, 0.5 mg/ml BSA, and 0.5 ml of 25% (w/v) TCA at 0°.

The TCA precipitate is collected by filtration on 24-mm Whatman (Clifton, NJ) GF/C filters, washed 3 times, 5 ml each, with ice-cold 5% TCA, rinsed with 5–10 ml of 95% ethanol, and dried. The filters are counted in a liquid scintillation counter at a known efficiency for ^{3}H and [14]C. The spillover of ^{14}C counts into the ^{3}H window should be determined and corrected for in all calculations unless the counter is equipped with automatic quench correction and dual-channel analysis that can directly report disintegrations per minute of ^{3}H and [14]C.

Calculation of Nucleotide Pool Size and Specific Activity

The ^{14}C and ^{3}H radioactivities incorporated in the RNA polymerase assay can be used directly for the calculation of the specific activity of the ^{3}H-labeled NTP pool in the embryo extract, on the basis of the known specific activity of the ^{14}C-labeled NTP precursor used in the assay.[2] The relationship is expressed in Eq. (1):

$$(^3H \text{ cpm}/^{14}C \text{ cpm}) \times k$$
$$= \text{specific activity } ^3\text{H-NTP/specific activity } ^{14}\text{C-NTP} \quad (1)$$

The conversion factor k represents the ratio of the counting efficiencies for ^{14}C and [3]H (i.e., it converts the observed count per minute ratio into the ratio of disintegrations per minute); the value of k must be determined

TABLE I

SIZE AND SPECIFIC ACTIVITY OF [³H]UTP AND [³H]ATP POOLS IN
EARLY MOUSE EMBRYOS[a,b]

Developmental stage	Total UTP (pmol/embryo)	[³H]UTP (%)	Total ATP (pmol/embryo)	[³H]ATP (%)
Day 0, 1-cell	0.048	0.06	1.11	2.6
Day 1, 2-cell	0.123	1.61	0.98	7.9
Day 2, 8-cell	0.323	3.53	0.82	13.3
Day 3, early blastocyst	0.464	12.6	1.08	33.0
Day 4, late blastocyst	0.538	27.3	0.99	52.3

[a] Mouse embryos were cultured in PLM for 2 hr in the presence of 5 μM [³H]uridine or 10 μM [³H]adenosine, extracted with perchloric acid and assayed for the specific activity of the UTP and ATP pools by an RNA polymerase assay. The amounts of [³H]UTP and [³H]ATP per embryo were derived from the total uptake of labeled nucleosides by the embryos and the percentage of label present in the form of [³H]UTP or [³H]ATP (determined chromatographically; results not shown). These values were used to derive the total UTP and ATP pools (the sum of endogenous and labeled NTPs) by means of Eq. (2). The percentage of [³H]UTP and [³H]ATP represents the fraction of labeled NTP in the embryo.

[b] Data taken from Clegg and Pikó.[2]

for individual counters and counting conditions if not corrected automatically. No correction needs to be made for dilution of the ¹⁴C specific activity in the reaction mixture by the endogenous ATP or UTP of the mouse embryos, since the assay conditions are designed such that the amount of [¹⁴C]ATP or [¹⁴C]UTP is in greater than 20-fold excess over the endogenous pool.

Pool sizes can be calculated from the ratio of the specific activity of the ³H-labeled NTP pool extracted from the embryos and that of the ³H-labeled nucleoside (N) originally used to label the embryos provided that the amount of label present in the form of NTP is known. The latter value is derived from the total soluble uptake of the ³H-labeled nucleoside and the percentage of label found in [³H]NTP.[2] The relationship is expressed in Eq. (2):

$$\text{Specific activity } {}^3\text{H-NTP}/\text{specific activity } {}^3\text{H-N}$$
$$= \text{pmol } {}^3\text{H-NTP}/\text{pmol } ({}^3\text{H-NTP} + \text{NTP}_{\text{endogenous}}) \quad (2)$$

The unlabeled endogenous portion of the NTP pool is determined by subtracting the labeled portion of the pool from the total pool size calculated for the embryos from Eq. (2). Table I shows the data obtained for the size and specific activity of the UTP and ATP pools in early mouse embryos that have been cultured *in vitro* in the presence of the corresponding labeled nucleoside.

It should be noted that the dual-labeling technique could also be adapted for direct measurement of the endogenous pool size in unlabeled embryo extracts, by using both ^{3}H- and ^{14}C-labeled NTPs of known specific activity in the RNA polymerase assay. By diluting a known amount of the limiting ^{3}H-labeled nucleotide (but not the driving ^{14}C-labeled nucleotide, which is in great excess) by an unknown sample, one can convert the change in specific activity to the amount of unlabeled NTP present [see Eq. (2)].

Choice of Labeled Precursor of RNA Synthesis

To measure rates of RNA synthesis as well as detect the earliest appearance of newly synthesized RNA in the mouse embryo, it is necessary to label one of the nucleotide pools to as high a specific activity as possible. The factors that affect the specific activity of the UTP and ATP pools in *in vitro* labeling experiments are the size of the endogenous NTP pool; the rate of uptake of the labeled nucleoside and its conversion to nucleoside triphosphate; and the length of the labeling period. For example, in the 1-cell zygote, the size of the UTP pool is about one-twentieth that of the ATP pool, but adenosine is taken up about 1000 times faster than uridine and is converted to ATP more efficiently; thus [^{3}H]adenosine is the label of choice in 1- and 2-cell mouse embryos.[2,10,11] From the 8-cell stage onward both the UTP and ATP pools can be efficiently labeled, and pool specific activities of up to 50% are attainable.[2]

The concentration of the labeled nucleoside in the culture medium has a considerable effect on the rate of uptake. For example, increasing the concentration of labeled adenosine from 10 to 50 μM increases the rate of adenosine uptake by the 1-cell mouse embryo about 5 times over a 2-hr period.[2,10] Although the rate of increase of nucleoside uptake is less than proportional at the later stages of development, it can still be substantial; for example, increasing the concentration of labeled uridine from 5 to 50 μM results in a 3- to 4-fold increase of uridine uptake by 8-cell embryos and early blastocysts over a 2-hr culture.[2] When the labeling period is extended to 5 hr or longer, the specific activity of the ATP and UTP pools rises nearly linearly over the first 80 min but increases at a progressively slower rate thereafter.[2,10]

Because of the higher rate of the uptake of adenosine, its efficient conversion to ATP (85–90%), and the relatively constant size of the ATP pool (see Table I), not only is labeled adenosine generally more sensitive for the study of RNA synthesis, but the specific activity of the labeled

[10] K. B. Clegg and L. Pikó, *Nature (London)* **295,** 342 (1982).
[11] K. B. Clegg and L. Pikó, *J. Embryol. Exp. Morphol.* **74,** 169 (1983).

ATP pool is more predictable. In these experiments, a reasonable estimate of the ATP pool specific activity can be made from the total soluble radioactivity in the embryos (assuming that 90% of this label was converted to ATP) and the known size of the endogenous ATP pool.[2,10-12]

The available evidence suggests that the intracellular ATP and UTP pools of early mouse embryos equilibrate relatively quickly, and, therefore, the average pool specific activity obtained for the whole embryo can be used for the calculation of RNA synthetic rates. However, there is some evidence that, after short incubation periods (40 min) in the presence of [³H]adenosine, the nuclear ATP pool may become preferentially labeled[11] and that, in blastocyst stage embryos, some labeled UTP may accumulate in the blastocoel fluid, where it is not available for RNA synthesis.[2]

Quantitative Analysis of RNA Synthesis

To obtain quantitative estimates of the rates of RNA synthesis, one needs to consider the length of the labeling period and the stability of the RNA species synthesized. Since the specific activity of the NTP pool does not change linearly over time (see above), for labeling periods longer than 1–2 hr one needs to determine the pool specific activity at several time points. The average specific activity of the labeled NTP pool can then be used for calculating the rates of synthesis of relatively stable RNAs, for example, ribosomal and transfer RNAs and stable messenger RNAs.[10] On the other hand, RNAs that are turning over rapidly, such as heterogeneous nuclear RNA, will reflect the specific activity of the precursor pool toward the end of the labeling period.

A strategy that can be used to provide relatively stable pool specific activities in labeling experiments is to pulse-label embryos with [³H]adenosine over a short period of culture (e.g., 40 min), then wash out the label from the culture medium and continue the incubation of the embryos. We have found that, after an initial pulse with [³H]adenosine, the specific activity of the [³H]ATP pool of 1-cell and 2-cell embryos remained constant at 7–8% for at least 5 hr, making it possible to calculate the time course of the molar accumulation of adenosine into different RNA fractions.[11] The stability of the [³H]ATP pool under these conditions is due to the large size of the ATP pool (about 1.1 pmol) and the low rate of RNA synthesis (less than 1% of the total ATP is incorporated into RNA over a 5-hr period[11]). The rate of RNA synthesis increases dramatically during succeeding cleavage, and, accordingly, the [³H]UTP pool was found to

[12] L. Pikó, M. D. Hammons, and K. D. Taylor, *Proc. Natl. Acad. Sci. U.S.A.* **81,** 488 (1984).

turn over with half-lives of about 5.5 hr in 8-cell embryos and 2.8 hr in early blastocysts.[2]

For an analysis of the absolute rates of RNA synthesis, an important safeguard is the inclusion of ^{14}C-labeled marker RNAs to monitor the recovery of embryo RNA fractionated according to the scheme described below. We have used ^{14}C-labeled 18 S ribosomal RNA isolated from mouse neuroblastoma cells[13] as a recovery marker for large RNAs and ^{14}C-labeled 4 S RNA from the same source for small RNAs (e.g., Clegg and Pikó[11]), but other labeled RNAs of an appropriate size could be used equally well. In our experience, RNA recovery is generally better than 90%, and thus a correction for recovery is rarely necessary.

Fractionation of Labeled RNA

To ascertain that the radioactive precursor has been incorporated into RNA and to identify the RNA species that have become labeled, one needs to fractionate and analyze further the labeled components. This is particularly important when labeled adenosine is used as a precursor, since adenosine may be incorporated by cytoplasmic enzymes into the poly(A) tracts of preexisting messenger RNA, may be incorporated into the CCA-OH termini of preexisting transfer RNA, or may be covalently attached to protein as mono- and oligo(ADP) ribose residues as well.[14] Even in the case of [^{3}H]uridine, we have found that it may be incorporated into high molecular weight components other than RNA in 1-cell mouse embryos, apparently as a result of covalent attachment to protein (on the basis of sensitivity to protease and banding with the protein band in a CsCl gradient; Ref. 2 and K. B. Clegg and L. Pikó, unpublished results).

An overall fractionation scheme that we have found satisfactory[10,11,15] is illustrated in Fig. 2, and the procedures used are described briefly below. Whenever possible, solutions used in the handling of RNA (other than Tris buffers[16]) should be rendered RNase free by treatment with 0.1% (v/v) diethyl pyrocarbonate followed by autoclaving, and routine precautions to prevent RNase contamination must be taken (see Ref. 16).

LiCl Precipitation of High Molecular Weight RNA. Labeled embryos are washed extensively and transferred to 300 μl RNA extraction buffer containing 20 μg each of unlabeled mouse liver ribosomal RNA (isolated from a microsomal fraction[13]) and yeast-soluble (4 and 5 S) RNA as

[13] L. Pikó and K. B. Clegg, *Dev. Biol.* **89**, 362 (1982).
[14] R. J. Young and K. Sweeney, *Biochemistry* **17**, 1901 (1978).
[15] K. B. Clegg and L. Pikó, *Dev. Biol.* **95**, 331 (1983).
[16] J. Sambrook, E. F. Fritsch, and T. Maniatis, "Molecular Cloning: A Laboratory Manual," 2nd Ed. Cold Spring Harbor Laboratory, Cold Spring Harbor, New York, 1989.

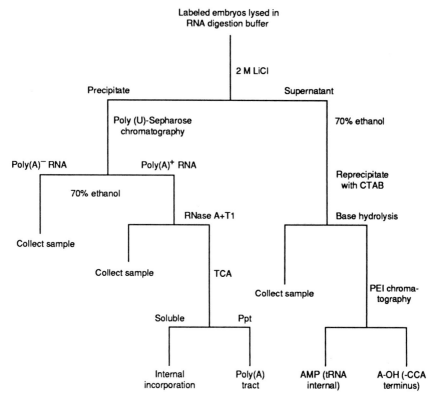

FIG. 2. Flow chart showing the fractionation of RNA extracted from 1-cell mouse embryos cultured in the presence of [³H]adenosine. The incorporation of labeled adenosine is followed into LiCl-precipitable (high molecular weight) poly(A)⁻ RNA; into the poly(A) tract and internal locations of poly(A)⁺ RNA; and into the -CCA terminus and internal locations of tRNA. The samples that are collected intact can be used for analysis by gel electrophoresis. For methodology, see text.

carriers. After digestion for 30 min at 37°, the extract is made 2 *M* in LiCl by the addition of an equal volume of 4 *M* LiCl and precipitated at 0°–4° for 18 hr to separate high molecular weight RNAs (that are precipitated by LiCl) from the 4–5 S soluble RNAs. The precipitate is collected by centrifugation at 20,000 *g* for 20 min at 4° and washed 3 times with ice-cold 2 *M* LiCl. The LiCl-soluble supernatants are pooled and the 4–5 S RNAs precipitated by the addition of 2.5 volumes ethanol at − 20°.[10,11]

Poly(U)-Sepharose Chromatography. The high molecular weight components containing ribosomal RNA, poly(A)⁺ and poly(A)⁻ messenger

RNA, and heterogeneous nuclear RNA are further fractionated by chromatography on poly(U)-Sepharose using a scaled-down batch procedure.[15,17] The LiCl-precipitated RNA is dissolved in 300 μl binding buffer and mixed with poly(U)-Sepharose (3–6 mg dry weight; freshly rehydrated in binding buffer) in a sterile conical microcentrifuge tube at 37°. (The rehydrated Sepharose occupies a volume of about 4 μl per mg dry weight.) The suspension is gently vortexed every 1–2 min for a period of 15 min, and the Sepharose beads are separated from the supernatant by centrifugation for 30 sec. The supernatant is carefully removed with a micropipette and the beads washed 3–4 times with 500 μl binding buffer/centrifugation cycles. The washes are pooled and the poly(A)$^-$ RNA precipitated with 2.5 volumes ethanol at $-20°$. The bound poly(A)$^+$ RNA is eluted from the Sepharose by three 300 μl washes/centrifugations with elution buffer at 45°, pooled, and precipitated with ethanol after the addition of 10 μg ribosomal RNA and one-fourth volume binding buffer. For an analysis of the distribution of label within the molecule, the poly(A)$^+$ RNA is redissolved in 200 μl of 10 μg/ml RNase A and 5 U/ml RNase T1 in RNase digestion buffer and incubated at 37° for 1 hr. Under these conditions poly(A) tracts are undigested and remain acid precipitable while nucleotides incorporated into internal (transcriptionally derived) sequences are rendered acid soluble. The digestion products can be analyzed by PEI chromatography.[15] Alternatively, the sample can be precipitated with ice-cold 10% TCA and the radioactivity in the precipitate and the soluble fraction determined by scintillation counting.[11]

CTAB Precipitation of Low Molecular Weight RNA. The low molecular weight LiCl-soluble RNAs are further purified to remove traces of contaminating nucleoside triphosphates precipitated by ethanol. The 4–5 S RNAs are collected by centrifugation and dissolved in 0.1 M sodium acetate, pH 5.0. The RNA is reprecipitated at room temperature for 30 min by the addition of one-tenth volume of 5% CTAB to 0.5% final concentration.[10,18] The CTAB precipitate is collected by centrifugation in a microcentrifuge and washed several times with 0.1% CTAB in 0.1 M sodium acetate, pH 5.0, to remove the soluble nucleotides. The CTAB counterion is removed from the RNA by exchange with sodium ions. This is accomplished by washing the CTAB–RNA precipitate 2–3 times with 70% ethanol/0.1 M sodium acetate, pH 5.0. Under these conditions the CTAB is removed while the 4–5 S RNA remains as an ethanol-insoluble precipitate. All the ethanol precipitates prepared above can be stored at $-20°$ until analyzed qualitatively or quantitatively.

[17] K. D. Taylor and L. Pikó, *Development (Cambridge, UK)* **101,** 877 (1987).
[18] A. R. Bellamy and R. K. Ralph, this series, Vol. 12B, p. 156.

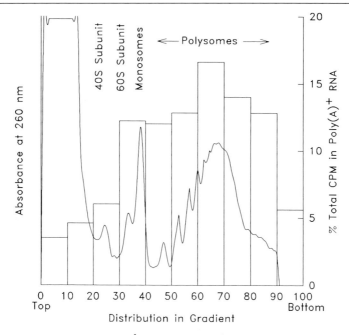

FIG. 3. Polysomal association of ³H-labeled poly(A)⁺ RNA from 8-cell mouse embryos. About 160 8-cell embryos were cultured *in vitro* for 5 hr in the presence of 50 μM [³H]adenosine (37 Ci/mmol). At the end of culture the embryos were washed free of label with PLM containing 20 μg/ml emetine hydrochloride and homogenized together with 25 mg freshly dissected mouse liver. After centrifugation to pellet nuclei and mitochondria, one-half of the homogenate (shown here) was sedimented on a 15–50% (w/v) sucrose gradient as described in the text; the other half of the homogenate was brought to 50 mM in EDTA and sedimented in the same way except that the MgCl₂ in the gradient was replaced with 5 mM EDTA. The continuous line is the A_{260} profile of the mouse liver carrier fractions read through a 2-mm Gilford flow cell (full scale = 0.2 OD reading). The gradient was collected from the bottom of the tube in 10 fractions, and the poly(A)⁺ RNA from each fraction was isolated by poly(U)-Sepharose chromatography. The bars represent the percentage of counts in each fraction in relation to the total (= 1860 cpm) recovered from the gradient as poly(A)⁺ RNA. Both the polysomal profile and the ³H radioactivity in the polysomal region were completely abolished by EDTA treatment (result not shown).

Gel Electrophoresis of RNA Fractions. For a qualitative analysis of labeled RNA, the RNA fractions are recovered by centrifugation, washed with cold 70% ethanol then 100% ethanol, and air dried. The RNA samples can then be run in polyacrylamide gels (4 and 5 S RNAs) or denaturing agarose gels (large RNAs). ¹⁴C-Labeled marker RNAs can be added to the sample, or unlabeled marker RNAs (e.g., RNA ladder, GIBCO-BRL, Gaithersburg, MD) can be run in a parallel lane and visualized by ethidium bromide staining.[19] The lanes are cut into about 1.5 mm slices and the radioactivity in the samples determined by scintillation counting.[10,15]

Polysomal RNA Fractions. The association of labeled poly(A)$^+$ RNA with polysomes is of interest for the study of the utilization of newly synthesized messenger RNA. We have found the procedure described below useful for the isolation of polysomal fractions from early mouse embryos since it appears to preserve the integrity of the polysomes while keeping nonspecific aggregation to a minimum.

Embryos are washed free of label with PLM containing 20 μg/ml emetine hydrochloride (Sigma) (which stops the elongation process on polysomes and prevents runoff during isolation) and transferred to a 0.5 ml glass homogenizer fitted with a Teflon pestle (Kontes, Vineland, NJ) containing about 25 mg freshly dissected mouse liver in 200 μl polysome isolation buffer. All solutions are kept in an ice bath, and subsequent treatments are carried out at 0°–4°. The embryos and liver are homogenized by 10 strokes of the hand-held homogenizer. The homogenate is transferred to a microcentrifuge tube, the homogenizer is rinsed with 200 μl polysome buffer, and the combined homogenate is centrifuged at 12,000 g for 10 min to remove nuclei and mitochondria. The supernatant is removed, adjusted to 1% in sodium deoxycholate (from a 10% stock), and immediately layered over a 3.5-ml 15–50% (w/v) sucrose gradient (in 100 mM NaCl, 50 mM Tris-HCl, pH 7.5, 10 mM MgCl$_2$) over a 0.2-ml 60% sucrose cushion in a 4.5 ml sterile polyallomer tube. The gradient is centrifuged for 70 min at 56,000 rpm, 4°, in a Beckman SW60 rotor. The gradient is drained from the bottom through a Gilford flow cell to monitor the absorbance profile at 260 nm. Fractions are collected directly in tubes containing 100 μl RNA extraction buffer for further processing and isolation of poly(A)$^+$ and poly(A)$^-$ RNA as described above. Figure 3 shows the A_{260} absorbance of a polysome gradient and the corresponding distribution of ^{3}H-labeled poly(A)$^+$ RNA from 8-cell mouse embryos. In this experiment about 75% of the total ^{3}H-labeled poly(A)$^+$ RNA migrated with the polysomal (postmonosomal) fractions.

Acknowledgments

This work has been supported by the Medical Research Service of the Veterans Health Administration and Public Health Service Research Grant No. HD19691 from the National Institute of Child Health and Human Development.

[19] Z. Gong, *BioTechniques* **12**, 74 (1992).

[16] Quantitative Analysis of Specific Messenger RNAs by Hybridization Using RNA Probes

By KENT D. TAYLOR and LAJOS PIKÓ

Introduction

Knowledge of the steady-state amounts of specific mRNAs in the course of oogenesis and early embryogenesis can provide insight into the developmental pattern of expression of a particular gene at the level of its primary transcript. The issues that one can attempt to address include the following: the time course of accumulation of the mRNA during oocyte growth; the amount of maternal mRNA stored in the fully grown oocyte and persisting during meiotic maturation and fertilization; and the time of degradation of maternal mRNA and accumulation of the mRNA during cleavage as a result of new synthesis by the embryonic genome. The quantitative determinations can be combined with preliminary subcellular and molecular fractionation [e.g., into nuclear and cytoplasmic fractions; mitochondrial and postmitochondrial fractions; polysomal and nonpolysomal fractions; poly(A)$^+$ and poly(A)$^-$ RNA]. If quantitative data on the synthesis and accumulation of the protein product are available, the overall rate of utilization and translational efficiency of the mRNA can be estimated.

We have found that dot-blot or slot-blot hybridization using single-stranded RNA probes can be used for the quantitative assay of small amounts of mRNA obtainable from 50–100 embryos, even in the case of rare class transcripts (500–1000 transcripts per egg). In the following we describe the experimental approach and the methods involved as we have used them in mouse oocytes and early embryos. For general techniques of molecular biology (including preparation of RNase-free glassware and solutions), the reader is referred to the standard textbook of Sambrook *et al.*[1]

Collection of Embryos

Oocytes and embryos can be recovered, cleaned of adhering follicle cells, and if necessary cultured *in vitro* as described.[2,3] After a rinse in a

---

[1] J. Sambrook, E. F. Fritsch, and T. Maniatis, "Molecular Cloning: A Laboratory Manual," 2nd Ed. Cold Spring Harbor Laboratory, Cold Spring Harbor, New York, 1989.

[2] L. Pikó and K. B. Clegg, *Dev. Biol.* **89,** 362 (1982).

[3] L. Pikó and K. D. Taylor, *Dev. Biol.* **123,** 364 (1987).

Copyright © 1993 by Academic Press, Inc.
All rights of reproduction in any form reserved.

buffered saline solution (e.g., 145 mM NaCl, 25 mM HEPES, pH 7.3), the eggs are transferred with a fine-bore pipette into 100–300 μl NTE buffer (100 mM NaCl, 50 mM Tris-HCl, 5 mM EDTA, pH 7.5) containing 0.5% sodium dodecyl sulfate (SDS) and 0.5 mg/ml pronase (nuclease-free; Calbiochem, La Jolla, CA) or proteinase K, allowed to lyse at room temperature for 15–30 min, and stored at $-20°$ until use. The egg lysate can be thawed and more eggs added until the desired number is reached.

Isolation of Embryo RNA

Before extraction of the RNA, the embryo lysate is thawed at 37° for 30 min. Radioactively labeled RNA should be added to the lysate at this point to monitor the recovery of the RNA during the extraction procedure and for the eventual quantitation of embryo RNA present in the samples used for hybridization. We generally use ^{3}H-labeled λ cRNA as a marker, because of its long shelf-life (up to 2 years, with about 30% of ^{3}H disintegrations resulting in a single-strand break[4]), the ready availability of pure λ DNA template, and because it does not seem to interfere with the hybridization reaction. About 10 ng λ cRNA with a specific activity of 2–3 $\times$ 10^4 disintegrations/min (dpm)/ng is sufficient even for relatively large batches of embryo extracts (about 2000 embryos per batch).

Synthesis of λ RNA Marker

We prepare λ cRNA essentially according to the procedures of Gall and Pardue.[5] The reaction is carried out in 100 μl assay buffer (0.15 M KCl, 40 mM Tris-HCl, pH 8, 10 mM MgCl$_2$, 2 mM MnCl$_2$, 70 μM EDTA, 10 mM 2-mercaptoethanol) containing 5 μg λ DNA (from bacteriophage λ strain cI857 Sam7; Promega, Madison, WI), 20 nmol each of rATP, rCTP, and rGTP, 5 nmol [^{3}H]rUTP [10–20 Ci/mmol; dried in the tube in a SpeedVac centrifuge (Savant Instruments, Farmingdale, NY) before the reaction mixture is added], and 4 units of *Escherichia coli* RNA polymerase (from a stock solution of 1000 U/ml made up in 50% glycerol, 100 mM NaCl, 10 mM Tris-HCl, pH 8, 0.1 mM EDTA, 0.1 mM dithiothreitol and kept at $-20°$). After incubation at 37° for 1–2 hr, RNase-free DNase (DPRF, Worthington, Freehold, NJ; stock solution of 1 mg/ml in distilled water kept at $-20°$) is added to 20 μg/ml, and incubation continues at 37° for 30 min to digest the DNA. The reaction mixture is brought to 0.1% in SDS, 100 μg/ml in proteinase K, then incubated at 37° for 30 min. After extraction with phenol–chloroform, chloroform, and ether, the RNA in

[4] P. N. Rosenthal and M. S. Fox, *J. Mol. Biol.* **54,** 441 (1970).
[5] J. G. Gall and M. L. Pardue, this series, Vol. 21, p. 470.

the water phase is recovered and unincorporated [³H]UTP removed by passage through a Nensorb-20 column (Du Pont–NEN; see below). (It is important to remove ether from the sample because it interferes with the binding of RNA to the column.) The eluate [in 20% (v/v) ethanol] is dried, redissolved in autoclaved distilled water, and stored at −70°. The final yield is 2–3 μg cRNA consisting of a heterogeneous population of molecules predominantly from 300 to 2000 nucleotides in length as determined by sucrose gradient centrifugation and migration on a formaldehyde–agarose gel (K. D. Taylor and L. Pikó, unpublished results).

Extraction and Purification of Embryo RNA

The embryo lysate containing the marker RNA is extracted with phenol–chloroform, chloroform, and several changes of water-saturated ether and the RNA isolated by chromatography on a Nensorb-20 column as described[6] (see also manufacturer's protocol). The addition of carrier RNA is not necessary in this procedure. Before loading the sample on the column, all traces of ether have to be removed, for example, by blowing a stream of nitrogen gas over the tube for 15–30 min; incubating the sample in a 68° water bath for 15–30 min; or evaporating the sample to about one-half volume in a SpeedVac centrifuge. The sample is then mixed with an equal volume of buffer A (0.1 *M* Tris-HCl, 10 m*M* triethylamine, 1 m*M* EDTA, pH 7.7) and loaded on the Nensorb column. The column is washed with 3 ml of buffer A, followed by two washes with 3 ml each of TE buffer (10 m*M* Tris-HCl, 0.1 m*M* EDTA, pH 7.5), and the nucleic acids are eluted with 0.5 ml of 20% ethanol. One has to be careful not to let the column run dry during these procedures. The eluate is dried in a siliconized, autoclaved microcentrifuge tube in a SpeedVac centrifuge under reduced pressure, redissolved in diethyl pyrocarbonate (DEPC)-treated[1] and autoclaved distilled water, and kept at −70° until use.

To avoid repeated freeze–thawing and possible loss of RNA owing to accidental contamination, it is advisable to divide large batches of RNA preparations into several aliquots. Recovery of the RNA by this method is generally better than 80%. It should be noted that the void volume of the column is about 170 μl, and thus a corresponding amount of TE buffer is carried over into the final sample. If it is necessary to digest the DNA, the sample is adjusted to 5 m*M* MgCl₂ and 20 μg/ml RNase-free DNase (DPRF, Worthington) and incubated at 37° for 15 min. Proteinase K is then added to 50 μg/ml, and the solution is incubated for an additional 15 min. The sample is extracted with an equal volume of chloroform, followed by ether, and the water phase is stored at −70°.

[6] K. D. Taylor and L. Pikó, *Development (Cambridge, UK)* **101,** 877 (1987).

For use in hybridization experiments and Northern blots, a suitable RNA preparation can also be isolated by ethanol precipitation of the water phase obtained after phenol–chloroform extraction of the embryo lysate in the presence of 30–50 μg/ml *E. coli* tRNA carrier. If the tRNA is added after the phenol–chloroform extraction, one has to make sure that it is free of RNase activity.

Synthesis of RNA Probe and RNA Standard

Subcloning of Template DNA

For the *in vitro* synthesis of RNA standards and radiolabeled RNA probes, the DNA of interest (cDNA or the appropriate sequence of genomic clones) can be conveniently cloned into commercially available plasmid vectors in which the polycloning site is bracketed by two different bacteriophage promoters arranged in the opposite orientation, such as SP6 and T7 (e.g., pGEM vectors, Promega) or T3 and T7 (e.g., Bluescript II and pBS vectors, Stratagene, La Jolla, CA). This arrangement allows synthesis of standard (sense) RNA using one promoter and the synthesis of the complementary probe using the other.

The factors that one should consider when subcloning the DNA into the vectors include the following: to reduce nonspecific binding of the probe, it is desirable to exclude long poly(A) stretches or other polynucleotide runs; longer DNA inserts generally increase the sensitivity of hybridization, thus requiring less RNA for assay; and assay of transcripts encoding specific protein isoforms generally requires probes complementary to the 5' or 3' untranslated segment of the mRNA (e.g., Ref. 7). For dot hybridization experiments, it is also highly desirable that the probe be completely homologous to the RNA assayed, so that high stringency hybridization can be used and possible nonspecific reactions can be eliminated by RNase treatment of the hybrids (see below). If only nonhomologous (nonspecies-specific) probe is available, Northern blot hybridization with total RNA should be done to check whether the probe reacts to any significant extent with sequences other than the mRNA assayed. If the nonspecific reaction is not eliminated by increasing the stringency of hybridization, quantitative dot hybridization cannot be done, and the mRNA should be assayed instead by Northern blot hybridization (which, although less sensitive and less quantitative, allows the specificity of the reaction to be ascertained more readily).

[7] K. D. Taylor and L. Pikó, *Mol. Reprod. Dev.* **16,** 111 (1990).

Synthesis of Probe RNA Strand

To be used for RNA synthesis, the recombinant plasmid clone is linearized on the side of the insert opposite to the promoter used. The use of restriction enzymes leaving 3′ protruding ends (e.g., *Pst*I and *Sph*I) should be avoided because these termini may initiate the production of long extraneous transcripts.[8] The digested DNA is extracted with phenol–chloroform, ethanol precipitated, and redissolved in TE buffer. The optimal conditions for *in vitro* transcription have been described[9] (see also Ref. 1), and a protocol is provided by the supplier of the RNA polymerase used. As an example, for the synthesis of high specific activity complementary RNA probes we use about 0.2 pmol linear DNA template (containing about 40 ng double-stranded insert DNA 300 bp in length), 200 μCi of [α-^{32}P]rUTP (1000 Ci/mmol, preferably no more than 2 weeks old), and 10 units of RNA polymerase in a 10-μl reaction volume, which is incubated for 1 hr at 37° (T7 RNA polymerase) or 40° (SP6). The yield is 100–200 ng probe RNA with a specific activity of about 1.7 × 10^6 cpm/ng. Template DNA is digested by incubation at 37° for 15 min with 20 μg/ml RNase-free pancreatic DNase I (DPRF, Worthington; we have found this enzyme the most consistently reliable). After the addition of distilled water to 100 μl, the sample is extracted with chloroform and ether and, after evaporation of the ether, passed through a Nensorb-20 column (see above) to remove unincorporated ribonucleotides from the RNA. We have found that removal of protein from the probe sample results in lower background with the GeneScreen Plus filter. Alternatively, the RNA can be purified by repeated ethanol precipitation from an ammonium acetate solution or by centrifugation through a Sephadex G-50 spun column (see Ref. 1).

The specific activity of the probe is calculated from its base composition and the specific activity of the labeled ribonucleoside triphosphate (rNTP) based on the information provided by the supplier. High specific activity probes labeled with ^{32}P should be used for hybridization within a few days to a week since each radioactive decay results in a single-strand break in the probe RNA,[4] and so the size of the probe decreases over time. The size of a ^{32}P-labeled single-stranded probe at a given time after its synthesis can be calculated from Eq. (1)[10]:

$$L_t = L_0/1 + N_x(1 - e^{-kt}) \qquad (1)$$

[8] E. T. Schenborn and R. C. Mierendorf, Jr., *Nucleic Acids Res.* **13**, 6223 (1985).

[9] D. A. Melton, P. A. Krieg, M. R. Rebagliati, T. Maniatis, K. Zinn, and M. R. Green, *Nucleic Acids Res.* **12**, 7035 (1984).

[10] C. P. Hodgson, R. Z. Fisk, and L. B. Willett, *BioTechniques* **6**, 208 (1988).

where L_t is the number-average strand length at time t, L_0 is the initial number-average length, N_x is the total number of labeled loci per initial number-average length, k is the decay constant for ^{32}P (0.0487/day), and t is the time in days. N_x can be calculated for a particular labeled base by multiplying the total number of that base in the probe by the specific activity of the base used in the synthesis reaction divided by the theoretical maximum specific activity for ^{32}P, 9131 Ci/mmol. For example, if a 500-base probe ($L_0 = 500$) contains 125 U's and is synthesized using a $[^{32}P]rUTP$ precursor with a specific activity of 1000 Ci/mmol, then $N_x = 125 \times 1000/9131$, and L_7, the number-average length of this probe after 7 days, would be 101 bases.

Synthesis of Sense RNA Standard

The synthesis of single-stranded sense RNA standard is carried out the same way as probe synthesis except that the reaction volume is scaled up to 100 μl, with a corresponding increase in the amounts of template DNA and RNA polymerase used. A total of 5 μCi of a tritiated rNTP (specific activity 1.0 Ci/mmol) is added to the reaction mixture. The expected yield is several micrograms of mostly full-length RNA, which is purified by chromatography on a Nensorb-20 column as above and kept at $-70°$ until use. We routinely check a sample (about 100 ng) of standard RNA for size and purity by electrophoresis on a 0.66 M formaldehyde–agarose gel and ethidium bromide staining. By adding the ethidium bromide (1 μl of a 5 mg/ml stock solution) directly to the sample to be loaded on the gel, the staining of the gel after electrophoresis is eliminated, and the sensitivity of RNA detection is greatly enhanced.[11,12] In our experience about 5 ng RNA in a band can be detected by this procedure. The quantity of standard RNA is determined from the base composition of the RNA and the amount of ^3H radioactivity incorporated (measured by scintillation counting of a sample of purified RNA) according to Eq. (2):

$$(\text{Counts min}^{-1} \text{ incorporated}/\text{counts min}^{-1} \text{ mol}^{-1} [^3H]rNTP) \times (100/\%N)$$
$$= \text{moles total nucleotides incorporated} \quad (2)$$

where $\%N$ is the mole percent of the labeled base in the RNA. Multiplying the above value with an average molecular weight of 330 per nucleotide, one obtains the grams RNA in the sample. The mass of RNA can also be estimated on the basis of its ultraviolet absorbance (assuming that the RNA is free of unincorporated rNTPs), using a value of 40 μg/ml per OD_{260} unit. We prefer to use the RNA concentrations derived from ^3H

[11] R. M. Fourney, J. Miyakoshi, R. S. Day III, and M. C. Paterson, *Focus* **10**, 5 (1988).
[12] Z. Gong, *BioTechniques* **12**, 74 (1992).

radioactivity and routinely monitor the standard RNA preparations used in hybridization experiments by scintillation counting.

Slot Hybridization Assay of RNA

In dot or slot hybridization experiments, the denatured RNA to be assayed is deposited on nitrocellulose or nylon filters and incubated with a hybridization solution containing the labeled probe. Slots are generally preferable to dots because they lend themselves more readily to quantitation by densitometric tracing. The general procedures for these experiments have been discussed.[1] Here we describe the methodology that we have used with good results, with an aim to increase the specificity and sensitivity of the assay and obtain reliable quantitation of small amounts of RNA.

Transfer and Immobilization of RNA to Filters

Several filtration devices are commercially available for loading the RNA samples on the filter. We prefer the Hybri-Slot apparatus manufactured by Bethesda Research Laboratories (Gaithersburg, MD) because its small slot size (4 × 0.8 mm) reduces the amount of RNA needed. Also, we have found that devices which use rubber gaskets tend to leave background marks on the filter. Before use, the apparatus is washed thoroughly in a detergent solution, soaked in a 0.1% solution of DEPC for 2 hr, and rinsed with autoclaved distilled water. It is advisable to repeat this treatment at least once a month. Between experiments, the apparatus is rinsed with autoclaved distilled water, blotted with autoclaved filter papers, and stored separately in an autoclaved metal tray.

One problem that may arise with the Hybri-Slot and other devices is that the samples leak on the membrane around the slots. Leakage can usually be eliminated by placing a piece of autoclaved Whatman (Clifton, NJ) No. 1 filter paper (cut to the size of the filter and wetted with 20× SSC 1 × SSC is 0.15 M NaCl, 15 mM sodium citrate) on the lower plexiglass block. A piece of GeneScreen Plus filter (Du Pont–NEN) that had been cut to the appropriate size is wetted with water and soaked in 20× SSC for 10 min. (We use GeneScreen Plus filters directly from the box but are careful in handling them and use an autoclaved ruler and blade to cut out the pieces.) The filter is then placed carefully over the Whatman No. 1 paper (avoiding air bubbles) with a pair of autoclaved forceps, and the apparatus is assembled and finger tightened over the filters. A mild suction is applied to the apparatus through a side-arm Erlenmeyer flask connected to a water aspirator. We prefer to use a charged nylon filter such as GeneScreen Plus because of its resistance to handling and because, when combined with UV irradiation (see below), it retains 60–80% of the RNA dotted (about 80 bases in length and longer) on completion of the hybridization procedure (but without RNase treatment of the filter). Although nitrocellulose filters tend to have fewer background problems, in our experience they retain less than 20% of the RNA.

Standard and embryo RNA samples to be dotted are denatured by heating in denaturation buffer containing 25 mM sodium phosphate, pH 7.0, 5 mM EDTA, 2.2 M formaldehyde, 50% formamide (deionized; Ref. 1) at 65° for 10 min and allowed to cool at room temperature. Because the embryo RNA contains ^{3}H-labeled λ cRNA recovery marker and the standard RNA is synthesized in the presence of a [^{3}H]rNTP (see above), an aliquot of the denaturation mixture is counted for determining the amount of RNA dotted. The denatured standard RNA sample is diluted further with denaturation buffer as necessary to prepare a 1 : 1 dilution series spanning the desired range for the experiment. Just before dotting, each denatured sample, generally 10–20 µl in volume, is mixed with 200 µl of 20 × SSC (at room temperature), vortexed briefly, and pipetted into the upper well of the assembled dotting apparatus while mild suction is being applied to give a flow-through rate of approximately 100 µl/min. Care should be taken to avoid trapping air bubbles at the bottom of the well when transferring the sample.

After dotting, the filter is removed from the apparatus, placed on an autoclaved Whatman No. 1 filter, and while moist the side of the membrane onto which the RNA was dotted is irradiated under a germicidal lamp (254 nm) at a distance of 15 cm for 2 min to UV cross-link the RNA to the filter.[13] The energy of irradiation at the level of the filter is about 720 µW/cm². We have determined empirically that this level of irradiation enhances the retention of RNA on GeneScreen Plus filters while at the same time allowing the maximum hybridization signal. The filter is then covered with two sheets of autoclaved Whatman 3MM paper to act as a weight and baked at 80° for 1 hr in a vacuum oven. If not used immediately for hybridization, the filter is kept at room temperature under vacuum until use.

Hybridization Reaction

Prehybridization and hybridization reactions with the filters are carried out in plastic sealer bags which are placed directly in a water bath at the appropriate temperature. We use polyethylene bags (4 mil, BelArt,

[13] G. M. Church and W. Gilbert, *Proc. Natl. Acad. Sci. U.S.A.* **81,** 1991 (1984).

purchased from Thomas Scientific, Swedesboro, NJ) taken directly from the package and sealed with a heat sealer. One should remove most of the air bubbles before sealing the bags. The filters are prehybridized for 2 hr at 65° in 1 M NaCl, 50 mM sodium phosphate, pH 7, 10 mM EDTA, 0.1% SDS, 10× Denhardt's solution (prepared according to Ref. 1); the bag is agitated from time to time, to make sure that the filter is completely wetted. This solution is exchanged with hybridization buffer (50–100 μl/cm^2 of filter) containing 50% (v/v) formamide, 1 M NaCl, 50 mM sodium phosphate, pH 7, 10 mM EDTA, 1% SDS, 5× Denhardt's solution, 200 μg/ml sonicated denatured herring sperm DNA, 20 μg/ml poly(A), and 50 μg/ml yeast tRNA, and the bag is resealed and incubated for 1 hr more at 65°. The ^{32}P-labeled RNA probe is then introduced into the bag to a concentration of 5–10 ng/ml and allowed to hybridize for 16–18 hr at 65–68°. To remove possible contaminating RNase activity, we treat the stock solutions of herring sperm DNA, poly(A), and tRNA with 200 μg/ml proteinase K for 1 hr at 37° followed by extraction with phenol–chloroform, chloroform, and water-saturated ether.

The above conditions work well for hybridization with RNA probes 100 bases in length or longer. They appear to result in saturation of the available sites, since increasing the concentration of the probe and/or the time of hybridization does not increase the hybridization signal but may result in higher background radioactivity. These conditions are similar to those found to be optimal for *in situ* hybridization using single-stranded RNA probes.[14] The factors affecting the rate of hybridization of single-stranded probes to target nucleic acids immobilized on solid supports have been discussed.[15] The optimum rate of hybridization occurs about 25° below the melting temperature (T_m) of the RNA–RNA hybrids formed, which can be derived from Eq. (3)[16] (see also Ref. 1):

$$T_m = 79.8 + 18.5(\log M) + 58.4(G + C) + 11.8(G + C)^2 - 0.35(\%\text{formamide}) - (820/l) \quad (3)$$

where M is the molarity of the monovalent cation (Na$^+$), (G + C) is expressed as a fraction, and l is the average length of the probe (number of bases).

On completion of hybridization, the filters are rinsed with wash A containing 1 M NaCl, 50 mM sodium phosphate, pH 7, 4 mM EDTA, 1% SDS. The filters may be treated at this time with 13–130 μg/ml RNase A and 6–60 U/ml RNase T1 (Boehringer-Mannheim, Indianapolis, IN) in 40

[14] K. H. Cox, D. V. DeLeon, L. M. Angerer, and R. C. Angerer, *Dev. Biol.* **101**, 485 (1984).
[15] J. Meinkoth and G. Wahl, *Anal. Biochem.* **138**, 267 (1984).
[16] D. K. Bodkin and D. L. Knudson, *Virology* **143**, 55 (1985).

ml of 0.375 M NaCl, 75 mM Tris-HCl, pH 7.5, for 5 min at room temperature, followed by two washes with wash A. The filters are then washed
for 2 hr at 68° in several changes of 250 ml each of wash B (25 mM sodium
phosphate, pH 7, 4 mM EDTA, 1% SDS) in a wide-mouth flask with
vigorous shaking (e.g., in a reciprocating water bath shaker; New Brunswick). After washing the filters are dried under a heat lamp and set up
for autoradiography at −70° using preflashed[17] X-ray film (Du Pont Cronex
No. 4 or Kodak, Rochester, NY, XAR-2) with intensifying screen (e.g., Du
Pont Cronex Lightning Plus). For a general discussion of autoradiographic
procedures, see Ref. 1).

Quantitative Evaluation of Hybridization Results

Quantitation of the hybridization reaction is usually through densitometric tracings of the slot images on the autoradiographic negatives. As
a rule, the experimental design is to dot a 1 : 1 dilution series of the standard
RNA on the same filter as the experimental RNA samples (Fig. 1B).
Exposures for various lengths of time may be necessary to enhance weak
slots and to bring the density of the experimental slots within the linear
range of the standard slots. If a small aperture is used for tracing (0.1 mm
in diameter or less), it is advisable to scan the slots in several different
areas because unevenness in the slots (e.g., due to small air bubbles during
dotting) may result in a false reading. From the densitometric readings of
the standard slots, a best-fit line is drawn by the least-squares method
and used to determine the number of RNA molecules hybridized in the
experimental slots. The mass of RNA in the standard slots is calculated
from the amount of ^{3}H radioactivity deposited on the filter and is converted
to number of RNA molecules according to the formula (grams RNA/
molecular weight RNA) (Avogadro's number) = number of molecules.
In calculating the number of mRNA molecules in the experimental sample,
a correction is applied if the probe length hybridized to the standard
RNA is greater (because of the presence of flanking genomic or vector
sequences) than that hybridized to the mRNA. When the amount of hybridization is sufficiently high, the slots can be excised and the ^{32}P radioactivity
(above the ^{3}H window) bound per slot determined by scintillation counting.
In our experience, the results obtained by direct counting and densitometry are in good agreement.

Using the above procedures, we find that approximately 10 fg standard
RNA in a slot can be reliably detected[7,18] (see Fig. 1), although it may
require that autoradiography be extended for up to 1 week. This means

[17] R. A. Laskey, this series, Vol. 65, p. 363.
[18] K. D. Taylor and L. Pikó, *Mol. Reprod. Dev.* **28**, 319 (1991).

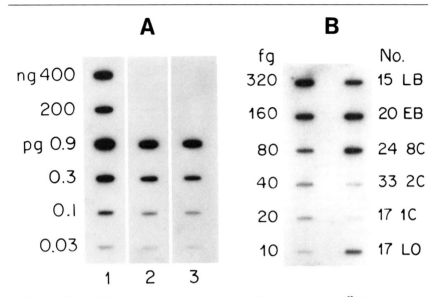

FIG. 1. Effect of RNase treatment on the slot hybridization reaction of ^{32}P-labeled single-stranded RNA probes (specific activity about 1.6×10^6 dpm/ng) with RNA immobilized on GeneScreen Plus filters. (A) Autoradiograph showing the reaction of a cardiac α-actin probe with total RNA from mouse plasmacytoma (nonspecific reaction, top two dots) and cardiac α-actin standard RNA (specific reaction, bottom three dots). Filter 1 is untreated, filter 2 was digested with 13 μg/ml RNase A plus 6 U/ml RNase T1, and filter 3 was digested with 130 μg/ml RNase A plus 60 U/ml RNase T1, for 5 min at 25° before washing of the filters and autoradiography for 2 days. (Reproduced from Ref. 7.) (B) Assay of mouse ribosomal protein S15 (formerly *rig*) mRNA in mouse embryo RNA using a complementary RNA probe. *Left*: A 1:1 dilution series of standard RNA. *Right*: Total RNA extracted from large oocytes (LO) and embryos at different stages of development from the 1-cell to late blastocyst (LB) stage. After hybridization, the filter was treated with 130 μg/ml RNase A and 60 U/ml RNase T1 as above and autoradiographed for 4 days. (Reproduced from Taylor and Pikó,[18] with permission from Wiley–Liss.)

that with an RNA probe 300 bases in length (about 6×10^3 molecules/fg), approximately 6×10^4 mRNA molecules can be assayed. Although this degree of sensitivity enables one to detect a rare class mRNA present in as few as 600 copies per egg in total RNA of about 35 ng from 100 eggs,[2] one has to ascertain that nonspecific reaction of the probe is reduced to 10^{-7} or less. This can be accomplished by treating the hybridized filter, after an initial rinse, with a mixture of RNase A and T1 as indicated above. Even at the lower dose range, nonspecific reaction is virtually abolished while the reaction of the probe with the complementary RNA slots is reduced only by about one-half (Fig. 1A). Because perfectly matched RNA–RNA hybrids are insensitive to RNase treatment under these conditions, the reason for the reduction in the standard RNA reaction may be that the RNase treatment promotes the loss of RNA from the filter during the wash procedure; however, the loss appears to be uniform and thus is unlikely to interfere with quantitation of the hybridization reaction.

Acknowledgments

This work has been supported by the Medical Research Service of the Veterans Health Administration and Public Health Service Research Grant No. HD19691 from the National Institute of Child Health and Human Development.

[17] Quantitative Analysis of Specific Messenger RNAs by Ribonuclease Protection

By Ross A. Kinloch, Richard J. Roller, and Paul M. Wassarman

Introduction

Analysis of gene expression during mammalian development is often hampered by difficulties in obtaining large quantities of RNA from tissues of interest. Although this problem can be overcome using techniques such as the polymerase chain reaction (PCR) or *in situ* hybridization, such analyses are not always amenable to quantification. Ribonuclease (RNase) protection assays provide an alternative and sensitive method for detecting RNAs ($>10^3$ copies per cell) that can be readily quantitated.

In RNase protection assays, a radiolabeled probe, complementary in part to an RNA species of interest, is mixed with tissue RNA under hybridization conditions that permit annealing of complementary transcripts. Following hybridization, the mixture is treated with a ribonuclease mixture that results in degradation of single-stranded, excess probe. The "protected" hybrid of labeled probe and complementary RNA fragment is resolved by polyacrylamide gel electrophoresis (PAGE). When probe is present in large molar excess over the RNA target, quantification is achieved by estimating the quantity of probe protected by the RNA species of interest from a standard curve. A standard curve is produced by determining amounts of probe protected by varying known amounts of synthetic target transcripts. Protocols describing the RNase protection assay have

Copyright © 1993 by Academic Press, Inc.
All rights of reproduction in any form reserved.

been described elsewhere,[1,2] and kits are available for both synthesis of single-stranded RNA probes[3] and the protection assay itself.[4] Such kits greatly facilitate the entire procedure and allow easy optimization of the assay.

Other uses for RNase protection assays include gene mapping[2,5] and mapping of mutation sites.[6] This chapter concentrates on the detection and quantification of individual RNA transcripts by RNase protection.

Synthesis of Single-Stranded RNA Probes

RNase protection assays depend on the use of ^{32}P-labeled, single-stranded, synthetic RNA transcripts as probes. The labeled probes must contain a region that is complementary to the RNA species of interest. It is most convenient to subclone a genomic fragment, containing at least part of an exon, into one of the commonly available transcription vectors that contain a bacteriophage promoter (e.g., T3, T7, or SP6 promoters). Consequently, antisense RNA probes of high specific activity can be generated using T3, T7, or SP6 polymerase from plasmid templates linearized at the 3' end of the genomic insert.

Probe length is an important consideration. In general, minimizing probe length will increase the chance of synthesizing full-length transcripts, thereby ensuring complete protection of the complementary sequences in target RNA. However, less label is incorporated with shorter probes, thereby reducing the sensitivity of the assay. It is convenient to synthesize probes that are longer than the length of the target sequences so that undigested probe is easily distinguished from protected fragment following PAGE. Considering these points, optimally, probes should be 200–500 nucleotides (nt) in length.

Materials and Sources

Solutions used are listed in order of their appearance in the text. Vectors should contain bacteriophage promoters, for example, the pGEM-

[1] F. M. Ausubel, R. Brent, R. E. Kingston, D. D. Moore, J. G. Seidman, J. A. Smith, and K. Struhl, "Short Protocols in Molecular Biology," p. 154. Wiley (Interscience), New York, 1987.

[2] F. J. Calzone, R. J. Britten, and E. H. Davidson, this series, Vol. 152, p. 611.

[3] Riboprobe Gemini System II, Promega Biotec. In vitro transcription kit, Ambion Inc.

[4] Ribonuclease Protection Assay Kit, Ambion Inc.

[5] R. A. Kinloch, R. J. Roller, C. M. Fimiani, D. A. Wassarman, and P. M. Wassarman, Proc. Natl. Acad. Sci. U.S.A. 85, 6409 (1988).

[6] E. Winter, F. Yanamoto, C. Almoguera, and M. Perucho, Proc. Natl. Acad. Sci. U.S.A. 82, 7575 (1985).

Zf series (Promega Biotec, Madison, WI). *In vitro* transcription systems are available from a number of commercial sources (e.g., Promega Biotec and Ambion Inc., Austin, TX). RNase protection kits are also commercially available (Ambion Inc.).

Reagents

Unlabeled ribonucleotides, 10 mM stock

[^{32}P]UTP, SP6/T7 grade, 20 mCi/ml, approximately 800 Ci/mmol aqueous (Amersham, Arlington Heights, IL)

Transcription buffer (5×) is 200 mM Tris-Cl, pH 7.5, 30 mM MgCl$_2$, 10 mM spermidine, 50 mM NaCl

Dithiothreitol (DTT), 100 mM stock

RNasin (30 U/μl) (Promega Biotec)

SP6, T3, or T7 RNA polymerase (15–20 U/μl) (Promega Biotec and New England Biolabs, Beverly, MA)

RQ1 RNase-free DNase (1 U/μl) (Promega Biotec)

TE-saturated phenol/chloroform is phenol saturated with 10 mM Tris-Cl (pH 7.5), 1 mM EDTA, and chloroform/isoamyl alcohol (25 : 24 : 1, v/v)

Chloroform/isoamyl alcohol (24 : 1, v/v)

Sephadex G-50 size-exclusion columns (Boehringer-Mannheim)

Hybridization buffer is 80% formamide, 40 mM MOPS (pH 6.6), 0.4 M NaCl, 1 mM EDTA

RNase digestion buffer is 10 mM Tris (pH 7.5), 0.3 M NaCl, 5 mM EDTA

RNase A and RNase T1 (Sigma, St. Louis, MO)

Proteinase K and yeast tRNA (Boehringer-Mannheim, Indianapolis, IN)

Gel loading buffer is 80% formamide, 0.1% xylene cyanol, 0.1% bromphenol blue, 2 mM EDTA

6% Polyacrylamide/8 M urea gels are typically 0.75 mm thick by 12 cm wide by 15 cm long

Aquasol (New England Nuclear, Boston, MA)

Radiolabeling of Probes

Procedures for synthesizing high specific activity radiolabeled probes are described in a number of standard methods manuals,[1,7] or are included with commercially available *in vitro* translation kits,[3] and are all quite similar. The following procedure is provided as a guideline. The individual

---

[7] G. M. Wahl, J. L. Meinkoth, and A. R. Kimmel, this series, Vol. 152, p. 572.

reaction components should be added in the order shown, at room temperature, in order to avoid precipitation of spermidine.

RNA probes are synthesized during a 60 min incubation at 37° in a 20-μl reaction volume containing the following: $1\times$ transcription buffer; 10 mM DTT; 24 units RNasin ribonuclease inhibitor; three unlabeled ribonucleotides (each at a final concentration of 500 μM); a fourth ribonucleotide added at 12 μM unlabeled and at 12 μM as a ^{32}P-labeled ribonucleotide (2.5 μl; 800 Ci/mM, 20 mCi/ml); 0.5–1.0 μg linearized template DNA; and 15–20 units bacteriophage RNA polymerase (SP6, T3, or T7). Following the *in vitro* transcription reaction, template DNA is destroyed by adding 1 unit of RQ1 RNase-free DNase and the incubation continued at 37° for an additional 15 min. The sample is extracted first with TE-saturated phenol/chloroform and then with chloroform/isoamyl alcohol, and newly synthesized RNA is separated from unincorporated radionucleotides by chromatography over Sephadex G-50 size-exclusion spin columns. The RNA probe is then recovered by ethanol precipitation. Removal of template DNA and unincorporated radionucleotides is important to ensure a low background and to increase the sensitivity of the RNase protection assay. Following the DNase step, RNA probe can also be recovered by purification following PAGE.[1,2] This alternative has the advantage that full-length transcripts can be isolated free from both prematurely terminated transcripts and unincorporated radionucleotides. By using this procedure, RNA probes will generally have specific activities of 6–9 $\times$ 10^8 counts/min (cpm)/μg.

Synthetic RNA target transcripts are used to construct a standard curve that is necessary to calibrate the assay. The transcripts are prepared using the same reaction conditions as above, except that all four unlabeled ribonucleotides are present at 500 μM and the single radiolabeled ribonucleotide is present at 0.8 μM (0.625 μl; 800 Ci/mM, 20 mCi/ml). Consequently, the amount of target RNA synthesized can be calculated from a knowledge of the percentage incorporation as follows:

Percentage incorporation (P):

$$(\text{cpm}_{\text{incorporated}}/\text{cpm}_{\text{total}}) \times 100 = P$$

Moles [^{32}P]UTP in reaction:

$$(1.25 \times 10^{-3} \text{ mCi}) \times (1 \text{ mmol}/8 \times 10^2 \text{ Ci}) \times (1 \text{ Ci}/10^3 \text{ mCi})$$
$$= 1.56 \times 10^{-3} \text{ nmol}$$

Moles cold UTP in reaction:

$$(5 \times 10^2 \ \mu\text{mol}/10^3 \text{ ml}) \times (20 \ \mu\text{l}) \times (1 \text{ ml}/10^3 \ \mu\text{l}) = 10 \text{ nmol}$$

Total UTP incorporated into reaction:

$$P \times (10 \text{ nmol} + 1.56 \times 10^{-3} \text{ nmol})$$

Nanograms RNA synthesized:

$$(4 \times 330 \times 10^9 \text{ ng})/(10^9 \text{ nmol}) \times (P \times 10 \text{ nmol})$$

The following assumptions are made in the calculation: RNA synthesized contains equimolar amounts of all four ribonucleotides; the average molecular weight of a nucleotide is 330; and the contribution of [^{32}P]UTP is negligible.

Synthetic "recovery targets" are used to estimate the efficiency of the RNA extraction procedure. These targets are sense transcripts, identical in sequence to part of the mRNA species of interest, but they protect a shorter portion of the probe than the authentic mRNA. Generally they can be transcribed from the same plasmid used to synthesize the labeled probe by employing the bacteriophage promoter located at the opposite end of the insert. The targets are prepared as described above except that all four ribonucleotides are present at 500 μM and no labeled ribonucleotide is necessary.

Experimental Design of Protection Assays

Detection and quantification of RNA transcripts can be carried out by RNase protection experiments. In these assays, the concentration of radiolabeled probe must be in large excess over complementary target sequence to ensure that the amount of probe protected from digestion is proportional to the amount of target sequence. Under such conditions, by using varying known amounts of a synthetic sense RNA as target, a standard curve can be constructed that relates the amount of protected probe (cpm) to the amount of target DNA (pg). Figure 1 provides an example of such a calibration, in which varying amounts of MOM-1 ("major oocyte messenger RNA-1") target transcript are protected by a constant amount of MOM-1 antisense probe.[8] When parallel assays are performed using both the RNA of interest and varying amounts of synthetic target RNA, the absolute amount of the RNA of interest can be determined from the standard curve.

RNA samples to be assayed (usually <10 μg) are combined with ^{32}P-labeled probe (2–10 $\times$ 10^5 cpm; 0.2–1.0 ng), precipitated with ethanol, and resuspended in 20 μl of hybridization buffer. RNA samples are dena-

[8] R. J. Roller, R. A. Kinloch, B. Y. Hiraoka, S. S.-L. Li, and P. M. Wassarman, *Development (Cambridge, UK)* **106,** 251 (1989).

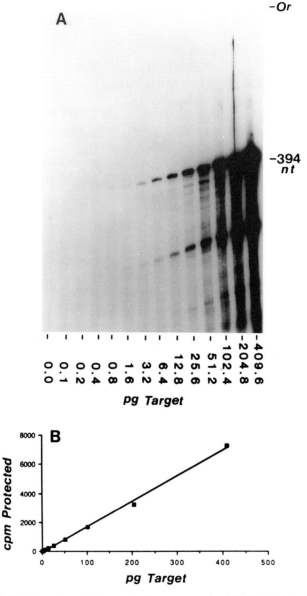

FIG. 1. Calibration of an RNase protection assay using MOM-1 antisense probe and synthetic target. Assays were performed as described, using a constant amount of MOM-1 probe and varying amounts of MOM-1 target (0–409.6 pg). (A) Autoradiogram of the poly-acrylamide/urea gel analysis of the RNase protection assay. The 394-nt MOM-1 protected fragment is indicated. Or, Origin. (B) Standard curve relating radioactivity (cpm) associated with the 394-nt protected fragment (A) with amount of target transcript added (pg). Appropriate bands were excised from the gel (A) and subjected to liquid scintillation counting as described in the text. The line drawn is based on least-squares fitting of the data.

tured at 85° for 2 min and hybridized overnight at 45°. Following the overnight incubation, unprotected probe is digested by the addition of 200 μl of RNase digestion buffer containing 40 μg/ml RNase A and 2 μg/ml RNase T1, then incubation at 37° for 60 min. Digestion is terminated by adding sodium dodecyl sulfate (SDS; to 0.7%) and proteinase K (to 150 μg/ml) and continuing the incubation for 15 min at 37°. Protected RNA fragments are purified by extraction with phenol/chloroform followed by precipitation with ethanol in the presence of carrier yeast tRNA (15 μg). Samples are resuspended in 5–10 μl of gel loading buffer, denatured by incubation at 90° for 3 min, and analyzed on 6% polyacrylamide/8 M urea gels. Protected fragments are visualized by autoradiography of dried gels. A ^{32}P-end-labeled *Hae*III digest of ϕX174 DNA provides convenient size standards.

Quantification is achieved by excising the appropriate bands from the dried gel, using the autoradiograph as a template, and subjecting the bands to liquid scintillation counting in Aquasol. In all experiments, a reaction containing yeast tRNA (10 μg) is included in order to correct for background. Yeast tRNA should not hybridize to any of the RNA probes used. A band corresponding in size to the protected fragments is excised from the yeast tRNA lane, processed as above, and used to control for background.

Analysis of Protection Assays

Figure 2 is an example of a typical quantitative analysis.[8] In this case, the mRNA of interest is MOM-1, a 4.6-kb species present in total RNA isolated from fully grown mouse oocytes. The antisense MOM-1 probe is complementary to 405 nt of authentic MOM-1 mRNA. The synthetic MOM-1 target is a 2.3-kb sense transcript containing 394 nt of MOM-1 cDNA sequence. Parallel assays are carried out using both oocyte RNA (~125 ng; 250 oocytes) and varying amounts of target RNA (equivalent to 0–80 pg of MOM-1 mRNA; note that MOM-1 mRNA is twice as long as the synthetic target, and thus 100 pg of target is equivalent in copy number to 200 pg of MOM-1 mRNA) and a constant amount of MOM-1 antisense probe (1 × 10⁶ cpm; ~800 pg). Protected fragments are resolved on denaturing gels, excised, and subjected to scintillation counting. As shown in Fig. 2, the target RNA protected a fragment of 394 nt, whereas MOM-1 mRNA protected a 405-nt fragment. To control for recovery of oocyte RNA, a 220-nt "recovery fragment" (MOM-1 cDNA sense transcript) was added to oocytes prior to RNA extraction, and an equivalent amount was assayed directly in the *Escherichia coli* ribosomal RNA control reaction. The ratio of the amount of recovery transcript protected in the reaction containing oocyte RNA to the amount protected in the

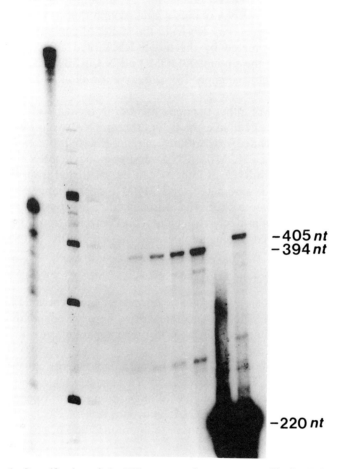

A B C D E F G H I J K

−405 nt
−394 nt

−220 nt

FIG. 2. Quantification of the RNase protection assays; specifically, MOM-1 mRNA steady-state levels in fully grown oocytes. An autoradiogram was prepared of the polyacrylamide/urea gel analysis of RNase protection assays of MOM-1 target transcripts (pg) and RNA from fully grown oocytes (see text for details). Positions of the 405-nt fragment protected by oocyte RNA, the 395-nt fragment protected by the MOM-1 target transcript, and the 220-nt fragment protected by the recovery transcript are indicated. Lane A, undigested probe transcript; lane B, undigested MOM-1 target transcript; lane C, ^{32}P-end-labeled size standards; lanes D–I, RNase protection assays performed with 0, 2.5, 5, 10, 20, and 80 pg of MOM-1 target transcript, respectively; lane J, *E. coli* ribosomal RNA (equivalent to using yeast tRNA), plus recovery transcript; lane K, RNA from 250 fully grown oocytes (~125 ng), plus recovery transcript. Or, Origin.

control reaction is used as a correction factor for the efficiency of the RNA extraction procedure. By comparing the amount of antisense probe protected by MOM-1 mRNA (minus a background that is assessed from the reaction containing *E. coli* RNA) to the standard curve derived from amount of probe protected by the target RNA, and adjusting for the percentage recovery, the amount of MOM-1 mRNA in fully grown oocytes is calculated to be approximately 185 fg/oocyte, or about 7.4×10^4 copies/oocyte.

Once an estimate of either the copy number or the mass of an mRNA species has been obtained, an estimate of other mRNA species within the same tissue can be made by performing parallel RNase protection assays for each message (e.g., MOM-1 mRNA versus an oocyte mRNA of interest). The amount of mRNA of interest can be calculated from the ratio of counts per minute associated with MOM-1 mRNA and the counts associated with another mRNA. The ratio must then be adjusted for differences in the lengths of protected fragments, the compositions of the probes with respect to labeled ribonucleotide, and the sizes of the mRNAs.

Optimization of Protection Assays

RNase protection assays tend to be template dependent and thus may be tricky to optimize. The protocols described here are adequate in general; however, certain steps can be taken when encountering difficulties with specific templates.

If sufficient sample RNA is available, pilot experiments can be set up to optimize the assay. These involve a series of assays in which the amount of probe used is kept constant and the following components or parameters are varied: amount of sample RNA; concentration of RNase cocktail; use of RNase T1 alone; hybridization temperature; and RNase digestion reaction time.

The choice of probe is important. The probe should not be self-complementary as intramolecular hybridization will result in numerous, smaller than expected fragments, in both experimental and control reactions. Probes containing A–U-rich regions result in probe/RNA hybrids that are susceptible to RNase digestion. This is due to local denaturation at these regions. Often this can be overcome by performing the RNase digestion step in the presence of RNase T1 alone, since RNase T1 cleaves 3' to guanosine residues. Furthermore, the probe might cross-hybridize to related mRNA species if they are part of a multigene family. Generally this is not a problem since the probe and related mRNAs contain sufficient mismatches that result in smaller protected fragments. The probe may share homology with yeast RNA, resulting in protected fragment(s) in the control reaction. Finally, for reasons discussed earlier, probe lengths should be 200–500 nt.

Occasionally, the amount of sample RNA will be limiting or the mRNA species of interest will be of low abundance, resulting in a weak signal. The sensitivity of the assay can be increased by increasing the amount of labeled ribonucleotide while omitting the corresponding unlabeled nucleotide. This produces a probe with a higher specific activity. It is important to maintain the final concentration of the labeled ribonucleotide at 12–24 μM to ensure synthesis of full-length probe.

[18] Analysis of Messenger RNA

By MAYI Y. ARCELLANA-PANLILIO and GILBERT A. SCHULTZ

Introduction

The period of early development in the mouse sees a multitude of changes in the embryo, culminating in the emergence of two distinct cell types in the blastocyst: the inner cell mass (ICM) and the trophectoderm. The molecular events leading up to and surrounding blastocyst formation are very interesting, and thus are the subject of avid research. This chapter focuses on the study of the qualitative as well as the quantitative aspects of mRNA prevalence through early mouse development. We describe procedures that we have used successfully in our laboratory to characterize patterns of mRNA expression.

RNA Extraction

Total RNA is extracted from mouse eggs and early embryos using an adaptation of the phenol–chloroform extraction method originally described by Braude and Pelham.[1] The necessity for ribonuclease-free handling practices through all manipulations involving RNA cannot be overemphasized, and ensuring that the laboratory provides such an environment is worthwhile.[2]

Pools of eggs/embryos (usually 100–300) are collected in a 500-μl microcentrifuge tube in the minimum volume (5 μl or less) of sterile M2 medium.[3] In a separate tube, the following are combined: 100 μl RNA

---

[1] P. R. Braude and H. R. B. Pelham, *J. Reprod. Fertil.* **56**, 153 (1979).
[2] D. D. Blumberg, this series, Vol. 152, p. 20.
[3] D. G. Whittingham, *J. Reprod. Fertil. Suppl.* **14**, 7 (1971).

METHODS IN ENZYMOLOGY, VOL. 225

Copyright © 1993 by Academic Press, Inc.
All rights of reproduction in any form reserved.

extraction buffer (0.2 M NaCl; 25 mM Tris, pH 7.4; 1 mM EDTA); 100 μl phenol (neutralized, stored under H_2O at $-20°$, and thawed to room temperature); and 100 μl Sevag's solution (24 : 1 chloroform–isoamyl alcohol). Ten micrograms *Escherichia coli* ribosomal RNA (rRNA; Boehringer-Mannheim, Indianapolis, IN), at 2–4 μg/μl, are added to the eggs/embryos to serve as carrier for the subsequent precipitation steps, as well as to allow the estimation of recovery through manipulations.

The buffer–phenol–Sevag's mix is added to the eggs/embryos and vortexed vigorously for two bursts of approximately 10 sec each. The phases are separated by centrifugation at 13,000 rpm in a microcentrifuge for 5–10 min. The supernatant aqueous phase is transferred to a slim 400-μl centrifuge tube and reextracted with 200 μl Sevag's solution. After centrifugation, the aqueous phase is carefully pipetted off and transferred to a fresh 500-μl tube, 2.5 volumes of cold 95% (v/v) ethanol is added, and the nucleic acids are allowed to precipitate at $-20°$ overnight. Typically, RNA is stored precipitated in ethanol at $-20°$.

When the RNA is required for analysis, it is recovered by centrifugation at 28,000 g for 30 min at 4°. The barely visible pellet is washed in 100 μl cold 70% (v/v) ethanol and recentrifuged for 15 min. The final pellet is dried under vacuum for approximately 5 min, dissolved in 5 μl cold diethyl pyrocarbonate-treated, autoclaved water (DEPC–water), and kept on ice.

Notes. In the event that it becomes necessary to store embryos prior to extraction, as when they have to be pooled from collections on different days, we have done so by freezing individual lots in sterile medium at $-70°$. Then extraction must proceed immediately on removal from storage.

It is wise to ascertain that the carrier RNA does not cross-react with whatever nucleic acid probes might be used later. Aside from rRNA, transfer RNA (tRNA), both from calf liver and from yeast, at similar concentrations, have been used in various instances in our laboratory. The recovery of RNA can be calculated by comparing the acid-precipitable radioactivity before and after extraction of embryos labeled *in vitro,* and in this manner we have estimated recovery to be 69 ± 2%.[4]

Northern Blots

The expression of particular mRNA species through development can be defined using the Northern analysis of RNA[5] from staged embryos. This RNA counterpart to the Southern analysis of DNA[6] involves size

[4] M. Y. Arcellana-Panlilio, unpublished observation, 1988.
[5] J. C. Alwine, D. J. Kemp, and G. R. Stark, *Proc. Natl. Acad. Sci. U.S.A.* **74,** 5350 (1977).
[6] E. M. Southern, *J. Mol. Biol.* **98,** 503 (1975).

fractionation of the nucleic acids by gel electrophoresis, transfer from the agarose gel matrix to a solid support, and hybridization to a labeled probe.

Fractionation of RNA

The maintenance of denaturing conditions during gel electrophoresis disrupts RNA secondary structure, thus promoting accurate separation of species by molecular weight. There are various ways by which denaturation of the RNA can be achieved, including the use of either formaldehyde[7] or methyl mercuric hydroxide[8] in the preparation of the samples prior to loading and in the gel itself. Another method is based on the glyoxylation of the RNA sample[9] prior to electrophoresis, which maintains RNA single-strandedness through later steps. Each method has its own strengths and weaknesses; a discussion comparing the different gel systems appears elsewhere.[10] The choice of which one to use depends very much on the objectives of the experiment. For the study of early embryos, where sensitivity of detection is key, particularly where amounts are limiting, the formaldehyde system is a good option, and this is the method detailed below.

Stock Solutions

10× MOPS: 0.2 M MOPS [3-(N-morpholino)propanesulfonic acid, Sigma, St. Louis, MO], 50 mM sodium acetate, 10 mM EDTA, pH to 7.0 with NaOH

Formaldehyde (BDH, Poole, UK): Supplied as 37% (w/v) solution in water (12.3 M)

Formamide (Fluka, Ronkonkoma, NY): Deionize prior to use by stirring 10 ml with 1 g AG-501 SA (Bio-Rad, Richmond, CA) for 1 hr at room temperature. Store filtered, single-use aliquots frozen at $-70°$

Buffer/denaturant/dye mix: Mix 100 μl of 10× MOPS, 160 μl formaldehyde, 500 μl formamide, 100 μl sterile 50% glycerol, plus 0.1% (w/v) bromphenol blue. Prepare fresh

Pouring Gel. To make a 1.5% agarose gel of dimensions 10 cm long by 15 cm wide by 6.6 mm deep with 20 wells (5 × 1 × 5 mm each), dissolve 1.5 g agarose in 74 ml water by boiling. Cool to 65°. Add 10 ml of 10× MOPS and 16 ml formaldehyde. Swirl to mix and pour. Allow to set. *Caution:* Mix and pour gels containing formaldehyde in a fume hood.

[7] H. Lehrach, D. Diamond, J. M. Wozney, and H. Boedtker, *Biochemistry* **16**, 4743 (1977).
[8] J. M. Bailey and N. Davidson, *Anal. Biochem.* **70**, 75 (1976).
[9] G. C. Carmichael and G. K. McMaster, this series, Vol. 65, p. 380.
[10] R. C. Ogden and D. A. Adams, this series, Vol. 152, p. 61.

The transfer can be accomplished successfully by a variety of methods, including by vacuum, by electroblotting, and by capillary action. The first two employ specially designed equipment which come with instructions for use. Capillary transfers have the simplest of requirements and when set up properly work extremely well. Procedures for capillary transfers are given below.

Transfer to Nitrocellulose. Take the gel right side up into a dish and trim away unused portions. Note that formaldehyde gels are more fragile than nondenaturing gels of equivalent agarose content and thus should be handled with care. Cut straight across the wells, in order to facilitate identification of the origin, or leave the wells intact and mark their position on the filter later.

With a new scalpel or a paper cutter, cut a piece of nitrocellulose the same size as the gel. Take care to handle filters with gloves or blunt-ended forceps at all times. Float the nitrocellulose filter on water to wet completely from underneath, then transfer to soak in $20 \times$ SSC ($1 \times$ SSC is 0.15 M NaCl, 0.015 M sodium citrate) for at least 5 min.

Fill a baking dish with $20 \times$ SSC. Take a glass plate of dimensions that would allow placing it on top of the baking dish and having some clearance on two opposite sides. Drape two layers of Whatman (Clifton, NJ) 3MM paper previously wet with $20 \times$ SSC over the straddled plate so that the short ends of the paper extend beyond the plate and into the liquid below. Thoroughly wet the paper and then smooth out any bubbles with a glass rod. Note that this is the wick on which the gel will be placed, and therefore its useful top surface must be wider and longer than the gel.

Place the gel upside down and centered on the wet Whatman 3MM paper. Ensure that there are no bubbles between gel and paper. Lay strips of Parafilm or Saran on the paper around the gel. This prevents the buffer from wetting the papers that will be stacked on top except via the gel and the nitrocellulose filter.

Take the wet nitrocellulose filter and, starting from one end, lower it onto the gel so that edges coincide. Doing it this way keeps bubbles from getting trapped between filter and gel, but should some be caught anyway, smooth these out by lightly rolling the glass rod on top of the nitrocellulose. It is important that the filter does not shift once it has come in contact with the gel.

Take two pieces of Whatman 3MM paper cut to the dimensions of the gel, wet them with $20 \times$ SSC, and place these on the filter. Take several *dry* pieces of similarly cut Whatman 3MM paper and set these on. Finally, take a stack of paper towels cut to gel dimensions about 8–10 cm high and place these on the Whatman 3MM papers. Position a glass plate on top of the stack, and place a 0.5-kg weight on it to compress the layers

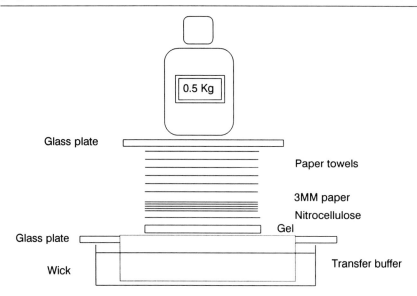

FIG. 1. Diagram of capillary blot setup for transfer of nucleic acids onto a solid support.

and aid capillary action. The aim is to set up the flow of buffer from the reservoir orthogonally through the gel so that RNA is deposited onto the nitrocellulose, producing a faithful replica of the gel. Thus, it is important that the stack is stable and that it is compressed evenly. (Figure 1 shows a diagram of the setup.)

Allow to transfer for 6 hr to overnight. When transfer is complete, dismantle the blot, peeling layers until the nitrocellulose filter is on top. With a soft lead pencil mark the filter (e.g., the positions of the wells). Snip the lower right-hand corner of the filter; this will correspond to the bottom of the first lane on the gel. Alternatively, this designation could be penciled in as well.

With blunt-ended forceps, peel the filter off the gel and soak it in 5× SSC for a few minutes. This allows the removal of any agarose that might have stuck to the filter. Drain the filter on a piece of Whatman 3MM paper and let air-dry for about 30 min. Place the filter between two fresh sheets of Whatman 3MM paper and bake at 80° under vacuum for 1–2 hr. Note that nitrocellulose is especially fragile when dry, and, if baked too long, it may become very brittle. Store the filter between the sheets of Whatman 3MM paper in aluminum foil at room temperature.

Transfer to Nylon Membranes. The makers of nylon filters usually accompany their product with protocols for use. These protocols invariably include one for capillary transfers very similar to that for nitrocellulose, with minor changes. For instance, with Nytran (Schleicher and Schuell), the blotting buffer is 10× SSC, and the use of vacuum during baking to fix the RNA is not necessary. UV cross-linking[15] is an alternative method of immobilizing the RNA onto the filter. However, as noted by Sambrook et al.,[13] there is an optimum amount of irradiation required, and this should be empirically determined.

Probes to RNA

We use two types of radiolabeled probes routinely: the oligolabeled cDNA probe[16] and the in vitro transcribed riboprobe.[17] Below are protocols for making these radiolabeled probes. We have started to use nonradioactive methods of detection, such as the digoxigenin system (Boehringer-Mannheim), based on the introduction into the probe of a specific hapten as an immunological marker coupled to a nucleoside triphosphate, which is detected after hybridization by enzyme-linked immunoassay using an antibody conjugate. Procedures for these methods are specific to the particular detection system.

Oligolabeling. The following protocol[16] optimizes the oligolabeling method originally described by Feinberg and Vogelstein[18] with respect to pH, enzyme, temperature, DNA, primer, and buffer components, thus enabling 70–90% incorporation of label into probe to specific activities as high as 5×10^9 disintegrations/min (dpm)/μg. Combine 60 ng of dp(N)$_6$ random primer (Pharmacia, Piscataway, NJ) with 200 ng template DNA in 4 μl sterile TE (10 mM Tris, pH 7.5; 1 mM EDTA) in a 500-μl tube. Immerse the tube in a boiling water bath for 2 min. Quench on ice. Add 2 μl sterile 10× oligolabeling buffer [10× OLB is 0.5 M Tris, pH 6.9; 0.1 M MgSO$_4$, 1 mM dithiothreitol (DTT), and 0.6 mM each of dATP, dGTP, and TTP]. Add 50 μCi of [α-^{32}P]dCTP (3000 Ci/mmol; 10 μCi/μl), 2–3 units Klenow enzyme (Boehringer-Mannheim), and DEPC–water to 20 μl volume. Incubate at 22° for 60 min. Remove unincorporated nucleotides by passing the reaction through a spun Sephadex G-50 column, packed at the bench[19] or obtained from a kit (Boehringer-Mannheim). Prior to hybridization, denature the probe in a boiling water bath for 5–10 min and then quench on ice.

[15] G. M. Church and W. Gilbert, *Proc. Natl. Acad. Sci. U.S.A.* **81**, 1991 (1984).
[16] C. P. Hodgson and R. Z. Fisk, *Nucleic Acids Res.* **15**, 6295 (1987).
[17] D. A. Melton, P. A. Kreig, M. R. Rebagliati, T. Maniatis, K. Zinn, and M. R. Green, *Nucleic Acids Res.* **12**, 7035 (1984).
[18] A. P. Feinberg and B. Vogelstein, *Anal. Biochem.* **132**, 6 (1983).
[19] J. L. Meinkoth and G. M. Wahl, this series, Vol. 152, p. 91.

Riboprobes. To avoid precipitating the DNA due to the spermidine present in the riboprobe buffer, set up the transcription reaction at room temperature and combine components in the order given. In a 500-μl tube, combine 4 μl of 5× riboprobe buffer (5× RB is 200 mM Tris, pH 8.0; 40 mM MgCl$_2$; 125 mM NaCl; 10 mM spermidine hydrochloride; 25 mM DTT; BRL), 0.5 μl RNase inhibitor (20 units), 4 μl of 5× ribonucleoside triphosphates (5× rNTPs is 2.5 mM each of rATP, rGTP, and rCTP; 60 μM UTP), 11 μl template DNA (100–1000 ng) plus [^{32}P]UTP label (e.g., 6 μl template plus 5 μl of 50 μCi label), and 0.5 μl RNA polymerase (either SP6, T7, or T3, about 10 units). Incubate at 38°–39° for 1–2 hr.

Purify as for oligolabeled probes or, alternatively, precipitate the reaction in ethanol. Incorporation of radiolabeled precursor as measured by trichloroacetic acid (TCA) precipitation should reach 70–80%. It is not necessary to destroy the DNA template with RNase-free DNase prior to hybridization. Because the probe is single-stranded, boiling the probe prior to use is not required.

Notes. RNase-free conditions must be maintained during probe preparation and use in hybridization to RNA. For riboprobes, it is imperative in order to yield any probe at all. Template preparation for both oligolabeled probes and riboprobes is an important factor for successful incorporation of radiolabeled precursor. We routinely gel-isolate and purify templates after restriction enzyme digestion, which either releases insert (for oligolabeling) or linearizes the plasmid (for riboprobes). Also, in the case of riboprobes, the orientation of the template must be such that an antisense strand is produced. Finally, we usually make probes from fresh label on the same day that we expect to use them. We also have stored the probes as a precipitate under ethanol at − 20° and recovered it by centrifugation the next day.

Northern Hybridization

Understanding the theoretical aspects of the hybridization of probes to nucleic acids immobilized on solid supports allows the optimization of conditions which will yield high signal-to-noise ratios. A discussion of these parameters appears elsewhere.[20] In our study of mRNAs in early mouse embryos, we have found the following protocols to work well.

Stock Solutions

20× SSPE: 3.6 M NaCl; 0.2 M sodium phosphate, pH 7.7; 1 mM EDTA

---

[20] G. M. Wahl, S. L. Berger, and A. R. Kimmel, this series, Vol. 152, p. 399.

50× Denhardt's reagent: 1.0% (w/v) each of Ficoll, polyvinylpyrrolidone, and bovine serum albumin
10% (w/v) Sodium dodecyl sulfate (SDS)
50 mg/ml Yeast tRNA, aliquoted and kept frozen at −20°
50% (w/v) Dextran sulfate
2% (w/v) Sodium pyrophosphate

Prehybridization. Prepare a sufficient volume of prehybridization buffer to ensure more than adequate wetting of the filter containing the immobilized RNA (e.g., for a 10 × 15 cm^2 blot, use 15 ml). The prehybridization buffer contains 2× SSPE, 10× Denhardt's, 0.2% SDS, 50 μg/ml yeast tRNA, 2.5% dextran sulfate, and 0.1% sodium pyrophosphate. Allow the filter containing the immobilized RNA to incubate in prehybridization buffer at 65° for 1–2 hr.

Hybridization. Prepare the probe as outlined previously. Add probe to the prehybridization buffer at 1 × 10^6 counts/min (cpm)/ml. Allow to hybridize overnight at 65°.

Washes. Immerse the filter in 2× SSPE, 0.1% SDS at room temperature for a few minutes. This allows the removal of most of the unhybridized radioactive material. Wash twice for 20 min, each time in 2× SSPE, 0.1% SDS at 65°. Wash twice for 10 min, each time in 0.1× SSPE, 0.1% SDS at 65°. Drain off excess liquid and wrap the wet filter in Saran. Place against Kodak XAR-5 film (Eastman Kodak Co., Rochester, NY) in a cassette and allow an overnight exposure at −70°.

Notes. The time, temperature, and salt conditions for the washes above allow the specific detection of fairly abundant mRNAs using probes of greater than 150 bp to total RNA from 150–300 preimplantation embryos. The detection of rarer messages is favored by the loading of more total RNA, or even of poly(A)$^+$ RNA, the isolation of which is described elsewhere.[21]

Slot/Dot Blots

RNA from early mouse embryos can be immobilized directly onto a solid support and analyzed by hybridization with an appropriate probe. Costanzi and Gillespie[22] review "fast blots" for DNA and for RNA onto nitrocellulose. Slot/dot blots can be set up for quantitation of a specific mRNA, and the procedure outlined below has been found useful.

Sample Preparation. Isolate RNA from embryos as described, using no more than 10 μg of carrier. Recover the pellet and dissolve in 50 μl

---

[21] A. Jacobson, this series, Vol. 152, p. 254.
[22] C. Costanzi and D. Gillespie, this series, Vol. 152, p. 582.

DEPC–water. Prepare two dilution series: (1) of the RNA sample and (2) of RNA standards containing known amounts of the particular mRNA. Have these in 50 μl volumes. Incubate at 90° for 1 min. Quench on ice. The RNA is now ready for loading.

Immobilization of RNA. Set up a minifiltration apparatus (such as Slot Blotter by Schleicher and Schuell) according to the manufacturer's specifications, hooking up to either the vacuum lines or a vacuum pump. Depending on how many slots will be used, cut to the appropriate dimensions a piece of nylon membrane (such as Nytran by Schleicher and Schuell) and a sheet of Whatman 3MM paper for backing. Prewet these in DEPC–water and place in the apparatus. Apply vacuum and load the RNA samples into the slots. The 50-μl volumes filter through almost instantly. Turn off vacuum about 1 min after the last sample has gone through. Air-dry the nylon membrane; then, if recommended by the manufacturer, bake at 80° for 1 hr.

Analysis. Prehybridize, hybridize, wash, and prepare autoradiographs of the slot/dot blots as for Northern blots. Analyze the autoradiograph by densitometry (Model 620 Video Densitometer, Bio-Rad). Construct one curve by plotting integration data from the RNA standards against the number of copies and another by plotting integration data from the RNA sample dilutions against number of embryos. Being careful to stay within linear limits of both curves, read off the number of copies of a particular mRNA species giving a signal equivalent to that of a given number of embryos.

Notes. As always, RNase-free conditions must be maintained throughout. The minifiltration apparatus can be rendered RNase-free by soaking in AbSolve (Du Pont, Wilmington, DE) for 30 min and rinsing with DEPC–water. The standard RNA can be *in vitro* transcribed RNA from cloned cDNA in a transcription vector. The procedures for preparing such standards are described in greater detail later, in the discussion of quantitative polymerase chain reaction (qPCR).

Consistency of technique is important, particularly when loading the RNA. Apply the sample while holding the pipette tip as close to the filter as possible without actually making contact and dispense smoothly. The pull of the vacuum should be such that the volume applied flows through quickly. Contrary to other quick blot methods[22] which invariably recommend high salt conditions during filtration, this protocol simply uses DEPC–water for prewetting the membrane as well as for carrying the RNA. Thus, no washing of the filter prior to drying is necessary.

The conditions of hybridization have to be quite stringent, maybe more so than for Northern blots. It is important that the signal seen is due only to hybridization to the specific message. Any cross-hybridization with other RNA molecules would increase the signal and cause an overestimation of the abundance of the species of interest. In Northern blots, the cross-hybridizing molecule is picked up on quite easily because the errant species is usually of a different molecular weight and therefore of a different mobility. It is good practice to work out stringency conditions for hybridization and washing through the use of Northern blots to ensure that the probe is specific.

Reverse Transcription–Polymerase Chain Reaction

The polymerase chain reaction (PCR) is an *in vitro* method which can amplify small amounts of a specific DNA fragment to levels that are readily detectable. The method involves the use of two synthetic oligodeoxynucleotide primers, 20 to 30 nucleotides in length, that hybridize to opposite strands of the target sequence, followed by amplification through repeated cycles of heat denaturation of DNA, annealing of the primers to complementary sequences, and extension by a thermostable DNA polymerase.[23] Because the extension product of each primer can serve as a template for the other primer, each cycle effectively doubles the amount of the target DNA sequence produced in the previous cycle, resulting in its exponential accumulation.

By combining the reverse transcription (RT) of RNA with the PCR amplification process, it is possible to detect RNA sequences present at low copy number in very small amounts of material. Because several mRNA species can be assayed within the same RNA preparation through the use of appropriate primer pairs, RT–PCR has been termed "mRNA phenotyping."[24] This is particularly applicable to the analysis of gene expression in oocytes and preimplantation mouse embryos, where it is not feasible to obtain large numbers for study.

Preparation of RNA Template

The RNA template for reverse transcription is isolated from mouse embryos in the manner described at the beginning of this chapter. Another method that has worked well for the purposes of RT–PCR is that described by Rappolee *et al.*,[25] which is a microadaptation of the guanidinium isothiocyanate/cesium chloride (GuSCN/CsCl) gradient ultracentrifugation

---

[23] R. K. Saiki, D. H. Gelfand, S. Stoffel, S. J. Scharf, R. Higuchi, G. T. Horn, K. B. Mullis, and H. A. Erlich, *Science* **239**, 487 (1988).

[24] D. A. Rappolee, D. Mark, M. J. Banda, and Z. Werb, *Science* **241**, 708 (1988).

[25] D. A. Rappolee, A. Wang, D. Mark, and Z. Werb, *J. Cell. Biochem.* **39**, 1 (1989).

method of Chirgwin *et al.*[26] Here, small numbers of cells, oocytes, or embryos are lysed by addition to 100 μl of 4 M GuSCN containing 5 to 10 μg *E. coli* rRNA (Boehringer-Mannheim) as carrier, layered over 100 μl of 5.7 M CsCl, and centrifuged in a TL-100A rotor of a Beckman T1-100 tabletop ultracentrifuge at 80,000 rpm for 2 hr or in a Beckman Airfuge at 95,000 rpm for 1 to 2 hr. The GuSCN and CsCl are aspirated from the pellet, which is then dissolved in DEPC–water, precipitated in the presence of 2.5 M ammonium acetate with 2.5 volumes of cold absolute ethanol, and kept at $-20°$. The RNA pellet recovered after centrifugation is washed twice in 80% ethanol to remove residual salts, then dried briefly under vacuum.

The RNA pellet (1 μg or less, with up to 10 μg *E. coli* rRNA carrier) is dissolved in 5 μl RNasin–water (1 unit RNasin per 10 μl water) and kept on ice. Immediately prior to reverse transcription, the RNA is placed in a heating block at 95° for 1 min and quenched on ice.

Reverse Transcription

First-strand cDNA synthesis can be achieved by reverse transcription of RNA primed with oligo(dT), random hexanucleotides, or a specific downstream sequence.[27] The use of primers other than the customary oligo(dT) appears to be helpful in improving efficiency of reverse transcription where unresolved secondary structure in the RNA is a problem.[28] However, in many instances, the choice of RT primer is not critical, and equal success can be achieved with the use of either primer. Figure 2 shows the amplification of a 400-bp region of a major histocompatibility complex gene cDNA after reverse transcription of mouse placenta total RNA with oligo(dT) (lane 1), random hexamers (lane 2), or specific downstream primer, in this case, the same downstream primer used in the subsequent PCR (lane 3).

Reverse transcriptases available commercially are of two types: cloned mouse Moloney leukemia virus (MMLV) reverse transcriptase and avian myeloblastosis virus (AMV) reverse trancriptase. Both MMLV and AMV enzymes, from BRL, Molecular Genetic Resources (Tampa, FL), and Boehringer-Mannheim, have yielded good results, and others from reputable suppliers would probably work as well.

[26] J. M. Chirgwin, A. E. Przybyla, R. J. MacDonald, and W. J. Rutter, *Biochemistry* **18**, 5294 (1979).
[27] E. S. Kawasaki, *in* "PCR Protocols: A Guide to Methods and Applications" (M. A. Innis, D. H. Gelfand, J. J. Sninsky, and T. J. White, eds.), p. 21. Academic Press, New York, 1990.
[28] D. A. Rappolee, *Amplifications* **4**, 5 (1990).

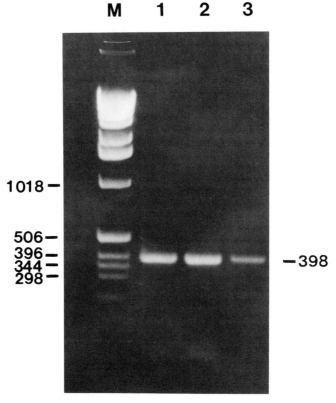

FIG. 2. PCR amplification of a mouse major histocompatibility complex cDNA fragment after reverse transcription of mouse placenta total RNA with oligo(dT) (lane 1), random hexamers (lane 2), or specific downstream primer used in the subsequent PCR (lane 3). The size of the amplified fragment is 398 bp and is shown in relation to a low molecular weight DNA ladder (lane M).

The reverse transcription reaction is usually carried out in a final volume of 10 or 20 μl and set up as outlined below. For a 20 μl reaction, combine the following:

4 μl of 5× RT buffer (250 mM Tris-HCl, pH 8.3 at 42°; 300 mM KCl; 15 mM MgCl$_2$)

0.4 to 2.0 μl primer [per reaction, add 0.1 to 0.5 μg oligo(dT), or 0.7 to 7.0 μg random hexanucleotides,[28] or 5 to 100 pmol specific downstream primer[27]]

2.0 μl of 10 mM dNTPs (Pharmacia)

0.2 μl of 1 M DTT

0.2 μl RNasin (40 units/μl, Promega, Madison, WI)
2.0 μl bovine serum albumin (BSA, nuclease-free, 1 mg/ml)
2.0 μl reverse transcriptase (20–25 units/μl, MMLV or AMV)
5.0 μl RNA template
RNasin–water to volume

Incubate at 42° for 1 hr. Heat the reaction at 95° for 5 min to denature the RNA–cDNA hybrid and inactivate the reverse transcriptase. Quench on ice. Make up to 50 μl with water and store at $-20°$.

Notes. When preparing several reverse transcription reactions, it is wise to prepare master mixes to minimize pipetting errors. Combining the RNA template with the RT primer, heating at 70° for 5 min, and cooling in ice, prior to the addition of the other reaction components, may improve the efficiency of reverse transcription.[29] If desired, a second cycle of reverse transcription can be carried out before making up to the 50 μl final volume. Prepare 4.0 μl of a second RT mix by combining 0.2 μl of 10 mM dNTPs, 0.2 μl BSA (1 mg/ml), 0.8 μl of 5× RT buffer, 1.0 μl reverse transcriptase, and RNasin–water to volume. Incubate for another 1 hr at 42°. The reverse transcription reaction yields enough material for approximately 10 PCR assays.

Polymerase Chain Reaction

Although the PCR is undoubtedly a very powerful method for the analysis of specific nucleotide sequences, each application must be optimized for the particular target sequence and its primer set.

Designing Primers. The design of primers is not trivial; it is a primary consideration well worth caring about. Although there are commercially available computer software packages (such as Primer Designer by Scientific and Educational Software, State Line, PA) to aid in primer selection, it is wise to be mindful nonetheless of several guidelines for good design.[25,28] Some of the more critical of these are listed below.

1. Oligonucleotide primers should be 20 to 30 nucleotides in length with a melting temperature (T_m) near 72° (55° to 75°), so that the PCR annealing step has optimal stringency. The G/C and A/T contents of the primers should be balanced, and the primers themselves should lack secondary structure. An empirical formula to define the temperature at which 50% of short duplexes (14–20 bp) dissociate (T_m) has been determined[30]:

[29] A. T. Garber, personal communication (1991).
[30] R. B. Wallace, J. Shaffer, R. F. Murphy, J. Bonner, T. Hirose, and K. Itakura, *Nucleic Acids Res.* **6**, 3543 (1979).

$$T_m = 4(G + C) + 2(A + T) \qquad (1)$$

where G, C, A, and T are the number of corresponding nucleotides, and denaturation is at 0.9 M NaCl. There are useful software packages, such as Primer Designer (Scientific and Educational Software) and Oligo (National BioSciences, Hamel, MN), which calculate T_m based on nearest-neighbor thermodynamic properties.

2. The primers should not have complementarity, and 3′ sequence overlaps should be avoided to prevent primer–dimer formation.

3. Generally, primers should bracket a sequence of 200 to 600 nucleotides in length, though longer targets have been amplified efficiently.

4. When possible, primers should be designed to bracket an intron so that amplification from contaminating genomic DNA can be distinguished from the cDNA of interest.

5. Primers should bracket a sequence with a diagnostic restriction site or a sequence that spans a cDNA clone possessed by the laboratory, so as to provide tools for the verification of the identity of the PCR product by restriction analysis or Southern blot.

PCR Amplification Protocol. Outlined below is an amplification procedure which works reasonably well in most cases; it can be used as a basis for developing a customized protocol. For a 50 μl reaction volume, combine in a 500-μl tube the following:

> 5.0 μl of 10× PCR buffer (100 mM Tris-HCl, pH 8.3; 500 mM KCl; 1 mg/ml BSA)
> 5.0 μl of 15 mM MgCl$_2$
> 5.0 μl of 2 mM dNTPs
> 0.2 μl *Taq* DNA polymerase (1 unit)
> 1.0 μl (100 pmol) of each primer (made up as 100 μM stock)
> 5.0 μl reverse-transcribed product
> Water to volume

Overlay mixed contents with an equal volume of light mineral oil (Sigma, M5904). Place in thermocycler programmed to give the following temperature profile: (i) 4 min at 94°; (ii) denaturation for 30 sec to 1 min at 94°, annealing for 1 to 2 min at 55° to 72°, and extension for 2 min at 72°; (iii) repeat (ii) for 25 to 40 cycles; (iv) incubate for 7 min at 72°; and (v) soak at 4°.

Optimization. As for reverse transcriptions, set up several PCRs from master mixes that combine common components. Run both negative and positive controls every time. Depending on the specific target, use between 0.5 and 1.5 units of *Taq* DNA polymerase (Cetus, Norwalk, CT) per 50-μl reaction. When working out optimum conditions, try concentrations between 0.2 and 3.0 units per 50-μl reaction. Note that excessive amounts of enzyme may lead to the accumulation of nonspecific products, while too little will decrease yields.

Make 10 mM stocks of dNTPs and store these at $-20°$. Make fresh working solutions of 1 or 2 mM, and use 50 to 600 μM of each deoxynucleotide per reaction. The fidelity and specificity of amplification are generally improved by using lower dNTP concentrations.[31] On the other hand, using dNTP concentrations at the high end of the range, and correspondingly higher magnesium, as recommended by Rappolee,[28] has worked very well, particularly for analyzing mouse embryos. In any case, keep the concentrations of dNTPs in the reaction mix balanced! *Taq* polymerase tends to misincorporate or even to terminate prematurely when any one nucleotide concentration is significantly different from the others.[31]

Empirically determine the optimum $MgCl_2$ concentration to use for each primer set and target sequence. The free magnesium ion (Mg^{2+}) concentration is critical, since it is required for *Taq* DNA polymerase activity. Note that Mg^{2+} can be bound by template DNA, primers, and dNTPs. Moreover, the presence of metal chelators like EDTA can decrease the free Mg^{2+} concentration. For most applications, we have found a range of concentrations from 1 to 4 mM $MgCl_2$ to be useful. Innis and Gelfand[32] recommend using 0.5 to 2.5 mM magnesium over the total dNTP concentration.

Use 12 to 120 pmol of each primer per 50-μl reaction. Note that adjusting the primer concentration will necessarily change the $MgCl_2$ optimum. Too high concentration of primers in the reaction may promote mispriming and subsequent accumulation of nonspecific product. Moreover, it may lead to the generation of the primer–dimer, an artifactual double-stranded fragment of DNA approximately the sum of the two primers in length that is formed when one primer is extended by the polymerase over the other primer.[33]

It is important that the denaturation step in each cycle accomplishes its purpose. Innis and Gelfand[32] note that incomplete denaturation of the target template and/or the PCR product is the most likely cause for failure of a PCR. On the other hand, excessively long denaturation times or too high temperatures lead to unnecessary loss of enzyme activity. The half-

[31] M. A. Innis, K. B. Myambo, D. H. Gelfand, and M. A. D. Brow, *Proc. Natl. Acad. Sci. U.S.A.* **85,** 9436 (1988).

[32] M. A. Innis and D. H. Gelfand, *in* "PCR Protocols: A Guide to Methods and Applications" (M. A. Innis, D. H. Gelfand, J. J. Sninsky, and T. J. White, eds.), p. 3. Academic Press, New York, 1990.

[33] R. K. Saiki, *in* "PCR Technology: Principles and Applications for DNA Amplification" (H. A. Erlich, ed.), p. 7. Stockton, New York, 1989.

life of *Taq* DNA polymerase activity is greater than 2 hr at 92.5°, 40 min at 95°, and 5 min at 97.5°, respectively.[32]

The annealing temperature is dependent on the T_m of the primers, which, in turn, is a function of length and of base composition. Higher annealing temperatures favor correct priming and therefore increase specificity. Test a range of annealing temperatures from 55° to 72°, up to about 5° below the T_m of the primers. The 72° extension temperature is optimum for *Taq* polymerase. The extension time is dependent primarily on the length of the target sequence, although the rate of nucleotide incorporation can be affected by buffer, pH, salt concentration, and the nature of the template. Estimates for the rate of incorporation at 72° vary from 35 to 100 nucleotides per second.[31]

The cycle number is dependent on the initial concentration of the template DNA. For the detection of messages of low copy number, intuitively one would try to execute as many cycles as possible. However, we observe higher background owing to nonspecific amplification at too many cycles (>40 cycles). In this case, it is often better to increase the initial template concentration. For example, if, for routine analysis, cDNA equivalent to 30 embryos is amplified per reaction, try doubling that amount for rarer messages. Optimizing cycle number becomes most important in quantitation (discussed later).

The final soak at 4° essentially inactivates the enzyme and is a good temperature at which to store the amplified product for the short term until analysis. For long-term storage, we routinely freeze samples at −20°.

Analysis of PCR Products. Following amplification, analyze 5 to 10 μl of each reaction by gel electrophoresis with appropriate molecular weight markers. To analyze products under 500 bp use 2% (w/v) agarose or, for improved resolution, use a composite gel made up of 2% agarose and 1–2% (w/v) NuSieve (FMC Corporation, Rockland, ME). Alternatively, run samples on a 6–10% (w/v) polyacrylamide gel. Use ethidium bromide staining for visualizing the PCR products, and determine that the bands migrate as expected for their predicted size.

In addition to sizing the PCR products, verify identity by cutting with a restriction enzyme that cleaves the amplified DNA at a known site and then analyzing the cut fragments on a gel. So as not to compromise the activity of the chosen restriction enzyme, ethanol precipitate the PCR reactions, centrifuge, wash the pellet with 70% ethanol, and dissolve in water or TE, prior to setting up the restriction digests. As a further verification, do a Southern blot[13] of the gel bearing the PCR products and hybridize it with an appropriate probe. Of course, the ultimate proof is to sequence the amplified product.[31]

Quantitative Polymerase Chain Reaction

It is relatively easy to amplify rare mRNA transcripts to detectable limits by RT–PCR. However, because PCR amplification is an exponential process, particularly in early cycles, small differences in any of the parameters that control the efficiency of the reaction can substantially affect the final yield of PCR product, making it difficult to quantitate the amount of mRNA in the original material.[34] Thus, the challenge of qPCR is to control these variables, which include concentrations of enzyme, dNTPs, DNA, Mg^{2+}, and primers; temperatures of denaturation, annealing, and extension; and cycle number. In our laboratory, we have employed two basic strategies for qPCR: (1) the use of an internal standard of known copy number, which employs the same primers as those of the cDNA of interest but can be distinguished from the target cDNA after coamplification; and (2) the comparison of sample PCR yields against a calibration curve of external standards. Protocols for each approach are given below.

Internal Standard Method

Standards. Regardless of the qPCR strategy, the standards largely determine the validity as well as the accuracy of the method. Thus, great care must be taken in the choice/design of the standard, its production, its purification, and its own quantitation.

Although cDNA standards can be utilized directly to coamplify with reversed-transcribed sample RNA,[34] we recommend the use of RNA standards, which can be added to the sample RNA, reverse transcribed together with the sample RNA, and then coamplified. When extrapolating to the number of copies of a transcript in the embryo, it is useful to add the RNA standard to the embryos *prior* to extraction of the embryo RNA, so that differences in efficiency even this early in the procedure are covered.

Internal standards must allow the distinction of amplified products in order to be useful. The standard can be designed to give an amplified product that is of a different size from the sample PCR product. Wang *et al.*[35] constructed a plasmid which can be transcribed *in vitro* to produce

[34] G. Gilliland, S. Perrin, and H. F. Bunn, *in* "PCR Protocols: A Guide to Methods and Applications" (M. A. Innis, D. H. Gelfand, J. J. Sninsky, and T. J. White, eds.), p. 60. Academic Press, New York, 1990.

[35] A. M. Wang, M. V. Doyle, and D. F. Mark, *Proc. Natl. Acad. Sci. U.S.A.* **86,** 9717 (1989).

a polyadenylated sense-strand RNA that after RT–PCR yields a product of about 300 bp for any of the 12 primer sets included. The 300-bp product is different in size from the true sample products and can thus be readily separated by gel electrophoresis. Another way to accomplish PCR product size differences is to clone into a transcription vector the genomic sequence containing the target, if it spans a small intron (e.g., 100 bp). Or a short *stuffer* fragment might be inserted into the cloned PCR amplified product so that the same primers could still be used but the product would be larger in size. Note that methods based on differences in the size of PCR products must be applied with caution, since the efficiency of RT–PCR may differ significantly, thereby invalidating quantitative comparisons. The variability in efficiency may not only be size dependent but sequence dependent as well. A standard as *like* the target as possible appears to work best.

Thus, we recommend distinguishing amplified products by designing the standard so that its PCR product contains a unique restriction site not found on the sample amplified product. Gilliland et al.[34] describe the use of a mutant cDNA that is identical to the target cDNA except that it contains a new restriction site introduced by site-directed mutagenesis.[36] Such a cDNA could be used to make mutant RNA for use as internal standard, whose PCR product can be differentiated from that of the sample RNA by gel electrophoresis of the restriction enzyme digest of the amplified products.

In our laboratory, we have taken advantage of natural minor sequence differences in the major histocompatibility complex class I genes between mouse strains. We obtain cDNA clones from one strain of mice and use these to make synthetic RNA for internal standards in the quantitation of transcript from another strain. As in the mutant template method, the PCR products are differentiated by restriction enzyme digestion. The added advantage of this particular application is that it is possible to choose enzymes so that in one case the standard product is cut while the sample is left intact, and in another case the cutting pattern is reversed.

The transcription vector of choice for making RNA standards has the following features: a multiple cloning region flanked upstream by an RNA polymerase promoter and downstream by a polyadenylated sequence followed by several unique restriction sites to facilitate linearization. *In vitro* transcription of the linearized plasmid should produce synthetic sense-strand RNA, with poly(A) sequences making the RNA amenable to purifi-

[36] R. Higuchi, B. Krummel, and R. K. Saiki, *Nucleic Acids Res.* **16**, 7351 (1988).

cation by oligo(dT)-cellulose chromatography[37,38] and later to reverse transcription with oligo(dT)priming.

Preparing RNA Standards. Given that the standard sequence has been cloned into a transcription vector, prepare the standard RNA using the riboprobe protocol with the following modifications: (1) Use an rNTP mix with equal amounts of rATP, rGTP, rCTP, and UTP (i.e., $5\times$ rNTPs would have 2.5 mM of each). (2) Do not use radioactive label. (3) After the 1- to 2-hr incubation at 38–39°, extract with phenol–chloroform and precipitate with ethanol at $-20°$ overnight. Recover by centrifugation as for sample RNA.

Purification of RNA Standards. Pool the synthetic RNA produced from several transcriptions, and purify through an oligo(dT)-cellulose column as described by Jacobson.[21] Note that, when calculating how much RNA to load on the column, one must subtract the contribution arising from the DNA template. Also, predicting the amount of oligo(dT)-cellulose resin (Collaborative Research, Bedford, MA) needed to bind the polyadenylated RNA is straightforward. Knowing the fraction of the RNA sequence that is poly(A) and the binding capacity of the resin [usually 4–5 mg pure poly(A) per gram resin], calculate the amount of pure poly(A) present and the amount of resin needed. Combine the elution buffer washes and recover the poly(A)$^+$ RNA by adding a one-tenth volume of 3 M sodium acetate and 2.5 volumes ethanol and precipitating at $-20°$ overnight.

Handling of Standards. Accurately quantitate the purified standards by spectrophotometry at 260 nm. To ensure that the standards are intact and homogeneous, run about 250 ng on a formaldehyde gel and visualize · the RNA by ethidium bromide staining. Prepare a precise dilution series ranging from 1 ng (equivalent to 3.6×10^9 copies of RNA 500 nucleotides long) to 1 fg per microliter in 500 μl total volumes. Store the dilutions at $-70°$ in aliquots (5–10 μl).

Quantitative RT-PCR. The method below is an adaptation of that described by Wang and co-workers.[35] To 250 ng embryo RNA (equivalent to about 250 blastocysts) add 3.6×10^3 to 10^7 copies of standard RNA and reverse-transcribe the mixture as described previously. Make up to 50 μl and label this RT^0. Take 5 μl, dilute to 50 μl, and label this RT^{-1}. Store at $-20°$.

[37] M. Edmonds, M. H. Vaughn, and H. Nakazoto, *Proc. Natl. Acad. Sci. U.S.A.* **68,** 1336 (1971).
[38] H. Aviv and P. Leder, *Proc. Natl. Acad. Sci. U.S.A.* **69,** 1408 (1972).

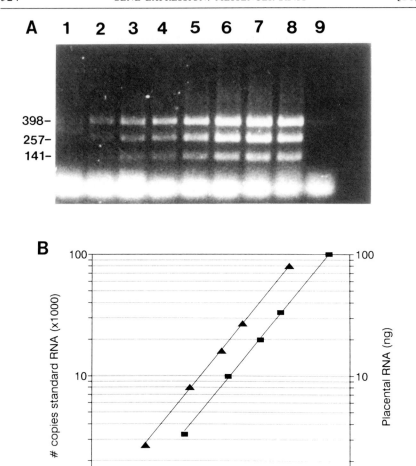

FIG. 3. Quantitative PCR by internal standards method. (A) One microgram total RNA from 13.5-day CD1 mouse (K^q) placenta was "spiked" with 8×10^5 copies of standard K^d RNA and reverse-transcribed. Dilutions were amplified using PCR primers to a 398-bp region of the class I major histocompatibility complex K gene mRNA in the presence of [^{32}P]dCTP. The PCR products were ethanol-precipitated and digested with *Rsa*I, which cuts the K^q product to yield 257- and 141-bp fragments and leaves the K^d product intact. The digests were electrophoresed on a 2% agarose–1.5% NuSieve gel. The bands were visualized by UV transillumination of the ethidium bromide-stained gel. Lanes 1 and 9 are negative controls. Lane 2 shows the PCR of 3.3 ng placental RNA spiked with 2.7×10^3 copies standard RNA; lane 3, 9.9 ng placental RNA with 8.0×10^3 copies standard RNA; lane 4, 19.8 ng placental RNA with 1.6×10^4 copies standard RNA; lane 5, 33.3 ng placental RNA

Set up the PCR to optimize cycle number. Amplify one-tenth aliquots of the RT^0 for 16, 18, 20, 22, 24, 26, and 28 cycles. Include 1 μCi of [^{32}P]dCTP per 50-μl reaction. Do *not* adjust the amount of unlabeled dCTP in the dNTP mix. After PCR, ethanol-precipitate the reactions and cut aliquots with the appropriate restriction enzyme. Run the products on a gel. Visualize by ethidium bromide staining. Excise bands from the gel and count the radioactivity. Plot the logarithm of the amount of radioactivity versus the cycle number. Do this for both the standard and sample products. Note whether the standard curve is parallel to that of the sample. This assures equivalent amplification efficiencies. See whether the amount of product for either curve has started to plateau, meaning that amplification has ceased to be exponential. Set up future PCRs so that reactions stay in the exponential phase. Note that, even with these data, the number of copies per nanogram total RNA is calculable. The number of copies (on the standard curve) giving the same amount of radioactivity as a given amount of total RNA (on the sample curve) can be read from the graph.

Amplify aliquots of the RT^0 and RT^{-1} in 50-μl reactions with label included. Try 2-, 4-, 6-, and 8-μl aliquots of each dilution. Run for the optimum number of cycles. Cut and analyze as above. Plot counts on the x axis versus copy number on one y axis (standard curve) and weight of total RNA on a secondary y axis (sample curve). Determine the number of copies of standard RNA with equivalent radioactivity to a given amount of total RNA. Figure 3 shows the results of an experiment using internal standards for determining the amount of major histocompatibility complex class I K^q mRNA in a sample of 13.5-day mouse placenta total RNA.

Notes. The amount of RNA to add to the original reverse transcription has to be determined empirically. It should approximate the amount of

---

2.7 $\times$ 10^4 with copies standard RNA; lane 6, 99.9 ng placental RNA with 8.0 $\times$ 10^4 copies standard RNA; lane 7, 198.0 ng placental RNA with 1.6 $\times$ 10^5 copies standard RNA; and lane 8, 333.3 ng placental RNA with 2.7 $\times$ 10^5 copies standard RNA. (B) The 398- and 257-bp bands were excised and placed in scintillation cocktail, and the radioactivity was measured. After correcting for background and normalizing for difference in size of bands (i.e., corrected counts for the 257-bp bands were multiplied by 398/257; corrected counts for the 398-bp bands were multiplied by 398/398), the standard RNA copy number (▲) and nanograms placental RNA (■) were plotted against radioactivity on log–log scales. Linear regression analysis was done on each data set, and the best fits were plotted as solid lines. To determine the number of copies of standard RNA corresponding to a given weight of placental RNA, such as 10 ng, a vertical line is drawn from the 10 ng point on the placental RNA curve to the standard RNA curve. The y coordinate of the point of intersection is the copy number ($\times$1000), in this case 2.0 $\times$ 10^4 copies. Note that the two points share the same x coordinate (i.e., they represent equivalent radioactivities).

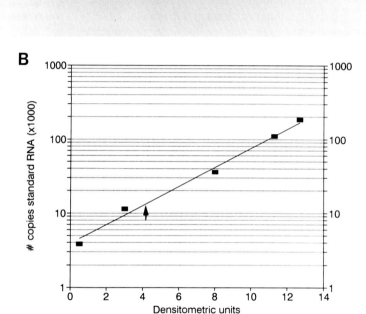

FIG. 4. Quantitative PCR by external standards method. (A) Total RNA was extracted from eggs/embryos as described. The egg/embryo preparations and dilutions of standard RNA were reverse-transcribed, and a 398-bp fragment from the class I major histocompatibility complex K gene message was amplified in the reverse-transcribed material by PCR, in the presence of $[^{32}P]dCTP$. The PCR products were electrophoresed on a 2% agarose–1.5% NuSieve gel. A "hot blot" was set up, as described, to transfer the amplified products onto Nytran, and Kodak XAR-5 film was placed against the dried filter. The autoradiograph is shown. Lanes 1 to 4 contain PCR product derived from the equivalent of 76 unfertilized eggs, 50 two-cell embryos, 44 eight-cell embryos, and 40 blastocysts, respectively. Lanes 5 to 9 contain PCR product derived from 3.6×10^3, 1.1×10^4, 3.6×10^4, 1.1×10^5, and 1.8×10^5 copies of standard RNA, respectively. Lane 10 is the negative control. (B) The

message in the original sample. When counting radioactivity, excise adjacent gel regions in negative control lanes for background. A rule of thumb for deciding if a quantitative experiment is worth counting at all is to look at the stained gel: if you can see it, you can count it quantitatively.

When amounts are limiting, setting up a "hot blot" works very well. Do not excise any bands. Instead, transfer the DNA onto a solid support. Do a rapid Southern blot onto Nytran (Schleicher and Schuell) using a transfer buffer with 0.4 M NaOH and 0.6 M NaCl and allow capillary transfer for 2–4 hr. After baking to immobilize the DNA, place the filter against XAR-5 film (Eastman Kodak). Exposure times vary depending on how "hot" the filter is. Take several exposures, to ensure that the signals are within the linear range of film response. Densitometric analysis will allow estimation of the amount of product in each band and construction of curves similar to the ones described. Other workers, such as Singer-Sam et al.,[39] do not use label in the amplification but do a Southern hybridization of the gel.

External Standards Method

Standards. The PCR amplified product itself cloned into a transcription vector like that described above makes a useful external standard which can be produced, purified, and quantitated in the same manner as the internal standards. A similar dilution series is made and kept in aliquots at $-20°$.

Quantitative RT–PCR. Set up reverse transcriptions so that reactions are equivalent in all respects except the increasing number of copies of standard RNA in the standard RT tubes. To correct for total nucleic acid content, include ribosomal RNA at the appropriate levels in each tube. Note that 200–300 eggs/embryos are extracted at a time, and therefore it is convenient to correct total content to the highest amount, usually the blastocyst preparation (which contains about 1 ng per embryo).

[39] J. Singer-Sam, M. O. Robinson, A. R. Bellvé, M. I. Simon, and A. D. Riggs, *Nucleic Acids Res.* **18,** 1255 (1990).

---

autoradiograph was analyzed by densitometry (Model 620 Video Densitometer, Bio-Rad). The area under each standard scan (lanes 5 to 9) was calculated, denoted by the arbitrarily chosen term "densitometric units," and plotted on the x axis against the known copy number corresponding to each standard dilution. The area under each sample scan (lanes 1 to 4) was calculated as well, and, from the standard curve, the copy number in each preparation was determined. The arrow indicates the point on the standard curve corresponding to the blastocyst preparation.

Do PCR on the reverse-transcribed products. Optimize cycle number as described, and use this optimum for future reactions. Construct a standard calibration curve and determine copy number for each sample from the curve. Figure 4 shows an example of a quantitation experiment using external standards to determine the number of copies of major histocompatibility complex class I K^q mRNA in a series of preimplantation mouse embryos.

Notes. The range of copy number of standard RNA to RT–PCR Must obviously encompass the range of actual copy numbers in the samples. The biggest concern when using this method is that tube-to-tube variability must be minimized. The use of master mixes in setting up both RT and PCR is important. Optimizing PCR conditions lowers tube-to-tube variability so that it is possible to obtain precisions of within 10%.[40] The parameters that are particularly important are magnesium ion concentration and proportions of primer to template. Increasing annealing times helps to decrease variability as well. Taking the time to work out the correct conditions for quantitative RT–PCR is well worth the effort.

Both methods of quantitation can work reliably. In terms of requirements for embryo RNA, the external method is much more economical. A single preparation of 200–300 embryos could be useful for multiple determinations by the external method. Moreover, an embryo preparation need not be dedicated to the quantitation of a single transcript. Once conditions are set, at least two other transcripts might be quantitated in the requisite triplicate determinations.

[40] K. K. S. Ng, M.Sc. Thesis, University of Calgary (1992).

[19] Quantitative Analysis of Specific Messenger RNAs in Small Numbers of Preimplantation Embryos

By MICHELLE F. GAUDETTE, QIU PING CAO,
and WILLIAM R. CRAIN

Introduction

The study of gene expression in preimplantation mouse embryos has the potential for uncovering and defining some of the earliest molecular events in the determination and differentiation of the mammalian embryo. One aspect of this analysis is to detect mRNAs from

Copyright © 1993 by Academic Press, Inc.
All rights of reproduction in any form reserved.

particular genes in early embryos to establish whether and when these genes are active. However, an understanding of the relationship between gene expression and early embryonic development may depend as much on a quantitative knowledge of mRNA levels as on the simpler, qualitative description of whether RNAs are present. In other words, the functional importance of a particular mRNA or its protein product in the early embryo may be determined largely by its prevalence. Thus, it is important to quantify the amount of specific mRNAs in early embryos. Although there are, in principle, a variety of ways to measure mRNA levels, the preimplantation mouse embryo presents a special and difficult case because of the limited number of embryos that can be obtained readily, and therefore the limited amounts of embryonic RNA available for analysis. With this constraint in mind, we have developed a method for quantifying specific transcripts in RNA preparations from small numbers of embryos using an assay based on the polymerase chain reaction (PCR), which is described here.

To detect specific sequences in RNA from small numbers of embryos, we combined reverse transcription, the polymerase chain reaction, and a ribonuclease protection assay (RT/PCR/RNase). The use of the RNase protection step to detect PCR-amplified DNA fragments by their ability to protect riboprobe from digestion increases the sensitivity of the assay and allows accurate determination of the amount of the amplified sequence. To quantify the signals detected from embryo RNA samples, different amounts of *in vitro* synthesized mRNA strand from the target sequence are also assayed by RT/PCR/RNase at the same time, to establish a standard curve. The signal from a known number of embryos is then superimposed on the standard curve to determine the number of transcripts present. Generally, the same oligonucleotide that is complementary to the mRNA sequence is used as both the primer for reverse transcription and the 3' primer for PCR. (The 5' primer, which is complementary to the opposite strand, should be located a convenient distance upstream.) By selecting the sequence of the 3' primer so that it is complementary to a region in the *in vitro* synthesized RNA that is used for the standard curve, the reverse transcription and PCR amplification are carried out on precisely the same sequences in the standard RNA and the target mRNA. This eliminates inaccuracies that might result from copying different sequences or lengths of sequences in the standard and target RNAs (a potential problem in some other quantitative PCR procedures). Furthermore, we have established conditions that allow the reverse transcription and PCR steps to be carried out sequentially in the same buffer and tubes, thereby simplifying the assay and reducing the number of manipulations of the samples.

TABLE I
COPY NUMBER OF FIVE DIFFERENT mRNAs IN
PREIMPLANTATION EMBRYOS

Transcript	Number of transcripts per embryo	Number of experiments	Ref.
Hox 2.1	18^a	3	1
β-Actin	$1.4 \times 10^{5\,a}$	4	1
594	$3.3 \times 10^{5\,b}$	6	c
MEA^g	350^a	1	d
Sry^e	4400–11,400^b	20	f

[a] Early blastocyst stage embryos.
[b] Late blastocyst stage embryos.
[c] Q. P. Cao and W. R. Crain, unpublished (1991).
[d] D. H. Robinson and W. R. Crain, unpublished (1990).
[e] For *Sry* the number of transcripts is per male embryo.
[f] Q. P. Cao, M. F. Gaudette, D. H. Robinson, and W. R. Crain, unpublished (1991).
[g] The clone for MEA, the male enhanced antigen gene, was obtained from Y. F. C. Lau, K. Chan, and R. Sparkes, *Proc. Natl. Acad. Sci. U.S.A.* **86**, 8462 (1989).

This procedure is sufficiently sensitive to measure as few as 100–200 transcripts in the RNA from one preimplantation embryo.[1] At the late blastocyst stage (about 110 cells), this is the equivalent of 1–2 transcripts/cell. If the amount of RNA used in the assay is increased to 10 embryo equivalents, as few as 10–20 transcripts/embryo can be measured accurately. So far, we have used this approach to determine the amounts of five different transcripts in preimplantation mouse embryos that range from about 18 transcripts/embryo up to 3.3×10^5 transcripts/embryo (Table I). A step-by-step description of the procedure and a discussion of some problems we have encountered follows. However, there are two overriding issues that we want to emphasize in general terms here. First, because of the exquisite sensitivity of PCR it is absolutely essential to demonstrate that all embryo and standard RNA preparations are free of DNA contamination. Second, the determination of the number of particular transcripts in embryo RNA, by comparison with a standard curve, must be carried out several times for each sequence because single measurements can vary severalfold.

Cloned Sequences

To use the methods described in this chapter, at least part of the sequence of interest must be available in a riboprobe vector. The cloned

[1] M. F. Gaudette and W. R. Crain, *Nucleic Acids Res.* **19**, 1879 (1991).

region should ideally be devoid of obvious inverted repeat sequences that might form hairpin structures and block transcription. Because both sense and antisense transcripts are used, it is advantageous to have the sequence in a vector that contains two different RNA polymerase promoters which flank a multiple cloning region in opposite orientations. Examples of such vectors are pBS or pBluescript from Stratagene (La Jolla, CA), or the pGEM series from Promega (Madison, WI). The *in vitro* transcription products for both the sense and antisense strands should be analyzed to determine the length of the RNAs that are produced.

Isolation of RNA from Preimplantation Mouse Embryos

When isolating RNA from embryos for use in a quantitative PCR experiment, it is important to consider several factors. First, the method used must yield intact RNA. Second, the RNA recovery must be determined. Finally, it is imperative that the RNA preparation be completely free of DNA contamination so that message levels are not overestimated. We have successfully used two different methods of RNA isolation.

In the first method,[2] the embryos (which are stored at $-70°$ in a minimal volume of Hank's balanced salt solution from the time of collection) are suspended in 31 μl of a solution containing $1 \times$ TNE (75 mM NaCl, 5 mM Tris-HCl, pH 8.0, 0.5 mM EDTA), 0.53% (v/v) Nonidet P-40 (NP-40), and 40 units of RNasin (Promega). The mixture is vortexed and the cells lysed by repeated rounds of flash freezing (in a dry ice/ethanol bath) and rapid thawing. An equal volume of $2 \times$ TNE is added to the cell lysate, plus sodium dodecyl sulfate (SDS) to 1% (w/v) (to lyse the nuclei) and 25 μg of carrier yeast tRNA. The mixture is then brought to 100 μl with water, extracted with phenol/chloroform, and precipitated with ethanol. After resuspending the pellet in 40 mM Tris-HCl, pH 7.9, 10 mM NaCl, and 6 mM MgCl$_2$, the sample is digested with 1 unit RQ1 DNase (Promega) at 37° for 15 to 30 min. The RNA is purified further by extraction with phenol/chloroform, precipitated by ethanol, and resuspended in water.

The second method of RNA isolation is that of Chomczynski and Sacchi,[3] with minor alterations. First, 100 μg of yeast tRNA is added as a carrier to the frozen embryos prior to cell lysis. Second, we routinely use 200 μl of solution D (see Ref. 3; rather than 1 ml per 100 mg of tissue) for all extractions, since only small quantities of embryos are generally available. The cells are disrupted by vortexing for 5 to 10 sec after the addition of solution D. The remainder of the protocol is followed as de-

---

[2] N. C. Nicolaides and C. J. Stoeckert, Jr., *BioTechniques* **8**, 154 (1990).
[3] P. Chomczynski and N. Sacchi, *Anal. Biochem.* **162**, 156 (1987).

scribed, using the recommended ratios for the volumes of water-saturated phenol, sodium acetate, and chloroform/isoamyl alcohol. Chomczynski and Sacchi[3] state that, after the first extraction step, DNA and proteins segregate into the phenol phase and RNA is retained in the aqueous phase. While this is generally true, we have observed variable amounts of genomic DNA contamination in some of these preparations of embryonic RNA. We speculate that DNA may be carried over during transfer of the aqueous phase away from the phenol phase. This problem can be overcome by digesting the RNA sample with RQ1 DNase, as described above.

The amount of RNA is determined by the number of embryos that were lysed, which must be counted carefully. Therefore, regardless of the RNA isolation method used, the percentage of RNA recovered must be established. This can be accomplished by adding labeled RNA to the embryos before cell lysis and monitoring recovery of the radioactivity throughout the course of the RNA isolation. In our recovery experiments, known quantities of [35]S- or ^{32}P-labeled RNA are added with the carrier tRNA to varying numbers of mouse embryos and total RNA isolated. The recovery of the tracer RNA for average method each was 65%.

Oligodeoxyribonucleotides

We routinely use oligonucleotide primers from our in-house facility without purification. The sequence of the region of the mRNA to be amplified must be known in order to select primer sequences. We generally choose oligonucleotides that are 20 to 25 nucleotides in length, although shorter sequences have been used successfully as primers for PCR amplification.

Sequences selected as PCR primers generally should have about 50–60% of G + C content.[4] In addition, the 5' and 3' primers should be matched as closely as possible for G + C content so that their optimal annealing temperatures are compatible. As a first approximation, the melting temperature of an oligonucleotide can be estimated by calculating 2° for every A or T residue and 4° for every G or C residue.[4] Primers in which the G + C contents are not well matched can be used together, but the PCR annealing temperature must be determined for the more AT-rich sequence. This may result in the synthesis of random sequences in addition to the amplified target sequence owing to annealing of the

[4] M. A. Innes and D. H. Gelfand, in "PCR Protocols: A Guide to Methods and Applications" (M. A. Innes, D. H. Gelfand, J. J. Sninsky, and T. J. White, eds.), p. 3. Academic Press, San Diego, 1990.

oligonucleotides to partially mismatched regions. The primer pairs also must not contain complementary sequences since they would then anneal to each other.

The locations of the oligonucleotide primers in the cloned sequence are also important. The region between the primers (the region to be amplified) should not contain sequences known to assume stable secondary structures that might prevent the synthesis of full-length cDNA and/ or amplified fragments. Furthermore, the amplified fragment should be a size that is conveniently detected. For all PCR experiments, the minimum size of the amplified fragment should be greater than that of the "primer–dimers" which sometimes form during the reaction. Among the practical considerations for deciding the maximum sizes is the fact that the duration of the elongation phase of the PCR cycle must be increased as the size of the amplified fragment increases. Because the method described in this chapter couples RT–PCR with RNase protection analysis, further constraints are imposed on the size of the amplified fragment. First, the maximum length of the amplified fragment should be smaller than the labeled riboprobe to allow discrimination between protected riboprobe fragments and the undigested, full-length riboprobe. Second, the amplified fragment should not be the same size as any RNase-resistant regions of the probe. Ideally, RNase digestion of the riboprobe alone produces only very small oligomers that are not retained on the assay gel. However, digestion of some riboprobes results in the appearance of discrete bands, apparently representing self-protected fragments. In addition, the digestion of some probes yields a heterogeneous population of oligomers that appears as a low molecular weight, diffuse smear on the assay gel.

Sometimes the target message is a member of a gene family. If this is the case, the presence of related sequences in the embryo RNA might interfere with the quantification by competing with the target for the primers during the amplification reaction. In addition, if the sequences are closely related, the amplified related sequences could hybridize to the labeled riboprobe, artificially elevating the number of messages detected. To avoid this problem, the oligonucleotide primers (at least one, preferably both) should be selected from regions of nonconserved sequences among the family members so that the specificity of amplification is retained. (If primer selection is insufficient to obtain the desired specificity, the annealing temperature in the PCR step and the hybridization temperature in the RNA–DNA hybridization may also be increased.)

For several reasons, we routinely test all new primers in an amplification reaction with the cloned sequence as template. First, we have found that occasionally a particular primer preparation is "bad." When this occurs, the test amplification of the cloned sequence yields little or no product. Subsequent preparations of identical oligonucleotide sequences, however, are usually functional and give a fragment of the expected size when tested. We have also found that some primer combinations simply do not work: test amplification yields no products, and subsequent preparations of either oligonucleotide sequence do not remedy the situation. However, when these primers are combined with different primers from the same gene, they can yield expected amplification products. Because the primer pairs are selected to be noncomplementary, there is no obvious explanation for this phenomenon. Finally, test reactions can be used to establish optimal amplification conditions for the specific primers used. This is usually accomplished by varying the concentration of Mg^{2+} ions in the reaction buffer and by altering the PCR annealing temperature, starting at 5° below the calculated melting temperature.

RNA Quality

Because the amount of RNA isolated from mouse embryos is low, standard methods for testing the quality of the RNA preparations cannot be used. Instead, we test the amplification of β-actin mRNA, a sequence that is known to be abundant in preimplantation embryos. Two samples are tested. In the first, cDNA is synthesized from an oligo(dT) primer, using the method of Gerard.[5] The second sample is identical to the first, except that the reverse transcriptase is eliminated. After purification of the cDNA, a region of the β-actin gene is amplified by PCR in both samples and detected by RNase protection analysis. If the quality of the RNA preparation is good, a protected fragment of the expected size will be detected in the sample containing reverse transcriptase. The appearance of a protected fragment in the mock-transcribed sample could indicate the presence of contaminating genomic DNA. In our experience, good RNA samples do not give signal in the absence of reverse transcriptase. However, there are reports that *Taq* DNA polymerase exhibits reverse transcriptase activity *in vitro*,[6,7] and it could produce a protected fragment in the absence of added reverse transcriptase. If this occurs, an additional control can be performed to verify whether there is DNA contamination by digesting the RNA sample with RNase before reverse transcription. No protected fragment should result if the RNA is free of contaminating DNA. If a protected fragment is produced despite predigestion of the sample, genomic DNA is present, and the RNA preparation should be treated with RQ1 DNase, as described above.

[5] G. F. Gerard, *BRL Focus* **10**, 12 (1988).
[6] M. D. Jones and N. S. Foulkes, *Nucleic Acids Res.* **17**, 8387 (1989).
[7] W. T. Tse and B. G. Forget, *Gene* **88**, 293 (1990).

It is convenient to attempt to amplify and detect target sequence when performing the test above. A crude estimate of the abundance of the target sequence can be made by comparing its signal to that of the β-actin sequence if the same cDNA template is used, and if both riboprobes are of the same specific activity and are used in equal amounts. The test for the presence of genomic DNA contamination (amplification with and without reverse transcriptase) should be repeated for the target sequence.

Synthesis of Message-Strand Standard RNA

To synthesize message-strand RNA *in vitro,* linearize the plasmid clone on the 3' side of the insert and transcribe it with the appropriate RNA polymerase in the presence of [^{32}P]UTP at a low specific activity [we synthesize the RNA at about 1×10^6 disintegrations/min (dpm)/μg]. Transcription reactions contain 0.5 μg linearized plasmid DNA template, 500 μM each ATP, GTP, and CTP, 12.2 μM UTP, 0.1 μCi [^{32}P]UTP (800 Ci/mmol, NEN, Boston, MA), 16 units RNasin, and 10 units of RNA polymerase in 40 mM Tris-HCl, pH 7.9, 10 mM NaCl, 10 mM dithiothreitol (DTT), 6 mM MgCl$_2$, 2 mM spermidine, and 50 μM EDTA. Reactions are performed in a final volume of 20 μl at 37° (T3 or T7 polymerase) or 40° (SP6 polymerase) for 1 hr. Template DNA is removed by digestion with 1 unit of RNase-free DNase (Promega) at 37° for 15 min (or longer). The synthesized RNA is purified by phenol/chloroform extraction, and unincorporated nucleotides are removed by centrifugation through a 1-ml bed of Sephadex G-50 resin. The RNA is then ethanol-precipitated in the presence of 5 μg of yeast tRNA and resuspended in water.

It is crucial that no contaminating DNA remains after isolation of the standard RNA, so an amplification/detection reaction (RT/PCR/RNase method described below) must be performed on at least the following three samples. In the first, the standard RNA serves as the template in a complete reaction mix. This should result in a protected fragment of the predicted size. A second sample uses the standard RNA in a reaction that is missing only the reverse transcriptase. As stated above, no protected fragment should result if the standard RNA is not contaminated with plasmid DNA. The detection of a protected fragment, however, suggests the presence of residual plasmid DNA and the necessity of further digestion with DNase. Finally, an aliquot of standard RNA is digested with RNase and then subjected to RT/PCR/RNase assay. If there is no plasmid contamination, the detected band will disappear; if DNA remains, the detected band will be RNase-resistant.

To determine how much standard RNA has been synthesized, the specific activity of the ^{32}P-labeled nucleotide is calculated for the day used, following the equations and decay tables provided by the manufacturer. If the entire sequence of the synthesized RNA is known, the actual fraction of the labeled nucleotide should be used in this calculation; if not, it is probably best to assume it to be one-fourth of the total nucleotides incorporated. By using the specific activity of the synthesized RNA, calculated in this way, and by determining the total radioactivity incorporated into trichloroacetic acid-precipitable material, it is possible to calculate precisely the total amount of RNA synthesized and recovered.

One-Step Reverse Transcriptase–Polymerase Chain Reaction Amplification

Once it has been demonstrated that the quality of both the standard and embryonic RNA preparations is good and that neither is contaminated by DNA, the quantitative reaction can be performed. The basic strategy used is reverse transcription of the target message from a sequence-specific oligonucleotide primer in the presence of *Taq* DNA polymerase, immediately followed by PCR amplification. Because the reactions occur in the same tube without any manipulations between the two steps, the reaction conditions for both enzymes must be compatible. The recommended conditions for the reverse transcriptase isolated from avian myeloblastosis virus (AMV) are quite different from those for *Taq* polymerase. Although Goblet *et al.*[8] report successful one-step amplification reactions using AMV reverse transcriptase, we instead chose to use the reverse transcriptase isolated from Moloney murine leukemia virus (MMLV). The buffer conditions for MMLV reverse transcriptase also differ somewhat from those for *Taq* polymerase [1 × reverse transcription (RT) buffer is 50 mM Tris-HCl, pH 8.3, 75 mM KCl, 3 mM MgCl$_2$, 10 mM DTT, whereas 1 × PCR buffer is 10 mM Tris-HCl, pH 8.3, 50 mM KCl, 2.5 mM MgCl$_2$, 100 μg/ml bovine serum albumin (BSA)], but we have shown that *Taq* polymerase functions equivalently under both conditions.[1] We therefore use reverse transcription buffer in these reactions. The source of the enzymes may also be important. We have used *Taq* DNA polymerase only from Perkin-Elmer Cetus (Norwalk, CT) and have found that both the native enzyme and the cloned enzyme, Ampli-Taq, work well. Different sources of reverse transcriptase seemed to give variable results. The most reproducible experiments were obtained when using MMLV reverse transcriptase from Stratagene.

To establish the standard curve, serial dilutions of the standard RNA are made, and a constant volume of each is aliquoted into separate reaction

[8] C. Goblet, E. Proust, and R. G. Whalen, *Nucleic Acids Res.* **17,** 2144 (1989).

tubes. It is convenient to begin with 10-fold dilutions, thereby amplifying the standard RNA levels over several orders of magnitude. Once the amount of message has been estimated in this fashion, all subsequent standard curves can more closely bracket this level (e.g., 3-fold dilutions over two orders of magnitude). The amount of embryonic RNA used will necessarily vary depending on the number of messages present. In our hands, the RNA from 5–10 embryos is sufficient to detect Hox 2.1 transcripts,[1] which are present at about 18 copies per embryo (Table I).

To perform the one-step reaction, embryonic RNA or varying amounts of standard RNA are mixed with 5 μg *Escherichia coli* rRNA and 1 μg each of the gene-specific 5′ and 3′ PCR primers. The RNA/oligonucleotide mixtures are heated to 70° for 5 min and cooled slowly by incubating first at 37° for 10 min and then at room temperature for 10 min. Each tube then receives reaction mix, such that each 50-μl reaction contains 150 μM each dNTP, 20 units MMLV reverse transcriptase (Stratagene), and 1 unit *Taq* DNA polymerase in $1 \times$ RT buffer. The samples are incubated at 37° for 1 hr in the automatic temperature cycler to allow reverse transcription to occur, and then the program is immediately looped into the appropriate PCR amplification cycle.

It is critical that great care be used in setting up the reactions described. To avoid contamination, we recommend the use of aerosol-resistant pipette tips and separate pipettors for PCR reagents. In addition, negative control reactions should be included with each experiment. These can be reactions that contain all components except the reverse transcriptase, or reactions containing RNase-digested template.

Ribonuclease Protection Analysis

After the RT/PCR amplification steps, DNA fragments are detected and quantified using an RNase protection assay. Antisense riboprobe is transcribed as described above (see Synthesis of Message-Strand Standard RNA) except that the amount of [^{32}P]UTP (800 Ci/mmol) added is calculated to yield a riboprobe specific activity between 6×10^7 and 3×10^8 dpm/μg. After purification, the probe is resuspended in hybridization buffer (80%, v/v, formamide, 40 mM PIPES, pH 6.7, 0.4 mM NaCl, 1 mM EDTA) and trichloroacetic acid-precipitable counts measured.

For quantification to be accurate, it is imperative that the entire amplification reaction is recovered. This can be accomplished by direct removal of the aqueous PCR reaction mixture from beneath the protective oil layer. Alternatively, Perkin-Elmer Cetus suggests extraction of the reaction with high-quality chloroform, so that the aqueous phase will float above the chloroform/oil mixture. The DNA is purified by phenol/chloroform extrac-

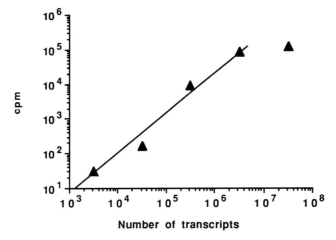

Number of transcripts

FIG. 1. Standard curve with limiting riboprobe. A one-step RT/PCR/RNase assay was performed on varying amounts of *in vitro* synthesized *Sry* RNA. *Sry* riboprobe (5×10^5 cpm, specific activity 6.3×10^7 dpm/μg) was used with each sample during the RNase protection step. Following gel electrophoresis and autoradiography, regions corresponding to the protected fragments were excised from the dried gel and counted. After correcting for background, the amount of radioactivity in each band was plotted against the number of transcripts originally present in each sample. The amount of radioactivity reached a maximum level when about 10^6 transcripts were present, increasing only slightly when 10^7 transcripts were amplified. The best-fit line is drawn only through the linear region of the curve.

tion and chloroform back-extraction, followed by ethanol precipitation. Next, each pellet is resuspended in 30 μl of probe [typically 5×10^5 counts/min (cpm) per sample] diluted in hybridization buffer. In addition to the experimental samples, a mock hybridization mixture should be used that contains only 5 μg of tRNA and 5×10^5 cpm of probe in hybridization buffer.

The amount of probe used in the hybridization step is an important variable. The probe must be in sufficient excess over the amplified DNA fragments in the standard curve samples so that it does not become limiting over the range of the experiment. At the point where the probe becomes limiting, the standard curve will stop rising linearly and begin to plateau. An example of this is shown in Fig. 1, where the standard curve signals increase linearly through about 10^6 transcripts and then level off. In our experience, the plot shown in Fig. 1 is reproducible for most probes when 5×10^5 cpm of probe with a specific activity of about 6.3×10^7 dpm/μg is used. The point of saturation can be shifted by changing the amount of probe used to optimize the standard curve in the range of the RNA of interest. For abundant RNAs, greater amounts of a lower specific activity probe might be used, and for somewhat rare RNAs lower amounts of higher specific activity probe might be employed.

Immediately after probe addition, the DNA pellets are resuspended, and the samples are heated to 85° for 5 min and allowed to hybridize overnight at 50° to 55°. Although this hybrization temperature range works well for most cases, it can be varied to alter the stringency of sequence detection. Higher hybridization temperatures increase the stringency of detection but may also decrease the signal. Optimum hybridization temperatures must be determined empirically for special cases.

Following hybridization, the samples are incubated with RNase to digest unannealed riboprobe while leaving the RNA–DNA hybrids intact. This should result in little radioactivity being detected in the mock hybridization sample and the presence of a single protected fragment of the predicted size in the standard curve reactions. An indication of underdigestion by RNase is the presence of full-length probe in all samples. The presence of heterogeneously sized probe fragments in samples, visible as a "smear" of radioactivity in the gel lanes, is another indication of poor or incomplete digestion. These should not be confused with self-protected fragments, which are usually discrete bands present in all samples, including the control. On the other hand, overdigestion should be suspected if the protected fragment is smaller than predicted, or there are multiple protected smaller fragments present in the standard curve samples. The sequences most susceptible to overdigestion seem to be AT-rich regions.[9] It is best to determine the optimal digestion conditions in pilot experiments.

We generally use three sets of RNase digestion conditions, although an infinite variety can be imagined. Under the most stringent conditions used ("high RNase"), 300 μl of reaction mix containing 40 μg/ml RNase A, 2.2 μg/ml RNase T1, 10 mM Tris-HCl, pH 8.0, 5 mM EDTA, and 0.3 M NaCl is added to each hybridization reaction. Under the least stringent conditions ("low RNase"), the reaction mix contains only 2.5 μg/ml RNase A and 0.3 μg/ml RNase T1, in 10 mM Tris-HCl, pH 8.0, 5 mM EDTA, and 0.9 M NaCl. The intermediate reaction condition contains 25 μg/ml RNase A and 1 μg/ml RNase T1, in 10 mM Tris-HCl, pH 8.0, 5 mM EDTA, and 0.5 M NaCl. Digests are incubated at either 37° for 30 min or at room temperature for 1 hr and stopped by incubation at 37° for 15 min with 50 μg of proteinase K and 0.6% (w/v) SDS. Undigested riboprobe is purified by phenol extraction and precipitated by ethanol in the presence of 5 μg carrier tRNA. Protected fragments are separated by electrophoresis through denaturing polyacrylamide gels and are visualized

---

[9] B. P. Bullock, P. E. Nisson, and W. R. Crain, *Dev. Biol.* **130,** 335 (1988).

by autoradiography. (Multiple exposures are helpful in analysis of the data.)

An example of such an experiment is shown in Fig. 2, in which levels of message detected by a particular cDNA clone (named 594) are quantified in early, mid, and last blastocyst stage preimplantation mouse embryos. In this particular experiment the points on the standard curve range over two orders of magnitude (3.2×10^3 to 3.2×10^5), and the embryo RNAs were from 0.1 embryo equivalents. The mock hybridization (lane d, Fig. 2A) demonstrates complete digestion, while the lane containing undigested probe (lane p, Fig. 2A) indicates the size of the full-length transcript. Complete digestion of unannealed riboprobe is further shown by the absence of full-length probe either in the standard curve or in the embryonic RNA samples. When loading the gel, it is often helpful to segregate the embryonic RNA sample away from the standard curve to prevent spillover of radioactivity from strong signals into lanes with lower signals.

Once the desired autoradiographic exposures are obtained, the regions of the dried gel containing the protected fragments as well as the corresponding regions of the control lane (digested probe background) are excised and counted in liquid scintillation fluor. At this point, it is suggested that another autoradiographic film be exposed to verify that the appropriate regions of the gel were removed. After subtracting background, the counts/minute present in each sample are plotted against the number of transcript molecules present in the sample before amplification. A best-fit straight line is drawn through the linear portion of the curve (see Figs. 1 and 2B). If riboprobe has become limiting and the standard curve has leveled off, data in the plateau region are not analyzed. The total number of transcripts present in the embryo sample is found by superimposing the counts data onto the standard curve and interpolating to determine transcript number. This number can be normalized to transcripts per embryo or transcripts per cell.

Variations

We think that the one-step assay procedure described above is best and should always be used, if possible. However, for one of the five messages that we have quantified in embryos so far (*Sry*) it was necessary to modify the approach somewhat. In the case of *Sry* we were unable to detect the expected amplified fragment reliably using the one-step assay. Primer extension analysis suggested that, although cDNA was synthesized from the 3' primers we tested, the major product was not long enough to serve as a PCR template because the cDNA did not include the 5' primer sequence. We have no good explanation for this.

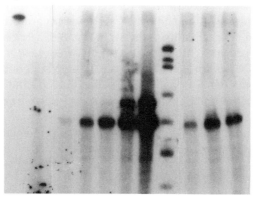

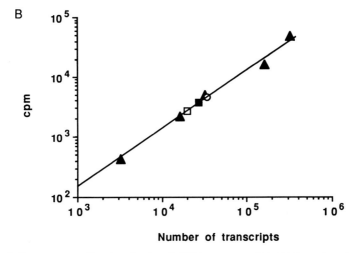

FIG. 2. One-step quantification of a cloned cDNA sequence, 594. (A) Detection of standard and natural 594 mRNAs by RNase protection of 594 riboprobe by RT–PCR amplified DNA fragments. The levels of 594 RNA from three different stages of preimplantation mouse embryos were determined in this experiment. The lanes contain the following: (p) undigested probe; (d) digested probe, no added DNA; (1) 3.2×10^3 copies of standard RNA; (2) 1.6×10^4 copies; (3) 3.2×10^4 copies; (4) 1.6×10^5 copies; (5) 3.2×10^5 copies; (M) markers; (e) RNA from the equivalent of 0.1 early blastocyst; (m) 0.1 mid blastocyst; (l) 0.1 late blastocyst. In this experiment, 1.3×10^6 dpm of riboprobe at a specific activity of 5.2×10^7 dpm/μg was hybridized to amplified DNA fragments. The film was exposed at $-70°$ for 1.5 hr. (B) Standard curve constructed after counting regions of the dried gel containing the 186-nucleotide protected fragments. ▲, Standard RNA; □, 0.1 early blastocyst; ○, 0.1 mid blastocyst; ■, 0.1 late blastocyst.

To quantify *Sry* RNA in embryos, we chose to modify our approach by using oligo(dT) as the primer for cDNA synthesis and by carrying out the reverse transcription and the PCR in two separate steps. An example of one of these experiments is shown in Fig. 3. The cloned *Sry* sequence, which was used to synthesize the standard curve RNA, had been sub-cloned into the vector pSP64[poly(A)]. This resulted in an *in vitro* transcription product that contained 30 A residues on the 3' end of the *Sry* sequence. To assure that all of the standard RNA transcripts were full-length and contained the poly(A) tail, they were passed over an oligo(dT)-cellulose column. Only the poly(A)-containing molecules were used as template in the standard curve. Known amounts of poly(A)-containing standard RNA and embryonic RNA were reverse-transcribed from an oligo(dT) primer, following the procedure of Gerard.[5] After addition of 5 μg of *E. coli* rRNA (carrier) to the standard and embryonic RNAs, the samples were denatured at 70° for 5 min followed by rapid cooling on ice. To each sample, reverse transcription mix was added so that each 30-μl reaction contained 1.5 μg oligo(dT), 250 μM each dNTP, and 30 units (Stratagene) or 300 units (BRL, Gaithersburg, MD) MMLV reverse transcriptase in 1× RT buffer. The mixtures were incubated at 37° for 1 hr, and the resulting cDNAs were purified by phenol extraction followed by ethanol precipitation. The pellets were resuspended in water and PCR components added so that each reaction contained 250 nM of each primer, 200 μM dNTPs, and 1 unit *Taq* DNA polymerase in 10 mM Tris-HCl, pH 8.3, 50 mM KCl, 1.5 mM MgCl$_2$, 100 μg/ml BSA, and 0.05% Tween 20.[10] Amplification cycles and subsequent detection followed standard protocols.

We believe the one-step protocol is more accurate in principle than the two-step protocol described here. In the one-step reaction, cDNA synthesis from a specific primer is the same with the standard RNA template and the embryonic RNA template. In the two-step procedure, the poly(A) tail of the standard RNA might be located closer to the target amplification sequence than it is in authentic mRNA. If this is the case, the length and the nucleotide composition of the region between the poly(A) tail and the target sequence will be different between the authentic and standard messages, and these factors might influence the efficiency of reverse transcription and, indirectly, the efficiency of PCR amplification.

[10] R. K. Saiki, *in* "PCR Protocols: A Guide to Methods and Applications" (M. A. Innis, D. H. Gelfand, J. J. Sninsky, and T. J. White, eds.), p. 13. Academic Press, San Diego, 1990.

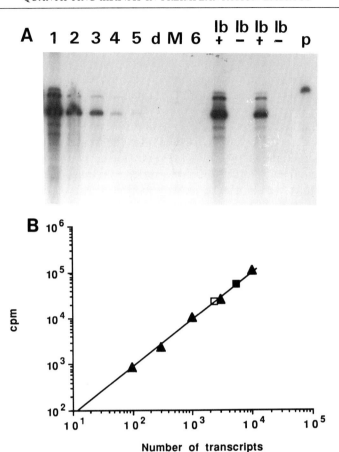

FIG. 3. Two-step quantification of *Sry* transcripts. (A) Detection of standard and natural *Sry* RNAs by RNase protection of *Sry* riboprobe by DNA fragments amplified by PCR in a two-step reaction. In this experiment, *Sry* message was detected in two separate RNA preparations from late blastocyst stage mouse embryos. In addition, amplification was shown to be dependent on reverse transcriptase by assaying in the presence ($+$) or absence ($-$) of reverse transcriptase. The lanes contain the following: (1) 10^4 copies of standard RNA; (2) 3×10^3 copies; (3) 10^3 copies; (4) 3×10^2 copies; (5) 10^2 copies; (d) digested probe, no DNA added; (M) markers; (6) 1×10^5 copies of standard RNA without reverse transcriptase; (lb$+$) RNA from 2 late blastocyst equivalents reverse-transcribed with MMLV reverse transcriptase; (lb$-$) RNA from 2 late blastocyst equivalents with no reverse transcriptase added during the cDNA synthesis step (two different blastocyst samples were tested); (p) undigested probe. In this experiment 5×10^5 cpm of riboprobe at a specific activity of 1×10^8 dpm/μg was hybridized to amplified DNA fragments. The protected fragment is 319 nucleotides. The film was exposed at room temperature for 2.5 hr. (B) Standard curve constructed after counting the regions of the dried gel containing the 319-nucleotide protected fragments. ▲, Standard RNA; ■, 2 late blastocyst equivalents, preparation 1; □, 2 late blastocyst equivalents, preparation 2.

Reproducibility

We have found that quantification of a specific message can vary up to 3- or 4-fold in different experiments using this approach. Part of this may be due to slight variations in embryo staging, particularly for messages such as *Sry* and 594 which exhibit a remarkable increase within a short time frame. Because of this, we routinely carry out at least three or four separate analyses and try to use several different embryo RNA preparations. It is also important to realize that the standard curve must be generated simultaneously with the reverse transcription and amplification of the experimental sample. Results from separate quantification reactions, although similar, are not necessarily superimposable on one another.

Acknowledgments

We would like to thank Douglas H. Robinson for the use of some of the unpublished results. Much of the work that resulted in this procedure was supported by a Basic Research Grant from the March of Dimes Birth Defects Foundation (to W.R.C.) and by a National Cancer Institute Cancer Center Grant. M.F.G. was a trainee on an institutional National Research Service Award from the National Institute of Child Health and Human Development.

[20] Quantitative Analysis of Messenger RNA Levels: Reverse Transcription–Polymerase Chain Reaction Single Nucleotide Primer Extension Assay

By J. SINGER-SAM and A. D. RIGGS

Introduction

Changes in absolute and relative amounts of specific RNAs can be measured by polymerase chain reaction (PCR) amplification following a reverse transcription step (RT–PCR).[1-3] Although Northern blotting has been used routinely to visualize specific RNAs, RT–PCR offers several advantages: (1) minimal purification of RNA is needed; (2) the amount of material required is much less (several hundred molecules of a given transcript are usually detectable); and (3) RT–PCR is quantitative even

---

[1] G. Veres, R. A. Gibbs, S. E. Scherer, and C. T. Caskey, *Science* **237**, 415 (1987).
[2] K. E. Noonan and I. B. Roninson, *Nucleic Acids Res.* **16**, 10366 (1988).
[3] E. S. Kawasaki, *in* "PCR Protocols" (M. A. Innis, D. H. Gelfand, J. J. Sninsky, and T. J. White, eds.), p. 21. Academic Press, New York, 1990.

Copyright © 1993 by Academic Press, Inc.
All rights of reproduction in any form reserved.

when the transcripts of interest are present in such low abundance that the signal from Northern blots is lost or irreproducible.

A number of laboratories have demonstrated the use of RT–PCR for quantitative detection of RNA. Two general types of methods have been used. The first method relies on adjustment of the amount of input RNA and the number of cycles of PCR, to assure that measurement is done in the "log phase" of PCR when the signal is proportional to the amount of input template. In its simplest form, this method is usable for measurement of relative changes in RNA levels when there are at least 2-fold differences.[4,5] To assure that saturation has not yet been reached, a standard curve may be generated by amplification of various dilutions of an external standard amplifiable by the same primer set as the target RNA. If the concentration of the external standard is known, the absolute amount of input RNA may be calculated.[6]

A second approach, involving use of an internal standard, becomes essential when one is measuring small changes in RNA levels, or when too little starting material is available for accurate determination of the amount of input RNA in each sample. The internal standard may be an endogenous ubiquitous RNA [i.e., mRNA for phosphoglycerate kinase (PGK-1) or β-actin], a known amount of a plasmid or an amplified PCR fragment, or *in vitro* transcribed RNA. The sequence of the internal standard may differ totally from that of the target RNA,[7–9] or it may be similar to the target RNA and amplifiable with the same primer set.[10–12]

After amplification and gel electrophoresis, a number of techniques can be used to measure the signal from each amplified product. Ethidium bromide-stained gels may be photographed onto film for densitometric measurement.[13] Alternatively, the radioactive signal may be measured

[4] J. Singer-Sam, M. O. Robinson, A. R. Bellvé, M. I. Simon, and A. D. Riggs, *Nucleic Acids Res.* **18**, 1255 (1990).

[5] R. A. W. Rupp and H. Weintraub, *Cell (Cambridge, Mass.)* **65**, 927 (1991).

[6] M. O. Robinson and M. I. Simon, *Nucleic Acids Res.* **19**, 1557 (1991).

[7] D. A. Rappolee, D. Mark, M. J. Banda, and Z. Werb, *Science* **241**, 708 (1988).

[8] K. E. Noonan, C. Beck, T. A. Holzmayer, J. E. Chin, J. S. Wunder, I. L. Andrulis, A. F. Gazdar, C. L. Willman, B. Griffith, D. D. Von Hoff, and I. B. Roninson, *Proc. Natl. Acad. Sci. U.S.A.* **87**, 7160 (1990).

[9] T. Hoof, J. R. Riordan, and B. Tümmler, *Anal. Biochem.* **196**, 161 (1991).

[10] A. M. Wang, M. V. Doyle, and D. F. Mark, *Proc. Natl. Acad. Sci. U.S.A.* **86**, 9717 (1989).

[11] M. Becker-André and K. Hahlbrock, *Nucleic Acids Res.* **17**, 9437 (1989).

[12] G. Gilliland, S. Perrin, and H. F. Bunn, *in* "PCR Protocols" (M. A. Innis, D. H. Gelfand, J. J. Sninsky, and T. J. White, eds.), p. 60. Academic Press, New York, 1990.

[13] J. Singer-Sam, T. P. Yang, N. Mori, R. L. Tanguay, J. M. LeBon, J. C. Flores, and A. D. Riggs, *in* "Nucleic Acid Methylation" (G. Clawson, D. Willis, A. Weissbach, and P. Jones, eds.), p. 285. Alan R. Liss, New York, 1990.

following incorporation of a ^{32}P-labeled nucleoside triphosphate (^{32}P] dNTP) during amplification,[6,14] or hybridization may be carried out to a ^{32}P-labeled oligonucleotide after transfer of the amplified products to nylon filters.[15] Similar methods may also be used in conjunction with further amplification of the signal by an *in vitro* transcription step following PCR.[14-16] An alternative method, the quantitative RT–PCR single nucleotide primer extension (SNuPE) assay, has been described[17] and is discussed below in further detail.

Selection of Internal Standard

Athough it is frequently convenient to use endogenous internal standards that differ totally in sequence (e.g., β-actin RNA), one runs the risk that the efficiency of amplification may be quite different between primer sets and may respond differently to tube-to-tube variation. In theory, therefore, and in practice, the surest way to obtain reproducible quantitative data is to include an internal standard that can be used with the same primer set as the template of interest. The standard must differ from the sample template either in size or in sequence, so that the ratio of the amplification products of the two templates can be determined.

Three methods have been used to distinguish the sample from the standard. (1) Sample and standard can differ in size. A deleted version of the template can be synthesized (also by PCR) by use of a "deletion primer" that loops out an internal segment of the sequence.[18] Alternatively, if a short intron (up to ~200 bases) is included in the region being amplified, the standard for amplification of the mRNA may consist of the cloned genomic template.[12] A potential disadvantage of standards prepared in this way is the possibility that the efficiency of hybridization to the primers may vary slightly owing to the size differences between the two templates. (2) The standard can be synthesized to contain a restriction site lacking in the template (or vice versa). This can be accomplished by site-specific mutagenesis,[19] although for mice a simpler route in many cases will involve exploitation of differences in sequence among different

[14] T. Horikoshi, K. D. Danenberg, T. H. W. Stadlbauer, M. Volkenandt, L. C. C. Shea, K. Aigner, B. Gustavsson, L. Leichman, R. Frösing, M. Ray, N. W. Gibson, C. P. Spears, and P. V. Danenberg, *Cancer Res.* **52,** 108 (1992).

[15] G. J. Murakawa, J. A. Zaia, P. A. Spallone, D. A. Stephens, B. E. Kaplan, R. B. Wallace, and J. J. Rossi, *DNA* **7,** 287 (1988).

[16] E. S. Stoflet, D. D. Koeberl, G. Sarkar, and S. S. Sommer, *Science* **239,** 491 (1988).

[17] J. Singer-Sam, J. M. LeBon, A. Dai, and A. D. Riggs, *PCR Methods Appl.* **1,** 160 (1992).

[18] J. Singer-Sam, J. M. LeBon, R. L. Tanguay, and A. D. Riggs, *Nucleic Acids Res.* **18,** 687 (1990).

[19] R. Higuchi, B. Krummel, and R. K. Saiki, *Nucleic Acids Res.* **16,** 7351 (1988).

strains. In this case only one base change is sufficient for comparison of both amplified products by measurement of the strength of the signal obtained before and after digestion with restriction endonucleases. A potential disadvantage of the method is the uncertainty of obtaining complete digestion.[12] (3) A standard differing by only a single base change can be used and quantified relative to the template of interest by use of the quantitative SNuPE assay. This assay, now used routinely in our laboratory, is described in detail below.

Quantitative Reverse Transcription–Polymerase Chain Reaction Single Nucleotide Primer Extension Assay

The single nucleotide primer extension assay was first described for diagnosis of allele-specific differences in DNA.[20] We have adapted it for quantitative measurement of specific RNAs[17] and find that it offers several advantages. (1) Any base difference should be usable, so that allelic sequence differences can be exploited for preparation of the internal standard, even when there are no differences at restriction sites. (2) The assay is quantitative even when the sample is present as 0.1–1% of the amount of the internal standard. (3) The assay lends itself to quantitative measurement of allele-specific transcripts so that imprinting or allelic exclusion can be studied. (Although only the assay of RNA is discussed here, the method also lends itself to studies of alterations in genomic DNA in a minor fraction of cells in a sample; in this way, the assay can be used to study genetic rearrangements and mutations.)

The steps of the assay are shown schematically in Fig. 1. (1) The RT–PCR is carried out with primers common to the target RNA and the internal standard. The standard, allele 2 in Fig. 1, which was shown to differ by a C to A transversion, may be added as RNA to the RT reaction or as DNA just prior to PCR.[12] For allele-specific measurement the internal standard is automatically present if heterozygous animals are used. The RT–PCR primers are designed so that the region of difference between the two alleles (or the target RNA and the internal standard) is amplified. (2) Amplified products of the expected size are gel-purified (both products are contained within the same band). (3) The sample is incubated with *Taq* polymerase for one round of extension of a third primer, the SNuPE primer. The 3′ end of the primer is just 5′ to the position of the difference between the amplified products. In separate tubes, the [^{32}P]dNTP appropriate for extension of each of the two amplified products is added. (4) The products of the two reactions are subjected to denaturing gel electro-

[20] M. N. Kuppuswamy, J. W. Hoffmann, C. K. Kasper, S. G. Spitzer, S. L. Groce, and S. P. Bajaj, *Proc. Natl. Acad. Sci. U.S.A.* **88,** 1143 (1991).

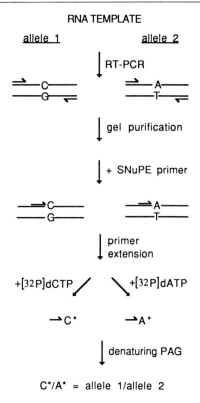

C*/A* = allele 1/allele 2

FIG. 1. Schematic outline of the quantitative RT–PCR SNuPE assay. Allele 2 may be an exogenously added standard or an allelic variant. PAG, Polyacrylamide gel electrophoresis. Arrows represent primers; asterisks denote ^{32}P-labeled products.

phoresis; the amount of radioactivity in each band of the right size is determined and the ratios calculated.

We have found that the RT–PCR SNuPE assay is quantitative over a wide range. We describe here the detailed procedure as used in our laboratory for measurement of allele-specific transcripts where at least one base pair difference between alleles has been identified. For measurement of the absolute amount of a given RNA, one should substitute an exogenously added internal standard as described above.

RNA Isolation

Total RNA is isolated from a strain heterozygous for the transcript of interest. RNA is extracted from small numbers of cells by addition of chloroform to a guanidinium thiocyanate–phenol mixture,[21] commercially

[21] P. Chomczynski and N. Sacchi, *Anal. Biochem.* **162**, 156 (1987).

available as RNAzol (Cinna/Biotecx Laboratories, Inc., Friendswood, TX). Addition of glycogen prior to precipitation is necessary for full recovery when the amount of starting material is low.

Procedure

1. Add RNAzol (100 μl) to sample in siliconized Eppendorf tube. (Samples may be stored at −70° at this point.)
2. Add chloroform (10 μl) and keep on ice for 5 min.
3. Centrifuge at 12,000 g for 15 min at 4°.
4. Transfer the aqueous phase to a fresh Eppendorf tube.
5. Add an equal volume of 2-propanol and 20 μg mussel glycogen (Sigma, St. Louis, MO) (stock solution, 20 mg/ml).
6. Keep on ice for 15 min; centrifuge as above.
7. Store precipitate in 75% (v/v) ethanol at −70°.

Reverse Transcription–Polymerase Chain Reaction

The RNA of interest is reverse-transcribed as described[3] (detailed instructions are also supplied in the protocol with the RT–PCR kit from Cetus, Norwalk, CT). A primer set should be selected which flanks the base(s) differing between alleles. Primers should be chosen which give a single strong band after amplification, especially in cases where the starting amount of material is low. The optimal concentration of $MgCl_2$ in both the RT and PCR steps should be determined empirically for each primer set. In some cases use of oligo(dT) or random primers in the RT step may improve the RT–PCR signal.[1,3,8] The amount of input RNA needed varies with the abundance of the target transcript and the efficiency of the primers. For a moderately expressed transcript (~20,000 molecules/cell), single-cell assays should be possible. A "hot start" of the PCR reaction[22] (see Step 3 below) can increase yields when the starting amount of material is low.

Procedure

1. Resuspend approximately 1 ng total RNA in RT buffer (10 mM Tris-HCl, pH 8.3, 50 mM KCl, 5 mM $MgCl_2$, 1 mM dNTPs, and 2 U/μl RNasin) in a final volume of 20 μl. Add downstream primer (1 μM), then incubate for 15–30 min at 42° with Moloney murine leukemia virus reverse transcriptase (BRL, Gaithersburg, MD), 2.5 U/μl.
2. Heat to 99° for 5 min.
3. Cool to a temperature allowing correct primer–template annealing (typically 40°–50°).

[22] R. T. D'Aquila, L. J. Bechtel, J. A. Videler, J. J. Eron, P. Gorczyca, and J. C. Kaplan, *Nucleic Acids Res.* **19**, 3749 (1991).

4. Add PCR master mix to give final concentrations of 10 mM Tris-HCl, pH 8.3, 50 mM KCl, 2 mM MgCl$_2$, 200 μM dNTPs, 0.2 μM downstream and upstream primers, and AmpliTaq polymerase (Cetus), 2.5 U/reaction, in a final volume of 100 μl.

5. Add 100 μl mineral oil.

6. Carry out PCR in a thermal cycler, typically about 40 cycles of denaturation (95°, 1 min), annealing (2 min at appropriate temperature), and elongation (72°, 1 min).

Purification of Amplified Products

Procedure

1. Extract with phenol–chloroform and precipitate with ethanol.[23]

2. Run samples on a low-melting agarose gel (BRL) in Tris-acetate or Tris-borate buffer.[23]

3. Extract bands of the appropriate size by use of Gelase (Epicentre Technologies, Madison, WI), following instructions of the manufacturer.

4. Precipitate by addition of 0.5 volume of 7 M ammonium acetate and 3 volumes of ethanol.

5. Resuspend the DNA in 10–20 μl of 10 mM Tris-HCl, pH 8, 1 mM EDTA and estimate the concentration by gel electrophoresis of an aliquot.

Quantitative Single Nucleotide Primer Extension Assay

Procedure

1. Add approximately 10 ng DNA to a master mix containing all ingredients except for the [32P]dNTP.[17,20] The final reaction mix contains the following in a total volume of 10 μl: amplified DNA, 1× SNuPE buffer (10 mM Tris-HCl, pH 8.3–8.5, 50 mM KCl, 2–5 mM MgCl$_2$, 0.001% gelatin, w/v), 1 μM primer (18-mer), 0.75 units of *Taq* polymerase (Cetus), and the appropriate [32P]dNTP. Briefly heat the primer to 90° prior to addition to the master mix to avoid artifacts due to repeated freeze–thawing.

2. Dilute [32P]dNTPs (3000 Ci/mmol, 10 μCi/μl) 5-fold in water; add 1 μl (2 μCi) of the appropriate [32P]dNTP to each tube just prior to incubation.

3. Incubate the samples at 95° for 1 min, 42° for 2 min, and 72° for 1 min for one cycle in a thermal cycler. As a control for background, incubate amplified products of each allele with the [32P]dNTP corresponding to the other allele.

[23] T. Maniatis, E. F. Fritsch, and J. Sambrook, "Molecular Cloning: A Laboratory Manual." Cold Spring Harbor Laboratory, Cold Spring Harbor, New York, 1982.

4. Electrophorese the samples on a 15% denaturing polyacrylamide gel.[23]

5. Determine the amount of radioactivity in the 19-mer bands by use of a radioisotope scanning system such as the AMBIS System II (Automated Microbiology System, Inc., San Diego, CA). Alternatively, the bands may be excised and counted in a scintillation counter. After subtraction of background (<1%), the ratio of the two radioactive products is proportional to the relative amount of each allelic transcript.[17]

Acknowledgments

This work was supported by a grant from the National Institutes of Health (AG08196) to A.D.R. We thank J. M. LeBon for valuable contributions.

[21] Microinjecting Antisense Sequences into Oocytes

By FERNANDO J. SALLÉS, WILLIAM G. RICHARDS, JOAQUIN HUARTE, JEAN-DOMINIQUE VASSALLI, and SIDNEY STRICKLAND

Introduction

In mice, primordial oocytes are arrested at the dictyate stage of meiosis I. On hormonal stimulation, they enter a growth phase during which a subset of mRNAs are synthesized and stored, translationally dormant, in the cytoplasm. These mRNAs, which are recruited for translation at later stages during meiotic maturation and early development while the genome is transcriptionally inactive, are called maternal mRNAs.[1]

The uncoupling of transcription and translation makes it possible to address the role of maternal mRNAs by allowing their specific and irreversible functional depletion. This is carried out by the microinjection of antisense RNAs or antisense oligodeoxynucleotides (ODNs) into the cytoplasm of primary oocytes. The antisense sequences presumably duplex with their complementary mRNA and target the region for attack by nucleases, thus preventing the later translational recruitment of the targeted stored maternal mRNA.[2-4]

[1] P. Wassarman, *in* "The Physiology of Reproduction" (E. Knobil, J. Neill, L. L. Ewing, G. S. Greenwald, C. L. Markert, and D. W. Pfaff, eds.), p. 69. Raven, New York, 1988.

[2] S. Strickland, J. Huarte, D. Belin, A. Vassalli, R. J. Rickles, and J.-D. Vassali, *Science* **241**, 680 (1988).

[3] R. S. Paules, R. Buccione, R. C. Moschel, G. F. Vande Woude, and J. J. Eppig, *Proc. Natl. Acad. Sci. U.S.A.* **86**, 5395 (1989).

Copyright © 1993 by Academic Press, Inc.
All rights of reproduction in any form reserved.

This chapter deals with the isolation of meiosis I arrested mouse primary oocytes, the cytoplasmic microinjection of antisense RNA transcripts and ODNs, and the subsequent analysis of the targeted mRNA. The principles defined are also applicable to the injection of translatable RNAs and RNA fragments for the purpose of analyzing mRNA regulatory elements.[5-7] Although we concentrate on the analysis of oocytes up to metaphase II of meiosis, we also briefly discuss ways of following the analysis through fertilization and other early developmental stages.

Oocyte Collection

Mice

We typically use 3-week-old B6D2F$_1$ [(C57BL/6 × DBA/2)F$_1$; Taconic Farms, Germantown, NY] or Swiss albino (BRL, Basel, Switzerland) female mice. The use of 3-week-old mice is beneficial for several reasons. (1) It takes advantage of the synchronized growth of a relatively large number of oocytes that occurs shortly after birth[1]; 40–80 fully grown primary oocytes can thus be obtained from each mouse. (2) Younger mice are less expensive, and the savings are significant when large numbers are used.

For the purposes of analyzing meiotic maturation of microinjected oocytes, we prefer to use denuded oocytes (oocytes devoid of adhering granulosa cells) from mice that have not been primed with pregnant mare serum gonadotropin (PMSG). These oocytes mature well and are less problematic during handling and injection than those from PMSG-primed mice. However, if microinjections are performed with the ultimate goal of analyzing *in vitro* fertilizations and early development, then PMSG-primed cumulus-enclosed oocytes should be used. A discussion on the isolation and use of cumulus-enclosed oocytes is provided elsewhere.[8]

Culture Media

A variety of culture media have been tested with respect to meiotic maturation and subsequent embryonic development.[9] For the purposes

---

[4] S. J. O'Keefe, H. Wofes, A. A. Kiessling, and G. M. Cooper, *Proc. Natl. Acad. Sci. U.S.A.* **86**, 7038 (1989).

[5] J.-D. Vassalli, J. Huarte, D. Belin, P. Gubler, A. Vassalli, M. L. O'Connell, L. A. Parton, R. J. Rickles, and S. Strickland, *Genes Dev.* **3**, 2163 (1989).

[6] J. Huarte, A. Stutz, M. L. O'Connell, P. Gubler, D. Belin, A. L. Darrow, S. Strickland, and J.-D. Vassalli, *Cell (Cambridge, Mass.)* **69**, 1021 (1992).

[7] F. J. Sallés, A. L. Darrow, M. L. O'Connell, and S. Strickland, *Genes Dev.* **6**, 1202 (1992).

[8] J. J. Eppig and E. E. Telfer, this volume [5].

[9] J. J. M. Van de Sandt, A. C. Shroeder, and J. J. Eppig, *Mol. Reprod. Dev.* **25**, 164 (1990).

of microinjection and analysis of meiotic maturation in overnight cultures, minimal essential medium (MEM) without L-glutamine or bicarbonate (GIBCO BRL, Grand Island, NY) supplemented with 10 mM HEPES, pH 7.0, 0.1 mg/ml sodium pyruvate, and 3 mg/ml polyvinylpyrrolidone (PVP) (MEM–PVP)[10] works well. This medium provides several technical advantages. (1) The absence of fetal calf serum (FCS) in the medium facilitates handling of oocytes during cleaning and microinjection. (2) PVP serves as a bulking agent that decreases "bleeding" of the oocytes immediately after injection, thereby increasing survival (typically, 70–80% of injected oocytes survive). (3) The addition of HEPES allows a more stable pH during handling and injection. (4) Oocytes efficiently undergo meiotic maturation in this medium, allowing the use of one medium for all manipulations.

If oocytes are to be fertilized *in vitro,* then the use of Waymouth's medium containing 5% (v/v) FCS[8,9] supplemented with 3 mg/ml PVP works well for injections and maturation.

Culture Dishes

Collecting dishes [35 mm dishes (Falcon 3001, Fisher, Pittsburgh, PA), 1 dish/mouse] are prepared with 2 ml of MEM–PVP containing 100 μg/ml dibutyryl-cAMP (db-cAMP). The db-cAMP is made as a 100× stock by dissolving 10 mg in 1 ml of MEM–PVP medium or deionized water, filter sterilizing, aliquoting, and storing at −20°. The isolation of oocytes from dissected ovaries will be performed in these dishes. The addition of db-cAMP to the medium maintains the oocytes in meiosis I during their isolation.[11]

Cleaning dishes (35 mm) are prepared with four 15-μl drops of MEM–PVP medium containing db-cAMP. The drops are completely covered with light paraffin oil (Fluka, Ronkonkoma, NY, or Fisher). Some lots of oil may contain soluble substances that are toxic to oocytes and inhibit meiotic maturation. All oil should be tested for toxicity and equilibrated with medium, if necessary, prior to use. The oocytes will be transiently cultured, prior to injection, in these dishes.

Culture dishes are prepared with six 15-μl drops of MEM–PVP medium either with db-cAMP (to prevent meiotic maturation) or without db-cAMP (to allow maturation) and covered with oil as before. After injection, the oocytes are washed by passage through several drops, then cultured in a clean drop. All dishes are equilibrated prior to use in a 37°, 5% (v/v) CO_2 incubator.

[10] W. T. Poueymirou and R. M. Schultz, *Dev. Biol.* **133,** 588 (1989).
[11] W. K. Cho, S. Stern, and J. D. Biggers, *J. Exp. Zool.* **187,** 383 (1974).

Pipettes

Mouth-handling pipettes for the purposes of collecting, denuding, and transferring oocytes are fashioned[12] using 50-μl disposable micropipettes (Fisher or Corning, Corning, NY). For denuding oocytes of adhering granulosa cells, the diameter of the pipette tip should be the same size as that of a zona pellucida-enclosed oocyte. Handling the oocytes by mouth pipette requires practice since even minute changes in pressure translate to large changes in medium volumes under the dissecting scope with the drawn capillaries.

Dissection of Ovaries

We have found the following procedure to be the easiest for preparing ovaries. Ovaries prepared by this method are essentially free of adhering fat; this simplifies subsequent oocyte isolation since fat droplets in the medium make visualization of the oocytes difficult.

1. Open the abdominal cavity of the mouse as described.[12] Secure the left half of the uterus with a pair of forceps (forceps B) and lift it to reveal the oviduct, ovary, fat pad, and kidney. Tear the uterine mesenteries to facilitate handling.

2. With a second forceps (forceps A) secure the fat pad and release forceps B.

3. With forceps B secure the uterus, oviduct, *and* fat pad, leaving the ovary resting on top of forceps B. Release forceps A.

4. With forceps A, using little pressure, gently pinch off the ovary. This action tears the bursa (membrane surrounding the ovary) and releases the ovary free of fat and connective tissue.

5. Carefully roll the ovary on a Kimwipe tissue or paper towel to remove any residual fat and place in a preequilibrated collecting dish.

6. Repeat procedure for the right ovary.

Isolation of Oocytes

Oocytes are isolated under a dissecting scope at 100–250× magnification.[13] Using a handling pipette, sift through the ovary follicular exudates, collecting oocytes. Oocytes are generally embedded in large granulosa cell masses, which can be dispersed by drawing them in and out of the

---

[12] B. Hogan, F. Costantini, and E. Lacy, "Manipulating the Mouse Embryo: A Laboratory Manual." Cold Spring Harbor Laboratory, Cold Spring Harbor, New York, 1986.

[13] M. L. DePamphilis, S. A. Herman, E. Martinez-Salas, L. E. Chalifour, D. O. Wirak, D. Y. Cupo, and M. Miranda, *BioTechniques* **7**, 662 (1988).

handling pipette. Collect only large, fully grown, zona pellucida-enclosed oocytes that clearly contain a germinal vesicle. Transfer groups of oocytes to the cleaning dishes. Oocytes are denuded of granulosa cells adhering to the zona pellucida by drawing them in and out of the handling pipette. Transfer oocytes free of granulosa cells to clean drops, and culture them in the incubator until ready for injection.

Oocyte Microinjection

Discussions of the equipment necessary for microinjection can be found elsewhere.[12,13] Because cytoplasmic injections do not require the precision of nuclear injections, a vibration-free table is not necessary but is helpful. In addition, phase-contrast optics are satisfactory.

Injection Chambers

We use either glass depression slides or a rectangular version of an aluminum chamber with a glass coverslip.[13] The aluminum chamber provides better optics than the slide if Nomarski differential interference contrast optics are used. A 4-μl culture drop of MEM–PVP containing db-cAMP is placed in the center of the siliconized depression slide or coverslip and overlayed with light paraffin oil or Voltalef 10S oil (Prolabo, Paris, France). This drop is generally sufficient to handle injection of up to 50 oocytes at one time.

Holding Pipettes

Holding pipettes are made from hand-drawn or mechanically (Sutter micropipette puller, see below) drawn glass capillaries as described using a De Fonbrune-type microforge (Technical Products International, Inc., St. Louis, MO).[12,13] These pipettes are used to stabilize the oocyte for injection.

Injection Needles

The most critical component of the microinjection setup is the injection needle. Injection needles can be produced quickly and reproducibly using a Flaming/Brown micropipette puller Model P-80/PC equipped with a box filament (Sutter Instruments, Co., Novato, CA). Filament-containing glass capillaries (6 inches with filament 1.0 mm, GC 120F-10, Clark Electromedical Instruments, Reading, UK; or 1B120F-6, World Precision Instruments, Inc., Sarasota, FL) are fitted on the puller such that the heating element is in the center of the capillary, yielding two usable microinjection needles.

TABLE I
PROGRAM USED FOR GENERATION
OF INJECTION NEEDLES[a]

Step	Heat[b]	Pull	Velocity	Time
1	R + 50	10	20	100
2	R − 25	70	20	10
3[c]	R − 130	200	30	—

[a] Glass capillaries are pulled on a Flaming/ Brown micropipette puller Model P-80/PC with box filament. Gas is set at 125. (All values in manufacturer's units.)

[b] R, Heat value derived from the ramp test on a particular lot of glass.

[c] It may be necessary to adjust the heat setting slightly at this step and/or the gas setting throughout the entire program to optimize needle length and strength.

The program used is set as shown in Table I. Using this program we find that needles can be pulled and immediately used; no pre- or posttreatment of the glass is necessary. Needles that are too long should be avoided; they generally are not rigid enough to pierce the membranes effectively, and they flow slowly, making it difficult to visualize the injection (see below). Alternatively, similar high-quality pipettes can be obtained using a BB-CH puller (Mecanex, Geneva, Switzerland) and the following parameters: Mode 4; Heat 1, 695; Heat 2, 610.

The injection needle is loaded immediately prior to injecting. Mouth pipettes are used to load the sample to be microinjected into the needle. These mouth pipettes are constructed the same way as the handling pipettes but are drawn out very long. When the sample is scarce, a small amount (0.2 μl) is allowed to enter by capillary action into the mouth pipette, which is then inserted through the back opening of the injection needle to deliver the sample close to the tip. Alternatively, when large samples are available, the back opening of the injection needle can be touched to a drop of sample and the needle will fill itself by capillary action owing to the presence of the filament. It is critical to avoid contaminating the sample with nucleases, especially if RNA transcripts are being injected.

If the flow is insufficient or if clogging occurs during injections, the tip of the needle can be broken slightly by first lowering it to the slide and breaking it by lowering the holding pipette over the tip. This can be repeated several times if necessary. However, if the diameter of the tip becomes too large or the needle becomes too jagged, oocyte survival may decrease, and a new needle should be loaded. These needles can also be used for pronuclear injections of fertilized eggs.

Injections

When all the oocytes are collected, transfer several (in groups of 20–40) into an equilibrated injection chamber. Load the injection needle with sample as described above. For the ideal injection, the needle should be mounted on the manipulator so that it is parallel to the slide and perpendicular to the surface of the oocyte. This minimizes oocyte trauma. Carefully inject each oocyte by maneuvering the needle through the zona pellucida and oocyte plasma membrane into the cytoplasm. Injection is performed by transiently increasing the pressure in the injection needle.[13] Continue the injection until a transient "pool" is seen forming in the cytoplasm (Fig. 1). Injecting too great a volume will decrease survival. Calculations of injected volumes can be carried out using a radiolabeled marker. Generally, injection volumes of 10 pl yield excellent survival.

If the injecting solution appears to "bubble" in the oocyte, the needle has not penetrated the cytoplasmic membrane, and the solution is being forced backward along the membrane and needle into the perivitelline space. In this event, push further inside the oocyte until the membrane breaks or withdraw the needle and repeat the injection.

Injections of cumulus-enclosed primary oocytes, secondary oocytes, and fertilized eggs are performed the same way. For the injection of cumulus-enclosed oocytes, the holding pipette should be fashioned with a slightly larger diameter to prevent clogging with cumulus cells. Greater care should be taken not to aspirate both cumulus cells and oocytes. With cumulus-enclosed oocytes it is also more difficult to visualize the injected pool.

Culture of Injected Oocytes

After injections are complete, sequentially wash the oocytes through three drops of a culturing dish and transfer them to a clean, numbered drop for overnight incubation. When washing the oocytes, minimize the volume of medium transferred between drops so that the db-cAMP is effectively removed and optimal maturation attained. Alternatively, oocytes can be flushed into 2.5 ml of medium in a 35-mm dish and then transferred to an organ culture dish (Falcon 3037) containing 0.5 ml of medium for overnight incubation. Sometimes it is necessary to prevent meiotic maturation in a group of injected oocytes as a control. This is accomplished by culturing the oocytes in db-cAMP-containing medium. Oocytes can then be scored and collected for analysis the following day.

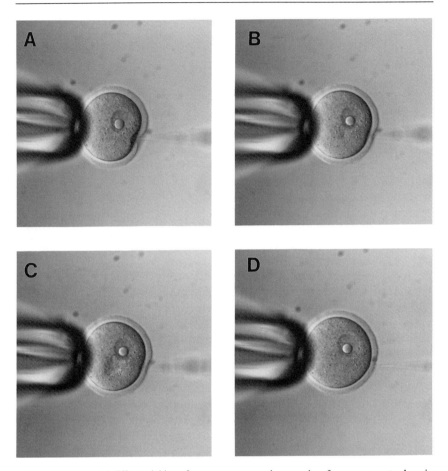

FIG. 1. Nomarski differential interference contrast micrographs of an oocyte cytoplasmic microinjection. A fully grown primary oocyte is shown secured by a holding pipette (left) and being microinjected by an injection needle containing an antisense ODN. The zona pellucida and germinal vesicle containing a prominent nucleolus are clearly visible. (A) The microinjection needle penetrates the zona pellucida and plasma membrane. (B) The needle has pierced the membranes. A small dimple can be seen in the cytoplasm at the tip of the injection needle where the solution is being injected. (C) A transient "pool" can be visualized as the solution is injected. (D) An injected oocyte immediately after withdrawal of the injection needle. Note that the injection "pool" is no longer visible. The diameter of the oocyte is approximately 80 μm.

Antisense Reagents

Injection of antisense sequences should always be accompanied by injections of a sense sequence control into a second group of oocytes. For inhibition of mRNA translation, either antisense RNA transcript[2] or ODNs[3,4] can be used.

It is important to ensure that the material to be injected is free of particulates that will clog the injection pipette. Therefore, the RNA or ODN is resuspended in 200 μl of deionized water and filtered through a SPIN-X centrifuge filter unit in a microcentrifuge. The filter unit (SPIN-X, low binding, 0.22-μm cellulose acetate filter; Costar, Cambridge, MA) is prewashed with 100 μl of deionized water and the flow-through discarded in order to remove particulates originating from the filter. The sample flow-through is lyophilized and resuspended in the appropriate sterile-filtered buffer using a washed pipette tip.

RNA Microinjections

RNA transcripts are synthesized, prepared, and resuspended in 150 mM KCl.[2] The inclusion of m^{7}G(5')ppp(5')G in the transcription reactions is essential to confer stability of the injected RNA in the cytoplasm. For antisense RNA, one must consider the number of injected transcripts required to prevent translation. A 400-fold excess abolishes tissue-type plasminogen activator (t-PA) translation, and a 4-fold excess inhibits by approximately 90%.[2] Because it is difficult to determine accurately the abundance of a specific mRNA in the oocyte, as concentrated a solution as possible should be injected. Also, it is important to note that different mRNA regions may be masked in the primary oocyte[2]; therefore, the most effective region targeted by the antisense RNA transcript needs to be experimentally determined. Sense RNA transcripts can also be used to study RNA processing and translation, by injection into the cytoplasm or nucleus.[5-7]

DNA Microinjections

Oligodeoxynucleotides can be used without HPLC purification. The ODNs are filtered as described above and resuspended at 1 μg/μl in 10 mM Tris, pH 7.2, 0.2 mM EDTA. This concentration represents approximately 10^9 molecules/10 pl for an ODN 25 nucleotides (nt) long. (*Note*: Although this buffer does not appear to inhibit meiotic maturation, it does appear to decrease the ability of *in vitro* fertilized eggs to cleave. A better buffer for this purpose is PBS.[14]) For the proper interpretation of antisense

---

[14] A. Hampl and J. J. Eppig, personal communication.

microinjections, several ODNs directed against different regions of the RNA should be used to ensure that phenotypes observed are not due to a nonspecific effect.[15]

ODNs as short as 15 nt have been used to prevent translation of a specific maternal mRNA in murine oocytes.[4] We have found this size to be ineffective in preventing t-PA mRNA translation. However, longer ODNs (25 nt) directed against the same regions abolish t-PA production.[16] The reasons for this observation are not clear. It is possible that different stretches of RNA molecules are more accessible for hybridization with ODNs. Because these differences are not known *a priori*, several considerations should be kept in mind when designing ODNs for antisense-mediated cleavage. (1) If a primer is too AT-rich it may not complement well in the cytoplasm at 37°. Therefore, ODNs with low melting temperatures (T_m values) should be avoided if possible. (2) Longer ODNs of 25 nt should be used to increase both specificity and RNase recognition. (3) ODNs with clear secondary structure should be avoided to prevent an intramolecular reaction.

Finally, the stability of ODNs in the cytoplasm of oocytes appears to be poor.[16] Therefore, the use of phosphorothioate ODNs, which are more stable in *Xenopus* oocytes, may increase the effectiveness of antisense injections. However, phosphorothioate ODNs are toxic at effective concentrations in *Xenopus* embryos and may not be useful for analysis of later time points. A discussion of antisense injections in *Xenopus* oocytes can be found elsewhere.[17]

Analysis of Antisense Targeted RNA

To ensure that the RNA or the ODN has recognized and effectively destroyed the mRNA, several methods can be employed. The mechanism of antisense-mediated translational inhibition appears to be a degradation of the RNA in the duplexed region by either an RNase III-like activity (in the case of antisense RNA)[2] or an RNase H-like activity (in the case of ODNs).[18] In both instances, sequences 5' of the duplexed region are stable, whereas sequences 3' are degraded.[2,7] The translation of the 5' portion of the mRNA is presumably prevented by the loss of the nuclear

[15] R. C. Smith, W. M. Bement, M. A. Dersch, E. Dworkin-Rastl, M. Dworkin, and D. G. Capco, *Development (Cambridge, UK)* **110**, 769 (1990).

[16] F. J. Salles, A. L. Darrow, and S. Strickland, unpublished observations, 1991.

[17] T. M. Woolf, *in* "Gene Regulation: Biology of Antisense RNA and DNA" (R. P. Erickson and J. G. Izant, eds.), p. 223. Raven, New York, 1992.

[18] P. Dash, I. Lotan, M. Knapp, E. P. Kandel, and P. Goelet, *Proc. Natl. Acad. Sci. U.S.A.* **84**, 7896 (1987).

polyadenylation signal AAUAAA that is required for cytoplasmic polyadenylation.[5] It should be noted that other mechanisms may exist for translational activation and that ODNs directed against the 3′ untranslated region may not abolish the translation of certain mRNAs.

The stability of the 5′ fragment of the cleaved mRNA allows the determination of appropriate antisense destruction by a variety of methods. (1) Northern analysis can be used to detect the targeted message in both antisense-injected and sense-injected control oocytes, where increased mobility of the antisense-injected oocyte RNA confirms cleavage.[2] This method is only useful for abundant transcripts where sensitivity is not a problem. (2) RNase protection can also be used by mapping the region of duplex formation. (3) For low-abundance mRNAs, it is possible to use a scheme based on the polymerase chain reaction (PCR).[7] This scheme, however, requires extended sequence information as well as the synthesis of several PCR primers.

These methods require the isolation of oocyte RNAs, which can be accomplished easily and efficiently.[19,20] Because there is a low DNA/RNA ratio in these samples, treatment with DNase is generally not required unless background is a problem in the PCR-based scheme. These RNA isolation methods can also be used in the analysis of zygotes and early embryos.

[19] J. Huarte, D. Belin, A. Vassalli, S. Strickland, and J.-D. Vassalli, *Genes Dev.* **1**, 1201 (1987).
[20] P. Chomczynski and N. Sacchi, *Anal. Biochem.* **162**, 156 (1987).

[22] Detection of Messenger RNA by *in Situ* Hybridization to Tissue Sections and Whole Mounts

By DAVID G. WILKINSON and M. ANGELA NIETO

Introduction

An important step in the characterization of genes with potential roles in development is the analysis of their temporal and spatial pattern of expression. *In situ* hybridization to mRNA is a rapid and convenient method for visualizing patterns of gene expression, and specific probes can be readily produced from cloned DNA or by the synthesis of oligonucleotides. The method involves a series of procedures: (a) synthesis of a labeled nucleic acid probe; (b) fixation and preparation of tissue; (c) hybridization of probe to the tissue and washing to remove unhybridized probe; and (d) visualization of the probe. Many alternatives exist for the

Copyright © 1993 by Academic Press, Inc.
All rights of reproduction in any form reserved.

type of probe, and for the methods of labeling and visualization. Single-stranded RNA probes[1] labeled with ^{35}S have been the most widely used for the hybridization of tissue sections because of their high sensitivity. The signal obtained with these probes has a good resolution (about a cell diameter) and, typically, can be visualized after an autoradiographic exposure of about 1 week. Recently, a method has been developed for the nonradioactive labeling of probe with digoxigenin and detection with a phosphatase-coupled antibody.[2,3] Although currently less sensitive than methods using radioactive probes, this method offers several advantages, in particular a single-cell resolution of signal after 1 day or less of detection. In addition, nonradioactive probes can be hydrized and detected in whole mounts, obviating the need to reconstruct serial sections in order to visualize patterns of gene expression in three dimensions.

Procedures are presented here for the *in situ* hybridization of ^{35}S-labeled probes to sections of mouse embryos and the hybridization of digoxigenin-labeled probes to sections and whole embryos; results obtained by the three methods are shown in Fig. 1. Precautions should be taken to avoid the degradation of cellular RNA prior to hybridization; we have not found it necessary to treat solutions with diethyl pyrocarbonate (DEPC) but do autoclave the phosphate-buffered saline (PBS), saline, and triethanolamine solutions. In addition, glassware is baked, and containers used for posthybridization steps (when ribonuclease is used) are kept separate from those used for prehybridization steps. The volumes given in the methods for the analysis of tissue sections are suitable for up to 9 slides in a 50-ml Stretton jar (BDH, Poole, UK) for the 40-ml washes and a 300-ml slide dish (BDH) for the 250-ml washes.

In Situ Hybridization of ^{35}S-Labeled Probes to Tissue Sections

Preparation of Single-Stranded RNA Probes

For the preparation of ^{35}S-labeled RNA probes, the required sequences are cloned into a plasmid vector containing RNA polymerase binding sites,[4] such as Bluescript (Stratagene, La Jolla, CA) or pGEM (Promega, Madison, WI). The plasmid is linearized by digestion with a restriction enzyme so that transcription will yield an RNA probe that is complemen-

[1] K. H. Cox, D. V. Deleon, L. M. Angerer, and R. C. Angerer, *Dev. Biol.* **101**, 485 (1984).
[2] D. Tautz and C. Pfeifle, *Chromosoma* **98**, 81 (1989).
[3] A. Hemmati-Brivanlou, D. Frank, M. E. Bolce, M. D. Brown, H. L. Sive, and R. M. Harland, *Development* (*Cambridge, UK*) **110**, 325 (1990).
[4] D. A. Melton, P. A. Krieg, M. R. Rebagliati, T. Maniatis, K. Zinn, and M. R. Green, *Nucleic Acids Res.* **12**, 7035 (1984).

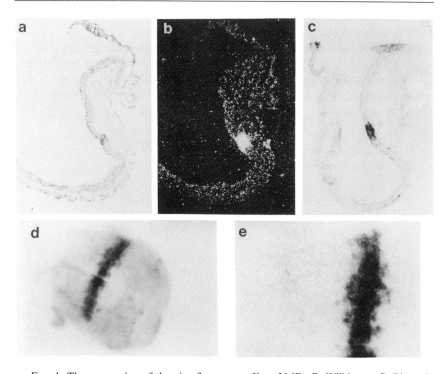

Fig. 1. The expression of the zinc-finger gene *Krox-20* [D. G. Wilkinson, S. Bhatt, P. Chavrier, R. Bravo, and P. Charnay, *Nature (London)* **337**, 461 (1989)] in the 8-day mouse embryo revealed by *in situ* hybridization. This gene is expressed in one domain in the hindbrain neural epithelium at this stage of development. (a, b) *In situ* hybridization analysis of a longitudinal section with an ^{35}S-labeled probe photographed under bright-field (a) and dark-field (b) illumination. The high density of silver grains indicates the site of *Krox-20* expression, but this method does not give single-cell resolution. (c) *In situ* hybridization analysis of a longitudinal section with digoxigenin-labeled probe. The dark stain indicates the site of gene expression at a single-cell resolution. (d) *In situ* hybridization analysis of a whole embryo with digoxigenin-labeled probe. A dorsolateral view is shown of the head, which has been dissected away from the trunk. (e) High-magnification view of the domain of expression shown in (d) after microdissection of the neural plate and flat-mounting under a coverslip. *Krox-20* transcripts are detected at a single-cell resolution in the cytoplasm.

tary (antisense) to the target mRNA and lacks plasmid sequences. Alternatively, DNA fragments amplified by the polymerase chain reaction (PCR) can be used for transcription by including an RNA polymerase binding site in one of the PCR primers.[5] Transcription is carried out in the presence of three unlabeled and one labeled ribonucleotide. The amount of radioac-

[5] M. Frohman and G. Martin, *Technique* **1**, 165 (1989).

tive ribonucleotide limits the transcription reaction, and the maximum yield is approximately 200 ng.

The length of the probe influences the signal strength: the greater the sequence representation, the greater the signal, but very long probes give lower signals because they penetrate the tissue less well. We find that probes of up to 1 kb in length give optimal results. If the probe is longer than 1 kb, it is partially degraded by limited alkaline hydrolysis as described in Step 5.

1. Mix in the following order at room temperature:

Sterile distilled water	2.5 μl
Nucleotide mix	2 μl
10 × Transcription buffer	2 μl
0.2 M Dithiothreitol (DTT)	1 μl
Linear template (1 μg/μl)	1 μl
[^{35}S]UTP (>1000 Ci/mmol)	10 μl
Placental ribonuclease inhibitor (100 U/μl)	0.5 μl
SP6, T7, or T3 RNA polymerase (10 U/μl)	1 μl

Nucleotide mix includes 2.5 mM GTP, 2.5 mM ATP, and 2.5 mM CTP; 10 × transcription buffer is 400 mM Tris-HCl, pH 8.25, 60 mM MgCl$_2$, and 20 mM spermidine.

2. Incubate at 37° for 2 hr.
3. Add 2 μl RNase-free DNase I (20 U/μl).
4. Incubate at 37° for 15 min.
5. If required, hydrolyze the RNA (starting length, L) to an average length of 750 bases by adding an equal volume of 80 mM NaHCO$_3$, 120 mM Na$_2$CO$_3$ and incubating at 60° for a time (t, in minutes) given by $t = (L - 0.75)/0.08L$.
6. Remove the unincorporated ribonucleotides by fractionation on a Sephadex G-50 column (e.g., in a Pasteur pipette) in 50 mM Tris-HCl, pH 8.0, 1 mM EDTA, 0.1% sodium dodecyl sulfate (SDS), and then ethanol-precipitate the RNA. Redissolve at 2 × 10^6 counts/min (cpm)/μl in 100 mM DTT and make hybridization mix by adding 1 volume to 9 volumes of 50% deionized formamide, 0.3 M NaCl, 20 mM Tris-HCl, 5 mM EDTA, pH 8.0, 10% dextran sulfate, 1 × Denhardts' solution, 0.5 mg/ml yeast RNA. Store at −70°.

Preparation of Tissue Sections

Tissue sections are prepared by fixing embryos, embedding in paraffin wax, cutting sections, and drying them onto subbed slides. The fixative is 4% paraformaldehyde in PBS, which is prepared on the day of use by heating at 65° until the paraformaldehyde has dissolved, followed by cooling on ice. Mouse embryos are dissected out in PBS, placed in 10 ml ice-cold fixative, and left at 4° overnight. The fixative is removed and the embryos dehydrated by washing for 30 min with gentle agitation in 10 ml saline at 4° (twice), 1 : 1 (v/v) saline–ethanol mix, 70% ethanol (twice), 85% ethanol, 95% ethanol, and absolute ethanol (twice). Washing times should be increased for embryos older than 12.5 days. Embryos can be stored at 4° in 70–100% ethanol. Embedding is carried out by incubating for 20 min with 10 ml toluene (three times), then with a 1 : 1 toluene–wax mix at 60°, followed by molten paraffin wax at 60° (three times). Paraplast (VWR, San Francisco, CA) or Fibrowax (BDH) give successful results. The embryos are then transferred to embryological dishes (preheated to 60°), oriented with a warmed needle under a dissection microscope, and the wax allowed to set. Paraffin wax blocks can be stored indefinitely at 4° until required for use.

Sections are cut as 6-μm ribbons that are then floated on a bath of distilled water at 50° until the creases disappear, and collected on subbed slides (prepared as described below). The sections are dried onto the slides at 37° overnight and are then stored desiccated at 4°. Subbed slides are prepared as follows:

1. Dip the slides in 10% HCl/70% ethanol, followed by distilled water and 95% ethanol, for 1 min each, and then air-dry.
2. Dip the slides in 2% TESPA (3-aminopropyltriethoxysilane) in acetone for 10 sec.
3. Wash twice with acetone and then with distilled water.
4. Dry at 37°.

Prehybridization Treatments and Hybridization of Tissue Sections

Before hybridization sections are dewaxed and subjected to a series of treatments. The proteinase K treatment increases signal significantly by digesting cellular protein, thus rendering the target RNA more accessible to the probe. It is essential to postfix the sections after this step, otherwise the tissue will disintegrate during hybridization and washing. The acetic anhydride step is intended to acetylate amino residues and prevent nonspecific binding of probe. This step is optional, however, since we find that it has no effect on the level of background under the conditions that we use for *in situ* hybridization.

1. Dewax the slides in 250 ml Histoclear (National Diagnostics), twice for 10 min, and then place them in 250 ml of 100% ethanol for 2 min to remove most of the Histoclear.

2. Transfer the slides quickly through 250 ml of 100% ethanol (twice), 95, 85, 70, 50, and then 30% ethanol. Wash the slides with 250 ml saline and then PBS for 5 min each.

3. Immerse the slides in 40 ml fresh 4% (w/v) paraformaldehyde in PBS for 20 min.

4. Wash the slides with 250 ml PBS, twice for 5 min.

5. Drain the slides and place horizontally on the bench. Overlay the sections with 20 μg/ml proteinase K (freshly diluted in 50 mM Tris-HCl, 5 mM EDTA, pH 8.0, from a 10 mg/ml stock) and leave for 5 min.

6. Shake off excess liquid and wash the slides with 250 ml PBS for 5 min.

7. Repeat the fixation (Step 3); the same solution can be used.

8. Place the slides in a slide dish with 250 ml of 0.1 M triethanolamine hydrochloride, pH 8.0, set up with a rapidly rotating stir bar and in a fume hood. Add 0.63 ml acetic anhydride (*caution:* acetic anhydride is toxic and volatile) and leave for 10 min.

9. Wash the slides with 250 ml PBS, and then saline for 5 min each. Dehydrate by passing through 250 ml of 30, 50, 70, 85, 95, and 100% ethanol, and air-dry. To avoid salt deposits on the slides, the slides should be left in 70% ethanol for 5 min; the other dehydration steps can be carried out quickly.

10. Apply the hybridization mix to the slide adjacent to the sections ($\sim$2.5 μl per cm^2 of coverslip is sufficient) and gently lower a clean coverslip so that the mix is spread over the sections.

11. Place the slides horizontally in a box containing tissue paper soaked in 50% formamide, 5$\times$ SSC (20 $\times$ SSC is 3 M NaCl, 0.3 M sodium citrate), seal the box, and incubate overnight at 60°.

Posthybridization Washing and Autoradiography

Following hybridization, slides are washed at high stringency and treated with ribonuclease to remove probe that has not annealed to the target RNA. The location of the radioactive probe is then revealed by autoradiography.

1. Remove the slides from the hybridization box and place in a slide rack in 250 ml of 5$\times$ SSC, 10 mM DTT (prewarmed) at 60° for 30–60 min until the coverslips fall off.

2. Place the slides in 40 ml of 50% formamide, 2$\times$ SSC, 20 mM DTT, at 65° for 30 min.

3. Wash the slides with 250 ml NTE buffer (0.5 M NaCl, 10 mM Tris-HCl, 5 mM EDTA, pH 8.0) at 37°, three times for 10 min each.

4. Treat the slides with 40 ml of 20 μg/ml ribonuclease A in NTE buffer at 37° for 30 min.

5. Wash the slides with 250 ml NTE buffer at 37° for 15 min.

6. Repeat the high-stringency wash of Step 2.

7. Wash the slides in 250 ml of 2× SSC, then 0.1× SSC for 15 min each.

8. Dehydrate the slides by quickly putting them through 250 ml of 30, 60, 80, and 95% ethanol, all including 0.3 M ammonium acetate, followed by absolute ethanol, twice. Air-dry the slides.

The slides are dipped into liquid photographic emulsion (Ilford K5, London, UK, or Kodak, Rochester, NY, NTB-2), exposed, and developed for the autoradiographic detection of signal as follows.

1. Under safelight conditions (e.g., Kodak safelight filter 6B) in a darkroom, melt the emulsion at 43° and mix gently with an equal volume of 2% glycerol that has been prewarmed to 43°. Aliquot approximately 15 ml into slide mailers (BDH) with a wide-mouthed pipette, and wrap in foil. Store at 4°. When required, melt an aliquot at 43°.

2. To remove any bubbles, dip clean slides in the emulsion until an even, bubble-free coating is obtained. Next, dip the experimental slides quickly into the emulsion, drain vertically for 2 sec, wipe the back of the slide and place horizontally in a light-tight (but not airtight) box.

3. Leave the slides for 2 hr, then add a packet of desiccant and leave for a further 2 hr.

4. Transfer the slides to a slide box containing a packet of desiccant, seal with tape, and place at 4° to expose. It is important that the slides are completely dry during the exposure. Typical exposure times are between 1 and 2 weeks.

5. About 1–2 hr before developing remove the box of slides from 4° and warm to room temperature.

6. Under safelight conditions in a darkroom, transfer the slides through Kodak D19 developer (80 g/liter) for 2 min, 1% glycerol, 1% acetic acid for 1 min, and then 30% sodium thiosulfate (Sigma, St. Louis, MO) for 2 min. The lights can now be turned on. At least 5 ml per slide should be used for each of these reagents.

7. Wash the slides twice in 250 ml distilled water for 10 min, quickly transfer through 70% and then 100% ethanol, and air-dry.

8. Stain in 0.02% toluidine blue for 1 min, then pass the slides quickly through water, 70% ethanol, and absolute ethanol. Repeat this if the staining is too weak, but do not overstain since this can obscure the signal. Once appropriate staining has been achieved, immerse the slides in Histoclear for several minutes, then briefly drain and mount the sections

under a coverslip using DPX (BDH) or Permount (Fisher Scientific, Pittsburgh, PA) mounting agent.

Autoradiography produces silver grains that can be observed on a microscope simultaneously with the stained tissue by using bright-field illumination. However, only intense signals can be observed under bright-field conditions, and a more sensitive means of visualizing the silver grains is to use dark-field illumination. For black-and-white photography, bright- and dark-field images of the stained tissue and silver grains, respectively, are photographed (Fig. 1a,b), but this can have the drawback of difficulty in precisely relating the signal to the morphology. With color photography, it is possible to present the signal and stained tissue simultaneously by double-exposure; a dark-field image of the silver grains is superimposed on a bright-field image of the stained tissue.

In Situ Hybridization of Tissue Sections with Digoxigenin-Labeled Probes

Many steps of the procedure for detecting transcripts with digoxigenin-labeled probes are identical or similar to those for the use of radioactive probes. The major difference is the signal detection method, which is achieved by the immunocytochemical detection of the probe with an alkaline phosphatase-conjugated antibody.

Probe Synthesis and Pretreatments of Tissue Sections

The method for probe synthesis is identical to that described for labeling with [^{35}S]UTP, except that 2 μl of 10 mM GTP, 10 mM ATP, 10 mM CTP, 6.5 mM UTP, 3.5 mM digoxigenin–UTP (Boehringer, Mannheim, Germany), pH 8.0, and 10 μl distilled water are used instead of the radioisotope and nucleotide mix. Because the labeled nucleotide is not limiting, up to 10 μg of RNA is synthesized. The amount made is estimated by running an aliquot (after Step 2) on an agarose gel containing 0.5 μg/ml ethidium bromide. Probe can be separated from unincorporated nucleotides on a Sephadex column; the fractions containing probe are revealed by spotting a small portion of each onto a nitrocellulose filter, followed by detection with phosphatase-conjugated antibody (Boehringer). However, it is more convenient to remove most of the nucleotides by adding sodium acetate to 0.25 M, then precipitating probe with 2.5 volumes of ethanol at $-20°$ for 30 min. After centrifugation for 10 min in a microcentrifuge, the pellet is washed twice with 70% ethanol, then dried and redissolved at approximately 0.1 μg/μl in distilled water. To make hybridization mix, 1 volume of probe is added to 99 volumes of 50% formamide, 5×

SSC, pH 5 (use citric acid to adjust pH), 50 μg/ml yeast RNA, 1% SDS, 50 μg/ml heparin (final concentration of probe is 1 μg/ml). Store at $-70°$.

The preparation of tissue sections and prehybridization treatments are carried out as described above except that a hydrogen peroxide step is performed after dewaxing of the sections (after Step 2): immerse the slides in 6% hydrogen peroxide for 30 min, then wash them twice for 5 min with PBS, and proceed with the fixation with paraformaldehyde. The peroxide treatment improves the background after signal detection, perhaps by inactivating endogenous phosphatases. Hybridization is carried out at 65° since we have observed higher signals under these conditions than after hybridization at lower temperatures.

Posthybridization Washing and Signal Detection

The protocol we use to remove unbound probe by high-stringency washing and ribonuclease treatment has been adapted from that used for whole mounts (see below) and is slightly different from that used for ^{35}S-labeled probes. Hybridized probe is detected with an alkaline phosphatase-coupled antidigoxigenin antibody (Boehringer). To prevent the nonspecific binding of this antibody, the embryos are preblocked with sheep serum (heat-treated at 70° before use), and the antibody is preabsorbed with embryo powder. After washing to remove excess antibody, the signal is produced by incubation with a substrate for alkaline phosphatase that yields an insoluble blue product (Fig. 1c). Levamisole is included during the color developing as it inhibits many alkaline phosphatases present in the embryo, but not the calf intestinal enzyme that is coupled to the antibody. NBT (nitro blue tetrazolium salt) and BCIP (5-bromo-4-chloro-3-indolyl phosphate, toluidinium salt) are obtained from Sigma or Boehringer.

1. Place the slides in a slide rack in 250 ml of 5× SSC at 65° until the coverslips fall off.
2. Wash the slides in 40 ml solution I (50% formamide, 5× SSC, pH 5.0, 1% SDS) at 65°, twice for 30 min.
3. Wash the slides in 250 ml solution II (0.5 M NaCl, 10 mM Tris-HCl, pH 7.5, 0.1% Tween 20) at room temperature, three times for 10 min.
4. Treat the slides with 40 ml solution II containing 25 μg/ml ribonuclease A for 45 min at 37°.
5. Wash the slides with 40 ml solution II for 5 min.
6. Wash the slides with 40 ml solution III (50% formamide, 2× SSC, pH 5.0) at 65°, twice for 30 min.
7. Wash with 40 ml PBT (PBS, 0.1%, v/v, Tween 20), twice for 10 min.

8. During Steps 7 and 9 preabsorb the antidigoxigenin antibody as follows. Weigh out 3 mg embryo powder (prepared as described below) into a microtube, add 0.5 ml PBT, and heat at 70° for 30 min. Cool on ice and add 5 μl sheep serum and 1 μl antidigoxigenin antibody. Shake gently at 4° for 2–3 hr, then spin in a microcentrifuge for 5 min. Dilute the supernatant to 2 ml with 1% sheep serum in PBT.

9. Preblock the sections (from Step 7) by overlaying with 10% sheep serum in PBT for 2–3 hr.

10. Quickly wash the sections with 40 ml PBT, overlay with the preabsorbed antibody (from Step 8), and incubate overnight in a moist chamber at 4°.

11. Wash the slides with 40 ml PBT three times for 5 min and then three times for 30 min.

12. Wash the slides with 40 ml NTMT (100 mM NaCl, 100 mM Tris-HCl, pH 9.5, 50 mM MgCl$_2$, 0.1% Tween 20, 2 mM levamisole; make from stocks on the day of use), three times for 5 min.

13. Overlay the sections with NTMT containing 4.5 μl NBT (75 mg/ml stock in 70% dimethylformamide) and 3.5 μl BCIP (50 mg/ml stock in dimethylformamide) per milliliter and incubate in the dark. Signal usually appears after 5–6 hr. Monitor the reaction and do not allow the background to become unacceptably high. To stop the color reaction wash with PBT. The slides can then be washed with distilled water and air-dried and the sections mounted under a coverslip with DPX (BDH).

Embryo powder is prepared as follows. Homogenize 12.5- to 14.5-day mouse embryos in a minimum volume of ice-cold PBS. Add 4 volumes of ice-cold acetone to the homogenate, mix, and incubate on ice for 30 min. Centrifuge at 10,000 g for 10 min and remove the supernatant. Wash the pellet with ice-cold acetone and repeat the centrifugation. Spread the pellet out and grind it into a fine powder on a sheet of filter paper. Air-dry the powder and store it at 4°.

Whole-Mount *In Situ* Hybridization with Digoxigenin-Labeled Probes

The whole-mount hybridization of mouse embryos[6] requires several modifications to the protocol described for the hybridization of sections with digoxigenin-labeled probe. Embryos are dissected free of extraembryonic membranes, and any cavities (e.g., the neural tube, amniotic cavity) are dissected open to avoid high backgrounds caused by the subsequent trapping of reagents. Fixation in 4% paraformaldehyde in PBS is carried

[6] D. G. Wilkinson, in *"In Situ* Hybridisation: A Practical Approach" (D. G. Wilkinson, ed.), p. 75. IRL Press, Oxford, 1992.

overnight, and then the embryos are pretreated before hybridization. Because younger than 9-day mouse embryos are quite delicate and small, the washes should be gentle and precautions taken to avoid damaging or losing them during these and later treatments. The washes are carried out in 10-ml tubes (Steps 1–9), which are not completely filled with solution and rocked gently (e.g., on a Denley, Billingshurst, UK, A600 rocker). To change the wash solutions, the embryos are allowed to settle to the bottom of the tube and some liquid is left above them, since otherwise the embryos will be flattened. Prehybridization and hybridization are carried out in microtubes at 70° by placing the tube in a heater block turned on its side and on a rocker, such that the tubes are tilted along their short axis.

1. After fixation, wash the embryos with PBT (PBS, 0.1% Tween 20), twice for 5 min.

2. Wash the embryos for 5 min each with 25, 50, 75% methanol in PBT, then twice with 100% methanol.

3. Rehydrate the embryos through the methanol/PBT series in reverse and then wash twice with PBT.

4. Incubate the embryos in 6% (v/v) hydrogen peroxide in PBT for 1 hr.

5. Wash the embryos with PBT three times for 5 min.

6. Treat the embryos with 10 μg/ml proteinase K in PBT for 15 min.

7. Wash the embryos for 5 min each with freshly prepared 2 mg/ml glycine in PBT and then twice with PBT.

8. Refix the embryos with fresh 0.2% (v/v) glutaraldehyde/4% (v/v) paraformaldehyde in PBT for 20 min.

9. Wash the embryos twice for 5 min with PBT.

10. Add 1 ml prehybridization solution (50% formamide, 5× SSC, pH 5, 50 μg/ml yeast RNA, 1% SDS, 50 μg/ml heparin) and transfer the embryos to a 1.5-ml microtube.

11. Replace the prehybridization solution with fresh mix and rock at 70° for 2–3 hr. The embryos can be stored indefinitely in this solution at −20°.

12. Add 0.4 ml hybridization mix including approximately 1 μg/ml digoxigenin-labeled RNA probe.

13. Incubate overnight at 70° with rocking.

Posthybridization Washes and Signal Detection

Following hybridization, the embryos are washed, and then probe that has not annealed to target RNA is removed by ribonuclease treatment and high-stringency washing. This is followed by incubation with antidigoxigenin antibody conjugated to alkaline phosphatase, washing, and production of the signal. All washes are with 1.7 ml solution in a 2-ml microtube and with rocking. A heater block placed on its side is used for the 37–70° washes, and the solutions are prewarmed. The histochemical reaction should be carried out in a glass container, since a precipitate can form if a plastic container is used.

1. Wash the embryos with solution 1 (50% formamide, 5× SSC, pH 5, 1% SDS) twice for 30 min each at 70°.

2. Wash the embryos with 1 : 1 solution 1–solution 2 for 10 min at 70°.

3. Wash the embryos three times for 5 min each with solution 2 (0.5 M NaCl, 10 mM Tris-HCl, pH 7.5, 0.1% Tween 20).

4. Wash the embryos twice with 100 μg/ml ribonuclease A in solution 2 for 30 min at 37°.

5. Wash the embryos with solution 2, then with solution 3 (50% formamide, 2× SSC, pH 5) for 5 min each.

6. Wash the embryos twice with solution 3 for 30 min at 65°.

7. Wash the embryos three times for 5 min each with PBT.

8. Preblock the embryos by incubating with 10% sheep serum in PBT for 2–3 hr.

9. During preblocking (Step 8), preabsorb the antibody with embryo powder (as described above for hybridization to tissue sections).

10. Remove the 10% serum from the embryos, replace with the preabsorbed antibody, and rock overnight at 4°.

11. Wash the embryos three times for 5 min with PBT.

12. Wash the embryos five times for 1 hr with PBT.

13. Wash the embryos three times for 10 min each with NTMT.

14. Incubate the embryos with NTMT including freshly added 4.5 μl NBT and 3.5 μl BCIP per milliliter (see above). Transfer the solution and embryos to a glass dish for easier observation and rock for the first 20 min of incubation. Keep in the dark as much as possible. Monitor the reaction at intervals under a dissecting microscope until the color has developed to the desired extent.

15. Wash the embryos twice for 10 min with NTMT and store at 4° in PBT.

Low-power photographs of the embryos can be taken on a dissecting microscope with overhead illumination (Fig. 1d). In addition, it can be useful to dissect out tissues and mount them under a coverslip in order to take photographs at high magnification (see Fig. 1e). The reaction product using NTB/BCIP is soluble in organic solvents such as ethanol and xylene, but following overnight fixation in 4% paraformaldehyde, the embryos can be passed quickly through these solvents to embed them in order to cut tissue sections.

Acknowledgments

We thank Phil Ingham, Barry Rosen, Richard Harland, and Ron Conlon for useful advice and protocols. We are also grateful to Romita DasGupta and Graham Alldus for the whole-mount *in situ* hybridization data shown in Fig. 1.

[23] Detection of Messenger RNA by *in Situ* Hybridization to Postimplantation Embryo Whole Mounts

By RONALD A. CONLON and BERNHARD G. HERRMANN

Introduction

The determination of the spatial distributions of specific mRNAs is important in the elucidation of the roles of genes in development. We describe procedures for the rapid and convenient visualization of mRNA distribution in early postimplantation mouse embryos. Standard *in situ* hybridization techniques utilize radiolabeled nucleic acid probes applied to sectioned tissue with subsequent detection by autoradiography. The procedures we have developed employ the hybridization of hapten-labeled probes to whole embryos with detection via enzyme conjugates that generate colored reaction products at sites of hybridization. Such techniques have been termed whole-mount *in situ* hybridization procedures. The procedures are derived from methods developed for embryos of other species.[1,2]

The visualization of RNA distribution in the whole embryo has a number of advantages over procedures utilizing sectioned material. The first and most obvious of these is the fact that the need for three-dimensional reconstruction of patterns of hybridization from serial sections is eliminated: the overall pattern of hybridization is immediately apparent. Second, the detection of small groups of cells that express the target RNA can be identified more readily, since individual cells are not partitioned over adjacent sections. Third, embryos can be more readily staged in whole-mount preparations. Fourth, the determination of the plane of section in gastrulating mouse embryos is occasionally difficult because there are few morphological landmarks. This problem is avoided in the whole-mount procedures. Finally, the topology of the mouse embryo changes

---

[1] D. Tautz and C. Pfeifle, *Chromosoma* **98**, 81 (1989).

[2] A. Hemmati-Brivanlou, D. Frank, M. E. Bolce, B. D. Brown, H. L. Sive, and R. M. Harland, *Development (Cambridge, UK)* **110**, 325 (1990).

Copyright © 1993 by Academic Press, Inc.
All rights of reproduction in any form reserved.

rapidly during the process of turning of the embryo on embryonic day 8, which makes it difficult to obtain comparable sections from different embryos. In general, the comparison of hybridization patterns of different genes is much easier in whole embryos than in sectioned material.

There are also advantages of hapten-labeled probes and histochemical detection over radiolabeled probes and autoradiography. First, the former provide greater safety, since radioisotopes are not involved. Second, histochemical detection provides greater resolution than autoradiography. Subcellular resolution is easily attainable with histochemical detection (e.g., Shermoen and O'Farrell[3]), whereas resolution at even the cellular level may require special methods when autoradiography is employed. In gastrulating mouse embryos, where the amounts of tissue are small, it can be difficult to obtain sufficient resolution with autoradiography to determine the germ layer in which expression occurs. The resolution of histochemical detection is more than adequate for this purpose. Third, the development of the histochemical color reaction can be monitored and controlled without loss of material, whereas the monitoring of autoradiography requires taking test slides. Fourth, histochemical detection is more rapid than autoradiography. Last, the hapten-labeled probes are stable almost indefinitely, whereas the stability of radiolabeled probes is limited by the half-life of the radioisotope. The combined advantages of the whole-mount procedures result in an increase in convenience and speed that allows for the analysis of more mRNA distributions than is possible with standard techniques.

As noted above, whole-mount *in situ* hybridization techniques have been developed for the embryos of a number of different species. Whole-mount procedures are applicable to embryos that are large enough to be handled easily, but not so large as to present problems with respect to the penetration of reagents. In the mouse embryo, this size range lies over embryonic days 6 to 9. We have used these protocols over these stages, but it is possible to use the procedures on preimplantation mouse embryos if great care is taken in handling the embryos to prevent losses. It is not possible as yet to use these procedures on whole embryos of later stages, but we have been successful with dissected organs from embryos of later stages, such as lungs from 12- and 13-day embryos.

Below we present two procedures that we have used successfully for whole-mount *in situ* hybridization to postimplantation mouse embryos. The procedures differ in the hapten and enzymatic detection systems as well as in numerous details throughout. Method 1 utilizes the hapten digoxigenin with antidigoxigenin–alkaline phosphatase detection, whereas

---

[3] A. W. Shermoen and P. H. O'Farrell, *Cell (Cambridge, Mass.)* **67**, 303 (1991).

Method 2 uses the hapten biotin with streptavidin–β-galactosidase detection.

Method 1: Whole-Mount *in Situ* Hybridization with Digoxigenin-Labeled Probes

The following methods have been adapted from similar procedures for embryos of other species which use digoxigenin-labeled probes and detection with alkaline phosphatase-conjugated antibody.[1,2] Modifications were necessary because of the large size of the later stages of mouse embryos, and because of the presence of endogenous alkaline phosphatases. The digoxigenin components are available from Boehringer-Mannheim (Mannheim, Germany).

Preparation of Digoxigenin-Labeled Probes

Single-stranded RNA probes labeled with digoxigenin are made by *in vitro* transcription in the presence of digoxigenin–UTP by the procedure provided by Boehringer Mannheim. After transcription is complete, the probe is precipitated with ethanol and dissolved in hybridization buffer directly. We do not routinely remove the DNA template or reduce probe size by limited hydrolysis. The amount of probe synthesized may be determined by comparison with labeled standards as suggested by the manufacturer. Some templates do not appear to be transcribed efficiently by SP6 RNA polymerase. In such cases, recloning for transcription by T3 or T7 RNA polymerase may be necessary. Digoxigenin probes are stable for periods of months to years when stored appropriately.

Preparation of Embryo Powder

An acetone powder of 13.5-day embryos is used to preabsorb the antidigoxigenin antibody to prevent binding to embryonic epitopes. The antibody is preabsorbed just before use. The following procedure was adapted from Harlow and Lane.[4]

1. Homogenize 13.5-day embryos (10–20) in a minimum of calcium- and magnesium-free phosphate-buffered saline (PBS) on ice.
2. Add 4 volumes of cold acetone and mix vigorously. Keep on ice for 30 min with occasional vigorous mixing. Collect the precipitate by centrifugation at 10,000 *g* for 10 min at 4°. Remove and discard the supernatant.

[4] E. Harlow and D. Lane, "Antibodies: A Laboratory Manual," p. 633. Cold Spring Harbor Laboratory Press, Cold Spring Harbor, New York, 1988.

3. Resuspend the pellet with cold acetone and mix vigorously. Keep on ice for 10 min. Spin at 10,000 g for 10 min at 4°. Transfer the pellet to a clean piece of filter paper, spread the precipitate, and allow to air-dry at room temperature. Spread and disperse the pellet as it dries. After the powder is dry, transfer it to an airtight container and store at 4°.

Fixation and Hybridization

Care must be taken throughout the steps preceding hybridization to preserve the integrity of the RNA. All solutions must be free of contaminating RNases. In addition, cleanliness throughout the procedure is important. Dirt accumulates and tends to stick to the embryos. Wash the embryos carefully after dissection with a view to removing hairs and lint, and use disposable plasticware where possible. We have found that screw-capped disposable 2-ml tube are ideal for the procedure.

1. Dissect 6.5- to 9.5-day embryos free from decidua in PBS, remove Reichert's membrane from the early embryos, and reflect the yolk sac and amnion from the later embryos. Day 9 embryos with closed anterior neuropores should have the brain punctured to allow flow of solutions in and out.

2. Fix embryos in fresh cold fixative (4% paraformaldehyde in PBS). Rock or rotate the tube gently for 2 hr at 4°.

    $10 \times$ PBS (100 ml): 8 g NaCl, 0.2 g KCl, 1.15 g Na_2HPO_4, 0.2 g KH_2PO_4

    Make 20% paraformaldehyde fresh before use: Heat 17 ml of water on a hot plate with stirring. Add 1 drop of 10 N NaOH, then 4 g of paraformaldehyde and stir until dissolved. Make up to 20 ml and filter.

    Fixative: 1 ml of $10 \times$ PBS, 2 ml of 20% paraformaldehyde, 7 ml water

3. Wash three times on ice with PBT (PBS containing 0.1% Tween 20), then wash once with PBT for 1 hr at 4° with rocking. Transfer the embryos directly into 100% methanol and store at $-20°$. Embryos may be stored for several months without degeneration.

4. Treat the embryos with 5 : 1 methanol/30% hydrogen peroxide for 4–5 hr at room temperature, followed by several washes in methanol. The embryos can be stored in methanol at $-20°$. The peroxide treatment lowers the background substantially, although the reasons are obscure. To resume the procedure, rehydrate the embryos through a graded series of methanol/PBT (75, 50, and 25%) at room temperature. Then wash through several changes of PBT.

5. Treat with 20 μg/ml proteinase K in PBT for 3–5 min at room temperature with rocking. The length of the protease treatment may need to be altered for a particular batch of protease.

6. Wash twice with freshly prepared 2 mg/ml glycine in PBT for 5 min each at room temperature with rocking. Wash twice with PBT.

7. Refix the embryos in fresh 0.2% glutaraldehyde/4% paraformaldehyde in PBS at room temperature for 20 min with rocking.

8. Wash the embryos through three changes of PBT. Treat with freshly prepared 0.1% sodium borohydride in PBT for 20 min. Do not cap the tubes tightly because hydrogen gas is produced.

9. Wash with three changes of PBT. Replace the PBT with hybridization buffer. The embryos become translucent in solutions containing 50% formamide; therefore, care must be taken in transferring and washing the embryos. Prehybridize for at least 1 hr at 63° with rocking.

Hybridization buffer:

Concentration (final)	To make 10 ml	Stock solution
50% formamide	5.0 ml	Formamide
0.75 M NaCl	1.5 ml	5 M NaCl
10 mM PIPES, 1 mM EDTA	1.0 ml	0.1 M PIPES, pH 6.8, 10 mM EDTA
100 μg/ml tRNA	0.1 ml	10 mg/ml tRNA
0.05% Heparin	0.1 ml	5% Heparin
0.1% Bovine serum albumin (BSA)	0.1 ml	10% BSA
1% Sodium dodecyl sulfate (SDS)	0.5 ml	20% SDS

10. Replace the hybridization buffer with enough fresh buffer to almost fill the tube, then add the digoxigenin-labeled probe to 0.5–2 μg/ml. Hybridize overnight at 63° with rocking.

11. Wash three times with Wash 1 (0.3 M NaCl, 10 mM PIPES, pH 6.8, 1 mM EDTA, 1% SDS). Wash twice with Wash 1 for 30 min each at 63° with rocking.

12. Wash two times with Wash 1.5 (50 mM NaCl, 10 mM PIPES, pH 7.2, 0.1% SDS), then twice with Wash 1.5 for 30 min each at 50° with rocking.

13. Wash twice with RNase buffer (0.5 M NaCl, 10 mM PIPES, pH 7.2, 0.1% Tween 20). Treat with 100 μg/ml RNase A and 100 U/ml RNase T1 in RNase buffer, twice for 30 min at 37° with rocking. Wash twice with RNase buffer.

14. Wash once with Wash 2 (50% formamide, 300 mM NaCl, 10 mM PIPES, pH 6.8, 1 mM EDTA, 1% SDS), then once for 30 min at 50° with rocking.

15. Wash once with Wash 3 (50% formamide, 150 mM NaCl, 10 mM PIPES, pH 6.8, 1 mM EDTA, 0.1% Tween 20), then once for 30 min at 50° (45° for AT-rich probes) with rocking. At this point inactivate the goat serum (see below) by heating to 70° for 30 min. The embryo powder for preabsorption of the antibody should also be inactivated by heating a few milligrams of powder in 1 ml of TBST (137 mM NaCl, 25 mM Tris-HCl, pH 7.6, 3 mM KCl, 0.1% Tween 20) to 70° for 30 min.

16. Wash twice with Wash 4 (500 mM NaCl, 10 mM PIPES, pH 6.8, 1 mM EDTA, 0.1% Tween 20), then place embryos in a heating block at 70° for 20 min to inactivate the endogenous alkaline phosphatases.

Detection of Hybridization

Hybridization of the digoxigenin-labeled probe is detected with an antidigoxigenin antibody coupled to alkaline phosphatase (Boehringer Mannheim). The alkaline phosphatases endogenous to the embryo and in the solutions used for blocking and preabsorption are inactivated by heating to 70°. In addition, levamisole, an inhibitor of certain alkaline phosphatases, is included in solutions.

17. Rock embryos in TBST containing 2 mM levamisole (0.5 mg/ml freshly added) and 10% heat-inactivated goat serum for at least 1 hr at room temperature. At this time preabsorb the antibody by dilution to 1/2000–1/5000 in cold TBST containing 2 mM levamisole, 1% freshly heat-inactivated goat serum, and the heat-inactivated embryo powder. Rock the tube for 30 min at 4°. Centrifuge the mixture at 10,000 g for 10 min at 4°. The preabsorbed antibody is in the supernatant.

18. Incubate the embryos with the preabsorbed antibody overnight at 4° with rocking.

19. Wash three times with TBST containing 2 mM levamisole (fresh), then wash five or six times with the same solution, 1 hr each, at room temperature with rocking.

20. Wash twice with freshly prepared NTMT (100 mM NaCl, 100 mM Tris-HCl, pH 9.5, 50 mM MgCl$_2$, 0.1% Tween 20) containing fresh 2 mM levamisole for 20 min each at room temperature with rocking.

21. Start the color reaction by washing the embryos into NTMT containing 2 mM levamisole, 4.5 μl/ml NBT (75 mg/ml nitro blue tetrazolium salt in 70% dimethylformamide), and 3.5 μl/ml BCIP (50 mg/ml 5-bromo-4-chloro-3-indolyl phosphate toluidine salt in 100% dimethylformamide). Protect from light. Rock the tubes for the first 5 min of the reaction, then stand the tubes in a rack. The purple reaction product should be visible within 5 min for prevalent messages and can be complete within 20 min. For rare messages the color reaction may need to proceed overnight.

22. Wash the embryos through several changes of TBST, then dehydrate through 30, 50, 70, and two changes of 100% methanol. The dehydration intensifies the pink-to-purple reaction products to dark blue. Rehydrate by going down the alcohol series to TBST. Clear the embryos by passing the embryos into 1 : 1 glycerol/PBT and then into 4 : 1 glycerol/PBT containing 0.02% sodium azide. The stained embryos may be stored in this for several months at 4°.

Method 2: Whole-Mount *in Situ* Hybridization with Biotin-Labeled Probes

The following procedure is based on the method of Wilkinson and Green[5] for *in situ* hybridization to sectioned mouse embryos using radioactively labeled probes, and on the whole-mount *in situ* hybridization protocols of Tautz and Pfeifle[1] and Hemmati-Brivanlou *et al.*[2] for the analysis of *Drosophila* and *Xenopus* embryos, respectively.[6] The procedure utilizes the strong binding of streptavidin to biotin-labeled probes and the low nonspecific binding of streptavidin–β-galactosidase complexes (Boehringer Mannheim). Under the conditions used for detection, the embryo does not exhibit endogenous β-galactosidase activity. Labeling of the probe is performed by incorporation of biotin-21–UTP (Clontech, Palo Alto, CA) into RNA synthesized *in vitro*.

Preparation of Biotin-Labeled Probes

Single-stranded RNA probes are synthesized by *in vitro* transcription from linearized DNA templates (0.5–1 μg) in the presence of biotin-21–UTP (Clontech), following standard procedures or the supplier's recommendation. Addition of unlabeled UTP to the reaction is not recommended. Preferentially, T7 RNA polymerase is used; T3 RNA polymerase was found to work less efficiently and to frequently produce prematurely terminated transcripts. Templates longer than 1 kb are preferred, since a larger proportion of the target transcript can be detected and this increases the signal.

After RNA synthesis, one-tenth of each of the reactions is taken for analysis by gel electrophoresis on a 2% RNase-free agarose gel containing ethidium bromide. The efficiency of probe synthesis can be estimated by the comparison of DNA to RNA staining intensity. It

[5] D. Wilkinson and J. Green, *in* "The Postimplantation Mouse Embryo: A Practical Approach" (D. Rickwood and K. L. Cockroft, eds.), p. 155. IRL Press, Oxford, 1990.
[6] B. G. Herrmann, *Development (Cambridge, UK)* **113**, 913 (1991).

should be considered that the RNA is stained much less intensely by ethidium bromide than DNA, and that very small amounts of probe are needed for hybridization. A normal synthesis reaction will produce an RNA band (or bands) of one-fifth to one-half the intensity of the DNA band. Less efficient reactions frequently are due to impurities in the DNA template preparations.

The RNA probe is hydrolyzed in 40 mM NaHCO$_3$/60 mM Na$_2$CO$_3$, pH 10.2, at 60° (40 min for RNA of 0.4–1 kb, 60 min for RNA of 1–2 kb), ethanol-precipitated, and dissolved in 50 μl Hyb($-$) solution (see below). The probe is stable for months at $-20°$ in this solution.

Fixation and Hybridization

All steps are carried out for 5 min at room temperature, except where indicated, in 1–1.5 ml solution in 1.5- or 2-ml round-bottomed Eppendorf plastic tubes. Wherever convenient, tubes are rotated around the long axis. Of course, solutions and glassware, etc., must be free of ribonucleases [i.e., diethyl pyrocarbonate (DEPC) treatment of aqueous solutions and baking of glassware].

1. Dissect embryos [6–9 days postcoitum (dpc)] in Dulbecco's modified Eagle's medium (DMEM, containing 10 mM HEPES, pH 7.4) or PBS. Remove Reichert's membrane and parietal endoderm or yolk sac and amnion.

2. Transfer to ice-cold PFA/PBS (made on day of use) and fix overnight at 4°.
   PFA/PBS: 4% paraformaldehyde dissolve at 65° in PBS (takes ~30 min)

3. Wash twice in PBS, replace PBS with methanol, and store at $-20°$ or proceed. (Embryos may be stored for many months.)

4. Rehydrate by consecutive washes in 75, 50, and 25% methanol and PBT (PBT is PBS containing 0.1% Tween 20).

5. Refix embryos in PFA/PBS for 20 min, then wash twice with PBT.

6. Treat with proteinase K (20 μg/ml in PBT) for 1 (6–6.75 dpc), 3 (7–7.75 dpc), or 5 min (7.75–9 dpc), quickly rinse with PBT, and wash again in PBT.

7. Postfix for 10 min at room temperature in PFA/PBS and rinse briefly in distilled deionized water.

8. Treat for 10 min with 0.1 M triethanolamine, pH 8, mixed just prior to use with 2.5 μl/ml acetic anhydride, then wash twice with PBT. (From here proceed without rotation.)

9. Replace the PBT with Hyb(−) solution and incubate at 55° (the embryos sink to the bottom).
   Hyb(−) solution:

Concentration (final)	To make 10 ml	Stock solution
50% Formamide	5.0 ml	Formamide (deionized)
5× SSC	2.5 ml	20× SSC
20 mM Tris-HCl, pH 8.0	0.2 ml	1 M Tris-HCl, pH 8.0
5 mM EDTA, pH 8.0	0.1 ml	0.5 M EDTA, pH 8.0
0.1% Tween 20	0.01 ml	Tween 20

10. Remove Hyb(−) solution, leaving a residual 20–50 μl, and add the same volume of Hyb(+) solution (a good estimate is sufficient).
    Hyb(+) solution:

Concentration (final)	To make 10 ml	Stock solution
50% Formamide	5.0 ml	Formamide (deionized)
5× SSC	2.5 ml	20× SSC
20 mM Tris-HCl, pH 8.0	0.2 ml	1 M Tris-HCl, pH 8.0
5 mM EDTA, pH 8.0	0.1 ml	0.5 M EDTA, pH 8.0
0.1% Tween 20	0.01 ml	Tween 20
0.2% polyvinylpyrrolidone	0.2 ml	10% Polyvinylpyrrolidone
0.2% Ficoll type 400	0.2 ml	10% Ficoll type 400
2 mg/ml Yeast RNA (RNase-free)	0.2 ml	100 mg/ml yeast RNA
2 mg/ml Heparin (e.g., Sigma, St. Louis, MO, H3393)	0.2 ml	100 mg/ml heparin
0.8 mM d-Biotin	1.0 ml	8 mM d-biotin
Water	0.39 ml	

11. Prehybridize at 55° for 2 hr or more.
12. Add the probe in Hyb(−) solution. Use probe amounting to 1/20 to 1/1000 the volume of the hybridization solution, and hybridize at 55° for 12–16 hr. [Before the first use, heat the probe in Hyb(−) solution to 80° for 3 min, chill on ice, and keep at −20°. Optimal probe concentration must be determined empirically.]

For convenience the following washes are done in a water bath without rotation.

13. Wash embryos in 50% formamide/2× SSC at 55° for 20 min.

14. Wash three times for 10 min each with NTET (0.5 M NaCl/10 mM Tris, pH 7.5/5 mM EDTA/0.1% Tween 20) at 37°.

15. Treat with RNase in NTET for 30 min at 37° (100 μg/ml RNase A, 100 U/ml RNase T1).

16. Wash with NTET for 10 min at 37°.

17. Wash consecutively with 50% formamide/2× SSC, 2× SSC/0.1% Tween 20, and 0.2× SSC/0.1% Tween 20 at 55° for 15 min each, then equilibrate the embryos with PBT.

Detection of Hybridization

Hybridization of the biotin-labeled probe is detected with bacterial β-galactosidase (β-Gal) coupled to streptavidin (available from Boehringer Mannheim), which binds tightly to biotin. There is no significant endogenous β-Gal activity in the embryo. Alternatively, a number of other detection systems in combination with avidin or streptavidin are available on the market and might be as useful.

The following steps are performed with rotation.

18. Incubate the embryos with streptavidin–β-Gal complex in PBT for at least 1 hr at room temperature or overnight at 4°. The working dilution giving acceptable background must be determined empirically but lies in the range of the dilution recommended by the supplier (about 1/2000).

19. Wash at least 4 times for at least 15 min each with PBT.

20. Equilibrate embryos with β-Gal buffer and stain at 37° for several hours to days in β-Gal buffer containing the following (per ml): 50 μl of 2% X-Gal in dimethylformamide, 10 μl of 0.5 M K$_3$Fe(CN)$_6$, 10 μl of 0.5 M K$_4$Fe(CN)$_6$, and 2 μl of 0.5 M EGTA, pH 8.0. It is recommended to check new batches of X-Gal for performance.

β-Gal Buffer: 0.1 M sodium phosphate, pH 6.8–7.5, 2 mM MgCl$_2$, 0.02% Nonidet P-40 (NP-40), 0.01% sodium deoxycholate

21. Stop the reaction by replacing the staining solution with PBT. The embryos may be stored in 70% ethanol or mounted (the stain is stable in water and ethanol).

Future Directions

As with all *in situ* hybridization protocols, there are many possible individual variations on the basic protocols described here, and we do not intend that the methods should be considered definitive. The techniques will continue to evolve in the future. For example, it may be possible to eliminate protease treatment by increased use of detergents (B. Rosen, personal communication, 1992), and many other variations can be envisioned. It is probable that techniques for the simultaneous detection of two different molecules within the same embryo will soon be developed. The simultaneous detection of two different RNAs should be possible through the combination of the two labeling/detection systems used here. In addition, it should be possible to combine the detection of transgene-directed β-galactosidase expression and detection of RNA, as has been done in *Drosophila* (e.g., Cubas *et al.*[7]). Finally, it might be possible to combine immunocytochemical detection of specific proteins with whole-mount *in situ* hybridization.

The great strength of whole-mount *in situ* hybridization lies in the fact that the overall three-dimensional pattern of hybridization is readily apparent. With some complex patterns of hybridization, however, it can sometimes be difficult to discern the exact cellular distribution of staining of internal tissues. This problem can be solved by subsequently sectioning the embryos to elucidate details of pattern.[6] An alternative solution would be to use fluorescent detection and optical sections obtained by confocal microscopy. Reagents for fluorescent detection of both biotin and digoxigenin are available, but it may be that the sensitivity of direct fluorescence is not sufficient for rare mRNAs. However, there are substrates for alkaline phosphatase that give products that are both fluorescent and colored.[8] If such dyes can be shown to be useful, the advantages of both sectioned and whole-mount *in situ* hybridization will be gained.

The whole-mount procedures we describe here allow for the convenient and rapid determination of many different mRNA distribution patterns. The ease of handling is such that the procedures could be used for the screening of large numbers of random or preselected cDNA clones for expression patterns of interest. It may be possible to streamline the procedures somewhat through modifications allowing the elimination of steps, thus permitting the analysis of a greater number of probes. Finally, it may be possible to automate some parts of the procedure, further increasing the number of genes that could be analyzed.

---

[7] P. Cubas, J.-F. de Celis, S. Campuzano, and J. Modolell, *Genes Dev.* **5**, 996 (1991).
[8] C. A. Ziomek, M. L. Lepire, and I. Torres, *J. Histochem. Cytochem.* **38**, 437 (1990).

[24] Detection of Messenger RNA by *in Situ* Hybridization

By DAVID SASSOON and NADIA ROSENTHAL

Introduction

The ability to localize mRNA transcripts on histological sections *in situ* is one of the most satisfying procedures for the cell biologist and embryologist. The field of molecular embryology has been revolutionized by the rapidity with which newly isolated genes can be characterized by their localization on single embryo tissue sections. It is astonishing how many sophisticated laboratories still consider this technique to be an obstacle despite its high visibility in the current literature. Many protocols are unreliable, incomplete, or dependent on tissue type or probe. In addition, protocols that are published in standard journal formats of materials and methods often leave out essential caveats that are journalistically too detailed but nonetheless critical for good results. Although numerous protocols have been devised using both radiolabeled and "cold" probes, a careful examination of these approaches reveals that specific steps and considerations are shared in common and are crucial for a successful outcome. First, tissue sections must be prepared that can withstand the pretreatment, hybridization, and vigorous washing procedures inherent in all techniques for *in situ* hybridization, retaining sufficient histological detail so that accurate localization to specific cells and tissues is possible. Second, the quality of the probe being used is of major importance. Although *in situ* hybridization of transcripts has long been a goal, it has become a practical approach with the advent of synthesis of high specific activity probes. In this respect, the term high specificity refers to a large percentage either of radiolabeled nucleotides or of modified nucleotides for nonradioactive techniques.

In this chapter, we consider one technique for the localization of cytoplasmic message using ^{35}S-labeled RNA polymerase-generated riboprobes. This technique is based largely on the techniques of Cox *et al.*[1] and Wilkinson *et al.*[2] with several modifications to improve signal-to-noise variables. We present below a detailed protocol that can be used directly in the laboratory and that may be modified for use in conjunction with other common histological techniques such as immunohistochemistry and

[1] K. H. Cox, D. V. DeLeon, L. M. Angerer, and R. C. Angerer, *Dev. Biol.* **101,** 485 (1984).
[2] D. G. Wilkinson, J. A. Bailes, J. E. Champion, and A. P. MacMahon, *Development (Cambridge, UK)* **99,** 493 (1987).

Copyright © 1993 by Academic Press, Inc.
All rights of reproduction in any form reserved.

histochemistry. In addition, a rationale is furnished for most steps that are specific to this technique.

In situ hybridization can only inform the investigator as to the cellular localization of steady-state levels of mRNA transcripts. It is generally more sensitive than other techniques such as Northern blot and RNase protection analysis, since *in situ* hybridization techniques can resolve a single positive cell among a field of negative cells. Polymerase chain reaction (PCR) amplification techniques can theoretically exceed *in situ* hybridization techniques, although much information is lost in the latter approach. For instance, a weak signal from a Northern or PCR analysis cannot distinguish between a large number of cells transcribing low levels of a particular mRNA and a small number of cells transcribing high mRNA levels. If an appropriate antibody is available, the investigator is encouraged to employ this first as it will answer the spatial localization question and also inform one as to whether the transcript of interest is translated. This latter point cannot be overemphasized. The appearance of a transcript in a particular cell is often a strong indicator as to the localization of the gene product. However, in the case of protein secretion and uptake by neighboring nontranscribing cells or where translation is uncoupled, *in situ* hybridization results cannot be interpreted to be firm proof of gene expression. Nonetheless, the term expression is often misused in this context.

There are situations when neither antibodies to the products of novel cDNA clones nor isoform-specific antibodies are available. This situation can be resolved by *in situ* hybridization approaches if specific cDNA-generated probes corresponding to the 5' or 3' untranslated regions of the transcript are available (see, e.g., Sassoon *et al.*[3]). The information obtained from a successful *in situ* hybridization assay gives a spatial and temporal description of gene transcription. Thus, it is not difficult to see the value of exploiting *in situ* hybridization to define the temporal and spatial distribution of a newly isolated gene during development. Such data serve as a critical first step in designing more functional experiments and ultimately assigning a putative role for a gene. Table I outlines several uses and benefits that *in situ* techniques afford. It is emphasized, however, that although the technique is essentially simple, it is time-consuming and requires both attention to molecular biological techniques as well as meticulous attention to proper histological techniques; both disciplines are not always available in a single laboratory, and both should be learned for a succcesful outcome.

[3] D. A. Sassoon, I. Garner, and M. Buckingham, *Development* (*Cambridge, UK*) **104,** 155 (1988).

TABLE I
In Situ STRATEGIES

Gene class	Strategy	Possible outcome	Major difficulties
Structural gene (e.g., actin, collagen)[a,b]	Tissues or embryos prepared for stages of interest	Gene may not be tissue-specific at all stages of development; may be localized to subset of cells that otherwise appear identical	Many structural genes are members of gene families; even if family is not identified, large portions of probe may encode motifs present in other genes, resulting in cross-reactivity
Regulatory gene (e.g., homeobox-containing gene, transcription factor)[c,d]	Tissues or embryos prepared for stages of interest, but since tissue distribution is usually less clear at onset of study, a large number of representative sections must be examined	Gene may appear to be present in gradient or over various regions of embryo, necessitating alternate planes of section for analysis; homeobox genes often display rostral–caudal gradient which is easily viewed in parasaggital plane	As for structural genes, families of related genes are abundant for regulatory transcription factors; specific probes are most prudent
Receptors and ligands (e.g., growth factors and receptors)[e]	If possible, it is useful to carry out study with both receptor and ligand to determine temporal linkage; here, antibodies are only way to determine localization of ligands	Thorough analysis may well lead to assignment of role for receptor and ligand	Same as above
Uncharacterized or novel clone	Strategy will depend on source of clone (stage, tissue, etc.) and degree of structural characterization; if untranslated regions are not identified, it must be remembered that pattern obtained may not be specific	Pattern such as temporal specificity or cell type or regional localization can often be only clue in designing future experiments and ultimately assigning role for gene	Same as above

[a] D. A. Sassoon, I. Garner, and M. Buckingham, *Development (Cambridge, UK)* **104**, 155 (1988).

[b] G. E. Lyons, M. Ontell, R. Cox, D. Sassoon, and M. Buckingham, *J. Cell Biol.* **111**, 1465 (1990).

[c] P. Dollé, J. C. Izpisua-Belmonte, H. Falkenstein, A. Renucci, and D. Duboule, *Nature (London)* **342**, 767 (1989).

[d] D. Sassoon, G. Lyons, W. E. Wright, V. Lin, A. Lassar, H. Weintraub, and M. Buckingham, *Nature (London)* **341**, 303 (1989).

[e] E. Keshet, S. D. Lyman, D. E. Williams, D. M. Anderson, N. A. Jenkins, N. G. Copeland, and L. F. Parada, *EMBO J.* **10**, 2425 (1991).

In Situ Hybridization Protocol

Tissue Preparation

Procedure. Remove tissue, working in cold phosphate-buffered saline (PBS) if possible, and place in fix (4% paraformaldehyde–PBS) at 4° overnight. A cardiac perfusion is done if the tissue is very large. Paraformaldehyde is prepared fresh. To prepare 100 ml, add 4 g paraformaldehyde powder to 50 ml distilled water and heat while stirring to 65° (the solution remains a bit cloudy). Add 5 to 10 μl of 10 N NaOH, which will make the solution clear. Filter through Whatman (Clifton, NJ) filter paper to remove debris and undissolved paraformaldehyde. Add 10 ml of 10× PBS and bring volume to 100 ml final with distilled water. The paraformaldehyde–PBS can be stored at 4° for no more than 24 hr. Do not fix more than about 0.5 cm^3 of tissue per 50 ml of fixing solution.

Rationale. Tissue must be quickly and well fixed in order to stabilize the RNA in the cells. The paraformaldehyde is prepared carefully and freshly; avoid using a polymerized formalin-type fix that will not penetrate the tissue fast enough to preserve the RNA.

Embedding in Paraffin

Procedure. Perform the following washes of fixed tissue:

1× PBS	4°	30 min
0.85% Saline (NaCl)	4°	30 min
1 : 1 Saline/100% ethanol	Room temperature	15 min
70% Ethanol (do twice)	Room temperature	15 min
85% Ethanol	Room temperature	30 min
95% Ethanol	Room temperature	30 min
100% Ethanol (do twice)	Room temperature	30 min

At this step the tissue can be stored at 4° overnight if time is short; otherwise continue. Sufficient paraffin should be ready, having been melted in a 60° oven and filtered. Wash tissue as follows:

100% Xylene (do twice, in fumehood)	Room temperature	30 min
1 : 1 Xylene/paraffin	60°	45 min
100% Paraffin (do three times)	60°	20 min

The third paraffin wash is performed with the tissue arranged and oriented in the container where it will be allowed to solidify at room temperature overnight. The following day, the embedded tissue can be either stored at 4° indefinitely or mounted on a microtome for sectioning. If the tissue large and difficult to infuse with the paraffin, a vacuum oven can be very useful.

Rationale. The paraffin used in modern histology contains both wax and plastic polymers, which are easier to cut (i.e., Paraplast brand). The paraffin should be melted ahead of time, preferably the night before or early the day of embedding, because it can be slow to melt and must also be filtered. For the washes, use volumes approximately 10 times that of tissue or more. The sequential washes are meant to remove the fixative by replacement with saline and finally to dehydrate the tissue by gradual replacement of saline with alcohol. These washes are normally done in 50-ml conical tubes and should be agitated periodically. Paraffin is a sufficiently hard compound that the different densities of the tissue will not affect the cutting thickness and quality. High-quality sections are essential for good results. A section with folds, bad fixation, or poor dehydration will not hybridize well! Times for each step may be increased 2- to 3-fold for larger tissues. Dehydration must be complete, or the paraffin will not penetrate the tissue, which might lead to uneven cutting.

Cultured Cell Preparation

Procedure. Cells grown in culture can also be processed for *in situ* analysis (see Fig. 1A,B). This is most useful in the case of mixed populations of cells from primary material or to verify that all cells are (or are not) behaving uniformly. Cells should be grown on a glass substrate to permit the use of solvents (such as xylene) and high temperatures, which will deform or alter the optical characteristics of plastic. Coverslips (treated with appropriate substrates) or commercially available Labtek glass slides (Nunc) are particularly useful. The washes performed on fixed cells are as follows:

1× PBS	4°	10 min
0.85% Saline (NaCl)	4°	10 min
1:1 Saline/100% ethanol	Room temperature	10 min
70% Ethanol (repeat twice)	Room temperature	5 min
95% Ethanol	Room temperature	10 min
100% Ethanol (repeat twice)	Room temperature	5 min

Preparation of Subbed Slides

Procedure. Because the slides will ultimately be handled in a darkroom, we recommend using single-sided frosted slides so that the side with the section can be easily determined. Ease of orientation is also advantageous during the hybridization procedures. It is essential that slides are clean. Place slides in metal slide racks and soak in hot tap water with detergent for glassware. Rinse 1 hr in hot running water followed by 3 changes of distilled water. Allow to air-dry or place in 40° oven. Slides are kept in racks and subbed.

To prepare subbing solution, boil 1 liter of distilled water. While stirring but with heat *off,* add 2 g of gelatin and 0.1 g potassium chromium (III) sulfate [$KCr(SO_4)_2 \cdot 12H_2O$]. When the solution temperature is approximately 35°, transfer to a slide staining tray. Dip the washed slides and allow 5 min to drain. Dip again and let drain at an angle such that the gelatin will drain toward the frosted ends of the slides. Dry overnight in a 35° oven and store in the original slide boxes (handle slides carefully by the edges only). Subbed slides are stable at room temperature for 1 year.

Rationale. This procedure ensures that tissue sections will remain attached to the slide. Gelatin will cross-link to tissue once fixed and does not react with the photographic emulsion (see last steps), nor will it react with the cRNA probe. Subbed slides are used only for tissue sections and not for cultured cells. Frozen sections may also be used and collected on subbed slides. This is necessary if an antibody sensitive to paraffin techniques is used in tandem.

Tissue Sectioning

A description of paraffin sectioning techniques is not included. There are many histological handbooks available; however, we recommend consultation with an experienced histologist or pathologist. There are many tricks for obtaining high-quality tissue sections. A sharp knife, a working microtome, and patience are the key ingredients. Whereas many histologists collect sections in long paraffin ribbons, we suggest collecting sections individually onto subbed slides.

Slides are placed on a heating tray at 50°. A large drop of water (0.2% ethanol) is placed on the slides, and the sections are carefully placed on the drops, where they are allowed to relax to their original configuration. This usually takes 3–5 min. The water drop is then blotted away with a corner of blotting paper (bibulous paper), and the section is pressed onto the slide by placing the moistened blotting paper over the section and pressing lightly with the fingers. Let dry overnight in racks. The slides can then be placed in slide boxes and stored indefinitely at 4° or room temperature.

Preparation of Probe

Rationale. We recommend the use of RNA probes. They are single-stranded, thus eliminating the presence of labeled sense probe that (1) may cross-react with other transcripts and (2) can duplex with the antisense probe, reducing the available concentration of probe for hybridization with the transcripts *in situ*. In addition, RNA probes form very stable RNA–RNA duplexes at high affinity, affording a higher sensitivity compared to DNA probes. Last, RNA probes can be generated readily with high specific activity.

The proper preparation of probe is the single most important step for a clean, low-background image. Free labeled nucleotides are the major cause of nonspecific background. For this reason, we do not recommend spin columns or repeated precipitations, which only reduce but do not eliminate unincorporated nucleotides. The probes can be transcribed with high efficiency from either the T3, T7, or Sp6 bacterial promoters. In principle these *in vitro* transcription reactions result in a high percentage of full-length material, which must then be alkaline hydrolyzed to an average length of approximately 100 bp to facilitate accessibility into the tissue sections. Because ^{35}S-labeled radionucleotides are used, it is important to use siliconized tubes when possible and also to maintain 10 mM dithiothreitol (DTT) in the solutions to minimize loss of probe caused by nonspecific sticking to the Eppendorf plastic tubes. Total yeast RNA is used as a carrier and to saturate contaminating RNases in several steps, so a substantial volume of clean yeast RNA stock should be prepared.

Transcription of RNA Probe. Prepare 5× transcription buffer:

Tris-HCl (pH 7.5)	200 mM
MgCl$_2$	30 mM
Spermidine	10 mM
NaCl	50 mM

The reaction is carried out in a final volume of 20 to 22 μl with about 300 ng of linear template. The radiolabeled nucleotide of preference is UTP (>1000 Ci/mmol). The reaction mix is as follows:

5× transcription buffer	4 μl
0.1 M DTT	2 μl
Cold CTP (10 mM)	1 μl
Cold ATP (10 mM)	1 μl
Cold GTP (10 mM)	1 μl
[^{35}S]UTP	5 μl
RNasin	1 μl
Polymerase	1.5 μl
DNA template (300 ng)	300 ng
Distilled water	To 22 μl

Incubate for 1–2 hr at 37°. Add 1 μl additional polymerase and incubate at least 30 min. Add 1 μl DNase RQ1 (RNase-free, to digest DNA template) and incubate at 37° for 15 min. After DNA digestion, add 3 μl total yeast RNA (10 mg/ml; this reagent is extracted with phenol and chloroform 3 times, then precipitated with ethanol several times to make a stock solution).

Hydrolysis (Size Reduction) of RNA Probe. Add 50 μl of hydrolysis solution A to newly synthesized probe:

Solution A:

DTT (1 M)	10 μl
NaHCO$_3$ (1 M)	80 μl
Na$_2$CO$_3$ (1 M)	120 μl
Distilled water	790 μl

The time of alkali hydrolysis depends on the length of the template used, and the reaction is carried out at 60°. We suggest that the final probe length be verified using an analytical acrylamide gel. As a guide, transcripts shorter than about 300 bases will need about 15–20 min of hydrolysis, whereas those over 300 bases will require about 30–45 min. Because the hydrolysis reaction randomly cuts the probe, a long transcript will quickly be reduced to fragments of around 300 bases in about 10 min. Once hydrolysis is complete, the reaction is neutralized by adding 50 μl of solution B.

Solution B:

Sodium acetate (1 M)	200 μl
Acetic acid	10 μl
DTT (1 M)	10 μl
Distilled water	780 μl

Separation of Probe from Unincorporated Nucleotides. The neutralized probe reaction mix is passed through a Sephadex G-50 column to separate the labeled probe from unincorporated nucleotides and small labeled fragments. To prepare the Sephadex G-50 column, use a 1 : 1 ratio of fine and medium matrix swelled in running buffer that does not contain DTT.

Running buffer:

Tris-HCl	10 mM
EDTA	5 mM
Sodium dodecyl sulfate (SDS)	0.1%
DTT	10 mM, added just prior to use

The column is run in a long Pasteur pipette which is plugged with a small piece of siliconized glass wool. The tip is broken off as close to the glass wool as possible. Pass through 1 ml running buffer to equilibrate the column followed by 50 μl total yeast RNA (10 mg/ml). Pass through 500 μl running buffer to get RNA into the column. The RNA in the column functions both as a carrier and to saturate RNases. Add 1 μl of saturated phenol red to the probe (to serve as a tracer for free nucleotides) and add the probe to the column. Allow about 500 μl to run through gently so as not to disturb the column bed, then collect fractions of about 400 μl until the phenol red is about 1/4 inch (0.635 cm) from the bottom. The fractions containing probe should be easily monitored by a sensitive Geiger counter.

Combine the radiolabeled fractions into the minimum possible number of siliconized Eppendorf tubes and precipitate with 1/10 volume of 3 M potassium acetate (filtered) and at least 2 volumes of 100% ethanol and place in dry ice for at least 1 hr. Spin 15–30 min, reduce the volume in a SpeedVac (Savant Instruments), but be sure not to dry the RNA completely as it is difficult to resuspend. Resuspend each fraction in 10 μl of 10 mM DTT. This is a crucial step because the probe is very sticky; one must be meticulous in pipetting up and down to get a good recovery. Combine the resuspended probe into one tube. Count a 1-μl aliquot in scintillation fluid; the radioactivity should exceed 40 million counts total. Dilute to a final concentration of 50,000 counts/min (cpm) in hybridization buffer (see below).

Prehybridization

Rationale. In the prehybridization step the tissue on the slide is prepared for hybridization by removal of the paraffin, leaving the section intact on the slide, followed by a deproteinization step to make the tissue more porous to the probe. In this and all subsequent procedures, glass slide racks (which hold about 19 to 20 slides oriented in a zigzag manner) and 250 ml glass staining trays are used for the various steps. We suggest that about 10 racks and staining trays be on hand, which can be emptied and rinsed in between steps. Alcohols are reused about 10 experiments (200 slides) and can be stored in bottles when not in use. To heat solutions or keep them hot, place staining trays with solution into a water bath and never directly on a heating plate.

Procedure. The following solutions should be ready:

4% paraformaldehyde–PBS (fresh)	500 ml
1× PBS	1 liter
0.85 NaCl (autoclaved)	1 liter
Proteinase K (20 mg/ml stock in distilled water, −20°)	Stock alloquated in 100 μl (−20°)
PK buffer:	
Tris (1 M)	10 ml
EDTA (0.5 M)	2 ml
Water	To 200 ml
0.1 M Triethanolamine (6.7 ml/500 ml, store at 4°), MW 149.2	
Acetic anhydride	

Bring the slides to be hybridized to room temperature. Place in clean racks. All washes that follow are at room temperature unless otherwise noted. The xylene is changed every 60 slides, but the xylene from the second wash can be used again as the xylene for the first wash for subsequent batches of slides.

Wash slides in xylene 2 times for 10 min each. Wash in 100% ethanol for 2 min (retain this ethanol for deparaffination only). Run quickly through the following rehydration series: 100, 95, 85, 70, 50, and 30% ethanol. Continue by washing as follows:

0.85% NaCl	5 min
1× PBS	5 min
4% Paraformaldehyde–PBS	20 min
1× PBS (do twice)	5 min

Remove the liquid from the tissue and wash in the presence of proteinase K for 7.5 min, using 18 μl of stock (20 mg/ml) per 15 ml of PK buffer, made freshly. The proteinase K wash is done on each slide individually.

Return slides to the rack and immerse in 1× PBS for 5 min. After treating the slides for 5 min in 4% paraformaldehyde, dip slides in water. Slides are then transferred into 250 ml of 0.1 M triethanolamine solution. Stir gently with a stir bar placed below the slide rack. Stirring is needed to mix in the acetic anhydride as follows. Add 625 μl of acetic anhydride to the staining tray with slides already immersed. Let the mixture stir for 10 min; the acetic anhydride will not immediately go into solution.

Wash slides in 1× PBS for 5 min followed by 0.85% NaCl for 5 min. Run quickly through a dehydration series: 30, 50, 70, 85, 95, 100% ethanol, 100% ethanol 2. The 100% ethanol 2 is used only at this step; this ethanol is never in contact with xylene. Air-dry slides for at least 30 min. Slides can be stored at room temperature for 2 ′ weeks or used the same day.

Hybridization

Procedure. The hybridization buffer and the coverslips must be prepared in advance. The coverslips are loaded into racks and washed successively in glass detergent, hot tap water, and distilled water, then allowed to dry. They are then dipped into a siliconization solution, dried, and stored in a glass petri dish. Dip in ethanol (100%) and dry just before use to remove dust and silicone residue.

The hybridization buffer is prepared to a concentration such that 9 parts of buffer are added to 1 part of the probe in 10 mM DTT. The buffer is stable for at least 6 months when stored at 4°.

Hybridization buffer:

Component	Concentration (final)	Amount (stock)
Formamide (deionized)	50%	15 ml (100%)
NaCl	0.3 M	2.25 ml (4 M)
Tris-HCl (pH 7.4)	20 mM	600 μl (1 M)
EDTA	5 mM	300 μl (.5 M)
$NaH_2PO_4 \cdot H_2O$ (pH 8.0)	10 mM	300 μl (1 M)
Dextran sulfate	10%	6 ml (50%)
Denhardt's solution	1×	600 μl (50×)
Yeast RNA (total)	0.5 mg/ml	1.5 ml (10 mg/ml)

This solution now represents 9 of 10 parts of the complete hybridization solution to which probe will be added. The probe has been resuspended in 10 mM DTT in the previous step. To arrive at the correct final concentration of hybridization buffer, the addition of probe should constitute the final one-tenth of the total volume; thus, water may be added with probe at this time. Adjust the hybridization buffer to 10 mM DTT prior to mixing with the probe, at which point the solution is stable for several months at $-20°$. The optimal concentration is between 25,000 and 50,000 cpm/μl. To dilute the probe (in hybridization buffer) to the final concentration, add hybridization buffer plus 1/10 volume of DTT in water.

The probe must be heated to 80° for at least 2 min prior to adding to the slides in order to denature any possible secondary structure, which would inhibit hybridization. Apply 10–20 μl (depending on size of tissue section) on each slide. Cover with a siliconized coverslip which has been rinsed in 100% ethanol to eliminate residues from the siliconizing solution (place one side down and slowly drop the coverslip in place). Incubate in a humid chamber at 50°–52° for about 16 hr. The slide box contains a towel soaked in a mixture of 50% formamide (5 ml of 100% formamide), 4× SSC (2 ml 20× SSC), and water (3 ml). The box, sealed with electrical tape to keep the slides from drying, is arranged in the oven such that the slides lay flat with the coverslips on top.

Washing

Procedure. It is important to keep the tissues moist. Therefore, transfer the slides from solutions into racks already immersed in solutions whenever possible. Because DTT is labile, it should be the last item added to solutions just prior to incubation. The wash in formamide–SSC–DTT is carried out in sealed plastic Coplin jars. Wash slides as follows:

5× SSC, 10 mM DTT	50°	30 min
50% Formamide, 2× SSC	65°	20 min
1× Washing solution (do twice)	37°	10 min
RNase A treatment in 1× washing solution	37°	30 min

A 10× stock of washing solution is made from 233.8 g NaCl, 100 ml of 1 M Tris (pH 7.5), 100 ml of 0.5 M EDTA, and distilled water to 1 liter. RNase A is kept as a 10 mg/ml stock made up in water. The RNase A stock is not boiled since any contaminating nucleases will not affect the quality of the hybridization and boiling will only decrease the activity of the enzyme. The final concentration of RNase A used is 20 μg/ml. It is also advisable to use RNase T1 at 2 μg/ml final concentration if background is a problem due to secondary RNA structures; however, this is a rare occurrence and usually is not responsible for high background. Continue with the following washes:

1× Washing solution	37°	5 min

Repeat the formamide wash if the background problem is prohibitive.

2× SSC	37°	15 min
0.1× SSC	37°	15 min

A dehydration step is now carried out as before, but 0.3 M ammonium acetate is used in all the ethanol solutions except the 100% ethanol. Naturally these solutions are kept separate for this step only. Wash quickly in 30, 60, 80, and 95% ethanol (with ammonium acetate), then 100% pure ethanol (no ammonium acetate). Allow to air-dry.

Autoradiography

Procedure. Safelight conditions are of extreme importance: this means no unplugging of switches, no phosphorescent watches, and gentle handling of the emulsion. Warm Kodak (Rochester, NY) NTB-2 in a 45° bath in the darkroom (this solution is stored carefully sealed from light at 4°). Pour an appropriate amount into a beaker and gently pass a clean, blank slide through it to remove bubbles. Dip slides and allow to air-dry for 1 hr. Box the slides in a light-tight box with a Drierite packet and store at 4°. The refrigerator used must be radioactivity-free. The emulsion is also sensitive to shock, so the slides must be treated gently. When it is time to develop the slides, they are first allowed to come to room temperature while still in boxes to avoid condensation, which causes image fading.

Develop with Kodak D-19 (stored at 4°) and then fix with Kodak Rapid Fix:

D-19	16°	3.5 min
Tap water	Room temperature	Dip
Rapid Fix (do twice)	Room temperature	5 min
Tap water	Cold (tap temperature)	10 changes

Scrape the emulsion off the back of the slide with a razor blade and stain lightly in an aqueous 0.2% toluidine blue staining solution. Rinse and destain in water followed by dehydration in graded ethanol solutions that are used exclusively for this part of the protocol. Dehydration in the 50 and 70% ethanol baths can be long to remove excess stain (i.e., 2 to 20 min each step). Staining is preferably light so that the silver grains will be clear when viewed with dark-field optics. If viewing is to be done only with bright-field optics, this is less important. Destaining is considered adequate when the photographic emulsion is no longer stained. A typical dehydration series for slides is as follows: 50, 70, 85, 95, 100% ethanol, then 100% xylene (repeat twice). Dehydration is followed by overlying coverslips with a mounting medium. For dark-field optics, slides should be coverslipped with an autoradiography-compatible mounting medium (consult the supplier) and allowed to dry flat for 1 week, although may be viewed and photographed immediately. The intensity of staining and choice of stain will ultimately reflect histological considerations.

Rationale. The time of exposure is variable and will depend on the abundance of the transcript and the level of detail required for the experiment. For instance, to prove that a transcript is absent (undetectable), exposure periods that are 2 to 3 times those normally required for tissues where it is easily detected are useful. Empirically, we find that a 1-week exposure is optimal for the majority of probes. Very abundant transcripts such as β-actin may require only a 12-hr exposure, whereas detection of a homeobox gene in an adult tissue (i.e., very low levels or very few cells) may require 2–3 weeks. It is our experience that if a signal cannot be detected after 2–3 weeks, longer exposures will not remedy the situation. The use of dark-field optics is also critical for sensitive detection. Signals that are normally not easily detected by standard light microscopy can be easily detected at low magnification using dark-field optics.

Applications in Mouse Embryology

The interpretation of *in situ* hybridization results with mouse embryo sections requires some knowledge of mouse embryogenesis. Although not all-encompassing the texts by Rugh (*The Mouse: Its Reproduction and Development,* Oxford Press), Theiler (*The House Mouse: Atlas of*

Embryonic Development, Springer-Verlag), and Kaufman (*The Atlas of Mouse Development,* Academic Press) are a good beginning. Often, the investigator can judge what stages of embryonic development are worth examining first. The important issue is where in the embryo to look. For instance, it is now clear that most homeobox (*Hox*) genes are expressed in the mouse limb bud. This fact was missed since the majority of studies of novel *Hox* genes were limited in scope to few tissue section levels. In particular, novices tend to hybridize exclusively sections that are easily recognizable in terms of structures that are near perfect mid-saggital sections. Clearly, many structures such as the limbs may not be present in these sections.

If *in situ* hybridization is going to be a central tool in a laboratory, it is worthwhile to assemble a collection of mouse embryos that are embedded in paraffin and already sectioned. Paraffin sections mounted onto glass slides will store indefinitely. Indeed, this is a major advantage of paraffin since the tissues are stable and sections are long-lived. This can be particularly useful if an additional gene is examined and the investigator wishes to compare directly with sections studied months (or years) before. If sections were meticulously collected on numbered slides, adjacent and serial sections will allow for a detailed comparison of several patterns of gene expression. In this section, we outline three applications of *in situ* hybridization: (1) localization of gene transcripts in the embryo and how this information ultimately can be used to assign putative gene function; (2) use in tissue culture, and (3) localization of reporter gene expression in transgenic mice.

Embryonic Development: In Situ Hybridization as Key to Understanding Gene Function

One widely published use of *in situ* hybridization has been to elucidate and interpret the temporal and spatial pattern of transcription of a regulatory gene whose function is incompletely characterized. There are many examples of this with regard to the isolation of murine homeobox-containing genes. Clearly, a pattern of gene expression gives insight into the possible role of a *Hox* gene.

In skeletal muscle development, much activity has recently centered around the recent characterization of the so-called myogenic factors, which include MyoD1, myogenin, myf5, and MRF4 (see Weintraub *et al.*[4]). Although the potential role of the myogenic factors as master regulatory genes was originally based on data obtained with cell lines, the tempo-

---

[4] H. Weintraub, R. Davis, S. Tapscott, M. Thayer, M. Krause, R. Benezra, T. K. Blackwell, D. Turner, R. Rupp, S. Hollenberg, Y. Zhuang, and A. Lassar, *Science* **251,** 761 (1991).

ral and spatial patterns of MyoD1, myogenin,[5] myf5,[6] and MRF4[7] have been described and represent the chief distinctions that can be attributed to each. In general, all four factors share in common a DNA-binding motif and a helix–loop–helix (HLH) domain which confers protein–protein binding activity with other HLH domain proteins. The *in situ* analysis has elucidated the following results and consequently the following interpretations. (1) Each factor, although capable of phenotypic conversion, has a distinct pattern of expression. (2) The first factor expressed temporally is myf5. Thus, the other three are not the initial myogenic lineage commitment gene. (3) Myf5 is expressed in the dermamyotome compartment of the early somites. Although this is the embryological "birth place" of skeletal muscle, *in situ* analysis failed to detect myogenic factor accumulation outside of the somite in structures such as the limb bud, which already possess committed myoblasts as demonstrated by classic embryological techniques. Thus, the true role of the factors is unclear, although their early expression suggests a role in development. An example of myogenic factor hybridization Fig. 1C–E) shows the localization to the somites of a 9.5 day postcoitum mouse embryo.

In a sense, the *in situ* analysis provides a molecular sleuthing that, if appropriately interpreted, can help elucidate the role(s) of a regulatory gene or gene family. Indeed, as myogenic factor knockouts are completed, it will be of critical interest to determine if the pattern of the other three factors are altered to compensate molecularly for the missing factor(s). Although detailed histological, fiber type, and muscle morphology analyses in gene disruption experiments must ultimately be performed, *in situ* hybridization using genetically manipulated mouse embryos will prove to be an important tool for understanding the molecular consequences of this approach.

Tissue Culture

The tendency to generalize gene expression data from relatively homogeneous sources of RNA, such as clonal cell lines grown in culture under identical conditions, and then to extrapolate this information to give number of transcripts per cell is probably one of the most misleading attempts to quantify data that the molecular biologist can make. One example is

[5] D. Sassoon, G. Lyons, W. Wright, V. Lin, A. Lassar, H. Weintraub, and M. Buckingham, *Nature (London)* **341,** 303 (1989).

[6] M.-O. Ott, E. Bober, G. Lyons, H. Arnold, and M. Buckingham, *Development (Cambridge, UK)* **111,** 1097 (1991).

[7] T. J. Hinterberger, D. A. Sassoon, S. J. Rhodes, and S. F. Koneiczny, *Dev. Biol.* **147,** 144 (1991).

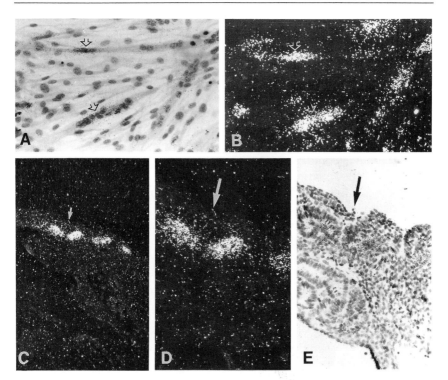

FIG. 1. Localization of myogenic regulatory gene trasncripts to cells in culture and to the mouse embryo. (A) Bright-field photomicrograph of F3 myotubes differentiated for 48 hr showing clusters of nuclei (arrows). (B) Dark-field photomicrograph of same field as in (A) showing hybridization signal over myotubes. In particular, note the dense accumulation of silver grains over the clustered nuclei [arrows in same position as for (A)]. (C) Low-magnification dark-field photomicrograph of a 9.5-day mouse embryo cut parasagittally at the level of the somites (arrow). This section has been hybridized with two cRNA probes, myf5 and myogenin. Note the dense signal over somites. (D) High-magnification dark-field photomicrograph of (C) showing degree of localization of grains to the central myotomal somite compartment. (E) Corresponding bright-field photomicrograph of same field in (D). Note that grains are visible but more difficult to resolve.

illustrated in this chapter. Myogenic cells can be grown in culture and induced to differentiate when serum concentrations are decreased. It had been assumed that MyoD1 was expressed in the myoblast and maintained in the multinucleated myotube. Northern analysis of many cell lines indicated that MyoD1 is abundant in both myoblasts and myotubes. We had investigated the implicit assumption that the situation at the cellular level was fairly uniform in a clonal cell line induced to differentiate in the same culture dish by hybridizing myoblasts (F3 cell line) and myotubes with a cRNA probe to MyoD1. We observed that only 10 to 20% of the myoblasts transcribe detectable levels of MyoD1 (data not presented) and that not all regions of the myotubes (nuclei) have detectable levels of MyoD1 (see Fig. 1A,B). In addition, we observe a localization of message to certain clusters of myonuclei, indicating that transcriptional heterogeneity exists in the syncytial myotube. This phenomenon had already been established by *in situ* analysis of acetylcholine receptor transcripts to the subjunctional myonuclei *in vivo* (see Fontaine *et al.*[8]).

This level of detailed analysis is simply not possible using alternate techniques and should be used to test whether a model for gene regulation during cell differentiation is not overgeneralized. An additional advantage of using *in situ* hybridization to track transcription in cultured cells is that only small quantities of cells are needed to obtain results. Thus, a single limb bud or somite can be cultured under various conditions and changes in gene transcription followed with minimal numbers of embryos and a greater yield of information.

Analysis of Exogenous Gene Expression in Stable Cell Lines and Transgenic Mice

The examples described above illustrate the appeal of *in situ* hybridization as a tool for a comprehensive analysis of endogenous gene expression in developing embryos, adult tissues, or cell cultures. A novel application of *in situ* hybridization involves the localization of transcripts from gene constructs introduced into cells in culture or into the germ line of transgenic mice. The choice of *in situ* hybridization for detecting transcripts from gene expression vectors depends on the specific biological question being addressed, but it often offers substantial advantages over more conventional methods of RNA analysis, as illustrated in the applications below.

Cell Culture. A wide variety of plasmid and retroviral eukaryotic vectors has facilitated the introduction and expression of virtually any gene in cultured cells. Alternatively, vectors in which reporter genes are linked to putative transcriptional regulatory sequences have uncovered cis-acting elements associated with numerous gene loci. In many cases, cell cultures transfected or infected with these vectors can be harvested and the exogenous gene transcript assayed by conventional means, such as Northern analysis or RNase protection. Reporter genes often encode gene products which can be detected with enzyme or radioimmunoassays. However, if

---

[8] B. Fontaine, D. Sassoon, M. Buckingham, and J.-P. Changeux, *EMBO J.* **7**, 603 (1988).

cultures contain multiple cell types, as is the case with primary cultures, only one of the cell types may express the exogenous gene. Moreover, transfected genes may not be homogeneously expressed, even in a clonal cell line, as illustrated above in muscle cells (Figs. 1A,B). By scoring exogenous gene transcripts in these cultures with *in situ* hybridization, heterogeneous expression patterns can be readily detected and assigned to specific cell types or stages of differentiation. As noted in the previous section on tissue culture, small cell populations can be analyzed, as opposed to standard RNA assays which require isolation and purification of transcripts from considerably larger quantities of starting material. Finally, the intracellular localization of exogenous gene transcripts is possible with this technique, providing an additional level of detail not achievable with more conventional methods.

Transgenic Mice. In transgenic animal models, it is often crucial to localize expression of the exogenous gene, either in a specific anatomical structure, during a developmental stage, or in response to a stimulus. *In situ* hybridization of transgene transcripts in tissue sections renders the whole animal accessible to analysis and in many cases yields much more information than multiple RNA assays of individual tissue samples. This feature is particularly useful in studies of transgene expression in the embryo, where dissection of specific tissues is less precise and where transgene expression may be initiated in individual cells. In this regard, *in situ* hybridization is one alternative among several techniques currently used for whole-body visualization of transgene expression, in which the transgene product is an enzyme such as β-galactosidase (e.g., Klarsfeld *et al.*[9]) that can be assayed *in situ*.

There are several advantages of *in situ* hybridization over other methods. First, both in the embryo and in the adult, expression of the transgene can be compared to that of the endogenous gene or other genes of interest by hybridizing adjacent sections to appropriate probes. Second, *in situ* hybridization measures accumulation of transgene transcripts, rather than enzyme products whose activity may be obscured by endogenous enzyme activity or may be subject to additional influences in the context of a given tissue. Third, an exclusive feature of *in situ* hybridization is the ability to localize expression of multiple genes in a specific tissue.

A representative example (Fig. 2A,B) shows transgene expression in the skeletal muscles of a neonatal mouse. In this case the transgene consisted of a chloramphenicol acetyltransferase (CAT) reporter driven by

---

[9] A. Klarsfeld, J. L. Bessereau, A.-M. Salmon, A. Triller, C. Babinet, and J.-P. Changeux, *EMBO J.* **10,** 625 (1991).

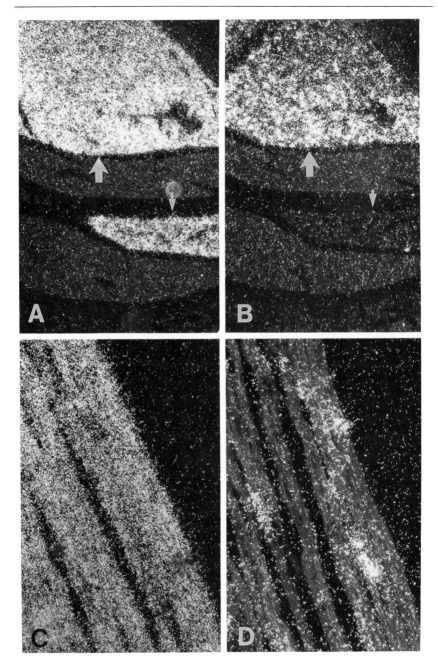

regulatory elements from a myosin light chain gene (MLC).[10] By probing adjacent sections with antisense RNA probes corresponding either to CAT or to MLC sequences, specific neck muscle groups expressing both the endogenous gene and transgene could be distinguished from those expressing only the endogenous gene. At higher magnifications (Fig. 2C,D), the same samples displayed significant differences in the subcellular distribution of the two gene transcripts. This example illustrates the power of *in situ* hybridization not only to localize exogenous gene transcripts to a specific tissue type, but also to reveal simultaneously potentially important information regarding the subcellular distribution of different transcripts.

Controls

This chapter provides a reproducible and reliable protocol that has been used to detect over 50 different gene transcripts. It is by no means the only approach, but it does work for all tissues and routinely has a sensitivity exceeding that of Northern blot and ribonuclease protection analyses by virtue of single-cell resolution. A common issue arising in the interpretation of *in situ* hybridization data concerns controls. When single-stranded nucleic acid probes are used, many wonder if a sense-strand probe should be employed as a control. This is reasonable only if a single probe is being used and important regions of expression are very low, and thus a comparison must be made to the sense hybridizations. It is not necessary if more than one probe has been used and each gives a different pattern of expression. For instance, as illustrated in Fig. 2, the MLC probe labels two muscles, whereas the CAT probe labels only one. Thus we know both muscles are capable of hybridizing to probe, and thus

[10] U. Grieshammer, D. Sassoon, and N. Rosenthal, *Cell (Cambridge, Mass.)* **69**, 79 (1992).

FIG. 2. Localization of exogenous and endogenous transcripts in the muscles of a transgenic mouse by *in situ* hybridization. Histological sections were prepared from the neck muscles of an 8-day neonatal transgenic mouse. In this line, the transgenic sequences comprise a CAT reporter gene driven by regulatory elements from the myosin light chain 1/3 locus, which restrict transgene expression exclusively to skeletal muscles.[10] (A, B) *In situ* hybridization of adjacent sections with ^{35}S-labeled cRNA probes to MLC and CAT sequences, respectively. Note that this method distinguishes between specific muscle groups which express both endogenous MLC and exogenous CAT genes (large arrows) and those which express only the endogenous MLC gene (small arrows). (C, D) Same samples at higher magnification, to show the subcellular distribution of each type of transcript. Note that MLC transcripts are distributed evenly in the muscle cells (C), whereas CAT transcripts are localized around a subset of nuclei in each muscle cell (D).

both probes serve as internal controls. The sense probe hybridizations are predicted to reveal no labeling. If labeling occurs, it must be due to specific hybridization of probe to another transcript since sense and antisense are quite different in base composition and sequence. Such a result may be interesting and may indicate antisense transcription or just spurious cross-hybridization.

Any research effort aimed at elucidating the role of regulatory, structural, or newly isolated genes during development can benefit greatly from establishing *in situ* hybridization techniques in the laboratory. This technique is indispensable to gain insight into critical embryological events such as cell–cell induction, patterning, and differentiation, and it is enjoyable and aesthetically pleasing as well.

Section V

Gene Expression: Reporter Genes

[25] Preparation of Injection Pipettes

By Miriam Miranda and Melvin L. DePamphilis

Introduction

Analysis of the replication and transcription of DNA injected into the nuclei of mouse oocytes and embryos necessitates reproducibly injecting the same amount of DNA into hundreds of ova. This is best done using a single injection pipette loaded with a single sample of DNA. In transient assays, the DNA injected into mouse oocytes and embryos is approximately 1000-fold more concentrated than DNA samples used in the production of transgenic animals (see [26], this volume). This appears to give rise to two problems. First, viscosity increases with DNA concentration, with the extent of the increase depending on the DNA conformation (linear > circular > superhelical). Therefore, a larger opening is required in the injection pipette tip to accommodate higher DNA concentrations. This, in turn, requires beveling the tip in order to penetrate the cell membrane and nucleus without lysing the cell. Second, in our experience, nuclear material tends to stick more easily to the tip of injection pipettes filled with high concentrations of DNA. This makes it more difficult to inject and reduces the fraction of embryos that survive injection. Therefore, injection pipettes are siliconized to reduce adhesion with cellular materials.

The concentration of DNA that can be injected, the number of embryos that can be injected with the same injection pipette, and the number of embryos that survive are determined by three parameters: the diameter of the injection pipette tip, whether the tip is beveled, and whether the glass is siliconized. Only those oocytes and embryos that survive injection are assayed for their ability to replicate or express the injected gene. Using the injection pipettes described here, relatively large amounts of DNA can be introduced into the cell nucleus of as many as 300 embryos using the same injection pipette. Survival is approximately 80% for 2-cell embryos, 65% for 1-cell embryos, and 90% for oocytes (see [26], this volume).

Pulling Pipettes

A detailed analysis of the pros and cons of micropipette size, shape, and applications to cell biology has been published.[1] First, set up a clean

[1] K. T. Brown and D. G. Fleming, in "Advanced Micropipette Techniques for Cell Physiology." Wiley, New York, 1986.

METHODS IN ENZYMOLOGY, VOL. 225

Copyright © 1993 by Academic Press, Inc.
All rights of reproduction in any form reserved.

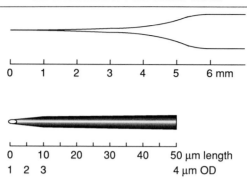

FIG. 1. Measurements for injection pipettes.

area where there is minimal traffic. Before beginning, wipe the area and equipment clean with damp paper towels. A speck of dust in the tip of an injection pipette significantly increases the chance that the pipette will pull out nuclear material when withdrawn from the injected nucleus, and this leads to a drastic decrease in embryo survival. Pipettes used to inject DNA into the nuclei of mouse oocytes and embryos are made from borosilicate glass capillaries, 1 mm (OD) × 0.78 mm (ID) × 15 cm, which can be purchased from Sutter Instrument Co. (40 Leveroni Court, Novato, CA 94949) or the Glass Co. of America (c/o Frederick & Dimmock, P.O. Box 230, Millville, NJ 08332). About 100 capillaries are first cleaned to remove particles and grease by soaking them for at least 3 hr in *aqua regia* (14 ml HNO_3 plus 66 ml HCl in a 100-ml graduated cylinder; work in a fume hood). Capillaries are rinsed 10 times with 2% sodium bicarbonate, then 10 times with double-distilled, deionized water. Both liquids are filtered through a 0.22-μm Millipore (Bedford, MA) filter to eliminate dust. Dry the capillaries at 200° for 1 hr and allow them to cool to room temperature.

Injection pipettes with tips less than 0.2 μm outside diameter are pulled using a programmable Sutter micropipette puller (Model P-80/PC, equipped with a 2 mm square box filament), which we find gives more reproducible results than the Narishige PN-3 (Medical Systems Corp., 1 Plaza Road, Greenvale, NY 11548). Once the desired program is established, about 100 uniform pipettes can be produced in 1 hr. Because each instrument varies somewhat, Sutter will help develop a program based on a drawing of the pipette shape desired.

Injection pipettes have a taper 6 mm long that is steep near the shoulder but shallow close to the tip (Fig. 1). The outside diameter at four locations along the first 50 μm of pipette tip is given for injection pipettes used in our laboratory (Fig. 1). It has a beveled open tip about 1 μm inner diameter.

Our program settings are as follows: heat, 670; pull, 40; velocity, 10; time, 1. The micrometer valve that controls the flow of N_2 gas cooling the filament is set at 60. With our instrument and glass capillary, the ramp value is 717. This value is related to the amount of heat necessary to melt the glass under standard conditions and thus serves as a way to standardize instrument conditions. The ramp value may change when a new batch of glass capillary is used to pull pipettes, or when the filament is changed. Therefore, adjustments in the heat setting may have to be made from time to time. In our experience, the change in ramp value equals the change required in the heat setting in order to return to the original pipette shape.

Pipettes to hold oocytes or embryos are made from thick-walled borosilicate capillaries, 1 mm (OD) × 0.58 mm (ID) × 15 cm. The capillaries are pulled into a long taper on the pipette puller (Sutter instruments settings: heat, 750; pull, 0; velocity, 8; time, 0; N_2, 100). Under a dissecting microscope, the glass is scored with a diamond pencil at a spot where the outer diameter is 50–100 μm, then bent until it breaks at that point. If the tip is not cleanly broken and flat, the holder is discarded (about 30%). Those with flat tips are fire-polished in a microforge (Narishige Model MF-83, Medical Systems) until the inner diameter is 14–20 μm. An eyepiece micrometer facilitates these measurements. The micrometer is calibrated with respect to the 100 μm diameter platinum filament in the microforge.

The best way to generate a flat, polished surface on the tip of the holding pipette is to melt a small bead of glass onto the filament of the microforge, place this bead adjacent to the segment of the pulled capillary that is 50–100 μm in diameter, and then touch them together. Immediately terminate the current in the filament, and the glass capillary wall will now be fused with the glass bead. A slight pull of the capillary away from the bead should generate a straight break in the capillary. The tip of the pipette is then placed near the hot filament. As the glass melts it is important to monitor the internal diameter of the holder. A holding pipette can be reused until it breaks or becomes clogged, if it is rinsed after completing injections by drawing filtered distilled water through it. Should the pipette become dirty, it can be cleaned by soaking it for 10 min in 1 M NaOH and then flushing copious amounts of water through it.

Beveling Tips of Injection Pipettes

Both the cytoplasmic and nuclear membranes of oocytes and embryos are extremely elastic. We found that beveling of the needles is required for easy penetration of the membrane and to increase embryo survival. Injection needles are beveled to produce a 0.5–1.5 μm opening with a clearly discernable point (Fig. 1). Using the Sutter micropipette beveler

(BV-10, base unit and micromanipulator), the needle is held at an angle of 25° in the micromanipulator and slowly lowered to the extrafine diamond abrasive plate, pointing in the direction of rotation of the plate. The beveling is monitored by coating the abrasive plate with a thin film of filtered water, forcing compressed N_2 through the needle (N_2 tank and regulator to deliver 50 psi), and observing the tip of the needle through a stereo zoom microscope (Olympus Model SZ-III with universal table stand, Model VS-4). A valve (Hamilton) should be located in the tubing between the nitrogen regulator and the needle so that the flow can be shut off, releasing pressure in the needle. Open the valve. Slowly lower the pipette and lightly touch the surface of the abrasive plate for a fraction of a second. A slight bending of the needle indicates that it has touched the plate. If you see a stream of bubbles flowing from the needle, the opening is too large. Pipettes with a shorter taper than the one recommended are difficult to bevel because they are more brittle.

Dry the tip and check that it is open by briefly immersing it in a vial of acetone. (Rinse the glass vial twice with fresh, clean acetone before filling it, and use fresh acetone each time a new batch of injection pipettes is prepared.) A fine stream of N_2 bubbles should be visible; otherwise, the opening is too small. The shape and opening of the injection pipette tip can be observed directly using an Olympus Model CHBS student microscope equipped with × 10 eyepieces and MSPlan5 and MSPlan100 objectives. An ocular scale in one of the eyepieces allows the size of pipette tips as small as 0.5 μm (OD) to be determined. The pipette is mounted on the movable slide holder using a small ball of soft clay.

Siliconization

Siliconization of the injection pipette is critical to prevent cellular components from adhering to the pipette tip. Cellular materials readily adhere to nonsiliconized pipettes. This results in cell lysis as the pipette tips become dull and eventually plug. Nonsiliconized pipettes last for 3 to 5 consecutive injections, whereas siliconized pipettes average 100 injections (some have even been used for 600 injections!). Unfortunately, siliconization cannot be done with capillary pipettes containing a central filament that facilitates sample loading by capillary action, because silicone eliminates the required surface adhesion. Therefore, samples are loaded into the back end of the injection pipette using a Hamilton syringe with a 3-inch blunt-tipped 26- to 28-gauge needle.

After beveling, injection pipettes are siliconized for 5 to 7 days. Place them on a rack like the one described below for storage. Place this rack

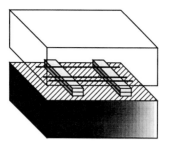

FIG. 2. Rack for storing pipettes.

in a large desiccator (220 cm ID) containing a 50-ml beaker with 10 ml of hexamethyldisilazane (H7300, Hüls America, P.O. Box 456, Piscataway, NJ 08855). Seal the desiccator with a thin layer of vacuum grease. Hexamethyldisilazane vapor will produce a monomolecular layer of silicone on glass surfaces. Other siliconizing agents, such as SurfaSil (Pierce, Rockford, IL), produce multiple layers of silicone that can eventually plug the pipette opening. A fresh sample should be used each time since hexamethyldisilazane absorbs water vapor and slowly inactivates. Purchase hexamethyldisilazane in small bottles, and, each time a bottle is used, purge the air in the bottle with dry nitrogen before closing. We recommend starting a new 100-ml bottle every 3 months. The injection pipettes are allowed to air-dry for at least 3 days before they are used. Shorter times for siliconization and drying result in pipettes more likely to stick to cellular materials. This siliconizing reagent is also good for plastic pipette tips and microtubes that one may wish to siliconize. Siliconizing agents that produce multiple layers tend to leave a residue that contaminates our DNA samples.

A convenient storage container for injection and holding pipettes can be made from a plastic rack in which plastic micropipette tips are commonly sold (Fig. 2). Two strips of foam mounting tape with adhesive on both sides are placed on the top surface of the box, parallel to each other and 5 to 6 cm apart. Two or three layers of tape provide sufficient height to handle the pipettes. The pipettes are then held in position by the adhesive on one side of the tape (Fig. 2). The object is to prevent either end of the pipette from touching a surface and thus prevent breaking the injection tip or accumulating "particles" around the back orifice into which the sample is loaded. Such particles can accidentally be knocked into the pipette when loading a sample and cause plugging. The box comes with a cover that keeps the pipettes free of dust.

Loading DNA into Injection Pipettes

DNA samples cannot be drawn up into the injection pipettes through the tip opening, because the siliconized surface eliminates capillary action. Therefore, DNA samples must be introduced through the rear of the injection pipette using a Hamilton syringe with a 3-inch blunt-tipped 26- to 28-gauge needle (Hamilton Co., P.O. Box 10030, Reno, NV 89520). The entire DNA sample must be placed near the tip of the injection pipette; otherwise, it is difficult to force it out the tip when pressure is applied.

First, blow any dust out of the syringe needle before loading the sample. All materials (pipette tips, microtubes, needles, etc.) that come in contact with the DNA should be dust-free. An aerosol duster (Ultrajet, Fisher Scientific, Pittsburgh, PA) is useful for this purpose. Then place 1.5 μl of sample onto a dust-free plastic surface (e.g., 60 mm tissue culture dish) and draw it slowly into the syringe needle using either a Hamilton glass syringe or a micropipette filler (VWR, 53518-007).[2] Lower the needle into the injection pipette as far as possible. Dispense the sample into the injection pipette while simultaneously withdrawing the syringe needle. Both dispensing the sample and withdrawing the syringe needle must be done slowly. The object is to transfer the liquid from the needle to the pipette without the liquid flowing upward into the space between the syringe needle and the injection pipette. If this problem does occur, then the pipette must be discarded, because once the inside of an injection pipette is wetted, it cannot be loaded. In addition, there must only be one drop (band) of liquid in the pipette. Air trapped between bands of liquid makes it impossible to force the sample into the injection pipette tip when it is attached to a microinjector.

[2] The micropipette filler works well when connected by latex tubing (5/16 inch OD × 3/16 inch ID) to a Hamilton Luer needle of the size and shape described.

[26] Application of Firefly Luciferase to Preimplantation Development

By Miriam Miranda, Sadhan Majumder, Maria Wiekowski, and Melvin L. DePamphilis

Introduction

DNA injected into the nuclei of mouse oocytes, 1-cell embryos, and 2-cell embryos responds to normal cellular signals which regulate DNA

Copyright © 1993 by Academic Press, Inc.
All rights of reproduction in any form reserved.

replication and gene expression in that it undergoes replication and transcription only in cells competent for that function, and only when unique eukaryotic regulatory sequences are present. For example, mouse oocytes can express some of their genes, but, because they are arrested in prophase of the first meiosis, they cannot replicate DNA. Accordingly, plasmid DNA does not replicate when injected into mouse oocytes, even if the injected DNA contains a viral origin and is provided with the appropriate viral proteins,[1,2] but injected eukaryotic promoters are active. The same sequence that provides oocyte-specific expression of zona pellucida protein-3 (ZP3) when integrated into the chromosomes of transgenic animals[3,4] also provides oocyte-specific expression when present on injected plasmid DNA.[5] However, whereas oocytes utilize some of the same promoter elements recognized by somatic cells, promoter activity in oocytes does not require enhancers.[2,6] One reason for this may be that oocytes produce unique trans-acting factors that mimic certain viral transcription factors.[7]

Oocytes mature into eggs that are then fertilized to produce a zygote, but transcription and translation of zygotic genes do not begin until the 2-cell stage of development.[8] Initiation of zygotic gene expression is governed by a "clock" that initiates transcription about 20 hr postfertilization, regardless of whether the 1-cell embryo (fertilized egg) has completed DNA replication or undergone mitosis.[9–11] Accordingly, promoters injected into the pronuclei of 1-cell embryos remain inactive until the "zygotic clock" initiates expression of the endogenous genes,[11,12] showing

[1] D. O. Wirak, L. E. Chalifour, P. M. Wassarman, W. J. Muller, J. E. Hassell, and M. L. DePamphilis, *Mol. Cell. Biol.* **5,** 2924 (1985).

[2] L. E. Chalifour, D. O. Wirak, P. M. Wassarman, and M. L. DePamphilis, *J. Virol.* **59,** 619 (1986).

[3] S. A. Lira, R. A. Kinloch, S. Mortillo, and P. M. Wassarman, *Proc. Natl. Acad. Sci. U.S.A.* **87,** 7215 (1990).

[4] M. Schickler, S. A. Lira, R. A. Kinloch, and P. M. Wassarman, *Mol. Cell. Biol.* **12,** 120 (1992).

[5] S. E. Millar, E. Lader, L.-F. Liang, and J. Dean, *Mol. Cell. Biol.* **11,** 6197 (1991).

[6] L. E. Chalifour, D. O. Wirak, P. M. Wassarman, U. Hansen, and M. L. DePamphilis, *Genes Dev.* **1,** 1096 (1987).

[7] T. Dooley, M. Miranda, N. C. Jones, and M. L. DePamphilis, *Development (Cambridge, UK)* **107,** 945 (1989).

[8] N. A. Telford, A. J. Watson, and G. A. Schultz, *Mol. Reprod. Dev.* **26,** 90 (1990).

[9] J. C. Conover, L. T. Gretchen, J. W. Zimmermann, B. Burke, and R. M. Schultz, *Dev. Biol.* **144,** 392 (1991).

[10] F. M. Manejwala, C. Y. Logan, and R. M. Schultz, *Dev. Biol.* **144,** 301 (1991).

[11] M. Wiekowski, M. Miranda, and M. L. DePamphilis, *Dev. Biol.* **147,** 403 (1991).

[12] E. Martínez-Salas, E. Linney, J. Hassell, and M. L. DePamphilis, *Genes Dev.* **3,** 1493 (1989).

that expression of injected genes is governed by the same mechanism that regulates expression of zygotic genes. This delay is observed only in 1-cell embryos whose morphological development is arrested. When injected 1-cell embryos develop to the 2-cell stage or beyond, expression of injected genes is reduced to less than 1% of levels observed in arrested 1-cell embryos. Similarly, DNA containing a viral origin replicates in the presence of its cognate recognition protein only in injected 1-cell embryos that spontaneously remain in appearance as 1-cell embryos; plasmid DNA replication is barely detectable in embryos that continue morphological development.[1] Apparently DNA injected into 1-cell embryos is repressed by changes in its physical state that occur during formation of a 2-cell embryo.

Replication and transcription of DNA injected into 2-cell embryos require the same regulatory sequences as DNA transfected into a variety of undifferentiated and differentiated mammalian cells. Plasmid DNA replication is not observed unless the injected DNA contains a known origin such as polyomavirus or SV40 and its cognate recognition protein, T-antigen.[1,2,13,14] Expression of a reporter gene is not observed unless it is fused to an appropriate eukaryotic promoter and embryo-responsive enhancer, and these cis-acting sequences must bind trans-acting factors in order to function (see Refs. 11, 12, 15, and 16).

The levels of promoter and origin activity in arrested 1-cell embryos are 12- to 500-fold greater than when injected into 2-cell embryos. The exact difference depends on the promoter or replication origin tested, the amount of DNA injected, and whether injected 2-cell embryos are arrested at the beginning of S phase in the 4-cell stage. For 2-cell embryos to produce levels of transcription or replication that are equivalent to or slightly greater (~2-fold) than those observed in arrested 1-cell embryos, the DNA injected into 2-cell embryos must carry an embryo-responsive enhancer. Enhancers that strongly stimulate promoter or origin activities in 2-cell embryos have no effect in arrested 1-cell embryos or oocytes. Thus, the requirements for replication and expression of genes in mammalian somatic cells appear to be established on formation of a 2-cell embryo. The differences in requirements for replication and transcription that are observed prior to this stage appear to reflect changes occurring in chroma-

[13] E. Martínez-Salas, D. Y. Cupo, and M. L. DePamphilis, *Genes Dev.* **2**, 1115 (1988).
[14] M. L. DePamphilis, S. A. Herman, E. Martínez-Salas, L. E. Chalifour, D. O. Wirak, D. Y. Cupo, and M. Miranda, *BioTechniques* **6**, 662 (1988).
[15] G. W. Stuart, P. F. Searle, H. Y. Chen, R. L. Brinster, and R. D. Palmiter, *Proc. Natl. Acad. Sci. U.S.A.* **81**, 7318 (1984).
[16] S. Majumder, M. Miranda, and M. DePamphilis, *EMBO J.* **12**, 1131 (1993).

tin structure and nuclear organization that are unique to preimplantation development.

The ability to dispense with enhancers is not a consequence of arresting 1-cell embryos in S phase, because enhancers are required for full promoter activity in either developing or S phase-arrested 2-cell and 4-cell embryos.[11] Likewise, it is not determined by the time of injection relative to the cell proliferation cycle or by the developmental history of the embryo. Expression of genes injected into arrested 1-cell embryos is tightly linked to zygotic gene expression, and the same promoter and origin components that are essential from the 2-cell stage to the adult animal are also essential in arrested 1-cell embryos.[11–13,15,16] Therefore, enhancers must provide a unique function *in vivo* that is dispensable in fertilized eggs. We suggest that this function is to relieve repression by chromatin structure.

Based on the above information, it appears that one can identify the requirements for DNA replication and gene expression in early mammalian development by injection of unique DNA sequences into oocytes and embryos. This chapter describes optimal conditions for quantitative evaluation of promoter activity from DNA injected into mouse oocytes and preimplantation embryos.

Measuring Promoter/Enhancer Activity in Mouse Ova

Oocytes and Embryos

Isolation and culture of mouse oocytes and embryos has been described.[11,13,14] Sigma (St. Louis, MO) bovine serum albumin (BSA) A4161 is superior to A7638 for development *in vitro* beyond the 2-cell stage. Oocytes are obtained from 14- to 15-day-old females, although oocytes from 10-day-old females may be more transcriptionally active.[5] Fertilized eggs are isolated from 7- to 8-week-old pregnant females 17 hr after human chorionic gonadotropin (hCG) is injected (Fig. 1). Two-cell embryos are either isolated from pregnant females at 40 to 42 hr post-hCG or allowed to develop *in vitro* from fertilized eggs. Both types of 2-cell embryos respond the same to injected DNA. Where indicated, embryos are cultured in 4 μg/ml aphidicolin (Boehringer-Mannheim, Indianapolis, IN) to arrest development as they enter S phase. Aphidicolin specifically inhibits DNA polymerases-α and -δ in eukaryotic cells.[17] Aphidicolin is prepared as a 2 mg/ml solution in dimethyl sulfoxide and stored in aliquots at $-20°$. When development is arrested by aphidicolin, it is included in the medium

[17] T. S.-F. Wang, *Annu. Rev. Biochem.* **60**, 513 (1991).

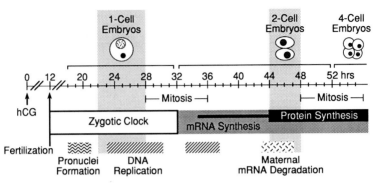

FIG. 1. The mouse developmental program is initiated at fertilization and proceeds in a series of easily identifiable steps [V. N. Bolton, P. J. Oades, and M. H. Johnson, *J. Embryol. Exp. Morphol.* **79**, 139 (1984); S. K. Howlett and V. N. Bolton, *J. Embryol. Exp. Morphol.* **87**, 175 (1985)]. Human chorionic gonadotropin (hCG) should be injected prior to the normal release of luteinizing hormone from the animal; release generally occurs between 3 and 8 PM in order to induce ovulation. We routinely inject hCG at 5 PM to provide convenient windows for injecting 1-cell and 2-cell embryos (shaded areas). Approximately 4 to 9 hr after fertilization, two pronuclei appear; the paternal pronucleus originates from the sperm (large stippled circle), and the maternal pronucleus originates from the oocyte germinal vesicle (small solid circle). The two haploid pronuclei each complete the first round of DNA replication before undergoing the first mitosis to produce a diploid zygotic nucleus. About 20 hr after fertilization, coinciding with formation of a 2-cell embryo [reviewed by N. A. Telford, A. J. Watson, and G. A. Schultz, *Mol. Reprod. Dev.* **26**, 90 (1990)], transcription (shaded bar) is triggered by the "zygotic clock" (open bar); nascent zygotic proteins (black bar) first appear 3 to 4 hr later, while the major synthesis of α-amanitin-sensitive proteins begins about 12 hr later. Prior to formation of a 2-cell embryo, most, if not all, newly synthesized proteins are made from maternally inherited mRNA, the bulk of which is degraded late during the 2-cell stage. Changes in protein patterns during the first cell cycle are due to posttranscriptional modifications [discussed by Howlett and Bolton, 1985 (see above)]. The times when 1-cell, 2-cell, and 4-cell embryos are visible are indicated (brackets). Cell division cycle asynchrony is greatest when fertilization occurs *in vivo* and least when fertilization is done *in vitro*.

in the injection dish as well as in the culture dish used to isolate the embryos.

Preparation of DNA for Injection

Circular plasmid or viral DNA (<10 kb) is commonly used for injection into nuclei, because it is convenient to prepare, stable after injection, and easy to inject. High molecular weight, linear DNA such as λ DNA is too viscous to inject through a 1-μm opening, although this problem may be overcome by using the circular, superhelical form. Plasmid DNA is prepared in dam$^+$ *Escherichia coli* strain DH5 (Bethesda Research Labs,

Gaithersburg, MD) or AG-1 (Stratagene, La Jolla, CA) to prevent re-arrangement and to produce *Dpn*I-sensitive DNA.

Plasmid DNA is isolated as described by the Brij/sodium dodecyl sulfate (SDS) method of Timmis *et al.*[18] This method is less efficient than the alkaline lysis method, but it avoids the potential problem of partially denatured superhelical forms of DNA. Bacterial cells are treated with lysozyme, gently lysed in the presence of 1% Brij and 0.4% SDS, and centrifuged to obtain a clear lysate. The lysate is treated further with pancreatic RNase (10 μg/ml, 37°, 30 min) and then with proteinase K (60 μg/ml in the presence of 0.5% SDS, 37°, 1 hr). The resulting plasmid DNA is finally purified through two consecutive CsCl–ethidium bromide equilibrium density gradients. Ethidium bromide is extracted with isobutanol that has been equilibrated with 10 mM Tris-HCl (pH 8.0), 1 mM EDTA, and 6 M CsCl, and the DNA is then dialyzed extensively against 10 mM Tris-HCl (pH 8.0) and 1 mM EDTA. The DNA concentration is measured at 260 nm (1 OD unit = 50 μg DNA/ml). Purified DNA has an absorbance ratio at 260/280 nm of 1.9 to 2.0, and 70 to 80% of the DNA should appear superhelical when fractionated by agarose gel electrophoresis.[19]

Purified DNA samples are stored as precipitates. Aliquots of DNA are adjusted to 0.3 M sodium acetate, combined with 2.5 volumes of absolute ethanol, and stored at $-20°$. Sodium acetate is soluble in ethanol and therefore will not precipitate along with the DNA. Before injection, DNA precipitates are collected using a microcentrifuge, and the pellet is washed once with cold 75% (v/v) ethanol, dried under vacuum, and then resuspended in either phosphate or Tris–EDTA injection buffer (Table I). All solutions that come in contact with purified DNA samples are filtered first through a 0.22-μm pore size Millipore (Bedford, MA) filter. In addition, all materials such as pipette tips, microtubes, and needles that come in contact with the DNA should be dust-free. An aerosol duster (Ultrajet, Fisher, Pittsburgh, PA) is useful for this purpose. Immediately prior to injection, 10 μl samples of DNA are centrifuged at 12,000 rpm in a microcentrifuge for at least 10 min to remove any insoluble material present that might clog the injection pipette.

Selecting Injection Buffer

Results with either injection buffer (Table I) are equivalent. Although phosphate injection buffer is more physiological than Tris–EDTA, it suffers from two problems. First, ethanol cannot be used to precipitate the

---

[18] K. N. Timmis, F. Cabello, and S. M. Cohen, *Mol. Gen. Genet.* **162**, 121 (1978).
[19] J. Sambrook, E. F. Fritsch, and T. Maniatis, "Molecular Cloning: A Laboratory Manual 2nd Ed. Cold Spring Harbor Laboratory, Cold Spring Harbor, New York, 1989.

TABLE I
INJECTION BUFFERS

Component	Concentration
Phosphate injection buffer for DNA[a]	
K_2PO_4	48 mM
KH_2PO_4	4.5 mM
NaH_2PO_4	14 mM
Final pH 7.2	
Isotonic phosphate buffer for embryos[b]	
NaCl	102 mM
KCl	2.7 mM
Na_2HPO_4	8 mM
KH_2PO_4	1.42 mM
$CaCl_2$	0.93 mM
$MgCl_2$	0.48 mM
Sodium pyruvate	13.2 mM
Glucose	1 mg/ml
Bovine serum albumin	4 mg/ml
(Sigma, A4161)	
Phenol red	10 μg/ml
Penicillin/streptomycin	0.5 mg/ml
(GIBCO, Grand Island, NY)	
Tris–EDTA injection buffer for DNA[c]	
Tris (pH 7.6)	10 mM
EDTA	0.25 mM

[a] M. Graessmann and A. Graessmann, this series, Vol. 101, p. 482.
[b] J. Mann, personal communication, 1991.
[c] R. L. Brinster, H.-Y. Chen, M. E. Trumbauer, A. W. Senear, R. Warren, and R. D. Palmiter, *Proc. Natl. Acad. Sci. U.S.A.* **82,** 4438 (1985).

DNA out of phosphate buffer without precipitating salt as well. Therefore, once DNA has been prepared in this buffer, it cannot be precipitated and resuspended at a higher concentration or in a different buffer. Second, the calcium present in embryo culture medium will cause calcium phosphate to precipitate at the tip of the injection pipette, reducing the ability of the injection pipette to penetrate the cytoplasmic and nuclear membranes and also reducing survival of injected embryos. This problem can be dealt with by placing embryos in isotonic phosphate buffer (Table I) for brief periods of no longer than 30 min, with no apparent ill effects on subsequent development. Second, materials that adhere to the injection pipette tip can be removed by inserting the tip into the end of the holding pipette, and then gently brushing it against the inside of the holding pipette as it is withdrawn. Medium is simultaneously drawn into the holding pipette to collect the debris.

Injection of DNA and Embryonic Development

DNA is routinely injected into one of the two pronuclei in 1-cell embryos between 22 and 28 hr post-hCG and into one of the two zygotic nuclei in 2-cell embryos between 44 and 48 hr post-hCG (Fig. 1). To arrest development at the beginning of S phase in 1-cell embryos, aphidicolin is included in the medium when eggs are isolated at 17 hr post-hCG. To arrest development at the 2-cell stage, aphidicolin is added at approximately 32 hr post-hCG, just after the first mitosis is completed (Fig. 1). To arrest development at the 4-cell stage, aphidicolin is added at 40 to 42 hr post-hCG when 2-cell embryos have completed DNA replication (Fig. 1). Maternal pronuclei are distinguished from paternal pronuclei by their smaller size and their proximity to the second polar body. However, injection of pronuclei is easier at later times (24–26 hr post-hCG) because nuclear material does not adhere to the injection pipette as readily and maternal nuclei present a larger target. Both factors improve the survival of injected embryos.

Injection is carried out as previously described.[14] About 100 embryos per hour can be injected. After all ova have been injected, they are transferred to culture medium and allowed to continue development *in vitro*. About 30 min later, injected ova are examined under a dissection microscope, and those that have survived injection (50–70%) are transferred to a new drop of equilibrated medium.[14] This step is important for good embryo development and for expression of the injected gene. Injection of DNA also limits embryonic development (Table II).

Firefly Luciferase Assay

The firefly luciferase gene cloned by DeWet *et al.*[20] provides a reporter gene with several advantages for evaluation of promoter activity in embryonic cells. Gene expression can be measured quantitatively over a 10^5-fold range in single cells.[21] The assay is relatively inexpensive and simple enough that several hundred a day may be performed. Cell lysates can be stored for long periods of time before being assayed. Firefly luciferase is at least 10-fold more sensitive than *E. coli* chloramphenicol acetyltransferase (CAT) as a reporter gene during transient assays of promoter activ-

---

[20] J. R. DeWet, K. V. Wood, M. DeLuca, D. R. Helinski, and S. Subramani, *Mol. Cell. Biol.* **7,** 725 (1987).

[21] S. J. Gould and S. Subramani, *Anal. Biochem.* **175,** 5 (1988).

TABLE II
DEVELOPMENT OF INJECTED AND UNINJECTED MOUSE EMBRYOS

Inject DNA into	Development 44 hr later[a] (%)				
	2-Cell	3-Cell	4- to 8-Cell	Morula[b]	Abnormal[c]
1-Cell embryos					
CD-1[d,e]	10		90		
CD-1 injected	65		15		20
B6SJL[e,f]			100		
B6SJL injected	65		15		20
2-Cell embryos					
CD-1[g]				95	5
CD-1 injected	15	10	20	20	35
B6SJL[g]				95	5
B6SJL injected	5	20	25	20	30

[a] Embryos are assayed routinely at 40 to 44 hr postinjection. Development data represent averages of three to four experiments.

[b] Morulae are defined as any compacted embryo that has not yet formed a blastocoel cavity.

[c] Abnormal embryos are defined as those that have undergone multiple "microcleavage" events, producing blastometeres of unequal size [D. O. Wirak, L. E. Chalifour, P. M. Wassarman, W. J. Muller, J. E. Hassell, and M. L. DePamphilis, *Mol. Cell. Biol.* **5,** 2924 (1985).]

[d] CD-1 Swiss albino mice are from Charles River Breeding Laboratories, Wilmington, MA.

[e] About 60% of CD-1 or 90% of B6SJL 1-cell embryos that had not been injected formed blastocysts in 96 hr.

[f] (C57BL/6J × SJL/J)F$_1$ (referred to as B6SJL) mice are from Jackson Laboratories, Bar Harbor, ME.

[g] From 90 to 100% of 2-cell embryos isolated from either mouse strain formed blastocysts within 72 hr.

ity in mammalian cells,[22] although CAT may be more stable than luciferase.[23] Luciferase also can be used for measuring promoter activity in transgenic mice,[3,24] and it can be visualized in intact cells by autoradiography.[21]

Luciferase (E) catalyzes the oxidative decarboxylation of luciferin (D-LH$_2$) to produce yellow-green light at pH 7.5 to 8.5 with a peak emission at 560 nm. When an excess of luciferin is added to an extract containing

[22] T. M. Williams, J. E. Burlein, S. Ogden, L. J. Kricka, and J. A. Kant, *Anal. Biochem.* **176,** 28 (1989).

[23] I. H. Maxwell and F. Maxwell, *DNA* **7,** 557 (1988).

[24] A. G. DiLella, D. A. Hope, H. Chen, M. Trumbauer, R. J. Schwartz, and R. G. Smith, *Nucleic Acids Res.* **16,** 4159 (1988).

luciferase, ATP, and Mg^{2+}, the resulting burst of light is proportional to the amount of luciferase present. Luciferase is inactivated in the process:

$$D\text{-}LH_2 + MgATP + E \rightarrow E \cdot LH_2AMP + PP_i$$
$$E \cdot LH_2AMP + O_2 \rightarrow E \cdot L + AMP + CO_2 + light$$

When luciferin is added to the luciferase reaction mixture, a flash of light is emitted that peaks 0.3 sec later (25°). This emission then decays in a biphasic manner to approximately 10% of peak level within 30 sec and declines gradually over the next 30 min.[25]

Luciferase can be used to measure the activity of promoter/enhancer regulatory sequences in mouse ova in the following manner. Plasmid DNA is prepared containing the firefly luciferase gene downstream of appropriate promoter/enhancer regulatory sequences (examples are found in Refs. 4, 5, 7, 11, 12, and 26). Spurious plasmid-initiated transcription of the reporter gene can be suppressed by placing polyadenylation signals upstream of the gene.[27,28] Plasmid DNA is injected into the nucleus of a mouse oocyte or preimplantation embryo, and the embryos are cultured *in vitro* for 1 to 2 days.[14] Individual ova are then harvested in a small volume of lysis buffer, frozen once and thawed, and then added to a reaction buffer containing MgATP. The sample is placed into a luminometer, and the reaction is initiated by injection of luciferin-CoA. Light output integrated over a preset period of time is proportional to luciferase concentration.

Luminometer

Because the reaction between luciferase and luciferin is nearly instantaneous, these measurements require a photometer designed to inject luciferin into a cuvette that is already inside the counting chamber. We use the Monolight 2010 luminometer (Analytical Luminescence Laboratory, 11760 Sorrento Valley Rd., Suite E, San Diego, CA 92121). The amount of light recorded varies with the luminometer used. For example, the Monolight 2010 records about twice the amount of light recorded by the Monolight 2001. Jago et al.[29] have evaluated the performance of 10 commercial luminometers. Luciferase assays can also be done using a scintillation counter.[30]

[25] S. Subramani and M. DeLuca, *Genet. Eng.* **10**, 75 (1988).

[26] A. R. Brasier, J. E. Tate, and J. F. Habener, *BioTechniques* **7**, 1116 (1989).

[27] I. H. Maxwell, G. S. Harrison, W. M. Wood, and F. Maxwell, *BioTechniques* **7**, 276 (1989).

[28] J. L. Fridovich-Keil, J. M. Gudas, I. B. Bryan, and A. B. Pardee, *BioTechniques* **11**, 572 (1991).

[29] P. H. Jago, *J. Biolumin. Chemilumin.* **3**, 131 (1989).

[30] V. T. Nguyen, M. Morange, and O. Bensaude, *Anal. Biochem.* **171**, 404 (1988).

TABLE III
PREPARATION OF REAGENTS

Reagents	Stock solution	Final concentration in assay
Assay components		
Firefly luciferase	1 mg/ml	Varied
D-Luciferin	1 mM	1 mM
Coenzyme A	100 mg/ml	0.2 mg/ml
Reaction buffer		
Glycylglycine (pH 7.8)	0.5 M	25 mM
Magnesium acetate[a]	1.0 M	10 mM
ATP (pH 7)	0.1 M	0.5 mM
Bovine serum albumin	40 mg/ml	100 μg/ml
Dithiothreitol	0.5 M	1 mM
Lysis buffer		
Reaction buffer + 0.1% Triton X-100		0.01% Triton X-100

[a] Acetate salts are generally less inhibitory of enzymatic activities than are chloride or sulfate salts.

Basically, all transparent materials such as glass, polystyrene, and polyethylene can be used to make cuvettes for the luminometer. However, some materials are more sensitive to electrostatic charging, which results in an increased and fluctuating background. Polystyrene cuvettes are available with antistatic additives (Analytical Luminescence Laboratory, No. 2050-10). However, we find that glass cuvettes give the most reproducible results. Do not expose cuvettes to shortwave light to prevent phosphorescence. Do not change the cuvette type or lot number within an assay.

Preparation of Reagents

Table III lists reagents and buffers needed for the luciferase assay. Firefly luciferase and D-luciferin are purchased from Analytical Luminescence Laboratory. Luciferase from Sigma varies significantly in specific activity and is difficult to dissolve. Luciferase powder is reconstituted in 0.5 M glycylglycine (pH 7.8) and 2 mg/ml BSA (make sure the powder is dissolved before freezing). Luciferase is a light-sensitive reagent that must be stored in single-use aliquots, in the dark, at −80°, where it is stable for at least 3 months. D-Luciferin is purchased as a sodium salt in 10-mg aliquots. Dissolve 10 mg in 34.6 ml deionized water, to make a 1 mM solution that is stable at 4° for up to 2 months when wrapped in aluminum foil. The solution should be pH 6 to 6.3 and visibly clear. At pH values greater than 7.5, luciferin turns yellow-green and forms dihydroluciferin, which racemizes to L-luciferin; both are competitive inhibitors of luciferase. Coenzyme A is purchased as the trilithium salt (grade II) from Boehringer-Mannheim, dissolved in deionized water, and stored as single-use aliquots at $-80°$. It is stable for several months.

Reaction buffer is freshly prepared from concentrated stock solutions at least every 2 weeks and stored at $4°$. MgATP slowly hydrolyzes. Glycylglycine and ATP stock solutions should be adjusted to indicated pH with 2 N NaOH. Do not overshoot the pH since the triphosphate hydrolyzes rapidly in alkali. One can make the ATP solution in 10 mM glycylglycine buffer to facilitate pH adjustment. Check that the pH of the final reaction buffer is 7.8. Lysis buffer is prepared just prior to use by adding 10% Triton X-100 to reaction buffer. Triton X-100 is prepared fresh every 2 months and is stored at $4°$ to prevent microbial contamination.

Lysis of Oocytes and Embryos

Place 50 μl of lysis buffer into a 1.5-ml microcentrifuge tube for each embryo to be assayed. We normally culture 30 to 60 embryos in a single drop of medium (1 μl/embryo). Wash the embryos through three drops of fresh medium. The third drop should contain about 100 μl of medium to allow for loss of medium during embryo transfer. Working under a dissection microscope, transfer one embryo at a time to each tube using a finely drawn out Pasteur pipette (inner diameter slightly greater than outer diameter of embryo). The pipette should be attached by rubber tubing to a saliva trap (tube containing glass wool and cotton) coupled to a Safe protective mouthpiece (Gelman Sciences, 600 S. Wagner Rd., Ann Arbor, MI 48106). Embryos will stick to the glass surface of the pipette unless it is first rinsed out with culture medium. To transfer an embryo, first allow capillary action to draw 2 or 3 μl of fresh medium into the pipette tip. Then pick up a single embryo, minimizing the amount of additional medium taken up. Place the tip of the pipette into the 50 μl of lysis buffer and expel the embryo (stop when air bubbles appear). Immediately freeze the tube in a dry ice/ethanol bath. Rinse out the pipette tip in fresh medium to prevent transfer of detergent to the remaining embryos. Enzyme activity is stable for at least 2 weeks at $-70°$.[22,26]

Assay Procedure

1. Turn on the luminometer at least 30 min before using, and equilibrate a water bath at ambient temperature for prewarming the reaction buffer and luciferin/CoA solution. Ambient temperature for the assays is between $22°$ ($71.6°$F) and $24°$ ($75.2°$F). Keep in mind that

significant changes in ambient temperature will affect the rate of an enzyme reaction.

2. Prepare luciferase standards by diluting stock solution in reaction buffer (e.g., 10^{-5}, 10^{-6}, 10^{-7}, 10^{-8}). These are protected from light with a sheet of aluminum foil and kept on ice. It is not necessary to work in a darkened room.

3. Warm 1 mM luciferin to ambient temperature. Add 90 μl coenzyme A stock per 10 ml luciferin. CoA is added to the luciferin solution rather than to the reaction buffer because CoA is unstable at pH values above 7.

4. Place the reservoir filled with luciferin/CoA in the appropriate compartment and fill the automatic injector supply line with luciferin/CoA solution. A cuvette must always be present in the counting chamber, or the photomultiplier tube can be damaged by contact with liquid.

5. To determine the background light emission in a Monolight 2010 luminometer, place 350 μl of reaction buffer into each of three cuvettes and warm to ambient temperature. The actual volumes used may vary depending on the luminometer. Dry the outside of the cuvette first with a tissue before placing the cuvette in the counting chamber. Press the start button. When the instrument is in the integration mode, photons are counted over a preset time period, and the final number recorded. The greatest signal-to-noise ratios are obtained by measuring the sum of the total light emissions (integration mode) for 10 sec.

6. To calibrate the luminometer, place 300 μl reaction buffer into each cuvette and warm to ambient temperature. Add 50 μl of the luciferase standard. Gently mix the tube contents, place it in the luminometer chamber, and press the start button. The Monolight 2010 automatic injector dispenses 100 μl of luciferin into the cuvette to initiate the reaction. Thus, the final reaction volume is 450 μl (300 μl reaction buffer, 100 μl luciferin, plus 50 μl sample).

7. To assay samples, first thaw at ambient temperature the number of embryo lysates that can be assayed in 1 hr ($\sim$50). Spin the lysates for approximately 10 sec in a desk-top microcentrifuge to ensure that the entire sample is at the bottom of the tube, and then place them on ice. Dispense 300 μl of reaction buffer into each luminometer cuvette and incubate them at ambient temperature. Add the embryo lysate to 300 μl of prewarmed reaction buffer, and measure luciferase activity as in Step 6.

8. When finished, flush the luciferin solution out of the automatic injector supply line with deionized water. The injector system

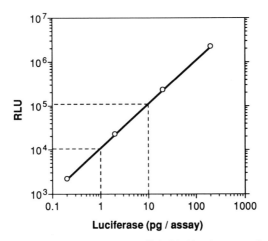

FIG. 2. Luciferase standard curve on a Monolight 2010 luminometer. Increasing amounts of purified luciferase were added to the standard assay. Each point is the average of five sets of duplicates run on different days. Total light emissions were integrated over the first 10 sec of the reaction. The standard error of the mean (SEM) was 3% of the mean value. Background readings are 130 to 180 relative light units (RLU). Luciferase activity from 10 pg of enzyme is approximately 100,000 RLU when CoA is included in the assay.

should be cleaned regularly by pumping approximately 0.5 ml of Monorinse (Analytical Luminescence Laboratory) through the tubing, followed by three wash cycles with deionized water. This prevents microbial growth in the tubing. Empty the tubing and leave it dry. It is wise to flush out tubing with deionized water prior to using the instrument. Remember to change cuvettes during the washes to keep them from overflowing into the reaction chamber and damaging the luminometer. Leave a clean cuvette in the chamber when finished.

Comments

Luciferase activity in relative light units (RLU) is proportional to the amount of luciferase enzyme present in the assay over a 10^5-fold range in enzyme concentration (Fig. 2).[31] As little as 0.1 pg of luciferase per assay is easily detected. Background readings for the Monolight 2010 luminometer are 130 to 180 RLU. The standard error of the mean (SEM) for 30 aliquots of a single sample of purified luciferase is 0.3% of the mean value.

[31] S. Gould, G. A. Keller, and S. Subramani, *J. Cell. Biol.* **105,** 2923 (1987).

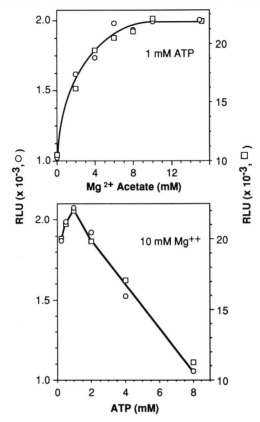

FIG. 3. Effects of Mg^{2+} ion (top) and ATP concentration (bottom) on luciferase activity. Luciferase was tested at low (0.5 pg, $\bigcirc$) and high (5 pg, $\square$) concentrations. CoA was not present in these assays.

The reaction tolerates high concentrations of magnesium acetate (Fig. 3, top) but is sensitive to ATP concentration (Fig. 3, bottom). Luciferase activity is optimal at 0.5 to 1 mM ATP in the presence of 10 to 15 mM Mg^{2+}. Similar results have been observed with $MgSO_4$.[26] Luciferase also is inhibited by AMP, one of the reaction products.[32] Luciferase is very pH sensitive; pH values below 7 inhibit the enzyme.

When luciferase is expressed in mammalian cells, it accumulates in peroxisomes, where it is not accessible.[31,33] Freezing and thawing is inef-

[32] A. Lundin and A. Thore, *Anal. Biochem.* **66,** 47 (1975).
[33] G. A. Keller, S. Gould, M. Deluca, and S. Subramani, *Proc. Natl. Acad. Sci. U.S.A.* **84,** 3264 (1987).

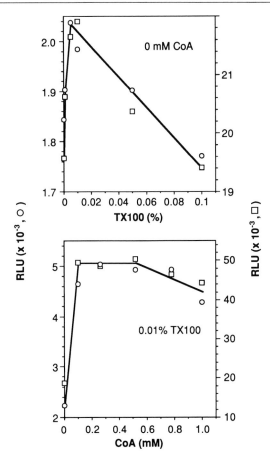

FIG. 4. Effects of Triton X-100 (TX100; top) and coenzyme A (CoA; bottom) on luciferase activity. Luciferase was tested at low (0.5 pg, ○) and high (5 pg, □) concentrations.

fective in breaking these vesicles, but Triton X-100 releases the luciferase activity into the soluble fraction. This problem can be severe for cells that have high numbers of peroxisomes (e.g., liver and kidney cells[26]). However, we did not observe a significant difference between freeze–thawing and Triton X-100 in the amount of luciferase activity released following injection of luciferase expression vectors into mouse oocytes, 1-cell embryos, or 2-cell embryos. Nevertheless, we include Triton X-100 in our lysis buffer as a precaution. With purified luciferase, low concentrations of Triton X-100 in the assay (0.01%) stimulate luciferase activity approximately 20%, but higher concentrations (>0.02%) begin

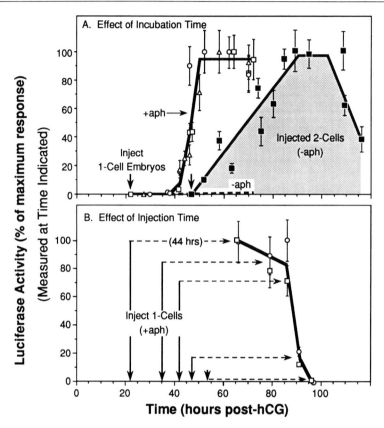

FIG. 5. Expression of firefly luciferase from plasmid DNA injected into mouse preimplantation embryos [E. Martínez-Salas, E. Linney, J. Hassell, and M. L. DePamphilis, *Genes Dev.* **3**, 1493 (1989); M. Wiekowski, M. Miranda, and M. L. DePamphilis, *Dev. Biol.* **147**, 403 (1991)]. Mouse 1-cell embryos were isolated at 17 hr post-hCG and either cultured in aphidicolin to arrest development at the G_1/S boundary (open symbols) or allowed to continue development *in vitro* (dashed line). Mouse 2-cell embryos were isolated from pregnant females at 40–42 hr post-hCG and allowed to continue development *in vitro*. (A) One-cell embryos were injected at 22 hr post-hCG with plasmid DNA containing the luciferase gene linked to the herpes simplex virus (HSV) *tk* promoter (○, ptkluc), to the HSV *tk* promoter and the polyomavirus F101 enhancer (□, pF101tkluc), or to the adenovirus EIIa promoter (△, pEIIaluc). Embryos were harvested at the indicated times and assayed for luciferase activity. Data are expressed relative to luciferase activity observed at 65 hr post-hCG. Two-cell embryos were injected with pF101*tkluc* (■) at 47 hr post-hCG and assayed for luciferase activity at the times indicated. Data are expressed relative to luciferase activity observed at 95 hr post-hCG. (B) Arrested 1-cell embryos were injected at the indicated times and then incubated for an additional 44 hr before assaying for luciferase activity. Data are expressed relative to activity observed when injected at 22 hr post-hCG. Each data point represents the average of 40 to 60 embryos. Error bars indicate the standard error of the mean.

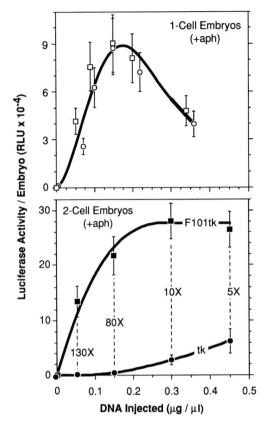

FIG. 6. Relationship between luciferase activity [relative light units (RLU)] and amount of DNA injected. The concentration of the plasmid DNA solution injected is given. An injected nucleus receives about 2 pl of DNA solution. Plasmids (~7 kb) injected were ptkluc (●, ○) and pF101tkluc (■, □) (see Fig. 5). Data for 1-cell embryos are from injection of paternal pronuclei. Injection of maternal pronuclei gives similar results.

to inhibit the enzyme (Fig. 4, top). Because Triton X-100 adheres to proteins, the optimal amount of detergent will depend on the type and number of cells assayed. Extracts of transfected cells may require higher concentrations of Triton X-100 for optimal stimulation (from 0.17%[22] to 0.25%[26]).

Recent studies reported that coenzyme A is also a substrate of luciferase. In the presence of CoA, oxidation occurs from luciferyl-CoA with greater efficiency than from luciferin.[34] We found that CoA stimulates luciferase activity 2.3- to 2.5-fold (Fig. 4, bottom).

[34] K. V. Wood, in "Bioluminescence and Chemiluminescence: Current Status" (P. E. Stanley and J. Kricka, eds.), p. 11. Wiley, Chichester, 1991.

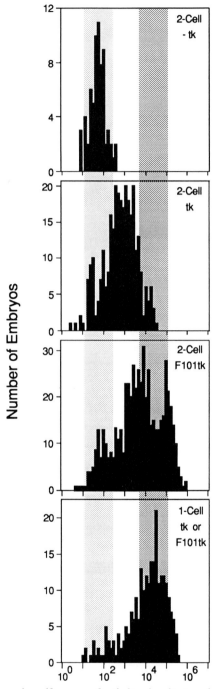

Number of Embryos

Luciferase Activity (units/embryo)

Parameters Affecting Expression of Genes Injected into Mouse Ova

The amount of gene expression observed depends on (1) stability of the injected DNA, (2) the time elapsed after injection, (3) the developmental state of the embryo, (4) the amount of DNA injected, and (5) the choice of promoter/enhancer sequences. The same results have been obtained with outbred CD-1 Swiss albino mice and with inbred (C57BL/ 6J × SJL/J)F_1 mice.

Stability of Injected DNA

Circular DNA is about 5 times more stable than linear DNA when injected into mouse nuclei. Once the injected DNA has been converted to chromatin (~5 hr in oocytes and 2-cell embryos), it is stable for up to 3 days. DNA injected into the cytoplasm of mouse ova is degraded almost completely (~4–6 hr in embryos and <1 hr in unfertilized eggs). The fraction of DNA that is degraded (30–80%) when injected into nuclei varies from one experiment to the next. Since any degradation that occurs is completed within 4 to 6 hr, it appears to result from DNA leaking into the cytoplasm. Leakage will depend on the volume injected into the nucleus, the holding pressure on the injection pipette, and the size of the opening in the injection pipette tip.

Time Course and Development

The time course for luciferase expression following transfection of mammalian cells in culture varies with the cell line, and it sometimes shows a sharp peak of activity, revealing that luciferase is unstable in certain cells.[23] In 2-cell mouse embryos (Fig. 5A), the greatest amount of luciferase activity is observed from 40 to 50 hr postinjection. At this time, the amount of luciferase in 4- to 8-cell embryos is approximately 5-fold greater than in morulae. Therefore, the total amount of luciferase observed depends on the extent of development. The highest levels of luciferase observed with injected 2-cell embryos occurs when the embryos are cultured in aphidicolin and arrested at the G_1/S phase of the 4-cell stage.[11]

Expression of genes injected into 1-cell embryos that subsequently develop to the 2-cell stage or further is less than 1% of the amount produced when DNA is injected into 2-cell embryos. However, when DNA is injected into 1-cell embryos arrested at the beginning of S phase by incuba-

FIG. 7. Distribution of responses by individual embryos to an injected luciferase expression vector. Plasmids injected were pluc, ptkluc, and pF101tkluc.

tion with aphidicolin, the level of gene expression is equivalent to the highest levels observed when 2-cell embryos are injected (Fig. 5A). This is true for several different promoters[11,12] as well as for the polyomavirus origin of DNA replication.[1,13] Aphidicolin arrests morphological development of 1-cell embryos at the G_1/S phase transition, but it does not arrest the zygotic clock that triggers the onset of transcription about 20 hr after fertilization (see Introduction and Fig. 1). However, by 90 hr post-hCG, arrested 1-cell embryos have lost their ability to express both injected genes (Fig. 5B) and endogenous genes.[11] Transfer of arrested 1-cell embryos to aphidicolin-free medium 30 hr post-hCG does not restore their ability to develop into 2-cell embryos.[11,35]

Individual oocytes are routinely assayed at 24 hr after injection.[2,5-7,11,12] The amount of promoter activity in oocytes is 6 to 10% the amount observed in arrested 1-cell embryos.

DNA Concentration

Diploid zygotic nuclei in 2-cell embryos tolerate more injected plasmid DNA than haploid pronuclei in 1-cell embryos (Fig. 6). Oocytes are similar in this respect to 2-cell embryos.[12] Haploid nuclei in parthenogenetically activated eggs or 2-cell embryos derived from activated eggs tolerate about half as much injected DNA as 1-cell embryos.[36] Therefore, the maximum response observed for the same promoter/enhancer sequence will depend on choosing the optimal DNA concentration. Analysis of the amounts of plasmid DNA present per embryo after injection indicates that approximately 2 pl of sample remains in the nucleus.

Stimulation of promoters by enhancers in 2-cell embryos is greatest when low concentrations of DNA are injected (Fig. 6) or when weak promoters are used because the basal level of promoter activity is reduced. The polyomavirus F101 enhancer is the strongest embryo-responsive enhancer found so far[12] (F. Melin, unpublished data, 1993). However, this enhancer has no effect on promoter activity in oocytes or 1-cell embryos. Promoter activity in aphidicolin-arrested 1-cell embryos is only 2- to 3-fold less than the maximum levels of enhancer-driven promoter activity observed when injected 2-cell embryos are incubated with aphidicolin, and 80% arrest development at the 4-cell stage.

Variation in Promoter/Enhancer Response among Embryos

The range of luciferase activities among individual embryos can vary as much as 1000-fold (Fig. 7). Nevertheless, the relative activities observed

[35] S. K. Howlett, Roux's Arch. Dev. Biol. 195, 499 (1986).

[36] M. Wiekowski, M. Miranda, and M. L. DePamphilis, Dev. Biol., in press (1993).

among different regulatory sequences and embryonic cells is highly reproducible, even when DNA injections are performed by different people. Sample size for a single data point is typically around 40 embryos with a standard error of the mean that varies from 15 to 25%. The mean value obtained from several independent experiments is reproducible to within 10 to 20%. Data from three or more independent experiments can be combined to give sample sizes that average about 100 embryos per data point in order to reduce the SEM to approximately 10%. The number of embryos responding to an injected promoter increases approximately 2-fold when the promoter is linked to an embryo-responsive enhancer such as the polyomavirus F101 enhancer. Furthermore, the distribution of responses appears bimodal, with much less variation among the strongest responding embryos (Fig. 7).

Transgenic Animals

Firefly luciferase reporter genes can also be used to identify tissues and cells in transgenic animals that utilize a particular promoter/enhancer sequence, and the extent of utilization can also be quantified. For example, the oocyte-specific upstream regulatory region of the mouse sperm receptor gene (ZP3) was identified by linking it to firefly luciferase and producing transgenic animals by pronuclear injection.[3,4] Ovaries from the transgenic animals contained approximately 175 pg of luciferase ($\sim 4 \times 10^5$ RLU/ovary under the assay conditions), and individual oocytes from these ovaries contained from 0.4 to 2 pg of luciferase ($1-5 \times 10^3$ RLU/oocyte). Tissue samples from transgenic mice containing the chicken α-skeletal actin promoter gave lower luciferase values (100–16,000 RLU).[24]

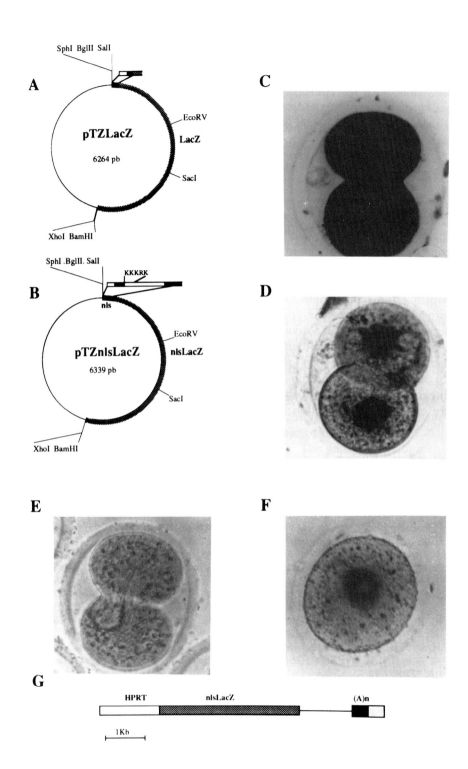

A

SphI BglII SalI

pTZLacZ
6264 pb

EcoRV

LacZ

SacI

XhoI BamHI

B

SphI .BglII. SalI

KKKRK

nls

pTZnlsLacZ
6339 pb

EcoRV

nlsLacZ

SacI

XhoI BamHI

C

D

E

F

G

HPRT   nlsLacZ   (A)n

1Kb

pMC 1871,[20] or, more conveniently, pTZLacZ (C. Bonnerot and J.-F. Nicolas, unpublished results, 1992) (Fig. 1A).[21,22] β-Galactosidase cleaves the substrate X-Gal to yield a reaction product that can be precipitated, leading to a blue staining in the entire cytoplasm (Fig. 1C). Cells with very low β-galactosidase activity show a punctate pattern.

The *LacZ* gene has been modified by an in-phase fusion with a sequence coding for a nuclear locating signal (*nlsLacZ*).[11,21] In *nlsLacZ*, the 5' region up to the first four amino acids is a fragment of the early region of the simian virus (SV40) genome (nucleotides 5221 to 5133).[22] It contains the initiation translational AUG codon of the t antigens. This segment is linked to a segment containing the nuclear locating signal (amino acids 127 to 147 of large T antigen of SV40), fused to β-galactosidase at the sixth amino acid. The source of *nlsLacZ* is L7RH β-Gal[11,21] or, more conveniently, pTZnlsLacZ (C. Bonnerot and J.-F. Nicolas, unpublished results, 1992) (Fig. 1B). The encoded β-galactosidase yields, when using the substrate X-Gal, a blue staining in the nucleus[11] (Fig. 1D).

One advantage of using *nlsLacZ* instead of *LacZ* for analysis in preimplantation embryo is that it permits direct visualization of the nuclei and, in 1-cell embryos, of the pronuclei.[17] In addition, the chimeric nuclear protein is dispersed in the cytoplasm during mitosis and relocated to the nucleus at the next interphase.[17] The nuclear staining is also compatible

[20] S. K. Shapira, J. Chou, F. V. Richaud, and M. Casadaban, *Genes* **25**, 71 (1983).

[21] D. Kalderon, B. L. Roberts, W. D. Richardson, and A. E. Smith, *Cell* (*Cambridge, Mass.*) **39**, 499 (1984).

[22] J. Tooze, "Molecular Biology of Tumor Viruses," 2nd Ed., Part 2. Cold Spring Harbor Laboratory, Cold Spring Harbor, New York, 1980.

FIG. 1. Reporter *LacZ* and *nlsLacZ* genes. (A) pTZLacZ is 6264 bp long. *LacZ* sequences are flanked by unique restriction enzyme sites. The 5' noncoding sequences and the first four amino acids (white box) are from SV40 (nucleotides 5122 to 5133).[22] It is followed by the linker sequence 5'-TTCCGGAGC-3', coding for amino acids 5 to 7 (black box) linked at the sixth amino acid of *LacZ*[18] (gray box). (B) pTZnlsLacZ is 6339 bp long. It is identical to pTZLacZ with the addition of the 5'-AAT TCC GCA AAA AAG AAG AGA AAG GTA GAA GAC CCC AAG GAC TTT CCT TCA GAA TTG CTA AGT TTT TTG AGT CCA-3' sequence between the sixth and seventh codons of the linker sequence (white box flanked by black boxes). The sequence contains a nuclear locating signal.[11,21] (C) Expression of a *LacZ* gene and (D) of a *nlsLacZ* gene in a 2-cell embryo. (E) Control 2-cell-embryos. (F) Expression of *nlsLacZ* in an oocyte, stained in the germinal vesicle. Fixation and X-Gal staining of the cells in (C)–(F) were performed as described in the text. (G) The HPRTnlsLacZ construct. The HPRT promoter (white box) is a 1.5-kb fragment of a human HPRT genomic clone. *nlsLacZ* (gray box) is identical to the *Sal*I–*Bam*HI fragment of pTZnlsLacZ. The polyadenylation signal (black box) is from Moloney murine leukemia virus.

with histological staining and immunochemistry (detailed in [28] in this volume).

HPRTnlsLacZ[14] (Fig. 1G) combines the promoter of the gene for hypoxanthine phosphoribosyltransferase (HPRT) that is expressed in all tissues and the *nlsLacZ* reporter (Fig. 1G). It is transcriptionally active in transient expression assays from the period of minor activation of the genome in early 2-cell embryos and in oocytes.[17] Therefore, it can be used as a positive control.

Preparation of DNA for Microinjection

DNA must be extensively purified to obtain reproducible results. Following large-scale preparation of plasmid DNA by techniques involving lysozyme–alkaline lysis or lysozyme–Triton X-100 lysis, two purifications by equilibrium ultracentrifugation in CsCl–ethidium bromide gradients are generally necessary.[1] When linearized DNA molecules are tested, a purification step must follow the digestion by restriction enzymes and electrophoresis in agarose of the plasmid or inserts. Impurities from the agarose may interfere with egg development and may generate false-negative results. Glass bead purification[23] is recommended.

State of Reporter Vector

Constructs can be tested as supercoiled DNA (plasmids) or as inserts deleted of plasmid DNA sequences. Both forms of DNA are adequate substrates for transcription in 1-cell and 2-cell embryos.[16,17] In contrast, linear DNA in oocytes is less active[24] than supercoiled DNA; therefore, before the injection, the inserts must be religated. Ligation is performed as follows. The DNA concentration is 1 to 6 nM; ligation buffer is 20 mM 1,4-dithiothreitol (DTT), 1 mM adenosine 5′-triphosphate (ATP), 10 mM MgCl$_2$, 50 μg/ml bovine serum albumin (BSA), 50 mM Tris-HCl, pH 7.8. Ligation is conducted for 18 hr at 15° with 1 unit of T4 DNA ligase. DNA in the ligation buffer can be directly injected into cells without purification. It is generally diluted 2- to 10-fold in 10 mM Tris-HCl, pH 7.4, 0.1 mM ethylenediaminetetraacetic acid (EDTA) to bring it to the right DNA concentration.

The DNA sequences of bacterial plasmids are not all neutral. We notice promoter activity in embryos injected with pGEM1–*nlsLacZ*, a construct lacking eukaryotic promoter sequences. It is therefore important to delete all plasmidic bacterial sequences before testing for the expres-

[23] B. Vogelstein and D. Gillespie, *Proc. Natl. Acad. Sci. U.S.A.* **76**, 615 (1979).
[24] L. E. Chalifour, D. O. Wirak, P. M. Wassarman, and M. L. DePamphilis, *J. Virol.* **59**, 619 (1986).

sion of the *LacZ* gene by the recombinant construct. In pTZLacZ- and pTZnlsLacZ-based vectors, useful restriction enzyme binding sites in the polylinkers flank the insert (Fig. 1A,B). Note that, in contrast to *Xenopus* and sea urchin eggs,[25,26] replication of plasmid DNA does not occur in mouse eggs.[27]

Glass Bead Purification of DNA

1. Electrophoresis of DNA is performed in 40 mM Tris–acetate, 1 mM EDTA. Cut the gel slice (preferably low-melting agarose) containing the fragment of DNA of interest under long-wavelength UV light (302 nm). Weigh and mince the slice.

2. Dissolve the gel slice in 6 M NaI, 20 mM Na_2SO_3 in water. The solution must be filtered and saturated with 0.5 g/100 ml Na_2SO_3 before use and be protected from light and stored at 4°. Use approximately 2 ml per gram of gel slice. Incubate at 37° until completely dissolved (about 15 min) in the dark (protect from light with an aluminum sheet).

3. Vortex the glass slurry until all the powder is in suspension. Glass slurry is prepared from silica 325 mesh as follows: Resuspend 100 ml of powder in 200 ml of distilled water; stir for 1 hr and let settle for 1 hr; take the supernatant and spin 10 min in a Sorvall centrifuge at room temperature; resuspend the pellet in 100 ml water; add nitric acid to 50% (v/v) in a chemical hood; bring close to boiling; and wash the pellet four times with water. Store as a 50% slurry in distilled water.

4. Add the slurry to the dissolved gel (1 μl for 1 μg of DNA). Leave for 2 to 5 hr in the tube on a rocker in the dark at 4°.

5. Spin for 20 sec in a microcentrifuge. Discard the supernatant.

6. Wash the pellet twice with 10 volumes of NaI, Na_2SO_3 solution in the cold. Resuspend the powder at each wash.

7. Wash the pellet twice with cold ethanol solution. Ethanol solution is 0.1 M NaCl, 1 mM EDTA, 50% ethanol, 10 mM Tris-HCl, pH 7.5 (stored at $-20°$). Remove as much of the supernatant as possible.

8. Air-dry the pellet briefly.

9. Elute the DNA for 30 min in 3 volumes of 0.1 mM EDTA, 10 mM Tris, pH 7.5, at 37°.

Notes. This procedure is modified from Vogelstein *et al.*[23] Fragments from 200 bp to 18 kb can be purified with nearly complete recovery using

[25] S. Rusconi and W. Schaffner, *Proc. Natl. Acad. Sci. U.S.A.* **78**, 5051 (1981).

[26] A. P. McMahon, C. N. Flytzanus, B. R. Hough-Evans, K. S. Katula, R. J. Britten, and E. H. Davidson, *Dev. Biol.* **108**, 420 (1985).

[27] D. O. Wirak, L. E. Chalifour, P. M. Wassarman, W. J. Muller, J. A. Hassel, and M. L. DePamphilis, *Mol. Cell. Biol.* **5**, 2924 (1985).

this method. DNA concentration can be determined by measuring the fluorescence of bisbenzimide H33258[28] or comparing the fluorescence intensity after electrophoresis. DNA can be stored at 4° for months.

Preparation of Embryos

When choosing the genetic origin of the cells, consider the ease of nuclear microinjection, the yield of eggs per female, and the ability to continue development *in vitro*. (C57BL/6J × DBA2)F_1 females mated with F_1 males of the same strain are convenient. They give a high number of eggs when treated for superovulation (30 per 7- to 10-week-old female), and fertilized eggs pass with a high efficiency to the 4-cell stage in *in vitro* culture. Other F_1 strains such as C57BL/6J × SJL/J and C57BL/6J × CBA are also convenient. Note that for the majority of mouse strains 1-cell embryos arrest development in culture at the late 2-cell stage (2-cell block).

There are two ways of looking at preimplantation embryos. *In vivo* material can be isolated at each stage (i.e., growing oocytes, fully grown oocytes, 1-cell, 2-cell, and 4-cell embryos) and analyzed immediately. Alternatively, the material can be isolated at one stage and cultivated *in vitro* until the eggs reach the desired stage.

The efficiency of *in vitro* development of 1-cell mouse embryos to the morula or blastocyst stage is very high (80%), as is the developmental potential when transferred back into foster mothers (35%). However, it should be noted that *in vitro* cultivated material may not exactly represent the *in vivo* material.[17] Moreover, the times of ovulation, insemination, and fertilization vary considerably *in vivo*. *In vitro* fertilization reduces these variations.[29]

Required Culture Media

Stock A solution (10× concentrated) for M16 and M2: NaCl 950 mM (5.534 g/100 ml), KCl 48 mM (0.360 g/100 ml), KH_2PO_4 11 mM (0.162 g/100 ml), $MgSO_4 \cdot 7H_2O$ 12 mM (0.294 g/100 ml), sodium lactate 60% syrup (3.2 ml/100 ml), glucose 55 mM (1 g/100 ml), penicillin (1 × 10^5 IU), and streptomycin (0.050 g/100 ml)

M16: Stock A solution diluted 10-fold (1×) plus $NaHCO_3$ 25 mM (2.101 g/liter), sodium pyruvate 0.32 mM (0.036 g/liter), $CaCl_2 \cdot 2H_2O$ 1.71 mM (0.252 g/liter), phenol red (0.01 g/liter), and BSA (4 mg/ml); adjust to pH 7.6

---

[28] C. Labarca and K. Paigen, *Anal. Biochem.* **101**, 339 (1980).
[29] S. K. Howlett and V. N. Bolton, *J. Embryol. Exp. Morphol.* **87**, 175 (1985).

M2: Stock A solution diluted 10-fold ($1\times$) plus NaHCO$_3$ 4 mM, sodium pyruvate 0.32 mM (0.036 g/liter), CaCl$_2$ · 2H$_2$O 1.71 mM (0.252 g/liter), HEPES (5 mg/ml), and BSA (4 mg/ml); adjust to pH 7.4

T6: NaCl 97.84 mM (5.719 g/liter), KCl 1.42 mM (0.106 g/liter), MgCl$_2$ · 6H$_2$O 0.47 mM (0.096 g/liter), Na$_2$HPO$_4$ · 12H$_2$O 0.36 mM (0.129 g/liter), CaCl$_2$ · 2H$_2$O 1.78 mM (0.262 g/liter), NaHCO$_3$ 25 mM (2.101 g/liter), sodium lactate 24.9 mM (2.791 g/liter), sodium pyruvate 0.47 mM (0.052 g/liter), glucose 5.56 mM (1 g/liter), penicillin (1×10^5 IU), streptomycin sulfate (0.05 g/liter), and phenol red (0.01 g/liter)

Phosphate-buffered saline (PBS): 138 mM NaCl, 2.7 mM KCl, 1.5 mM KH$_2$PO$_4$, 8.1 mM Na$_2$HPO$_4$, pH 7.3

Stock solutions can be stored at $-20°$ for months. Solutions are stored at 4° for no more than 2 weeks.

Isolation of Mouse Oocytes

Fully grown oocytes are obtained from ovaries of 8- to 10-week-old F$_1$ females. Such oocytes have acquired meiotic competence.[30] Each female yields approximately 20–30 oocytes.

1. The ovaries are dissected and placed in M2 medium.
2. The oocytes (80 μm in diameter) are released from follicles by puncturing them with forceps.
3. Oocytes are washed three times (by transfer) in M2 containing 4 mg/ml BSA, an operation which eliminates follicular cells of the zona radiata and debris by pipetting. Keep the denuded oocytes.
4. Oocytes are cultured in M16 containing 4 mg/ml BSA in 5% CO$_2$ in air.

If the manipulations have been carried out in the absence of $N^6,2'O$-dibutyryladenosine 3',5'-cyclic monophosphate (dbcAMP, Sigma), the breakdown of the germinal vesicle occurs within 1–2 hr. Therefore, nuclear injections are done immediately after preparation of oocytes. Only denuded oocytes are used because visualization of the germinal vesicle is very difficult with oocytes still in the corona radiata. At 18–20 hr after isolation in culture, this material yields 11% oocytes with intact germinal vesicles (GV), 37% oocytes at metaphase I, and 52% oocytes that had completed maturation as demonstrated by the emission of the first polar body.[17]

---

[30] R. M. Schultz, *in* "Experimental Approaches to Mammalian Embryonic Development" (J. Rossant and R. A. Pedersen, eds.), p. 195. Cambridge Univ. Press, Cambridge, 1986.

Notes. Injection in the cytoplasm does not disturb the maturation process. Injection in the germinal vesicle reduces the number of eggs which complete maturation (32% instead of 52%). To prevent spontaneous germinal vesicle breakdown, 100 μg/ml dbcAMP is added in M2 medium from the step of the dissection of the ovaries.[31]

Growing oocytes are obtained from ovaries of mice less than 15 days of age (usually 12- to 14-day-old females). They are smaller (40 to 60 μm in diameter) than fully grown oocytes. They do not mature spontaneously and fail to resume meiosis when placed in a suitable culture medium. The ovaries are dissected manually in M2 medium and then treated exactly as the fully grown oocytes. The oocytes are also cultured in M16 containing 4 mg/ml BSA in 5% (v/v) CO_2 in air. DNA is injected 1–4 hr after isolation.

In Vivo Fertilized 1-Cell and 2-Cell Embryos

Superovulated females are obtained by injecting 6- to 8-week-old mice with 5 IU of pregnant mare serum gonadotropin (PMSG; folligon) diluted in 0.9% NaCl or in PBS (<100 μl total volume) in the abdominal cavity and 42–48 hr later with 5 IU of human chorionic gonadotropin (hCG; chorulon) diluted in 0.9% NaCl or in PBS (<100 μl total volume) in the abdominal cavity. The females are bred with males immediately after hCG injection. Vaginal plugs are observed the next morning (vaginal plugs are still visible 12 hr after mating). To obtain 1-cell embryos, fertilized oocytes are isolated from the ampulla tubae of superovulated females 20 hr post-hCG injection.

1. The female is sacrificed by cervical dislocation and dissected.

2. Oviducts are placed in M2 containing 0.5 mg/ml hyaluronidase. The ampulla is manually dissected with forceps. This operation liberates the cumulus oophorus. Eggs are released from cumulus cells within 2 min.

3. Eggs are rinsed three times in M2, transferred to M16 medium, and incubated at 37°. Note that the male pronucleus starts forming 16–19 hr post-hCG and is clearly visible 20 hr post-hCG. The female pronuclei is visible after 19 hr post-hCG. Amphimixis occurs 30–32 hr post-hCG (Fig. 2).

The 2-cell embryos are isolated from the oviducts of superovulated females 43 hr post-hCG. The oviducts are dissected in M2 and flushed with the same medium with a 5-ml syringe and a 30-gauge eroded needle.

Both 1-cell and 2-cell embryos are incubated in M16 medium at 37° in 5% CO_2 in air until microinjection. Dulbecco's modified Eagle's medium

---

[31] W. K. Cho, S. Stern, and J. D. Biggers, *J. Exp. Zool.* **187,** 383 (1974).

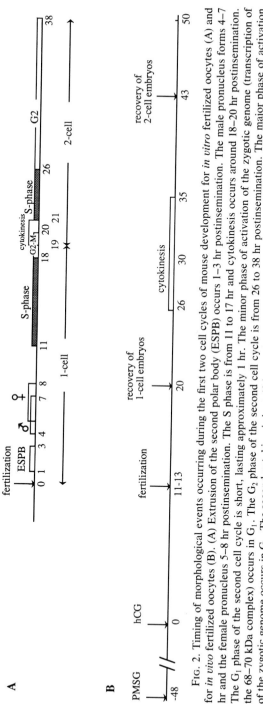

Fig. 2. Timing of morphological events occurring during the first two cell cycles of mouse development for *in vitro* fertilized oocytes (A) and for *in vivo* fertilized oocytes (B). (A) Extrusion of the second polar body (ESPB) occurs 1–3 hr postinsemination. The male pronucleus forms 4–7 hr and the female pronucleus 5–8 hr postinsemination. The S phase is from 11 to 17 hr and cytokinesis occurs around 18–20 hr postinsemination. The G₁ phase of the second cell cycle is short, lasting approximately 1 hr. The minor phase of activation of the zygotic genome (transcription of the 68–70 kDa complex) occurs in G₁. The G₂ phase of the second cell cycle is from 26 to 38 hr postinsemination. The major phase of activation of the zygotic genome occurs in G₂. The second cytokinesis is at 40 hr postinsemination. Correspondence to hours post-hCG injection is indicated. The timing is valid for F₁ hybrid embryos. PMSG, Pregnant mare serum gonadotropin; hCG, human chorionic gonadotropin.

(DMEM) containing 10% (v/v) fetal calf serum (FCS) is also adequate, but only for 2-cell embryos. A more complete description of these manipulations can be found in Hogan *et al.*[3]

In Vitro Fertilization

Variation in the time of division of 1-cell embryos after *in vivo* fertilization is great (from 26 to 35 hr post-hCG, Fig. 2B). It is the consequence of variation both in the time of ovulation[32] and in the interval between coitus and fertilization.[33,34] In contrast, the time of division of 1-cell embryos after *in vitro* fertilization is reduced (from 19 to 21 hr postinsemination, Fig. 2A) especially if the source of spermatozoa is limited to the cauda epididymides.[29] Therefore, in experiments where timing is important, this kind of material is recommended.

1. Superovulated 6- to 8-week-old females are prepared by injection of 5 IU of folligon followed exactly 46 hr later by injection of 5 IU of hCG. Collection of the sperm is performed 12 hr after hCG injection, and oocytes are prepared exactly 14 hr after hCG injection.

2. The male is sacrificed by cervical dislocation, and sperm suspensions are prepared from the two cauda epididymides of an F_1 male of at least 7 weeks of age. Sperm is collected in 1 ml of T6 medium by pressing the two cauda epididymides.

3. Sperm is next incubated for 20 min at 37° in preequilibrated warm T6 (in 5% CO_2 in air) and counted.

4. Sperm is diluted in preequilibrated warm T6 containing 15 mg/ml BSA (Sigma, St. Louis, MO, fraction V) so as to give a sperm concentration ranging from 1×10^6 to 2×10^6 cells/ml. In general, one male yields 2×10^7 spermatozoa. The sperm is incubated for 1.5 hr to allow for capacitation.

5. Oviducts are prepared 14 hr after hCG injection from swollen ampulla of oviducts of the superovulated females. The cumulus masses are directly released in 1 ml of fully capacitated sperm in T6. This step is performed 2 hr after collection of the sperm.

6. Samples are incubated for 3–4 hr. Fertilized oocytes are washed three times by transfer in M16 containing 4 mg/ml BSA.

7. The fertilized oocytes are incubated in M16 containing 4 mg/ml of BSA at 37° in 5% in air until microinjection.

[32] R. G. Edwards and A. H. Gates, *J. Endocrinol.* **18**, 292 (1959).
[33] H. Krzanowska, *Folia Biol.* **12**, 231 (1964).
[34] A. Nicol and A. McLaren, *J. Reprod. Fertil.* **39**, 421 (1974).

Notes. Timing of developmental events in mouse oocytes fertilized *in vitro* is summarized in Fig. 2A. DNA synthesis occurs from 11 to 17 hr postinsemination. Cleavage is at 20 hr postinsemination. The fertilization protocol presented is slightly modified from Wood *et al.*[35]

Microinjection of DNA

Volume. It is possible to inject 1 to 2 pl in pronuclei (1-cell embryos), 0.4 to 0.8 pl in the nuclei of 2-cell embryos, and 1 to 4 pl in germinal vesicles. Injection volumes are visually estimated by the increase in nuclear diameter. The diameter of the nucleus expands by 20% of the original value. Exact measurement of injected volumes can be obtained using [^{3}H]thymidine. For instance, a solution of [^{3}H]thymidine at 2.5×10^9 disintegrations/min (dpm)/ml is microinjected and ^{3}H is directly measured after transfer of the embryos in 1 ml of scintillator liquid (ACS2, Amersham). Radioactivity is assayed in spectrometer counter (Rack-β 12II, LKB). With this protocol, we have estimated that we microinject 1 to 2 pl in male pronuclei, 6 to 10 pl in the cytoplasm of 1-cell and 2-cell embryos, and 2 to 3 pl in the cytoplasm of growing oocytes.

Quantity. At 5 ng/μl a 5-kb construct is present at 1000 copies/pl. Up to 100 ng/μl DNA solutions are viscous, and the percentage of surviving embryos falls in consequence. DNA has toxic effects from approximately 0.1 to 1 μg/μl. The DNA is diluted in 10 mM Tris-HCl, 0.1 mM EDTA at pH 7.4.

In Vitro Culture of 1-Cell and 2-Cell Embryos

Incubation medium is M16 (containing 4 mg/ml of BSA) at 37° in 5% CO_2 in air. Embryos are placed in plastic petri dishes treated for tissue culture. As embryos are particularly sensitive to low temperature or pH, manipulations are strictly performed with preequilibrated and warm medium at 37°.

Inhibiting Embryo Development

Inhibition of DNA Replication. Inhibition of DNA replication in 1-cell embryos causes a failure of the first cleavage. In 1-cell arrested embryos, synthesis of the 68–70 kDa complex normally detected in early 2-cell

[35] M. J. Wood, D. G. Whittingham, and W. F. Rall, *in* "The Low Temperature Preservation of Mouse Oocytes and Embryos in Mammalian Development" (M. Monk, ed.), p. 255. IRL Press, Oxford and Washington, D.C., 1987.

embryos[36,37] occurs, but the synthesis of the polypeptide chains indicative of the major activation of the zygotic genome normally detected in late 2-cell embryos[36,37] does not occur.[38] Therefore, arrested 1-cell embryos are not indicative of normal late 1-cell embryos. It remains to be analyzed to what extent they correspond to "normal" early 2-cell embryos.

If 1-cell embryos are maintained in the presence of aphidicolin, an inhibitor of DNA polymerase α,[39] or 5-fluorodeoxyuridine[40] (FUdR), a thymidine analog, at any time between 8 and 16 hr postinsemination (i.e., during the S phase), they fail to divide. The effect is reversible if the embryos are cultured in the presence of the drug for less than 2 hr. Inhibition of DNA synthesis is obtained by culturing eggs in M16 containing 4 mg/ml BSA and either 2 μg/ml aphidicolin or 100 μM FUdR. Stock solutions of aphidicolin are at 2 mg/ml in dimethyl sulfoxide (DMSO) stored at $-20°$, and those of FUdR are at 10 mM in water stored at $-20°$. Two-cell embryos maintained in aphidicolin of FUdR from 21 to 25 hr postinsemination fail to cleave. The effect is reversible. Similarly, inhibition of DNA synthesis in 4-cell and 8-cell embryos prevents the following cleavage.[41]

Inhibition of Cleavage. Methyl [5-(2-thienylcarbonyl)]-1,4-benzimidazole 2-carbamate (nocodazole) causes arrest of 1-cell embryos in metaphase by inhibiting spindle formation.[42] This microtubule-disrupting drug causes arrest in metaphase in second meiosis when eggs are fertilized in presence of the drug and in metaphase at first mitosis when the drug is added to fertilized eggs from 15 hr postinsemination. The effect is reversible when the drug is removed. It is used at 10 μM in the culture medium of the egg. The stock solution of Nocodazole is 10 mM in DMSO. Cytochalasin D inhibits cytokinesis but not karyokinesis. Late 1-cell embryos treated with cytochalasin D (from 15 hr postinsemination) fail to divide into 2-cell embryos, but the nuclei divide and the DNA replicates. The embryos synthesize late 2-cell polypeptides.[37] Therefore, the drug has no effect on transcriptional activation. Cytochalasin D is used at 0.5 to 1 μg/ml in the culture medium of the egg.

[36] G. Flach, M. H. Johnson, P. R. Braude, R. A. S. Taylor, and V. N. Bolton, *EMBO J.* **1,** 681 (1982).

[37] V. N. Bolton, P. J. Oades, and M. H. Johnson, *J. Embryol. Exp. Morphol.* **79,** 139 (1984).

[38] S. K. Howlett, *Roux's Arch. Dev. Biol.* **195,** 499 (1986).

[39] S. Ikegami, T. Taguchi, M. Ohashi, M. Oguro, H. Nagano, and Y. Mano, *Nature (London)* **275,** 458 (1978).

[40] N. R. Cozzarelli, *Annu. Rev. Biochem.* **46,** 461 (1977).

[41] R. K. W. Smith and M. H. Johnson, *J. Embryol. Exp. Morphol.* **89,** 133 (1985).

[42] J. Hoebeke, G. Van Nigen, and M. De Brabander, *Biochem. Biophys. Res. Commun.* **69,** 319 (1976).

Survival and Development of Injected Embryos

Uninjected control embryos develop to 2-cell embryos in culture at a frequency of 90 to 98% and to fetuses (after transfer to foster mothers) at a frequency of 75%. DNA-injected 1-cell embryos develop to 2-cell embryos at a frequency of 70 to 82%. Similarly, 80% of the blastomeres which survive injection at the 2-cell stage cleave normally.[4,43]

The timing of cleavage of injected 1-cell embryos and of injected blastomeres is retarded a few hours relative to that of controls.[17,43] This is an important parameter to take into account during the interpretation of results, as biochemical events may be independent of morphological events and, in particular, of cleavage.

X-Galactoside Staining

1. Cultured oocytes or 1-cell, 2-cell, or 4-cell embryos are rinsed in PBS and then transferred in the fixative medium for 5 min. Fixative medium is 1% (v/v) formaldehyde and 0.2% (v/v) glutaraldehyde in PBS containing 1% (v/v) serum.

2. Cells are then rinsed three times (2 min each, by successive transfer) in PBS containing 1% serum.

3. Cells are transferred to the histochemical reaction mixture. The histochemical reaction mixture is 1 mg/ml 4-chloro-5-bromo-3-indolyl-β-galactoside (X-Gal), 4 mM K$_4$Fe(CN)$_6$ · 3H$_2$O (potassium ferrocyanide), 4 mM K$_3$Fe(CN)$_6$ (potassium ferricyanide), and 2 mM MgCl$_2$ in PBS. Stock solutions are 40 mg/ml in DMSO for X-Gal (storage at $-20°$) and 0.2 M in water for potassium ferricyanide and potassium ferrocyanide. Stock solutions must be protected from light and made fresh every 2 weeks.

4. Cells are incubated at 37° in an humidified chamber (usually a petri dish). The histochemical reaction proceeds slowly. In general, the first positive eggs can be seen after 1–3 hr of incubation, but the positive cells are scored only after 20 hr of incubation. There is no background activity in control cells (see Fig. 1E).

Notes. For best results, it is important to use a fixative medium with both formaldehyde and glutaraldehyde and to respect the 5-min fixation time. Using only one aldehyde or a shorter fixation time is inadequate. The eggs appear ghostly in the histochemical reaction mixture. To minimize the stickiness of the embryo to glass pipettes during transfer, it is essential

---

[43] K. Ueno, Y. Hiramoto, S. Hayashi, and H. Kondoh, *Dev. Growth Differ.* **30**, 61 (1987).

to have proteins (provided by the serum) in all media and to transfer the cells rapidly in groups of no more than 5 to 10. Siliconized pipettes can also be used. To avoid background activity, the pH (7.3) of the reaction mixture is important. Also note that enzyme activity decreases rapidly in acidic media.

Reference Scale of X-Galactoside Stained Embryos

The generation of quantitative data from experiments with *LacZ* reporter genes is complicated by several factors. Visual observation does not allow precise quantification of the reaction, and, in particular, it is not possible to see differences above a certain level of precipitate. In addition, there are large variations in the response of an individual egg to a given construct. Nevertheless, it is possible to estimate roughly semi-quantitatively the strength of constructs by visual observation of the eggs in reference to a scale of X-Gal staining. The reference scale is best generated by microinjection of known quantities of purified β-galactosidase (Sigma). Differences in the intensity of staining are observed in eggs microinjected with increasing amounts of β-galactosidase from 2×10^{-10} to 2×10^{-8} units and stained with X-Gal for 18 hr at 37°. One β-galactosidase unit corresponds to the hydrolysis of 1 μmol of *o*-nitrophenyl-β-D-galactopyranoside to *o*-nitrophenol and D-galactose per minute at pH 7.3 and 37°. Under these conditions, we estimate the lower limit of sensitivity of β-galactosidase detection at approximately 10^3–10^4 molecules per egg.

1. Dissolve 1000 units of β-galactosidase (EC 3.2.1.23) from *E. coli* (Sigma) in 1 ml of 50 m*M* Tris-HCl, pH 7.3. Prepare serial 10-fold dilutions.

2. Prepare siliconized micropipettes for microinjection as follows. Place the micropipettes in a 15-ml Falcon tube and fill it with Sigmacote (Sigma); discard the Sigmacote and let dry for 1 hr. Rinse 6 times with distilled water. Dry and autoclave the pipettes.

3. Prepare oocytes or 1-cell or 2-cell embryos as described above.

4. Inject the solution (from 1 to 4 pl) into the nucleus. Injection into the nucleus allows better control of the delivered volume (judged by swelling of the nucleus) compared to injection into the cytoplasm.

5. Let the eggs recover for 30 min in M16 containing 4 mg/ml BSA at 37° in 5% CO_2 in air.

6. Stain the eggs with X-Gal. β-Galactosidase-positive eggs are obtained at 2×10^{-10} to 2×10^{-9} units per nucleus after an overnight incubation. The staining is in the nucleus.

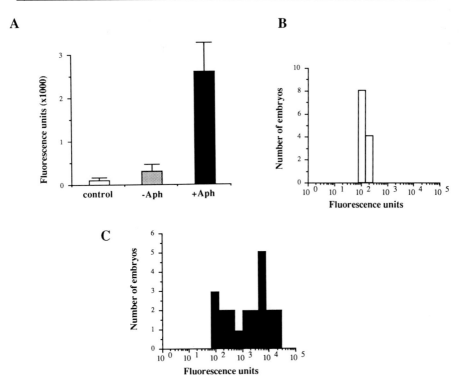

FIG. 3. Assay of β-galactosidase with MUG. (A) Fertilized eggs were microinjected with 4000 copies of SVnlsLacZ inserts and cultured for 20 hr in the absence ($-$ Aph) or in presence ($+$ Aph) of 2 μg/ml aphidicolin, a drug which blocks the first segmentation of the eggs. Control corresponds to noninjected eggs. Values are the mean per embryo calculated from the measure of the β-galactosidase activity of pooled eggs. (B, C) Enzymatic reaction measured on individual eggs. The experiment was the same as in (A), on control embryos (B) and aphidicolin-treated embryos (C).

Quantitative Analysis: 4-Methylumbelliferyl-β-D-galactoside Assay

4-Methylumbelliferyl-β-D-galactoside (MUG) is a fluorogenic substrate of β-galactosidase. The reaction product, 4-methylumbelliferone, can be measured by its emission at 355 nm (excitation at 480 nm). MUG permits detection of β-galactosidase in single embryos. Quantitative indication of the enzyme activity can be obtained by measuring methylumbelliferone fluorescence by a Fluoroskan fluorimeter (Flow Labs, McLean, VA). Figure 3 illustrates the effect of blockage of DNA replication on expression of SVnlsLacZ, a construct in which the promoter of SV40 drives *nlsLacZ*. Fertilized oocytes are injected with 4000 copies of the insert and incubated for 20 hr in the presence or absence of 2 μg/ml aphidicolin. β-Galactosidase is then measured in individual embryos with MUG (Fig. 3C). Noninjected control embryos have a mean of 50 fluorescence units (FU), whereas microinjected 2-cell embryos cultured in the absence of aphidicolin have a mean of 300 FU. The blockage of egg development at the 1-cell stage results in a 10-fold increase of β-galactosidase activity (Fig. 3A). The result is consistent with the idea of DePamphilis and co-workers that some negative regulatory factor first appears as a component of zygotic nuclear structure in 2-cell embryos.[8]

MUG Assay

1. Cultured eggs are transferred individually in 50 μl of lysis buffer in a 96-well microplate. Lysis buffer is 60 mM Na$_2$HPO$_4$, 40 mM NaH$_2$PO$_4$, 10 mM KCl, 1 mM MgSO$_4$, pH 7.2, and 0.5% Triton X-100. It is a modified Z buffer.

2. The reaction is started by the addition of 0.5 mM MUG. MUG stock solution is 5 mM in Z buffer. Solubilization is achieved at 90° for a few minutes. Incubation of the eggs is allowed at 37° for 20 hr.

3. The reaction is stopped by the addition of 50 μl of glycine 100 mM, pH 10.

4. Fluorescence of methylumbelliferone is quantitated by a Fluoroskan fluorimeter.

Notes. Another useful substrate for β-galactosidase is a phenylgalactose-substituted dioxetane. Chemiluminescence of the enzymatic reaction product is measured with a luminometer. Conditions of the luminometric assay are described by Beale *et al.*[44]

Transgenic Mice Expressing *LacZ* in Preimplantation Embryos

To date only two mouse transgenic lines that express *LacZ* at the 2-cell stage have been reported in the literature. HPRTnlsLacZ-4 has been obtained by microinjection of a position-dependent transgene.[14] Therefore, its expression at the preimplantation stages is probably due to cis complementation by a genomic control element of a transcriptional unit active at this stage. CMZ12-LacZ was obtained by microinjection of a *LacZ* construct driven by the promoter of cytomegalovirus.[45] Its expression is subject to variegation and is dependent on maternal inheritance.

[44] E. G. Beale, E. A. Deeb, R. S. Handley, H. Akhavan-Tafti, and A. P. Schaap, *BioTechniques* **12**, 320 (1992).
[45] M. A. Surani, R. Kothary, N. D. Allen, and P. B. Singh, *Development (Cambridge, UK) Suppl.*, 89 (1990).

These and other *LacZ* transgenic mice derived from position-dependent or position-independent constructs will certainly provide 2-cell embryos with markers indicative of the minor and major activations of the zygotic genome. In the near future, this material will be of help in nuclear transplantation experiments[46,47] and in studies designed to compare *in vivo* material and *in vitro* cultured embryos.

Species Other Than Mouse

Comparative analysis of the biochemistry of egg development in several species could indicate the importance of observations made in one species. It is relatively easy to prepare and to microinject embryos from rabbits, pigs, and cattle. Details and references for induction of superovulation and collection and *in vitro* culture of rabbit,[48-50] pig,[51] and bovine[52,53] embryos have been published.

Acknowledgments

We thank Iris Tong for comments and C. Tran for typing the manuscript. C.B., P.B., and J.F.N. are from Institut National de la Santé et de la Recherche Médicale.

[46] M. A. H. Surani, S. C. Barton, and M. L. Norris, *Cell (Cambridge, Mass.)* **45**, 127 (1986).
[47] D. Solter, J. Aronson, S. F. Gilbert, and J. McGrath, *Cold Spring Harbor Symp. Quant. Biol.* **50**, 45 (1985).
[48] C. Delouis, C. Bonnerot, M. Vernet, and J. F. Nicolas, *Exp. Cell Res.* **201**, 284 (1992).
[49] M. Techakumphu, S. Wintenberger-Torres, and C. Sevellec, *Anim. Reprod. Sci.* **12**, 297 (1987).
[50] A. K. Voss, A. Sandmöller, G. Suske, R. M. Strojek, M. Beato, and J. Hahn, *Theriogenology* **34**, 813 (1990).
[51] A. J. Clark, A. L. Archibald, M. McClenaghan, J. P. Simons, C. B. A. Whitelaw, and I. Wilmut, *Proc. N. Z. Soc. Anim. Prod.* **50**, 167 (1990).
[52] K. Saeki, M. Hoshi, M. L. Leibfried-Rutledge, and N. L. First, *Biol. Reprod.* **44**, 256 (1991).
[53] T. Greve and V. Madison, *Reprod. Nutr. Dev.* **31**, 147 (1991).

[28] Application of *LacZ* Gene Fusions to Postimplantation Development

By CLAIRE BONNEROT and JEAN-FRANÇOIS NICOLAS

Introduction

LacZ can be used as an *in situ* enzyme reporter gene for the visualization of gene activity or as a marker of cells during embryogenesis. The

Copyright © 1993 by Academic Press, Inc.
All rights of reproduction in any form reserved.

ease with which this gene can be used is clear from the work of Sanes *et al.*,[1] Price *et al.*,[2] Bonnerot *et al.*,[3] and Goring *et al.*[4] The technical advantages of using *LacZ* compared to other reporter genes are based on the unique possibilities to detect expression in single cells,[1] to stain whole mouse embryos until very late stages (E13–E14),[5–7] and to purify viable cells by fluorescein-activated cell sorting (FACS).[8]

LacZ fusion genes have many uses in the study of mammalian development. Use of *LacZ* allows (1) following the pattern of expression of an endogenous gene tagged with the *LacZ* gene, which is inserted via homologous recombination,[9] (2) determination *in vivo* in *LacZ* transgenic mice of the activity of cis-acting elements of a particular transcriptional unit,[5,10] (3) examination of patterns and mutations by using position-dependent *LacZ* genes[6,7,11,12] or promoter and enhancer *LacZ* traps[13–15] in transgenic animals, and, finally, (4) progenies of a single infected cell of an embryo may be marked with *LacZ* defective recombinant retroviruses for clonal analysis.[1,2] Data on cell lineages in the cerebral cortex,[16–18] retina,[19] skin and yolk sac,[1] and muscle[20] have been obtained.

[1] J. Sanes, J. Rubenstein, and J.-F. Nicolas, *EMBO J.* **5**, 3133 (1986).

[2] J. Price, D. Turner, and C. Cepko, *Proc. Natl. Acad. Sci. U.S.A.* **84**, 156 (1987).

[3] C. Bonnerot, D. Rocancourt, P. Briand, G. Grimber, and J.-F. Nicolas, *Proc. Natl. Acad. Sci. U.S.A.* **84**, 6795 (1987).

[4] D. R. Goring, J. Rossant, S. Clapoff, M. L. Breitman, and L. C. Tsui, *Science* **235**, 456 (1987).

[5] J. Zakany, C. K. Tuggle, M. D. Patel, and M. C. Nguyen-Huu, *Neuron* **1**, 679 (1988).

[6] J.-F. Nicolas, C. Bonnerot, C. Kress, H. Jouin, P. Briand, P. Grimber, and M. Vernet, *in* "Vectors as Tools for the Study of Normal and Abnormal Growth and Differentiation" (H. Lother, *et al.*, eds.), NATO **34**, 33 (1989). ASI Ser: Vol. 34, p. 33. Springer-Verlag, Berlin and New York, 1989.

[7] N. D. Allen, G. Cran, S. C. Barton, S. Hettle, W. Reik, and M. A. Surani, *Nature (London)* **333**, 852 (1988).

[8] G. P. Nolan, S. Fiering, J.-F. Nicolas, and L. A. Herzenberg, *Proc. Natl. Acad. Sci. U.S.A.* **85**, 2603 (1988).

[9] A. L. Joyner, *BioEssays* **13**, 649 (1991).

[10] C. Kress, R. Volgels, W. de Graaf, C. Bonnerot, M. Hameleers, J.-F. Nicolas, and J. Deschamps, *Development (Cambridge, UK)* **109**, 775 (1990).

[11] C. Bonnerot, G. Grimber, P. Briand, and J.-F. Nicolas, *Proc. Natl. Acad. Sci. U.S.A.* **87**, 6331 (1990).

[12] N. D. Allen, E. B. Kerverne, and M. A. Surani, *Development (Cambridge, UK)* **55**, 181 (1990).

[13] A. Gossler, A. L. Joyner, J. Rossant, and W. C. Skarnes, *Science* **244**, 463 (1989).

[14] R. Kothary, S. Clapoff, A. Brown, R. Campbell, A. Peterson, and J. Rossant, *Nature (London)* **335**, 435 (1988).

[15] G. Friedrich and P. Soriano, *Genes Dev.* **5**, 1513 (1991).

[16] M. B. Luskin, A. L. Pearlman, and J. R. Sanes, *Neuron* **1**, 635 (1988).

[17] G. E. Gray, J. C. Glover, J. Majors, and J. R. Sanes, *Proc. Natl. Acad. Sci. U.S.A.* **85**, 7356 (1988).

[18] C. Walsh and C. L. Cepko, *Science* **255**, 434 (1992).

Many of these experimental approaches generate new autonomous cell markers for cell types, cell states, and cells from developmental compartments. These markers are precious in tissue transplantation and graft experiments, as the donor cells can be unambiguously identified by the expression of the *LacZ* marker, and are useful in genetic crosses to describe in a novel way the manner in which mutants affect the positioning of cells.

Reporter *LacZ* Molecules

LacZ and *nlsLacZ* (containing a nuclear locating signal) plasmid vectors are discussed elsewhere in this volume.[21] Other noteworthy vectors combine a fused protein with both APH[3′]II phosphotransferase (*neo*) activity[15,22] or phleomycin-binding activity[23] and β-galactosidase activity (Fig. 1A,B). Phosphotransferase and phleomycin-binding activities are both used for selection in geneticin- or phleomycin-containing medium, respectively, of stably transfected or retrovirally infected embryonal stem (ES) cells, and β-galactosidase is used for description of the spatial and temporal expression of the construct. The markers can be combined with appropriate control elements to generate vectors for trap experiments and for homologous recombination.

More specialized *LacZ* constructs include position-dependent reporters,[7,11,24] enhancers, and promoters (*LacZ* traps)[13,15] (Fig. 1B–D). To mark cells, *LacZ* or *nlsLacZ* defective recombinant retroviruses (dRRV) designed for ubiquitous expression can be harvested from LZ1[1] and LZ2 cell lines (C. Bonnerot and J.-F. Nicolas, unpublished results, 1992), which produce *LacZ* retroviruses with an SV_{40} internal promoter [Moloney murine leukemia virus (M-MuLV)SVLacZ] (Fig. 1E), or from ψ2.21c,[25] which produces *nlsLacZ* retroviruses with an SV_{40} internal promoter (M-MuLVSVnlsLacZ) (Fig. 1F). The retroviruses mark cells regardless of the tissue type. Other recombinant retroviruses are described by Price *et al.*,[2] Savatier *et al.*,[26] and Reddy *et al.*[27]

[19] D. L. Turner and C. L. Cepko, *Nature (London)* **328**, 131 (1987).

[20] S. M. Hughes and H. Blau, *Cell (Cambridge, Mass.)* **68**, 659 (1992).

[21] M. Vernet, C. Bonnerot, P. Briand, and J.-F. Nicolas, this volume [27].

[22] P. J. Southern and P. Berg, *J. Mol. Appl. Genet.* **1**, 327 (1982).

[23] P. Mulsant, A. Gatignol, M. Dolens, and G. Tiraby, *Somatic Cell Mol. Genet.* **14**, 243 (1988).

[24] J. Weis, S. M. Fine, C. David, S. Savarirayan, and J. R. Sanes, *J. Cell Biol.* **113**, 1385 (1991).

[25] N. Savatier, D. Rocancourt, C. Bonnerot, and J.-F. Nicolas, *J. Virol.* **26**, 229 (1989).

[26] N. Savatier, J. Morgenstern, and R. S. P. Beddington, *Development (Cambridge, UK)* **109**, 655 (1990).

[27] S. Reddy, J. V. DeGregori, H. Von Melchnier, and H. E. Ruley, *J. Virol.* **65**, 1507 (1991).

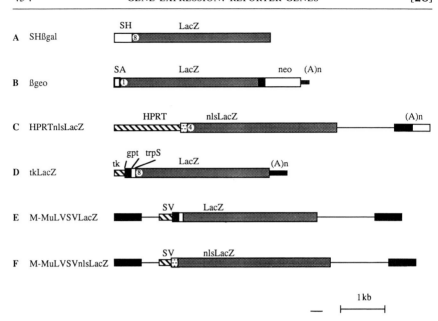

FIG. 1. Examples of *LacZ* vectors. (A) SH-β-gal encodes a hybrid protein with both phleomycin-inactivating[23] (white box) and β-galactosidase (gray box) activities at the eighth amino acid. (B) β-geo encodes a hybrid protein with both APH[3′]II phosphotransferase[22] and β-galactosidase (gray box) activities. The black box corresponds to linker sequences. SA is from the adenovirus major late splice acceptor. The hybrid protein includes β-galactosidase (amino acids 1 to 1021), 13 amino acids from linker sequences, and *neo* (amino acids 2 to the end).[15] Both constructs (A, B) are used as gene traps. (C) HPRTnlsLacZ[11] and (D) tkLacZ[7] are position-dependent transgenes. tkLacZ uses the gpt–trpS–LacZ fusion protein of pCH110. (E) M-MuLVSVLacZ[1] and (F) M-MuLVSVnlsLacZ[3] are defective murine retrovirus constructs. They are used to create induced genetic mosaics during embryogenesis (see Fig. 2).

Genetic Mosaicism Induced by Retroviral Infection

Defective recombinant retroviruses[28,29] are produced in a packaging cell line to provide, in trans, the factors for the assembly and maturation of the viral particles. Engineered helper-free cell lines that constitutively

[28] J.-F. Nicolas and J. Rubenstein, *in* "Vectors: A Survey of Molecular Cloning Vectors and Their Uses" (R. Rodriguez and D. Denhardt, eds.), p. 493. Butterworth, Boston, 1987.

[29] R. Weiss, N. Teich, H. Varmus, and J. Coffin, "Molecular Biology of Tumor Viruses," 2nd Ed., Supplements and Appendixes. Cold Spring Habor Laboratory, Cold Spring Harbor, New York, 1985.

express the viral genes from a unique[30,31] or several nonpackageable mR-NAs.[32,33] are available. M-MuLVSVLacZ and M-MuLVSVnlsLacZ (Fig. 1E,F) are produced in ψ2 (in LZ1 and LZ2, respectively) and in ψ2.21c, an ecotropic packaging help-free cell line. Using a mixture of two retroviruses, the clonal origin of cells within a cluster of labeled cells can be clearly defined.[34,35] A mixture of *LacZ* and *nlsLacZ* retroviruses is adequate, as the two β-galactosidases can be easily distinguished by their staining (Fig. 2A–C). Retrovirus encoding alkaline phosphatase can also be used.[36]

Harvesting of Virus

The retrovirus is produced by budding from the cell surface; its half-life is 6 hr at 37°, but the virus is stable for several days at 4°. Cells of the helper-free cell line ψ2 producing the *LacZ* recombinant retrovirus are plated at $10^5/cm^2$. The culture medium [Dulbecco's modified Eagle's medium, (DMEM) containing 10% of either fetal calf serum or newborn serum][37] is changed 36 hr later, when the cells are subconfluent. The supernatant containing virus is harvested after an additional overnight incubation. It is immediately filtered through a 0.22-μm Millipore (Bedford, MA) filter to remove debris and cells. The filtration is stopped when it becomes difficult to pass liquid through the filter, as viruses will stick to the membrane. Aliquots can be stored in liquid nitrogen with no apparent loss of titer. Under these conditions of harvesting, the titers (tested on NIH 3T3 cells) of ψ2.21c and LZ2 are 1 to 5 × 10^6 blue colony-forming units (bcfu)/ml.

Concentration of Virus

For small aliquots (50 to 170 μl), a quick and efficient concentration method involves centrifuging in an Airfuge (Beckman).

[30] R. Mann, R. C. Mulligan, and D. Baltimore, *Cell (Cambridge, Mass.)* **33**, 153 (1983).
[31] A. D. Miller, M.-F. Law, and I. M. Verman, *Mol. Cell. Biol.* **5**, 431 (1985).
[32] D. Markowitz, S. Goff, and A. Bank, *J. Virol.* **167**, 400 (1988).
[33] O. Danos and R. C. Mulligan, *Proc. Natl. Acad. Sci. U.S.A.* **85**, 6460 (1988).
[34] J.-F. Nicolas and C. Bonnerot, *in* "Cellular Factors in Development and Differentiation: Embryos, Teratocarcinomas and Differentiated Tissues," p. 125. Alan R. Liss, New York, 1988.
[35] D. S. Galileo, G. E. Gray, G. C. Owens, J. Majors, and J. R. Sanes, *Proc. Natl. Acad. Sci. U.S.A.* **87**, 458 (1990).
[36] S. C. Fields-Berry, A. L. Halliday, and C. L. Cepko, *Proc. Natl. Acad. Sci. U.S.A.* **89**, 693 (1992).
[37] H. Jakob and J.-F. Nicolas, this series, Vol. 151, p. 66.

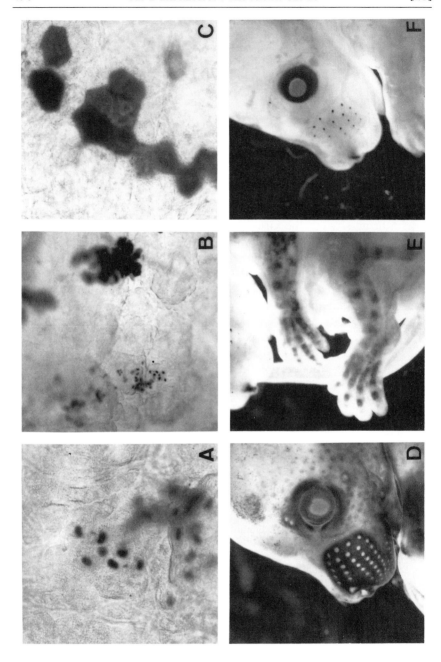

1. The filtered medium from an overnight culture of producer cells is centrifuged at 30 psi for 5 min (165,000 g) at 4° in 175-μl tubes to pellet the virus.

2. After the centrifugation carefully remove as much of the supernatant as possible with a Pasteur pipette pulled by hand.

3. Resuspend the pellet in the remaining medium (~10 μl). Nearly complete recovery of viral activity is achieved.

For large-scale preparation, ultracentrifuge the supernantant containing virus in a SW28 rotor.

1. Harvest supernatant containing virus of overnight cultures.

2. Centrifuge in a clinical centrifuge at 2000 rpm for 10 min at 4° (to pellet cells and debris). Keep the supernatant.

3. Filter the supernatant through a 0.22-μm Millipore filter into polyallomer tubes. Polyallomer tubes are sterilized with ethanol [70% (v/v) in distilled water].

4. Centrifuge at 11,500 rpm (25,000 g) at 2° for 16 hr using a SW28 rotor.

5. Discard the supernatant very slowly and resuspend the pellet in 1% of its original volume in culture medium.

Aliquots of concentrated virus can be stored for months in liquid nitrogen.

Titration of Viral Stocks

Titers are measured by the ability of the dRRV to confer β-galactosidase activity to cells in culture (and are expressed in bcfu/ml).

1. BRL (Bufalo rat liver) or NIH 3T3 cells are used to titrate the virus. BRL cells are easier to grow in culture and give very reproducible results. Replate BRL cells at 3×10^4 cells in 35-mm plastic petri dishes (treated for tissue culture) 1 day before titration.

2. Dilute the viral stock (serial 10-fold dilutions) in culture medium.

3. Aspirate the supernatant of the tester cells. Add 2 ml of culture

---

FIG. 2. Examples of X-Gal staining of embryos and organs. (A–C) Retrovirally induced genetic mosaics in skin. (B) Low-magnification of a whole mount; note the presence of a *LacZ* clone and an *nlsLacZ* clone in the skin of an E17 embryo injected with a mixture of both *LacZ* viruses (Fig. 1E,F) at E9. (A, C) High magnification view, showing the *nlsLacZ* and *LacZ* clones, respectively (photographs kindly provided by Iris Tong). (D–F) *In toto* X-Gal analysis of E14 *LacZ* transgenic embryos. (D) With β-LacZ-7, the expression pattern includes the ectodermal placodes, the mesenchyme surrounding the vibrissae, and the mesenchyme in the head orbita. (E, F) With I_1-I_2-LacZ1, the expression pattern includes skeletal elements (E) and keratinocytes in the vibrissae (F).

medium containing 10 $\mu g/ml$ of Polybrene. (Polybrene is a positively charged molecule that increases the probability of adsorption of enveloped viruses by neutralizing the negatively charged cell surface. Other polycations such as DEAE-dextram are also convenient.)

4. Add 10 μl of the viral dilutions.

5. Incubate the cells for 24 to 48 hr. The maximum number of clones is reached 24 hr after infection.[25]

6. Stain the cells with X-Gal (see below).

Notes. It is not necessary to change the culture medium before X-Gal staining. The cells are apparently insensitive to Polybrene. For both BRL and NIH 3T3 cells, the mean number of cells per clone 24 hr after infection is 1.5 and 1 day later, 4.[25] The number of β-galactosidase-positive clones is directly proportional to the volume of viral supernatant up to 1 ml for BRL and NIH 3T3 cells. For a multiplicity of infection above 2 per cell, all BRL or NIH 3T3 cells are positive. Failure to obtain proportionality and/or complete infection of the culture indicates potential problems with the producer cell line (i.e., presence of material interfering with infection).

Injection of Defective Recombinant Retrovirus into Embryos

Virus is injected through the uterine wall of the pregnant female. The earliest stage amenable to injection is E7 (E0 is the day vaginal plugs are detected).[1] The critical point of the experiment is the micropipette.

1. To prepare the micropipettes, pipettes (CG 100-10, 0.8 mm internal diameter from Clark Electro Medical Instruments) are pulled with a vertical electrode puller from electrode glass (Model 720, David Kopf Instruments) and sealed. The internal diameter of the pulled micropipettes is approximately 40 μm.

2. To avoid plugging of the micropipettes when going through the uterine wall, a hole is made laterally in the pipette as follows. Air pressure is applied by hand to the end of the sealed pipette through a 50-ml injection syringe. The micropipette is moved laterally near the filament of the microforge. A lateral glass bubble forms and bursts, leaving a lateral hole. The filament is immediately turned off. Bumps are removed by scraping the micropipette against the cold filament.

3. Pregnant mice are anesthetized by intraperitoneal injection of 0.1 ml of Avertine solution per 10 g of body weight. Avertine is tribromoethyanol in tertiary amyl alcohol. A stock solution is prepared by dissolving 5 g of tribomoethanol in 5 ml of tertiary amyl alcohol and stored at 4° in the dark. The stock solution is dissolved in 3% (v/v) in phosphate-buffered saline (PBS) on use. PBS is 138 mM NaCl, 2.7 mM KCl, 1.5 mM KH_2PO_4, and 8.1 mM Na_2HPO_4, pH 7.3.

4. Vertical incisions are made in the skin and body wall musculature, and the uterine horns are withdrawn from the ventral side.

5. The micropipette filled with the viral solution is then driven in between the veins of the white portion of the decidual swelling on the antimesometrial side of the embryo. Approximately 0.2 μl of viral solution is injected per embryo.

6. The body wall and skin are then closed with sutures.

Notes. The choice of anesthesic is important. Avertine does not provoke embryo resorption of C57BL/6J × CBA/J or C57BL/6J × DBA/2J pregnant females. Do not use mice from strains that harbor infective endogenous retroviruses.[38] The C57BL/6J, CBA/J, and 129/Sv strains are convenient. From E12 on, it is possible to target the injection into organs, which are visible through the uterine wall (use fiber optics to visualize the embryo more easily). Additional information can be found in other texts.[6,28,29,34,39]

LacZ Transgenic Mice

Several genes have been tagged via homologous recombination using a *LacZ* insertion vector, including *Hox 1-3*.[40] In all cases the β-galactosidase expression pattern strictly follows the pattern of expression of the endogenous gene. Neutrality of *LacZ* and *nlsLacZ* in the corresponding tissues and also outside the lineages is suggested by the lack of β-galactosidase expression in regions where the endogenous gene is not also expressed.

Determinations of the activity of promoters and of cis-acting elements by combination with a *LacZ* or *nlsLacZ* reporter gene have been done with various cloned genes. This type of analysis is particularly useful in the determination of boundaries of expression of *Hox* genes[5,10] and of segmentally restricted genes.[41,42] In these analyses frequent expression of the reporter gene in patterns unique for each transgenic line is noticed. These patterns of expression are probably due to position effects (Fig. 3F,G).

---

[38] R. Weiss, N. Teich, H. Varmus, and J. Coffin, "Molecular Biology of Tumor Viruses." Cold Spring Harbor Laboratory, Cold Spring Harbor, New York, 1982.

[39] P. Soriano, T. Gridley, and R. Jaenisch, "Retroviral Tagging in Mammalian Development and Genetics." American Society for Microbiology, Washington, D.C., 1989.

[40] H. Le Mouellic, Y. Lallemand, and P. Brulet, *Proc. Natl. Acad. Sci. U.S.A.* **87**, 4712 (1990).

[41] J. M. Greenberg, T. Boehm, M. V. Sofroniew, R. J. Keynes, S. C. Barton, M. L. Norris, M. A. Surani, M. G. Spillantini, and T. H. Rabbits, *Nature (London)* **344**, 158 (1990).

[42] S. S. Tan, *Dev. Biol.* **146**, 24 (1991).

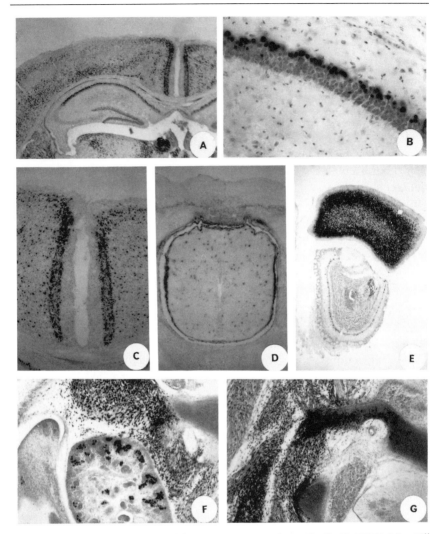

FIG. 3. Examples of X-Gal staining on cryostat sections. (A–C, E) HPRTnlsLacZl[11] (P11). (A) Brain coronal section through the cortex, hippocampus, and thalamus. Only neurons in certain laminar structures express the transgene. (B) High-magnification view of CA1. Note the expression in only the most dorsally located pyramidal neurons. (C) High-magnification view of the cingulate cortex. (E) More rostral section of the same brain, showing expression in the cortex (area frontalis), in mitral cells of the olfactory bulb, and in neurons of the anterior olfactory nucleus. (D) HPRTnlsLacZ.3[11] (E14). The pia mater and glial cells in the spinal cord are positive for β-galactosidase. (F, G) *Hox 2.3/LacZ* transgenic line.[10] (F) Transgene-dependent expression in the epithelium of collecting tubules in the kidney. (G) Transgene-independent expression in muscle and cartilage. In (A–E) the sections were counterstained with neutral red and those in (F, G), with phloxine.

Targeting of *LacZ* in specific tissues has been successfully obtained with the neuron-specific enolase promoter[43] (in postmitotic neurons), the serum retinol-binding protein promoter[42] (in liver cells), the cellular retinoic acid-binding protein promoter,[44] the β-amyloid protein promoter[45] (in neurons), the platelet factor four promoter[46] (in megakaryocytes), and the promoter of the receptor of acetylcholinesterase[47,48] (in muscles). Strategies to search for patterns and mutations have been devised. Position-dependent transgenes with their own promoters are presumably sensitive reporters of enhancers and locus control region elements. They are generally not associated with the disruption of a gene. Powerful reporters include HPRTnlsLacZ,[11] tkLacZ,[7] and hsp68LacZ (p3LSN)[14] constructs (Fig. 1C,D). Gene trap constructs without promoters are gene reporters[13,15] (Fig. 1B). They are activated by fusion with a transcriptional unit and presumably report its pattern of expression with fidelity. They disrupt the gene. Plasmid vectors and retroviruses[15,49] are both used. Trapping experiments are now almost exclusively done in embryonic stem (ES) cells to facilitate selection of clones of interest and conservation of mutations. The varieties of *LacZ* activation patterns have been illustrated[7,11,13,15] (Figs. 2 and 3).

Preparation of Embryos for *in Toto* X-Galactoside Staining

To follow the spatiotemporal expression pattern of *LacZ* (Fig. 2D,E) and to localize clones of virally infected cells (Fig. 2A–C), the expression of the *LacZ* reporter gene is detected in whole embryos. In addition to this important technical advantage, the sensitivity of the detection of β-galactosidase is such that single β-galactosidase-positive cells can be identified; it allows direct visualization of the shape and boundary of a clone (Fig. 2B). Before E14 the embryo can be directly analyzed in whole mounts (Fig. 2D–F). From E14 to newborn, a simple dissection such as a sagittal section of the fetus greatly improves internal staining. Almost all tissues of the animal can be stained with X-Gal. Finally, endogenous β-galactosidase activity is not detected by the assay because of its optimum pH of 4.

[43] S. Forss-Petter, *et al.*, *Neuron* **5**, 187 (1990).
[44] L.-N. Wei, G.-J. Chen, Y.-S. Chu, J.-L. Taso, and M. C. Nguyen-Huu, *Development (Cambridge, UK)* **112**, 847 (1991).
[45] D. O. Wirak, C. Bayney, C. A. Kundel, A. Lee, G. A. Scangos, B. D. Trapp, and A. J. Unterbeck, *EMBO J.* **10**, 289 (1991).
[46] K. Ravid, D. L. Beeler, M. S. Rabin, H. E. Ruley, and R. D. Rosenberg, *Proc. Natl. Acad. Sci. U.S.A.* **88**, 1521 (1991).
[47] A. Klarsfeld, *et al.*, *EMBO J.* **10**, 625 (1991).
[48] J. R. Sanes, Y. R. Johnson, P. T. Kotzbauer, J. Mudd, T. Hanley, J. C. Martineau, and J. P. Merlie, *Development (Cambridge, UK)* **113**, 1181 (1991).
[49] A. Choulika and J.-F. Nicolas, in press (1992).

Preparation of Embryos until E14

1. A pregnant female is sacrificed by cervical dislocation or by suffocation in a carbon dioxide-enriched atmosphere obtained by placing dry ice at the bottom of a beaker. Interpose wood shavings between the dry ice and the mouse.

2. Make an incision in the ventral skin and body wall musculature and withdraw the uterine horns.

3. Liberate the embryos from the uterus by an incision of the uterus wall on the antimesometrial side. In general the embryo remains in the splanchnopleure. Transfer the embryo into cold PBS.

4. Transfer the embryos into 4% (v/v) paraformaldehyde in PBS at 4° for 20 min on a rocker. Rocking increases the quality of the fixation. Paraformaldehyde solution is prepared as follows. Dissolve 8 g paraformaldehyde in 100 ml water at 80° under constant stirring (in a chemical hood). While stirring add drops of 1 M NaOH to dissolve the paraformaldehyde and equilibrate the pH to 7.4. Store the 8% stock solution at 4°. Before use, make a fresh 4% solution in PBS (using a 10× stock solution of PBS).

5. Wash three times in PBS, 5 min each.

6. Transfer to X-Gal reaction mixture (see below) and incubate at 30° overnight (longer incubation times are possible).

7. Observe the embryo with a binocular microscope.

Notes. After E13, the fixation of the embryos is as follows:

4a. Transfer the embryos in 4% paraformaldehyde in PBS at 4° for 10 min on a rocker.

4b. Make a sagittal section of the embryo with a razor blade (maintain the embryo in hand during this operation).

4c. Transfer the embryos in the fixation medium for 15 min and go to Step 5.

Preparation of Fetuses and Newborn Animals

Fetuses at E15 and later are intracardially perfused in order to improve penetration of the fixative.

1. Mice are anesthetized by intraperitoneal injection of Avertine (0.1 ml per 10 g of body weight) and attached to a dissecting board.

2. The rib cage is opened and maintained with a clamp. An incision is made in the right ventricle of the animal, and an eroded 30-gauge needle adapted to a peristaltic pump is introduced.

3. The right auricle is cut.

4. Five milliliters of warm PBS is perfused at low velocity (5 min).

5. Thirty milliliters of 4% paraformaldehyde in PBS is perfused at low velocity (15 min for an adult mouse).

6. Five milliliters of PBS is perfused at low velocity.

7. If the perfusion is successful, the mice become stiff. The organs are then dissected, washed in PBS for 10 min, and placed either in X-Gal reaction mixture at 30° overnight or in 30% sucrose in PBS for cryostat sectioning.

Notes. Adapt the volume of the perfusion to the size of the animal. For certain tissues, it may be necessary to perform a permeabilization step after Step 4 (Preparation of Embryos until E14) or Step 7 (Preparation of Fetuses and Newborn Animals): incubate the anatomic piece in 0.01% (v/v) deoxycholate, 0.02% (v/v) nonidet P-40 (NP_{40}) in PBS for 10 min. Fixation of the embryos or anatomic pieces for more than 20–40 min decreases the X-Gal staining.

X-Galactoside Staining

1. Whole embryos or anatomic pieces are transfered from the fixative medium to PBS for rinsing.

2. They are then transferred into histochemical reaction mixture. The histochemical reaction mixture is 1 mg/ml 4-chloro-5-bromo-3-indolyl-β-galactoside (X-Gal), 4 mM $K_4Fe(CN)_6 \cdot 3H_2O$ (potassium ferrocyanide), 4 mM $K_3Fe(CN)_6$ (potassium ferricyanide), 2 mM $MgCl_2$ in PBS. Stock solutions are 40 mg/ml in dimethyl sulfoxide (DMSO) for X-Gal (storage at $-20°$) and 0.2 M in water for potassium ferricyanide and potassium ferrocyanide; both stocks must be protected from light and made fresh every 2 weeks.

3. Samples are incubated at 30° in an humidified incubator. The histochemical reaction proceeds slowly. The reaction can be prolonged for 1 or 2 days at 30°. If the embryos were correctly fixed, no background activity develops.

4. The embryos are then transferred to PBS for a few hours and then to 30% sucrose (see below) at 4°. Owing to the clearance effect of the tissue, the X-Gal staining is intensified at this step.

Notes. To avoid background activity, the pH (7.3) and temperature (30° or below) of the reaction mixture are important. At 37°, numerous tissues exhibit cytoplasmic background activity. Also note that enzyme activity decreases rapidly in acidic media. If the fixation of the anatomic piece is insufficient or has not been done properly, leakage of the reaction product stains the reaction mixture blue.

Tissue Sections

Material prepared for *in toto* analysis is adequate for sectioning. Four routine histology, tissues are sectioned in a cryostat. Cryostat sections can be restained with X-Gal (Fig. 3). This is almost always necessary for embryos older than E11, as direct staining of whole embryos or organs limit the staining of the cells to the periphery of the anatomic piece (in general, 100 to 200 μm depth). Paraffin sections cannot be successfully restained, as β-galactosidase is heat-sensitive. Therefore, sectioning of paraffin-embedded tissues is restricted to cases when X-Gal restaining is not necessary.

Cryostat Sectioning

1. Equilibrate the anatomic piece in 30% sucrose (30 g/100 ml) in PBS, pH 7.3, overnight at 4°. Sucrose is a cryoprotector.

2. Embed the anatomic piece in Tissue-tek (OCT medium, Miles Scientific) and immerse slowly in cold isopentane. Cold isopentane is obtained by placing it in a beaker on dry ice for 10 min.

3. Place the cold anatomic piece on the cryostat holder and let it equilibrate at −20° in the cryostat for 1 to 4 hr.

4. Prepare gelatinized glass slides for microscopy. Heat to 65° 5 g of gelatin (Merck) and 1 g of chromium potassium sulfate dodecahydrate (chrome alum) in 1 liter of distilled water under constant stirring until completely dissolved. Filter the hot solution through Whatman (Clifton, NJ) 3 MM paper. Immediately immerse clean microscopy slides for 1 min in the solution at 45°. Dry the slides at 37°, overnight. Store the glass slides at 4°.

5. Sections are cut 100 to 5 μm thick, depending on tissue type and purpose.

6. Eventually, postfix the sections in 4% paraformaldehyde for 5 min.

7. Wash twice in PBS.

8. Stain with X-Gal at 30° for 4 to 18 hr.

9. Wash twice in PBS and then once in distilled water.

10. Stain with the appropriate chemical.

11. Dehydrate rapidly and successively in ethanol: 70, 90, 99.9, and 99.9% (v/v).

12. Wash twice in pure toluene (Eukitt). This last operation must be performed in a chemical hood (toluene is hazardous).

Counterstaining. The following stains are compatible with identification of β-galactosidase-positive cells by X-Gal staining: red neutral (1% in 50 mM sodium acetate, pH 3.3), phloxine (5 g/liter in water), hematein

(2.6 g/liter in 100 g/liter KAlSO$_4$ in water), Safran (2 g/200 ml ethanol), cresyl violet, toluidine blue, and hematoxylin–eosin. Counterstaining of the tissue section is performed at Step 10 of the protocol.

Notes. After X-Gal staining, small embryos (until E11) or small anatomic pieces can be embedded in paraffin. Rinse the X-Gal-stained tissue in PBS. Then dehydrate by rapid transfer in graded alcohols and clear in xylene. Embed in paraffin wax. Counterstain the tissue section with various chemicals, but not with X-Gal.

Immunohistochemistry on X-Galactoside-Stained Cells

A great variety of antibodies are available to determine the presence of specific markers to identify cell types precisely. For immunoperoxidase detection the following method[50] can be used.

1. Fix the cells in 4% paraformaldehyde, 0.2% picric acid in PBS, pH 7.3.
2. Rinse twice with PBS.
3. Stain with X-Gal for a few hours at 30° (until blue staining is evident).
4. Rinse twice with PBS.
5. Block nonspecific antibody binding sites by incubation with 3% normal goat serum in 0.3% Triton-X 100 in PBS for 1 hr.
6. Incubate cells with the primary antibody (in general, rabbit antibodies) in the presence of 3% normal goat serum in 0.3% Triton-X 100 in PBS.
7. Visualize the labeling using the Vectastain ABC kit followed by reaction with 0.025% 3,3'-diaminobenzidine tetrahydrochloride and 0.013% H$_2$O$_2$ in 50 mM Tris.

For tissue sections, rinse the slides in PBS after X-Gal staining (Step 3) and incubate in 1% H$_2$O$_2$ to reduce peroxidase activity. Then go on to Step 5.

Notes. Excellent results are obtained with nuclear localized β-galactosidase (*nlsLacZ*) because the cytoplasm is free of the reaction product precipitate. Adapt the time of incubation in X-Gal reactive-medium to these conditions.

In Vivo Staining of β-Galactosidase-Positive Cells in Suspension

A fluorogenic assay for detection of β-galactosidase as a quantitative marker molecule in single cells has been developed.[8,11] It is based on the

---

[50] B. E. Wojick, F. Nothias, M. Lazar, H. Jouin, J.-F. Nicolas, and M. Peschanski, *Proc. Natl. Acad. Sci. U.S.A.* **90,** 1305 (1993).

use of fluorescein di-β-D-galactopyranoside (FDG),[51] which is cleaved by β-galactosidase to yield fluorescein. Under the conditions described below, fluorescein is retained within the cells and is not transferred to β-galactosidase-negative cells. The major factor affecting the assay is temperature, which must be below 16° during the enzymatic reaction. Remember that at 37° fluorescein passes the cell membrane 200 times faster that at 5°.[52] To load cells with FDG,[52] a very short hypotonic shock in a medium containing a high concentration of the substrate is applied. Cells that exhibit β-galactosidase activity are isolated by fluorescence-activated cell sorter (FACStar or FACStar Plus, Becton Dickinson) analysis.

Excellent discrimination between β-galactosidase-positive and β-galactosidase-negative cells is obtained[6,53,54] with the protocol described below, but, owing in part to the rapid exhaustion of the substrate during the loading at 37°, it does not give reliable quantitative data. To obtain quantitative data, loading of FDG is conducted at 14° for 5 min, and enzymatic reaction is also done at 14° (or below). The increase in fluorescein is linear for at least several hours, even for cell exhibiting high β-galactosidase levels.

1. Prepare a cell suspension at 10^7 cells/ml in RPMI 1640 medium (GIBCO BRL, Grand Island, NY) containing 2% fetal calf serum and 10 mM HEPES, pH 7.3.

2. Bring 200 μl of the cell suspension to 37°.

3. Add 200 μl of prewarmed 2 mM FDG in water. The stock solution of FDG (Molecular Probes, Eugene, OR) is 200 mM in 50% (v/v) DMSO in water.

4. Mix gently with a Pipetman for 2 min.

5. Cool cells immediately to 4° and add 3600 μl of cold RPMI medium containing 2% fetal calf serum and 10 mM HEPES, pH 7.3.

6. Let the enzymatic reaction proceed for 15 to 60 min.

7. Stop the reaction by adding 1 mM phenylethyl-β-thiogalactoside. The stock solution is 50 mM in water.

8. Use FACS analysis to measure cellular fluorescence and to sort positive and/or negative cells.[55,56]

[51] B. Rotman, J. A. Zderic, and M. Edelstein, *Proc. Natl. Acad. Sci. U.S.A.* **50**, 1 (1963).

[52] K. Wallenfels and O. P. Malhotra, *in* "The Enzymes," Vol. 4, p. 409. Academic Press, New York 1960.

[53] D. Rocancourt, C. Bonnerot, H. Jouin, M. Emerman, and J. F. Nicolas, *J. Virol.* **64**, 2660 (1990).

[54] D. Milan and J. F. Nicolas, *J. Virol.* **65**, 1938 (1991).

[55] D. R. Parks, L. L. Lanier, and L. A. Herzenberg, *Handb. Exp. Immunol.* **4**, 29.1 (1986).

[56] S. Alberti, D. R. Parks, and L. A. Herzenberg, *Cytometrics* **8**, 114 (1987).

Notes. Positive cells can be observed with a microscope equipped for fluorescence. Cells that have been successfully sorted using this method include multipotential stem cells [embryonal carcinoma (EC) and embryonal stem (ES) cells][6]; cell types of various origins such as NIH 3T3 and derivatives in particular retroviral packaging helper-free lines established in culture, such as ψ2, PA-12, ψCRIP; SPE/0 (a B hybridoma) and PW5147 cells (a T-cell line)[8]; HeLa cells (an epithelial human cell line)[53]; NS20Y cells (a neuronal line); G26 cells (a glial line); BRL cells (a hepatocyte line); and FLKBLV[54] cells (a fibroblastic bovine line).

Several cell types have also been purified from transgenic embryos and fetuses after a short *in vitro* culture. These cell types include a number of fibroblastic types, epithelium types such as cells of the metanephric tubules (in the kidney),[10] and glial cells from HPRTnlsLacZ-3 animals.[11] With the exception of neuronal cells, which are irreversibly damaged with currently available techniques of dissociation, the majority of other cell types can probably also be purified this way.

We found that a critical point for cell culture following the enrichment of a given cell population by FACS is the density of plating. Low plating density is deleterious. Also, it is important to change the culture medium of cells a few hours after plating to remove fluorescein and substrate.

Electron Microscopy

The 5-bromo-4-chloro-3-indolyl precipitate obtained in the reaction medium containing $K_4Fe(CN)_6$, $K_3Fe(CN)_6$, and $MgCl_2$ is electron-dense. Therefore, it can be detected by direct examination of β-galactosidase-positive cells processed for electron microscopy (EM). *nlsLacZ* yields precipitated product that is essentially found associated with the nuclear membranes and with the endoplasmic reticulum of the nuclear periphery.[3] The cytoplasm is free of patches. Examination by EM can give information on the ultrastructure of the cell together with information on *LacZ* expression.

1. For cells in culture, fixation is in 2% (v/v) glutaraldehyde in PBS for 5 min. For tissues of transgenic animals, intracardial perfusion of the mice with 4% paraformaldehyde and 0.2% (v/v) glutaraldehyde in PBS is necessary.

2. Rinse the material twice with PBS and stain with X-Gal or Bluo-Gal as described above. Bluo-Gal is 5-bromo-3-indolyl-β-D-galactosidase (BRL Life Technologies, Gaithersburg, MD).

3. After staining wash twice with PBS and dissect the piece of interest.

4. Incubate overnight in 2% paraformaldehyde and 2% glutaraldehyde in PBS.

5. Incubate for 1 hr in 1% OsO_4 in PBS to refix tissues.

6. Wash in PBS and then in distilled water.

7. Stain in 1% uranyl acetate in distilled water.

8. Dehydrate using ethanol: 10, 90, and 99.9% (v/v) ethanol.

9. Expose to propylene oxide and to propylene oxide plus accelerated Epon–Araldite.

10. Embed in accelerated Epon–Araldite.

11. Make sections.

Notes. A mild silver staining is compatible with the detection of the X-Gal or Bluo-Gal reaction products.[3,34] X-Gal and Bluo-Gal give different results in EM.[24] The Bluo-Gal reaction product forms a fine, very electron-dense precipitate that binds to internal membranes. The X-Gal reaction product forms a less dense and sometimes displaced precipitate.

Rapid Screening of Animals for *LacZ* and β-Galactosidase

Detection of transgenic founders among newborn animals from fertilized ova microinjected with DNA and identification of their *LacZ*-positive offspring are greatly facilitated by using unpurified DNA extracts from tails or ears for the polymerase chain reaction (PCR).

1. Cut a small piece of ear or tail from the animal and place it in 1.5 ml microtube.

2. Digest overnight at 55° in 400 μl of 50 mM KCl, 1.5 mM MgCl$_2$, 0.1 mg/ml gelatin, 0.45% (w/v) NP-40, 0.45% (w/v) polyoxyethylene sorbitan monolaurate (Tween 20), 10 mM Tris-HCl, pH 8.3, containing 1 μg/μl of proteinase K (lysis buffer).

3. Transfer 25 μl of the digest to a microtube with a pierced cap (made with a 30-gauge needle).

4. Heat at 95° exactly for 15 min (to denature proteins).

5. Cool on ice.

6. Add 2 μl of the heated digest to 25 μl of *Taq* polymerase reaction medium and 1 unit of *Taq* polymerase in a 750-μl microtube. The composition of 25× *Taq* reactional medium is 1.25 M KCl, 250 mM Tris-HCl, pH 8.3, 37.5 mM MgCl$_2$, 2.5 mg/ml gelatin, 5 mM each of the four deoxyribonucleoside triphosphates (dNTPs), and 6.25 mM each of the four oligonucleotide primers (see below). Store at −20°.

7. Add a drop (25 μl) of mineral oil.

8. Place the tubes in a DNA thermal cycler with the following parameters: 35 cycles of amplification with denaturation at 95° for 1 min, annealing and elongation at 72° for 4 min.

9. The products are electrophoresed in a 1.5% agarose gel.

Oligonucleotide primers used to detect LacZ[57] are

5′ GCATCGAGCTGGGTAATAAGGGTTGGCAAT 3′

and

5′ GACACCAGACCAACTGGTAATGGTAGCGAC 3′

which amplifies a 822-bp fragment. Oligonucleotide primers to detect an endogenous mouse gene, RAP SYN,[57] are

5′ AGGACTGGGTGGCTTCCAACTCCCAGACAC 3′

and

5′ AGCTTCTCATTGCTGCGCGCCAGGTTCAGG 3′

which amplifies a 590-bp fragment.

Another convenient way to distinguish transgenic from wild-type animals is by biopsies from an organ in which cells are β-galactosidase-positive. For instance, HPRTnlsLacZ-1 and HPRTnlsLacZ.3[11] express *LacZ* in mesenchymal cells detectable in biopsies of a digit at postnatal ages. HIV transgenic animals express *LacZ* in keratinazed cells of the skin[58] and can therefore be identified by finger biopsies. Transgenic lines for β-*LacZ* (a construct combining HPRTnlsLacZ and the locus control region of β-globin)[59] can be identified by the presence of hematopoietic β-galactosidase-positive cells. In addition to detecting *LacZ* animals, this approach gives information on variegated phenotypes.

Acknowledgments

We thank J. C. Benichou, S. Gasca, H. Jouin, C. Laugier, M. Peschanski, D. Rocancourt, and B. Wojcik for participation in elaboration of the methods, Iris Tong for comments, and C. Tran for typing the manuscript.

[57] T. Hanley and J. P. Merlie, *BioTechniques* **10**, 56 (1991).

[58] C. Cavard, A. Zider, M. Vernet, M. Bennoun, S. Saragosti, G. Grimber, and P. Briand, *Am. Soc. Clin. Invest.* **86**, 1369 (1990).

[59] C. Bonnerot and J.-F. Nicolas, *C. R. Acad. Sci.* **316**, in press (1993).

Section VI

Gene Expression: Proteins

[29] Two-Dimensional Gel Analysis of Protein Synthesis

By KEITH E. LATHAM, JAMES I. GARRELS, and DAVOR SOLTER

Introduction

Two-dimensional (2D) protein gel electrophoresis has been used extensively to investigate changes in gene expression during mouse embryogenesis.[1-11] The value of this technique lies in its ability to provide information about the synthesis of a large number of gene products with comparatively little starting material. Using high-resolution 2D gel electrophoresis, it is possible to detect over 2000 individual polypeptides in cultured fibroblast whole-cell lysates.[12] Quantitative gel image analysis combined with computer software for constructing protein databases has allowed more than 1200 individual polypeptides to be detected in mouse embryo lysates and their relative rates of synthesis to be quantified throughout preimplantation development.[10,11] Depending on the stage in question, between 3 and 10 preimplantation stage embryos and a single early postimplantation stage embryo can provide sufficient incorporated radiolabel for analysis. With the resolution that can be achieved with 2D gels, it is possible to address a variety of issues including transcriptional regulation, mRNA utilization and stability, posttranslational modification, and subcellular localization.[1-9]

The combination of high-resolution gel formats and quantitative gel image analysis has increased the usefulness of this approach. This technology has been used to construct a 2D gel protein database for the mouse

[1] O. Bensaude, C. Babinet, M. Morange, and F. Jacob, *Nature (London)* **305,** 331 (1983).

[2] P. Braude, H. Pelham, G. Flach, and R. Lobatto, *Nature (London)* **282,** 102 (1979).

[3] G. Flach, M. H. Johnson, P. R. Braude, R. A. S. Taylor, and V. N. Bolton, *EMBO J.* **1,** 681 (1982).

[4] C. C. Howe and D. Solter, *J. Embryol. Exp. Morphol.* **52,** 209 (1979).

[5] S. K. Howlett, *Cell (Cambridge, Mass.)* **45,** 387 (1986).

[6] S. K. Howlett and V. N. Bolton, *J. Embryol. Exp. Morphol.* **87,** 175 (1985).

[7] J. Levinson, P. Goodfellow, M. Vadeboncoeur, and H. McDevitt, *Proc. Natl. Acad. Sci. U.S.A.* **75,** 3332 (1978).

[8] J. Van Blerkom, *Proc. Natl. Acad. Sci. U.S.A.* **78,** 7629 (1981).

[9] J. Van Blerkom, *in* "Cellular and Molecular Aspects of Implantation" (S. R. Glasser and D. W. Bullock, eds.), p. 155. Plenum, New York, 1979.

[10] K. E. Latham, J. I. Garrels, C. Chang, and D. Solter, *Development (Cambridge, UK)* **112,** 921 (1991).

[11] K. E. Latham, J. I. Garrels, C. Chang, and D. Solter, *Appl. Theor. Elect.* **2,** 163 (1992).

[12] J. I. Garrels and B. R. Franza, *J. Biol. Chem.* **264,** 5283 (1989).

Copyright © 1993 by Academic Press, Inc.
All rights of reproduction in any form reserved.

embryo.[10,11] This chapter describes the techniques used to culture and label embryos for this analysis, and it offers practical suggestions for selecting an appropriate labeling and sampling strategy to detect changes in gene expression in genetically altered or experimentally manipulated mouse embryos.

Methods

Embryo Isolation and Culture

Embryos are isolated from superovulated mice at the 1-cell stage and cultured *in vitro*. Embryos can also be isolated at later stages to reduce the length of time spent in culture, although such embryos may differ subtly from *in vitro* cultured embryos. Adult female mice at least 6 weeks of age are superovulated with 5 IU of pregnant mare serum gonadotropin (PMSG) followed 46 hr later by 5 IU of human chorionic gonadotropin (hCG). Superovulated mice are placed in mating cages overnight and the embryos isolated approximately 20 hr post-hCG from the ampullae in HEPES-buffered Whitten's medium[13] supplemented with 100 μM EDTA,[14] treated for 2 min with hyaluronidase to remove cumulus oophorous cells, and washed 4 times in Whitten's medium. Fertilized embryos bearing two pronuclei are selected and incubated at 37° in bicarbonate-buffered Whitten's medium supplemented with 100 μM EDTA[14] under an atmosphere of 5% CO_2 and 5% O_2. Alternatively, embryos may be cultured in CZB medium[15] until compacted and then switched to Whitten's medium.

Embryo Labeling and Lysis

A summary of the procedure used for labeling and lysis of embryos is given (Table I). For labeling with L-[35S]methionine, embryos are washed through Whitten's or CZB medium containing 1 mCi/ml high specific activity (>1100 Ci/mmol) L-[35S]methionine and then labeled in 40-μl droplets of the same medium under oil. Incorporation of L-[35S]methionine is similar with the two media under these labeling conditions [e.g., 23,100 and 21,300 disintegrations/min (dpm)/embryo incorporated after 3 hr labeling in Whitten's and CZB medium, respectively, for the 2-cell stage]. For labeling of phosphoproteins, 2 mCi/ml of ortho[32P]phosphate (carrier-

[13] W. K. Whitten, *Acta Biosci.* **6,** 129 (1971).
[14] J. Abramczuk, D. Solter, and H. Koprowski, *Dev. Biol.* **61,** 378 (1977).
[15] C. L. Chatot, C. A. Ziomek, B. D. Bavister, J. L. Lewis, and I. Torres, *J. Reprod. Fertil.* **86,** 679 (1989).

TABLE I
EMBRYO LABELING AND LYSIS PROCEDURE FOR ANALYSIS OF L-[35S]METHIONINE- AND
ORTHO[32P]PHOSPHATE-LABELED PROTEINS FROM PREIMPLANTATION STAGES

1. Isolate embryos at 1-cell stage and culture *in vitro*
2. Synchronize embryos by controlling time of fertilization or by using pick-off method
3. Wash embryos in medium containing 1 mCi/ml L-[35S]methionine (>1100 Ci/mmol) or 2 mCi/ml ortho[32P]phosphate (carrier-free) and label for 2–3 hr in same medium under oil
4. Wash embryos once in PBS containing 0.4% (w/v) PVP
5. Transfer embryos in a minimal volume to preheated (100°) SDS lysis buffer in a siliconized microcentrifuge tube
6. Heat for 30 sec in boiling water bath and cool on ice for 1 min
7. Digest with 1/10 volume of DNase/RNase on ice for 1 min
8. Freeze in liquid nitrogen and store at −70°
9. Lyophilize, resuspend in an equal volume of 2D gel sample buffer, and quantitate incorporated radiolabel by TCA precipitation

free) in phosphate-free Whitten's medium is used. This labeling procedure yields significantly more incorporated radiolabel per embryo than was observed in some studies and permits the use of fewer embryos for analysis (Table II). Comparisons of several independent studies of protein synthesis in the mouse embryo (Table II) indicate that the use of concentrations of L-[35S]methionine greater than 1 mCi/ml or labeling times longer than 3 hr may be detrimental, as evidenced by reduced L-[35S]methionine incorporation.

After labeling, embryos are washed in phosphate-buffered saline (PBS), pH 7.4, containing 0.4% (w/v) polyvinylpyrrolidone (PVP) to remove excess label and bovine serum albumin (BSA) and then lysed. The embryos must be lysed in such a way as to maximize solubilization and dissociation of proteins while preventing proteolysis. Lysis is performed in hot (100°) sodium dodecyl sulfate (SDS) buffer containing 0.3% SDS, 1% (v/v) 2-mercaptoethanol, and 50 mM Tris-HCl (pH 8.0).[12] This buffer is prepared in advance and stored in aliquots at −70°. Lysis in SDS efficiently solubilizes most proteins and dissociates most protein complexes, thereby improving separation during electrophoresis. For preimplantation stage embryos, a small volume of lysis buffer (13–30 μl) is heated in a siliconized Eppendorf tube for 30 sec in a boiling water bath. The lysis volume is adjusted according to the number of embryos available and the number of gels (10 μl for each gel) to be run, allowing for an additional 2–3 μl for TCA (trichloroacetic acid) precipitation. The embryos are transferred to the hot buffer in a minimal volume and the lysis buffer heated for another 30 sec in the boiling water bath. If embryos are not lysed in hot SDS buffer, significant proteolysis can occur. After cooling

TABLE II
COMPARISON OF TWO-DIMENSIONAL GEL ANALYSES OF MOUSE EMBRYOS

	Latham et al. (1991)[10]	Howe and Solter (1979)[4]	Levinson et al. (1978)[7]	Van Blerkom (1979)[9]
Disintegrations/min (dpm)/embryo ($\times 10^4$)				
1-Cell	2.99[a]	0.09[b]	0.13	—
2-Cell	2.31[a]	0.084[b]	0.18	—
4-Cell	2.74[a]	0.476[b]	—	—
8-Cell	3.88[a]	—	0.18	—
Morula	8.0[a]	—	—	1.28[c]
Blastocyst	15.7[a]	0.57[b]	0.47	—
Number of embryos/gel	4–22 (12)[d]	170–420[e]	10–119	Single blastomere
Exposure time	25–81 days (40)[d]	1–4 months	40–44 days	8 months
Number of spots resolved	Up to 855[f]	—[g]	Up to 600	—[h]
Number of regulated spots (2-cell stage)	502	105	34	n.a.[i]
Labeling				
Specific activity	>1100 Ci/mmol	>400 Ci/mmol	600–1000 Ci/mmol	—[h]
Concentration	1 mCi/ml	0.5 mCi/ml	3–5 mCi/ml	—[h]
Time	3 hr	5 hr	1.5 hr	—[h]

[a] Average dpm/embryo over entire stage.
[b] Calculated by summing cytoplasmic and nuclear fractions.
[c] Calculated from 800 dpm per 16-cell stage blastomere.
[d] Numbers in parentheses show average values.
[e] Divided into cytoplasmic and nuclear fractions.
[f] High-quality spots only.[15]
[g] Samples were divided into cytoplasmic and nuclear fractions.
[h] Not available.
[i] n.a., Not applicable.

on ice for 1 min, 1/10 volume of a solution containing protease-free DNase I and RNase A (1.0 and 0.5 mg/ml, respectively; Worthington, Freehold, NJ), 1.5 M Tris-HCl (pH 7.0), and 1.0 M MgCl$_2$ is added and the reaction incubated on ice for 1 min. The DNase/RNase solution is also prepared in advance and stored at $-70°$. After digestion, the sample is frozen in liquid nitrogen and stored at $-70°$. The entire lysis procedure should require less than 3 min to complete. The samples should be kept on ice after boiling to prevent proteolysis. Prior to electrophoresis, the sample is lyophilized, redissolved in an equal volume of 2D gel sample buffer, and an aliquot removed for quantitation of incorporated radiolabel by precipitation with TCA as described.[16]

Two-Dimensional Gel Electrophoresis

Selecting Gel Format

Reproducibility and resolution are the primary concerns for selecting a gel format. No single gel format is capable of resolving every protein. The mouse embryo database used a pH 4–8 ampholine (British Drug House, Poole, UK) for isoelectric focusing and 10% acrylamide for the second dimension.[10,11] This format gives reproducible spot resolution over the range of pI and M_r of approximately 4.4–6.7 and 20,000–127,000, respectively. A broader range ampholine (pH 3.5–10, LKB) can extend slightly the range of pI values resolved.[12,16] More acidic proteins (e.g., SPARC) have been resolved using the pH 3.5–10 ampholines and 0.1 M phosphoric acid as the anode solution.[17] Additionally, nonequilibrium gels can be run as well as the standard equilibrium gels to visualize more basic and low-M_r proteins,[18] and altering the acrylamide concentration in the second dimension can reveal proteins of greater or lesser M_r.

Amount of Material Required for Analysis

Gel electrophoresis and fluorography are performed as described.[19,20] In general, analyses should be performed on duplicate samples in order to exclude the possibility that any observed differences are due to gel artifacts. To obtain sufficiently well-exposed gel images within a reasonable period of time, between 2×10^5 and 1×10^6 dpm of L-[^{35}S]methionine-labeled protein should be applied to each gel in approximately 10 μl. For L-[^{35}S]methionine-labeled samples, the best results are obtained with at least 4×10^5 dpm/gel which allows exposure times of approximately 4 weeks for the longest exposure (typically, 3–4 exposures of different lengths are obtained for each gel). For the 1-, 2-, and 4-cell stages, the mean incorporated radioactivity per embryo is relatively constant (29,900, 23,100, and 27,400 dpm, respectively),[10] and between 16 and 20 embryos provide sufficient material for a single gel. This value increases to approxi-

[16] J. I. Garrels, *J. Biol. Chem.* **264**, 5269 (1989).

[17] K. E. Latham and C. C. Howe, *Roux's Arch. Dev. Biol.* **199**, 364 (1990).

[18] R. Bravo, J. V. Small, S. J. Fey, P. M. Larsen, and J. E. Celis, *J. Mol. Biol.* **154**, 121 (1982).

[19] J. I. Garrels, this series, Vol. 100, p. 411.

[20] W. M. Bonner and R. A. Laskey, *Eur. J. Biochem.* **46**, 83 (1974).

mately 8×10^4 and $1-1.4 \times 10^5$ dpm/embryo for the morula and blastocyst stages, respectively. One study produced 2D gel data for a single, isolated 16-cell stage blastomere, although this required a prolonged exposure of 8 months.[9] Early postimplantation stage embryos incorporate enough activity so that single embryonic or extraembryonic regions or 3–4 isolated germ layers provide enough material for analysis.

Using pH 4–8 ampholines (British Drug House) in the first dimension and 10% acrylamide in the second dimension, an average of approximately 700 high-quality spots (designated as high quality based on their shapes, intensities, and degree of overlap with neighboring spots[16]) are reproducibly detected for samples of preimplantation stage embryos. A total of more than 1200 proteins have been detected as high-quality spots and monitored from fertilization through the blastocyst stage using this gel format.[11] Spots accounting for as little as 20 parts per million (ppm) of incorporated radiolabel can be detected and reliably quantified. For ortho-[^{32}P]phosphate-labeled samples, sufficiently dark exposures can be obtained within 1–3 weeks with approximately 300 embryos. With good reproducibility in the gel system, it is possible to align gels of embryos labeled with these different isotopes in order to study posttranslational modification.

Quantification of Gel Images

In many cases, simple visual comparison of two gel images may be sufficient to meet experimental needs, particularly where one or a few particular proteins of known location are to be examined. In other cases, however, gel images must be compared in their entirety in order to define both qualitative and quantitative differences between cell or tissue types, between embryos of different stages, or between normal and genetically altered or experimentally manipulated embryos. The details of gel image analysis using the QUEST system have been described.[16] One method of quantitation utilizes gel calibration strips that contain known amounts of radiolabeled protein. These are processed for fluorography and exposed in parallel with each gel. Using these calibration strips, film densities are converted to disintegrations per minute per unit of area. Each detected spot is thus quantified in terms of dpm and, dividing by the total TCA-precipitable radioactivity applied to the gel, as a fraction of total incorporated activity (ppm).[16]

One point that is worth considering for analysis of embryonic material, however, is the method used to detect radiolabeled proteins. Exposure of X-ray films by fluorography typically requires between a few weeks and several months. Recently, photostimulable phosphor imaging plates (e.g., Fuji Photo Film Co., Tokyo, Japan) have been developed that can significantly shorten the requisite exposure times.[21] Application of this technology to the analysis of embryonic material may significantly reduce the exposure times required and possibly the amount of labeled material that is required as well.

The resolution, format, and sensitivity obtained for embryonic samples was similar between several independent studies. Until recently, however, the technology to perform in-depth, quantitative analyses on gel images was not available, and only a relatively small number of developmentally regulated proteins were described (Table II). Quantitative gel image analysis combined with computer software for direct matching of multiple gel images greatly increases the efficiency and speed of gel image comparisons and has allowed many more developmentally regulated proteins to be revealed than was possible previously (Table II).[10]

Experimental Design

Reproducibility and Selection of Time Points for Analysis

When devising a strategy for using 2D gels to analyze embryonic gene expression, care must be taken to select the appropriate time point(s) and controls. Protein synthesis patterns of mouse embryos can change a great deal. An extensive reprogramming occurs during the 2-cell stage, for example[10] (Figs. 1 and 2), and many proteins can undergo substantial quantitative changes in their rates of synthesis within as little as 3 hr.[10] The amount of change that can occur within 3 hr of the 2-cell stage, for example, can exceed the difference observed between proliferating and quiescent fibroblasts.[10,22] Furthermore, proteins that are synthesized during the 2-cell stage include a set of proteins that are induced transiently during the mid 2-cell stage (Fig. 3). This set includes some proteins that are apparently synthesized almost exclusively during the 2-cell stage, appear and disappear within as little as 9–12 hr, and can change by 2-fold or more in synthesis within as little as 3 hr. Such transient expression illustrates a potential difficulty that can be encountered when analyzing embryonic protein synthesis patterns, namely, the selection of the proper time points for analysis. Meaningful interpretation of alterations that result from genetic differences between embryos or experimental manipulations may require that labeling be performed within a narrow period of time in order to visualize changes in the expression of such transiently expressed,

[21] R. F. Johnston, S. C. Pickett, and D. L. Barker, *Electrophoresis* **11**, 355 (1990).
[22] J. I. Garrels and B. R. Franza, *J. Biol. Chem.* **264**, 5299 (1989).

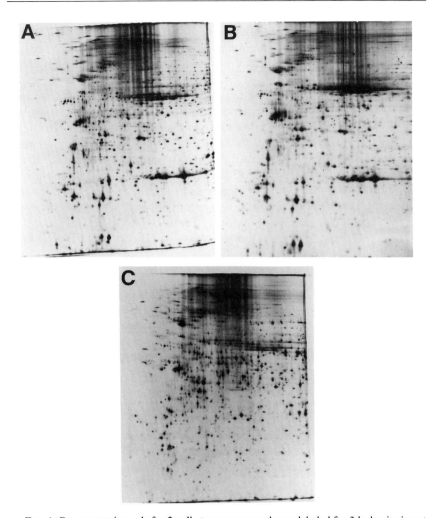

Fig. 1. Representative gels for 2-cell stage mouse embryos labeled for 3 hr beginning at (A) 3 hr, (B) 6 hr, and (C) 21 hr postcleavage. Note that even within as little as 3 hr (A versus B) the relative rates of synthesis of some proteins can change significantly and that the overall pattern of proteins synthesized changes a great deal over the course of the 2-cell stage.

stage-specific proteins. Additionally, for stages where rapid changes in embryonic protein synthesis patterns occur, comparisons should be performed between samples of embryos that, to the greatest extent possible, are contemporaneous.

For certain periods of development, direct comparisons of embryos labeled in different experiments can be complicated by variability in the relative timing of protein synthetic changes. For example, comparisons between duplicate samples of mouse embryos, prepared in two independent experiments, at the mid 2-cell stage (9 and 12 hr postcleavage) when the rate of change in protein synthesis is greatest, revealed 2-fold or greater differences in 10–13% of the proteins analyzed, even though the patterns of proteins were essentially identical between duplicate samples (K. E. Latham, J. I. Garrels, and D. Solter, unpublished observations, 1992). Duplicate samples of cultured cell lines or mouse embryos at other stages (e.g., 4-cell stage) typically show approximately 2% 2-fold or greater differences.[22] Thus, for some stages of development, comparisons might produce potentially misleading results in the form of apparent quantitative differences in protein synthesis that are actually due to slight age differences between embryos. To minimize the chance of this occurring, comparisons should be made between samples of embryos that have been isolated at the same time and cultured identically.

Synchronization of Embryos

In addition to the rapidity of changes in protein synthesis, asynchrony among mouse embryos can also complicate analyses. Asynchrony arises from variation in the time of ovulation, insemination, and fertilization.[23] As a consequence, the first cleavage division occurs in a population of eggs over a period of 9 hr or more. This asynchrony can potentially obscure the transient appearance of minor spots (like spots DF190 and EP60, Fig. 3) as well as the relative timing between different events. Thus, the use of synchronously developing embryos whenever possible is preferable. Synchronous cohorts can be collected by the pick-off method,[23] which consists of selecting at each cleavage division those embryos that cleave within a given time period (e.g., 1 hr). Typically, a group of approximately 500 embryos will produce 10–12 synchronous 1-hr cohorts of 25–40 embryos within a period of 4–5 hr. For experiments where synchronous 1-cell stage embryos are to be analyzed, either *in vitro* fertilization or delayed matings can be used. With the latter method, superovulated females are placed with male mice for 1 to 2 hr beginning at 13–14 hr after injection of hCG. A few mice can be sacrificed as early as 3 hr postmating to yield a few embryos bearing second polar bodies for immediate labeling. The remaining mice are sacrificed beginning at 6 hr postmating, and fertilized embryos are identified by the presence of two pronuclei. In this way,

[23] V. N. Bolton, P. J. Oades, and M. H. Johnson, *J. Embryol. Exp. Morphol.* **79**, 139 (1984).

2 Cells   4 Cells

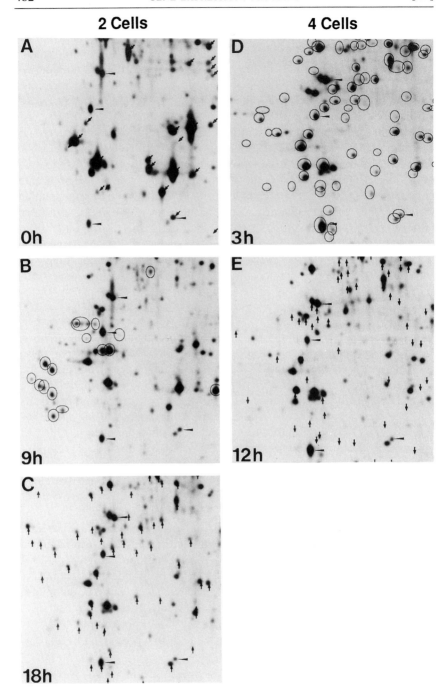

embryos can be synchronized at fertilization to within 2 hr of each other. As many as 75% of the embryos obtained from delayed matings will cleave within a 2-hr period.

Frequency of Sampling

A third consideration in designing a labeling strategy is the need to distinguish between discrete, qualitative alterations in the array of proteins synthesized and more subtle alterations in the temporal pattern of synthesis of a common set of proteins. Genetic alterations or experimental manipulations might, as their primary defect, lead to loss of expression or inappropriate expression of a particular gene. Alternatively, genetic or experimental manipulations of the embryo might affect the regulatory mechanisms that govern time- or stage-specific inductions and repressions, thereby altering the temporal pattern of gene expression. Mechanistically, these two types of effects are, of course, very different.

The possibility that an alteration in the temporal pattern of gene expression is involved can be at least partially addressed by examining the normal pattern of synthesis for a given protein. Androgenetic embryos (two paternal genomes) constructed with DBA/2 eggs by nuclear transplantation, for example, exhibit 11 proteins with quantitatively different rates of synthesis in comparison to unoperated or gynogenetic (two maternal genomes) embryos.[24] The normal patterns of synthesis of all 11 proteins reveal induction or repression between the 8-cell and blastocyst stages. Thus, a partial interruption or delay in the normal sequence of regulatory events occurred in the androgenones, possibly as a secondary consequence of some other effect of genomic imprinting. To discriminate between specific primary defects that alter the array of genes expressed and more subtle alterations in the time of expression, genetically altered or experimentally manipulated embryos should be sampled at as many time points as feasible

[24] K. E. Latham and D. Solter, *Development (Cambridge, UK)* **113,** 561 (1991).

FIG. 2. Enlarged regions of gels from the 2-cell and 4-cell stage. Each panel shows the same region of gels obtained for L-[^{35}S]methionine-labeled 2-cell (A–C) and 4-cell (D, E) stage embryos labeled at the indicated times after the first and second cleavage divisions, respectively. For the 2-cell stage, proteins that decline (downward pointing arrows), appear transiently (ovals), or increase (upward pointing arrows) during the 2-cell stage are indicated. For the 4-cell stage, proteins that are synthesized at a constant rate (ovals) or increase or decrease (arrows) during the 4-cell stage are indicated. Horizontal chevrons mark four proteins that are present in all five gels to facilitate alignment. Note that most of the major proteins of the 2-cell stage change significantly in rates of synthesis, whereas most of the major 4-cell stage proteins are synthesized at relatively constant rates.

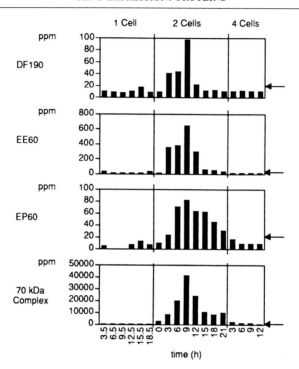

Fig. 3. Transiently induced proteins of the 2-cell stage mouse embryo. Graphs show the rates of synthesis of four representative proteins that are synthesized at elevated rates during the mid 2-cell stage. Individual bars show the rate of synthesis expressed as the fraction (parts per million) of incorporated radiolabel at sequential 3-hr time intervals beginning at the times indicated. Standard spot numbers, assigned to each protein using the protein database,[10] are given to the left of each graph. Arrows positioned at 20 ppm denote the minimum intensity (20 ppm) that could be reliably quantitated. Note that these proteins are essentially stage-specific.

and compared to a carefully staged series of normal embryos. Labeling at multiple time points will also facilitate the observation of transient changes in protein synthesis as well as comparisons between independent experimental series.

Analysis of Postimplantation Stages

Two-dimensional gel electrophoresis has also been applied to the analysis of postimplantation stages of mouse embryogenesis.[25] Analysis of more advanced stages by 2D gel electrophoresis becomes more difficult as multi-

[25] K.-F. Murach, M. Frei, D. Gerhauser, and K. Illmensee, *J. Cell. Biochem.* **44,** 19 (1990).

ple cell and tissue types become established. Characterization of protein synthetic patterns must account for both stage and cell or tissue specificity. The analysis of protein synthesis patterns of isolated tissues or embryonic regions offers a useful means of identifying tissue- or region-specific differences in protein synthesis as well as time-dependent changes in gene expression. Additionally, proteins that are selectively synthesized at enhanced rates in specific tissues or regions should be more easily visualized in lysates of the isolated tissue fragments than in lysates of whole embryos. Indeed, several proteins are more easily detected in lysates of isolated germ layers from 6.5 and 7.5 day embryos than in lysates of whole embryonic or extraembryonic regions (K. E. Latham, J. I. Garrels, and D. Solter, unpublished observations, 1993). Detection and accurate quantitation of such proteins will be facilitated by isolation or enrichment of that cell or tissue type. Thus, 2D gel analysis of isolated tissues or spatially distinct regions (e.g., anterior versus posterior) offers a sensitive means of detecting developmentally significant differences in gene expression.

Protein Identification

Two potentially very useful applications of 2D gel electrophoresis in embryology are to identify developmentally regulated genes and to monitor the expression of a particular known protein in different tissues or at different stages. The former requires the identification and/or characterization of a specific protein spot that exhibits a tissue- or stage-specific pattern of expression. The latter application requires the localization of a particular known protein of interest within the 2D gel pattern so that the synthesis of that protein can be followed. Limitations in the amount of embryonic material available may preclude the application of methods such as direct protein microsequencing and immunoblotting, particularly for proteins that are synthesized transiently and do not accumulate to high abundances. An indirect method of spot identification, based on the construction of the mouse database, is to align a 2D gel image containing spots to be identified with a 2D image containing previously identified spots. Twelve proteins have been identified in mouse embryo gels by alignment with 2D gels of rat fibroblast in which these proteins had been previously located.[22] This approach offers the advantage that tentative identifications can be assigned to embryonic proteins based on identifications made using more readily available material. For all but the most easily located, well-characterized proteins (e.g., actin), identifications made in this manner must be confirmed directly. Immunoprecipitation provides a convenient means for this purpose, and for locating spots that correspond to particular known proteins in gels of embryonic lysates. For example,

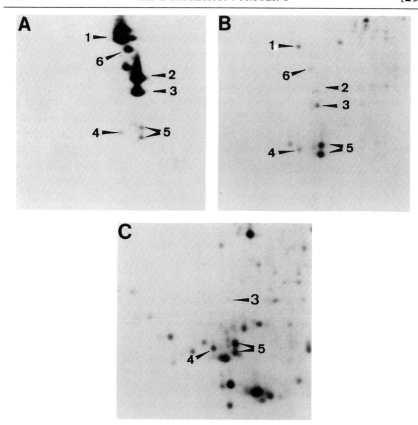

FIG. 4. Identification of tropomyosin isoforms by immunoprecipitation. Blastocysts were labeled and subjected to immunoprecipitation as described in the text using a rabbit polyclonal antiserum to tropomyosin.[22] (A) Immunoprecipitate from labeled mouse 3T3 cells, (B) immunoprecipitate of labeled mouse blastocysts, and (C) control blastocyst whole-cell lysate. The positions of tropomyosins 1–6 are indicated.

a rabbit polyclonal antitropomyosin serum[22] was used to verify the identification of tropomyosin isoforms in blastocyst stage lysates prepared from approximately 450 embryos (Fig. 4).

An effective immunoprecipitation protocol for mouse embryos is as follows. Embryos are labeled as described above and lysed in 100 μl of immunoprecipitation buffer [20 mM Tris-HCl, pH 8.0, 1% (v/v) Triton X-100, 0.5% sodium deoxycholate, 0.15 M NaCl, 5 mM $MgCl_2$, and 1 mM freshly added phenylmethylsulfonyl fluoride (PMSF)]. To this lysate are added 5 μl of 10% SDS and 10 μl of RNase/DNase (see above). After a

1-min incubation on ice, 10 μl are frozen immediately as a whole-cell lysate. An additional 100 μl lysis buffer is added and the lysate centrifuged (13,000 g) for 5 min at 4°. The lysate is precleared by incubation with 10 μl of normal serum and 50 μl protein A-conjugated Sepharose (25% suspension in lysis buffer) for 30 min on ice with agitation every 5 min. The supernatant is incubated with 1–5 μl of primary antiserum on ice for 1 hr. The final incubation is for 30 min on ice after addition of 50 μl protein A-conjugated Sepharose with agitation every 5 min. The Sepharose beads are centrifuged for 30 sec at 13,000 g. If an antibody for which protein A has low affinity is to be used, an additional incubation with the appropriate secondary antibody can be included. An aliquot of the supernatant is frozen in liquid nitrogen and stored at − 70°. The precleared (from above) and antigen-containing pellets are washed 6 times with 0.5 ml lysis buffer. During the final wash, the material is transferred to a new tube. After washing, the pellets are boiled for 1 min in 30 μl of SDS buffer (above) and centrifuged. The supernatants are frozen in liquid nitrogen and stored at − 70°.

The whole-cell lysate, precleared and antigen-containing pellet fractions, and supernatants are lyophilized, resuspended, and processed for 2D gel electrophoresis. Gels are obtained for the whole-cell lysate, immunoprecipitate alone, mixed whole-cell lysate and immunoprecipitate, precleared fraction, and supernatant fraction. Immunoprecipitated proteins are visualized on the gel receiving immunoprecipitate alone (Fig. 4) and on the gel receiving the mixture of whole-cell lysate and immunoprecipitate (not shown). The latter allows location and identification of antigen(s) within a gel of a whole-cell lysate as a spot(s) that is significantly enhanced in intensity relative to the whole-cell lysate alone. Proteins that coprecipitate with the antigen will also be enhanced. The supernatant gel should exhibit a specific reduction in the relative intensity of the same spot(s). The precleared sample provides a control for specificity.

This procedure should be generally applicable to identification of specific proteins in 2D gels of preimplantation stage mouse embryos. It is also possible to use purified proteins obtained from alternate sources to identify spots in 2D gel patterns.[17] Purified proteins are coelectrophoresed with labeled whole-cell lysates and later visualized either by staining (e.g., with silver or Coomassie blue) or by autoradiography (e.g., iodinated proteins). The position of the purified protein within the pattern of spots obtained from the whole-cell lysate is thus determined by comigration. If, however, different posttranslational modifications or alternatively spliced isoforms are expressed between the embryo and the source from which the purified protein was prepared, then confirmation by immunoprecipitation of labeled embryonic proteins is necessary. Immunoprecipitation can also reveal additional isoforms of some proteins that are not easily detected in gels of whole-cell lysates (Fig. 4C).

Protein Databases

Numerous 2D gel studies have been described for the mouse embryo.[1-9] Variability in gel conditions, resolution, and embryo staging and insufficient quantitation can limit the amount of information that can be gained from 2D gels. These problems can largely be overcome by adopting a protein database approach.[10] This approach combines standardized gel electrophoresis with a system for quantifying and matching gel images.[12,22,26-30] The advantage of this approach is that each detected protein spot can be quantified and followed through a series of gel images representing samples collected in a number of independent experiments. Thus, the data are cumulative, and most of the data contained within the gel images can be extracted. The database approach also allows for spot annotation to record such information as protein names, known modifications, subcellular localization, and regulatory patterns.[16] The details of constructing a protein database are described elsewhere[12,16,22] and will not be duplicated here.

Conclusion

The ability of high-resolution 2D gel electrophoresis to resolve so many gene products for analysis makes this technique an especially valuable one for the embryologist. Continued improvement in spot resolution, detection, and quantification should increase the value of this technique further, particularly when combined with computer software for accumulating and managing diverse types of data in the form of a 2D gel database. Additionally, it should be possible to integrate 2D gel data with data obtained through other approaches such as molecular cloning. Immuno-

[26] J. E. Celis, K. Dejgaard, P. Madsen, H. Leffers, B. Gesser, B. Honore, H. H. Rasmussen, E. Olsen, J. B. Lauridsen, G. Ratz, S. Mouritzen, B. Basse, M. Hellerup, A. Celis, M. Puype, J. Van Damme, and J. Vandekerckhove, *Electrophoresis* **11**, 1072 (1990).

[27] J. E. Celis, B. Gesser, H. H. Rasmussen, P. Madsen, H. Leffers, K. Dejgaard, B. Honore, E. Olsen, G. Ratz, J. B. Lauridsen, B. Basse, S. Mouritzen, M. Hellerup, A. Andersen, E. Walbum, A. Celis, G. Bauw, M. Puype, J. Van Damme, and J. Vandekerckhove, *Electrophoresis* **11**, 989 (1990).

[28] R. A. VanBogelen, M. E. Hutton, and F. C. Niedhardt, *Electrophoresis* **11**, 1131 (1990).

[29] R. D. Appel, D. F. Hochstrasser, M. Funk, J. R. Vargas, C. Pelligrini, A. F. Muller, and J.-R. Scherrer, *Electrophoresis* **12**, 722 (1991).

[30] I. Ali, Y. Chan, R. Kuick, D. Teichrow, and S. M. Hanash, *Electrophoresis* **12**, 747 (1991).

precipitation offers a good means of identifying spots on 2D gels that correspond to the products of specific cloned genes. Antisera can be generated from cloned DNA sequences using synthetic oligopeptides or polypeptides generated with expression vectors as immunogens. The combination of molecular cloning and 2D gel protein database approaches offers an excellent means of identifying developmentally regulated genes, characterizing in detail their patterns of expression, and examining the mechanisms that regulate these patterns of expression.

Acknowledgments

This work was supported in part by U.S. Public Health Grants HD-17720, HD-23291, and HD-21355 from the National Institute of Child Health and Human Development and CA-10815 from the National Cancer Institute. The QUEST Center at Cold Spring Harbor Laboratory was supported by a grant (P41RR02188) from the National Institutes of Health Biomedical Research Technology Program. K.E.L. was supported in part by a training grant from the NCI (CA 09171-14).

[30] One-Dimensional Gel Analysis of Histone Synthesis

By MARIA WIEKOWSKI and MELVIN L. DEPAMPHILIS

Introduction

All eukaryotic DNA is organized into a nucleoprotein complex referred to as chromatin. The basic unit of chromatin is the nucleosome, consisting of approximately 200 bp of DNA wrapped around a histone octamer. The nucleosome consists of 146 bp of "core" DNA tightly associated with two copies each of histones H2A, H2B, H3, and H4 and about 60 bp of loosely associated "linker" DNA. Histone H1 associates with linker DNA and thereby pulls nucleosomes together into a regular repeating array called the 30-nm fiber.

Histone composition can change during development. For example, in *Drosophila*, the core histones are always present, but histone H1 does not appear until the blastula stage.[1] In other animals, such as the sea urchin, sea worm, mud snail, and annuran amphibians (e.g., *Xenopus*), embryo-specific H1 variants are present.[2–6] In *Xenopus*, histone H1 first

[1] S. C. R. Elgin and L. E. Hodd, *Biochemistry* **12**, 4985 (1973).
[2] R. R. Franks and F. C. Davis, *Dev. Biol.* **98**, 101 (1983).
[3] A. M. Flenniken and K. M. Newrock, *Dev. Biol.* **124**, 457 (1987).
[4] D. Poccia, J. Salik, and G. Krystal, *Dev. Biol.* **82**, 287 (1981).

Copyright © 1993 by Academic Press, Inc.
All rights of reproduction in any form reserved.

TABLE I
COMMON HISTONE MODIFICATIONS

Histone	Acetylation[a] (on lysine position)	Species	Phosphorylation (number of serines)	Species
H4	16, 8, 12, 5	Mammals	1	Duck[b]
H3	14, 23, 18, 9	Mammals	?	—
H2A	5, 9	Mammals	1	Human[c]
H2B	12, 15, 20, 5	Mammals	4	Mammals[c]
H1	None	Vertebrates	13 (5 subtypes)	Mouse[d]
H1°	—	—	1	Mouse[d]

[a] Data are from B. M. Turner, *J. Cell Sci.* **99**, 13 (1991), and amino acid positions are listed in order of acetylation.

[b] A. Ruiz-Carrillo, L. J. Wangh, and V. G. Allfrey, *Science* **190**, 117 (1975).

[c] C. von Holt, W. F. Brandt, H. J. Greyling, G. G. Lindsey, J. D. Retief, J. de A. Rodrigues, S. Schwager, and B. T. Sewell, this series, Vol. 170, p. 431.

[d] R. W. Lennox and L. H. Cohen, *J. Biol. Chem.* **258**, 262 (1983).

appears with the onset of zygotic gene transcription.[5–7] Embryo-specific variants of the core histones H2A and H2B also were identified in early cleavage embryos of sea urchins[8] and anuran amphibians.[6]

All histones are subject to a variety of posttranscriptional modifications such as acetylation, phosphorylation, and methylation that also change during embryonic development and gametogenesis. Because interactions of histones with DNA are mediated through electrostatic interactions, acetylation of specific lysine residues or phosphorylation of serine residues (Table I) must affect the conformation and stability of chromatin. For example, histone modification during spermatogenesis and following fertilization appears to facilitate condensation and decondensation of paternal chromatin. During spermatogenesis in rainbow trout, replacement of histones by protamines coincides with the appearance of hyperacetylated histone H4.[9] Sperm-specific variants of histone H1 and H2B are dephosphorylated in sea urchins and then rephosphorylated following fertilization, presumably to allow replacement by embryo-specific histones.[10] Similarly, the acetylation state of histones H4 and H3 changes during oocyte

[5] R. C. Smith, E. Dworkin-Rastl, and M. B. Dworkin, *Genes Dev.* **2**, 1284 (1988).

[6] K. Ohsumi and C. Katagiri, *Dev. Biol.* **147**, 110 (1992).

[7] J. M. Flynn and H. R. Woodland, *Dev. Biol.* **75**, 222 (1980).

[8] D. Poccia, T. Greenough, G. R. Green, E. Nash, J. Erickson, and M. Gibbs, *Dev. Biol.* **104**, 274 (1984).

[9] M. E. Christensen and G. H. Dixon, *Dev. Biol.* **93**, 404 (1982).

[10] C. S. Hill, L. C. Packman, and J. O. Thomas, *EMBO J.* **9**, 805 (1990).

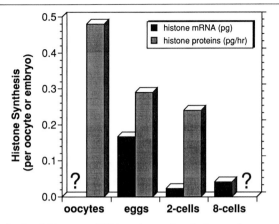

FIG. 1. Synthesis of histone mRNA and proteins during early mouse development. [Calculated from data in R. M. Schultz, *in* "Experimental Approaches to Mammalian Embryonic Development" (J. Rossant and R. A. Pedersen, eds.), p. 195. Cambridge Univ. Press, Cambridge, 1986; G. A. Schultz, *in* "Experimental Approaches to Mammalian Embryonic Development" (J. Rossant and R. A. Pedersen, eds.), p. 239. Cambridge Univ. Press, Cambridge, 1986.]

maturation in *Xenopus*.[11] The acetylated state of core histones is also correlated with gene expression (reviewed by Turner[12]).

Little is known about the composition and modification of histones synthesized during early mouse development. About 0.05% of the total protein synthesis of mouse oocytes and unfertilized eggs is dedicated to histone H4, an amount sufficient to support two or three cell divisions.[13] All four core histones and their mRNAs are synthesized at the beginning of mouse development, with the greatest amounts found in oocytes and unfertilized eggs (Fig. 1). Histone mRNA is reduced 10-fold by the middle of the 2-cell stage.[14,15] All five major histones are synthesized at the 2-cell stage, and histone synthesis is independent of DNA synthesis[16] (M. Wiekowski, unpublished data).

As with other animals, it seems likely that mammalian histones will undergo modification during preimplantation development. Fertilization of a mouse egg triggers completion of meiosis followed by formation of

[11] H. R. Woodland, *Dev. Biol.* **68,** 360 (1979).
[12] B. M. Turner, *J. Cell Sci.* **99,** 13 (1991).
[13] P. M. Wassarman and S. C. Mrozak, *Dev. Biol.* **84,** 364 (1981).
[14] D. H. Giebelhaus, J J. Heikkila, and G. A. Schultz, *Dev. Biol.* **98,** 148 (1983).
[15] R. A. Graves, W. F. Marzluff, D. H. Giebelhaus, and G. A. Schultz, *Proc. Natl. Acad. Sci. U.S.A.* **82,** 5685 (1985).
[16] P. L. Kaye and R. B. Church, *J. Exp. Zool.* **226,** 231 (1983).

two pronuclei. The maternal genome in oocytes carries a normal complement of core histones and forms the maternal pronucleus, but the paternal genome arrives packaged in protamines that must be replaced with histones provided by the egg.[17,18] Paternal DNA is depleted of protamines before core histones appear, whereas core histones are always associated with maternal DNA.[19] Both maternal and paternal nuclei undergo changes in morphology during this period, although changes in the paternal nucleus are more dramatic.[20] When remodeling is complete, chromatin structure is further modified in order to repress expression of all promoters not intended to be active at the beginning of zygotic gene transcription.[21]

In an effort to analyze histone synthesis in mouse eggs and embryos, published methods for analysis of histones have been optimized for application to small numbers of mouse oocytes and preimplantation embryos. Emphasis is placed on using the minimum number of steps that would still give good, reproducible results. To identify acetylated forms of histones in oocytes and embryos, mouse fibroblasts are cultured and radiolabeled in the presence of sodium butyrate. Sodium butyrate is an inhibitor of the histone deacetylase and induces the appearance of acetylated histones.[22]

Preparation of Histones

Siliconization of Tubes

Siliconized plastic tubes are used to reduce adsorption of basic histones to surfaces. Place tubes and pipette tips overnight in a beaker resting in a large desiccator (220 cm ID) containing a 50-ml beaker with 10 ml of hexamethyldisilazane (CH7300EG, Dynamil Nobel Microelectronics, 2570 Pearl Buck Rd., Bristol, PA 19007, or No. 84769, Pierce Chemical Co., Rockford, IL). Hexamethyldisilazane vapor will produce a monomolecular layer of silicon on glass or plastic surfaces. A fresh sample should be used each time since hexamethyldisilazane absorbs water vapor and slowly inactivates. Purchase hexamethyldisilazane in small bottles, and, once opened, purge the air in the bottle with dry nitrogen before closing.

[17] B. R. Zirkin, S. D. Perreault, and S. J. Naish, "The Molecular Biology of Fertilization," (H. Schatten and G. Schatten, eds.), Cell Biology Ser., p. 91. Academic Press, New York, 1989.
[18] S. Nonchev and R. Tsanev, *Mol. Reprod. Dev.* **25**, 72 (1990).
[19] T. C. Rodman, F. H. Pruslin, H. P. Hoffmann, and V. G. Allfrey, *J. Cell Biol.* **90**, 351 (1981).
[20] P. G. Adenot, M. S. Szöllösi, M. Geze, J. P. Renard, and P. Debey, *Mol. Reprod. Dev.* **28**, 23 (1991).
[21] M. Wiekowski, M. Miranda, and M. L. DePamphilis, *Dev. Biol.* **147**, 403 (1991).
[22] J. Covault and R. Chalkley, *J. Biol. Chem.* **255**, 9110 (1980).

We recommend starting a new 100-ml bottle every 3 months. Autoclave the tubes before using.

Mouse Fibroblast Extract

Because the number of mouse ova available for histone analysis is relatively small (50 to 100 embryos), radiolabeled embryonic histones are isolated in the presence of an excess of unlabeled mouse histones to reduce the fraction of labeled histones lost in any step and to provide internal histone standards. Carrier histones are isolated from mouse fibroblasts (3T3 or C127) grown in 15-cm dishes until they reached about 80% confluency. The cells are washed twice with isotonic phosphate buffer (GIBCO, Grand Island, NY) to eliminate culture medium, and then the dishes of cells are stored at −70° until needed. To prepare an extract, cells are thawed at room temperature, scraped from the dishes, and centrifuged in a Sorvall HB-4 rotor for 5 min at 6000 rpm (4°). The pellet is resuspended in 0.5 ml lysis buffer per dish (Table II). After breaking the cells in a Dounce homogenizer (7 strokes with pestle B), 100 μl of cell extract is added to 100 embryos or oocytes. This is about one-fifth of a 15-cm dish of cells or 2.5 × 10^6 fibroblasts per 100 mouse embryos, which provides enough histones to produce a visible pellet following centrifugation (see below).

Radiolabeling Preimplantation Mouse Embryos and Oocytes

Oocytes and 1-cell or 2-cell embryos are isolated from CD-1 Swiss albino mice (Charles River Breeding Laboratories, Wilmington, MA) or (C57BL/6J × SJL/J)F$_1$ hybrids (referred to as B6SJL; Jackson Laboratories, Bar Harbor, ME). From 20 to 25 fertilized eggs are obtained from one superovulated and mated female. About 300 oocytes are obtained from ten 14-day-old females. Ova are cultured under oil in M16 medium[23] containing 40 mg/ml bovine serum albumin (BSA; Sigma, St. Louis, MO, A4161) as previously described.[21,24,25]

[³H]Lysine and [³H]arginine are most appropriate to label nascent histones since these amino acids are abundant in all five histone types (>20 mol %).[26] ³H-Labeled amino acids are purchased as L-[4,5-³H]lysine

[23] B. Hogan, F. Constantini, and E. Lacy, "Manipulation of the Mouse Embryo." Cold Spring Harbor Laboratory, Cold Spring Harbor, New Yorik, 1986.

[24] M. L. DePamphilis, S. A. Herman, E. Martinez-Salas, L. E. Chalifour, D. O. Wirak, D. Y. Cupo, and M. Miranda, *BioTechniques* **6,** 662 (1988).

[25] E. Martinez-Salas, E. Linney, J. Hassell, and M. L. DePamphilis, *Genes Dev.* **3,** 1493 (1989).

[26] C. von Holt, W. F. Brandt, H. J. Greyling, G. G. Lindsey, J. D. Retief, J. de A. Rodrigues, S. Schwager, and B. T. Sewell, this series, Vol. 170, p. 431.

TABLE II
SOLUTIONS USED IN HISTONE ISOLATION

Buffer	Stock solution		Final concentration	
Embryo wash buffer				
NaCl	5	M	140	mM
Tris-HCl (pH 7.5)	1	M	10	mM
EDTA	0.5	M	1	mM
Bovine serum albumin	40	mg/ml	0.2	mg/ml
Lysis buffer[a]				
Tris-HCl (pH 7.5)	1	M	10	mM
Sucrose	1	M	0.25	M
Triton X-100	100%		1%	(v/v)
Sodium butyrate	1	M	5	mM
MgCl$_2$	1	M	10	mM
Phenylmethylsulfonyl fluoride (PMSF)	0.1	M	0.1	mM
Sodium bisulfite[a]	—		5.2	mg/ml
Nuclear disruption buffer[a]				
Tris-HCl (pH 7.5)	1	M	10	mM
EDTA	0.5	M	13	mM
Sodium butyrate	1	M	5	mM
Sodium bisulfite[a]	—		5.2	mg/ml
Chromatin wash buffer				
Tris-HCl (pH 7.5)	1	M	10	mM
EDTA	0.5	M	13	mM
Sodium butyrate	1	M	5	mM
NaCl	5	M	140	mM
Sodium bisulfite[a]	—		5.2	mg/ml

[a] These recipes are from L. Sealy, R. R. Burgess, M. Cotten, and R. Chalkley, this series, Vol. 170, p. 612. Sodium bisulfite is added just before the buffer is used. This changes the pH to 6.0. Sodium butyrate (Sigma B5887) is prepared in 10 mM Tris-Hcl (pH 7.5) and, if necessary, adjusted to pH 7.5 with 1 N HCl. Aliquots of sodium butyrate can be stored at −20°.

monohydrochloride in aqueous solution (specific activity 70–100 Ci/mmol) and L-[2,3,4,5-³H]arginine monohydrochloride (specific activity 35–70 Ci/mmol) in aqueous solution containing 2% ethanol (Amersham, Arlington Heights, IL). Methionine is absent in histones H2A and H1 but present in histones H2B, H3, and H4 (1.5 mol %).[26] Therefore, [³⁵S]methionine can be used to label the latter proteins.[13] ³⁵S is approximately 10-fold more energetic than ³H, and it has a specific activity (>1000 Ci/mmol) that is 10- to 30-fold greater than ³H-labeled amino acids (Amersham). However, ³⁵S has a shorter half-life than ³H (87.4 days compared to 12.4 years).

Optimum incorporation of amino acids into histones occurs with no more than 50 oocytes or embryos per 20-µl drop of culture medium. To this drop is added 2.5 μl of each radiolabeled amino acid to give a final concentration of 100 μCi/ml [^{3}H]arginine and 100 μCi/ml [^{3}H]lysine. Embryos are generally incubated for 6 hr and oocytes for 24 hr at 37° in 5% CO_2. Cells are then washed free of radiolabel by transferring them through two 100-μl drops of embryo wash buffer (Table II); cells are subsequently transferred to a microcentrifuge tube containing 50 μl lysis buffer (Table II), frozen in a dry ice/ethanol bath, and stored at $-70°$.

Radiolabeling Mouse Fibroblasts

Mouse fibroblasts (3T3 or C127) are cultured in 15-cm dishes to approximately 50% confluency. Dishes are washed with minimal essential medium without lysine (GIBCO, 410-2400) and than cultured in minimal essential medium containing 10% dialyzed fetal bovine serum (GIBCO, 220-6300). Radiolabeled amino acids are added to give a final concentration of 1.25 μCi/ml [^{3}H]lysine and 1.25 μCi/ml [^{3}H]arginine. Mouse fibroblasts are cultured for 20 hr in the presence or absence of 10 mM butyrate and are then harvested as described above.

Isolation of Histones

Embryos or oocytes are thawed at room temperature and then immediately placed on ice. Freshly prepared mouse fibroblast extract is added and the nuclei pelleted in an Eppendorf centrifuge (14,000 rpm, 5 min, 4°). The nuclear pellet is resuspended in 200 μl nuclear disruption buffer (Table II) using a vortex device, and the chromatin is recovered by centrifugation at 14,000 rpm for 10 min at 4°. Chromatin is resuspended in 200 μl of chromatin wash buffer (Table II) and again centrifuged to remove proteins soluble in 140 mM salt. The chromatin is pelleted again, resuspended in 50 μl nuclear disruption buffer, and then sonicated using a microtip probe placed directly into the sample (seven 1-sec pulses). Histones are extracted into acid for 1 hr on ice by addition of 5.5 μl of 3.6 N H$_2$SO$_4$ to each sample (final concentration 0.4 N H$_2$SO$_4$). Of the different acids that can be used, H$_2$SO$_4$ is the most efficient.[26] Cell debris is then removed by centrifugation (14,000 rpm, Eppendorf Microfuge, 30 min, 4°). Histones in the supernatant are precipitated by adding 10 volumes of acetone, mixing thoroughly using a vortex device, and then incubating overnight at $-20°$. Recover the histone precipitate by centrifugation (Microfuge, 14,000 rpm, 30 min, 4°) and discard the supernatant. Dry the histone pellet under vacuum until it appears white and powdery. This should require no more than 10 min. If the pellet appears dark (i.e., wet), continue drying. A wet pellet will smear during gel electrophoresis.

Dissolve the pellet in deionized water and store at $-20°$ for not more than 1 week to avoid histone charing.[26]

Isolation of Fibroblast Histones

The method used to isolate histones from radiolabeled mouse fibroblasts follows the same protocol as for the isolation of embryo histones, but the volumes used in each step are increased. After harvesting the cells, the cell pellet is resuspended in 0.5 ml lysis buffer (Table II) per dish. After breaking the cells in a Dounce homogenizer, the nuclei are pelleted by centrifugation in an Eppendorf centrifuge (14,000 rpm, 10 min, 4°). The nuclei are resuspended in 500 μl nuclear disruption buffer (Table II) and the chromatin pelleted (14,000 rpm, Microfuge, 10 min, 4°). The chromatin is washed once with 500 μl chromatin wash buffer (Table II) and pelleted again. After adding 200 μl nuclear disruption buffer, the sample is sonicated as described above. Histones are extracted with a final concentration of 0.4 N H_2SO_4 for 1 hr on ice. After centrifugation (14,000 rpm, Microfuge, 10 min, 4°), histones in the supernatant are precipitated with 10 volumes of acetone overnight at $-20°$. Histones are collected by centrifugation (10 min in Sorvall HB-4 rotor at 9000 rpm) and the pellet dried under vacuum. Histones are resuspended in water (200 μl per dish) and kept frozen at $-20°$.

Analysis of Histones by Gel Electrophoresis

SDS–Polyacrylamide Electrophoresis

SDS–polyacrylamide electrophoresis separates histones according to their molecular weight. Calf thymus histones migrate the same as mouse histones and have the following molecular weights: H2A, 14,500; H2B, 13,700; H3, 15,300; H4, 11,300; and H1, 22,000. However, basic proteins bind SDS in an anomalous fashion, with the result that histones migrate aberrantly in SDS gels.[26] Optimum separation of all five histones, including subtypes of histone H1, is achieved in 12% polyacrylamide gels (Table III). Gels are 20 cm long and 0.5 mm thick. Electrophoresis is carried out at 250 V (constant voltage) for 4 hr at room temperature using a Bio-Rad (Richmond, CA) Protean II xi system with a circulating water-cooled jacket. The negative pole is at the top of the gel, the positive pole at the bottom. Detailed procedures for handling gels and electrophoresis devices are described by Sambrook et al.[27] and Hames and Rickwood.[28]

[27] J. Sambrook, E. F. Fritsch, and T. Maniatis, "Molecular Cloning: A Laboratory Manula," 2nd Ed. Cold Spring Harbor Laboratory, Cold Spring Harbor, New York, 1989.
[28] B. D. Hames and D. Rickwood, "Gel Electrophoresis of Proteins: A Practical Approach," 2nd Ed. IRL Press, Oxford 1990.

TABLE III
SDS–POLYACRYLAMIDE GEL ELECTROPHORESIS[a]

Gel	Stock	Amount added	
12% Polyacrylamide resolving gel			
Tris-HCl (pH 8.8)	1.5 M	7.5	ml
Sodium dodecyl sulfate (SDS)	10%	300	μl
Acrylamide (IBI, New Haven, CT)	40%	9	ml
Bisacrylamide (IBI)	2%	4.65	ml
Water	—	8.25	ml
Deaerate for 5 min in a 50-ml side-arm flask			
Ammonium persulfate (freshly prepared)	10%	225	μl
TEMED (Bio-Rad)	—	10	μl
Pour the gel immediately at room temperature			
5% Polyacrylamide stacking gel			
Tris-HCl (pH 6.8)	1.5 M	1.26	ml
Sodium dodecyl sulfate	10%	100	μl
Acrylamide (IBI)	40%	1.25	ml
Bisacrylamide (IBI)	2%	0.75	ml
Water	—	6.8	ml
Deaerate for 5 min in a 25-mL side-arm flask			
Ammonium persulfate (freshly prepared)	10%	100	μl
TEMED (Bio-Rad)	—	10	μl
Pour the gel immediately			
Electrophoresis buffer (1 liter of 1× buffer)			
Tris (base)	—	6.06	g
Glycine	—	28.5	g
Sodium dodecyl sulfate	—	1	g
Add water to 1 liter			
Loading buffer (8 ml of 2× buffer)			
Tris-HCl (pH 6.8)	1.5 M	0.33	ml
Glycerol	—	0.8	ml
Sodium dodecyl sulfate	10%	1.6	ml
2-Mercaptoethanol	—	0.4	ml
Bromphenol blue	0.1 %	0.2	ml

[a] Reagents are added in the order given.

Polyacrylamide solutions are deaerated for 5 min before addition of TEMED. After the resolving gel has been poured, the polyacrylamide solution is overlaid with 0.1% SDS. Polymerization is monitored by formation of a sharp interphase between the SDS solution and the gel. Remove the SDS solution and pour the stacking gel solution on top of the resolving gel. Insert a 15-well comb (avoid gas bubbles between the comb and stacking gel solution). After polymerization, assemble the gel electrophoresis apparatus per the manufacturer's instructions. Fill the upper and lower chambers with electrophoresis buffer (Table III) before removing the comb in order to avoid disturbing the wells.

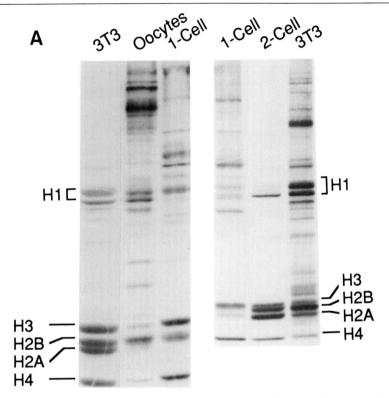

FIG. 2. Analysis of newly synthesized histones from 1-cell and 2-cell mouse embryos and from mouse 3T3 fibroblasts. (A) SDS gels. (B) Triton/acetic acid/urea gels.

Rinse the unpolymerized acrylamide from the bottom of the wells by bubbling air through a flat 0.3-mm gel loading pipette tip (Marsh Biomedical Products, Rochester, NY).

Histones from 50 embryos (3000–8000 counts/min) are diluted 1 : 1 in loading buffer (Table III). Sometimes the histone samples reduce the pH of the loading buffer. In that case, readjust the pH by adding 1.5 M Tris-HCl (pH 6.8) until the color has turned from yellow to blue. Cap the microcentrifuge tube and lock the lid in place with a Microlock (Marsh). The samples are incubated in boiling water for 5 min and then centrifuged briefly to collect all the sample at the bottom of the tube. Load the entire sample into a single well.

When electrophoresis is completed, stain the gel in 30% methanol, 10% acetic acid, and 0.1% Coomassie blue for 15 min at room tempera-

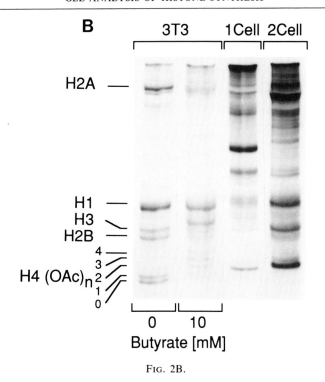

FIG. 2B.

ture with gentle agitation. Destain in 30% methanol and 10% acetic acid. The carrier histones purified from somatic mouse cells should be visible at this point. To visualize the nascent ^{3}H-labeled histones, incubate the gels in En^{3}Hance (Du Pont) for 30 min to permit fluorography (do this in a ventilated hood). This procedure destains the gel to some extent. Aspirate the En^{3}Hance solution into a trap and dispose of properly. Incubate the gel in water for an additional 30 min. This precipitates the En^{3}Hance and turns the gel opaque. The gel is quite fragile at this point. Slip a sheet of Bio-Rad filter paper backing under a gel floating in water and lift the gel out to avoid distorting the gel. Lay this over a second sheet of filter paper resting on a Hoefer gel dryer. Cover the gel with a sheet of Saran wrap. Cover this with the Mylar sheet provided with the gel dryer. Dry the gel onto the filter paper at 60° to 70° under vacuum. Make sure that the gel is completely dry before releasing the vacuum, or it will crack. Expose gel to X-OMAT AR film (Kodak, Rochester, NY) at −80°. Nascent histones should be visible in 3 days.

TABLE IV
TRITON/ACETIC ACID/UREA POLYACRYLAMIDE GELS[a]

Solution	Stock	Addition	Final concentration
15% Polyacrylamide gel			
Urea	—	4.8 g	8 M
Acrylamide/bisacrylamide (19 : 1)	40%	3.75 ml	15%
Triton X-100	10%	375 μl	0.375%
Acetic acid	Glacial	0.5 ml	5%
Water		10 ml final volume	
Incubate at $37°$ until urea is dissoved ($\sim$10 min)			
Deaerate for 15 min at *room temperature* in a 25-mL side-arm flask			
Ammonium persulfate	10%	60 μl	
(freshly prepared)			
TEMED (Bio-Rad)		50 μl	
Pour the gel immediately at room temperature			
Loading buffer			
Urea	—	4.8 g	8 M
Acetic acid	Glacial	0.5 ml	5%
2-Mercaptoethanol	—	0.5 ml	5%
Pyronin Y (Sigma)	0.2%	1 ml	0.02%
Protamine sulfate (Sigma)	—	12.5 mg	1.25 mg/ml
Water		10 ml final volume	
Incubate at 37° until urea is dissolved ($\sim$10 min)			
Well supporting buffer			
Urea	—	4.8 g	8 M
Triton X-100	10%	375 μl	0.375%
Acetic acid	Glacial	0.5 ml	5%
Water		10 ml final volume	
Incubate at 37° until urea is dissolved ($\sim$10 min)			

[a] Reagents are added in the order given.

Triton–Acetic Acid–Urea Gels

Electrophoresis through polyacrylamide gels containing acetic acid and urea separates proteins according to their mass and their charge.[29] Protein mobility depends on both the pH and the urea concentration of the gel and may vary for histones obtained from different species. Addition of Triton X-100 results in a reduction of the mobility of individual histones that is proportional to their hydrophobic character. This allows separation of histone subtypes that differ in only a few amino acids and is used particularly to separate the various subtypes of core histones. The relative mobilities of individual histones depend on the concentrations of urea and detergent as well as histone posttranslational modifications. What

[29] R. W. Lennox and L. H. Cohen, this series, Vol. 170, p. 532.

appears as a single histone band in SDS gels can appear as multiple bands in Triton/acetic acid/urea Gels (Fig. 2). Each acetylation of a lysine residue reduces the number of positive charges on the histone by 1. Each phosphorylation of a serine or threonine residue also neutralizes the number of positive charges by one.

Optimum separation of embryonic mouse histones is achieved in 15% polyacrylamide, 8 M urea, and 0.375% Triton X-100 (Table IV). The ratio of acrylamide to bisacrylamide is critical for the appearance of straight bands. Gels are 9 cm long and 0.75 mm thick. Electrophoresis is carried out at 200 V (constant voltage) for 4 hr at room temperature using a Mini-Protean II Dual Slab Cell (Bio-Rad). The positive pole is at the top of the gel, and the negative pole is at the bottom. The fastest migrating form of histone H4 reaches the bottom of the gel in 4 hr; the dye (pyronin Y) runs off the gel within 1 hr.

Pour the gel (Table IV) and add the comb immediately, since a loading gel is not used. Polymerization is complete after 1 hr at room temperature. Pull out the comb and rinse the sample wells with deionized water. Use a flat-tipped sample loading pipette tip (Marsh) attached to an aspirator to remove all liquid from the wells. Fill the well with well supporting buffer (Table IV) to prevent dialysis of urea from the gel, and then fill the chamber with 5% (v/v) acetic acid. Prerun the gel for 1 hr at 200 V or overnight at 30 V to establish a constant current and to remove ammonium persulfate from the gel, which can interfere with protein mobility. Histone preparations in water are diluted 1 : 1 in loading buffer (Table IV). Protamine sulfate in the loading buffer produces sharper bands of histones.[29] Remove the acetic acid from the reservoir, rinse wells with water, and load the samples into the dry wells. Carefully overlay samples with 5% acetic acid, and then fill the reservoir with 5% acetic acid without disturbing the samples. The gels are stained overnight in 500 ml of 30% methanol, 10% acetic acid, and 0.1% Coomassie Blue. It is important to dialyze out all of the urea, or the gel will crack during drying. Process the gel as described for SDS–polyacrylamide gels.

[31] Whole-Mount Immunohistochemistry

By Claytus A. Davis

Introduction

Whole-mount immunohistochemistry is the localization of antigens in unsectioned tissues using specific antibodies.[1,2] It can be successfully applied to whole mouse embryos from zygote to 10.0 days of gestation and to dissected tissues at any stage. The intention of this chapter is to provide enough information for the development of a working protocol; to discuss procedure variations; and to indicate possible solutions for the most common problems.

Although the whole-mount technique is similar in outline to immunohistochemistry of sectioned material and shares some of the problems, it has several advantages. Most importantly, it quickly and clearly conveys the spatial relationship between antigen distribution and the rest of the embryo.[3] It also requires less effort and working time to complete and permits many samples to be easily processed in parallel. There are also several disadvantages. The elapsed time, from the addition of primary antibody to getting the results, is longer; cellular level detail is poorer; and it does not work for older embryos.

The design of a working procedure and its successful application depend on the particular antigen–antibody combination, so there is no one protocol that will work in all situations. The protocol presented in Table I serves primarily as an example and as a reference point for discussing the individual steps.

Whole-Mount Procedures

General Points

Whole-mount procedures largely consist of a series of incubations, during which reagents must fully penetrate the sample and react, and

---

[1] M. Costa, Y. Patel, J. B. Furness, and A. Arimura, *Neurosci. Lett.* **6,** 215 (1977).

[2] M. Costa, R. Buffa, and E. L. Solcia, *Histochemistry* **65,** 157 (1980).

[3] People who are able to visualize well in three dimensions will be able to reconstruct this relationship from serial section data, but this information is very difficult to convey to anyone else. Several computer programs are available that automate the reconstruction and presentation of the data; however, they currently require an inordinate amount of time and effort.

METHODS IN ENZYMOLOGY, VOL. 225

Copyright © 1993 by Academic Press, Inc.
All rights of reproduction in any form reserved.

TABLE I

SAMPLE WHOLE-MOUNT IMMUNOHISTOCHEMISTRY PROTOCOL[a]

Step	Procedure
Dissection	Dissect out the embryos in cold phosphate-buffered saline (PBS), removing all extraembryonic membranes
Fixation	Fix in methanol–dimethyl sulfoxide (DMSO) (4 : 1) overnight at 4°
Pretreatment	Blocking endogenous peroxidase: Transfer to methanol–DMSO–30% H_2O_2 (4 : 1 : 1) for 4–5 hr at room temperature
	Storage: Embryos may then be stored in 100% methanol at $-20°$; rehydrate embryos through methanol series diluted in water: 50%, 20%, PBS, for 30 min each
	Blocking background binding: Incubate twice in PBSMT (PBS plus 2% instant skim milk powder and 0.1%, v/v, Triton X-100) for 1 hr each at room temperature
Primary antibody incubation	Incubate overnight at 4° with primary antibody diluted in PBSMT
Washes	Wash 2 times in PBSMT at 4° and 3 times at room temperature for 1 hr each
Secondary antibody incubation	Incubate overnight at 4° with secondary antibody diluted in PBSMT
Washes	Repeat the above washes, adding a final 20-min wash in PBT at room temperature
Color development	Incubate embryos in 0.3 mg/ml DAB (diaminobenzidine, e.g., Sigma, St. Louis, MO, D-5637; *possibly carcinogenic*) +0.5% $NiCl_2$ in PBT (PBS plus 0.2% bovine serum albumin, e.g., Sigma, A-4378, and 0.1%, v/v, Triton X-100) at room temperature for 20 min; add H_2O_2 to 0.03% and incubate at room temperature until color density looks good, usually about 10 min
	Rinse several times in PBT and dehydrate through methanol series: 30, 50, 80, 100% for 30 min each; embryos may be stored in methanol
Clearing	Incubate embryos in BABB [benzyl alcohol–benzyl benzoate (1 : 2)]–methanol (1 : 1) for 20 min; transfer to BABB

[a] Throughout the entire protocol, embryos should be gently rocked.

subsequent washes, during which excess reagents must be completely removed. To facilitate both, the embryos should be gently rocked or rolled throughout the procedure.

Frequent solution changes increase the chances of damaging or losing embryos. To avoid this, transfer the solution rather than handling the embryos. Postimplantation embryos may be processed in 15-ml disposable polypropylene tubes. To transfer, stand the tubes upright for a few minutes to let the embryos sink to the bottom, then gently decant the solution. For small embryos, monitor the process over a light box. Check that no embryos have stuck to the inside of the tube cap. In aqueous solutions without detergent, the embryos tend to stick to the tube walls. The incubation of small postimplantation embryos in primary antibody may be done in 1.5-ml cryovials or Reactivials (Pierce, Rockford, IL) to minimize the amount of reagent used. For preimplantation embryos, perform the incubations and washes in a depression slide or microtiter plate. Remove and add reagents by pipette. Avoid drawing up the embryos, however, since they may stick to the inside of the pipette. Monitor the transfers under a microscope. Before using a mouth pipette, remember that some of the reagents are poisonous. A few embryos will be lost during the course of the procedure.

Some of the transfers involve a large change of osmotic potential, for example, between aqueous and organic solutions and between different clearing agents. Even when the embryos are fully permeable, a sudden transfer may degrade the appearance. Take the embryos through a graded series of two or three intermediate mixes (e.g., 80, 50, 20%).

Mouse embryos of vastly different sizes may produce acceptable whole mounts. As the size increases, the minimum duration of each step also increases. The times indicated in Table I are usually long enough for 10.0-day mouse embryos. Observed background staining is proportional to the size of the specimen. It may determine the upper age limit for which the procedure works. For some reason, background is variable between individual large specimens, and so it may be necessary to process many before a few suitable for good photographs are found. Smaller embryos yield uniform results.

Dissection

The goal of dissection is to get the embryos from the uterus to the fixative in good condition and in a reasonably short time. A period of 30 min should be fine. Dissect in PBS (phosphate-buffered saline, without Ca^{2+}/Mg^{2+}) and keep the embryos on ice as much as possible. Good dissections, especially for the early stages, require practice, steady hands, a well-adjusted dissecting microscope with a good field depth, and undamaged forceps.

The stages between egg and late blastocyst are flushed out of the fallopian tubes and uterus. Between implantation and approximately 6.5 days of gestation, the embryos are extremely difficult to find and dissect

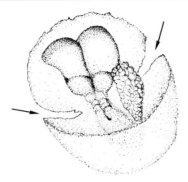

FIG. 1. Flattening an early neurulation embryo. A dorsal view of an 8.0-day embryo is shown. The covering amnion has been removed and two vertical cuts made in the lateral amnion (indicated by arrows). Once the cuts are made, the embryo will flatten out.

away from the mass of decidual tissue. Stages between 6.5 and 7.5 days are manageable but require practice. Later stages are easier. Hogan *et al.*[4] give detailed instructions for removing the different stages; however, if you are unfamiliar with the dissections, it is best to get a demonstration from someone knowledgeable.

During gastrulation and early neurulation (~6.5–8.0 days), the embryo is U-shaped and is bound in the amnion and yolk sac. Although whole-mount staining will work on the entire structure, it may be difficult to photograph the results. The embryo will flatten if, before fixing, the amnion cap is dissected off and two cuts are made in the lateral amnion (Fig. 1).

After approximately 9.0 days of gestation, the neural tube closes. The lumen of the tube sometimes nonspecifically binds both primary and secondary antibodies, causing high background. To both reduce this and improve penetrability, the thin roof of the hindbrain may be opened and the embryos left to gently rock in PBS at 4° for 10 min prior to fixing. Torn edges and crushed tissues often show high background staining.

Fixation

A perfect fixative would preserve antigenicity and morphology, while leaving the embryo completely open to the passage of antibody and other reagents. Fixatives fall into two groups: those that work by covalently cross-linking proteins, forming a meshwork (e.g., aldehyde fixatives), and

---

[4] B. Hogan, F. Costantini, and E. Lacey, "Manipulating the Mouse Embryo: A Laboratory Manual." Cold Spring Harbor Laboratory, Cold Spring Harbor, New York, 1986.

those that precipitate the cellular components (e.g., organic solvents). Sternberger discusses their preparation and relevance to immunohisto-chemistry in more detail.[5] Recipes for two good starting fixatives, one from each group, are given below. The same fixatives used for sectioned material may be tried with whole-mount procedures, although a fixative that works for one may not work for the other. In particular, glutaralde-hyde-containing fixatives may generate an almost impenetrable mesh of tightly cross-linked proteins.

4% Paraformaldehyde in PBS. Add 4 g of paraformaldehyde to 50 ml of water plus 1 drop of 1 N NaOH. Dissolve at 37° with shaking for 1 hr. Adjust the pH to approximately 7 with HCl, add an equal volume of 2 × PBS and then filter through Whatman (Clifton, NJ) No. 1 paper. Store at 4° and use within 1 day of preparation. Although less convenient, paraformaldehyde is more likely to preserve antigenicity than any other fixative.

Dent's Fixative. Dent's fixative is 80% methanol, 20% dimethyl sulfox-ide (DMSO).[6] If antigenicity is preserved, this may give a lower back-ground than aldehyde fixatives, and requires less work. Fix on ice in a volume more than 20-fold greater than the sample volume, for between 1 hr and overnight, depending on the size of the embryos. Transfer the embryos to fresh fixative after the first 15–30 min.

Pretreatment

Ideally, the pretreatments leave the embryo in such a condition that antibodies and other reagents can freely diffuse through them and bind to only the correct molecules. Which pretreatments are necessary depends on the fixative and the detection scheme used.

Blocking Free Aldehydes. Aldehyde-fixed embryos require an incuba-tion in 0.1 M glycine in PBS for 30–60 min at room temperature to block any free aldehyde groups.

Blocking Endogenous Enzyme Activity. The fixation protocol may not abolish all endogenous peroxidase activity. At the risk of reducing antige-nicity in some instances, the remaining activity may be eliminated by incubation in either PBS plus 0.3–1.0% H_2O_2, or methanol or Dent's fixative plus 0.5–5% H_2O_2, for 3–4 hr at room temperature. The methanol treatment is preferable because the aqueous reaction may produce destruc-tive oxygen bubbles within the embryo, especially at higher H_2O_2 concen-trations. Most endogenous phosphatase activities are inhibited by includ-

[5] L. A. Sternberger, "Immunocytochemistry." Wiley, New York, 1986.
[6] J. A. Dent, A. G. Polson, and M. W. Klymkowsky, *Development (Cambridge, UK)* **107**, 35 (1989).

ing 1–2 mM levamisole prior to and during the incubation with the enzyme substrate.[7]

If the embryos are still in methanol after completing these blocking steps, they should be rehydrated into PBS through a series of methanol dilutions in water (e.g., 50%, 20%, PBS, for 30 min each).

Blocking Nonspecific Antibody Binding. The final pretreatment, antibody dilutions, and washes are all done in PBS plus 0.1–0.2% detergent plus 1–5% protein. The protein may be bovine serum albumin (BSA) or nonfat milk powder (both used at 1–3%) or nonimmune serum from the animal in which the secondary antibody was raised (5–20%, v/v). Nonimmune serum can be heat-treated for 30 min at 56° to inactivate complement. In our hands, milk powder worked the best. The detergent added is usually nonionic: Nonidet P-40, Triton X-100, or Tween 20. Both the protein and the detergent decrease background staining. The excess protein blankets sites in the embryo that interact nonspecifically with the antibodies. The detergent may weaken nonspecific antibody binding. Which protein and detergent will work best needs to be determined empirically for each antigen–antibody combination.

Antibody Incubations

Dilute the primary antibody in the last blocking solution (PBS + protein + detergent) and incubate the embryos for a few hours to overnight, depending on their size. Keep the embryos gently rocking or rolling. Long incubations with a low concentration of antibody at 4° will give lower background staining than short incubations with high concentration of antibody.

The success of a whole-mount protocol depends greatly on the characteristics of the primary antibody. Antibodies vary, among other things, in their specificity (how well they recognize the correct antigen to the exclusion of others), their affinity (how stable the antigen–antibody interaction is), and their source (polyclonal antisera or monoclonal). An antibody with poor specificity will generate increased background, which, because of the thickness of the material, will readily obscure specific staining.

Because the antigen–antibody complex is noncovalent, and bound antibody is in equilibrium with free antibody, some of the bound antibody will be lost during the washes. In sectioned material, where 30 min of washing is usually sufficient, most of the antibody will remain bound, even if it does not have a high affinity. In the whole-mount procedures,

[7] B. A. Ponder and M. M. Wilkins, *J. Histochem. Cytochem.* **29,** 981 (1981).

especially for large embryos, the wash steps may span most of a day, and there is the possibility that much of the signal may be lost if the antibody has a low affinity.

A polyclonal antiserum usually contains antibodies recognizing more than one epitope. Consequently, its apparent affinity (acidity) is often higher than that of a monoclonal antibody, which will recognize only one epitope. There are specific problems associated with using mouse monoclonal antibodies on mouse tissues. Self-tolerance makes the production of monoclonal antibodies against epitopes present in the mouse less likely. For this reason, a mouse monoclonal antibody raised against a conserved non-mouse protein may recognize the protein in a wide variety of species except the mouse, since the mouse immune system may preferentially respond to regions of the injected protein that are different from the endogenous version. Also, embryonic or maternal mouse immunoglobulin G (IgG) present in the embryo will bind to any anti-mouse secondary antibody used and so increase background. To keep the embryos as free of mouse immunoglobulin as possible, dissect away maternal and extraembryonic tissues, layer by layer, in different changes of PBS, and wash them thoroughly in PBS before fixing. Remaining immunoglobulins may be blocked by incubating the embryos with unlabeled secondary antibody and then washing prior to the addition of primary antibody. This problem may be avoided by labeling the primary antibody, although this will reduce the sensitivity. A preferable alternative is to use rat monoclonal antibodies. Anti-rat immunoglobulin secondary antibodies which do not cross-react with mouse immunoglobulin are commercially available. Problems arising from self-tolerance will also be reduced.

Polyclonal antisera also have disadvantages. Unlike monoclonal antibodies, they contain a majority of nonspecific antibodies, which may cause background. To remedy this, the antiserum may be either diluted, affinity-purified against the antigen, or preadsorbed. If the specific antibody titer is high, then it may be possible to dilute the antiserum to such an extent that the background due to nonspecific antibodies is negligible. Affinity purification produces the cleanest antibody solution but requires 1 week to perform and a large amount of purified antigen; in addition, antibodies that have a very high affinity for the antigen may not elute well, thus stripping the antiserum of its best antibodies. Preabsorption to an acetone powder of embryonic mouse tissue that does not contain any of the desired antigen may remove nonspecifically binding antibodies. Harlow and Lane describe these techniques.[8]

---

[8] E. Harlow and D. Lane, "Antibodies: A Laboratory Manual." Cold Spring Harbor Laboratory, Cold Spring Harbor, New York, 1988.

Washes

Five consecutive washes in the blocking solution following antibody incubations are typical. Keep the embryos gently rocking or rolling at either 4° or room temperature. The duration must be determined empirically, but each wash should not need to be longer than 1 hr. If washes are too short, there will be increased background; if too long, there may be a loss of signal.

The length of the washes is dictated by diffusion rates and the intensity of nonspecific antibody binding. In general, the rate of diffusion decreases for larger molecules, for larger embryos, and for the denser tissue meshworks generated by some fixatives (e.g., glutaraldehyde). Improving the blocking steps will most often reduce the background better than greatly increasing the wash times.

Detection

The binding of primary antibody may be monitored, and the signal amplified, by a number of different methods. For whole-mount techniques, enzyme-linked and fluorochrome assays are usual. Briefly, the complex of antigen and primary antibody is detected by binding a secondary antibody coupled to either a dye, which is then observed by fluorescence microscopy, or to an enzyme, which generates an insoluble colored reaction product from a soluble substrate. Although the fluorochrome assays take less time and localize antigens more precisely, the enzyme-linked assays are preferable because they are more sensitive, permanent, and easier to photograph. Both horseradish peroxidase[9] (HRP) and alkaline phosphatase[10] are used. Of the two, HRP is better, because it is smaller and its diaminobenzidine (DAB) reaction product is extremely stable. (Sodium azide, which is often added to protein solutions to prevent bacterial or fungal growth, must not be used during the procedure, since it is a potent inhibitor of HRP.)

A variant technique, the avidin–biotin complex[11,12] (ABC) method, also works well. The secondary antibody is tagged with biotin and bound to the primary antibody–antigen complex. This in turn is bound to a complex of avidin or streptavidin[13] and biotinylated peroxidase. Because of the additional amplification step, the technique is reputedly more sensitive (although this has been questioned),[5] permitting background arising

---

[9] P. K. Nakane and G. B. Pierce, Jr., *J. Histochem. Cytochem.* **14,** 929 (1966).
[10] A. S. Bulman and E. Heydermann, *J. Clin. Pathol.* **34,** 1349 (1981).
[11] J. L. Guesdon, T. Ternynck, and S. Avarameas, *J. Histochem. Cytochem.* **27,** 1131 (1979).
[12] S.-M. Hsu, L. Raine, and H. Fanger, *J. Histochem. Cytochem.* **29,** 577 (1981).
[13] Streptavidin is preferable because of its greater affinity for biotin.

from nonspecific antibodies to be reduced by further diluting the primary antiserum. Endogenous biotin may be a source of increased background. It may be blocked in the pretreatment step by first incubating the embryos with unlabeled streptavidin or avidin, then with biotin, followed by washing away the excess before adding the primary antibody.[14]

Coupled secondary antibodies and ABC kits are available commercially. Irrespective of the chosen detection scheme, each incubation step should be followed by a thorough set of washes in PBS plus detergent plus protein.

Color Development

Enzyme-linked assays require the addition of enzyme substrate. If an HRP detection scheme has been chosen then DAB is the best choice. DAB is a potential carcinogen. Treat all waste DAB with bleach. Following the final wash, incubate the embryos in PBS plus 0.3 mg/ml DAB for 30 min. Add H_2O_2 to 0.03% and follow the color development under a dissecting microscope. If the development takes longer than 15 min, keep the embryos in the dark to avoid light-catalyzed DAB polymerization. To stop the reaction rinse the embryos three times in PBS plus 0.1% (v/v) Triton X-100.

If the signal is strong, the color development may be the shortest step in the entire procedure and the H_2O_2 may not completely permeate the embryo before the formation of background staining prompts termination of the reaction. To avoid this, try altering the antibody or reagent concentrations such that the color reaction takes 30 min or longer in the large embryos. The DAB color reaction can be enhanced by including 0.5% $NiCl_2$ in the PBS plus DAB solution. On adding the nickel, a precipitate occasionally forms, which may be removed by filtration through Whatman No. 1 paper.

Clearing

To see the staining pattern clearly, the embryos can be made more transparent or more opaque by transfer into solutions of different refractive indices. They are increasingly transparent in the following: methanol < PBS < glycerol < BABB. BABB (Murray's solution), is a mix of benzyl alcohol and benzyl benzoate (1:2). It is an obnoxious solution that dissolves polystyrene, but it works very well. Because BABB is nonmiscible with aqueous solutions, embryos should first be dehydrated

---

[14] G. S. Wood and R. Warnke, *J. Histochem. Cytochem.* **29**, 1196 (1981).

through an alcohol series (20, 50, 100%) and then soaked in alcohol–BABB (1 : 1) before being transferred into BABB.

To illustrate the relationship between the staining and the embryo, greater transparency is not necessarily better. Large embryos will need more clearing, but staining in small embryos or near the surface may be best observed if the tissues are more opaque. Background is also less apparent in more opaque embryos.

Microscopy and Photography

The embryos are usually photographed submerged in an open dish of clearing medium such that they can be manipulated with forceps into different orientations. Whole mounts are more difficult to photograph than slides. Normal microscopes used for viewing slides are not suitable, since they usually have a shallow field depth. A good dissecting microscope with a camera or one of the microscopes designed for lower magnifications [e.g., Leitz (Wetzlar, Germany) Apozoom, Photomakroskop M400; M10] is preferable. Use a stable viewing platform to avoid vibrations. If the illumination starts to heat the mounting medium, then the resulting convection currents may degrade the image quality by moving the embryo or causing diffraction differences in the medium.

Vary lighting conditions to get the best pictures (Fig. 2). Try reflected or dark-field illumination on more opaque embryos when the relation between the staining and the surface of the embryo is to be emphasized. Sometimes a combination of lighting works well. For more transparent embryos, transillumination is best. To photograph deep staining, large embryos must exhibit low background and be made almost completely transparent. Unfortunately, this results in a low contrast between the embryo and the clearing medium in which it sits (Fig. 2B). Staining is usually more obvious in color prints than in black and white, since the eye uses color cues. As journals typically prefer black-and-white figures, try an enzyme-linked detection assay that gives a dark-colored reaction product (e.g., nickel-enhanced DAB staining) for making black-and-white prints.

Whole-Mount Immunohistochemistry and Scanning
   Electron Microscopy

Scanning electron microscopy (SEM) is the best technique for observing surface detail and morphology. Surface antigens have been detected during SEM by using secondary antibodies coupled to a morphologically

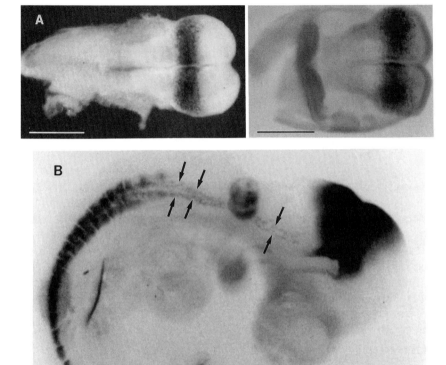

FIG. 2. Whole-mount immunohistochemistry on different stage embryos. All embryos were processed according to the example protocol given in Table I, using an affinity-purified polyclonal antiserum against the mouse *engrailed* proteins as primary antibody and a commercial affinity-purified goat anti-rabbit IgG coupled to HRP (Jackson Immunoresearch, West Grove, PA) as secondary antibody, and photographed using a Leitz macroscope and Ektachrome (Kodak, Rochester, NY) color slide film. Anterior is to the right. (A) Dorsal views of early 8.0-day embryos showing staining in the midbrain/hindbrain neuroepithelium. (Left) The DAB reaction was enhanced with NiCl₂. The embryo was photographed in methanol using reflected illumination. (Right) The DAB reaction was not enhanced. The embryo was cleared in BABB and transilluminated. (B) Lateral view of a 10.0-day embryo showing expression in the midbrain/hindbrain region, the somites, the ventral limb buds, and in the ventral spinal cord and hindbrain (arrows). The DAB reaction was not enhanced. The embryo was cleared in BABB and transilluminated. Bars: 0.3 mm.

distinct marker[15] or by a detecting secondary electrons emitted from osmium deposited preferentially at the site of DAB polymerization.[16]

We have reported a different method that may have some advantages.[17] Nickel ions, added to enhance the HRP/DAB reaction, are incorporated into the growing DAB polymer network. If the nickel is on or near the surface, it is detected during SEM by monitoring for element-specific X-ray emissions. Machines that can do this are typically found in metallurgy or geology departments. Although untried, it should be possible to localize several different antigens simultaneously on the SEM image by doing serial whole-mount reactions, incorporating different metals (e.g., nickel and cobalt) at each step.

There are some practical considerations to this technique. First, the nickel deposition must be on or within 1 μm of the surface, since the electron beam does not penetrate far. Second, the embryo surface must be conductive. To accomplish this, coat the surface with as thin a layer of carbon as possible. A thicker coat, or a metal coat, will greatly reduce the sensitivity. Third, the procedure is slow, requiring a few hours to build up an adequate map. Increasing the beam intensity helps. Fourth, nickel emits characteristic soft $K\alpha$ X-rays at 7.427 keV, which may be partially absorbed by the bulk of the embryo. Ensure that the detector is capable of surveying the same region of the embryo as the electron beam.

Storage

Embryos may be stored after fixation, after the first blocking steps (e.g., after aldehyde blocking and/or endogenous peroxidase inactivation), and after staining is complete. Embryos in most organic solvents may be safely left for a few days at 4° and for a few months at −20° if first transferred into methanol. Aldehyde-fixed embryos may be stored in methanol as well if antigenicity is retained. If not, it may be possible to freeze and store the fixed embryos in liquid nitrogen without severely damaging the morphology.

Once stained, the embryos are best stored at −20° in methanol. The DAB reaction product is completely stable in methanol and is stable for at least a few days in BABB, but we have noticed variable loss of staining in embryos stored in BABB for several months at 4°.

[15] R. S. Molday, *in* "Techniques in Immunochemistry" (G. R. Bullock and P. Petrusz, eds.), Vol. 2, p. 117. Academic Press, San Diego, 1983.

[16] A. L. Hartman and P. K. Nakane, *in* "Techniques in Immunochemistry" (G. R. Bullock and P. Petrusz, eds.), Vol. 2, p. 103. Academic Press, San Diego, 1983.

[17] C. A. Davis, D. P. Holmyard, K. J. Millen, and A. L. Joyner, *Development (Cambridge, UK)* **111**, 287 (1990).

TABLE II
TROUBLESHOOTING GUIDE

Problem	Possible causes	Relevant section of text
Poor morphology	Damage during dissection	Dissection
	Damage during incubations or transfer	General Points
	Poorly fixed embryos	Fixation
Background	Embryos poorly penetrable	Fixation; Pretreatment; Dissection
	Inadequate blocking	Pretreatment
	Nonspecific antibody binding	Pretreatment; Antibody Incubations
	Wash times too short	Washes
	Wash volumes too small	Washes
	Concentration of antibody or other detection reagent too high	Antibody Incubations; Detection
	Enzyme exposed to substrate for too long	Color Development
	Excess enhancement	Color Development
Poor signal	Antigenicity not retained	Fixation
	Rare antigen	Detection
	Poor primary antibody	Antibody Incubations; Developing Working Protocol
	Antibody or other detection reagents too dilute	Antibody Incubations; Detection; Developing Working Protocol
	Embryos poorly penetrable	Fixation; Pretreatment; Dissection
	One of reagents is bad	(Include positive controls)
Variable results	Deep interior staining faint in large embryos	Dissection; General Points; Color Development
	Variable background in large embryos	General Points

Sectioning

For a more detailed examination at the tissue level, embryos stained with DAB can be sectioned. The DAB reaction product is stable in xylene and wax, so follow normal paraffin wax embedding and sectioning procedures. Both staining and background will appear much fainter in the sections than in the whole mounts. Therefore, if the embryos are to be sectioned, it may be desirable to let the color development reaction proceed until the background starts to look quite bad.

Developing Working Protocol

The whole-mount procedures have a large number of parameters. Faced with an untried antibody–antigen combination, the task is to find the set of conditions that give the best results. There are a few practical points that can make this less arduous.

For the initial tests of a new antibody, use the smallest embryos that express the antigen well, since they will exhibit cleaner whole-mount results. Try the HRP indirect immunohistochemical detection method, since it is reasonably sensitive and straightforward. Fix some embryos in 4% paraformaldehyde and some in Dent's fixative. For each group test 7–8 different serial dilutions of the primary antibody. Working dilutions are typically between 1/20 and 1/1000 for polyclonal antisera or monoclonal antibodies from ascites, and between undiluted and 1/20 for hybridoma culture supernatants. If possible, include another primary antibody known to produce good whole mounts as a positive control, and use it to find a dilution range for the secondary HRP-coupled antibody (typically between 1/200 and 1/1000). Start with overnight antibody incubations at 4°, and five 1-hr washes. Do not initially enhance the HRP–DAB reaction, since even a small amount of enhanced background may conceal specific staining.

Following the initial trials, use the troubleshooting guide to help choose modifications to the procedure. It would take months of work to explore all the possible variations of the whole-mount procedure. If there is no staining or if there is unacceptable background after the first few trials of a new antigen–antibody combination, then it may be worthwhile to start raising more antibodes. Meanwhile, further alterations of the procedure can be tried.

Troubleshooting

Table II lists the most common difficulties that may be encountered during the whole-mount procedure, their possible causes, and portions of the text relevant to their solution.

Controls

Artifactual staining is common in immunohistochemical techniques. The same controls are required in whole-mount procedures as in others (e.g., no antibodies, no primary antibody, primary antibody preadsorbed to the antigen). They are fully discussed elsewhere.[18]

False-negative results are more problematic. They may be due to an insufficiently sensitive procedure, masking or loss of epitopes, or insufficient penetration of the specimen by the antibodies or detection reagents. The last is especially applicable to whole-mount protocols, where the increased thickness and possibly more impermeable basement membranes of older embryos may make exposure of the entire sample

---

[18] Chr. W. Pool, R. M. Buijs, D. F. Swaab, G. J. Boer, and F. W. Van Leeuwen, *in* "Immunohistochemistry" (A. C. Cuello, ed.), p. 1. Wiley, Chichester, UK, 1983.

less likely. Finally, no set of controls is perfect, so consistent results using other techniques (e.g., Western blot or RNA *in situ* hybridization) are desirable.

Acknowledgments

I thank Drs. I. Gitelman, A. L. Joyner, and N. Bonini for helpful comments.

[32] Techniques for Localization of Specific Molecules in Oocytes and Embryos

By CALVIN SIMERLY and GERALD SCHATTEN

Introduction

The fields of cell, developmental, molecular, and reproductive biology and genetics have tremendously benefited from studies on eggs at fertilization.[1,2] These early studies relied on systems like frogs and sea urchins in which fertilization *in vitro* and embryo culture were performed using simple solutions like seawater or pond water at room temperature. The design of methods for routinely and reliably obtaining excellent *in vitro* fertilization of many mammals now permits detailed experimentation on molecular and structural features of development in mammals. These investigations have led to many important and unexpected basic discoveries including genomic imprinting[3–5]; gametic recognition involving unique receptors and galactosyltransferases[6,7] atypical maternal inheritance patterns of the centrosome in mice[8]; both paternal and maternal inheritance

[1] T. Boveri, "Zellen-Studien: Ueber die Natur der Centrosomen," Vol. 4. Fisher, Jena, Germany, 1901.

[2] D. Mazia, *Exp. Cell Res.* **153,** 1 (1984).

[3] J. McGrath and D. Solter, *Cell (Cambridge, Mass.)* **37,** 179 (1984).

[4] M. A. H. Surani, S. C. Barton, and M. L. Norris, *Cell (Cambridge, Mass.)* **45,** 127 (1986).

[5] C. Sapienza, A. C. Peterson, J. Rossant, and R. Balling, *Nature (London)* **328,** 251 (1987).

[6] P. M. Wassarman, *Annu. Rev. Cell Biol.* **3,** 109 (1987).

[7] B. D. Shur, in "The Molecular Biology of Fertilization" (H. Schatten and G. Schatten, eds.), Cell Biology Ser., p. 38. Academic Press, San Diego, 1989.

[8] G. Schatten, C. Simerly, and H. Schatten, *Proc. Natl. Acad. Sci. U.S.A.* **88,** 6785 (1991).

Copyright © 1993 by Academic Press, Inc.
All rights of reproduction in any form reserved.

of mitochondria[9]; and unexpected signal transduction pathways for fertilization and cell cycle regulation.[10]

The justification for performing initial mammalian studies on the mouse model is quite strong. There is a superb and accurate literature, huge numbers of genetic strains are available, and, for fertilization researchers, the clarity and malleability of the oocytes are excellent. In addition, mouse oocytes are easily maintained, and development occurs with good synchrony. Notwithstanding these virtues, mammalian oocytes and embryos provide unusual challenges for the researcher interested in detecting cytoplasmic or nuclear structural features: they are large cells (80–150 μm), some oocytes are nearly opaque (e.g., bovine, porcine, canine), and the extracellular zona pellucida and cumulus cells can cause problems in introducing the imaging probe (e.g., antibody, calcium-sensitive dye) and in removing excess probe for later detection.

This chapter considers the methods used in the detection of cytoskeletal and nuclear architectural structures in mouse oocytes during fertilization and the methods used in exploring their regulation by intracellular calcium ion imaging. We first describe the methods used for static immunocytochemical localization of structural molecules in fixed specimens. Next we consider the advantages of and methods for microinjecting probes to detect rarer structural components such as centrosomal γ-tubulin[11] and for the investigation of the function of particular proteins such as motors in the kinetochore.[12] The detection of cytoskeletal dynamics in living oocytes using fluorescence recovery after photobleaching (FRAP)[13] or fluorescence-activated cytochemistry is also discussed. Finally, calcium ion imaging is reviewed using both conventional and confocal microscopy.[14,15]

[9] U. Gyllensten, D. Wharton, A. Josefsson, and A. C. Wilson, *Nature (London)* **352**, 255 (1991).

[10] R. S. Paules, R. Buccione, R. C. Moschel, G. F. Vande Woude, and J. J. Eppig, *Proc. Natl. Acad. Sci. U.S.A.* **86**, 5395 (1989).

[11] M. J. Palacios, H. C. Joshi, C. Simerly, and G. Schatten, *J. Cell. Sci.* in preparation (1993).

[12] C. Simerly, R. Balczon, B. R. Brinkley, and G. Schatten, *J. Cell Biol.* **111**, 1491 (1990).

[13] G. J. Gorbsky, P. J. Sammak, and G. G. Borisy, *J. Cell Biol.* **104**, 9 (1987).

[14] R. M. Tombes, C. Simerly, G. G. Borisy, and G. Schatten, *J. Cell Biol.* **117**, 799 (1992).

[15] S. A. Stricker, V. E. Centonze, S. W. Paddock, and G. Schatten, *Dev. Biol.* **149**, 370 (1992).

Methods

Superovulation, Collection, and Handling of Mouse Oocytes and Embryos

Routine embryological techniques for the superstimulation, *in vitro* fertilization or *in vivo* mating, and recovery of oocytes and embryos from mice have been employed for these studies. Readers interested in this methodology are referred to a number of excellent reviews.[16-19]

Culture media used for the collection and handling of oocytes and embryos is a modified Krebs–Ringer solution containing 10 mM HEPES (M2).[20] *In vitro* fertilization and culturing is performed in M16, a CO_2-buffered medium.[21] Culture medium employed for attaching embryos to polylysine-coated coverslips and subsequent permeabilization is a modified version of M2 without calcium or bovine serum albumin (BSA) and containing 250 μM EGTA (0 Ca^{2+}/0 protein M2). Details of the preparation these solutions can be found in the reviews by Whittingham[21] and Pratt.[18]

Removal of Cumulus Cells and Zona Pellucida

Immature ovarian oocytes and mature ovulated oocytes often are surrounded by follicular and/or cumulus cells that must be removed prior to fixation to achieve good visualization of oocyte structure. This can be accomplished by mechanical means using a small-bore pipette drawn out in a flame, by incubating in 0.1% (w/v) hyaluronidase prepared in warm (37°) M2 culture medium, or using a combination of both techniques.[16] If using hyaluronidase, the treatment should be kept short (5–7 min) to prevent artificially activating the oocytes. It is important to remove adhering cumulus cells before removal of the zona pellucida to prevent these cells from sticking to the oocyte surface.

Removal of the zona pellucida is usually necessary in order to attach oocytes to substrates and subsequently stain intracellular proteins. Removal of this extracellular layer facilitates better fixation, antibody penetration during immunostaining, and imaging during microscopic analysis.

[16] K. A. Rafferty, "Methods in Experimental Embryology of the Mouse." Johns Hopkins Univ. Press, Baltimore, Maryland, 1970.

[17] J. D. Biggers, W. K. Whitten, and D. G. Whittingham, *in* "Methods in Mammalian Embryology" (J. C. Daniel, Jr., ed.), p. 86. Freeman, San Francisco, 1971.

[18] H. P. M. Pratt, *in* "Mammalian Development: A Practical Approach" (M. Monk, ed.), p. 13. IRL Press, Oxford, 1987.

[19] B. D. Bavister, *Gamete Res.* **23**, 139 (1989).

[20] B. P. Fulton and D. G. Whittingham, *Nature (London)* **273**, 149 (1978).

[21] D. G. Whittingham, *J. Reprod. Fertil. Suppl.* **14**, 7 (1971).

Several methods have been described to remove the zona pellucida, including enzymatic digestion with 0.5% proteinase prepared in 37° M2 culture medium (5 min), dissolution in warm acidified M2 culture medium (pH 2.5; 15–30 sec), or mechanical stripping with a small-bore pipette.[18] Care should be exercised with any method employed to remove the zona, as damage to cell membranes can result in lysis of oocytes or embryos. Incubation times with enzymes and acidified culture medium should be kept to a minimum to reduce intracellular damage. Monitor zona pellucida swelling and thinning by microscopy, and remove the oocytes from the enzyme or acid M2 just before complete dissolution. It is recommended that recovery periods of 30 min or longer in normal culture medium be permitted after zona removal.

Details of the routine methods employed to remove cumulus cells and the zona pellucida from mouse oocytes in preparation for immunocytochemical processing are given in Table I.

Static Immunocytochemistry Protocols

A number of excellent reviews on immunocytochemical (ICC) techniques are recommended to acquaint readers with the theories and methodology of immunofluorescent staining.[22–24] The application of ICC technology to mammalian oocytes and embryos can be complicated, however, since these large, yolk-filled cells can contain significant amounts of soluble proteins and cytoplasmic inclusions that can degrade image quality by increasing nonspecific antibody staining. In addition, the small numbers of oocytes or embryos generated and the fact that they are not attached to substrates makes it cumbersome to process them by standard ICC protocols.

In this section, we consider some of the methods used for preparing mouse oocytes for immunocytochemical detection of intracellular proteins. Table I lists the detailed steps for processing mouse oocytes and embryos for immunostaining following detergent extraction in a microtubule stabilization buffer and methanol fixation, a protocol we have found useful for the localization and detection of a wide assortment of cytoskeletal and nuclear proteins.

Attaching Oocytes and Embryos to Glass Coverslips. Affixing oocytes and embryos to substrates facilitates ICC processing. Poly(L-lysine), a charged polyamino acid, is a particularly useful compound with which to

[22] B. R. Brinkley, S. H. Fistel, J. M. Marcum, and R. L. Pardue, *Int. Rev. Cytol.* **46,** 59 (1980).

[23] M. Osborn and K. Weber, *Methods Cell Biol.* **24,** 97 (1982).

[24] E. Harlow and D. Lane, "Antibodies: A Laboratory Manual," Chap. 10. Cold Spring Harbor Laboratory, Cold Spring Harbor, New York, 1988.

TABLE I
IMMUNOCYTOCHEMICAL PREPARATION OF UNFERTILIZED MOUSE OOCYTES
BY BUFFER M PERMEABILIZATION AND METHANOL FIXATION

A. Removal of cumulus cells and zona pellucida
  1. Place unfertilized oocytes in 0.1% (w/v) hyaluronidase (Sigma; type I-S) in M2 culture medium kept at 37°. Incubate for 5 min
  2. Remove any adhering cumulus cells by rapid pipette agitation using small-bore pipette with diameter slightly larger than zona-intact oocytes
  3. Rinse oocytes 3 times in warm M2 culture medium
  4. Transfer oocytes to warm, acidified M2 (pH 2.5, 37°) for 15–30 sec. Monitor zona pellucida swelling and thinning by microscopic observation and remove from acid M2 just before complete dissolution
  5. Wash 3 times in M2 culture medium to neutralize the pH and remove remaining zona remnants
  6. Allow zona-free oocytes a minimum of 30 min to recover in M2 or M16 culture medium
B. Attaching mouse oocytes to polylysine-coated coverslips
  1. Place a polylysine-coated 22-mm^2 coverslip in bottom of 6-well flat-bottom assay plate (Falcon, 3046)
  2. Add 5 ml of Ca^{2+}- protein-free M2 culture medium at 37°
  3. Carefully pipette washed zona pellucida-free oocytes onto polylysine-coated coverslips; incubate at 37° for 2–3 min
C. Permeabilization in detergent-containing microtubule-stabilizing buffer M
  1. Remove most of 0 Ca^{2+}/0 protein M2 culture medium from plate; do not expose attached oocytes to liquid–air interface
  2. Carefully pipette into plate ~6 ml buffer M containing 1% Triton X-100, 1 mM 2-mercaptoethanol, and 0.2 mM PMSF (pH 6.8, 37°)
  3. Incubate 10 min at 37°
D. Affixing extracted, permeabilized oocytes to polylysine-coated coverslips and methanol fixing
  1. Prepare new polylysine-coated coverslip; place in 4-well petri dish (Falcon, 1009), rinse 1 min in 0.1 M PBS + 0.1% Triton X-100 (PBS–TX), and aspirate completely dry
  2. Using a flame-drawn pipette with opening slightly larger than diameter of extracted oocytes, transfer permeabilized oocytes to dry polylysine-coated coverslip; transfer minimum amount of fluid with oocytes. Continually move pipette across glass surface as oocytes are expelled; do not introduce air bubbles. Step works best when performed with dissecting microscope at low magnification using dark-field optics
  3. Immediately add 3 ml warm buffer M without detergent or methanol additives. Do not air-dry attached, permeabilized oocytes
  4. Slowly add − 10° absolute methanol into buffer M until saturation. Incubate for 10 min at room temperature
  5. Remove methanol by rinsing in PBS–TX. Permeabilized, fixed oocytes can be stored overnight before immunostaining
E. Immunostaining permeabilized, methanol-fixed oocytes for detection of microtubules, centrosomes, and DNA
  1. Blot excess PBS–TX off coverslip with Kimwipe; add 70 μl of 0.1 M PBS + 0.1% Triton X-100 + 3 mg/ml BSA (PBS–TX–BSA) to block nonspecific secondary antibody-binding sites. Incubate at 37° for 30 min in humidified chamber

TABLE I (*continued*)

2. Blot off excess blocking solution; add 70 μl human polyclonal anticentrosomal antibody SPJ diluted 1 : 200 in 0.1 *M* PBS + 3 mg/ml BSA (PBS–BSA). Incubate at 37° for 40 min in a humidified chamber
3. Remove anticentrosome antibody; rinse 3 times in PBS–TX–BSA at room temperature 10 min each (30 min total)
4. Blot off excess rinse solution; add 70 μl biotin-conjugated goat anti-human IgG diluted to 10 μg/ml in PBS–BSA
5. Rinse as in Step E.3
6. Blot off excess rinse solution; add 70 μl fluorescein-conjugated streptavidin diluted 1 : 40 in PBS–BSA. Incubate at 37° for 40 min in humidified chamber
7. Rinse as in Step E.3
8. Remove rinse solution and add 70 μl mouse monoclonal β-tubulin antibody E-7 diluted 1 : 10 in PBS–BSA. Incubate at 37° for 40 min in humidified chamber
9. Rinse as in step E.3
10. Blot off excess rinse solution; add 70 μl rhodamine-conjugated goat anti-mouse IgG diluted 1 : 40 in PBS–BSA. Incubate at 37° for 40 min in a humidified chamber
11. Rinse as in step E.3
12. Remove rinse solution; add 70 μl of 2.5 μg/ml 4′,6-diamidino-2-phenylindole (DAPI) DNA stain (10 min, room temperature)
13. Blot off excess DAPI; rinse once in distilled water to remove salt
14. Mount coverslip with oocytes face down in 30 μl glycerol–DABCO solution. Do not introduce air bubbles. Blot off any excess mounting solution and seal with nail polish

attach living or fixed mouse oocytes to coverslips.[25] A stock solution of 2 mg/ml poly(L-lysine) (PL) hydrobromide (MW >300,000; Sigma Chemical, St. Louis, MO) is prepared in distilled water, aliquotted, and stored frozen at −20°.

Before treating coverslips (22 mm² or others) with the PL solution, clean the glass by boiling for 20 min in distilled water containing a laboratory detergent to remove dirt and greases that interfere with good polylysine adherence. Rinse the detergent-cleaned coverslips well with distilled water and store in 95% (v/v) ethanol.

To coat coverslips with PL, remove coverslips from the alcohol and wipe surfaces clean with a lint-free cloth. Soak coverslips for 5 min in the stock solution of PL. The PL solution should not form a bead on the coverslip but should smoothly coat the entire surface. After 5 min, rinse in distilled water and aspirate dry or let air-dry.

The procedure to attach zona-free oocytes or embryos to the PL-treated coverslips is given in Table I. Good egg attachment will occur only if the culture medium or phosphate-buffered saline (PBS) solution

[25] D. Mazia, G. Schatten, and W. Sale, *J. Cell Biol.* **64,** 198 (1975).

does not contain proteins such as BSA or serum. It is advisable to first rinse zona-free oocytes once in Ca^{2+}-free and protein-free M2 in an agar-coated dish (to prevent zona-free oocytes from sticking to dish) immediately before transferring to the PL-coated coverslip. Agar dishes can be prepared by adding 2.5 ml of a 1% Bacto-agar solution dissolved in distilled water to a 10×35 mm petri dish. Plates can be prepared in advance and stored at $4°$ for up to 1 week.

Other compounds used in a similar manner as PL to adhere oocytes to glass coverslips include protamine sulfate,[15] concanavalin A, or phyto-hemagglutinin.[18]

Permeabilization before Fixation. Permeabilization of oocytes and embryos in detergent-containing buffers which stabilize intracellular proteins have several practical advantages: they increase antibody penetration, lower nonspecific antibody binding, reduce monomer concentration for polymer detection, and allow greater sensitivity for the detection of small cytoplasmic structures such as centrosomes and kinetochores.

Table I demonstrates the technique for permeabilizing and extracting mouse oocytes in buffer M, a glycerol-based microtubule-stabilizing solution shown to preserve cytoplasmic microtubules while extracting about 80% of the soluble cellular proteins in cultured cells.[26] Prepare buffer M by mixing the following constituents and heating to $37°$: 25% (v/v) glycerol, 50 mM KCl, 0.5 mM MgCl$_2$, 0.1 mM EDTA, 1 mM EGTA, and 50 mM imidazole hydrochloride, pH 6.8. Before use, add 1 mM 2-mercaptoethanol. To permeabilize and extract oocytes and embryos, add 1% nonionic detergent such as Triton X-100 or Nonidet P-40 (NP-40) and 0.2 mM phenylmethylsulfonyl fluoride (PMSF; prepared as a 0.1 M stock in absolute methanol).

Several observations are worth noting in employing this technique. Buffer M permeabilization/extraction will not be complete unless zona-free oocytes are used. Also, it is advisable not to have extracellular calcium present during the permeabilization step to prevent calcium-induced alterations in the cytoskeletal system. Low concentrations of detergent (<0.5%) in buffer M can lead to insufficient extraction and cell lysis, probably as a result of a differential osmolarity gradient between the external buffer and the egg cytoplasm. Attachment of oocytes to polylysine-coated coverslips before detergent permeabilization is crucial to prevent the oocytes from floating in the buffer, which can greatly hinder recovery and increase structural damage to the extracted cells. For reat-

---

[26] A. D. Bershadsky, V. I. Gelfand, T. M. Svitkina, and I. S. Tint, *Cell Biol. Int. Rep.* **2**, 425 (1978).

taching detergent-permeabilized oocytes to new PL-coated coverslips following buffer M treatment, keep the volume of oocytes/buffer M to a minimum. It is also important not to expel the oocytes and buffer fluid all at once at the center of the coverslip, or the cells may not stick. Gently sweep the pipette across the surface of the coverslip to get the best reattachment. When using cold organic solvents like methanol to fix the cells, add the solvent slowly, or detachment and/or damage to the extracted eggs can occur. Never let detergent-extracted cells become exposed to the air–liquid interface.

In selecting a cell fixation method following buffer M permeabilization, it is crucial to take into account the nature of the protein(s) being targeted for detection intracellularly. It is necessary to avoid permeabilization and/or fixation steps that may alter antigenicity and redistribution of the proteins in vivo.[27] Fixatives that have been successfully used after buffer M extraction of mouse oocytes and embryos include methanol, glutaraldehyde, and other cross-linking reagents like dimethyl 3,3'-dithiobispropionimidate (DTBP; Pierce, Rockford, IL) or ethylene glycol bis(succinimidylsuccinate) (EGS; Pierce).[28]

To fix extracted oocytes in cold methanol, slowly dilute the organic solvent into buffer M until saturation and remove after 10 min by rinsing in PBS plus 0.1% Triton X-100 detergent as outlined in Table I.

Glutaraldehyde as a fixative is not recommended for immunofluorescence because it often destroys the intracellular antigenicity of proteins or increases nonspecific fluorescent background in the wavelengths commonly employed for ICC labeling. However, glutaraldehyde can be used as a fixative for immunoelectron microscopy observations. Electron microscopy (EM) grade glutaraldehyde prepared in buffer M at 1–2% concentrations will satisfactorily fix extracted eggs within 30 min. Formaldehyde is not recommended as a fixative for mouse oocytes and embryos following buffer M detergent extraction because of the extensive cytoskeletal lysis and damage that occurs when aqueous solutions such as PBS are applied during subsequent rinse stages.

Cross-linking reagents like DTBP and EGS can be employed in situations where antigenicity is deleteriously affected by organic solvents or by aldehyde fixation. A stock solution of 100 mM DTBP or EGS is prepared in dimethyl sulfoxide (DMSO) just before use and diluted 1 : 10 into buffer M.[28] Adjust the pH of the buffer solution to 7.8 to 8.1, and allow the cross-linking reaction to proceed for 3–6 hr at room temperature to get the best results with these reagents.

[27] M. A. Melan and G. Sluder, J. Cell Sci. 101, 731 (1992).
[28] R. Miake-Lye and M. W. Kirschner, Cell (Cambridge, Mass.) 41, 165 (1985).

Other investigators have employed variations of the microtubule-stabilizing buffer PHEM (60 mM PIPES, pH 6.9/25 mM HEPES/1 mM MgCl$_2$ · 6H$_2$O/10 mM EGTA)[29] to permeabilize and extract mouse oocytes and embryos before fixation. These simple, nonglycerinated buffers sometimes include microtubule-stabilizing substances such as taxol (0.6–1.0 μM) and/or deuterium oxide (50%) to secure the cytoplasmic microtubule populations; details of the use of these buffers for permeabilizing mouse oocytes can be found by consulting the original articles.[30,31]

Fixation before Permeabilization. In certain instances, soluble proteins or proteins with transient or weak associations with intracellular structures may not be preserved using detergent-based permeabilization buffers. In these cases it is advisable to fix oocytes or embryos before detergent extracting.

Zona-denuded oocytes attached to polylysine-coated coverslips are fixed in fresh 2% formaldehyde (EM grade, methanol free; Polysciences, Warrington, PA) prepared in protein-free M2 culture medium or 0.1 M PBS at 37° for 1 hr. After fixation, rinse in PBS and permeabilize/ extract in 0.1 M PBS containing 0.5% Triton X-100 for an additional 30 min. Be aware that nonionic detergents such as Triton X-100 can reverse the cross-linking property of formaldehydes, resulting in poorly preserved oocytes.[24]

Oocytes and embryos fixed with glutaraldehyde and formaldehyde can have unreacted aldehyde groups, which will increase background staining.[32] It is necessary, therefore, to use blocking steps to limit degradation of image quality following antibody application. If dilute solutions of glutaraldehyde have been employed in the ICC protocol, 0.5 mg/ml sodium borohydride prepared fresh in 0.1 M PBS can be employed to reduce any unwanted background signal.[32] For formaldehyde, any of a number of blocking solutions will reduce background staining. Included in this list are 0.1 M PBS solutions containing 150 mM glycine and 3 mg/ml BSA,[33] 0.26% ammonium chloride,[34] or 2% BSA/2% (w/v) powdered milk/2% (v/v) normal goat serum/0.1 M glycine.[31] Incubation times in blocking solutions range from 10 min to 3–4 days.

[29] M. Schliwa, U. Euteneuer, J. C. Bulinsky, and J. G. Izant, *Proc. Natl. Acad. Sci. U.S.A.* **78,** 1037 (1981).
[30] E. Houliston and B. Maro, *J. Cell Biol.* **108,** 543 (1989).
[31] S. M. Mesinger and D. F. Albertini, *J. Cell Sci.* **100,** 289 (1991).
[32] M. Osborn, *Tech. Cell. Physiol. Part 1–2 (Tech. Life Sci.: Physiol.)* **P107,** 1 (1981).
[33] T. Ducibella, E. Anderson, D. F. Albertini, J. Aalberg, and S. Rangarajan, *Dev. Biol.* **130,** 184 (1988).
[34] S. J. Pickering, M. H. Johnson, P. R. Braude, and E. Houliston, *Hum. Reprod.* **3,** 978 (1988).

Simultaneous Fixation/Permeabilization. Concurrent permeabilization and fixation protocols that allow for the localization of soluble and insoluble intracellular proteins in their native distribution can be a useful protocol for processing oocytes and embryos. Among the reagents to employ, protein-precipitating fixatives such as methanol or the combination of paraformaldehyde/lysolecithin will preserve cellular constituents while allowing cytoplasmic antibody penetration during immunostaining.

Oocytes affixed to PL-coated coverslips can be directly immersed in −20° methanol for 10 min followed by rehydration in PBS.[35] No detergent extraction is employed to further antibody penetration, since methanol solubilizes the plasma membrane. Intracellular structural damage and a loss in antigenicity of targeted proteins can be greater if oocytes are fixed in other organic solvents such as acetone or ethanol.

Alternatively, oocytes attached to PL-coated coverslips can be directly fixed and permeabilized by incubating in 2% paraformaldehyde, 50 μg/ml lysolecithin prepared in 0 Ca^{2+}/0 protein M2 culture medium at 37° for 1 hr.[36] Oocytes are rinsed in PBS after fixation and blocked using 150 mM glycine, 3 mg/ml BSA prior to immunostaining.

Fluorescent Immunostaining Protocols. The procedure to immunostain permeabilized and fixed mouse oocytes is given in Table I. All antibodies should be diluted into an appropriate aqueous buffer solution like 0.1 M PBS containing a protein carrier such as BSA (3 mg/ml) or 1% immunoglobulin derived from the same species as the detection reagent (i.e., normal goat, human, or rabbit serum) before applying to fixed, permeabilized, and properly washed cells. The dilution of primary antibody will depend on the specific antibody being used (monoclonal, polyclonal, or ascites fluid) and needs to be empirically determined by testing several dilutions. Secondary antibodies purchased from any number of commercial companies work well when diluted 1 : 10 to 1 : 100. It may be necessary to experiment with the concentration of the antibodies in order to obtain a detectable signal while keeping nonspecific background fluorescence to a minimum. It is advisable to spin each diluted solution briefly in an Eppendorf microcentrifuge for 2 min to remove insoluble particulates.

To apply the antibodies, use a 4-well petri dish (Falcon, No. 1009) and place the coverslip with attached oocytes in the center of the petri dividers, with a small amount of 0.1 M PBS buffer added into each well for humidification purposes. Other chambers for immunostaining can be employed.[24]

---

[35] E. Houliston, M. Guilly, J. Courvalin, and B. Maro, *Development (Cambridge, UK)* **102**, 271 (1988).

[36] G. Schatten, H. Schatten, I. Spector, C. Cline, N. Paweletz, C. Simerly, and C. Petzelt, *Exp. Cell Res.* **166**, 191 (1986).

For 22-mm^2 coverslips, pipette 70 μl of appropriately diluted primary antibody onto the coverslip and incubate for 40–90 min at 37° or overnight at 4° to label intracellular antigens. After staining, rinse in 0.1 M PBS with 0.1% Triton X-100 and 3 mg/ml BSA (PBS–TX–BSA) for an additional 30 min to remove excess unbound primary antibody. If using the indirect labeling technique, apply the diluted secondary antibody with fluorochrome after this rinse step in an analogous manner as the primary antibody. Follow secondary antibody staining with a wash in PBS–TX–BSA as stated previously.

The number of proteins one can detect following immunostaining will depend on the number of species of primary antibodies used and the number of fluorochromes that can be imaged by the microscope system. Table I lists a typical scenario for imaging three different wavelengths in the mouse. Best results are achieved if each antibody is added sequentially, but it is possible to mix primary antibodies together for a single-step application.

Some intracellular proteins such as kinetochores are present in extremely low copy numbers, and in these cases it may be advisable to amplify the detectable signal by labeling with the biotin/streptavidin system (see Schatten et al.[36a]). After primary antibody application, add 10 μg/ml biotin-conjugated secondary antibody and incubate for 60 min at 37°. Rinse in PBS–TX–BSA for 30 min and then apply a 1 : 40 dilution of fluorochrome-conjugated streptavidin prepared in 0.1 M PBS plus 3 mg/ml BSA. Because higher background fluorescence is generally produced with this labeling and detection system, it is advisable to employ the protocol only after extensive cell permeabilization.

Mounting Immunostained Cells for Fluorescent Observation. Because fluorochromes are susceptible to fading under fluorescent irradiation, stained oocytes are mounted in a buffer containing an oxygen scavenger reagent. There are several varieties of aqueous or nonaqueous buffers and antifade reagents that will work well with fluorescent samples.[24] In some instances, the method of imaging the stained cells will dictate the type of quenching reagent to use. Such is the case when viewing immunostained cells in the confocal microscope, where p-phenylenediamine appears to be a superior antifade reagent. For routine epifluorescence microscopy, however, 1,4-diazobicyclo[2.2.2.]octane (DABCO; Aldrich Chemical Co., Milwaukee, WI) can be used: it is less toxic and has a longer shelf-life.

[36a] G. Schatten, C. Simerly, D. J. Asai, E. Szöke, P. Cooke, and H. Schatten, *Dev. Biol.* **130**, 74 (1988).

After the final rinse step, coverslips containing stained cells are removed from 0.1 M PBS–TX–BSA and briefly rinsed in distilled water to prevent salt crystallization on the surface of the coverslip. Blot excess water off the coverslip, invert, and mount in a glycerol–PBS (9 : 1) mixture containing the antifade compound DABCO (100 mg/ml), being careful to avoid air bubbles. Seal the coverslip to a glass slide with nail varnish. Store samples at 4° or lower in a light-tight slide box until the photography session.

Photography. Immunostained specimens can be viewed by conventional epifluorescence or by confocal laser scanning microscopy.[23,37] Results of cell staining should be photographed as soon as possible after mounting to obtain a permanent record of the data before fading or redistribution of the fluorochromes occurs. To reduce exposure of immunostained samples to UV excitation and minimize potential bleaching problems, it is recommended that fast speed black-and-white professional films that can be push-processed in high-contrast developer be employed to obtain publication-quality photographs. Color films of ASA 160 or greater can be used to obtain presentation-quality slides.

Black-and-white images presented in this chapter were recorded on Kodak (Rochester, NY) Tri-X film exposed at 1600 ASA and developed in Diafine according to the manufacturer's instructions. Color images were taken on Ektachrome (Kodak) 200 ASA daylight color film and commercially processed.

Controls. The localization of intracellular antigens by any immunofluorescence method must be confirmed by careful scrutiny against a number of controls, as outlined by Brinkley *et al.*[22] and Osborn.[32] Staining patterns observed after replacing the primary antibody with preimmune sera, after preabsorption of primary antibody with excess antigen, or after staining with secondary antibody alone should be routinely performed. Similar dilutions of each control solution should be used for comparison with the staining patterns obtained with the primary antibody and similar exposure/development times used when photographing and processing are done.

Detection of Cellular Antibodies Following Microinjection of Primary Antibodies

Focal concentrations of antigens present in low copy numbers, such as centromere-associated proteins in kinetochores and γ-tubulin in centrosomes, are often more difficult to stabilize *in vivo* during permeabilization and fixation steps or to detect against background signal following standard

[37] S.J . Wright and G. Schatten, *J. Electron Microsc. Tech.* **18**, 2 (1991).

immunofluorescent staining. To improve the detectability of such proteins, primary antibodies can first be microinjected into living mouse oocytes to bind the target antigen before processing for immunocytochemistry. If the microinjected antibody can recognize the intracellular protein in its native form, this technique can markedly improve detection of rarer structures, since fixation damage to antigen-binding sites prior to antibody application can be avoided. In addition, microinjection of antibodies serves as a valuable method to explore the possible functions of particular intracellular proteins *in vivo* (see, e.g., Ref. 2).

Several good reviews of microinjection technology are available to help design and perform these manipulations in mouse oocytes.[38–40] Micropipettes used for injecting antibodies into mouse oocytes should be pulled and beveled (Sutter Instruments, Novato, CA) to create tips with outer diameters of 1–2 μm.[38] Egg-holding pipettes with outer diameters of about 20 μM can be constructed using a microforge (deFonbrune, Technical Products International, St. Louis, MO).[39] The useful life of an injection micropipette can be considerably extended if the pipettes are siliconized for 2–4 days by exposure to hexamethyldisilazane vapor in a vacuum desiccator.[40]

To prepare the oocytes for microinjection, remove the cumulus cells with hyaluronidase as described earlier but leave the zona pellucida intact. Executing the injection in M2 culture medium under oil at room temperature will prevent excessive cell lysis caused by puncturing the cell membrane at 37°. Antibody can be front-loaded into the pipette from a 1–2 μl drop adjacent to the oocytes. While securing an oocyte with the holding pipette, microinject the solution into the egg by puncturing the zona, drawing in a small amount of cytoplasm, and expelling the antibody/cytoplasmic contents back into the oocyte. Do not exceed 5% of the cell volume (~10 pl) or cell lysis will occur. About a dozen oocytes should be injected over a period of 20–30 min to limit exposure time to room temperature. Allow injected oocytes to recover at 37° for at least 30 min to monitor cells for lysis and to allow the antibody to bind the intracellular antigen.

Microinjected oocytes can be permeabilized, fixed, and processed for immunocytochemical detection of intracellular antigens by the methods outlined in Table I, except that the primary antibody staining step is omitted. Advisable controls to perform to confirm the specificity of the

---

[38] T. Uehara and R. Yanagimachi, *Biol. Reprod.* **15**, 467 (1976).

[39] V. M. Thadani, *J. Exp. Zool.* **212**, 435 (1980).

[40] M. L. DePamphilis, S. A. Herman, E. Martínez-Salas, L. E. Chalifour, D. O. Wirak, D. Y. Cupo, and M. Miranda, *BioTechniques* **6**, 662 (1988).

antibody–antigen reaction include microinjection of nonimmune sera or immunodepleted sera and staining noninjected oocytes with secondary antibodies only. Additionally, sham microinjections and use of other antibodies that recognize the protein(s) of interest should be included in the controls if the functions of the intracellular antigens *in vivo* are being examined.

Studies in Living Mouse Oocytes: Time-Lapse Video Microscopy and Fluorescent Cytochemistry

Time-Lapse Video Microscopy. The sequence of movements during fertilization responsible for sperm incorporation, pronuclear formation and apposition, and meiosis/mitosis can be documented *in vitro* using video microscopy employing differential interference contrast optics (e.g., Prather *et al.*[41]; Simerly *et al.*[42]).

Fertilized oocytes should have the zona pellucida removed to obtain the best imaging quality. Denuded embryos are placed in 50-μl droplets of M16 culture medium in a 10 × 35 mm glass-bottomed microwell petri dish (Cat. No. P35G, MatTek Corporation, Asland, MA) and overlaid with 4 ml of mineral oil. Alternatively, growth chambers such as the Dvorak-Stotler controlled environment culture system (Nicholson Precision Instruments, Gaithersburg, MD) can be used. Temperature maintenance is crucial for proper development and can be controlled using an air curtain heater in a plexiglass enclosure which surrounds the stage and imaging objective. Culture the samples under prewarmed, humidified 5% CO_2 gas directed across the petri dish or chamber using a laminar flow device (Model 5010, Precision Digital, Watertown, MA) modified to fit to the microscope stage.

Images are captured with a Newvicon video camera (Panasonic WV1350), processed on an Image 1/AST computer, and recorded on either videotape or optical disc at 30-sec intervals. Embryos develop best if not continually illuminated by the halogen light source; therefore, a computer-controlled shutter system should be used to block out light exposure when not recording (Uniblitz; Vincent Associates, Rochester, NY).

Tracing Dynamic Properties of Microtubules in Living Mouse Oocyte by Fluorescent Chemistry. The dynamic properties of intracellular proteins such as microtubules can be explored by microinjecting chromophore-derivatized native protein into oocytes and determining fluorescent

---

[41] R. Prather, C. Simerly, G. Schatten, D. R. Pilch, S. M. Lobo, W. F. Marzluff, W. L. Dean, and G. A. Schultz, *Dev. Biol.* **138**, 247 (1990).

[42] C. Simerly, N. B. Hecht, E. Goldberg, and G. Schatten, *Dev. Biol.* in preparation (1993).

recovery after photobleaching (FRAP; see Gorbsky *et al.*[43]). Readers interested in the possible applications of fluorescent analog cytochemistry in living cells to explore dynamic processes are referred to the review by Wang.[44]

Polymerization-competent porcine brain tubulin is derivatized with X-rhodamine isothiocyanate (XRITC) according to the protocols of Sammak and Borisy[45] and stored in liquid nitrogen until used. A 5-μl aliquot is centrifuged in an airfuge at 90,000 g for 10 min at 4° immediately before microinjection. Injection into unfertilized oocytes is performed as previously described.

To perform FRAP, microinjected oocytes are transferred to 50 μl of M2 culture medium without phenol red indicator in a Sykes-Moore (Bellco Glass, Vineland, NJ) tissue culture chamber. Temperature is maintained on an inverted Zeiss IM 35 microscope stage at 37° by an air curtain heater. The instrumentation and protocols for FRAP analysis have been described by Gorbsky *et al.*[13] Briefly, focusing and alignment of meiotic spindles are performed with 0.1-sec attenuated mercury arc illumination and computer-enhanced SIT imaging, keeping exposure times under 1 sec. After proper registration, live images before and after laser photobleaching are captured with a colled CCD camera following a 1-sec exposure with green mercury arc light from a 100-W source attenuated to 10% by neutral density filters. Laser photobleaching is accomplished by directing laser light into the microscope and focusing the beam to 4 × 57 mm at the specimen focal plane by means of a 200-mm cylindrical lens. Images are stored on an Image 1/AST 386 image processor for analysis according to published protocols.[43]

The introduction of derivatized protein into oocytes may alter the functional properties of microtubules *in vivo,* and the potential for introducing structural damage to cytoplasmic microtubules following UV exposure of fluorescent tubulin protein is a concern which must be addressed experimentally.[46] Oocytes microinjected with derivatized tubulin and exposed to UV illumination should be artifically activated by exposure to 7% (v/v) ethanol in M2 for 7 min at room temperature, washed 3 times in M2, and cultured in M16 at 37° to confirm normal resumption of meiosis, pronuclear development, and cleavage *in vitro.* In addition, total microtubule patterns in the bleached regions following FRAP should be analyzed by standard microtubule immunofluorescence staining as outlined in

[43] G. J. Gorbsky, C. Simerly, G. Schatten, and G. G. Borisy, *Proc. Natl. Acad. Sci. U.S.A.* **87,** 6049 (1990).
[44] Y. L. Wang, *Methods Cell Biol.* **29,** 1 (1989).
[45] P. J. Sammak and G. G. Borisy, *Nature (London)* **332,** 724 (1988).
[46] G. P. A. Vigers, M. Coue, and J. R. McIntosh, *J. Cell Biol.* **107,** 1011 (1988).

Table I to confirm that microtubules have not been structurally altered by UV illumination.

Fluorescent Calcium Imaging. Quantitative radiometric calcium dyes such as Fura-2 are powerful tools for investigating changes in intracellular ionic signals responsible for initiating oocyte activation.[14] Fura-2/am, the acetoxymethyl ester form of the calcium indicator, is prepared by first dissolving in 2% (w/v) pluronic F-127 (Molecular Probes, Eugene, OR) in anhydrous DMSO followed by diluting 1000-fold into M2 culture medium at pH 7.3 to a final concentration of 1 μM. Dye loading is achieved in oocytes or embryos by incubating at 37° for 1 hr followed by two rinses in M2.

Imaging of Fura-2/am-treated oocytes and embryos is accomplished using an intensified silicon target video camera (ISIT; Model 66; Dage-MTI, IN) coupled to a Nikon Diaphot inverted microscope (Nikon, NY) equipped with a UV-transmitting phase-contrast objective (20× Fluor, 0.75 NA; Nikon).[14] Image acquisition and processing are performed with the Image-1 calcium analysis software (Universal Imaging Corp., Media, PA). Quantitation of intracellular calcium is ratiometrically determined at dual-excitation spectra of 340 and 380 nm. Illumination of the 75-W xenon power source is attenuated 1000-fold using a computer-controlled neutral density filter wheel (EMPIX Imaging, Inc., Mississauga, Ontario).

Fura-loaded oocytes are placed in 50-μl drops of M2 culture medium in a 10 × 35 mm glass-bottomed microwell petri dishes (Cat. No. P35G, MatTek Corporation) and kept at 37° using the Diaphot incubator with temperature control unit. Images are collected at 10-sec intervals and stored on the Image 1/AST hard drive for later analysis.

It is important that the concentration of calcium indicator selected allow just enough signal to be imaged, or dye toxicity may artificially influence the behavior of the oocytes.[14] Internal compartmentalization of calcium dyes into intracellular vesicles should also be monitored for potential false results of ratiometric imaging.[47] Analyzing the configurations of the microtubules, microfilaments, and chromosomes in dye-treated oocytes by the immunocytochemical methods as outlined in Table I is useful to confirm that the cytoskeletal architecture has not been compromised by the chelators and/or imaging techniques.

Detection of Cytoskeletal and Nuclear Proteins in Mouse Oocytes and Embryos

Static Immunocytochemistry

It is accepted knowledge that motility and cytoskeletal rearrangements are essential for the successful completion of mouse fertilization (reviewed

[47] A. Malgaroli, D. Milani, J. Meldolesi, and T. Pozzan, *J. Cell Biol.* **105,** 2145 (1987).

by Longo[48] and Yanagimachi[49]). Our understanding of cytoskeletal organization and changes in nuclear architecture during oocyte maturation, fertilization, and development has benefited tremendously by applying static immunocytochemistry technology (reviewed by Schatten and Schatten.[50]) These investigations have led to important understandings of the basic features that prime and position the parental genomes for merging during fertilization, a step which, if inaccurate, can lead to the disastrous consequences of chromosome imbalance. The power of ICC technology will continue to help lay the foundations for future investigations on crucial intracellular events important to understanding fully mammalian fertilization.

As first recognized by Boveri,[1] the sperm typically introduces the microtubule-organizing center (MTOC), or centrosome of the cell at fertilization. Modern ICC technology, however, was instrumental in demonstrating that the mouse oocyte violates this dogma. Antitubulin immunostaining of the unfertilized oocytes after buffer M permeabilization and methanol fixation showed unusual astral microtubule patterns present in the cytoplasm in addition to the microtubules existing in the second meiotic spindle (see review by Schatten and Schatten[50]). As shown in Fig. 1A, immunodetection of microtubules, centrosomes, and DNA within the same oocyte reveals that the MTOCs are found at the spindle poles as expected[51] and also as multiple punctate foci (cytasters) scattered throughout the cytoplasm. During fertilization (Fig.

[48] F. J. Longo, *Am. J. Anat.* **174**, 303 (1985).
[49] R. Yanagimachi, *in* "Fertilization in Mammals" (B. Bavister, J. Cummins, and E. R. S. Roldan, eds.), p. 401. Serno Symposia, Norwell, Massachusetts, 1990.
[50] G. Schatten and H. Schatten, *Curr. Top. Dev. Biol.* **23**. 23 (1987).
[51] P. C. Calarco-Gilliam, M. C. Siebert, R. Hubble, T. Mitchison, and M. Kirschner, *Cell (Cambridge, Mass.)* **35**, 621 (1983).

FIG. 1. Detection of microtubules, centrosomes, and DNA in an unfertilized (A) and early pronuclear (B) mouse oocyte. (A) In the mature unfertilized mouse oocyte arrested at metaphase of second meiosis (blue), centrosomes are found as cytoplasmic foci and at the meiotic spindle poles (green). The microtubules extend from the centrosomal material, forming the barrel-shaped, anastral spindle and cytoplasmic asters (red). (B) In the fertilized mouse oocyte containing the male and female pronuclei (blue), an extensive matrix of microtubules (green) is present, nucleated by the maternal centrosomes (orange) scattered throughout the cytoplasm. Note that no single large sperm aster structure is associated with the incorporated sperm axoneme (arrow). Images were triple-labeled for affinity-purified rabbit antitubulin antibody, 5051 centrosomal antibody, and the DNA stain 4′,6-diamidino-2-phenylindole (DAPI). Bar: 10 μm. [Figure 1A reprinted, with permission, from G. Schatten, C. Simerly, D. J. Asai, E. Szöke, P. Cooke, and H. Schatten, *Dev. Biol.* **130**, 74 (1988).]

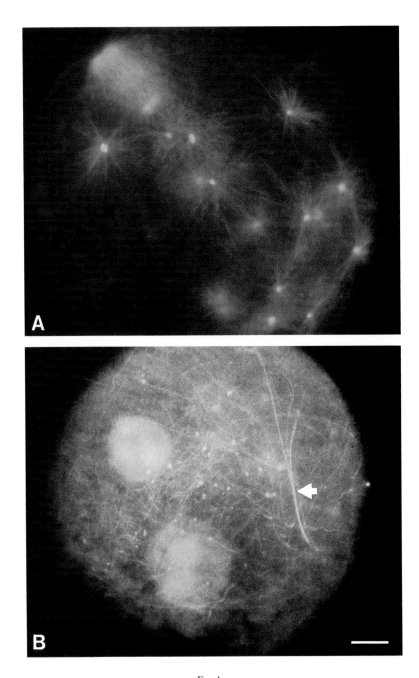

Fıg. 1

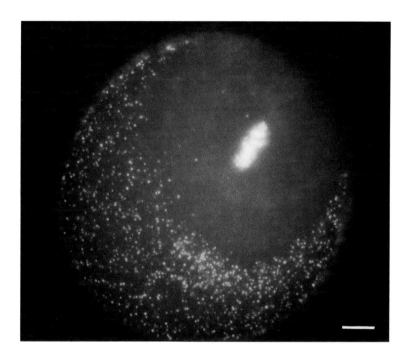

FIG. 4

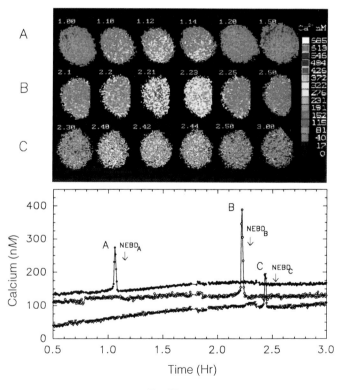

FIG. 13

1B), the egg microtubules responsible for pronuclear apposition are organized by these multiple cytasters instead of the expected finding of a single sperm aster. These observations led to the hypothesis that the dominate centrosome is maternally inherited in the mouse, a theory that was proved correct by examining the events which occur during parthenogenesis and polyspermy.[8]

Microtubule stability has been correlated with a posttranslational acetylation of α-tubulin.[52] In the mouse, new patterns of microtubules selectively appear and disappear after activation and/or fertilization of the arrested oocyte at second meiosis, making this an ideal system to test any modifications arising in cytoplasmic microtubules during early development. Figure 2 demonstrates how acetylated microtubules can be localized in mouse oocytes after buffer M permeabilization and methanol fixation. The binding patterns of acetylated α-tubulin antibody, when compared to total microtubule staining, permit the tracing of the behavior and disappearance of a uniquely modified microtubule subset during completion of second meiosis and fertilization.[53] The acetylated microtubule form is found predominately at the spindle poles at metaphase (Fig. 2A); the cytoplasmic asters do not bind the acetylated antibody (Fig. 2B,C). During meiotic anaphase (Fig. 2D), the staining intensity of the acetylated antibody increases (Fig. 2E,F). At the completion of second meiosis (Fig. 2G), when cytasters (arrows, Fig. 2H) and the midbody connecting the

[52] J. C. Bulinski, J. E. Richards, and G. Piperno, *J. Cell Biol.* **106**, 1213 (1988).
[53] G. Schatten, C. Simerly, D. J. Asai, E. Szöke, P. Cooke, and H. Schatten, *Dev. Biol.* **130**, 74 (1988).

FIG. 4. Detection of cortical granules in unfertilized mouse oocytes. In the mature mouse oocyte arrested at metaphase of second meiosis, numerous cortical granules can be detected at the cell cortex except at the site overlying the meiotic spindle region. The oocyte was double-labeled with DAPI stain for DNA (blue) and fluoresceinated *Lens culinaris* agglutinin (LCA). Bar: 10 μm.

FIG. 13. The first mitotic nuclear envelope breakdown is frequently coupled to a Ca^{2+} transient. (Top) Calibrated Ca^{2+} images from three eggs (A–C) undergoing NEBD (times are indicated in fractions of hours after imaging had begun). Scale represents nanomolar free Ca^{2+} ion. Ca^{2+} increases were global in nature. Egg B was at the edge of a field of view. (Bottom) The Ca^{2+} signal was averaged over the entirety of the three eggs and plotted on the same time scale. In these eggs, NEBD occurred within 5 min of the Ca^{2+} transient, as marked. Cytokinesis occurred approximately 1 hr after NEBD, but neither it nor anaphase, whose timing was not precisely determined, was associated with a detectable transient. (Reproduced from R. M. Tombes, C. Simerly, G. G. Borisy, and G. Schatten, *Journal of Cell Biology*, 1992, Volume 117, pp. 799–811, by copyright permission of the Rockefeller University Press.)

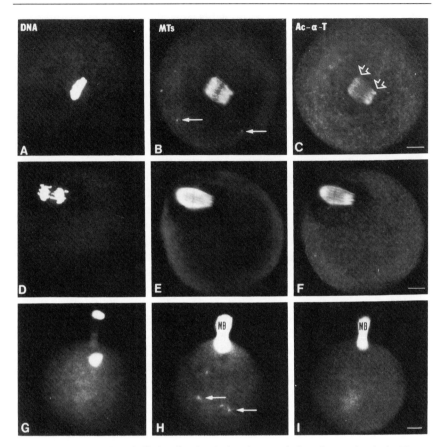

FIG. 2. Acetylated α-tubulin localization during the completion of second meiosis. Mouse oocytes are ovulated-arrested at metaphase (A). Microtubules are in the anastral barrel-shaped spindle and cytoplasmic asters (B, arrows). The acetylated α-tubulin antibody recognizes microtubules at the centrosomes (C, open arrows) of the spindle. At anaphase (D), the entire meiotic spindle (E) is more heavily acetylated (F). By late telophase (G), the meiotic midbody is acetylated, although the cytasters are not (I). All images were triple-labeled for DNA with Hoechst dye 33258 (A, D, G), total tubulin (MTs, either rabbit affinity-purified antibodies to porcine brain tubulin or rat anti-α-tubulin YL1/2; B, E, H), and acetylated α-tubulin antibody (1-6.1; C, F, I). Bars: 10 μm. [Reprinted, with permission, from G. Schatten, C. Simerly, D. J. Asai, E. Szöke, P. Cooke, and H. Schatten, *Dev. Biol.* **130**, 74 (1988).]

oocyte with the second polar body are apparent, only the midbody structure remains acetylated (Fig. 2I). These results demonstrate the microtubules in the mouse oocyte are posttranslationally acetylated in a cell-cycle-specific manner.

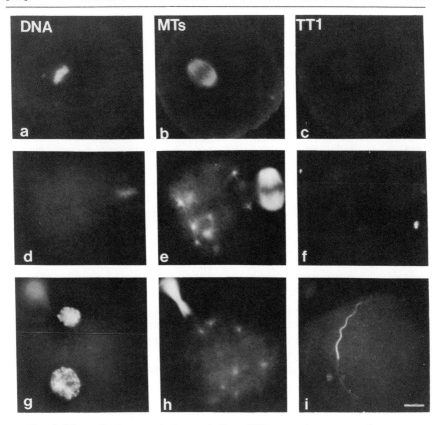

FIG. 3. The antibody to testicular α-tubulin, α-TT1, recognizes sperm tail components but not oocyte spindle or cytoplasmic microtubules. The unfertilized oocyte arrested at metaphase of second meiosis (a and d) has a barrel-shaped, anastral spindle and numerous cytoplasmic asters recognizable by E-7 mouse monoclonal antibody to β-tubulin (b and e). The α-TT1 antibody generally does not detect any class of microtubules in the unfertilized oocyte (f) but occasionally very weakly detects the polar microtubules of the meiotic spindle (c). Following fertilization, by 12 hr postinsemination, the male and female pronuclei have developed after completion of second meiosis (g). E-7 antitubulin antibody identifies the midbody spindle remnent and numerous asters in the cytoplasm of embryos (h). The α-TT1 antibody clear detects the incorporated sperm midpiece and axoneme (i) but does not stain other cytoplasmic microtubules. All images were triple-labeled with DAPI for DNA (a, d, g), E-7 β-tubulin antibody (b, e, h), and α-TT1 (c, f, i). Bars: 10 μm. [Reprinted, with permission, from Simerly et al.[42]]

Multiple tubulin isotypes are known to exist for α- and β-tubulins. Employing the ICC protocols as outlined in Table I, it is possible to demonstrate the presence of an isotype of α-tubulin present in sperm, but not oocyte, microtubules.[42] As shown in Fig. 3, the testicular α-tubulin

(α-TT1) antibody applied to methanol-fixed unfertilized oocytes detects no egg-derived microtubules (Fig. 3a–f). After fertilization, α-TT1 reacts strongly with the sperm axoneme (Fig. 3i) but not the cytoplasmic egg microtubules (Fig. 3h). Such observations have led to the tracing, at the molecular level, of the fate and dispersion of the sperm axonemal α-tubulin.[42]

ICC technology has also provided significant contributions to our understanding of the global organization of the mouse cytocortex region, an area of particular importance to gametic fusion during the process of fertilization. The dynamic structural and functional changes that occur during meiotic maturation establish a uniquely polarized oocyte with regard to surface microvilli, submembranous actin, and underlying cortical granules (reviewed by Longo[54]), which prime the mature oocyte for the reductional divisions necessary to restore the diploid condition on sperm–egg fusion. In Fig. 4, an example of this regionalization of the mouse cortex is demonstrated: the oocyte arrested at metaphase of second meiosis does not contain cortical granules overlying the meiotic spindle, as directly imaged by the fluorescent lectin, *Lens culinaris* agglutinin.[33,55]

In most lower animals, microfilaments assemble at the site of sperm–egg fusion and actively participate in the incorporation of the successful sperm (reviewed in Schatten[56]). Immunocytochemical examination of mouse sperm incorporation, however, has led to some surprising observations on the requirement for microfilament assembly in the mouse. The presence and need of assembled actin in acrosome-reacted mouse sperm remains unclear (reviewed by Schatten and Schatten[50]), whereas egg–actin assembly appears to be required for entrance of the sperm tail, but not the sperm head.[42] In Fig. 5, mouse oocytes were fertilized *in vitro* in the presence of 2.6 μM latrunculin, a microfilament inhibitor, and simultaneously fixed/permeabilized with formaldehyde and lysolecithin.[36] Sperm head penetration and pronuclear formation were able to occur in the presence of the inhibitor, although second polar body formation was blocked, resulting in a trinucleate oocyte (Fig. 5A). Polymerized microfilaments were present in the egg cortex and in the pseudocleavage furrow as detected after rhodamine–phalloidin staining (Fig. 5B,C), but the fertilization process was arrested from any further development. These results, similar to the observations found with the microfilament inhibitor cytochalasin B (reviewed by Maro *et al.*[57]), indicate that mammals may use differ-

[54] F. J. Longo, *in* ''The Cell Biology of Fertilization'' (H. Schatten and G. Schatten, eds.), p. 108. Academic Press, San Diego, 1989.
[55] G. N. Cherr, E. Z. Drobnis, and D. F. Katz, *J. Exp. Zool.* **246,** 81 (1988).
[56] G. Schatten, *Int. Rev. Cytol.* **79,** 59 (1982).
[57] B. Maro, S. K. Howlett, and E. Houliston, *J. Cell Sci. Suppl.* **5,** 343 (1986).

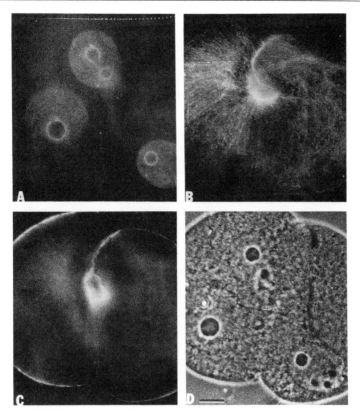

FIG. 5. Mouse fertilization *in vitro*. Sperm incorporation and pronuclear development occur in mouse oocytes inseminated in the presence of 2.6 μM latrunculin A. Pronuclear centration and the formation of the second polar body are arrested, resulting in a trinucleate oocyte (A), which undergoes pseudocleavage (B–D), as revealed by DNA fluorescence (A), rhodamine–phalloidin detection of microfilaments (B, C), and phase-contrast microscopy (D). Bar: 10 μm, magnification ×700. [Reprinted, with permission, from G. Schatten, H. Schatten, I. Spector, C. Cline, N. Paweletz, C. Simerly, and C. Petzelt, *Exp. Cell Res.* **166**, 191 (1986).]

ent mechanisms for sperm incorportaion and the events leading to pronucler apposition during the fertilization process.

The technology of ICC can also be applied to the characterization of nuclear constituents. Mammals like the mouse are fertilized as oocytes rather than eggs, and considerable modifications to chromosomes and developing pronuclei must occur after egg activation if the parental genomes are to be successfully united during the fertilization process. Static immunocytochemistry techniques have beem employed to follow the dra-

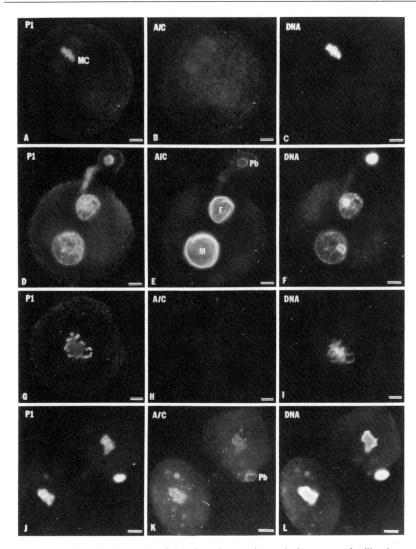

FIG. 6. Nuclear lamins and peripheral nuclear antigens during mouse fertilization and early development. (A–C) Unfertilized oocyte. (A) The P1 peripheral antigens ensheathe the surface of each meiotic chromosome (MC). (B) Lamin staining is lost in the ovulated oocyte, which is arrested at the second meiotic metaphase (lamins A/C). (C) Hoechst DNA fluorescence. (D–F) Pronucleate egg. (D) The peripheral antigens are associated with the rims of the male and female pronuclei and with the polar body nucleus. (E) Lamins A/C reassociate with the nuclear surface, and characteristically the polar body nucleus (Pb) stains only weakly. (F) Hoechst DNA fluorescence. (G–I) Mitotic egg. (G) At prophase, the P1 antibody against the peripheral antigens is redistributed from the pronuclear surfaces to

matic rearrangements that occur in two proteins crucial to the architecture of the nuclear surface, namely, the lamins and the peripheral proteins (reviewed by Stricker *et al.*[58]), as well as the changes correlated with nucleolar structure and function during early embryogenesis.[41]

In the unfertilized mouse oocyte permeabilized in buffer M and fixed in methanol, peripheral antigens are found to ensheathe each meiotic chromosome (MC in Fig. 6A), but the lamins A/C are not detected (Fig. 6B). After sperm incorporation and during pronuclear formation, the peripheral antigens relocalized from the chromosomes to the nuclear periphery (Fig. 6D), and lamins are reacquired at the nuclear surface (Fig. 6E). The polar body nucleus is barely labeled by the lamin A/C antibody (Pb in Fig. 6E). At first mitosis, the peripheral antigens again condense around each chromosome (Fig. 6G), but the lamins become diffuse throughout the cytoplasm (Fig. 6H). Following first division, the peripheral antigens recycle to the circumference of each daughter nucleus (Fig. 6J), as do the lamins (Fig. 6K). These dramatic changes in the architecture of the nuclei during the first cell cycle may be crucial for later events leading to development and cell differentiation.[59]

Localization by ICC technology of the proteins thought to be involved in ribosomal RNA processing are presented in Fig. 7. In the preovulatory oocyte, anti-U3 small nuclear ribonucleoprotein (snRNP) antibody detects the predominate nucleolus within the intact germinal vesicle (Fig. 7B). During meiotic maturation and following metaphase arrest at second meiosis, this snRNP is dispersed throughout the cytoplasm and does not associate with chromosomes (Fig. 7D). After sperm insemination, pronuclear formation initiates nucleolar formation, and U3 antigen is distinctly detected at the multiple nucleolar bodies within the developing male and female pronuclei (Fig. 7E–G). At subsequent interphase stages at the two-cell stage (Fig. 7H,I) and morula (Fig. 7J,K), U3 protein again associates

[58] S. Stricker, R. Prather, C. Simerly, H. Schatten, and G. Schatten, *in* "The Cell Biology of Fertilization" (H. Schatten and G. Schatten, eds.), p. 225. Academic Press, San Diego, 1989.

[59] G. G. Blobel, *Proc. Natl. Acad. Sci. U.S.A.* **82**, 8527 (1985).

cover each chromosome. (H) The lamins dissociate from the mitotic chromosomes. (I) Hoechst DNA fluorescence. (J–L) Cleavage. (J) As the daughter nuclei reform after first division, the peripheral antigens dissociate from the decondensing chromosomes and reassociate with the nuclear periphery (P1 antigen in J). (K) The lamins associate with the reformed nuclear envelope. (L) Hoechst DNA fluorescence. Bars: 10 μm. [Reprinted, with permission, from G. Schatten, G. G. Mual, H. Schatten, N. Chaly, C. Simerly, R. Balczon, and D. L. Brown, *Proc. Natl. Acad. Sci. U.S.A.* **82**, 4727 (1985).]

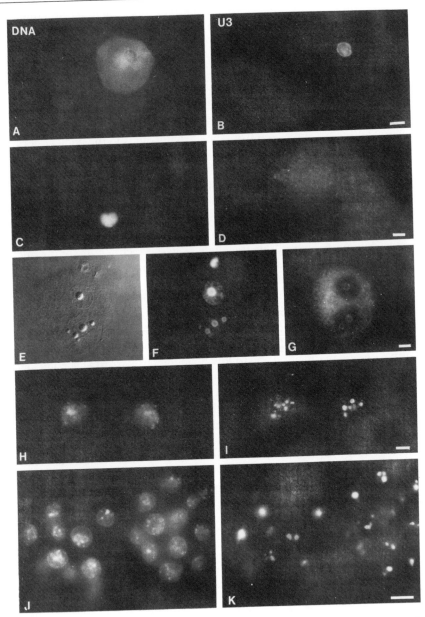

FIG. 7. Anti-U3 snRNP localization in mouse oocytes, zygotes, and embryos, revealed by anti-U3 snRNP antibody staining (B, D, G, I, K) and corresponding DNA localization (A, C, F, H, J). (A, B) Intact germinal vesicle in a preovulatory oocyte. U3 snRNP is

with the nucleoli in each blastomere nucleus. These observations correlate well with the observed biochemical and morphological changes indicative of rRNA synthesis in this mammal.[60]

Detection and Functional Assays of Microinjected Antibodies by Immunocytochemical Methods

Enormous samples like mouse oocytes and embryos can present significant problems for the detection of rarer structural constituents despite the marvelous detection probes and imaging techniques (i.e., confocal microscopy) that have been developed to trace such intracellular proteins. Difficulties can also be exacerbated by the limitations inherit with static immunocytochemistry; detection of soluble or weakly bound antigens can be difficult if protein modifications induced by ICC destroy or block antigen accessibility.

One useful technique to overcome these concerns is to microinject the primary antibody into the living oocyte before processing the cell for immunodetection of the intracellular antigen. The antibody–antigen complex is often more stable to the rigors of ICC processing, and the signal-to-background ratio is vastly increased compared to conventional staining methods.

An example of the ability to localize intracellular proteins that are otherwise poorly preserved by standard immunocytochemical techniques is demonstrated in Fig. 8. Microtubules in the unfertilized mouse oocyte are nucleated from maternal-derived centrosomes that do not contain centrioles.[61] γ-Tubulin, a component of the centrosome/centriole complex in somatic cells that may nucleate growing microtubules and define their

---

[60] L. Pikó and K. B. Clegg, *Dev. Biol.* **89**, 362 (1982).
[61] D. Szöllösi, P. Calarco, and R. P. Donahue, *J. Cell Sci.* **11**, 521 (1972).

---

localized in the nucleolus. (C, D) Unfertilized oocyte arrested at second metaphase of meiosis. U3 is not detected in association with the chromosomes, although diffuse cytoplasmic fluorescence is noted. (E–G) Early pronuclear stage, 18 hr post-hCG (human chorionic gonadotropin). After insemination the U3 antibody detects nucleolar particles within the enlarging male and female pronuclei. At the end of first interphase the U3 staining becomes patchy and punctate as it disperses throughout the nucleus. (E) Differential interference contrast microscopy. (H, I) Mid two-cell stage, 41 hr post-hCG. Nucleoli strongly bind the U3 antibody. (J, K) Morula stage, 88 hr post-hCG, shows U3 distribution restricted exclusively to the nucleolus of each interphase blastomere nucleus. Bars: 10 μm. [Reprinted, with permission, from R. Prather, C. Simerly, G. Schatten, D. R. Pilch, S. M. Lobo, W. R. Marzluff, W. L. Dean, and G. A. Schultz, *Dev. Biol.* **138**, 247 (1990).]

γ–Tub   ß-Tub   DNA

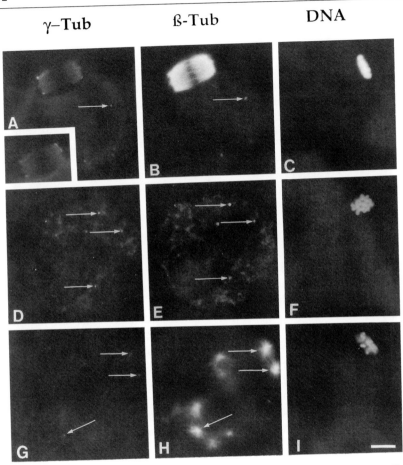

FIG. 8. γ-Tubulin in the centrosomes of unfertilized mouse oocytes. (A–C) Unfertilized oocyte. γ-Tubulin is localized exclusively at the centrosomal foci found at the poles of the metaphase II-arrested meiotic spindle and at the center of the cytoplasmic asters (arrow, A). (A, inset) Second meiotic spindle at a different focal plane showing γ-tubulin at spindle poles. The microtubule pattern is shown in (B), and the chromosomes are aligned on the metaphase plate in (C). (D–F) Microtubule disassembly induced by cold depolymerization (0° for 90 min). γ-Tubulin remains associated with the centrosomes after the microtubules are disrupted with cold or nocodazole treatment (data not shown). (G–I) Microtubule recovery from cold treatment. γ-Tubulin remains at the centrosomal foci in the spindle poles and cytastral centers after microtubule reassembly. (A, D, G) γ-Tubulin localization. (B, E, H) Microtubules. (C, F, I) DNA detection. Bar: 10 μm. [Reprinted, with permission, from Palacios et al.[11]]

polarity[62,63] is detected at the poles of the acentriolar second meiotic spindle (Fig. 8A, inset) and at the center of the cytoplasmic asters (arrow, Fig. 8A) following microinjection of γ-tubulin antibody. γ-Tubulin is still found in association with the MTOCs in the mouse unfertilized oocyte cold-shocked to disrupt the microtubule patterns (Fig. 8D–F). Unlike the findings in cultured mammalian cells,[64] however, this antibody does not block the assembly of microtubules in the mouse oocyte following recovery from cold-induced depolymerization (Fig. 8G–I), and normal meiotic spindle formation can occur in the presence of the antibody.[65] These observations demonstrate that γ-tubulin is a centrosomal component in the mouse oocyte, where it may participate as an obligatory constituent of the MTOC responsible for the assembly of microtubules that control the motions crucial to the events of fertilization and early development.

Another benefit of the microinjection of primary antibodies into living oocytes and embryos is the ability to explore critical protein functions *in vivo*. Figure 9 demonstrates the application of this technique in the study of centromere-associated proteins in the mouse, which have been probed with human autoimmune antisera containing antibodies directed against antigens detected in the centromere/kinetochore complex (CREST sera).[12] The mouse oocyte and embryo represent an excellent model system for investigating the functions of centromere proteins *in vivo* since arrest occurs at metaphase of second meiosis, permitting an analysis of the events of chromosome segregation independently of the events of chromosome alignment. In addition, meiotic maturation and fertilization can be blocked at stage-specific sites by drugs allowing the events of chromosome congression at metaphase to be explored without overlapping with the events of anaphase onset.

When prometaphase-arrested mitotic mouse oocytes were microinjected with CREST autoimmune antibody and allowed to recover, normal chromosome alignment (Fig. 9B) and segregation (Fig. 9D) were prevented, resulting in lagging chromosomes in the interzonal microtubules of the cleaving embryo (Fig. 9F). Control oocytes recovered from drug inhibition completed these events normally (Fig. 9A,C,E). If the CREST antibody was introduced into metaphase-arrested oocytes at either meiosis or mitosis, normal chromosome segregation occurred. These observations were among the first to demonstrate that a centromere-specific antibody interferes with a definitive chromosome motion *in vivo*, namely, congres-

[62] B. R. Oakley, *Trends Cell Biol.* **2,** 1 (1992).
[63] C. A. Alfa and J. S. Hyams, *Nature (London)* **352,** 471 (1991).
[64] H. C. Joshi, M. M. Palacios, L. McNamara, and D. W. Cleveland, *Nature (London)* **356,** 80 (1992).
[65] M. M. Palacios, H. C. Joshi, C. Simerly, and G. Schatten, *J. Cell Sci.* in press (1993).

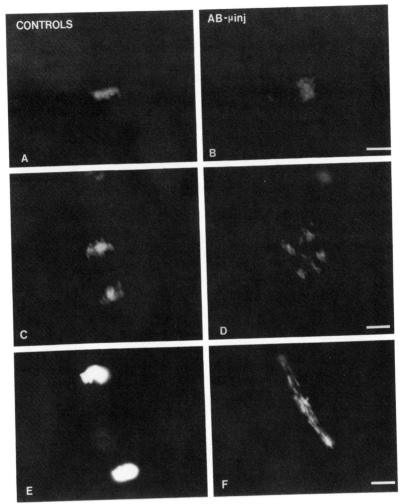

FIG. 9. Antikinetochore/centromere antibodies microinjected into zygotes at first mitosis interfere with prometaphase congression. To explore the possibility that the antibodies were affecting kinetochore maturation at the interphase–mitotic transition versus congression, E.K. CREST antibody was microinjected into mitotic zygotes arrested at prometaphase with 5 μM nocodazole. (A, C, and E) Uninjected controls recovering from nocodazole block for 1 hr. (B, D, and F). Recovery in the presence of antikinetochore/centromere antibody E.K. Whereas control cells can undergo proper chromosome alignment at metaphase (A), microinjected oocytes are unable to complete congression (B). At anaphase controls segregate chromosomes in an orderly fashion (C), but microinjected cells do not (D). After first mitosis, controls display two well-separated nuclei (E), whereas the injected cells have many chromosomes that remain trapped in the interzonal area of the mitotic spindle (F). Cells were double-labeled for DNA and microtubules (not shown). Bars: 10 μm. (Reproduced from C. Simerly, R. Balczon, B. R. Brinkley, and G. Schatten, *Journal of Cell Biology*, 1990, Volume III, pp. 1491–1504, by copyright permission of the Rockefeller University Press.)

sion (see also Bernat et al.[66]), and they serve to underscore the practical application of this powerful immunocytochemical technique.

Dynamic Studies in Living Mouse Oocytes

For fertilization to be successful, several dynamic movements must occur. For instance, the sperm must propel itself to the surface of the oocyte, bind and penetrate the zona pellucida, fuse with the egg plasma membrane, and be drawn into the cytoplasm. Following sperm penetration, the sperm and egg nuclei must be repackaged and transported during interphase so that the male and female pronuclei are adjacent to one another at the onset of mitosis (reviewed by Schatten[56]). Application of conventional immunocytochemistry technology as outlined in the preceding section can offer remarkable two-dimensional images of such events. However, some limitations must be realized with static ICC: specimen preparation can alter some structural features, fixation artifacts can be introduced during processing, and invasive permeabilization techniques can extract material of interest.

To appreciate fully the multitude of interacting intracellular proteins and their regulation, which occurs in a spatial and temporal fashion, the living cell should be investigated. Advances in imaging technology using conventional video-enhanced microscopy or laser-scanning confocal microscopy and the availability of new detection probes such as photoactivatable tubulin and calcium-sensitive fluorescent dyes have significantly improved the detectability and resolution of cellular events in living cells.[15,37,67,68] These newer technologies are now beginning to be employed to investigate mammalian oocytes and embryos.

Conventional computer-enhanced time-lapse video microscopy was employed to investigate the dynamic process of nucleolar formation in pronuclear stage mouse oocytes (Fig. 10).[41] Oocyte activation by the incorporating sperm is confirmed by the presence of a sperm tail bound to the vitelline membrane and the second polar body (Fig. 10a). One hour later, during early pronuclear formation, small nucleolar granules form within the nucleoplasm (arrows, Fig. 10b). Over the next 30 min, as pronuclear enlargement occurs, several of these small granules fuse to form a single predominate nucleolus within each pronucleus (Fig. 10c–i, arrows). Investigation of the mechanism of nucleolar formation by the

[66] R. L. Bernat, G. G. Borisy, N. F. Rothfield, and W. C. Earnshaw, J. Cell Biol. 111, 1519 (1990).
[67] T. J. Mitchison, J. Cell Biol. 109, 637 (1989).
[68] J. Holy, C. Simerly, S. Paddock, and G. Schatten, J. Electron Microsc. Tech. 17, 384 (1991).

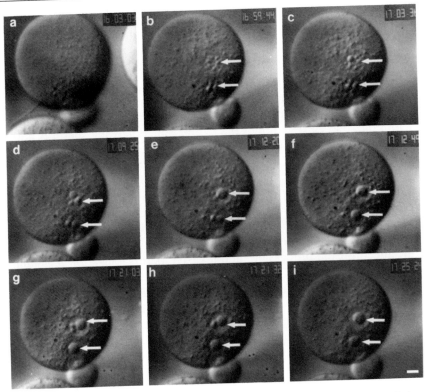

FIG. 10. Time-lapse video microscopy of nucleolus formation. Sperm–oocyte fusion, penetration, and second polar body extrusion occur within 3 hr of insemination (a). Within the next hour numerous small nucleolar bodies appear within the developing pronuclei (b, arrows). Over the next 30 min, several of these small structures fuse with one another to form a single large nucleolus (c–i, arrows point to unifying nucleolar bodies). Time (hr) post-hCG is noted. Magnification: ×160. [Reprinted, with permission, from R. Prather, C. Simerly, G. Schatten, D. R. Pilch, S. M. Lobo, W. R. Marzluff, W. L. Dean, and G. A. Schultz, *Dev. Biol.* **138**, 247 (1990).]

application of living video microscopy and static ICC has led to the correlation of the appearance of U3 snRNP antigen with nucleolar formation. This demonstrates the usefulness of studying in tandem living and fixed cell populations to investigate architectural changes in the mouse nucleus.[41]

Although documentations of the motility events associated with mammalian fertilization by video microscopy can yield relevant information on a global scale, one would like to understand these events at the molecular level in the living cell. Fluorescent analog cytochemistry can be a phenomenal mechanism to perform such investigations. Living fluorescent cytochemistry studies of tissue culture cells have shown that microtubules are not static protein structures but instead rapidly exchange subunits in both interphase arrays and mitotic metaphase spindles.[69-72] In the unfertilized mouse oocyte, the microtubules of the metaphase spindle at second meiosis might have unusual dynamics given the unique properties of this structure: it is arrested, contains a large proportion of acetylated α-tubulin which has been correlated with microtubule stability, and contains no centrioles (reviewed by Schatten et al.[73]).

As shown in Fig. 11, polymerization-competent biotinylated tubulin protein microinjected into the living mouse oocyte will incorporate into the microtubules of the second meiotic spindle and cytasters (arrows, Fig. 11C) and the midbody structure following egg activation (Fig. 11G) as detected by fluorescent streptavidin staining following buffer M permeabilization and methanol fixation. These observations demonstrate the derivatized tubulins can serve as useful probes for the exploration of tubulin turnover in microtubules of the mouse oocyte.

However, it is impossible to process oocytes microinjected with biotin–tubulin quickly enough by standard ICC techniques to explore accurately rates of tubulin exchange. Therefore, FRAP analysis was employed to measure microtubule turnover following microinjection of XRITC-derivatized tubulin protein. Figure 12 demonstrates the rapid recovery of a fiduciary bar bleached across the XRITC-labeled spindle in the living mouse oocyte (Fig. 12A–D). The recovery half-time was measured at 77 sec, a remarkable turnover rate for a spindle considered arrested. In contrast, the meiotic midbody structures in interphase oocytes exhibited limited recovery, averaging only 22% after 4 min (Fig. 12E–H). These observations indicate that most microtubules within the arrested metaphase spindle undergo rapid cycles of assembly and disassembly, whereas the microtubules of the midbody structure are more stable.

Notwithstanding the power of observing dynamic processes in living mouse oocytes, these techniques have potential technical problems that must be addressed. For instance, Vigers et al.[46] demonstrated the destruction of fluorescent microtubules in some instances where unattenuated exposure to photobleaching was performed. Likewise, certain dynamic

[69] Y. Hamaguchi, M. Toriyama, H. Sakai, and Y. Hiramoto, Cell Struct. Funct. 12, 43 (1987).

[70] E. D. Salmon, R. J. Leslie, W. M. Saxton, M. L. Karow, and J. R. McIntosh, J. Cell Biol. 99, 2165 (1984).

[71] P. J. Sammak, G. J. Gorbsky, and G. G. Borisy, J. Cell Biol. 104, 395 (1990).

[72] W. M. Saxton, D. L. Stemple, R. J. Leslie, E. D. Salmon, M. Zavortink, and J. R. McIntosh, J. Cell Biol. 99, 2175 (1984).

[73] H. Schatten, C. Thompson-Coffe, G. Coffe, C. Simerly, and G. Schatten, in "The Molecular Biology of the Cell" (H. Schatten and G. Schatten, eds.), p. 189. Academic Press, San Diego, 1989.

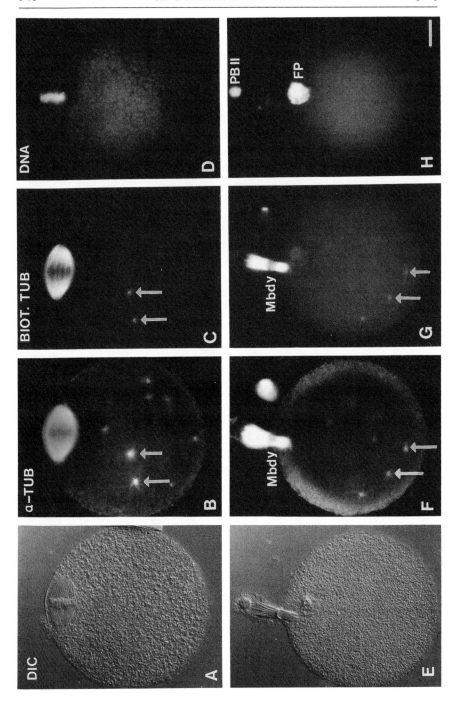

molecular events within living cells may be sensitive to the imaging technology. For example, recent investigations into microtubule behavior in tissue culture cells have suggested that tubulin subunits move poleward from kinetochore sites during mitosis, an observation not detected by FRAP experiments.[74] New probes consisting of a "caged" derivative of fluorescein which is activated by exposure to UV light after microinjection and incorporation into microtubules supposedly give better imaging resolution since the signal is a bright mark against a dark background.[67] It will be interesting to apply such new technological advances to the study of dynamic microtubule processes in mouse oocytes and embryos.

In addition to structural investigations by living fluorescent analog cytochemistry, it is also possible to explore ionic regulation events within the living cell using new quantitative radiometric calcium dyes like Fura-2.[75] In the mouse, calcium ions have been implicated in the regulation of meiotic maturation, oocyte activation, and nuclear envelope breakdown (NEBD) at the onset of first mitosis.[14,76,77] The majority of newer evidence for investigating the role of calcium as an intracellular ionic regulator has come from the fluorescent imaging studies on living oocytes and embryos following treatment with calcium indicator dyes. For instance, investigations on living fertilized mouse embryos loaded with Fura-2 indicator dye and imaged over several hours detected a single short, rapid calcium release shortly before NEBD (Fig. 13A–C).[14] By plotting the average calcium signal versus time, the kinetics of the amount of free calcium released in each embryo was determined as a roughly 2.5-fold increase over a 2-min period (Fig. 13, bottom). Additionally, it was determined

---

[74] T. J. Mitchison and K. E. Sawin, *Cell Motil. Cytoskeleton* **16**, 93 (1990).
[75] G. Grymkiewicz, M. Ponie, and R. Tsien, *J. Biol. Chem.* **260**, 3440 (1985).
[76] M. De Felici, S. Dolci, and G. Siracusa, *J. Exp. Zool.* **260**, 401 (1991).
[77] D. Kline and J. T. Kline, *Dev. Biol.* **149**, 80 (1992).

FIG. 11. Incorporation of microinjected biotin–tubulin into mouse meiotic spindles, midbodies, and cytoplasmic microtubules. (A–D) Metaphase-arrested unfertilized oocyte processed 48 min after microinjection with biotin–tubulin. Spindle and cytoplasmic microtubules incorporate biotin–tubulin (C), although incorporation into the cytastral microtubules is primarily detected at the centrosomes (arrows, B and C). (E–H) Biotin–tubulin incorporation in a microinjected oocyte that was subsequently activated to proceed to telophase. A meiotic midbody is produced after activation (F, Mdby) as the female pronucleus (H, FP) and second polar body nucleus form (H, PbII). The microtubules of the midbody (G, Mdby) and cytaster (arrows, F and G) contain biotin–tubulin. Cells were visualized by differential interference contrast (DIC; A and E); staining with α-TUB, an affinity-purified rabbit antitubulin antibody (B and F); streptavidin staining of biotin–tubulin (BIOT.TUB; C and G); and DAPI staining of DNA (D and H). Bar: 10 μm. [Reprinted, with permission, from G. J. Gorbsky, C. Simerly, G. Schatten, and G. G. Borisy, *Proc. Natl. Acad. Sci. U.S.A.* **87**, 6049 (1990).]

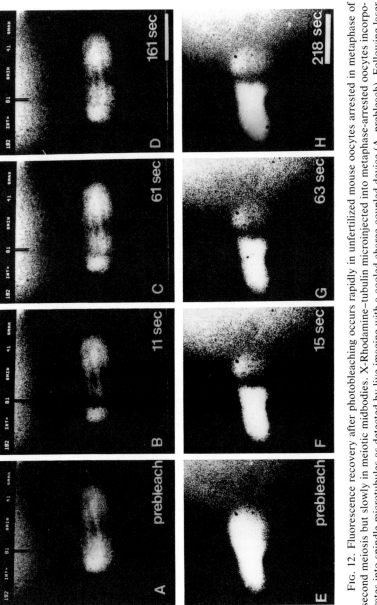

Fig. 12. Fluorescence recovery after photobleaching occurs rapidly in unfertilized mouse oocytes arrested in metaphase of second meiosis but slowly in meiotic midbodies. X-Rhodamine–tubulin microinjected into metaphase-arrested oocytes incorporates into spindle microtubules as detected by live imaging with a cooled charge-coupled device (A, prebleach). Following laser photobleaching of a cylindrical bar across the fluorescent spindle (B), a progression of images after irradiation shows that fluorescence recovery is rapid in the bleached zone (C and D). A prebleach image of a meiotic midbody is shown in (E). Following laser photobleaching of a cylindrical bar across the midbody (F), subsequent images (G and H) demonstrate that fluorescence recovery is slow. Times after photobleaching are indicated in the lower right-hand corners of B–D and F–H. Bar: 10 μm. [Reprinted, with permission, from G. J. Gorbsky, C. Simerly, G. Schatten, and G. G. Borisy, *Proc. Natl. Acad. Sci. U.S.A.* **87**, 6049 (1990).]

that intracellular chelators such as BAPTA [bis-(o-aminophenoxy)-ethane-N,N,N',N'-tetraacetic acid] and EGTA were sufficient to block this calcium release and NEBD. These findings were compared and contrasted to those found for germinal vesicle breakdown during resumption of meiotic maturation, where a role for intracellular calcium was not found (Tombes et al.,[14] but see De Felici et al.[76]). These results underscore the tremendous advantage of employing living cells with ratiometric indicators to investigate the dynamic events associated with diffusible ions such as calcium for single-cell analysis of phenomena like NEBD.

Understanding the sites of calcium release and action in addition to understanding the kinetics of intracellular calcium release will be important to appreciating the dynamic role of this ion in mammalian oocytes. It is often difficult to achieve the necessary resolution to discern reliable spatial information with conventional microscopic techniques because out-of-focus light tends to obscure details of subcellular organization, particularly from large cells such as oocytes. Technological innovations in the development of laser-scanning confocal microscopy (LSCM) have been introduced as a means for obtaining information from thin optical sections in living cells. Along with the ability to produce high-resolution images, confocal microscopy can produce stacks of optical sections that can be reconstructed to yield highly informative three-dimensional renderings.[37,68]

This technological innovation has been used to explore the dynamics of the calcium wave in the sea urchin egg using the calcium-sensitive dye fluo-3 (Molecular Probes, Eugene, OR).[15] Among the interesting new findings from this study was the determination of the extent of calcium wave propagation following fertilization. Previously, it had been unclear if the calcium wave was cortically localized or whether the free ion transversed throughout the entire egg cytoplasm. Volumetric renderings of confocal sections taken over time revealed that fertilization led to a global elevation of free intracellular calcium, with the greatest concentration occurring at the cortex. In addition to this discovery, a transient intranuclear calcium rise was first demonstrated in an egg. The detection of this fertilization-induced calcium elevation would have been difficult without the power of confocal imaging. Although the biological significance of changes in nuclear calcium levels is not known yet, its discovery by this remarkable technology may serve to unfold a new field of ionic regulation effects in nuclear activities.

Prospectives

Imaging structural dynamics in mouse oocytes and embryos has led to extraordinary discoveries with regard to organization of cytoskeletal elements, nuclear architecture, rearrangements in vivo, and their regulation. Many of the findings seem to violate long-held notions on the way fertilization systems behave. For instance, static ICC investigations have shown that the mouse oocyte acquires its dominant MTOC from maternal sources, rather than being a paternal contribution as noted for most lower animals from coelenterates to invertebrates, including fish and amphibians. Assembled microfilaments are not required for sperm head penetration but are needed for pronuclear apposition, in direct contrast to cases in other animals where actin plays a crucial role in sperm incorporation but not intracellular pronuclear movements. In addition, mouse oocytes are highly polarized with regard to subcellular organization when compared to most animal eggs, and this arrangement appears critical for achieving proper fertilization.

The uniqueness of the mouse oocyte as a model mammalian fertilization system to explore biological phenomena has also been exploited. With the mouse oocyte, it is possible to investigate meiotic resumption, fertilization, and early developmental events within a single species. Thus, for investigation of the in vivo effects of CREST antibodies on chromosome alignment and segregation patterns in the mouse, the natural and drug-induced arrest points were utilized to demonstrate that anticentromere antibodies affected the alignment of chromosomes during congression and not on chromosome segregation. Likewise, the ability to employ similiar methodologies on a single system led to the discovery of the differential role of calcium ions for the breakdown of the germinal vesicle versus the pronuclear envelopes.

The mouse oocyte has been an unrivaled model for investigating the dynamic properties of microtubules in mammalian oocytes. Second meiotic spindles in mouse oocytes have been shown to be highly dynamic structures and to assemble/disassemble tubulin subunits at rates equivalent to mitotic somatic cells, despite being arrested, posttranslationally modified, and acentriolar. The functions of such rapidly cycling microtubules at this stage are not known, but they may be important to proper chromosome segregation.

With the foundation of applying immunocytochemical methods to living and fixed mouse oocytes now firmly established, expansion of our investigations to other mammalian species is now feasible. Emerging data from studies of other mammals by static ICC investigations seem to indicate that other species do not necessarily follow the rodent model for the inheritance of centrosomes or organization of cytoplasmic microtubules. Information derived from the mouse, therefore, cannot be simply extrapolated to higher animals, including the human. As we have seen for investigations in the mouse, many exciting and fundamental properties of cell, developmental, and reproductive biology can be revealed through the study of mammalian taxa. With the explosion in new imaging techniques and probes, the time may be ripe indeed to explore cytoskeletal and nuclear molecules, and their intracellular dynamics, in other mammals.

Acknowledgments

We are indebted to our many wonderful colleagues and collaborators who have contributed greatly to various aspects of this research. We would like to mention particularly Drs. Ron Balczon, Barry Bavister, Dottie Boatman, Gary Borisy, Bill Brinkley, Gary Gorbsky, Jon Holy, Sara Steffen, and Robert Tombes. Our research reviewed here was funded by research grants from the National Institutes of Health.

Section VII

Gene Expression: Methylation

[33] Polymerase Chain Reaction for the Detection of Methylation of a Specific CpG Site in the *G6pd* Gene of Mouse Embryos

By MAURIZIO ZUCCOTTI, MARK GRANT, and MARILYN MONK

Summary

This chapter describes the polymerase chain reaction (PCR) detection of methylation changes at specific CpG sites in DNA isolated from minute quantities of biological material, such as single preimplantation mouse embryos or small numbers of stem cells, or germ cells. We have concentrated on refining these techniques to monitor specific sites in X-linked genes for methylation changes associated with X-chromosome inactivation. The general principles of the *Hpa*II-sensitive PCR assay described here should be applicable and adaptable to specific CpG sites in other genes of interest.

Introduction

The methylation of the pyrimidine base cytosine is the only known modification of vertebrate genomic DNA. In mammals, about 2–7% of cytosine residues are methylated and the majority (90%) of these occur in the dinucleotide sequence 5'CpG3'.[1] Studies on local methylation patterns reveal that tissue-specific genes are heavily methylated in most tissues but are unmethylated in their tissue of expression.[2] On the other hand, the CpG islands[3] of housekeeping genes (excepting those on the inactive X chromosome) are unmethylated in every tissue and at all stages of development.[4] This inverse correlation between gene expression and DNA methylation levels suggests that methylation may function in controlling gene activity.

We have studied the role of methylation in X-chromosome inactivation in female mouse embryos. During development, inactivation of an X chromosome occurs first in the extraembryonic lineages as they differentiate,

---

[1] M. Ehrlich and R. Y. H. Wang, *Science* **212**, 1350 (1982).
[2] J. Yisraeli and M. Szyf, *in* "DNA Methylation and Its Biological Significance" (A. Razin, H. Cedar, and A. D. Riggs, eds.), p. 353. Springer-Verlag, New York, 1984.
[3] A. P. Bird, Nature (*London*) **321**, 209 (1986).
[4] R. Stein, N. Sciaky-Gallili, A. Razin and H. Cedar, *Proc. Natl. Acad. Sci. U.S.A.* **80**, 2422 (1983).

METHODS IN ENZYMOLOGY, VOL. 225

Copyright © 1993 by Academic Press, Inc.
All rights of reproduction in any form reserved.

and then in embryonic tissue at approximately the time of implantation in the uterus.[5,6] In the early extraembryonic lineages, the paternally-inherited X chromosome is preferentially inactivated[7,8] suggesting that the paternal and maternal X chromosomes must be imprinted in some way so as to be distinguishable. In the female germ line the inactive X chromosome is reactivated.[9] We have asked whether changing patterns of methylation might be the molecular basis for inactivation and reactivation of X-linked genes.

Previously, the DNA methylation methods available were not sufficiently sensitive to study the role played by X-linked gene methylation in the initiation of X-inactivation in development. However, recently, Singer-Sam et al.[10] developed a procedure which uses the PCR to detect changes in methylation of specific CpG sites in early embryos. The CpG sites studied are in the 5' regions of X-linked genes; they have been identified as critical sites whose methylation is strictly correlated with the inactive state of the gene on the inactive X chromosome.[11,12] Singer-Sam et al.[10] developed procedures to study the methylation of the *Pgk* (phosphoglycerate kinase) gene. In this chapter, we present as an example the procedures we developed for the *G6pd* (glucose phosphate dehydrogenase) gene.

Methylation of the single informative X-linked CCGG site is assayed by PCR amplification of a sequence containing the site, before and after digestion with the methylation-sensitive restriction enzyme, *Hpa*II. When the site is unmethylated (in male embryos with a single active X chromosome or in female embryos with two active X chromosomes), amplification is not possible after *Hpa*II has cut the CCGG site in the sequence. When the site becomes methylated on the inactive X during female embryonic development, amplification of the sequence becomes resistant to *Hpa*II digestion. The *Hpa*II-sensitive PCR approach is illustrated in Fig. 1.

Methods

Tissue Isolation

Embryonic and adult tissues from which DNA is derived for the *Hpa*II-sensitive PCR assay are isolated according to procedures described else-

[5] M. Monk and M. I. Harper, Nature (*London*) **281**, 311 (1979).
[6] N. Takagi, O. Sugawara and M. Sasaki, Chromosoma **85**, 275 (1982).
[7] N. Takagi and M. Sasaki, Nature (*London*) **256**, 640 (1975).
[8] M. I. Harper, M. Fosten and M. Monk, J. Embryol. Exp. Morphol. **67**, 127 (1982).
[9] M. Monk and A. McLaren, J. Embryol. Exp. Morphol. 63, 75 (1981).
[10] J. Singer-Sam, M. Grant, J. M. LeBon, K. Okuyama, V. Chapman, M. Monk and A. D. Rigs, Mol. Cell. Biol. **10**, 4987 (1990).
[11] L. F. Lock, N. Takagi and G. R. Martin, Cell (*Cambridge, Mass*) **48**, 39 (1987).
[12] D. Toniolo, M. Filippi, R. Dono, T. Lettieri and G. Martini, Gene **102**, 197 (1991).

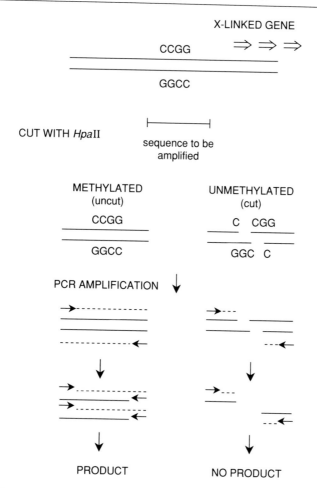

FIG. 1. Schematic representation of the polymerase chain reaction detection of CCGG methylation.

where.[13] Particular care should be taken to ensure that the instruments and media used to isolate these samples are clean and, wherever possible, sterile. Ideally, the isolation should take place in an area remote from the PCR amplification laboratory.

[13] M. Monk (ed.), "Mammalian Development: A Practical Approach," IRL Press, Oxford, 1987.

TABLE I

PROTEINASE K–GUANIDINE HYDROCHLORIDE LYSIS BUFFER

Reagent[a]	(Stock)[b]	Volume[c] (μl)
Guanidine hydrochloride	6 M	140
Ammonium acetate	7.5 M	10
Sarkosyl	20%	10
Proteinase K[d]	20 mg/ml	1.6
Total volume per sample		161.6

[a] Reagents are obtained from BDH, with the exception of proteinase K (Sigma).

[b] Stock solutions are prepared in a location remote from the PCR amplification laboratory, filter-sterilized in a laminar flow hood, aliquotted and stored at room temperature.

[c] Lysis buffer is made up freshly as required. For multiple samples, a master mix is prepared and aliquotted as appropriate.

[d] Proteinase K is stored in solution (with sterile water) at −20°. After use, thawed aliquots should be discarded.

DNA Preparation

DNA may be prepared from adult tissue using the standard phenol–chloroform procedure,[14] and resuspended in KGB (see below). Small-scale DNA preparations from embryonic samples are prepared using the proteinase K–guanidine hydrochloride method,[15] which has been modified to give maximal recovery of DNA from minute quantities of tissue (down to 100 cells) in a short period of time;[16] using this protocol, DNA can be PCR-ready in one working day. Moreover, the restriction endonucleases readily digest DNA isolated in this way [i.e., in the absence of sodium dodecyl sulfate (SDS) in the lysis buffer].

Tissue samples are isolated directly into 161.6 μl of freshly prepared lysis buffer (see Table I) in a sterile, preferably siliconized, 1.5-ml Eppendorf tube. Included in the mix is 1.5 g of ϕX174 RF (replicative form) (BRL, Gaithersburg, MD) or M13 RF DNA (BCL) to act as carrier and internal control for complete digestion by HpaII (see below). The sample may be stored at −70° for several months in this form. To prepare for PCR, the sample is incubated at 60° for 60–90 min to lyse the cells in the

[14] J. Sambrook, E. F. Fritsh and T. Maniatis, "Molecular Cloning: A Laboratory Manual," 2nd Ed. Cold Spring Harbor Laboratory, Cold Spring Harbor, New York, 1989.

[15] M. Jeanpierre, Nucleic Acids Res. 15, 964 (1987).

[16] S. Singer-Sam, J. M. LeBon, J. M. Tanguay and A. D. Riggs, Nucleic Acids Res. 18, 687 (1990).

presence of proteinase K, the lysate is spun briefly, loaded into a 1-ml syringe (sterile), and passed through a hypodermic needle (25-gauge, 5/8 inch) (Becton Dickinson, San Jose, CA) 10–20 times in order to shear the high molecular weight genomic DNA into smaller fragments. Glycogen (20 mg/ml) (BCL) is added (as carrier) to the resulting suspension along with 3 volumes of 95% (v/v) ethanol. Care must be taken to ensure that ethanol and lysate are thoroughly mixed by inversion. When DNA is present in small amounts, the yield may be improved by further DNA precipitation at −70° for 30 min or overnight (the ethanolic DNA solutions may be stored at this stage at −70°). The precipitated DNA is collected by centrifugation at ~14000 g for 30 min at room temperature in a microfuge. The pellet is washed twice with 2 volumes of 70% ethanol and resuspended in 40–50 μl of 0.5 × potassium glutamate buffer [1 in 4 dilution of 2× KGB stock: 200 mM potassium glutamate (Sigma, St. Louis, MO), 50 mM Tris–acetate (pH 7.5), 20 mM magnesium acetate (BDH, Poole, UK), 100 g/ml bovine serum albumin (BSA) (Sigma), 1 mM 2-mercaptoethanol (Sigma)].[14,17] Resuspension by storage overnight at 4° is sometimes required.

HpaII treatment

Restriction endonucleases may be added directly to a sample resuspended in KGB. Prior to HpaII treatment, DNA samples ideally should be digested with a restriction endonuclease which cuts outside of the sequence of interest to further reduce the genomic DNA fragments of a size readily more available to PCR amplification.[18]

The DNA suspension is divided into two aliquots; one is treated with the methylation-sensitive restriction endonuclease HpaII (New England Biolabs, Beverly, MA) and the other is the non-digested control [an equivalent volume of TE8 (Tris–EDTA, pH 8) is added]. The samples are incubated at 37° for 60–90 min. Two further enzymes additions followed by incubation are made (the third may be left at 37° overnight) to ensure completeness of digestion at the CpG site. Completion of digestion is monitored by examining the restriction pattern of carrier DNA (e.g., ϕX174 RF DNA) in a small aliquot of the reaction mix run on a 2% agarose–TBE (Tris–borate–EDTA buffer) gel. If the carrier DNA is cut to completion by HpaII, the samples are incubated for an additional 15 min at 100° to inactivate the enzyme. At this stage, samples may be stored at 4° until required for PCR assay. An MspI digest may be included as control to examine the result for complete digestion. The restriction enzymes MspI and HpaII are isoschizomers: they both recognize the se-

[17] M. McClelland, J. Hanish, M. Nelson and Y. Patel, Nucleic Acids Res. **16**, 364 (1988).
[18] B. N. Beck and S. N. Ho, Nucleic Acids Res. **16**, 9051 (1988).

quence CCGG, but only *Msp*I will cut the site if the internal cytosine is methylated.[19]

Primer Selection for PCR

Primers flanking a single *Hpa*II (CCGG) site of interest should be chosen according to the general guidelines (i.e., ~50–60% G + C composition and low complementarity to each other, particularly in the 3' region).[20] The sequence to be amplified is limited in size by the proximity of other CCGG sites in GC-rich sequences. More than one CCGG site could be included in the target sequence, but in this case interpretation of the methylation pattern for each individual site will not be possible since, in the absence of methylation of either site, amplification is prevented by *Hpa*II digestion. An additional set of primers is included in the same reaction as a control for successful PCR; the control primers direct amplification of a sequence that does not contain an *Hpa*II site. Control primers should be chosen with GC : AT ratio matching that of the target primers so that the control sequence is amplified under the same annealing conditions as the experimental sequence. So long as they meet these basic requirements, the internal control primers may be located in any part of the genome.

Polymerase Chain Reaction

In our experiments with mouse embryos, primordial germ cells, and gametes, we use 50 μl of PCR reaction mix (RM) for each sample. The RM contains: 10× buffer (Perkin-Elmer Cetus, Norwalk, CT, or Advanced Biotechnologies, London, UK), dNTPs (Perkin-Elmer Cetus), primers (Oswel DNA Service, Edinburgh, UK), *Taq* polymerase in the supplied buffer (Perkin-Elmer Cetus or Advanced Biotechnologies), autoclaved double-distilled water, and an aliquot of template estimated to contain 600 pg of DNA (equivalent to ~100 mouse nuclei). The volume and concentration of each of the components is shown in Table II. The RM is aliquoted in sterile Eppendorfs and overlaid with 50 μl mineral oil (BDH). The sample DNA is amplified through successive rounds of denaturation, annealing, and elongation in a PCR thermal cycler; the actual conditions depend on the nature and concentration of the starting DNA template.[21]

---

[19] C. Waalwijk and R. A. Flavell, *Nucleic Acids Res.* **5**, 3231 (1978).

[20] M. A. Innis and D.H. Gelfand, *in* "PCR Protocols: A Guide to Method and Applications" (M. A. Innis, D. H. Gelfand, J. J. Sninsky and T. Y. White, eds.), p. 3. Academic Press, San Diego and London, 1990.

[21] R. K. Saiki, D. H. Gelfand, S. Stoffel, S. J. Scharf, R. Higuchi, G. T. Horn, K. B. Mullis and H. A. Ehrlich, *Science* **239**, 487 (1988).

TABLE II
PREPARATION OF PCR MIX[a]

Component[b]	Volume	Final concentration
Sterile, double-distilled water	$(x - y)^c$ μl	
10× PCR buffer	5 μl	1 × [10 mM Tris-HCl (pH 8.3); 15 mM MgCl$_2$; 500 mM KCl; 0.01% gelatin]
dNTP mix (1,25 mM each)	16 μl	400 μM each dNTP
5' primer, 50 pmol	1 μl	50 pmol
3' primer, 50 pmol	1 μl	50 pmol
Taq polymerase, 5 units/μl	0.5 μl	2.5 units
Overlay with 50 μl mineral oil	—	
Add DNA (under oil)[d]	y μl	
Total volume per sample	50 μl	

[a] Add in sequence shown. The conditions shown here have been optimized for amplification of mouse G6pd and Pgk-1 genes. Primers sequences are shown in Fig. 2.

[b] Reagents are obtained from Perkin-Elmer Cetus. The 10× PCR buffer, dNTPs, primers, and Taq polymerase are stored at −20°. After use, thawed aliquots are discarded.

[c] Volume of water x required (depends on volume of DNA sample y) to give final volume of 50 μl.

[d] DNA is added in an area remote from the PCR setup.

Experience suggests that amplification of GC-rich DNA sequences by PCR requires prolonged denaturation at high temperature, for example, 2 cycles of amplification which include denaturation at 95° for 4 min followed by a further 35–40 cycles which include denaturation at 95° for 1 min. For particularly GC-rich sequences (i.e., 70% GC : 30%AT), denaturation temperatures of 96° are necessary.

For small quantities of DNA (e.g., DNA from embryos) it is necessary to use a large number of cycles of amplification (e.g., up to 40). In our experience, the Taq polymerase remained sufficiently active throughout the amplification cycles although it is possible to exhaust the Taq polymerase activity through prolonged periods of denaturation at high temperature over many cycles. In some cases, it may be advisable to add a second aliquot of Taq polymerase (although this increases the risk of contamination). Certain Taq polymerases have enhanced thermostability, for example, AmpliTaq DNA polymerase, Stoffel fragment (Perkin-Elmer Cetus) or Vent DNA polymerase (New England Biolabs), with half-lives of 20 min at 97.5° and 95 min at 100°, respectively.[21] Elevated levels of magnesium may be required for some of these enzymes.

A simpler and less expensive method of overcoming the problems associated with GC-rich templates, without elevating denaturation temperatures, may be to use cosolvents such as dimethyl sulfoxide (DMSO)

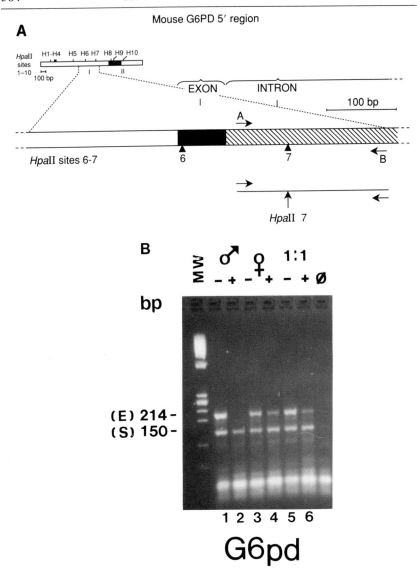

FIG. 2. (A) Schematic diagram of the 5′ regions of the mouse *G6pd*. *Hpa*II sites are represented by triangles. The sequence amplified by PCR contains one critical *Hpa*II site. Primers flanking this site are represented by horizontal arrows. Unshaded bar, 5′ untranslated region; shaded bar, exon; hatched bar, intron. Primers used in the *Hpa*II-sensitive PCR for the amplification of the *G6pd* sequence are: 5′-GCCCATGAGGACTAGACCTT; 3′-ACATCCACTGTGGGGCAGCTA (product size 214 bp).[12] The control primers are located in the 5′ untranslated region on the mouse *G6pd* gene (not shown in map in Fig. 2A) and have the following sequences: 5′-ACTCGCCCCATTTTCAAGGC; 3′-AGCTGCTAGTTTG-

or glycerol that lower the temperature required for DNA denaturation. Optimum concentrations of cosolvent have to be determined empirically for each set of primers.[22]

Precautions against Contamination

PCR products from previous experiments can be a major source of contamination of the following PCR amplifications. Consequently, a number of precautions should be observed in the preparation of DNA samples for use in the assay and when setting up the *Hpa*II-sensitive PCR (see also Refs. 14, 23): (1) Prepare PCR reaction mixes in a setup area remote from the laboratory where the final amplification products are analyzed. All the equipment and reagents used in the assay should be stored in this area. (2) Wear gloves at all stages of the procedure and ensure that they are changed before every entry into the setup area. (3) Use a set of dedicated micropipettes for setting up the PCR experiments. A different set of micropipettes should be used for pipetting the final amplification products and these should be kept in the amplification laboratory. (4) Aliquot all PCR reagents, stocks, and dilutions, where appropriate (i.e., primers, $10\times$ buffer, and dNTPs). (5) Autoclave all tips and tubes and the double-distilled water. With these precautions, reaction mixtures can be prepared on the open bench in the PCR setup area. Several reaction mixtures that do not contain DNA should be included in every PCR as control for contamination.

Gel Electrophoresis and Quantitation

Following amplification, a 20-μl aliquot of the reaction mix (with added bromphenol blue marker dye or equivalent loading buffer) is electropho-

---

[22] G. Sarkar, S. Kapelner and S. S. Sommer, *Nucleic Acids Res.* **18,** 7465 (1990).
[23] C. Holding and M. Monk, *Lancet* **ii,** 532 (1989).

GCTTCGG (product size 150 bp).[24] (B) PCR analysis of X-linked *G6pd* CpG methylation in male (♂) (lanes 1 and 2) and female (♀) (lanes 3 and 4) adult somatic DNA, and a 1:1 mixture (lanes 5 and 6) of these DNAs. Starting DNA (600 pg) is subjected to 40 rounds of PCR amplification: 2 cycles of 95° for 4 min, 60° for 1 min and 73° for 1 min, followed by 38 cycles of 95° for 1 min, 60° for 1 min and 73° for 1 min. The DNA was then electrophoresed in an ethidium bromide-stained 3% agarose–TBE gel. Band sizes are shown in base pairs (bp). E, Sequence containing specific CpG site of interest; S, internal standard; –, mock-digested DNA; +, DNA sample treated with *Hpa*II; φ, negative control (water added to reaction mix in place of DNA); MW, molecular weight markers (*Hae*III fragments of φX174 RF DNA).

TABLE III
METHYLATION OF SPECIFIC CpG SITES ON
X-LINKED GENES[a]

	G6pd	Pgk-1
Gametes		
Mature sperm	−	−
Mature oocyte	−	−
Pre-implantation embryos		
1-cell–morulae	−	−
Blastocysts (3.5 dpc)[b]	−	+/−
ICM (3.5 dpc)	−	+/−
Post-implantation embryos		
5.5–7.5 dpc EMB[b]	+	+
6.5–7.5 dpc EEE[b]	+	+
Female germ cells		
7.0–18.5 dpc oocytes	−	−
Newborn oocytes	−	−

[a] Data from Refs. 10 and 24.
[b] dpc, Days' post coitum; EEE, extra-embryonic portion; EMB, embryonic portion.

resed on a 3–4% agarose–TBE gel, containing 2 μg/ml ethidium bromide. The amplified DNA sequences are viewed directly under shortwave ultraviolet (UV) light. The presence of a band in the *HpaII* track indicates that the site is methylated; conversely, the absence of a band on the gel in the *HpaII* track indicates that the site is unmethylated. The gel may be photographed under UV light onto technical film (Polaroid 665), and band intensities measured from the negative by densitometry. The degree of methylation is determined by comparing band densities of the signals from control and *HpaII* digested DNA (see example below).

Results

An example of the *HpaII* sensitive PCR procedure is shown in Fig. 2. A sequence of the *G6pd* gene containing a critical CCGG site is amplified in adult male and female DNA and a 1 : 1 mixture of male and female DNA. The undigested DNA shows two bands on the gel resulting from amplification of the sequence containing the critical site of interest (E) and the control sequence (S). Amplification of the *G6pd* sequence containing the critical site is prevented by prior *HpaII* digestion (+) of male DNA (single active X chromosome unmethylated at this site) and is halved by prior *HpaII* digestion (+) of female DNA (one active chromosome unmethylated at this site, one inactive X chromosome methylated at this site). For the 1 : 1 mixtures of male and female DNA, the expected 30% of the signal remains (as determined by densitometry).

The use of this very sensitive procedure has allowed us to examine methylation of specific CpG sites of the X-linked *Pgk-1* and *G6pd* genes throughout female mouse development.[10,24] The sites become methylated at the time of X-inactivation. The results of this study and the earlier study by Singer-Sam *et al.*[10] are summarized in Table III. Methylation occurs at the time of inactivation of the X chromosome, but earlier for the *Pgk-1* gene which is closer to the X-inactivation center.[25] In female primordial germ cells, these two genes escape methylation at these sites, a condition which may be necessary for the reversibility of X inactivation at meiosis.

Acknowledgments

We thank Judy Singer-Sam for her guidance in developing the *Hpa*II-sensitive PCR techniques for *Pgk-1* and Daniela Toniolo for providing us with the mouse *G6pd* sequence ahead of publication. M. Z. is supported by Grants from the Company of Biologists (Cambridge), the European Science Foundation, and a Travelling Fellowship from the Wellcome Trust.

[24] M. Grant, M. Zuccotti and M. Monk, *Nature Genet.* **2**, 161 (1992).
[25] B. Borsani *et al., Nature* (London) **351**, 325 (1991).

[34] Analysis of Methylation and Chromatin Structure

By G. P. PFEIFER, J. SINGER-SAM, and A. D. RIGGS

Introduction

Changes in methylation and chromatin structure are an important part of mouse development. We describe here several polymerase chain reaction (PCR)-based assays that provide greatly improved sensitivity for the study of methylation and chromatin structure. We first describe a genomic sequencing method that uses ligation-mediated PCR (LM-PCR)[1,2] and can reveal both methylation and chromatin structure information. The original method of genomic sequencing[3] was very difficult to perform and required large amounts of material. LM-PCR is the most sensitive of the improved

[1] P. R. Muller and B. Wold, *Science* **246**, 780 (1989).
[2] G. P. Pfeifer, S. D. Steigerwald, P. R. Mueller, B. Wold, and A. D. Riggs, *Science* **246**, 810 (1989).
[3] G. M. Church and W. Gilbert, *Proc. Natl. Acad. Sci. U.S.A.* **81**, 1991 (1984).

Copyright © 1993 by Academic Press, Inc.
All rights of reproduction in any form reserved.

A

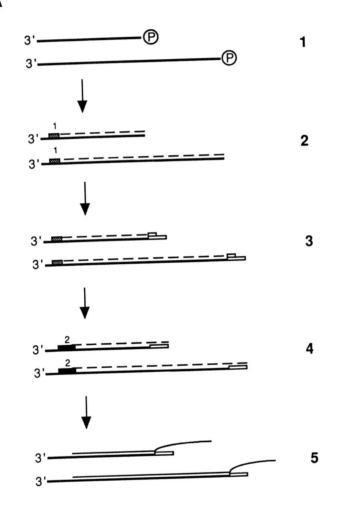

1

2

3

4

5

B

procedures for genomic sequencing,[4] and it requires only about 1 μg of total genomic DNA per lane of a sequencing gel for good-quality sequence-type ladders containing information on *in vivo* methylation, DNA structure, and DNA-bound proteins. We describe here the standard LM-PCR method of methylation analysis and *in vivo* DNase I footprinting as routinely performed in our laboratories. We also describe an assay for methylation, the *HpaII*–PCR method,[5] that can be used for quantitative measurement of methylation at a specific site in samples as small as several hundred cells.

Ligation-mediated PCR assays are based on the ligation of an oligonucleotide linker onto the 5′ end of each target DNA molecule. This provides a common sequence at the 5′ ends, and, together with a gene-specific primer, a family of fragments of different length can be exponentially amplified. The LM-PCR procedure is outlined in Fig. 1. The first step is cleavage of DNA to produce 5′-phosphorylated molecules. This can be achieved, for example, by chemical DNA sequencing (β-elimination) or by cutting with the enzyme DNase I. Primer extension of a gene-specific oligonucleotide (primer 1) generates molecules having a blunt end on one side. Linkers are ligated to the blunt ends, and then an exponential PCR amplification of the linker-ligated fragments is done using the longer oligonucleotide of the linker (linker–primer) and a second, nested gene-specific primer (primer 2). The PCR-amplified fragments are separated on a sequencing gel, electroblotted onto nylon membranes, and hybridized with a gene-specific single-stranded probe. Single-stranded hybridization probes can be made from PCR products that flank the LM-PCR primers (Fig. 1). By rehybridization, several gene-specific ladders can be sequentially visualized from one sequencing gel if several primer sets are included in the LM-PCR reactions.[2]

[4] H. Saluz, K. Wiebauer, and A. Wallace, *Trends Genet.* **7**, 207 (1991).
[5] J. Singer-Sam, J. M. LeBon, R. L. Tanguay, and A. D. Riggs, *Nucleic Acids Res.* **18**, 687 (1990).

Fig. 1. (A) Outline of the LM-PCR procedure. Step 1, Cleavage and denaturation of genomic DNA; step 2, annealing and extension of primer 1 with Sequenase; step 3, ligation of the linker; step 4, PCR amplification of gene-specific fragments with primer 2 and the 25-mer linker–primer; step 5, visualization of the sequence ladder by hybridization with a single-stranded probe which abuts on primers 1 and 2. (B) Arrangement of LM-PCR primers to sequence both strands of a 200-bp region (checkered area). Primer 1, Sequenase primer; primer 2, PCR primer. Primers 3 are used to make initially a double-stranded PCR product; then, one of primers 3 is used to make a single-stranded hybridization probe from the double-stranded template.

The LM-PCR has been used for sequencing genomic DNA and for determination of complete DNA cytosine methylation patterns.[2] Methylated cytosines are recognized by their failure to react with hydrazine in the Maxam–Gilbert[6] DNA sequencing reactions, which results in a gap in the C-specific sequencing lane. LM-PCR requires a minimum of 0.5 to 1 μg of genomic DNA for reproducible amplification of sequence ladders. The theoretical basis of these limitations has been discussed.[2] However, when only one specific site (instead of sequence ladders) is analyzed by LM-PCR, the amount of DNA required is significantly lower. For example, LM-PCR can be used in DNA methylation studies to amplify restriction-cut DNA fragments.[7] DNA is cleaved simultaneously with two restriction enzymes, one sensitive and one insensitive to methylation. After cleavage, a gene-specific oligonucleotide primer is used for primer extension, followed by linker ligation and then PCR amplification of both fragments. The ratio of the two amplified fragments provides a measure for the methylation state of the analyzed restriction enzyme site. With this LM-PCR technique, DNA from 100 cells can be qualitatively analyzed for methylation, and 50 ng of DNA can be analyzed quantitatively.[7]

Chromatin structure and *in vivo* protein–DNA contacts can be studied by footprinting experiments done on intact cells using dimethyl sulfate (DMS), a small molecule which penetrates the cell membrane.[1,8] However, DMS does not reveal all protein–DNA contacts. Nucleosomes, for example, are undetectable with DMS. LM-PCR provides adequate sensitivity to map rare DNA adducts, like those formed after UV irradiation,[9,10] and it can be used in conjunction with *in vivo* photofootprinting.[10] Reasoning that bulky agents such as enzymes would be quite useful in chromatin studies, we adapted DNase I footprinting for use with LM-PCR and obtained informative genomic DNase I footprints using cells permeabilized with lysolecithin.[11]

First, we describe the chemical DNA sequencing method used for DNA methylation analysis. Then, a description of chromatin structural studies by genomic DNase I footprinting of permeabilized cells is presented. The reader is referred to Ref. 8 for details on *in vivo* footprinting with DMS. We then present the details of the LM-PCR technique that is used to amplify fragments derived from the various cleavage methods.

[6] A. M. Maxam and W. Gilbert, this series, Vol. 65, p. 499.
[7] S. D. Steigerwald, G. P. Pfeifer, and A. D. Riggs, *Nucleic Acids Res.* **18**, 1435 (1990).
[8] G. P. Pfeifer, R. L. Tanguay, S. D. Steigerwald, and A. D. Riggs, *Genes Dev.* **4**, 1277 (1990).
[9] G. P. Pfeifer, R. Drouin, A. D. Riggs, and G. P. Holmquist, *Proc. Natl. Acad. Sci. U.S.A.* **88**, 1374 (1991).
[10] G. P. Pfeifer, R. Drouin, A. D. Riggs, and G. P. Holmquist, *Mol. Cell. Biol.* **12**, 1798 (1992).
[11] G. P. Pfeifer and A. D., Riggs, *Genes Dev.* **5**, 1102 (1991).

In a later section we describe the *Hpa*II–PCR assay, which can be used for the quantitative detection of DNA methylation at specific methylation-sensitive restriction sites. The assay can be used even when as few as several hundred cells are available for analysis, making it useful for the study of individual embryos or highly purified oocytes.

Methylation and Chromatin Studies by Ligation-Mediated Polymerase Chain Reaction

Methylation Analysis by Chemical Cleavage

The method described here is based on chemical DNA sequencing.[6] 5-Methylcytosines fail to react with hydrazine in the C-specific reaction. This results in a gap in the sequence ladder at positions of methylated cytosines. Unmethylated cloned DNA diluted to the single-copy gene level with bacterial carrier DNA can be used as a control. 5-Methylcytosines can also be detected by UV irradiation of DNA since they do not form pyrimidine (6-4)–pyrimidone photoproducts at 5'-CCG and 5'-TCG sequences.[9] The UV analysis is limited to these trinucleotide sequences, which comprise about 50% of all potential methylation sites in mammalian DNA. An example for methylation analysis by hydrazine cleavage is shown in Fig. 2A.

Produce

1. Isolate DNA by standard procedures using phenol/chloroform extraction and ethanol precipitation. The DNA preparation should be mostly free of RNA.

2. Digest the DNA with a restriction enzyme that does not cut within the region to be sequenced. This is done to reduce the viscosity of the solution. After digestion, extract the DNA once with phenol/chloroform and once with chloroform and then precipitate with ethanol. Dissolve the DNA in water and adjust the concentration to 2–10 $\mu g/\mu l$. The conditions below work well for 10 to 50 μg DNA.

3. Mix the following on ice: 5 μl genomic DNA (10 to 50 μg), 15 μl of 5 M NaCl, and 30 μl hydrazine (Aldrich, Milwaukee, WI). Hydrazine is a highly toxic chemical and should be handled in a well-ventilated hood. Hydrazine waste (including plastic material) is detoxified in 3 M ferric chloride. Store hydrazine under nitrogen at 4°. Replace the bottle at least every 6 months.

4. Incubate at 20° for 20 min.

5. Add 200 μl of 0.3 M sodium acetate (pH 7.5), 0.1 mM EDTA.

A   **B**

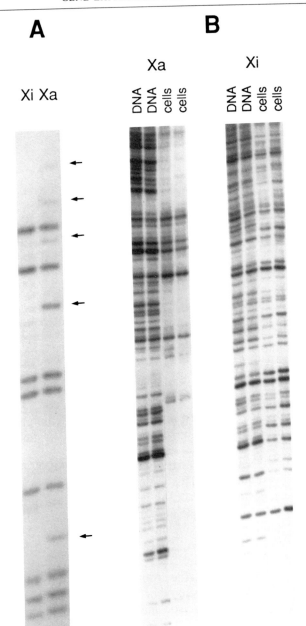

6. Add 750 μl precooled ethanol ($-70°$), keep in a dry ice/ethanol bath for 20 min, and centrifuge at 4° at 14,000 g, 10 min.

7. Remove the supernatant, dissolve the pellet in 225 μl water, and reprecipitate the DNA by adding 25 μl of 3 M sodium acetate and 750 μl of ethanol.

8. Wash with 1 ml of 75% ethanol and dry the pellet in a SpeedVac (Savant Instruments).

9. Dissolve the DNA in 100 μl of 1 M piperidine (Fluka, Ronkonkoma, NY, freshly diluted).

10. Heat at 90° for 30 min in a heat block (lead weight on top).

11. Add 10 μl of 3 M sodium acetate (pH 5.2) and 2.5 volumes of ethanol.

12. Keep on dry ice for 20 min.

13. Spin at 14,000 g, 15 min; remove the supernatant.

14. Wash with 1 ml of 75% ethanol.

15. Remove all traces of remaining piperidine by drying the sample overnight in a SpeedVac. Dissolve DNA in water to a concentration of about 1 μg/μl.

16. Determine the cleavage efficiency by running 1 μg of the samples on a 1.5% alkaline agarose gel. There should be fragments below the 200-nucleotide (nt) size range. Only these will be detected by LM-PCR.

17. Process the samples by LM-PCR (see below).

Genomic Footprinting with DNase I

A cell permeabilization system[12] is used for *in situ* digestion of chromatin with DNase I. Specific DNase I footprints of single-copy gene sequences are then obtained from genomic DNA by LM-PCR amplification.[11] Digestion of isolated nuclei with DNase I has consistently resulted in less

[12] L. Zhang and J. D. Gralla, *Genes Dev.* **3,** 1814 (1989).

FIG. 2. Chromatin analysis with LM-PCR-aided genomic sequencing. (A) Analysis of cytosine methylation in the CpG island of human *PGK1*. After reaction of genomic DNA with hydrazine and piperidine cleavage, fragments of the promoter of the X-linked *PGK1* gene were amplified by LM-PCR. Reactivity with hydrazine distinguishes methylated and unmethylated cytosines; unreactive positions (no band) are indicative of methylation and are marked with an arrow. The example shows data for hamster–human hybrid cell lines containing either an active human X chromosome (Xa) or an inactive human X chromosome (Xi). (B) DNase I footprinting of the *PGK1* promoter in permeabilized cells. Cells were rendered permeable by treatment with lysolecithin and then treated with DNase I. Naked DNA controls treated with DNase I are shown for comparison. Three footprints are seen for the active and none for the inactive X chromosome.[11]

clear footprints, presumably owing to detachment of transcription factors from their binding sites.[11] Cell permeabilization can be used for both monolayer and suspension culture cells,[13] and, with minor modifications, the method should be adaptable for homogenized tissue samples.

Procedure

1. Grow cells as monolayers to about 80% confluency.

2. To permeabilize the cells, treat the cell monolayers ($\sim$4 $\times$ 10^6 cells) with 0.05% lysolecithin (type I, Sigma, St. Louis, MO) in prewarmed solution I [150 mM sucrose, 80 mM KCl, 35 mM HEPES, 5 mM K$_2$HPO$_4$, 5 mM MgCl$_2$, 0.5 mM CaCl$_2$ (pH 7.4)] for 1 min at 37°.[12]

3. Remove the lysolecithin and wash with 5–10 ml of solution I. Incubate the cells with DNase I (10–50 μg/ml, grade I, Boehringer-Mannheim, Indianapolis, IN) in solution II [150 mM sucrose, 80 mM KCl, 35 mM HEPES, 5 mM K$_2$HPO$_4$, 5 mM MgCl$_2$, 2 mM CaCl$_2$ (pH 7.4)] at room temperature for 5 min. DNase I concentrations and incubation times may have to be adjusted for different cell types. During DNase I treatment, less than 10% of the cells should become detached from the plastic surface.

4. Stop the reaction and lyse the cells by removal of the DNase I solution and addition of 2.5 ml stop solution [20 mM Tris-HCl (pH 8.0), 20 mM NaCl, 20 mM EDTA, 1% sodium dodecyl sulfate (SDS), 600 μg/ml proteinase K]. Add 2.5 ml of 150 mM NaCl, 5 mM EDTA (pH 7.8) and incubate the solution for 3 hr at 37°.

5. Purify the DNA by phenol/chloroform extraction and ethanol precipitation. Remove RNA by digestion with RNase A (50 μg/ml in TE buffer, 1 hr at 37°). Extract with phenol/chloroform and precipitate with ethanol.

6. Naked DNA controls are obtained by DNase I digestion of purified DNA. Digest 40 μg of genomic DNA in 400 μl of 40 mM Tris-HCl (pH 7.7), 10 mM NaCl, 6 mM MgCl$_2$, with 0.2 to 1.6 μg/ml DNase I for 10 min at room temperature.

7. To prevent nonspecific priming from the 3'-OH groups of DNase I-cut genomic DNA fragments, block the 3' ends by addition of a dideoxynucleotide. This step can significantly increase the signal-to-noise ratio of the procedure. Heat-denature the DNase I-cleaved DNA (10 μg in 50 μl) and incubate with 5 units of Sequenase 2.0 (USB, Cleveland, OH) in 5 μM ddNTPs, 40 mM Tris-HCl (pH 7.7), 25 mM NaCl, 6.7 mM MgCl$_2$ for 20 min at 45°. Add sodium cacodylate (pH 7.4) to 200 mM and 2-mercaptoethanol to 1 mM, denature the DNA again, and incubate with 30 units terminal transferase (BRL, Gaithersburg, MD) at 37° for 30 min.

---

[13] R. Contreras and W. Fiers, *Nucleic Acids Res.* **9**, 215 (1981).

Extract with phenol/chloroform and selectively precipitate DNA fragments at room temperature by addition of ammonium acetate (to a final concentration of 2 M) and 2 volumes of ethanol. Centrifuge at 10,000 g at room temperature, wash the pellets in 75% ethanol, and dissolve the DNA in water at about 1 $\mu g/\mu l$.

8. Check the size of the DNA fragments on alkaline 1.5% agarose gels.
9. Process the sample by LM-PCR (see below).

The 3'-end blocking step can be omitted if an extension product capture procedure is used to enrich for specific fragments.[13a]

Identification of *in vivo* protein–DNA contacts requires a comparison of the *in vivo* treated sample with a naked DNA control. Purified DNA samples might be different depending on cell type. This is due to heterogeneous cytosine methylation patterns of genomic DNA samples, which can give altered DNase I cleavage patterns.[11] Therefore, it is imperative to compare all *in vivo* samples with an appropriate *in vitro* control from the same cell type.

For optimum molecule usage, the average fragment size should be between 100 and 250 nt. DNase I digestion often results in a broader distribution of fragment sizes (50–1500 nt). Estimate the amount of DNA to be used in the Sequenase reaction (see below) from the relative amount of DNA fragments within the lower size range to obtain similar band intensities on the sequencing gel in all lanes. Genomic footprinting of permeabilized cells with DNase I has given very clear footprints at the *PGK1* promoter.[11] An example is shown in Fig. 2B.

Ligation-Mediated Polymerase Chain Reaction

First Primer Extension. A gene-specific primer is used on genomic DNA fragments for primer extension with Sequenase. The primers we have used as Sequenase primers are 17- to 20-mer oligonucleotides with a calculated melting temperature (T_m) of 50° to 56°. Calculation of T_m is done with a computer program.[14]

1. Mix the following in a siliconized tube: 0.5–5 μg of cleaved genomic DNA, 0.6 pmol of primer 1, and 3 μl of 5× Sequenase buffer [250 mM NaCl, 200 mM Tris-Cl (pH 7.7)] in a final volume of 15 μl.
2. Denature at 95° for 3 min, then incubate at 45° for 30 min.
3. Cool on ice, then spin 5 sec.

[13a] V. T. Törmänen, P. M. Swiderski, B. E. Kaplan, G. P. Pfeifer, and A. D. Riggs, *Nucleic Acids Res.* **20**, 5487 (1992).
[14] W. Rychlik and R. E. Rhoads, *Nucleic Acids Res.* **17**, 8543 (1989).

4. Add 7.5 μl cold, freshly prepared Mg–DTT–dNTP mix [20 mM MgCl$_2$, 20 mM dithiothreitol (DTT), 0.25 mM of each dNTP] and 5 units of Sequenase 2.0 (USB).

5. Incubate at 48° for 15 min, then cool on ice.

6. Add 6 μl of 300 mM Tris-Cl (pH7.7).

7. Heat-inactivate the Sequenase at 67° for 15 min.

8. Cool on ice, then spin 5 sec.

Incubation with Sequenase 2.0 at 48° instead of 37° not only increases the processivity of the enzyme through sequences that can form secondary structures, but also decreases the terminal transferase activity associated with Sequenase. At 48°, Sequenase adds an extra nucleotide to 30–40% of the molecules, but this activity does not appear to be very sequence-selective (G. P. Pfeifer, J. Singer-Sam, and A. D. Riggs, unpublished results, 1992).

Ligation. The primer-extended molecules that have retained a 5'-phosphate after chemical sequencing or DNase I cleavage are ligated to an unphosphorylated synthetic double-stranted linker. The linker primers are gel-purified on 20% polyacrylamide gels. All oligonucleotides including linker primers are then dissolved in water as a stock solution of 50 pmol/μl. Linkers are prepared in 250 mM Tris-Cl (pH 7.7) by annealing a 25-mer (5'-GCGGTGACCCGGGAGATCTGAATTC, final concentration 20 pmol/μl) to an 11-mer (5'-GAATTCAGATC, final concentration 20 pmol/μl).[1] This mixture is heated to 95° for 3 min and gradually cooled to 4° over a time period of at least 3 hr. Linkers are stored at −20°. They are always thawed and kept on ice.

1. Add 45 μl of freshly prepared ligation mix [13.33 mM MgCl$_2$, 30 mM DTT, 1.66 mM ATP, 83 μg/ml bovine serum albumin (BSA), 3 units/reaction T4 DNA ligase (Promega, Madison, WI), and 100 pmol linker/reaction (=5 μl linker)].

2. Incubate overnight at 18°.

3. Heat-inactivate at 70° for 10 min.

4. Precipitate the DNA by adding 8.4 μl of 3 M sodium acetate (pH 5.2), 10 μg *Escherichia coli* tRNA, and 220 μl ethanol; cool samples on dry ice for 20 min.

5. Centrifuge for 15 min at 4° at 14,000 g.

6. Wash pellets with 1 ml of 75% ethanol.

7. Dry samples in a SpeedVac.

8. Dissolve pellets in 50 μl water and transfer to 0.5-ml siliconized tubes.

Polymerase Chain Reaction. Gene-specific fragments are amplified with a second, nested gene-specific primer and the common linker–primer, the longer oligonucleotide of the linker. It is preferable to gel-purify the amplification primers. The primers used in the amplification step (primer 2) are 21- to 28-mers. They are designed to extend 3' to primer 1. Primer 2 should overlap several bases with primer 1. The annealing temperature in the PCR is chosen to be at the T_m of the gene-specific primer (calculated T_m between 60° and 68°, Ref. 14).

1. Add 50 μl of freshly prepared 2× *Taq* polymerase mix [20 m*M* Tris-HCl (pH 8.9), 80 m*M* NaCl, 0.02% gelatin, 4 m*M* MgCl$_2$, 0.4 m*M* of each dNTP]. This mix also contains 10 pmol of the gene-specific primer (primer 2), 10 pmol of the linker–primer (5'-GCGGTGACCCGGGAGAT-CTGAATTC), and 3 units *Taq* polymerase (AmpliTaq, Perkins-Elmer, Cetus, Norwalk, CT) for each sample.

2. Cover the samples with 50 μl mineral oil and spin briefly.

3. Cycle 16 to 20 times at 95° for 1 min, 60° to 68° for 2 min, and 76° for 3 min.

4. To extend completely all DNA fragments and uniformly add an extra nucleotide, an additional *Taq* polymerase extension step is performed. Add 1 unit of fresh *Taq* polymerase per sample together with 10 μl reaction buffer. Incubate for 10 min at 74°.

5. Stop the reaction by adding sodium acetate (pH 5.2) to 300 m*M* and EDTA to 10 m*M*, then add 10 μg tRNA.

6. Extract with 70 μl of phenol and 120 μl chloroform (premixed).

7. Add 2.5 volumes of ethanol and put on dry ice for 20 min.

8. Centrifuge samples for 15 min at 14,000 g at 4°.

9. Wash pellets in 1 ml of 75% ethanol.

10. Dry pellets in a SpeedVac.

Gel Electrophoresis. The PCR-amplified fragments are separated on 0.4 mm thick, 60 cm long sequencing gels consisting of 8% polyacrylamide and 7 *M* urea in 0.1 *M* TBE. The gel is run until the xylene cyanole marker reaches the bottom. Fragments below the xylene cyanole dye hybridize only very weakly.

Dissolve pellets in 1.5 μl of water and add 3 μl formamide loading dye [94% formamide, 2 m*M* EDTA (pH 7.7), 0.05% xylene cyanole, 0.05% bromphenol blue]. Heat the samples to 95° for 2 min prior to loading, and load one-half of the sample with an elongated, flat tip (National Scientific, San Rafael, CA).

Electroblotting. We have been using electroblotting and hybridization instead of directly extending a ^{32}P-labeled primer followed by gel electrophoresis[1] because longer single-stranded probes provide a higher specific activity than end-labeled oligonucleotides. In addition, nylon membranes can easily be rehybridized after multiple primer sets have been included in the Sequenase reaction and PCR amplification (multiplexing).[2]

Electroblotters for transfer of sequencing gels are available from Hoefer Scientific (San Francisco, CA) and Owl Scientific (Cambridge, MA). We have used a simple homemade apparatus. Stainless steel plates from a gel drier are used as electrodes (obtainable from Hoefer Scientific). They are connected directly with a platinum wire and a cord to a Bio-Rad (Richmond, CA) 200/2.0 power supply. A homemade plastic container is the buffer chamber. The size of the chamber is 50 by 40 by 14 cm (length × width × height). Opening the lid interrupts the current (safety precaution).

1. After the run, transfer the lower part of the gel (length 40 cm) to Whatman (Clifton, NJ) 3MM paper and cover it with Saran wrap.

2. Pile 12 layers of Whatman 17 paper, 43 × 19 cm, presoaked in 90 mM TBE, onto the lower electrode, which is resting on four plastic incubation racks (height ~3 cm). Squeeze the paper with a rolling bottle to remove air bubbles between the paper layers.

3. Place the gel piece covered with Saran wrap onto the paper and remove the air bubbles between gel and paper by wiping over the Saran wrap with a soft tissue.

4. When all air bubbles are squeezed out, remove the Saran wrap and cover the gel with a GeneScreen (Du Pont, Boston, MA) nylon membrane cut somewhat larger than the gel and presoaked in 90 mM TBE.

5. Put 12 layers of Whatman 17 paper presoaked and cut as above onto the nylon membrane. Papers can be reused several times except for the two papers closest to the gel and closest to the upper electrode.

6. Place the upper electrode onto the paper and put 2 kg of lead weights on top of it.

7. Fill the electroblotting chamber with 90 mM TBE until the buffer level is about 5 layers of paper below the gel. The electroblotting is performed at 1.6 A and 30 V.

8. After 1 hr, remove the nylon membrane and mark the DNA side.

Hybridization

1. Dry the nylon membrane briefly at room temperature and then bake it at 80° for 20 min in a vacuum oven. Then irradiate the membrane at a UV dose of 1000 J/m^2 (~1 min at a distance of 20 cm from five germicidal tubes).

2. The hybridization is performed in rotating 250-ml plastic or glass cylinders in a hybridization oven. Wet the nylon membrane in 90 mM TBE, and roll it into the cylinders so that the membrane sticks completely

to the walls of the cylinders. This can be easily done by rolling the membrane first onto a 25-ml pipette and then unspooling it into the cylinder.

3. Prehybridize with 15 ml hybridization buffer [0.25 M sodium phosphate (pH 7.2), 1 mM EDTA, 7% SDS, 1% BSA] for 10 min at 60° to 68°.

4. Remove the prehybridization solution. Dilute the labeled probe into 5 ml hybridization buffer. Hybridize overnight at 60° to 68° depending on the G + C content of the probe.

5. After hybridization, wash each nylon membrane with 2 liters of washing buffer [20 mM sodium phosphate (pH 7.2), 1 mM EDTA, 1% SDS] at a temperature 5° lower than the hybridization temperature. Perform several washing steps (5 min each) with prewarmed buffer. Dry the membranes briefly at room temperature, wrap them in Saran wrap, and expose for 0.5 to 8 hr to Kodak (Rochester, NY) XAR-5 films. Nylon membranes can be rehybridized if several sets of primers have been included in the primer extension and amplification reactions. Probes can be stripped by soaking the filters in 0.2 M NaOH for 30 min at 45°.

Preparation of Single-Stranded Hybridization Probes. Initially, we used single-stranded DNA probes made from RNA templates by reverse transcription.[15] Perhaps the most convenient method to prepare labeled single-stranded probes is repeated primer extension by *Taq* polymerase with a single primer (primer 3) on a cloned, double-stranded template DNA,[16] which can be a plasmid or PCR product. The length of these probes can be easily controlled by an appropriate restriction cut. The primer that is used to make a probe (primer 3) should be on the same strand just 3′ to the amplification primer (see Fig. 1) and should have a T_m of 60° to 66°. It can overlap a few bases with primer 2.

1. Mix 50 ng of the respective restriction-cut plasmid DNA (or 5 ng of a PCR product) with 20 pmol of primer 3, 100 μCi of [^{32}P]dCTP, 20 μM of the other three dNTPs, 10 mM Tris-Cl (pH 8.9), 40 mM NaCl, 0.01% gelatin, 2 mM MgCl$_2$, and 3 units of *Taq* polymerase in a volume of 100 μl. Cover with mineral oil.

2. Linear amplification is done in a thermocycler using 35 cycles at 95° (1 min), 60–66° (2 min), and 75° (3 min).

3. Recover the probe by phenol/chloroform extraction, addition of ammonium acetate to a concentration of 0.7 M, precipitation with ethanol at room temperature, and centrifugation. Alternatively, the probe may be

[15] F. Weih, A. F. Stewart, and G. Schütz, *Nucleic Acids Res.* **16**, 1628 (1988).
[16] M. Stürzl and W. K. Roth, *Anal. Biochem.* **185**, 164 (1990).

recovered by centrifugation through a Sephadex G-50 spun column (5 prime–3 prime, Inc., Boulder, CO).

Quantitative *Hpa*II–Polymerase Chain Reaction Assay to Measure Methylation of DNA from Small Number of Cells

The *Hpa*II–PCR assay is based on the fact that a DNA template must be intact in order to be amplified efficiently by the PCR. If, prior to PCR, the DNA sample is treated with *Hpa*II (or another methylation-sensitive restriction endonuclease) only DNA molecules methylated at sites between the two PCR primers will be amplified in a subsequent PCR reaction. A comparison of the amount of amplified product present with and without prior digestion with *Hpa*II reflects the percent methylation at that site.

By limiting the number of PCR cycles, one can determine the percent DNA methylation of a given *Hpa*II site without the use of internal standards in the PCR reaction.[17] However, to obtain quantifiable data with limiting amounts of sample (several hundred cells), we include an internal standard. The standard is amplified by the same primers as the template DNA, but it can be distinguished from it by size or sequence.[5,18] Use of the internal standard is advantageous for several reasons: (1) the results become independent of the number of PCR cycles; (2) it is not necessary to know the amount of DNA in each sample beforehand; and (3) comparison of the amplified signal in tubes containing different amounts of the internal standard provides a control for the quantifiability of the results of each experiment.

Purification of DNA

We have used two methods to prepare DNA suitable for the *Hpa*II–PCR assay.[19–21] Both methods work well when there are about

---

[17] J. Singer-Sam, T. P. Yang, N. Mori, R. L. Tanguay, J. M. LeBon, J. C. Flores, and A. D. Riggs, in "Nucleic Acid Methylation" (G. Clawson, D. Willis, A. Weissbach, and P. Jones, eds.), p. 285. Alan R. Liss, New York, 1990.

[18] G. Gilliland, S. Perrin, and H. F. Bunn, in "PCR Protocols: A Guide to Method and Applications" (M. A. Innis, D. H. Gelfand, J. J. Sninsky, and T. J. White, eds.), p. 60. Academic Press, New York, 1990.

[19] M. Jeanpierre, *Nucleic Acids Res.* **15**, 9611 (1987).

[20] J. Singer-Sam, M. Grant, J. M. LeBon, K. Okuyama, V. Chapman, M. Monk, and A. D. Riggs, *Mol. Cell. Biol.* **10**, 4987 (1990).

[21] S. A. Miller, D. D. Dykes, and H. F. Polesky, *Nucleic Acids Res.* **16**, 1215 (1988).

1000 cells or fewer in each sample. For samples containing more cells, it may be necessary to increase the volume to obtain digestible DNA. A phenol extraction step may also be necessary.

Procedure 1

1. Mix fresh or frozen tissue (up to 1000 cells) with 163 μl of freshly prepared lysis solution [6 M guanidine hydrochloride (ultrapure, BMB), 140 μl; 7.5 M ammonium acetate, 10 μl; 20% (w/v) sodium sarkosyl, 10 μl; 20 mg/ml proteinase K (Boehringer-Mannheim, Indianapolis, IN), 1.5 μl; 0.75 mg/ml M13 RF DNA, 1.3 μl] in a 1.7-ml siliconized Eppendorf tube. Seal the tube with Parafilm and mix gently.[19,20]

2. Incubate at 60° for 1 hr.

3. Cool to room temperature; centrifuge for 1 min.

4. Shear DNA by passage 10 times through a 26-gauge needle. Transfer the sample to a 1.7-ml Eppendorf tube prior to this step if the sample is in a smaller tube. Shearing is done to reduce the DNA in size, increasing its solubility in later steps.

5. Precipitate the DNA by addition of 350 μl of 95% ethanol at room temperature. As carrier, add 20 μg of mussel glycogen (Sigma, stock solution, 20 mg/ml). Mix well and centrifuge for 30 min at room temperature in a microcentrifuge.

6. Remove the supernatant. Wash the DNA/glycogen pellet with 500 μl of 70% (v/v) ethanol, vortex, centrifuge for 10 min at room temperature (10,000 g), and remove the supernatant. Repeat the 70% ethanol wash. The samples may be stored in 70% ethanol at this point if desired. Remove supernatant, even the last few microliters, by use of a micropipette. (Drying under vacuum makes later resuspension irreproducible.)

Procedure 2

1. Add lysis buffer (75 mM NaCl, 25 mM EDTA, pH 8, 1% SDS, 200 μg/ml freshly dissolved proteinase K) to the cells (1 ml for ~1000 cells) and incubate overnight at 37°.[21]

2. Add prewarmed saturated NaCl (~6 M) to a final concentration of about 1.5 M (350 μl/ml).

3. Shake vigorously for 15 sec and centrifuge the sample (10,000 g for 10 min at room temperature); extract the supernatant with 1 volume of chloroform.

4. Reduce the DNA in size by passage 10 times through a 26-gauge needle. Add glycogen (20 μg) and M13 DNA (1 μg) as carriers, then precipitate the DNA with 1 volume of 2-propanol.

5. Follow Step 6 of Procedure 1.

Enzymatic Digestion and Polymerase Chain Reaction

Procedure

1. Resuspend the DNA in 31.5 μl of 0.1× TE (Tris-HCl, pH 8.0, 1 mM; EDTA 0.1 mM). Heat at 37° for 15 min, then at 60° for 2 min, to dissolve. Cool and centrifuge at 10,000 g for 2 min at room temperature.

2. Add 10.5 μl of 2× KGB[22] (potassium glutamate, 200 mM; Tris–acetate, pH 7.6, 50 mM; magnesium acetate, 20 mM; Bacto-gelatin, 0.02%). We find the KGB buffer system to be convenient, as many enzymes retain activity, and the buffer does not interfere with subsequent steps of the assay as long as it is diluted to 0.1× KGB in the final 100 μl PCR mix.

3. Reduce the DNA in size by incubation with a restriction endonuclease that does not cut within the sequence flanked by the PCR primers. Typically, enzymes such as *Bam*HI or *Xba*I are added (1 μl, 10–20 units), and the samples are incubated at 37° for 1.5 hr.

4. Remove two 19-μl aliquots of the reaction mix to 0.65-ml siliconized Eppendorf tubes (save remainder for Step 7).

5. To first tube add 1 μl of *Hpa*II (10 units/μl, BRL). To the second tube ($-$*Hpa*II control) add 1 μl TE. Incubate at 37° for 1.5 hr.

6. Repeat Step 5.

7. Quick-spin; remove a 2-μl aliquot to 8 μl TE; run this aliquot and 2 μl of undigested DNA on an agarose gel to check for completeness of *Hpa*II digestion (M13 bands should be visible).

8. Heat-inactivate remainder of *Hpa*II digest in a heatblock at 100° for at least 15 min (seal tubes with Parafilm). This is an important step, since incomplete inactivation will result in no PCR products.

9. Chill on ice, then centrifuge for 5 min.

10. Dilute the sample (up to 19 μl) into the PCR mix (final volume 100 μl). The PCR mix typically contains, together with primers flanking the *Hpa*II site of interest (0.2 μM) and the internal standard (see below), 10 mM Tris-HCl, pH 8.3, 50 mM potassium chloride, 1.5 mM MgCl$_2$, 200 μM dNTPs, 0.001% gelatin, and 2.5 units of AmpliTaq polymerase (Cetus). For several hundred molecules of target DNA, PCR amplification is done for approximately 40 cycles, typically at 95° for 1 min, 60° for 2 min, and 72° for 2 min.

Notes on Assay

Parameters for PCR Amplification. The optimum pH and magnesium chloride concentration should be determined empirically for each primer

[22] M. McClelland, J. Hanish, M. Nelson, and Y. Patel, *Nucleic Acids Res.* **16,** 364 (1988).

set. We find that the reproducibility and range of linearity of the assay are increased if the first two cycles of PCR are done with a 4-min heat-denaturation step at 95–96°.

Internal Standard. The internal standard may differ in size or sequence from the template DNA to be amplified. Methods for synthesizing such standards are discussed elsewhere.[5,18] To assure that each experiment is done in a quantifiable range of the assay, at least two PCR reactions with different amounts of the internal standard should be carried out for each determination. (For example, if we estimate 500 molecules of template DNA to be present, we divide the sample in half, and do two PCR reactions, with 250 molecules and 500 molecules of internal standard, respectively.) The ratio of template to internal standard should vary with the amount of internal standard added. Alternatively, a standard curve may be generated by amplification of a known amount of mouse genomic DNA with varying amounts of the internal standard.

Quantitation. We describe here methods based on size differences between the sample DNA and the internal standard. Methods of quantitation based on sequence differences are discussed elsewhere.[18,23] The PCR products are analyzed by electrophoresis in agarose gels containing ethidium bromide and quantified by densitometry of photographic negatives. Alternatively, after electrophoresis the products may be transferred to nylon filters, probed with a ^{32}P-labeled oligonucleotide, and quantified by use of a radioisotope scanning system.[24,25] For each tube, the number of DNA molecules initially present in the PCR mix is calculated by comparison of the amplified signal produced by the sample DNA to that produced by the known amount of added internal standard. A comparison of two tubes (with or without prior *Hpa*II digestion) then reveals the percent methylation of a given *Hpa*II site.

Acknowledgments

We thank R. L. Tanguay, J. M. LeBon, A. Dai, S. Tommasi, S. D. Steigerwald, and V. T. Törmänen for contributions. This work was supported by National Institutes of Health Grant (AG08196) to A.D.R.

[23] J. Singer-Sam and A. D. Riggs, this volume [20].
[24] G. J. Murakawa, J. A. Zaia, P. A. Spallone, D. A. Stephens, B. E. Kaplan, R. B. Wallace, and J. J. Rossi, *DNA* **7**, 287 (1988).
[25] J. Singer-Sam, M. O. Robinson, A. R. Bellvé, M. I. Simon, and A. D. Riggs, *Nucleic Acids Res.* **18**, 1255 (1990).

Section VIII

Gene Identification

[35] Construction of Primary and Subtracted cDNA Libraries from Early Embryos

By JAY L. ROTHSTEIN, DABNEY JOHNSON, JOEL JESSEE,
JACEK SKOWRONSKI, JULIE A. DELOIA, DAVOR SOLTER,
and BARBARA B. KNOWLES

Introduction

Mammalian preimplantation development is characterized by a period following fertilization when the embryo grows, differentiates, and develops the extraembryonic cell types necessary for implantation and maternal communication. Little genetic information is available about this period, primarily because of the inaccessibility of cells and the lack of molecular probes to distinguish these cell types. Immunohistochemistry, two-dimensional (2-D) gel electrophoresis, enzyme assays, dot blotting, and Northern blotting were the only methods available for examining transcripts or proteins in the preimplantation mouse embryo. These methods were restricted to the examination of moderate to abundantly expressed gene products, failing to provide information regarding genes expressed at relatively low levels during embryogenesis.

More recently, the polymerase chain reaction (PCR) has been used in the detection of growth factor gene expression in the preimplantation mouse embryo.[1] Although the PCR provides a method to detect rare messages from small amounts of starting material, quantification is difficult and not without ambiguity. In addition, to isolate new members of a gene family by PCR, primers in a known conserved region must amplify a DNA segment unique to the new family member. An alternative but less reliable method, namely, anchored PCR,[2] relies on a single primer in a region of known sequence.

Clearly, the molecular analysis of early embryogenesis would benefit greatly from the construction of large and representative stage-specific cDNA libraries which can in turn be used as a means to isolate rare transcripts of known genes by subtractive-enrichment hybridization. Previous attempts to generate such libraries from early embryos were largely unsuccessful. In these cases the libraries were primarily com-

[1] D. A. Rappolee, C. A. Brenner, R. Schultz, R. D. Mark, and Z. Werb, *Science* **241**, 1823 (1988).
[2] M. A. Frohman, M. K. Dush, and G. R. Martin, *Proc. Natl. Acad. Sci. U.S.A.* **85**, 8998 (1988).

Copyright © 1993 by Academic Press, Inc.
All rights of reproduction in any form reserved.

posed of clones containing inserts from the carrier RNA,[3] PCR artifacts,[4] or ribosomal RNA,[5] or were composed of less than 100 clones.[6] In addition, these previous cDNA libraries were constructed from RNA derived from a single stage of development and, therefore, were not useful in the analysis of changes in gene expression which likely occur during embryonic development. In this regard, only cDNA libraries constructed from serial stages of embryogenesis would provide a useful resource.

The PCR has been utilized in the construction of embryonic cDNA libraries[4,7–10]; however, there are many shortcomings to this approach. For example, large numbers of PCR cloning artifacts pose a significant hindrance to screening and subtractive hybridization. In addition, PCR amplification may progressively select for smaller products with each subsequent cycle, limiting the possibility of obtaining full-length cDNAs. Finally, PCR-synthesized libraries are poor substrates for subtraction since very short DNA segments compromise the degree of overlap and therefore the amount of complementary sequence available for hybridization. To avoid these limitations yet still be able to work with the small amounts of poly(A)$^+$ mRNA available from the egg and preimplantation embryo, we devised a cloning scheme for the generation of conventional, directionally cloned stage-specific embryonic cDNA libraries (Fig. 1). Each step in the cloning strategy is first optimized to yield the highest quality and quantity of product so as to improve the overall efficiency of embryonic RNA extraction, cDNA synthesis, and cloning. In addition, cDNAs are directionally cloned into a plasmid vector that provides T3 and T7 polymerase promoters for use in subtractive hybridization. A bacterial transformation protocol that uses electroporation in lieu of the standard chemical transformation protocols[11] is used to increase the number of independent clones and improve the likelihood that the libraries would contain rare and/or stage-specific embryonic genes.

[3] J. McConnell and C. J. Watson, *FEBS Lett.* **195,** 199 (1986).

[4] J. Welsh, J.-P. Liu, and A. Efstratiadis, *Genet. Anal. Tech. Appl.* **7,** 5 (1990).

[5] D. E. Weng, R. A. Morgan, and J. D. Gearhart, *Mol. Reprod. Dev.* **1,** 233 (1989).

[6] K. D. Taylor and L. Pikó, *Development (Cambridge, UK)* **101,** 877 (1987).

[7] A. Belyavsky, T. Vinogradova, and K. Rajewsky, *Nucleic Acids Res.* **17,** 2919 (1989).

[8] M. S. H. Ko, *Nucleic Acids Res.* **18,** 5705 (1990).

[9] J. L. R. Rubinstein, A. E. J. Brice, R. D. Ciaranello, D. Denney, M. H. Porteus, and T. B. Usdin, *Nucleic Acids Res.* **18,** 4833 (1990).

[10] C. Timblin, J. Battey, and W. M. Kuehl, *Nucleic Acids Res.* **18,** 1587 (1990).

[11] D. Hanahan, J. Jessee, and F. R. Bloom, this series, Vol. 204, p. 63.

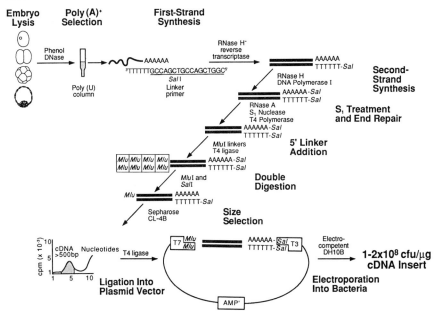

FIG. 1. Scheme for the generation of cDNA libraries. Embryo RNA was isolated and the poly(A)$^+$ RNA extracted. First-strand cDNA synthesis was primed with an oligo(dT) linker–primer containing two SalI restriction enzyme sites using Superscript RNase H$^-$ reverse transcriptase. Following second-strand DNA synthesis, the double-stranded cDNA was treated with RNase followed by S$_1$ nuclease to remove hairpin loops. The ends of the nuclease-treated cDNA were then repaired with T4 polymerase followed by the ligation of 5' MluI linkers. cDNA was double-digested using SalI and MluI, size-selected on a Sepharose CL-4B column to recover cDNA longer than 500 bp, ligated into a Bluescribe plasmid cloning vector, and electroporated into electrocompetent bacteria.

Preparation for cDNA Synthesis

Preparation of Chemicals and Reagents: RNA Shadowing

All distilled, deionized water (obtained from a Millipore, Bedford, MA, Milli-Q apparatus), chemicals, and reagents used for the isolation of RNA and subsequent synthesis of first-strand cDNA are screened for RNase contamination by the RNA shadowing method. In this procedure a small sample (1–5 μl) of the chemical or reagent is mixed with 1 μg of synthetic 7.5-kb poly(A)$^+$ RNA (GIBCO/BRL, Grand Island, NY) in a total volume of 10 μl and incubated at 37° for 30 min. In most cases samples are then directly mixed with 10 μl deionized formamide, 3.5 μl formaldehyde (37%, w/v), and 1 μl of 10× MOPS buffer (0.2 M MOPS, pH 7.0, 50 mM sodium acetate, 10 mM EDTA) and loaded onto a 1.2% (v/v) agarose formaldehyde vertical gel as described.[12] Reagents containing Tris-HCl in concentrations greater than 10 mM are precipitated after incubation with 1/10 volume of 3 M sodium acetate and 2.5 volumes of absolute ethanol, microcentrifuged, and resuspended in water prior to loading. Samples are electrophoresed for 1–2 hr at 100 V, after which the gel is removed and placed on an intensifying screen (Kodak, Rochester, NY) and illuminated from above with a hand-held UV lamp. RNase-free chemicals or reagents do not affect the 7.5-kb RNA, and a full-length band is visible. Degradation resulting from RNase contamination is evident by band smearing or the complete absence of the 7.5-kb band. Any reagent or chemical that shows degradation is remade, purchased from an alternative source, or, if possible, as in the case of *Escherichia coli* tRNA, treated with 5 μg proteinase K (Boehringer-Mannheim, Indianapolis, IN) for 15 min at 65°, phenol/chloroform-extracted and ethanol-precipitated.[13] In general, chemicals are found to be RNase-free when purchased from Fluka Biochemicals (Ronkonkoma, NY), Boehringer-Mannheim, or GIBCO/BRL (Ultra-pure).

Unfertilized Egg and Embryo Isolation

Embryos are collected from 6- to 8-week-old B6D2 mice (Jackson Laboratories, Bar Harbor, ME) superovulated and mated to B6D2 male mice. Unfertilized eggs are treated with hyaluronidase to remove cumulus cells and subsequently with pronase to remove the zona pellucida, but cleavage stage embryos and blastocysts are treated with pronase alone.[12] Eggs and embryos from all stages are repeatedly washed in modified Whitten's medium,[14] and pools of 500–1000 are placed in 200 μl of embryo lysis buffer [ELB; 100 mM NaCl, 50 mM Tris-HCl, pH 7.5, 5 mM EDTA, 0.5% sodium dodecyl sulfate (SDS), 5 μg *E. coli* tRNA (Boehringer-Mannheim)] which has been pretreated with 5 mg/ml proteinase K at 37° for 30 min. Two-cell stage embryos are harvested at 40–42 hr post-human chorionic gonadotropin (hCG), 8-cell stage at 68–70 hr, and blastocyst stage at 92–95 hr. RNA from embryos is extracted and purified immediately following embryo isolation to ensure a high yield of intact RNA.

[12] B. Hogan, F. Constantini, and E. Lacy, "Manipulating the Mouse Embryo." Cold Spring Harbor Laboratory, Cold Spring Harbor, New York, 1987.

[13] J. Sambrook, E. F. Fritsch, and T. Maniatis, *in* "Molecular Cloning: A Laboratory Manual" (C. Nolan, ed.), 2nd Ed. Cold Spring Harbor Laboratory, Cold Spring Harbor, New York, 1989.

[14] J. Abramczuk, D. Solter, and H. Koprowski, *Dev. Biol.* **61,** 738 (1977).

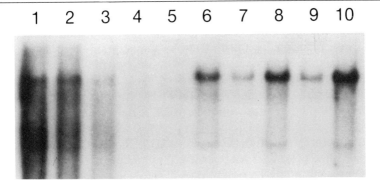

FIG. 2. RNA quantification. Total embryonic RNA and total RNA from an epithelial cell line standard were subjected to Northern blot analysis. Total standard cellular RNA was loaded as follows: lane 1, 50 ng; lane 2, 25 ng; lane 3, 12.5 ng; lane 4, 6.25 ng; lane 5, 3.125 ng. Lanes 6–10 were loaded with samples of embryonic RNA from early embryos (1% of five different samples). Following electrophoresis (100 V, 2 hr) the gel was transferred to an MSI nylon membrane and hybridized with a [α-³²P]dCTP-labeled 28/18 S ribosomal RNA probe. Autoradiograph shown is from an 18-hr exposure at −80°.

Total RNA Quantification

Total RNA is quantified by estimating the amount of ribosomal RNA present in each sample as determined by Northern blot analysis. RNA representing 0.5–1% of the amounts in individual frozen samples is prepared as in the RNA shadowing technique (above) and run on a 1.2% agarose formaldehyde gel as described.[13] Serial dilutions of a known amount of RNA standard derived from a mouse epithelial cell line are run on the same gel as the embryonic RNA (Fig. 2). The amount of embryonic RNA is estimated by visual comparison of the band intensities in the embryo RNAs with those in the RNA standards after hybridization of the blot with a random-primed [α-³²P]dCTP-labeled 28/18 S ribosomal probe.[15] Any aliquots of RNA that show degradation are discarded; all other RNAs originating from the same embryonic stage are pooled and stored either in RNase-free water or as a pellet in 70% (v/v) ethanol at −80°.

Poly(A)⁺ RNA Isolation

After a 60-min incubation at 37°, the embryo/ELB solution is extracted twice with 1:1 phenol/chloroform and ethanol-precipitated. Aliquots of embryo RNA are centrifuged and the pellets redissolved in 80 μl of RNase-free water, 20 μl of 5× DNase buffer (250 m*M* Tris-HCl pH 7.5, 1 *M*

---

[15] B. G. Herrmann, D. P. Barlow, and H. Lehrach, *Cell (Cambridge, Mass.)* **48**, 813 (1987).

NaCl, 50 mM MgCl$_2$, 25 mM CaCl$_2$), and 1.5 μg RNase-free DNase I (Worthington Biochemicals, Freehold, NJ) and incubated at 37° for 30 min. DNase reactions are terminated by the addition of 10 μl of 0.25 M *trans*-1,2-diaminocyclohexane-N,N,N',N'-tetraacetic acid monohydrate (CDTA), 5 μl of 10% SDS, and 2 μl proteinase K (5 mg/ml) followed by incubation at 56° for 15 min. The solution is extracted twice with phenol/chloroform and ethanol-precipitated as before.

Poly(A)$^+$ mRNA is selected using poly(U)-Sephadex as directed by the manufacturer (GIBCO/BRL). Poly(U) columns are used instead of oligo(dT) since poly(U) is more efficient in binding the short stretches of poly(A) present in mRNA of embryos at the 2-cell stage.[13] Briefly, 20–50 mg poly(U)-Sephadex beads are resuspended in 1 ml NTS (20 mM Tris-HCl, pH 7.5, 1 mM EDTA, 0.2% SDS, 0.4 M NaCl) in a 1.5-ml microcentrifuge tube and spun briefly to pellet the beads. The swollen gel is washed 3 times with 1 ml NTS, followed by the addition of a volume of NTS equal to the volume of the swollen beads pelleted.

DNase-treated total embryonic RNA is resuspended in 20–25 μl RNase-free water. RNA aliquots are pooled and added to an equal volume of 2× NTS. The NTS–RNA solution is added to packed prewashed Sephadex beads and agitated lightly for 10–20 min. Unbound RNA is removed by three 1-ml washes of NTS followed each time by a brief spin to pellet the beads. Nonspecifically bound RNA is further washed as described above with low-salt NTS (NTS with 0.1 M NaCl). Bound mRNA is eluted from the beads by the successive addition of 50 μl EL (0.1% SDS, 20 mM Tris-HCl, pH 7.5, 1 mM EDTA, 90% deionized formamide) followed by incubation at room temperature with gentle agitation for 10 min. After a 30-sec spin in the microcentrifuge to pellet the beads, the supernatants are pooled into a new tube containing 200 μl chloroform, 1/10 volume of 3 M sodium acetate, and 5 μg *E. coli* tRNA carrier. Two volumes of ethanol are added to the aqueous phase, and the solution is precipitated on dry ice or at −80° for several hours and pelleted by centrifugation for 60 min. After washing with 70% ethanol, the pellet is air-dried, resuspended in 8.3 μl RNase-free water, and stored at −80°.

cDNA Library Construction

First- and Second-Strand Synthesis

Because the first strand of cDNA determines the length and quantity of cDNA in the final library, the conditions for first-strand synthesis are optimized using a 7.5-kb synthetic mRNA molecule (GIBCO/BRL). After analyzing several first-strand buffers and reverse transcriptase enzymes,

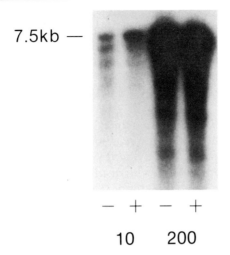

7.5kb —

− + − +

10   200

FIG. 3. First-strand synthesis optimization. The 7.5-kb mRNA first-strand reaction products were analyzed on an alkaline agarose gel. The 7.5-kb mRNA (GIBCO/BRL) was added (10 or 200 ng) to first-strand reaction mix in a total volume of 33 μl (see text for details) in the absence (−) or presence (+) of 5 μg E. coli tRNA. First-strand products were resolved on a 1% alkaline agarose gel, which was fixed in 10% trichloroacetic acid, dried, and exposed for 18 hr to X-ray film at −80°.

the conditions for optimal length and quantity of first-strand cDNA are found that allow the synthesis of cDNA from as little as 10 ng of input mRNA (Fig. 3). The addition of E. coli tRNA as carrier (5 μg/reaction) improves the recovery of full-length cDNA 2- to 5-fold (Fig. 3). Glycogen as carrier (5 μg/reaction) is inhibitory to first-strand synthesis and resulted in a reduced yield of full-length first-strand product (data not shown). Given a constant concentration of reaction components, the volume of the first-strand reaction does not dramatically affect the efficiency of the reaction; however, consistent results are obtained in a total volume of 33 μl. In addition, an oligo(dT) primer and the oligo(dT)/SalI linker–primer are equally effective in priming first-strand reactions (data not shown). Thus, to simplify the procedure and increase directional cloning efficiency, first-strand cDNA synthesis is performed using an oligo(dT)/SalI linker–primer.

Poly(A)$^+$ RNA is heated in 8.3 ml water to 65° for 15 min to remove secondary structure and immediately placed on ice. The first-strand reactants added are 6.6 μl of 5× reverse transcriptase buffer (GIBCO/BRL), 1 μl dithiothreitol (DTT, 0.1 mM), 3.3 μl of 10 mM dNTPs (Pharmacia, Piscataway, NJ), 5.0 μl (500 ng) oligo(dT)/SalI linker–primer (containing two SalI sites separated by 2 nucleotides: 5' pCGGTCGACCGTCGAC-CG(T)$_{15}$ 3'), 1.3 μl bovine serum albumin (BSA, 2.4 mg/ml, Boehringer-Mannheim), 1 μl (1 unit) of human placental RNase inhibitor (Boehringer-Mannheim), 5 μl (50 μCi) of [α-^{32}P]dCTP (3000 mCi/mol, Amersham), and 1.5 μl (200 units) Superscript RNase H$^-$ Moloney murine leukemia virus (MMLV) reverse transcriptase (GIBCO/BRL); incubation is at 37° for 60 min.

The amount of RNA converted to cDNA after incubation is quantified by placing 1 μl of the first-strand reactions onto a small (1 cm^2) piece of MSI nylon membrane (Magna nylon, MSI, Fisher Scientific, Pittsburgh, PA) which has been prewetted with distilled water. The samples are then soaked in three changes of 200 ml of 0.5 M sodium phosphate, pH 7.2, for 15 min each. Incorporation of [α-^{32}P]dCTP is calculated by comparing the counts per minute (cpm) on the filter to the total counts per minute in the sample (obtained by directly spotting 1 μl each from 1 : 10 and 1 : 100 dilutions of the first-strand reaction onto a similar nylon filter, air-dried without washing) and subtracting the background (obtained from spotting 1 μl of the first-strand reaction onto a nylon filter before incubation and washing as above) by the following formula: [(cpm incorporated − background cpm)/(average total cpm − background cpm)] × 85,800 = ng single-stranded cDNA, where the constant is used for a 33-μl reaction containing 2 mM of each dNTP.[13] The percent conversion of RNA to cDNA using standard amounts of cellular mRNA is consistently 30 ± 8% (data not shown). Thus, the amount of poly(A)$^+$ RNA collected is calculated for each of the embryonic states from the percent conversion of RNA to cDNA and ranges between 45 and 175 ng per stage (Table I).

A standard second-strand synthesis reaction is performed in the same tube in a total volume of 200 μl. After the amount of cDNA produced during first-strand synthesis is determined, the remaining 30 μl of first-strand reaction is combined with 20 μl of 10× second-strand buffer[13] [0.7 M Tris-HCl, pH 7.4, 100 mM MgCl$_2$, 100 mM (NH$_4$)$_2$SO$_4$], 2 μl (2 units) RNase H (Pharmacia), 14 μl (70 units) endonuclease-free DNA polymerase I holoenzyme (Boehringer-Mannheim), 20 μl dNTPs (10 mM each), and sterile, nuclease-free water, then incubated first at 15° for 1 hr and then at room temperature for 1 hr. The reaction is terminated by the addition of CGSK [2.5 μl of 0.25 M CDTA (3 mM final concentration), 5 μl glycogen (5 μg, Boehringer-Mannheim), 7 μl of 10% SDS (3.5% final concentration), and 5 μl proteinase K (5 μg)], incubated at 56° for 15 min, and phenol/chloroform-extracted and ethanol-precipitated using 200 μl of 5 M ammonium acetate as salt. Using 5 M ammonium acetate as salt at this step instead of 3 M sodium acetate preferentially enhances the precipitation of

TABLE I

EFFICIENCY OF EMBRYONIC cDNA CLONING

Stage of cDNA library	Poly(A)$^+$ RNA (ng isolated)[a]	Amount of ds cDNA (ng)		Number of clones[d]	Efficiency[e] (cfu/μg)
		Synthesized[b]	Yield (%)[c]		
Unfertilized egg	175	53	15 (28)	2×10^6	1.3×10^8
2-Cell	46	14	7 (53)	1×10^6	1.4×10^8
8-Cell	87	26	16 (63)	2×10^6	1.3×10^8
Blastocyst	45	14	5 (38)	1×10^6	2.0×10^8

[a] Calculated based on the percent conversion of mRNA to ds (double-stranded) cDNA (30 ± 8%).

[b] Determined by the amount of [α-^{32}P]dCTP incorporation into first-strand cDNA.

[c] Based on recovery of [α-^{32}P]dCTP labeled cDNA over 500 bp from the Sepharose CL-4B column.

[d] Number of independent transformants plated.

[e] Number of clones/μg of cDNA recovered from Sepharose CL-4B column.

cDNA over small pieces (<100 bp) of cDNA, primers, and unincorporated nucleotides.[13]

S_1 Nuclease Treatment and End Repair

The formation of hairpin loops following second-strand synthesis prevents the ligation of 5' linkers and subsequent insertion of cDNA molecules into plasmid vectors.[13] Addition of an S_1 nuclease step before ligation of the 5' linker greatly decreases the hairpin loops that form during the second-strand synthesis reaction when β-globin mRNA (600 bp) is either treated with S_1 nuclease following second-strand synthesis (Fig. 4, lane 1) or not treated (Fig. 4, lanes 2 and 3). Hairpin loops are observed in the alkaline agarose gel as a band at 1200 bp. Accordingly, S_1 nuclease is used to increase the amount of clonable cDNA in all cDNA synthesis reactions.

S_1 Nuclease treatment is performed in a total volume of 100 μl using 200 units of S_1 nuclease (Boehringer-Mannheim) in S_1 buffer (0.1 M sodium acetate, 0.8 M NaCl, 2 mM ZnCl$_2$) at 37° for 20 min. The amount of S_1 nuclease used in the reaction has previously been optimized to 2 units/μl of reaction volume and for the batches used is determined by treating 1 μg of HaeIII-digested pSP64 DNA (Promega, Madison, WI) with serial dilutions of S_1 nuclease diluted from 0.5 to 4 units/μl in S_1 buffer. The 2 units/μl dilution is chosen as the amount of S_1 nuclease that best degrades the ends of the digested fragments, as determined by smearing of low

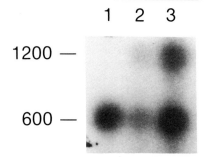

Fig. 4. S_1 nuclease digestion of β-globin mRNA. After synthesis into double-stranded DNA as described, 200 ng (lanes 1 and 3) or 50 ng (lane 2) of poly(A)$^+$ β-globin mRNA was treated with 200 units of S_1 nuclease for 20 min at room temperature in S_1 nuclease buffer (lane 1) or not (lanes 2 and 3). Reaction products were resolved on a 1% alkaline agarose gel, which was fixed, dried, and exposed to X-ray film for 18 hr at $-80°$. Full-length β-globin is 600 bp; β-globin with hairpin loops migrates as a dimer of 1200 bp.

molecular weight fragments on a 1.2% agarose gel, and is used in cDNA synthesis. This amount of nuclease does not significantly affect the *Sal*I cloning sites incorporated into the 3′ end of cDNA molecules.

Following S_1 nuclease treatment, Tris-HCl, pH 8.3, is added to a final concentration of 20 mM, the reaction is terminated by adding CGSK, and the cDNA is phenol-extracted and ethanol-precipitated in the presence of 3 M sodium acetate. The ends of the nuclease-treated cDNA are repaired by resuspending the pellets in 11 μl of sterile, nuclease-free water, 4 μl of 5× T4 polymerase buffer (0.2 M Tris-HCl, pH 7.5, 50 mM MgCl$_2$, 10 mM EDTA, 40 mM DTT, 1 mg/ml BSA), 4 μl dNTPs (10 mM each), and 1 μl (1 unit) of T4 DNA polymerase (Boehringer-Mannheim). The mixture is incubated at 37° for 15 min, phenol/chloroform-extracted and ethanol-precipitated as before.

5′ Linker Ligation and Double Digestion

The 5′ *Mlu*I linkers are ligated to the cDNA in 30 μl using 1.5 μl (1.5 Weiss units) of T4 ligase (GIBCO/BRL), 3 μl ligase buffer (50 mM Tris-HCl, pH 7.6, 10 mM MgCl$_2$, 10 mM DTT), 3 μl of 0.2 mM ATP, and 3 μl of 5′-phosphorylated *Mlu*I linkers (3 μg, Pharmacia) and incubated at 15° for 16–18 hr. After incubation, the ligase is inactivated at 65° for 10 min, and the cDNA is double-digested with 500 units of the restriction enzymes *Sal*I and *Mlu*I (New England Biolabs) in a total volume of 400 μl for 5–6 hr at 37° using NEB buffer 3 supplied by the manufacturer. Digestion reactions are terminated with CGSK, phenol/chloroform-extracted, and ethanol-precipitated as described above. The source of T4 ligase is important for the highest efficiency of ligation. We observe as much as a 10-fold variation in the number of transformants resulting from the use of different T4 ligase preparations (data not shown). In our hands, T4 ligase from GIBCO/BRL is consistently superior to several others tested.

Size Fractionation and Vector Ligation

MluI- and SalI-digested cDNA is resuspended in 15 μl of sterile, nuclease-free water, 10 μl saturated urea, and 1 μl bromphenol blue tracer dye (1 mg/ml) and loaded onto a column of Sepharose CL-4B (Pharmacia) made in a 1-ml disposable pipette and prewashed in column buffer (20 mM Tris-HCl, pH 7.5, 0.2 M sodium acetate, 4 mM EDTA, 0.1% SDS). The cDNA is loaded onto the column bed, and 100- to 200-μl fractions are collected. Elution of cDNA molecules is monitored using a hand-held Geiger counter. The first radioactive peak contains the cDNAs greater than 500 bp, whereas the second peak contains small cDNAs and unincorporated nucleotides. When the bromphenol blue dye reaches the end of the column, sample collection is stopped, the radioactivity in each fraction is determined by Cerenkov counting, and the quantity of cDNA in each is calculated by the following formula: [(cpm after column fractionation)/ (total cpm of first-strand cDNA)] × ng first-strand cDNA × 2 = ng double-stranded cDNA. Generally, if 2000 cpm represents less than 5% of the total pooled radioactivity, then this amount is examined on an alkaline agarose gel (Fig. 5). Fractions containing cDNA with an average size greater than 500 bp are pooled and ethanol-precipitated using 5 μg glycogen as a carrier. The yield of cDNA of at least 500 bp from the column determined for the egg and embryonic stages varies from 28 to 63% of the sample loaded (Table I).

Precipitated cDNA is resuspended in water at a concentration of 0.25–1 ng/μl and ligated into a 10-fold molar excess of MluI- and SalI-digested Bluescribe vector (Stratagene, La Jolla, CA) modified so that the EcoRI site is converted to an MluI site and a HindIII site is converted to a SalI site. Ligations are performed in a total volume of 25 μl using 2 Weiss units of T4 ligase (GIBCO/BRL), 0.2 mM ATP, 20 mM Tris-HCl, pH 7.6, 10 mM MgCl$_2$, and 10 mM DTT, with incubation overnight at 15° as described above. Following incubation the ligations are phenol/chloroform-extracted, ethanol-precipitated with 5 mg of glycogen as carrier and resuspended in 10 μl TE (10 mM Tris-HCl, pH 7.5, 0.1 mM EDTA).

Purification of the digested vector is critical to increasing the frequency of cDNA-containing clones in the final library. We propagate the vector

1 2 3

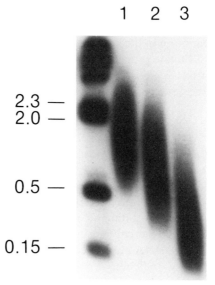

2.3 —
2.0 —

0.5 —

0.15 —

FIG. 5. Size of cDNA fractions collected from the Sepharose CL-4B column. cDNA was purified over a small Sepharose CL-4B column prepared in a 1-ml disposable pipette, and the cDNA fractions that eluted in the first radioactive peak were arbitrarily divided into three groups and resolved on a 1% alkaline agarose gel. Lane 1 contains cDNAs with an average size of 1.5 kb; lane 2 contains cDNAs with an average size of approximately 0.8 kb; and lane 3 contains cDNAs with an average size below 500 bp. Size markers are HindIII-digested [α-^{32}P]dCTP-labeled λ DNA.

with a 500-bp "stuffer" sequence, which is large enough to allow efficient separation of double-digested vector from insert and linear vector. The vector is digested overnight and then gel-purified away from the stuffer fragment by electrophoresis through at least two vertical agarose gels and spin column elutions. The vector background (vector ligated without cDNA insert) is no greater than 1 × 10^6 colony-forming units (cfu)/μg of vector when the electroporation efficiency is 2 × 10^{10}/μg for super-coiled plasmid.

Bacterial Electroporation and Plating

Escherichia coli DH10B bacteria are grown and made electrocompetent as described.[8] All electroporations are performed using a Cell Porator (GIBCO/BRL) set at 400 V and 4000 Ω, resulting in line voltages of 2.4–2.5 kV. In all electroporations, 10 pg pUC19 is used as a control, yielding transformation efficiencies of 3–6 × 10^{10} cfu/μg plasmid. Electroporation of cDNA libraries routinely results in transformation efficiencies of 1–2 × 10^8 cfu/μg input cDNA (Table I).

cDNA is electroporated in 1-μl aliquots using 20 μl of electrocompetent bacteria. Each electroporated sample is grown in 1 ml of SOC medium[13] at 37° for 60 min; samples are pooled and spun at 400 g (3500 rpm) for 10 min. Cell pellets for each library are resuspended in 4 ml of SOC and each 1 ml plated on a single 8.5 × 8.5 inch MSI nylon membrane (Fisher Scientific) placed on an Luria broth (LB) agar plate (Nuncleon 8.5 × 8.5 inch cell culture plates) containing 70 μg/ml ampicillin (LB/amp) and incubated overnight at 37°.

The next day filters are replicated using modifications of methods described previously[15] (Fig. 6). The master filter is removed from the agar surface and placed, colony side up, on Whatman (Clifton, NJ) 3MM blotting paper dampened with LB/amp medium. The first replica filter is then placed on top of the master followed by another piece of dampened Whatman 3MM blotting paper. Firm pressure is applied using a glass plate for 1 min, and then the filter paper is removed and 10–15 orientation marks made through both master and replica using an 18-gauge needle. The two filters are then separated and placed onto agar plates. The master filter is reincubated at 37° for 1 hr, and the second and third replicas are made from the master in the same fashion.

Replica filters are incubated for 4–6 hr or until the bacterial colonies are of similar size to those on the master. The third replica is grown overnight at room temperature and scraped with Hogness modified freezing medium[15] [10× HMFM is 6.3 g K_2HPO_4, 0.45 g sodium citrate, 0.09 g $MgSO_4 \cdot 7H_2O$, 0.9 g $(NH_4)_2SO_4$, 1.8 g KH_2PO_4, 44 ml glycerol, and water to a final volume of 100 ml] diluted in LB/amp, aliquotted, and stored at −70°. The master filter is placed on Whatman 3MM paper saturated with HMFM for 10 min, placed on a plexiglass plate (8.5 × 8.5 × 1/8 inch), and covered with a transparency of standard graph paper. Orientation marks on the filter are transferred to the transparency using a permanent marking pen, and both are covered by another plexiglass plate and stored at −80°. A copy of the transparency containing the marks is kept for later orientation of autoradiographs. Replica filters are denatured, neutralized, baked, and UV-treated as described.[13] All hybridizations are performed using both replicas to authenticate colony hybridization with specific probes.

General Screening

All probes used for hybridization are labeled with [α-^{32}P]dCTP by the primer extension method,[16] and 1–2 × 10^6 cpm/ml is hybridized to MSI nylon filters in Church buffer[17] (7% SDS, 1 mM EDTA, 0.5 M sodium

---

[16] A. P. Feinberg and B. Vogelstein, *Anal. Biochem.* **132**, 6 (1983).

[17] G. M. Church and W. Gilbert, *Proc. Natl. Acad. Sci. U.S.A.* **81**, 1991 (1984).

A

-250,000 clones plated on MSI nylon membrane placed on agar
surface 2 min prior to plating

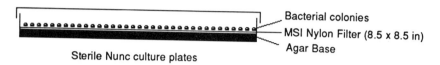

B

-Master plate grown overnight prior to replication

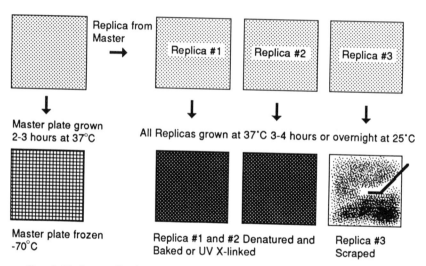

FIG. 6. Plating, replication, and storage of cDNA libraries. (A) Orientation of bacterial colonies plated on MSI nylon filters placed on top of 8.5 × 8.5 inch LB/amp plates. (B) Replication procedure for cDNA libraries. The master plate contains bacterial colonies plated the previous day on nylon filters. Three replicas are made and grown at 37° for 3–4 hr or for 18 hr at room temperature. The master plate is frozen, bacterial colony side up, beneath a copy of a transparency grid. A replica transparency is kept for orientation of autoradiographs. Replica filters are denatured, UV-cross-linked and/or baked, and stored for later use. The final replica is scraped into small aliquots and stored frozen at −80° in LB/amp-containing HMFM (see text for details).

phosphate, pH 7.2) at 65° for 18–20 hr. Filters are washed the following day first in Church wash[17] (1% SDS, 50 mM sodium phosphate buffer, pH 7.2) at 65° and then under high-stringency conditions (0.1× SSC,[13] 0.5% SDS) at 65°. Probes for known genes such as 28/18 S ribosomal DNA, cytochrome-c oxidase I and II, β-actin, intracisternal A-type particle (IAP), tissue plasminogen activator (t-PA), and B1/B2 repeats are used to screen library filters generated by replica plating to determine cDNA library quality and representativeness.[18]

Using such an approach, the unfertilized egg and embryonic cDNA libraries are shown to contain sequences that are representative of the poly(A)$^+$ mRNA in the corresponding unfertilized egg and embryonic stages.[18] Colonies that hybridize with a probe on two replicated filters are considered positive. Figure 7 shows an example of this analysis for IAP. Expression values for a given gene are determined as the number of positive colonies per 250,000 screened. The number of poly(A)$^+$ transcripts of a given gene per embryo is calculated by multiplying the frequency of its occurrence in the cDNA library by the total number of poly(A)$^+$ RNAs per embryo at each corresponding stage. The total number of poly(A)$^+$ mRNA molecules for the unfertilized egg, 2-cell, 8-cell, and early blastocyst stages are 1.4×10^7, 7×10^6, 1.3×10^7, and 3.6×10^7, respectively.[19]

Size Analysis of cDNA Clones

Aliquots of each cDNA library are diluted and plated on small (100 mm) LB/amp plates to obtain 50–200 single colonies. Individual colonies are randomly picked from each library using a sterile pipette tip, placed into 30 μl of PCR buffer [10 mM Tris-HCl, pH 8.3, 50 mM KCl, 2.5 mM MgCl$_2$, 0.1 mg/ml gelatin, 0.45% Nonidet P-40 (NP-40), 0.45% Tween 20], and denatured at 100° for 15 min. After denaturation 20 μl of PCR mix [5′ T7 primer, 0.25 μg; 3′ T3 primer, 0.25 μg; 0.5 mM dNTPs; 2 units Thermalase (IBI); and 2 ml of 10× PCR buffer (IBI)] is added, the tubes are placed in a thermal cycler for 35–45 cycles of 94° for 30 sec, 50° for 30 sec, and 72° for 1.5 min, and the products are resolved on a 2.0% agarose gel. Primers for T7 and T3 polymerase promoters are synthesized in the Wistar Institute (Philadelphia, PA) Oligonucleotide Synthesis Facility using the sequence published by Stratagene.

The average insert size of each cDNA library is determined using this random PCR approach; representative results from the 2-cell stage cDNA

---

[18] J. L. Rothstein, D. Johnson, J. A. Deloia, J. Skowronski, D. Solter, and B. Knowles, *Genes Dev.* **6,** 1190 (1992).

[19] K. B. Clegg and L. Pikó, *Dev. Biol.* **95,** 331 (1983).

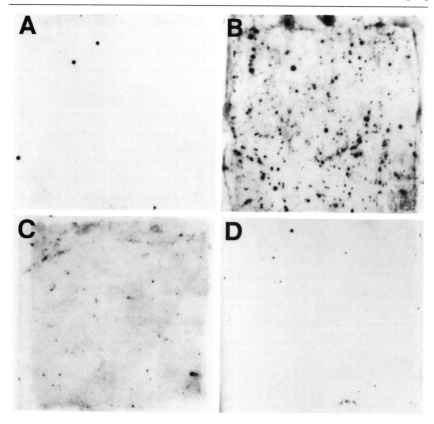

FIG. 7. Expression of IAP in egg and embryonic cDNA libraries. Library filters show the result of [α-^{32}P]dCTP-labeled IAP probe hybridization to 250,000 colonies of (A) unfertilized egg, (B) 2-cell, (C) 8-cell, and (D) blastocyst cDNA libraries. Filters were hybridized for 18 hr at 65° in Church buffer, washed, and exposed to X-ray film for 18 hr at −80°.

library are shown in Fig. 8. For each library no fewer than 50 independently isolated cDNA clones are analyzed to demonstrate that the average insert size of the unfertilized egg, 2-cell, 8-cell, and blastocyst cDNA libraries is 1.0, 1.3, 0.7, and 1.0 kb, respectively.

Gene-Specific Polymerase Chain Reaction Analysis

To determine whether the embryonic cDNA libraries contained specific genes important for early growth and differentiation, the libraries are subjected to PCR analysis using primers for several cytokines. In cells of the hematopoietic system cytokines such as interleukins 1–7 are expressed

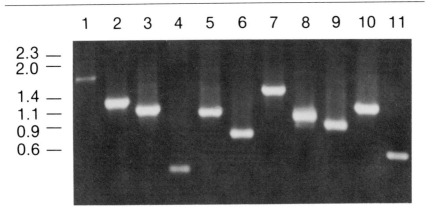

FIG. 8. PCR analysis of cDNA inserts from a 2-cell stage cDNA library. Randomly chosen cDNA clones from the 2-cell stage cDNA library were analyzed on an agarose gel. Lanes 1–11 are PCR-amplified reaction products resolved on a 1.5% TAE-agarose gel and stained with ethidium bromide.

in a lineage-specific fashion, making them appropriate probes for the analysis of genes potentially regulated in the egg and preimplantation embryo.

For PCR analysis, bacterial aliquots from each library are plated at high density (2×10^6 cfu) on Nuncleon 8.5 × 8.5 inch plates. Plates are incubated at 37° for 16 hr, scraped into 50-ml centrifuge tubes, and spun at 2500 g, and plasmid DNA is isolated by the alkaline lysis method.[20] Plasmid DNA purified by CsCl gradient centrifugation[13] is digested using MluI and SalI as described by the manufacturer (New England Biolabs). Insert cDNA is purified from vector sequences by 1% agarose gel electrophoresis and separated from agarose using spin columns.[13] For each PCR reaction 10–50 ng of purified insert cDNA is used as the starting template. PCR reactions for cytokine primers (Clonetech, Palo Alto, CA) are performed as recommended by the manufacturer. After amplification, PCR products are analyzed on 2% agarose gels and, when necessary, transferred to MSI nylon membranes and hybridized with appropriate probes.

Using this approach we have shown that several cytokine genes are expressed in a stage-specific fashion in the egg and embryo.[15] For example, screening the egg library showed that 8 of 250,000 clones (0.0003%) hybridized with the mouse interleukin 7 (IL-7) probe, suggesting that IL-7 is expressed as a maternal transcript in the mouse egg. Since the mouse egg contains 1.4×10^7 poly(A)$^+$ mRNA molecules, this suggests that 448 transcripts in the mouse egg encode IL-7. No expression of IL-7 is detected

[20] H. C. Birnboim and J. Doly, Nucleic Acids Res. 7, 1513 (1979).

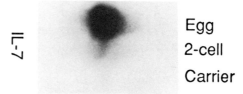

IL-7

Egg

2-cell

Carrier

FIG. 9. IL-7 expression in the unfertilized egg and 2-cell stage embryos. Eggs and embryos were collected (25 each) and subjected to RT–PCR as previously described.[2] PCR reaction products were resolved on a 1.5% TAE-agarose gel, transferred onto an MSI nylon membrane, and hybridized with an [α-^{32}P]dCTP-labeled murine IL-7 cDNA probe (kindly provided by L. Park, Immunex, Seattle, WA). The blot was exposed to X-ray film for 18 hr at −80°.

in the 8-cell or blastocyst libraries (data not shown). These data are verified by reverse transcription (RT)–PCR of mRNA isolated from mouse eggs (Fig. 9). Although a very low signal of IL-7 is present in the RT–PCR from 2-cell stage mRNA, IL-7 is not detected in the 2-cell stage cDNA library. These data demonstrate the sensitivity of this approach and its capacity to detect low-level gene expression during early embryogenesis.

Subtractive cDNA Library Construction

The embryonic cDNA libraries have served as the starting point for generation of the first embryonic stage-specific subtraction libraries, and several novel genes expressed at the 2-cell stage of mouse preimplantation development have been identified using these libraries. Traditionally, subtractive hybridization has been accomplished by hybridizing two different single-stranded DNA populations together and removing the double-stranded hybrids from the single-stranded unique DNAs using a hydroxy-apatite column,[13] a technique requiring large amounts of nucleic acid. The subtraction technique adopted exploits the cRNA synthesis capabilities of the Bluescribe plasmid cloning vector (Stratagene) and uses the high-affinity interaction between biotin and avidin along with phenol/chloroform extraction as a means to separate biotinylated molecules from nonbiotinylated molecules.[21] Using this method the unique nonbiotinylated poly(A)$^+$ sense RNA remains in the aqueous phase, and the biotinylated antisense single- and double-stranded (sense and antisense) molecules are bound to streptavidin and partitioned to the interface during the phenol/chloroform extraction. This extraction procedure ensures efficient removal of streptavidin–biotin complexes and purification of single-stranded sense RNA.[21]

[21] H. H. Sive and T. St. John, *Nucleic Acids Res.* **16**, 10937 (1988).

RNA Transcription

Templates for the transcription of sense RNA are generated by overnight digestion of CsCl-purified DNA with *Sal*I in buffer supplied by the manufacturer, followed by treatment with 5 μg proteinase K (Boehringer-Mannheim) for 15 min at 56° to remove RNase, phenol/chloroform extraction, and ethanol precipitation. The template for antisense RNA is similarly prepared, except *Mlu*I in buffer supplied by the manufacturer is used instead of *Sal*I. RNA synthesis is performed using T7 or T3 RNA polymerase diluted in reaction buffer supplied by the manufacturer (Promega) to generate sense and antisense transcripts, respectively. For a single transcription reaction 1 μg of linearized DNA template (*Sal*I-digested for T7-generated sense transcripts and *Mlu*I-digested for T3-generated antisense transcripts) is mixed with 1 mM each of ATP, CTP, GTP, and either UTP alone for sense strand synthesis or, for antisense RNA, biotin-UTP (Sigma) and UTP together (10:1, w/v, respectively) and 10 units of the appropriate RNA polymerase in reaction buffer supplied by manufacturer. As a tracer, [α-^{32}P]UTP (Amersham) is added at 1–2 μCi/ reaction. After incubation at 37° for 30 min, the template is removed by treatment with RNase-free DNase I (Worthington, 1 mg/ml for 30 min at 37°), phenol/chloroform extraction, and ethanol precipitation. RNA synthesized is quantified spectrophotometrically or by calculating the incorporation of [α-^{32}P]UTP as described above for first-strand cDNA.

Subtractive Hybridization

To determine the efficiency of the subtraction protocol, an initial control experiment is performed on two distinct cDNA clones (p17 and p44) randomly isolated from the blastocyst library. In the first part of this experiment, 1 μg of clone p17 DNA (830 bp) is transcribed in the sense orientation in the presence of 10 μCi of [α-^{32}P]UTP, and the resulting RNA is hybridized to RNA transcribed in the antisense orientation from 10 μg of p44 DNA (1400 bp) in the presence of biotinylated UTP in 4.5 μl of hybridization buffer with SDS (250 mM HEPES, pH 7.5, 10 mM EDTA, 1% SDS) and 0.5 μl of 5 M NaCl under 50 μl of mineral oil at 65° for 48 hr. After hybridization, the mineral oil is removed, and 50 μl of hybridization buffer (without SDS) and 5 μg of streptavidin (GIBCO/ BRL) are added, followed by incubation at room temperature for 5 min and phenol/chloroform extraction. The streptavidin and phenol/extraction procedure is repeated 5 times to separate any biotinylated molecules from the [α-^{32}P]UTP-labeled sense RNA. Streptavidin and streptavidin–biotin complexes partition to the interface between the organic and aqueous phases.[21]

TABLE II
CONTROL EXPERIMENT: SINGLE SUBTRACTIVE
HYBRIDIZATION CYCLE

	Counts per minute	
cDNA hybridization pairs	Aqueous	Organic
p17–[^{32}P]rUTP[a] versus p44–biotin[b]	248,550	4650
p17–[^{32}P]rUTP versus p17–biotin	12,150	106,200

[a] p17 cDNA clone, 830 bp.
[b] p44 cDNA clone, 1400 bp.

Clone p17 DNA (1 μg) is also transcribed in the sense orientation in the presence of 10 μCi of [α-^{32}P]rUTP, and the resulting RNA is hybridized to antisense RNA transcribed in the presence of biotinylated rUTP from 10 μg of the same plasmid DNA at 65° for 48 hr. Streptavidin treatment and phenol/chloroform extraction are described above. Finally, samples are removed from the pooled aqueous phases and pooled organic phases for counting (Table II). Little hybridization occurs between dissimilar cDNA clones (p17 versus p44), with the result that 98% of the counts are recovered in the aqueous phase. On the other hand, the majority of the counts are retained in the organic phase when p17 is hybridized to itself (Table II).

To obtain a 2-cell-specific subtraction library, hybridizations are performed between 2-cell, egg, and 8-cell stage cDNA libraries. The ratio of sense and antisense RNA used is 5 : 1 for unfertilized egg versus 2-cell stage; following hybridization and phenol/chloroform extraction as above, the RNA remaining in the aqueous phase is hybridized to at least a 100-fold excess of antisense 8-cell RNA. For hybridization of 2-cell and egg RNA, 200 ng of sense RNA transcribed from the 2-cell library is coprecipitated with 1 μg of biotinylated antisense RNA from the unfertilized egg library by the addition of 1/10 volume of 3 M sodium acetate and 2.5 volumes of ethanol. The pellet is resuspended in 4.5 μl of hybridization buffer with SDS and 0.5 μl of 5 M NaCl. Fifty microliters of mineral oil is placed on top of the mixture, and hybridization is carried out at 65° for 48 hr as described above.

After incubation, the hybridization mixture is treated as above to separate single-stranded sense RNA from biotin–streptavidin-complexed molecules. The 2-cell sense RNA remaining in the aqueous phase is hybridized to at least a 100-fold excess of 8-cell biotinylated antisense RNA at 65° for 48 hr followed by streptavidin treatment and phenol/chloroform extraction as described above. The 2-cell sense RNA remaining in the aqueous

phase after the two sequential subtractive hybridizations is ethanol-precipitated and the RNA pellet washed with 70% ethanol. The RNA is resuspended in 8.3 μl of water, reverse-transcribed using the *Sal*I linker–primer,
and cloned into the Bluescribe cloning vector (Stratagene) as described
above (see first- and second-strand synthesis). All subtractions are generally performed at least twice to ensure the removal of any common sequences,[19] and a control experiment such as that shown in Table II should
be performed prior to any subtractions involving valuable RNA (see
Fig. 10).

Subtraction cDNA Library Analysis

The average insert size of the cDNA in the first 2-cell subtraction
library (2CSL-I) is 1.0 kb (data not shown). For library analysis, 2CSL-I
is subtracted by repeating the procedure (i.e., digestion with *Sal*I and
transcription using T7 polymerase to generate sense transcripts), substitut-
ing 2CSL-I for the original 2-cell library. The second 2-cell-specific subtraction library (2CSL-II) has an average insert size of approximately
300–400 bp (data not shown). Hybridization of both libraries with a probe
to IAP, a gene abundantly expressed at the 2-cell stage[15] (Fig. 7), reveals
no IAP-reactive cDNAs in either library. Similarly, no cDNAs for β-actin
and none containing B1/B2 repeat sequences, also abundantly expressed
at the 2-cell stage,[15] are detected in either subtraction library (data not
shown). Thus, abundant gene products common to the egg, 2-cell, and
8-cell stages are removed by this subtraction protocol.

Stage-Specific Embryonic Clone Analysis

Clones are picked at random from the first and second 2-cell stage
subtraction libraries, since cDNAs in the two libraries are highly enriched
for transcripts expressed at the 2-cell stage of embryogenesis; only those
with insert sizes of at least 500 bp are analyzed further. Such clones are
sequenced from the 5' end with the T7 primer and from the 3' end with the
T3 primer using the Sequenase kit (USB, Cleveland, OH) and [^{35}S]dATP as
described by the manufacturer. Sequencing reactions are run on a 10%
polyacrylamide, 6% urea gel at 2000 V for 6–8 hr and exposed to X-ray
film for 12–18 hr at $-80°$. All sequences are compared to those listed in the
GenBank/EMBL databases using the FASTA command of the UWGCG
sequence analysis program.[22] In addition, all clones are hybridized to
replica filters of the original 2-cell cDNA library. Clones expressing at
levels greater than or equal to 1 clone/250,000 screened are analyzed

---

[22] J. Devereux, P. Haeberli, and O. Smithies, *Nucleic Acids Res.* **12**, 387 (1984).

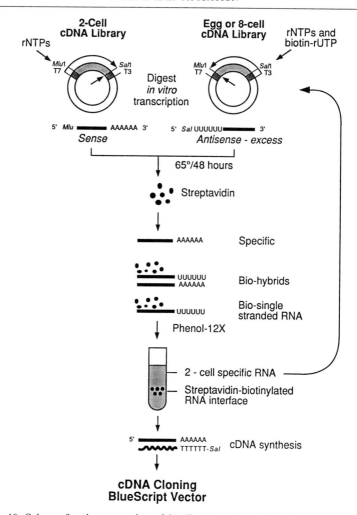

Fig. 10. Scheme for the generation of 2-cell subtractive cDNA libraries. Sense RNA was synthesized from the linearized 2-cell stage cDNA library and hybridized to a 5-fold excess of antisense biotinylated RNA from the unfertilized egg cDNA library. The mixture was treated with streptavidin and extracted with phenol/chloroform. The nonbiotinylated single-stranded RNA from the aqueous phase was then hybridized to greater than 100-fold excess of biotinylated RNA transcribed from the 8-cell library. Streptavidin treatment and phenol/chloroform extraction of the mixture left the 2-cell-specific RNA in the aqueous phase, which was used to generate the first 2-cell subtraction library, 2CSL-I (see text for details).

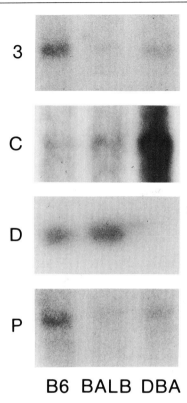

B6 BALB DBA

FIG. 11. SSEC clones 3, C, D, and P as probes on Southern blots of *Bam*HI-digested mouse genomic DNA (10 μg) from C57BL/6, BALB/c, and DBA/2 mice. Note that there is no DBA/2 lane on the blot which was probed with SSEC-D. Four independent 1% TAE-agarose gels were blotted overnight in 10× SSC onto a nylon filter and hybridized with [α-³²P]dCTP-labeled SSEC clones 3, C, D, and P in Church buffer. After washing, filters were exposed to X-ray film for 1–7 days at −80°.

further for stage-specific expression by hybridization to the egg and 8-cell libraries.

Of the 20 randomly chosen clones, we identified 5 stage-specific embryonic cDNA clones (SSEC) present predominantly or exclusively in the 2-cell stage library[15] with no significant homology to sequences listed in the GenBank/EMBL databases. The remaining 15 clones include 9 novel genes that demonstrate expression levels of below 1 clone/250,000 screened at the 2-cell stage and 6 that prove to be of bacterial origin traced to the lot of tRNA used in the construction of the subtraction library. Southern analysis of 4 of the stage-specific cDNA clones indicates they are single-copy mouse genes (Fig. 11).

Summary

By modifying current cDNA cloning[23] and electroporation[11] methods, large and representative murine cDNA libraries were synthesized from 10 to 100 ng mRNA isolated from unfertilized egg and preimplantation mouse embryos. High cloning efficiency is essential for complete representation of genes expressed in egg and preimplantation embryos and for the isolation of stage-specific genes using subtractive hybridization. Because the mouse embryo contains no more than 50 pg of poly(A)$^+$ mRNA at any stage of preimplantation development,[17] approximately 5000–10,000 embryos are required to obtain enough mRNA to synthesize libraries using current methods. To obtain a representative library that also includes rare transcripts, the size of the library should be at least 10^6 clones. The average percent conversion of mRNA to single-stranded cDNA was 20–40%, so that a cloning efficiency of nearly 2×10^8 cfu/μg cDNA is required for such a cDNA library. No previous methods have provided directional cloning of cDNA into plasmids with these high efficiencies. The advent of electroporation methods for the introduction of nucleic acids into bacteria has made possible the use of standard plasmid vectors for high-efficiency cDNA cloning.[13] Plasmid vectors are currently available that can accommodate the directional cloning of cDNA such that T7 and T3 RNA polymerase promoter sequences can be used to generate sense and antisense transcripts for subtractive hybridization and riboprobe synthesis.

The cDNA libraries we derived using this methodology are a reusable and abundant source of genetic information about the control of preimplantation development. Specialized subtractive cDNA libraries enriched for genes expressed exclusively at a predetermined time in development give access to genes expressed in a stage-specific manner. The ability to construct new cDNA libraries from limited amounts of starting material ensures the provision of new and important resources for the identification and study of novel genes or gene families, and it is an important new tool for understanding the molecular control of mammalian development.

Acknowledgments

We thank D. Hanahan for advocating the use of directional plasmid cloning methods in lieu of the more widely used λ cloning methods for cDNA synthesis, which simplified cDNA library handling and subsequent subtractive hybridization.

[23] U. Gubler and B. J. Hoffman, *Gene* **25,** 263 (1983).

[36] Construction of cDNA Libraries from Single Cells

By GERARD BRADY and NORMAN N. ISCOVE

Introduction

The development of the polymerase chain reaction[1,2] (PCR) has expanded the application of traditional molecular biology to allow examination of both DNA and RNA in samples as small as a single cell.[3,4] The methods developed have generally focused on amplification of known sequences with specific primers, and there has been little attention given to global amplification of expressed genes.[5,6] Here we describe a simple approach, referred to as poly(A) PCR,[6] which was designed for unbiased amplification of cDNA representing all polyadenylated RNA present in a sample as small as a single cell.

General Principle of Poly(A) Amplification

The basic principle of the PCR requires that the target sequence be bracketed by known sequences to which the amplification primers can anneal and initiate polymerization. To prepare total cDNA for poly(A) PCR, known sequences are added in two steps (Fig. 1). One end is initially defined through a cDNA reaction using reverse transcriptase and an oligo(dT) primer that will prime via the poly(A) tail present at the 3' end of most mRNA molecules (Fig. 1, step 1). A homopolymer dA tract is then added to the 3' end of the first-strand cDNA using terminal transferase (Fig. 1, step 2).

Because the PCR reaction is most efficient for relatively short sequences, amplification of full-length cDNA would result in disproportionate amplification of smaller cDNAs. This bias can be avoided by limiting the length of the initial cDNA strand to around 100–700 bases regardless of the size of the original RNA template.

[1] R. K. Saiki, F. A. Faloona, K. B. Mullis, C. T. Horn, H. A. Erlich, and N. Arnheim, *Science* **230**, 1350 (1985).
[2] K. B. Mullis and F. A. Faloona, this series, Vol. 155, p. 335.
[3] L. H. Gyllensten, X. Cui, R. K. Saiki, H. A. Erlich, and N. Arnheim, *Nature (London)* **335**, 414 (1988).
[4] D. A. Rappolee, A. Wang, D. Mark, and Z. Werb, *J. Cell. Biochem.* **39**, 1 (1989).
[5] A. Belyavsky, T. Vinogradova, and K. Rajewsky, *Nucleic Acids Res.* **17**, 2919 (1989).
[6] G. Brady, M. Barbara, and N. N. Iscove, *Methods Mol. Cell. Biol.* **2**, 17 (1990).

Copyright © 1993 by Academic Press, Inc.
All rights of reproduction in any form reserved.

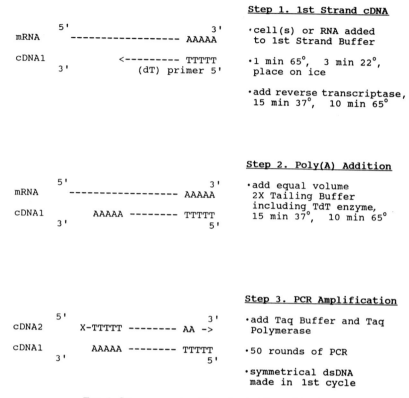

Step 1. 1st Strand cDNA

```
       5'                    3'      •cell(s) or RNA added
mRNA   ------------------ AAAAA       to 1st Strand Buffer

cDNA1             <--------- TTTTT    •1 min 65°,  3 min 22°,
       3'         (dT) primer 5'      place on ice

                                     •add reverse transcriptase,
                                      15 min 37°,  10 min 65°
```

Step 2. Poly(A) Addition

```
       5'                    3'      •add equal volume
mRNA   ------------------ AAAAA       2X Tailing Buffer
                                      including TdT enzyme,
cDNA1        AAAAA -------- TTTTT      15 min 37°,  10 min 65°
       3'                    5'
```

Step 3. PCR Amplification

```
       5'                    3'      •add Taq Buffer and Taq
cDNA2    X-TTTTT -------- AA ->        Polymerase

cDNA1        AAAAA -------- TTTTT     •50 rounds of PCR
       3'                    5'
                                     •symmetrical dsDNA
                                      made in 1st cycle
```

FIG. 1. Summary and outline of poly(A) amplification

PCR amplification of the dA/dT bracketed cDNA is carried out using a single oligonucleotide, known as (dT)-X, which consists of a 5' sequence containing a restriction enzyme site and a 3' dT stretch. Priming of the cDNA in the PCR is initiated via annealing of the dT region of (dT)-X to the homopolymer dA regions present at the termini of the cDNA molecules. The unique 5' sequence provides a convenient means of cloning the amplified product using the included restriction enzyme site, and it increases the stability and precision of primer annealing during PCR.

Purification steps have been avoided to minimize loss of starting material and opportunities for contamination. The final protocol is carried out in a single tube and involves only three pippetting steps prior to PCR amplification. The amplified product can be cloned into bacterial vectors and renewed indefinitely through further rounds of the PCR.

Poly(A) Amplification Protocol

The present protocol has been used for several years to amplify many hundreds of samples. During this period only slight modifications and additions have been made to the original method.[6] Most attempted "improvements" have been counterproductive, and for this reason even minor deviations from the given protocol should be avoided. Unless otherwise stated, it is advisable to store stock solutions as small aliquots at $-70°$.

cDNA Preparation

First-strand cDNA synthesis is carried out using reverse transcriptase in first-strand buffer which is made up from three stock components:

cDNA/lysis buffer: 52 mM Tris-HCl, pH 8.3, 78 mM KCl, 3.1 mM $MgCl_2$, 0.52% Nonidet P-40 (NP-40)

RNase inhibitors: 1 : 1 mixture of RNAguard (Pharmacia, Piscataway, NJ) and Inhibit Ace (5'–3' Inc.)

cDNA primer mix: mixture of nucleotides and dT primer [12.5 mM each dNTP, 6.125 OD_{260}/ml dT_{24} primer]

First-strand buffer should be prepared fresh and used within a few hours. For 100 μl mix 96 μl cDNA/lysis buffer, 2 μl RNase inhibitors, and 2 μl cDNA primer mix freshly diluted 1 : 24 with water.

To 4 μl of the first-strand buffer add 1 to 40 cells or up to 1 μg total RNA (see below) in no more than 0.5 μl. Heat for 1 min at 65° to promote unfolding of the mRNA, cool to room temperature for 3 min to allow annealing of the (dT) primer, and place on ice. To initiate first-strand synthesis, add 0.5 μl of a 1 : 1 mixture of Moloney murine leukemia virus and avian myeloblastosis virus reverse transcriptases (GIBCO–BRL, Grand Island, NY, or Boehringer-Mannheim, Indianapolis, IN), mixed directly as supplied by the manufacturer. (Both reverse transcriptase enzymes are used to avoid any sequence specificity attributible to either one alone.) Incubate for 15 min at 37°. Stop the reaction by heating to 65° for 10 min.

Nucleic Acid Preparation for 1–40 Cells: Direct Addition. Provided that the volume does not exceed 0.5 μl, samples of 1–40 cells may be added directly to 4 μl of first-strand buffer and stored on ice for a maximum of 1 hr prior to initiating the reverse transcriptase reaction. All biochemical reactions leading to a final poly(A) PCR product can be carried out without further purification. The direct addition procedure relies on NP-40 to lyse the outer cellular membrane and RNase inhibitors to protect the RNA from degradation. Because the NP-40 used does not lyse the nuclear membrane, genomic DNA may be recovered by pelleting the nuclei by a 50-sec centrifugation (12,000 g) at 4°. The cytoplasmic RNA present in the supernatant and genomic DNA recovered in the pellet can be processed separately. If there is no requirement for genomic DNA, the entire unfractionated sample can be processed following the poly(A) protocol without any interference from nuclear DNA.

Nucleic Acid Preparation for 40–10⁶ Cells: Mini-GT Protocol. Often it is desirable to process and to store large samples such as sections of tissue and cell cultures prior to poly(A) processing. RNA produced by any of the standard methods is readily amplified using the poly(A) PCR procedure. The following protocol provides a convenient method for preparing samples larger than 40 cells.

Mini-GT Protocol for Preparation of Total Nucleic Acids

Lyse 40–10⁶ cells either as a pellet, tissue, or in a suspension of no more than 10 μl by adding to 50 μl GT solution (5 M Guanidine thiocyanate, 0.5% Sarkosyl, 25 mM Sodium citrate, pH 7.0, 20 mM 1,4-Dithioerythritol). Store stock GT solution at room temperature (stable for several months). Vortex the cell lysate for 30 sec (at this point the sample may be processed directly or stored at $-70°$ for several months). Recover total nucleic acids from the sample by ethanol precipitation using glycogen as a carrier, add 25 μl 7.5 M ammonium acetate, 20 μg glycogen (Boehringer-Mannheim), and 150 μl ethanol. Mix and leave on wet ice a minimum of 30 min. For samples containing excess extracellular matrix, it is advisable to remove insoluble material after addition of ammonium acetate by centrifuging at 12,000 g at 4° for 5 min and transferring the supernatant. Pellet the ethanol precipitate by centrifuging at around 12,000 g for 30 min in a refrigerated microcentrifuge. To remove residual salts, wash the pellet three times at room temperature with 70% ethanol (each time briefly centrifuge to ensure the pellet is not lost), dry, and resuspend in 5–50 μl of 0.5% NP-40 containing a 1 : 100 dilution of the RNase inhibitor Inhibit Ace (5'–3' Inc.). Both DNA and RNA extracted in this procedure can be used as substrates for reverse transcriptase and genomic PCR. Although genomic DNA is present, it does not appear to contribute measurably to the poly(A) PCR product.

Timing of the cDNA Reaction

The incubation time of the reverse transcriptase reaction is kept deliberately short to produce abbreviated cDNA molecules that will be efficiently amplified in the final PCR step. Longer incubation times will result in longer overall cDNA molecules and consequently poor and nonrepresentative PCR amplification.

Oligonucleotide and Nucleotide Concentrations. The final concentrations of dNTPs and oligonucleotides in the cDNA reaction are 0.01 mM and 0.005 OD_{260} units/ml, respectively. The low dNTP concentration effectively limits the rate of reverse transcriptase polymerization and allows convenient addition of a predominantly homopolymer dA tail in a subsequent tailing reaction by the addition of excess dATP. The low dT primer concentration also plays a role in limiting first-strand synthesis and use of the primer as a substrate in the following tailing and amplification reactions.

Possible Contaminants. Amplification of cells added directly to the first-strand buffer appears to be relatively insensitive to the cell-suspension medium. Samples have been successfully processed directly from a variety of growth media that have included 10% serum, 1% methylcellulose, as well as buffered salt solutions (including phosphate buffers). Although we have never observed cell-related amplification from growth media and serum alone, a cDNA product has been obtained on occasion from filtered supernatants of dense cell cultures. For this reason appropriate controls should always be included, particularly for single-cell work. Nucleic acids prepared using the mini-GT protocol have invariably proved to be excellent substrates for both genomic and cDNA-directed PCR. Possible contamination from guanidine and ammonium salts is avoided by stringent washing with 70% ethanol.

Application. The direct addition approach has led to successful amplification of single primary hemopoietic cells, hemopoietic colonies grown in liquid culture or methylcellulose, fluorescence-activated cell sorted (FACS) populations, various cell lines, single mouse eggs, 2- to 8-cell embryos, blastocysts, and fetal tissue from various stages of development. The mini-GT protocol has proved to be invaluable for examining gene expression and genomic DNA from a multitude of sources. Combined with the poly(A) PCR amplification procedure, it is routinely being used as a simple and more economical alternative to Northern analysis.

Tailing

After allowing the heat-inactivated first-strand cDNA reaction to cool on ice for approximately 2 min, add an equal volume 2× tailing buffer [200 mM potassium cacodylate, pH 7.2, 4 mM $CoCl_2$, 0.4 mM dithiothreitol (DTT), and 1.5 mM dATP] containing 10 units Terminal transferase. (*Note:* The above formula for 2× tailing buffer is based on the 5× terminal transferase buffer supplied by Gibco–BRL and can be conveniently made up by mixing 800 μl of 5× BRL terminal transferase buffer, 30 μl of 100 mM dATP, and 1.17 ml water.) Incubate the tailing reaction for 15 min at 37° and heat-inactivate at 65° for 10 min.

Storage. At this point tailed cDNA preparations from 10 or more cells can be stored at −70° for at least 1 month. However, preparations from single cells noticeably deteriorate with storage at −70°, and it is advisable to continue with them directly to the PCR step.

Timing. Unlike the cDNA reaction, the exact timing of the terminal transferase reaction is not critical, and extended incubation up to 1 hr does not adversely affect the amplified product.

Primary Polymerase Chain Reaction

Optimization of Amplification Oligonucleotide and Salt Concentrations. The concentrations of both MgCl$_2$ and the amplification oligonucleotide in the primary PCR are critical for efficient amplification of samples equivalent to 1–50 cells. A given set of PCR conditions may allow efficient amplification from 100 cells but fail to produce material from a single cell. Establish optimal PCR conditions by amplifying aliquots equivalent to 1–2 cells from a scaled-up cDNA/tailing reaction. To do this, add 1 μl containing 50–100 cells to 100 μl first-strand buffer and process to yield 200 μl of tailed cDNA following the protocol above. Divide into 4-μl samples and amplify in a final PCR volume of 20 μl. For MgCl$_2$ titration, prepare a series of 10× *Taq* polymerase buffers consisting of 100 mM Tris-HCl, pH 8.3, 500 mM KCl, 15–50 mM MgCl$_2$ in increments of 5 mM, 1 mg/ml bovine serum albumin (BSA, Boehringer/Mannheim molecular grade), and 0.5% Triton X-100. For each MgCl$_2$ concentration, test the amplification oligonucleotide at final concentrations of 1, 2, and 4 OD$_{260}$ units/ml.

PCR Amplification. For every microliter of tailed cDNA add 4 μl of PCR mixture having the following proportions:

10× *Taq* polymerase buffer (optimized for MgCl$_2$)	4 μl
25 mM each dNTP	1.5 μl
(dT)-X amplification oligonucleotide (see optimzation above)	x μl
Taq polymerase (Perkin-Elmer Cetus, Norwalk, CT)	5 units
Water	Adjust to a final volume of 32 μl

Overlay the sample with mineral oil (if required) and amplify using the following cycle profile: 25 cycles consisting of 1 min at 94°, 2 min at 42°, and 6 min at 72° linked to a second 25 cycles each having 1 min at 94°, 1 min at 42°, and 2 min at 72°.

Estimate of Overall Amplification. The efficiency of overall amplification cannot be judged by ethidium bromide staining alone since control samples generated in the absence of added RNA or DNA template frequently yield visible DNA. This is believed to be due to prokaryotic DNA present in commercially supplied *Taq* polymerase. However, this nonspecific product has not hybridized to any of more than 20 specific eukaryotic probes tested so far. Successful amplification is judged by hybridization to two or more specific cDNA probes.

Primer Sequence. The (dT)-X amplification oligonucleotide sequence is 5′ ATG TCG TCC AGG CCG CTC TGG ACA AAA TAT GAA TTC 3′ plus 24 dT residues. Apart from the *Eco*RI restriction enzyme site prior to the dT stretch, the unique sequence of the amplification primer given above was chosen arbitrarily from primers available in the laboratory. Several other sequences have also worked, although most of the single-cell amplifications have been carried out with the given sequence.

Oligonucleotide Preparation. Different batches of the same oligonucleotide sequence vary markedly in their ability to amplify small samples (<10 cells). The reason for this is unclear, but oligonucleotides that failed to amplify single-cell samples have frequently been "rescued" by deionization and, in one instance, by adjusting the PCR to include a final concentration of 5 mM EGTA. Commercially available oligonucleotides appear to vary considerably in the purity and percentage of full-length product. Although cost has precluded a systematic analysis of differing sources, oligonucleotides provide by Oligos Etc. Inc. (Wilsonville, OR) have worked well in the single-cell poly(A) PCR protocol without further purification.

Deionization. Wash 1 g of mixed-bed resin three times with 30 ml water. Filter to remove excess water and add approximately 0.5 ml of the packed filtered resin directly to the oligonucleotide (in around 0.5–2 ml water). Heat at 65° for 5 min and swirl at room temperature for 30 min. Filter out the resin using a 0.22-μm minifiltration unit and measure the absorption ($OD_{260/280}$). Make suitable aliquots, lyophilize to dryness, and store at −70°. Reconstitute when needed with an appropriate volume of water.

Buffer Composition. The PCR buffer described appears to be critical for amplification since changes to a higher Tris-HCl concentration or slight changes in the pH of the 2× tailing buffer and the *Taq* polymerase buffer have led to dramatic reductions in PCR yield (S. Varmusa and F. Billia, personal communication).

Storage of PCR Samples. Prolonged storage (>2 months) of unpurified poly(A) PCR samples at 4°, −20°, and −70° can lead to deterioration to the point where they can no longer be amplified by additional rounds of poly(A) PCR. It is recommended for precious samples that a portion be stored as a dried ethanol precipitate. A small aliquot of each sample can also be expanded through reamplification and cloning (see below), providing an inexhaustable supply of working material.

Reamplification of Poly(A)-Amplified Material

Conditions for successful amplification of a previously amplified sample appear to be less stringent than those used in the primary PCR. Both *Taq* polymerase, oligonucleotides, and dNTPs may be reduced to around one-fifth of the initial levels. However, as with the primary PCR, it is advisable to titrate the oligonucleotide to establish efficient and economical working conditions. Typically reamplify five samples:

Water	440 μl
10× *Taq* polymerase buffer	50 μl
25 mM dNTPs	4 μl
Amplification oligonucleotide	1 OD_{260}/ml
	(final concentration)
Taq polymerase	5 units

Add 100 μl of this mix to 0.5 μl of original sample and amplify through 25 cycles: 1 min at 94°, 1 min at 42°, and 2 min at 72°.

Reamplification Oligonucleotides. A variety of oligonucleotides may be used during reamplification, including the original primer used for primary PCR, or an oligonucleotide consisting solely of the unique component of the (dT)-X oligonucleotide, or any oligonucleotide having a 3′ stretch of dT linked to another sequence. The latter type of reamplification oligonucleotide provides a useful way of introducing new restriction sites, bacteriophage promoter sequences, etc.

Dilution of Original Sample. Efficient reamplification appears to be inhibited in part by a product of the primary PCR. This can be overcome by diluting the original sample at least 100 times into the fresh PCR.

Sample Analysis

Successful amplification will produce samples containing a heterogeneous collection of 3′ cDNA fragments representing all of the poly(A) mRNA added to the starting cDNA reaction. On examination by agarose gel electrophoresis (1.5–2% agarise/Tris–borate), 3–5 μl of the completed PCR sample will produce a strong ethidium bromide-stained distribution from 100–700 base pairs in length. Frequently, a biphasic distribution will be apparent that is reduced if the sample is treated with single-stranded nucleases, suggesting that it may reflect the presence of both single- and double-stranded DNA. Genes expressed in the sample can be conveniently examined by conventional Southern blotting procedures as well as dot or slot-blot techniques. However, in our hands, Southern hybridization of amplified material is 10–50 times more sensitive than dot/slot blotting of the same poly(A) amplified sample. This appears to be due to an interfering component of the PCR that separates from the amplified cDNA during electrophoresis.

Bidirectional Southern Transfer

A convenient way of providing multiple filters for hybridization analysis is the following adaptation of the bidirectional transfer[7] protocol. Add 3–5 μl of each poly(A) sample directly after PCR to 10 μl of a suitable gel loading buffer containing bromphenol blue (BPB). Load a 1.5–2% agarose/Tris–borate gel together with size markers and run until the BPB has migrated approximately 3–4 cm. After photographing the ethidium bromide-stained gel, soak it for 30 min in 0.5 M NaOH/1.5 M NaCl, rinse in water, and soak in 1 M ammonium acetate for a further 30 min. Soak two sheets of GeneScreen Plus (Du Pont, Wilmington, DE) for a few minutes in 1 M ammonium acetate. Transfer the DNA by layering on a level surface the following in order: a 3-cm pile of paper towels, one sheet of Whatman (Clifton, NJ) 3MM filter paper wetted with 1 M ammonium acetate, one of the soaked GeneScreen Plus filters, the processed gel, the second GeneScreen Plus filter, another Whatman 3MM sheet, and a 3-cm pile of paper towels. Compress the entire sandwich by placing a 0.5 kg weight on top. Although over 90% of the DNA will be transfered within 1–2 h, the transfer setup may be left overnight. The same transfer procedure has been used successfully for nitrocellulose and a variety of commercial nylon membranes. For Hybond-N$^+$ (Amersham), the DNA was efficiently transferred directly after denaturation in 0.5 M NaOH, 1.5 M NaCl. DNA was fixed to the membrane following the manufacturer's recommendations. Because the proportion of DNA transferred to top and bottom filters is variable, estimation of relative hybridization intensities must be compared to standards included on the same filter.

Nylon Membrane Hybridization

This is based on the procedure recommended for GeneScreen Plus. Prehybridize the filter 2–20 hr at 42° in a minimum volume of the following

[7] G. E. Smith and M. D. Summers, *Ann. Biochem.* **109,** 123 (1980).

mixture: 50% deionized formamide, 1 M NaCl, 1% sodium dodecyl sulfate (SDS), 10% dextran sulfate, 5 μg/ml denatured oligo(dA)/(dT) (optional, see below), and 25 μg/ml denatured herring sperm DNA. Add denatured radiolabeled probe to a final concentration of approximately 2 × 10^6 counts/min (cpm)/ml and hybridize for 12–48 hr.

Washing Conditions. For each probe the stringency of washing necessary to produce strong, clean signals is variable owing to the small size of hybridized fragments and varying A/T richness. Generally, washing for 1 hr at 65° in 2× SSC followed by a second wash for 1 hr further in 0.5× SSC at 65° will produce clear signals. If background is excessive, increase the temperature to 70° and/or decrease the final wash to 0.1× SSC.

General Background. The most frequent cause of "spotty" background is the presence of particulate material in the hybridization buffer or from gloves worn when handling the filter (J. Jongstra, personal communication). This can be eliminated by filtering either the complete buffer or each of the individual components and by avoiding powdery gloves. Diffuse darkening of the filter is usually associated with preparation of individual probes and is often eliminated by reisolating the problem probe.

Oligo(dA)/(dT). Because 3' sequences are frequently A/T-rich, some probes exhibit an A/T-dependent background. This can be reduced by adding a 20-base oligonucleotide consisting of random additions of either dA or dT at each position to the hybridization mix.

Choice of Probes. Because of the 3' nature of the amplified product, detection of an expressed sequence requires that the probe include sequences close to the extreme 3' end of the native transcript. The ideal probe would consist of 400 bases of 3' cDNA sequence directly abutting the polyadenylation site. Where 3' probes are not readily available, the following adaption of the rapid amplification of cDNA ends (RACE) protocol[8] provides a means of generating a usable probe provided some 3' sequence information is available.

Rapid Amplification of cDNA Ends

Note that all oligonucleotides used in the procedure are adjusted to 50 OD$_{260}$/ml. Prepare the cDNA reaction containing the following:

cDNA lysis buffer (see cDNA section)	46 μl
RNA containing gene of interest	1 μl (10 ng–1 μg)
25 mM dNTPs	1 μl
(dT)$^+$ oligonucleotide	1 μl

---

[8] M. A. Frohman, M. K. Dush, and G. R. Martin, *Proc. Natl. Acad. Sci. U.S.A.* **85**, 8998 (1988).

Heat for 5 min at 65°, cool at room temperature for 5 min, add 200 units Moloney reverse transcriptase (GIBCO–BRL), incubate for 30 min at 37°, and heat-inactivate for 10 min at 65°. Set up the PCR as follows:

cDNA reaction	10 μl
10× *Taq* polymerase buffer	10 μl
25 m*M* dNTPs	2 μl
Unique sequence of the (dT)$^+$ oligonucleotide	1.5 μl
24-Base specific-sense oligonucleotide	1 μl
Water	75 μl

Boil for 5 min, add 5 units *Taq* polymerase, overlay with mineral oil if necessary, incubate at 72° for 30 min, and transfer a thermocycling machine for 25 rounds of 1 min at 95°, 1 min at 42°, and, depending on the size of fragment, 1–5 min at 72°.

Annealing Temperature and Elongation Time. If a large amount of nonspecific material is apparent after amplification, increase the annealing temperature until only specific products remain. Since extended elongation times can also contribute to nonspecific products, the 72° step in the PCR should be kept to the minimum time required for the target sequence.

Cloning into Bacterial Vectors

Additional Polymerization Step

After the final round of PCR a varying amount of denatured single-stranded material may be left. To maximize the proportion of clonable double-stranded material, an additional round of polymerization is advisable. For 10 μl of the original PCR material add 50 μl fresh *Taq* buffer containing additional nucleotides and oligonucleotide at the same concentration as the starting PCR. Boil the mixture for 2 min, anneal to the primer for 4 min at 42°, add 5 units *Taq* polymerase, and incubate for 30 min at 72°. Extract the sample with ether or chloroform to remove residual mineral oil and ethanol-precipitate by adding 30 μl 7.5 *M* ammonium acetate and 200 μl ethanol. After 30 min on ice, spin down the DNA and wash two times with 70% ethanol, dry, and resuspend in 10 μl TE (10 m*M* Tris-HCl pH 7.5, 1 m*M* EDTA).

Restriction Digestion

Bring the entire sample to 100 μl in restriction buffer (50 m*M* Tris-HCl, pH 8.0, 100 m*M* NaCl, 10 m*M* MgCl$_2$) and digest with 100 units *Eco*RI at 37° for 2 hr to overnight. Complete digestion will be apparent by a slight reduction in overall size and the appearance of a 36-bp fragment equivalent to the unique component of the amplification oligonucleotide. Because this small fragment contains *Eco*RI ends, it will compete with the cDNA sequences during the vector ligation step and therefore must be removed either by a sizing column or by elution of the cDNA from an agarose gel. (*Note:* Taq polymerase is still present in the sample and may interfere with some enzymatic modifications, making it necessary to purify further by proteinase K digestion and phenol–chloroform extraction.)

Ligation and Library Construction

For the construction of poly(A) cDNA libraries, it is advisble to titrate the amount of insert against a constant amount of dephosphorylated vector. Under these conditions the absolute number of colonies/plaques produced will increase with increasing amount of insert and will reach a plateau as the number of vector molecules become limiting. Libraries consisting of primarily single inserts will be produced if an insert concentration is used where vector molecules are not limiting. The ligation buffer used for both λ and plasmid cloning is a dilution of $5\times$ LIP [250 mM Tris-HCl (pH 7.6), 50 mM MgCl$_2$, 2.5 mM ATP, 5 mM DTT, and 25% (w/v) polyethylene glycol 8000].

Plasmid Vectors. A typical titration series would be as follows:

Insert	0, 1, 2, 4, 8 ng
Dephosphorylated plasmid	50 ng
$5\times$ LIP	5 μl
Water	Final volume of 50 μl
T4 DNA ligase	40 units

Ligate for 1 hr with the tubes in a beaker of 22° (room temperature) water. Place the entire beaker and samples in a sealed Styrofoam container and transfer to 4° overnight. Transform 10 μl into competent recombination-defective bacteria and count the resulting colonies to establish a suitable insert concentration for library construction.

Lambda Vectors. Set up the equivalent of the following:

Insert	0, 1, 2, 4, 8 ng
Dephosphorylated λ vector	400 ng
$5\times$ LIP	0.6 μl
Water	Final volume of 6 μl
T4 DNA ligase	20 units

Ligate as described for plasmid cloning and package 1 μl of the ligation into λ phage particles (commercial kits are available and suitable). Plate out a range of dilutions (1 : 1000 to 1 : 10) and estimate a suitable insert concentration for library construction.

Screening and Analysis of Clones

Standard molecular procedures can be used for both screening and examination of libraries and individual clones generated from poly(A) PCR material. Because of the abbreviated nature of the amplified material, specific probes must include the extreme 3′ sequences in order to detect positive clones in a poly(A)-derived library.

Differential Library Screening

Poly(A)-derived PCR material can also be used as a labeled probe to screen either poly(A) or conventional cDNA libraries and as such form the basis of a differential screening approach. Briefly, this would entail producing two replica filters for a cDNA library and hybridizing one with labeled poly(A) cDNA from source A and screening the other with labeled material from source B. Individual clones showing strong hybridization with one and not the other probe represent genes whose expression differed in the two starting samples.

Acknowledgments

We thank M. Barbara for excellent technical assistance, P. A. Johnson, L. Addy, F. Billio, L. Trumpner, X. Q. Yan, and M. Barbara for critical reading of the manuscript, and B. Vennstrom, R. Gronostajski, and M. Minden for helpful discussions and suggestions. The work was supported by operating grants to N.N.I. from the Medical Research Council and National Institute of Canada. G.B. is a Special Fellow of the Leukemia Society of America.

[37] Construction and Characterization of Yeast Artificial Chromosome Libraries from the Mouse Genome

By Zoia Larin, Anthony P. Monaco, Sebastian Meier-Ewert, and Hans Lehrach

Introduction

The generation of yeast artificial chromosome (YAC) libraries[1] with large DNA inserts provides a powerful tool for contributing to the long-range physical, genetic, and functional mapping of the murine genome. It should now be feasible to construct murine genetic maps that contain

[1] D. T. Burke, G. F. Carle, and M. V. Olson, *Science* **236**, 806 (1987).

Copyright © 1993 by Academic Press, Inc.
All rights of reproduction in any form reserved.

defined markers 1–2 cM (2000–4000 kb) apart, and the ability to clone large DNA fragments by YAC vectors (100–1500 kb) will bridge the gap between the genetic and physical map. The development of this technology offers advantages over other cloning systems and will be of significant value in relating these maps to the functional map and ultimately the molecular analysis of genes.

YACs can be easily modified within the yeast host by homologous recombination procedures. New markers or regulatory sequences can be incorporated in the vector or insert DNA, termed "retrofitting," for selection and functional analysis of YACs in mammalian systems.[2,3] Overlapping YACs can be recombined in yeast to generate fragments spanning extensive regions, which is especially useful with large and complex gene loci.[4] Sublocalization of genes within large frgments could be defined by high-resolution restriction maps, fragmentation,[5] and exon mapping[6,7] strategies. In addition, YACs are sufficiently large to encompass both the coding and regulatory regions of genes, and it is envisaged that they will be of value in constructing functional maps in the mouse by complementation approaches.[8]

In mouse genomes, important areas of biological function have included regions of extensive chromosomal rearrangements such as inversions, deletions, or translocations giving rise to interesting and often developmental stage-specific mutations. Some of the rearrangements such as deletions have arisen either spontaneously in laboratory inbred mice or as induced mutations by exposure to X rays or to chemical mutagens. YACs will be useful in mapping such areas (the c locus at 6–11 cM of chromosome 7 or the t complex at 12–15 cM of chromosome 17),[9,10] where existing genetic and physical mapping strategies have been frustrated because of the size of these regions. YACs will also be used to span

[2] W. J. Pavan, P. Heiter, and R. H. Reeves, Mol. Cell. Biol. 10, 4163 (1990).

[3] V. Pachnis, L. Pevny, R. Rothstein, and F. Costantini, Proc. Natl. Acad. Sci. U.S.A. 87, 5109 (1990).

[4] E. D. Green and M. V. Olson, Science 250, 94 (1990).

[5] W. J. Pavan, P. Heiter, and R. H. Reeves, Proc. Natl. Acad. Sci. U.S.A. 87, 1300 (1990).

[6] G. M. Duyk, S. Kim, R. M. Myers, and D. R. Cox, Proc. Natl. Acad. Sci. U.S.A. 87, 8995 (1990).

[7] A. J. Buckler, D. D. Chang, S. L. Graw, J. D. Brook, D. A. Haber, P. A. Sharp, and D. E. Houseman, Proc. Natl. Acad. Sci. U.S.A. 88, 4005 (1991).

[8] Z. Larin and H. Lehrach, Genet. Res. 56, 203 (1990).

[9] E. M. Rinckik and L. B. Russell, in "Genome Analysis: Genetic and Physical Mapping" (K. E. Davies and S. M. Tilghman, eds.), Vol. 1, p. 121. Cold Spring Harbor Laboratory, Cold Spring Harbor, New York, 1990.

[10] B. G. Herrmann, D. P. Barlow, and H. Lehrach, Cell (Cambridge, Mass.) 48, 813 (1987).

distances from flanking genetic markers to regions containing genes that, owing to point mutations, give rise to other interesting phenotypes.

The generation of YAC libraries containing large insert DNA is extremely important for this goal to be realized. YAC libraries have been constructed from both prokaryotic and eukaryotic genomes (reviewed in Hieter et al.[11]), with several from the mouse genome.[8,12-14] However, the generation of large insert libraries has been difficult to achieve owing to a number of technical problems involved in preparation and manipulation of large DNA fragments and transformation of the yeast host.

In this chapter we discuss methods to construct large insert mouse YAC libraries where DNA in agarose is size-fractionated by pulsed-field gel electrophoresis (PFGE). We also describe a procedure for rapid screening of YAC libraries by colony replication and hybridization, as well as initial characterization of insert DNA.

Preparation of Large DNA Fragments

Reports of YAC library constructions have indicated that size fractionation of DNA is essential to eliminate small fragments.[12,15,16] Size-fractionation procedures have been conducted before ligation, to decrease the number of small insert molecules which could ligate to vector ends, and after ligation, to remove unligated and self-ligated vector material which may cause background after transformation of the yeast host.

DNA in aqueous solution has been prepared and size-fractionated after partial digestion by using sucrose gradient density centrifugation prior to cloning by the vector[1] (reviewed by Burke and Olson[17]). This method subjects DNA to shear stress, producing YACs with an average insert size of 225 kb.[18] DNA is better protected when prepared in agarose, or

[11] P. Hieter, C. Connelly, J. Shero, M. K. McCormick, and R. Reeves, in "Genome Analysis: Genetic and Physical Mapping (K. E. Davies and S. M. Tilghman, eds.), Vol. 1, p. 83. Cold Spring Harbor Laboratory, Cold Spring Harbor, New York, 1990.

[12] M. K. McCormick, J. H. Shero, M. C. Cheung, Y. W. Kan, P. A. Hieter, and S. E. Antonarakis, Proc. Natl. Acad. Sci. U.S.A. 86, 9991 (1989).

[13] D. T. Burke, J. M. Rossi, J. Leung, D. S. Koos, and S. M. Tilghman, Mammal. Genome 1, 65 (1991).

[14] Z. Larin, A. P. Monaco, and H. Lehrach, Proc. Natl. Acad. Sci. U.S.A. 88, 4123 (1991).

[15] R. Anand, A. Villasante, and C. Tyler-Smith, Nucleic Acids Res. 17, 3425 (1989).

[16] H. M. Albertsen, H. Abderrahim, H. C. Cann, J. Dausset, D. Le Paslier, and D. Cohen, Proc. Natl. Acad. Sci. U.S.A. 87, 5109 (1990).

[17] D. T. Burke and M. V. Olson, this series, Vol. 194, p. 251.

[18] B. H. Brownstein, G. A. Silverman, R. D. Little, D. T. Burke, S. J. Korsmeyer, D. Schlessinger, and M. V. Olson, Science 244, 1348 (1989).

agarose beads,[19] and size-fractionated by PFGE.[12,15,16] PFGE allows greater flexibility in selecting DNA of different size ranges.

Two size-fractionation procedures by PFGE have produced YACs with an average insert size of 430 kb, which is an improvement on the size of YACs generated by a single size selection.[16] However, using the same procedures we experienced difficulties in routinely constructing YACs containing DNA fragments greater than 400 kb, and we observed a discrepancy between size-fractionated (PFGE) DNA (>400 kb) and the size of insert DNA found in YACs (<250 kb). On the basis of this size discrepancy, we determined that degradation of DNA during the cloning procedure had occurred as a result of damage to DNA.[14]

Using yeast chromosomes in mock cloning procedures, degradation of DNA occurred after the first size fractionation by PFGE when agarose containing DNA was melted at 68° prior to ligation to the vector, and it occurred after the second size-fractionation step when agarose containing ligated DNA was melted before transformation of the yeast host. The cause of degradation may have been nonspecific cleavage or denaturation of DNA when heated to 68°.[14]

The degradation of yeast chromosomes when heated to 68° was prevented in the presence of the divalent cation Mg^{2+} (>15 mM) and the polyamines (i.e., triamines) spermidine and spermine [1 and 10 mM, respectively, and in combination to 1.05 mM[12] (0.3 mM spermine, 0.75 mM spermidine)]. On the basis of this finding, polyamines were incorporated into the cloning procedure by equilibrating DNA in buffers containing polyamines, and DNA remained intact after both size-fractionation steps and when heated to 68°. The size discrepancy previously observed was eliminated; that is, when DNA was size-selected to greater than 400 and 600 kb, the insert size of YAC clones generated ranged between 500 and 800 kb. By following this protocol, a mouse YAC library (C3H, male) containing three genome equivalents (15,000 clones) with an average insert size of 700 kb was constructed.[14] In addition, we have recently generated a further 5000 YAC clones with an average insert size of 720 kb (C57BL/6, female; S. Meier-Ewert and H. Lehrach, unpublished).

Library Construction

Preparation of DNA

Many YAC libraries have been prepared from genomic DNA isolated from different sources, including cultured cells from a homogeneous back-

[19] T. Imai and M. V. Olson, *Genomics* **8,** 297 (1990).

ground, somatic cell hybrid lines containing the chromosome of interest on a heterogeneous background, or fresh tissue cells. Whatever the source, the amount of starting material required (microgram quantities in comparison to nanogram quantities required to transform a bacterial host) is important because of the poor transforming efficiency of the yeast host with high molecular weight DNA.

Protocol. In our experiments high molecular weight DNA in agarose blocks is prepared from fresh mouse spleen tissue according to published protocols,[9] at a concentration of approximately 2×10^6 cells/ml (equivalent to ~15 μg of DNA).

1. Lyse cells in agarose blocks in 0.4 M EDTA, 1% N-lauroylsarcosine, and 2 mg/ml proteinase K (BDH, Poole, UK) or pronase (Boehringer-Mannheim, Mannheim, Germany, ~50 blocks per 50 ml) and incubate at 50° for 48 hr. The blocks can be stored in this solution or washed several times and stored in 10 mM Tris-HCl, pH 7.5–8, 500 mM EDTA at 4°.

2. Prior to enzyme digestion remove the proteinase K or pronase either by washing blocks extensively in 1× TE (at least 4 times in 10 mM Tris-HCl, pH 7.5; 1 mM EDTA) or by inactivating the proteases with the addition of phenylmethylsulfonyl fluoride (PMSF; 40 μg/ml prepared in 2-propanol) in 1× TE at 50°. Remove PMSF by washing blocks in 1× TE 2 times at 50° and then 2 times at room temperature.

Partial Digestion of DNA Using EcoRI and EcoRI Methylase

The extent of partial digestion within and between batches of agarose blocks can be inconsistent, probably because of inhomogeneity in DNA concentration within the blocks. To achieve a better range of size fragments after partial digestion, and to eliminate some of the variability between blocks, a combination of *Eco*RI and *Eco*RI methylase enzymes are used. Recognition sites methylated by *Eco*RI methylase will be rendered insensitive to *Eco*RI. The extent of partial digestion can be controlled by varying the concentrations of the two enzymes with respect to one another, with the ratio of each being determined by the extent of ethidium staining of DNA after PFGE. Partial digestion of DNA can be controlled by varying either the concentration of *Eco*RI to constant *Eco*RI methylase or by varying the concentration of *Eco*RI methylase to constant *Eco*RI.

Protocol

1. Place individual agarose blocks in a restriction enzyme digest mixture containing 100 mM NaCl, 100 mM Tris-HCl, pH 7.5, 2 mM MgCl$_2$,

80 μM S-adenosylmethionine, 0.5 mg/ml bovine serum albumin (BSA), 2.6 mM spermidine trihydrochloride, and 1 mM dithiothreitol (DTT) in a final volume of 500 μl. Digest at least 10 blocks ($\sim$10–20 μg/block) in individual reactions. Each new batch of agarose blocks containing genomic DNA should be standardized to determine the optimal concentration for both enzymes.

2. Add both enzymes in a predetermined ratio (empirically determined by PFGE, but usually in the range 1 unit of EcoRI to 50–200 units of EcoRI methylase) to the restriction digest mix. Equilibrate blocks on ice for 1 hr prior to incubation at 37° for 4 hr.

3. Terminate the reaction by adding EDTA to 20 mM and incubating with proteinase K (0.5 mg/ml) at 37° for 30 min.

First Size Fractionation by Pulsed-Field Gel Electrophoresis

DNA fragments can be size-fractionated by PFGE according to the switch time selected. In our experiments we use a 30- or 50-sec switch time for selection of fragments 400 or 600 kb and above, respectively.

Protocol

1. Prior to size fractionation by PFGE, pool agarose blocks (containing partially digested DNA) into a 50-ml Falcon tube (Becton Dickenson, Lincoln Park, NJ) and wash 2 times in 10 mM Tris-HCl, pH 7.5, and 50 mM EDTA.

2. Place blocks in a trough prepared in a 1% (v/v) low-melt agarose gel (Sea Plaque GTG, FMC, Rockland, ME) which has been set at 4°. Place a similar genomic digest in gel slots either side of the trough and yeast chromosomes (as size markers) on the outside of these lanes. Seal the gel trough with the remainder of the low-melt agarose. The gel is subjected to electrophoresis at 160 V, 30- or 50-sec switch time, for 18 hr at 15° in a CHEF (clamped homogeneous electric fields) apparatus (Bio-Rad, Richmond, CA; or EMBL Workshop, Heidelberg, Germany).

3. Cut the marker lanes away from the main portion of the gel and stain with ethidium bromide. Store the rest of the gel in 10 mM Tris, pH 7.5, and 50 mM EDTA at 4°. Notch the gel on one side in the marker lanes (the side adjacent to the main portion of the gel) to indicate the position of the limiting mobility (the unresolved DNA which is focused into a thin band). Reassemble the whole gel, carefully aligning the notched marker lanes adjacent to the main portion of the gel, and excise the limiting mobility of the sample DNA from the gel using the notched marker lanes as guidelines. Restain the whole gel and visualize under UV light.

4. Equilibrate the slide (usually 1–2 ml) in four changes (30 min each) of ligation buffer containing 50 mM Tris-HCl, pH 7.5, 10 mM MgCl$_2$, 30

mM NaCl, and 1× polyamines (1× polyamines being 0.75 mM spermidine trihydrochloride, 0.3 mM spermine tetrahydrochloride; prepare in
distilled water as 100× and store at −20° in frozen aliquots), prior to
ligation.

Vector DNA Preparation and Ligation Reaction

The partial digest libraries are constructed with the artificial chromosome cloning vector pYAC4 containing an *Eco*RI cloning site described
previously.[1] The vector pYAC4 is linearized by digesting with *Eco*RI
enzyme, and the telomeric sequences present at the ends of both vector
arms are released by digesting with *Bam*HI. The vector DNA is then
dephosphorylated with calf intestinal phosphatase to prevent self-ligation
of the vector arms. The extent of the phosphatase reaction can be assayed
by adding T4 polynucleotide kinase to an aliquot of the ligation reaction
to check that the vector arms retain the ability to self-ligate. The prepared
vector DNA is then added to molten agarose solution containing partially
digested genomic DNA in a 1 : 1 ratio by weight in the presence of ligase
buffer, ATP, DTT, and T4 DNA ligase. The reaction is allowed to proceed
at 37° for approximately 0.5–1 hr to allow diffusion of ligase enzyme
through the molten agarose before transferring the reaction to room temperature for approximately 16 hr.

Vector DNA. Before proceeding with a large-scale purification of
pYAC vector DNA, check several minipreparation DNAs to ensure that
the pYAC vector is intact. A common problem occurring with these vectors is deletion of the telomeric sequences. Vector DNA digested with
*Hin*dIII (assayed on a 1% agarose gel) should produce four fragments of
the expected size 3.5, 3.0, 1.9 kb, and a 1.4-kb doublet. The presence of
a smaller fifth fragment indicates that telomeric sequences have deleted,
and this DNA should not be used in the cloning procedure.

1. Digest approximately 100–200 μg of purified vector DNA with
*Eco*RI and *Bam*HI. Again, to ensure the reaction has proceeded to completion, check that the vector is intact by analyzing a small aliquot on a 1%
agarose gel. Three fragments should be detected, 6.0, 3.7, and 1.7 kb.
Terminate the reaction by heating to 68° for 5 min.

2. Dephosphorylate vector ends by adding calf intestinal phosphatase
(CIP; 0.03–0.06 units/μg) directly to the restriction buffer and incubating
at 37° for 30 min. Inactivate the CIP by adding trinitriloacetic acid (BDH;
prepared by dissolving in distilled water) to 15 mM at 68° for 15 min.

3. Purify the DNA by extracting 2 times with phenol, precipitating
with ethanol, and resuspending in 1× TE at a final concentration of 1
μg/μl. Set up a small-scale ligation reaction (1–2 μg in a final volume of
20 μl) at 25° with and without the addition of T4 polynucleotide kinase.

Visualize the products on a 1% agarose gel to check the efficiency of the phosphatase reaction and that the vector arms can religate together after adding terminal phosphates.

Ligation Reaction

1. Place the agarose slice containing size-selected DNA in an Eppendorf tube and add vector DNA in a ratio of 1 : 1 by weight. Melt at 68° for 10 min (gently invert the tube to mix) and cool to 37°.

2. Add ligase buffer (not ice-cold), containing 5 μl of ligase enzyme (400 units/μl) and ATP (pH 7.5) and DTT to 1 mM each, to the molten agarose and mix initially by stirring extremely slowly with the pipette tip and then by gently inverting the tube 2 or 3 times.

3. At this stage remove two 10-μl aliquots of the reaction for ligation controls, and in a separate reaction add approximately 0.5 μl (<0.5 unit) of T4 nucleotide kinase to one aliquot. Incubate both aliquots with the ligation reaction at 37° for 0.5–1 hr prior to room temperature overnight.

4. Terminate the reaction by adding EDTA, pH 8.0, to 20 mM.

Second Size Fractionation

Protocol

1. Melt the ligation reaction at 68° for 10 min and cool to 37°.

2. Pipette the molten agarose with a tip (bore diameter not less than 4 mm) into a gel trough prepared in a low-melt gel as described above. Pipette a small aliquot (~60 μl) of the reaction into gel slots adjacent to the trough as marker lanes. In addition, use yeast chromosomes as size standards.

3. Use the same electrophoretic conditions for the CHEF apparatus as in the first size-fractionation procedure. Excise the limiting mobility from the gel using the marker lanes as guidelines as described above.

4. Wash the agarose slice (in a volume of about 2–3 ml) 4 times (30 min each) in 10 mM Tris-HCl, pH 7.5, 1 mM EDTA, 30 mM NaCl, and 1 × polyamines. Score the slice into pieces (volume not greater than ~600 μl), place in individual Eppendorf tubes, melt at 68°, and cool to 37°.

5. Digest the molten agarose with 150–200 units of agarase (Sigma, St. Louis, MO) for 2–3 hr at 37° prior to transformation of the yeast host.

Transformation of Yeast Host

Yeast cells from the host strain AB1380[1] are spheroplasted and transformed according to Burgers and Percival[20] (these procedures are also

---

[20] P. M. J. Burgers and K. J. Percival, *Anal. Biochem.* **163,** 391 (1987).

reviewed by Burke and Olson[17]). The yeast URA3 CEN plasmid YCp50,[21] which is stable in the yeast host, is used as a control of transformation efficiency (10 ng produces $\sim 1–3 \times 10^3$ transformants). In our experiments the desired density of cell culture for AB1380 prior to transformation is between 1 and 1.5 OD_{600} units, equivalent to approximately 5×10^7 cells/ml, and we use the enzyme lyticase (Sigma) to form spheroplasts. Inducing an extent of spheroplast formation between 80 and 90% takes approximately 15–20 min at 30°, which is checked by spectrophotometer readings after lysis in distilled water and observation by phase-contrast microscopy of percent lysis ("ghost formation") in sodium dodecyl sulfate (SDS). The two readings do not correlate exactly, since lysis in SDS always proceeds faster than lysis in water. We also observe that temperature has a significant effect on the efficiency of transformation. When temperatures are above 26° the transformation efficiency is reduced 50-fold; the optimum temperature is 20°.

Transformation efficiency is optimal when the ratio of DNA concentration to cell volume is between 1 and 5 μg (in no more than 50 μl of agarose solution) to 150 μl of spheroplasted cells at a final concentration of 4.5–5×10^8 cells/ml. The efficiency is reduced when the transformation reaction is scaled up. Cells from individual transformations can be plated directly onto 10-cm agar plates (prepared with selective medium lacking uracil),[22] or several transformations can be pooled in the recovery or SOS stage such that cells containing 10 μg of exogenous YAC DNA are plated onto a 22×22 cm agar plate in approximately 50 ml of top agar (kept warm at 48°). Agar plates should be prewarmed at 37° for 2–3 hr prior to plating. Plates are incubated at 30° for 4 days.

The efficiency of transformation of ligated DNA varies between 100 and 500 clones/μg of insert DNA. In our experiments we do not include polyamines in the transformation procedure,[12,23] since this can reduce the transformation efficiency; the absence of polyamines in the cells does not reduce the size of insert clones generated.

Protocol

1. Inoculate a single colony of AB1380 into 10 ml YPD (1% yeast extract, 2% Bacto-peptone, 2% dextrose, Difco, Detroit, MI) and let sit for 16 hr at 30°. Inoculate 200 ml of YPD with 200 μl of the 10-ml culture. Let shake for 16 hr at 30°.

[21] M. D. Rose and J. R. Broach, this series, Vol. 194, p. 195.
[22] R. Rothstein, *in* "DNA Cloning Volume II" (D. M. Glover, ed.), p. 45. IRL Press, Oxford, 1985.
[23] C. Connelly, M. K. McCormick, J. H. Shero, and P. Hieter, *Genomics* **10**, 10 (1991).

2. When the OD_{600} of a 1/10 dilution is between 0.1 and 0.15, split the culture into 50-ml Falcon tubes. Check some of the culture for bacterial contamination. Spin tubes at 3000 rpm for 5–10 min at 20°.

3. Decant media and resuspend pellets in 20 ml of distilled water for each tube. Spin at 3000 rpm for 5 min at 20°.

4. Decant media and resuspend pellets in 1 *M* sorbitol. Spin at 3000 rpm for 5 min at 20°.

5. Decant the sorbitol and resuspend pellets in 20 ml of SCE (1 *M* sorbitol, 0.1 *M* sodium citrate, pH 5.8, 10 m*M* EDTA, pH 7.5, and 30 m*M* 2-mercaptoethanol added fresh during transformation). Add 46 μl of 2-mercaptoethanol and take out 300 μl of one tube for a prelyticase control. Add 400–800 units of lyticase enzyme (L5263 Sigma), mix gently, and incubate at 30°. At 5, 10, 15, and 20 min, test the extent of spheroplasting of one tube by two independent methods:

   a. On the spectrophotometer measure the OD_{600} of a 1/10 dilution in distilled water.

   b. Check cells under the microscope. Mix 10 μl of cells with 10 μl of 2% SDS. When cells are dark, they are spheroplasted. Monitor cells until approximately 85% of cells have become spheroplasts. This should take between 15 and 20 min.

6. At this point cells should be immediately spun at 1000 rpm for 5 min at 20°. It is important not to allow the enzyme to remain with the cells any longer than necessary.

7. Decant the SCE and resuspend the pellets gently in 20 ml of 1 *M* sorbitol. Spin at 1000 rpm for 5 min at 20°.

8. Decant the sorbitol and resuspend pellets gently in 20 ml of STC (1 *M* sorbitol, 10 m*M* Tris-HCl, 10 m*M* $CaCl_2$). Count cells by making a 1/10 to 1/50 dilution of cells in STC. Spin cells for 5 min at 20°, then resuspend in the calculated volume of STC for the desired final concentration.

9. Aliquot 50 μl of genomic DNA into individual 15-ml clear Falcon tubes. Add 150 μl of spheroplasted cells so that the final cell concentration is $3.0–6.0 \times 10^8$/ml. For transformation controls use 10 ng YCp50,[21] 100 ng of digested and phosphatased pYAC4 or pYACRC vector, and a tube containing cells without DNA. Let DNA and cells sit for 10 min at 20°.

10. Add 1.5 ml of polyethylene glycol (PEG) (20% PEG 6000, 10 m*M* Tris-HCl, pH 7.5, 10 m*M* $CaCl_2$, prepared fresh prior to transformation and filter-sterilized) and mix gently by inverting the tubes. Let sit for a further 10 min at 20°.

11. Carefully pipette off the PEG without disturbing the pellet. Gently resuspend the pellet in 225 μl of SOS (1 *M* sorbitol, 25% YPD, 6.5 m*M*

$CaCl_2$, 10 μg/μl tryptophan, 1 μg/μl uracil; prepare fresh prior to transformation and filter-sterilize). Place at 30° for 30 min.

12. Add 5 ml of regeneration top agar [1 M sorbitol, 2% (w/v) dextrose, 0.67% (w/v) yeast nitrogen base without amino acids, amino acid supplements (as 10× stock: adenine, 200 μg/ml; arginine, 200 μg/ml; isoleucine, 200 μg/ml; histidine, 200 μg/ml; leucine, 600 μg/ml; lysine, 200 μg/ml; methionine, 200 μg/ml; phenylalanine, 500 μg/ml; tryptophan, 200 μg/ml, light-sensitive, filter-sterilize and store at 4°; valine, 1.5 mg/ml; tyrosine, 300 μg/ml, needs NaOH to go into solution), and 2% agar, kept at 48°] lacking uracil. Invert the tube quickly to mix and pour onto the surface of a prewarmed (37°) regeneration agar plate (solutions as for top agar). Incubate inverted plates at 30° for 3–4 days.

Rapid Colony Replication and Preparation of Filters

Another major problem in YAC library construction has been the inaccessibility of YAC clones after transformation, since regenerating yeast cells require a supportive agar matrix for optimal growth. In most laboratories, two procedures have been adopted to retrieve clones. First, to avoid amplification, clones have been individually picked into microtiter dishes in ordered arrays and transferred to filters for screening by colony hybridization.[18] Second, as an alternative to colony hybridization, pools of colonies containing YACs (prepared from mixing colonies grown on agar plates) have been screened by a polymerase chain reaction (PCR) procedure involving several rounds of purification.[24,25]

Our laboratory has circumvented these approaches by developing a rapid method of screening YAC clones embedded in the top agar. A multipin transfer device, consisting of an aluminum plate (22 × 22 cm) with 40,000 closely spaced machined pins [EMBL Workshop, and now available from Imperial Cancer Research Technology (ICRT), Sardinia House, Sardinia St., London, UK], can transfer the majority of YAC clones from within the agar matrix to the surface of another agar plate for eventual screening by colony hybridization. About 75% of the clones are transferred, but some clones that are in close proximity are replicated together and require a secondary screen by colony hybridization. High-density nylon filter lifts are prepared directly from the surface of these selective agar plates. For the large insert mouse library,[14] each filter contains clones approximately equivalent to three-fourths of genome cover-

[24] E. Heard, B. Davies, S. Feo, and M. Fried, *Nucleic Acids Res.* **17**, 5861 (1989).
[25] E. D. Green and M. V. Olson, *Proc. Natl. Acad. Sci. U.S.A.* **87**, 1213 (1990).

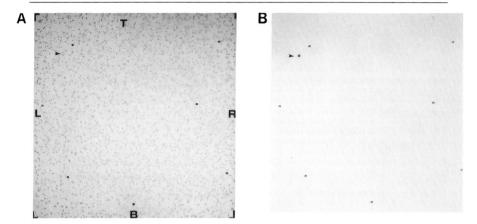

FIG. 1. (A) Random array of mouse YAC clones from the large insert *Eco*RI partial digest library replicated to the surface of a selective agar plate (lacking uracil and tryptophan) by the multipin transfer device. Arrow indicates the position of a positive clone. The plate contains about 5000 clones, equivalent to a one genome coverage with an average insert size of 600 kb. (B) Colony hybridization of a replicated filter with a single-copy probe from the mouse Y chromosome. An arrow indicates the positive clone that was picked from the corresponding plate (A).

age, such that a 3-fold coverage of the genome is represented on four filters (Fig. 1).

As an additional backup resource, YAC clones can be picked from the surface of the selective plates into microtiter dishes containing YPD and 15% glycerol for permanent storage at −70°. In our laboratory mouse YAC clones from the microtiter dishes have been spotted at high density onto filters by a robotic device in an ordered array[26] (compared to the replicated filters which have a random pattern). Each filter can contain either 9216 clones, equivalent to almost a 2-fold coverage of the genome, or 18,000 clones, equivalent to almost a 4-fold coverage of the genome (see this volume [38]). Pools of the same clones have also been prepared for screening by the polymerase chain reaction.

Protocol. The following procedures must be carried out in a hood to prevent fungal contamination.

1. Prepare several selective medium agar plates (2% dextrose, 0.67% yeast nitrogen base without amino acids, and amino acid supplements but

[26] D. Nizetic, G. Zehetner, A. P. Monaco, L. Gellen, B. D. Young, and H. Lehrach, *Proc. Natl. Acad. Sci. U.S.A.* **88,** 3233 (1991).

lacking tryptophan and uracil[22]); as many as eight copies of one pattern can be made by the device. Dry thoroughly under sterile conditions before use.

2. Dry the transformation plates (bottom agar plate inverted on top of its lid) while keeping cool at 4° for about 2 hr (cover the plates to prevent fungal contamination). This helps to compact the top surface by reducing the moisture content from the top agar, and it enables some clones to come closer to the surface.

3. Before use, sterilize the device by flaming in ethanol, then leave to cool in a bacterial hood. Press several transformation plates (as many as six) firmly onto the surface of the metal grid, taking care not to transfer agar pieces. Then place each selective plate in turn on top of the grid and press lightly and evenly over the entire plate such that the random pattern of clones is transferred to the surface only. It is important not to be heavy-handed so that clones grow on the surface only. This will ensure even colony lifts and enable clones to be more accessible for picking.

4. Incubate plates at 30° for 2 days. Do not allow clones to grow too large.

Colony Lifts

1. Mark each filter (Hybond N⁺, 22 × 22 cm; Amersham) at the top (T), bottom (B), right (R), and left (L) and mark the corresponding under-side of the replicated agar plate. Place the filter onto the plate in the correct orientation. Make marker holes and leave for about 10 min.

2. Lift the filter slowly off the plate using forceps, one at each diagonal corner. Flip the filter so that it is placed colony side up onto Whatman (Clifton, NJ) 3MM paper soaked in SCE containing Novozym (Nova Biolabs, Copenhagen, Denmark, 8 mg/ml final) and DTT to 10 mM (it is necessary to remove the yeast cell wall before processing the filters for hybridization). Leave filters in this condition in dishes (wrap to avoid evaporation) for 4–16 hr at 37°.

3. Filters can also be frozen for permanent storage at −70° in YPD and 30% glycerol between two polycarbonate 22 × 22 cm plates. Take a master filter containing colonies (marked with holes from the agar plate) and align on top a clean filter dampened in YPD and 30% glycerol. Place between glass plates and press firmly. This will give a duplicate of the desired colony. Copy marker holes from the master filter onto the other filter. In addition, prepare a transparent grid and record the position of the marker holes from the respective agar plate. This will assist in aligning positive YACs from the autoradiograph to the frozen filter. In a hood, retrive clones from the frozen filter (kept on dry ice) by cutting out the area of the filter containing the positive clone. Place both sides of the filter, colony side up, onto a YPD plate, and incubate at 30° for 2 days.

Store colonies from one filter at $-70°$ in YPD and 15% glycerol. The duplicate clones on the other filter should be diluted in YPD and replated on selective plates (lacking uracil and tryptophan) for a secondary screen by colony hybridization.

Filter Processing

1. Place filters containing lysed colonies onto Whatman 3MM paper soaked in denaturant (0.5 M NaOH, 1.5 M NaCl) for 5 min, then allow to air-dry for 5 min on fresh Whatman 3MM paper to remove the remaining denaturant. This helps to dry the colonies down onto the filter.

2. Float filters on top of neutralization solution (1 M Tris-HCl, pH 7.6, and 1.5 M NaCl) for 5 min and then for a further 5 min on 0.1 M Tris-HCl and 0.15 M NaCl before briefly submerging filters for a final 2 min. Place filters in fresh 0.1 M Tris-HCl, 0.15 M NaCl, and 0.25 mg/ml proteinase K and incubate at 42° (rocking) for 60 min.

3. Wipe off any remaining colony debris with a soft tissue soaked in 0.1 M Tris-HCl and 0.15 M NaCl, then rinse in this solution before air-drying on Whatman 3MM paper. Bake filters in a vacuum oven at 80° for 10–20 min and cross-link under UV light for 2 min (for Hybond N$^+$ filters this is optional).

Filter Hybridization and Screening

Filters containing YAC colonies and yeast DNA from PFGE transfers are hybridized in formamide buffer (50% formamide, 2× SSC, 50 mM sodium phosphate, pH 7.2, 1 mM EDTA, 8% dextran sulfate, 1% SDS, 10× Denhardt's solution[27]) at 42°, in the presence of yeast RNA (0.025 mg/ml) and salmon sperm DNA to avoid cross-hybridization to yeast RNA and yeast DNA on the filters.

1. Radiolabel approximately 20–40 ng of insert DNA (free of contaminating pBR322 plasmid) by random priming[28] and hybridize to filters at a concentration of 0.5–1 $\times$ 10^6 counts/min (cpm)/ml in formamide buffer 42° for at least 16 hr.

2. Wash filters once in cold 0.1× SSC and 1% SDS and 2 times at 65° while rocking for 20 min. Expose filters to autoradiographic film (Kodak, Rochester, NY, X-AR) with an intensifying screen for 2–4 days, depending on the signal-to-noise ratio and the number of hybridizations previously done on that set of filters. Generally, the filters give good signals for up

[27] T. Maniatis, E. F. Fritsch, and J. Sambrook, "Molecular Cloning: A Laboratory Manual." Cold Spring Harbor Laboratory, Cold Spring Harbor, New York, 1982.
[28] A. P. Feinberg and B. Vogelstein, *Anal. Biochem.* **132,** 6 (1984).

to 15 hybridizations after exposure to autoradiographic film for 3 days. Any further hybridizations tend to require a longer exposure.

3. Identify positive clones by aligning the autoradiographic film containing marker holes from the filter to the same marker holes on the respective agar plate. Pick positive clones directly from the selective agar plates either onto selective agar plates (lacking uracil and tryptophan) or, after 2 months, onto YPD agar plates before transferring to the selective plates (a greater nutrient source is needed since after time clones lose the ability to grow directly under selection).

4. Screen colonies from the secondary plates either by colony hybridization to detect the correct positive clone or by preparing cultures from single colonies grown in selective media for yeast chromosomal DNA minipreparations (see this volume [38]). DNA is subsequently analyzed by PFGE. Recombinant YACs are transferred to nylon filters and hybridized under the conditions described above to specific or mouse repeat sequence (e.g., B1/B2) probes to detect the presence and size of YACs.

[38] Genome Mapping and Cloning of Mutations Using Yeast Artificial Chromosomes

By ROGER D. COX, SEBASTIAN MEIER-EWERT,
MARK ROSS, ZOIA LARIN, ANTHONY P. MONACO,
and HANS LEHRACH

Introduction

The large insert size of yeast artificial chromosomes (YACs) makes it feasible to clone a mutation within 1 cM ($\sim 2 \times 10^6$ bp) of a DNA marker within two to three walks (assuming an average insert size of 700 kb), thus making YAC cloning a very attractive system. The Imperial Cancer Research Fund (ICRF) YAC reference library system (described below) allows distribution of libraries to many laboratories for screening and accumulates the data generated on probe distribution on each individual clone. Furthermore, the probe distribution pattern generated for each YAC will identify overlapping YACs and thus the clones that form a contiguous stretch of DNA over a whole chromosome. The consequence of this is that physical mapping information can be rapidly generated. The aim of this chapter is to describe the Imperial Cancer Research Fund

Copyright © 1993 by Academic Press, Inc.
All rights of reproduction in any form reserved.

YAC reference library system,[1] together with a set of protocols allowing the characterization of clones isolated from this library, and their use in the positional cloning of mutations.

Yeast Artificial Chromosome Library Arraying

YAC clones can be picked into and grown in microtiter plates, frozen at −70°, and used for arraying the clones on membranes for screening by colony hybridization. The method for analysis by hybridization is the same as described elsewhere (this volume [37]). Identified clones can then be picked from the corresponding microtiter plate wells and used for further analysis. The arrayed libraries provide a permanent source of clones.

Colony Picking

Clones can be picked either from replicated plates, the preparation of which was described previously (see this volume [37]), or from primary library plates. It is necessary for YAC clones to the grown under double selection [in medium lacking uracil and tryptophan (− Ura, − Trp)] at some point before long-term storage in microtiter plates, to select against clones containing only one vector arm. For this reason the two protocols for picking YAC clones are not the same. In our laboratory clones are picked using a twelve-pin wheel, mounted on a short handle, so that a colony can be picked onto each pin and then inoculated into one row of a microtiter plate (see Fig. 1A).

1. Prepare replica library plates on double selective medium (− Ura, − Trp), as described in [37].

2. Pick individual clones into separate 96-well microtiter plate wells, each filled with 100 μl growth medium per well. When picking from replicated plates grow in YPD medium (1% yeast extract, Difco, Detroit, MI, 1% Bacto-peptone, Difco, 2% glucose BDH), and when picking from primary plates grow in double selective medium (− Ura, − Trp).

3. Incubate at 30° for 24–48 hr.

4. For clones grown in double selective medium, inoculate all clones into a fresh microtiter plate containing 100 μl YPD medium per well and incubate at 30° for a further 24–48 hr. The inoculation is done using a transfer device which consists of 96 pins mounted on a plate so that the

---

[1] H. Lehrach, R. Drmanac, J. Hoheisel, Z. Larin, G. Lennon, A. P. Monaco, D. Nizetic, G. Zehetner, and A.-M. Poustka, *in* "Genome Analysis, Volume 1: Genetic and Physical Mapping," p. 39. Cold Spring Harbor Laboratory, Cold Spring Harbor, New York, 1990.

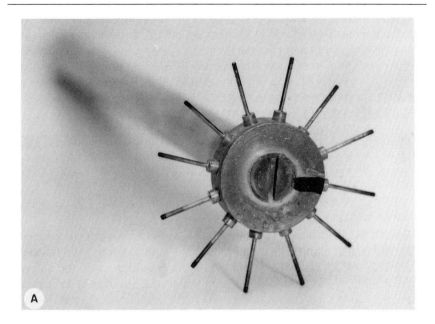

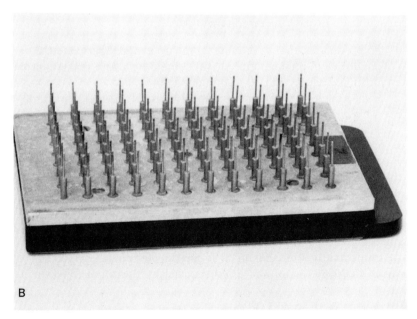

FIG. 1. (A) Twelve-pin inoculating wheel; (B) 96-pin transfer device.

pin array matches that of the microtiter plate wells (see Fig. 1B). The transfer device is sterilized by immersing in 70% (v/v) ethanol followed by drying on a hot plate. This step is included because clone regeneration after freezing is more successful when cells are stored in YPD medium.

5. Add 100 μl YPD plus 40% glycerol to all wells, mix into the grown culture, and place on dry ice until solid.

6. Store microtiter plates at $-70°$.

A duplicate copy of the microtiter plates should be made, to avoid contamination and decreased viability of original plates during repeated handling. This can be done in the same way as in Step 4.

Arraying Clones

In our laboratory we use a custom-made robotic machine that arrays 10,000 or 20,000 clones per membrane filter, simultaneously on up to six membranes of 22 × 22 cm size (see Fig. 2). The filters form the basis of the ICRF reference library system, by which the libraries in the system can be screened.[1,2] Presently there is a mouse YAC library in the system.[3] The following steps are performed by the robotic machine, requiring the operator to change the 96-well plates and 96-pin inoculating device after the inoculation of each membrane with each plate.

1. To produce each arrayed filter, prepare a large agar plate (22 × 22 cm) of single selection medium ($-$Ura) containing 0.25% calcium propionate (antifungal agent) and 50 μg/ml ampicillin.

2. Soak a piece of nylon membrane (Hybond N$^+$, Amersham) and two pieces of Whatman (Clifton, NJ) 3MM paper in single selection medium ($-$Ura) containing 0.25% (w/v) calcium propionate. Place the nylon membrane on top of the two Whatmann 3MM filters to help retain the moisture for the time taken to array all clones. Fix the membranes onto the supports on the bed of the robotic machine (see Fig. 2F). The membranes are held in place at the edges with metal frames (see Fig. 2F).

3. Shake the microtiter plates, sufficiently to resuspend the yeast cells but not to cross-contaminate the wells, using a vibrational shaker (Wesbart Engineering Ltd., IS89).

4. Immediately after shaking, a 96-pin transfer device, which has been sterilized in 70% ethanol and dried, is used by the robot to inoculate a

[2] M. T. Ross, J. D. Hoheisel, A. P. Monaco, Z. Larin, G. Zehetner, and H. Lehrach, *in* "Techniques for the Analysis of Complex Genomes" (R. Anand, ed.), p. 137. Academic Press, San Diego, 1992.

[3] Z. Larin, A. P. Monaco, and H. Lehrach, *Proc. Natl. Acad. Sci. U.S.A.* **88,** 4123 (1991).

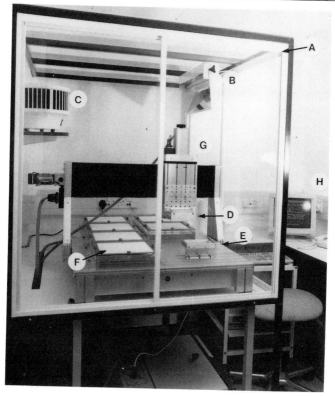

FIG. 2. (A) Perspex-enclosed aseptic cabinet; (B) UV lights; (C) air filter; (D) 96-pin transfer device holder mounted on the robot arm; (E) 96-well microtiter plate holder; (F) membrane supports on the bed of the robot; (G) robot arm capable of moving back and forth and side to side (x and y axes) as well as lifting the 96-pin transfer device up and down on its arm (z axis); (H) computer terminal controling the robot.

small number of cells from each well of a microtiter plate onto each of the prepared nylon membranes (Fig. 2D,E).

5. Repeat the process as many times as required to array all desired YAC clones. For 10,000- and 20,000-clone filters, 96- and 216-well microtiter plates, respectively, are arrayed onto each membrane by the robot (see Fig. 3). This process takes up to 5 hr (in the case of 20,000-clone filters), during which time the membranes are exposed to air. To help reduce fungal contamination we have built a closed cabinet (Fig. 2A) to house the spotting robot, which is UV-treated (Fig. 2B) for 10 min prior to beginning each spotting cycle.

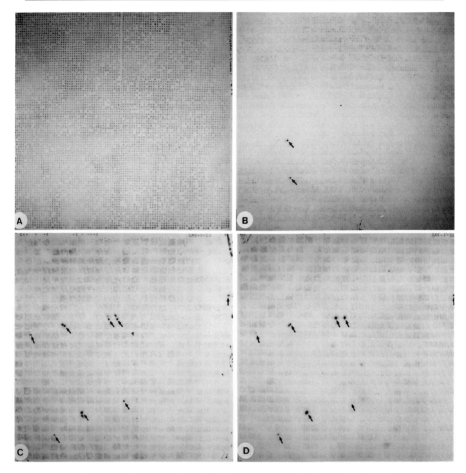

Fig. 3. (A) Autoradiograph of the 10,000-clone array hybridized with YAC vector. (B) Autoradiograph of 10,000-clone array hybridized with a single-copy DNA probe; two positive clones are marked by arrows. (C, D) Autoradiographs of 20,000-clone arrays hybridized with a single-copy probe; the filters were hybridized in duplicate, and duplicating positive clones are arrowed. Note the background hybridization to clones arranged in squares of 36 clones.

6. Transfer the membrane, inoculated side up, onto a single selection media plate (– Ura) prepared in Step 1, ensuring that there are no air bubbles between filter and plate.

7. Incubate the plate and membrane at 30° sufficiently long so that each inoculated clone has formed a colony, but has not merged with neighboring colonies (~2 days).

In Situ DNA Preparation on Arrayed Filters

1. Transfer membrane, colony side up, onto two pieces of Whatman 3MM paper soaked in SCE [1 M sorbitol; 100 mM sodium citrate, pH 5.8; 10 mM EDTA; 10 mM dithiothreitol (DTT)] plus 4 mg/ml NovoZym 234 (Novo Biolabs), avoiding any air bubbles between the membrane and Whatman 3MM paper.

2. Incubate at 37° for 16 hr. To prevent drying during incubation, seal the membrane and Whatman 3MM paper inside a plastic tray (e.g., Nunc bioassay plate). Good spheroplasting of yeast cells can be seen as the colonies change from semispherical and matte to flattened and shiny in appearance.

3. Transfer the membrane onto Whatman 3MM paper soaked in denaturant solution (0.5 M NaOH, 1.5 M NaCl) and leave at room temperature for 20 min.

4. Air-dry the membrane on fresh Whatman 3MM paper for 10 min, to absorb excess denaturant solution and dry the colonies to the membrane.

5. Float the membrane on neutralizing solution (1 M Tris-HCl, pH 7.6; 1.5 M NaCl) for 5 min.

6. Float the membrane on 0.10× neutralizing solution (0.1 M Tris-HCl, pH 7.6; 0.15 M NaCl) for 5 min.

7. Submerge the membrane in 0.10× neutralizing solution plus 0.25 mg/ml proteinase K (BDH, Poole, UK) for 1 hr at 37°, shaking gently at approximately 20 rpm.

8. Submerge the membrane in 50 mM sodium phosphate for 2 min.

9. Air-dry on Whatman 3MM paper.

10. Cross-link the DNA onto a nylon membrane by UV irradiation according to manufacturer's specification. Filters are now ready for prehybridization (see [37]).

Preliminary Characterization of Yeast Artificial Chromosomes

After a YAC clone of interest has been identified by hybridization with a probe, the well coordinate containing the clone is identified. Clones are picked from microtiter plates and distributed in YPD stabcultures. These should be streaked out on selective medium (−Ura, −Trp) as soon as possible (allowing 1–2 days growth at 30° from picking). If any difficulty is encountered in growing the cells, they can be streaked onto nonselective YPD medium first. After a few days of growth on plates (colonies should be a 2–4 mm in diameter) filter lifts are made and processed for hybridization as described above.

After screening with the probe used for isolation, single clones are picked for agarose plug DNA minipreparation. The use of plugs preserves the high molecular weight of the DNA. Clones are then analyzed by electrophoresis on pulsed-field gels (PFGE) to determine the size of the YACs and to ensure each clone contains only a single YAC. If cotransformation has occurred, then replate the YAC at clonal density (by serial dilution), rescreen by hybridization, and grow up several positive clones. If this fails to separate the YACs, then extract DNA from plugs and transform into yeast as described in [37]. To confirm that the YAC contains the region of interest, plugs should be digested with frequently cutting restriction enzymes, resolved by electrophoresis on agarose gels alongside digested mouse genomic DNA from the same strain, blotted, and hybridized with the probe used for screening (and any others from the region).

Agarose Plug Minipreparation

Protocols differ in the way in which the plugs are processed. We currently use two methods which employ NovoZym 234 (Novo Biolabs) to spheroplast the cells and either proteinase K and sarcosyl[4,5] or lithium dodecyl sulfate to extract the DNA.[4,6] Both methods are effective, although the former may be more expensive because of the use of proteinase K.

1. Inoculate 10–100 ml of −Ura, −Trp medium and incubate at 30° until the culture is saturated (5 × 10^7–10^8 cells/ml), about 24–48 hr.

2. Prepare a frozen aliquot of cells for permanent storage. Remove 0.5 ml and mix with 0.5 ml of 30% glycerol in YPD, freeze on dry ice, and store at −70°. Whenever fresh cells are required, scrape a small amount out of the frozen tube with an inoculating loop and streak onto a YPD plate. Return to −Ura, −Trp selection as soon as possible.

3. Centrifuge the remaining culture at 3000 rpm (Beckman, J6-B) for 10 min at room temperature.

4. Decant the supernatant and resuspend the cell pellet in 10 ml of TE50 (10 mM Tris-HCl and 50 mM EDTA at pH 7.5).

5. Centrifuge the cells at 3000 rpm for 10 min at room temperature.

6. Decant the supernatant and resuspend the pellet in 10 ml of SCE (1.0 M sorbitol; 0.1 M sodium citrate, pH 5.8; 10 mM EDTA, pH 7.5).

7. Centrifuge the cells at 3000 rpm for 10 min at room temperature.

[4] D. C. Schwartz and C. R. Cantor, *Cell* (*Cambridge, Mass.*) **37,** 67 (1984).

[5] Z. Larin and H. Lehrach, *Genet. Res.* **56,** 203 (1990).

[6] E. M. Southern, R. Anad, W. R. A. Brown, and D. S. Fletcher, *Nucleic Acids Res.* **15,** 5925 (1987).

8. Decant the supernatant and resuspend the pellet in SCE at 40 μl/ ml of original culture (5 × 10^7 cells). For more concentrated blocks increase this to 40 μl for 2 ml. The final yields will be approximately 1–2 μg of DNA per plug.

9. At this point have ready molten agarose and block formers. The molten 1.5% low melting point agarose (SeaPlaque GTG, FMC Bioproducts, Rockland, ME) is prepared in SCE with 8 mg/ml NovoZym 234 freshly added (it is easier to dissolve the enzyme in SCE before adding to the molten agarose). Keep at 50°. Block formers consist of 7.5 × 7 × 2 mm slots drilled through a Perspex sheet cut to convenient sizes (12 slots per former). It should be possible to purchase such formers from the suppliers of the PFGE system. Put plastic tape across one side of the block formers to seal the holes and place on a glass plate sitting on ice to facilitate setting of the plugs.

10. Divide the cell suspension into 500-μl aliquots in 1.5-ml Eppendorf tubes. Then in turn add an equal volume of agarose and Novozym solution, rapidly invert to mix, and then pipette 80 μl into each slot of the block formers. Allow to set on ice for 30 min.

11. Remove the tape from the block formers and gently push out the plugs, using a large inoculating loop, into 50 ml of SCE plus 10 mM DTT (sufficient for up to 100 blocks) in a 50-ml Falcon tube. To allow the cells to spheroplast by incubating at 37° for 1 hr, inverting occasionally.

12. To lyse the cells transfer the plugs into 50 ml of solution containing proteinase K or lithium dodecyl sulfate.

a. Proteinase K method. The lysis solution is 0.4 M EDTA, pH 7.5, 1% sarkosyl, and 2 mg/ml proteinase K. Incubate overnight at 50° with gentle rocking.

b. Lithium method. The lysis solution contains 1% lithium dodecyl sulfate, 100 mM EDTA, 10 mM Tris-HCl, pH 8. Incubate at 37° for 1 hr. Replace with fresh solution and incubate overnight at 37°.

13. Proceed to washing of the plugs.

a. Proteinase K method. The plugs can be stored in the proteinase K solution at 4° indefinitely. Before use the plugs must be washed and the proteinase K inactivated. Wash the plugs twice in TE50 for 30 min at 50°, then twice again with 0.04 mg/ml phenylmethylsulfonyl fluoride in TE50 (PMSF, make up as a 1000× stock in ethanol). (*Caution:* PMSF is extremely toxic), and finally again twice in TE50 at room temperature.

b. Lithium method. Wash the plugs four times in TE50 at room temperature.

Before the plugs can be used they should be washed (2 times for 30 min each) into TE (10 mM Tris-HCl, pH 7.5; 1 mM EDTA). Wash only the number required since it is better to store the plugs in a higher EDTA concentration (at least 50 mM).

Determining Sizes of Yeast Artificial Chromosomes

1. We use the Bio-Rad (Richmond, CA) clamped homogeneous electric fields (CHEF) system for PFGE. Size fractionation is carried out on a 1% agarose (SeaKem GTG, FMC Bioproducts) gel in 0.5× TBE (5× is 54 g Tris base, 27.5 g boric acid, and 20 ml of 0.5 M EDTA, pH 8.0). Size-fractionate the YACs with *Saccharomyces cerevisiae* (commercially available YP148 is useful because it contains pBR322 sequences in two of its chromosomes, allowing detection on blots) and λ DNA ladders (FMC Bioproducts) as size markers. Switching times of 40 sec for 16 hr, followed by 80 sec for 12 hr, and then 110 sec for 10 hr at 5 V/cm will yield a good size separation (90–1580 kb).

2. Transfer the DNA by Southern blotting to a nylon membrane (following the manufacturer's instructions) and prepare for hybridization.

The blot should be hybridized with the probe used to isolate the YAC. This is to determine the size of the YAC, confirm that it is a single chromosome, and judge the quality of the plugs. The blot should then be probed with labeled mouse genomic DNA, or mouse repeats (B1, B2, and R elements), in order to detect any additional YACs (cotransformants) present in the clone.

Walking with Yeast Artificial Chromosomes

The resolution of the mouse genetic map is about 3 cM[7,8] and increasing, and it should reach 1 cM soon. At this level of resolution, genetic mapping complements physical mapping by PFGE and YAC cloning, making it feasible to walk to any genetically mapped mutation from the nearest DNA marker. It is clear that long-range cloning techniques are going to make a major contribution to the cloning of developmentally important mutations.

Having isolated a YAC in the region of interest, one has to orientate it relative to the mouse centromere, generate new probes for isolating more overlapping YAC clones, and generate polymorphic markers for genetic mapping on crosses carrying the mutation of interest. It is important to be aware of the problem of chimerism when using YACs; this may arise by a process of homologous recombination in yeast when two or

[7] R. D. Cox and H. Lehrach, *BioEssays* **13**, 193 (1991).
[8] N. G. Copeland and N. A. Jenkins, *Trends Genet.* **7**, 113 (1991).

more DNA fragments are cotransformed into the host yeast strain,[9] or it may be due to coligation. Consequently it is useful to have several YACs from a region and to map any probes isolated using genetics and somatic cell hybrids.

Partial Restriction Digest Mapping

Partial restriction digest mapping of YACs with rare cutting restriction enzymes allows correlation with the physical map constructed from genomic DNA, and it allows probes to be physically mapped onto the YAC. Yeast DNA is unmethylated; therefore, more sites are cut in YAC DNA than in methylated genomic DNA isolated from mouse tissues, but it is still possible to make correlations with genomic DNA. The pYAC4 vector contains different parts of pBR322 in each of its arms: the right arm (URA3) can be detected with a 1.4-kb *Pvu*II–*Sal*I digest DNA fragment of pBR322, and the left arm (TRP, AMP[r]) can be detected with the 2.3-kb *Pvu*II–*Eco*RI digest DNA fragment of pBR322.[10]

Any rare cutting restriction enzyme can be used, but a certain amount of optimization must be carried out for each, and for each batch of enzyme. The following method uses *Bss*HII as and example.

1. Equilibrate one plug in TE.
2. Cut the block into quarters and place in separate 1.5-ml Eppendorf tubes. The digests are carried out in a final volume of 200 μl which includes the volume of the agarose plug (20 μl). Add to each tube 180 μl of buffer composed of 20 μl of 10$\times$ *Bss*HII buffer (New England Biolabs or the buffer recommended by the enzyme supplier) and 50 μg of acetylated bovine serum albumin (BSA), plus sterile distilled water.
3. Allow to equilibrate on ice for 30 min.
4. Add the enzyme diluted in 1$\times$ restriction buffer to each tube and mix. For *Bss*HII use 20, 0.5, 0.15, and 0.05 units. A range of concentrations must be used in order to detect all of the partial digest products efficiently. Allow to equilibrate for 30 min on ice.
5. Transfer the tubes to a water bath at the recommended digest temperature, 50° for *Bss*HII, for 1 hr.
6. Add 0.5 μl of 0.5 M EDTA and mix. Add 1 ml of TE 50 and place the tubes on ice.
7. Load the digests onto an agarose gel for PFGE as soon as possible.

The following enzyme concentrations will yield partial digests in the above protocol but should be optimized by a titration of enzyme concentra-

[9] E. D. Green, H. C. Riethman, J. E. Dutchik, and M. V. Olson, *Genomics* **11,** 658 (1991).
[10] D. T. Burke, G. F. Carle, and M. V. Olson, *Science* **236,** 806 (1987).

tions: *Mlu*I, 0.3, 1.0, 10, and 20 units; *Nru*I, *Sal*I, *Sac*II, and *Not*I, all at 1, 0.1, and 0.05 units. The enzyme *Sfi*I gives only partial digestion even at 20 units in the above protocol, cutting very little at lower concentrations.

Two pulsed-field gels (14 × 12.7 cm) should be run for accuracy, one to fractionate DNA between 90 and 1580 kb and one to fractionate DNA between 10 and 300 kb. Both should be 1% agarose gels in 0.5× TBE, run at 14°. For the former, switching times on the Bio-Rad CHEF system are 40 sec for 16 hr, 80 sec for 12 hr, followed by 110 sec for 10 hr at 5 V/cm. For the latter a linear ramp of switching times from 0.47 to 26.29 sec for 21 hr at 6 V/cm is used. In addition to λ ladders and *S. cerevisiae* markers, λ *Hin*dIII markers should be included on the latter gel.

After blotting, radiolabeled right and left pBR322 arm probes are hybridized to the filters. Maps are easily constructed from the sizes of the fragments detected. We have found that there are strain differences in the partial maps, reflecting the fact that the majority of the sites *in vivo* are methylated and therefore prone to mutation. Other probes can be positioned on the map by probing these blots. In addition to generating fragments by partial digestion of YAC DNA, fragments produced by complete digestion (this can be achieved by digesting with 20 units of enzyme for 6 hr or preferably overnight) are useful for positioning probes on the map of the YAC, especially when there is a high density of probes in the region.

Generation of Probes from Yeast Artificial Chromosomes

One of the first aims in chromosome walking projects is to isolate probes from the ends of the YAC in order to screen and isolate the next set of overlapping clones. Below are described two common methods used to achieve this, followed by a useful method for generating probes at random. Other methods are discussed at the end.

It is worth noting at this point that probes often require competition with genomic DNA in order to suppress the hybridization of repetitive elements. A simple protocol is as follows.[11,12]

1. To a radiolabeled probe in 100 μl TE add 100 μg of sonicated (average size ~300 bp) mouse genomic DNA, 100 μg of yeast tRNA, and TE to a final volume of 200 μl (or less).

2. Denature at 100° for 15 min.

3. Add 28 μl of 1 *M* sodium phosphate buffer, pH 7.2 (final concentration of 0.12 *M*).

[11] P. G. Sealey, P. A. Whittaker, and E. M. Southern, *Nucleic Acids Res.* **13**, 1905 (1985).
[12] M. Litt and R. L. White, *Proc. Natl. Acad. Sci. U.S.A.* **82**, 6206 (1985).

4. Incubate at 65° for 1–2 hr.
5. Add competed probe to hybridization buffer of your choice.

Plasmid Rescue Method

The pYAC4 vector has the pBR322 origin of replication and ampicillin resistance gene in its left arm.[10] The telomere sequence can be digested off this arm by *Xho*I digestion, and, providing that there is another such site in the insert within a reasonable distance of the vector, a plasmid can be "rescued" by ligating the ends of the DNA fragment carrying the vector and some insert DNA together and transfecting into a bacterial host.

1. Equilibrate one plug in TE.
2. Digest the plug with 50 units of *Xho*I in a 200 μl reaction (including the 80 μl plug) with the buffer recommended by the enzyme supplier. Digest at 37° for 3 hr.
3. Wash the plug in 50 ml of TE for 30–60 min.
4. Transfer to a clean 1.5-ml Eppendorf tube and equilibrate with 1 ml of 1× T4 DNA ligase buffer (as per enzyme supplier, but without ATP) at 4° for 1 hr.
5. Then add 100 μl of fresh 1× ligation buffer and melt the block at 68° for 15 min.
6. Mix and cool to 37° and then add 1 μl of ATP (100 m*M* stock) and 400 units of T4 DNA ligase (1 μl of high activity, New England Biolabs). Incubate at 37° for 1 hr.
7. Heat at 68° for 15 min and then cool to 37°.
8. Add 20 units of agarase (Sigma, St. Louis, MO) and incubate at 37° for 1 hr.
9. Purify the DNA by phenol/chloroform extraction, followed by a chloroform extraction.
10. Ethanol-precipitate the DNA and wash the pellet in 70% ethanol.
11. Dry the pellet in a vacuum dryer and resuspend in 4 μl of sterile distilled water.
12. Transfect 2 μl of DNA by electroporation (we use the Bio-Rad Gene Pulser apparatus) into electrocompetent cells such as *Escherichia coli* XL1-Blue. After 1 hr of incubation in 1 ml of growth medium at 37°, plate onto ampicillin (50 μg/ml) plates.

The yield of colonies is very low (reflecting the low concentration of YAC ends). (*Note*: Yield could be increased by carrying out plasmid rescue on DNA from minipreparations of YAC clones; see end rescue in Other Methods for Isolating Probes and Mapping Yeast Artificial Chromosomes section, below.) Therefore, an efficient transformation system must be used. An alternative to agarase purification (Steps 7–10) is the Gene-clean II method (from Bio 101). To analyze the clones, DNA should be prepared and digested with EcoRI (the YAC cloning enzyme), XhoI (used to digest and rescue the plasmid), and HindIII (which produces a character-istic YAC vector band of about 1.9 kb). The probes generated should be mapped back to the YAC by hybridization to filters containing the YAC (with appropriate negative control YACs), and it should also be ensured that the probe maps to the correct mouse chromosome.

Vectorette Probes

Preparation of vectorette probes is described in detail by Riley et al.,[13] and a kit is available from Cambridge Research Biochemicals. This effective method uses a linker ligatable to a number of frequently cutting restriction enzyme sites. A YAC plug is digested and ligated with such a linker, and then vector-specific primers are used in polymerase chain reaction (PCR) amplifications with a primer specific to the linker. The design of the linker is such that priming from the linker cannot occur until the linker is copied by a vector-primed strand, ensuring that only the vector ends are amplified. One advantage of this method is that both ends of the YAC can be obtained, but it generally yields less DNA from the YAC insert than the rescue method.

We routinely prepare our own vectorettes. When doing this it is neces-sary to treat the top strand of the vectorette oligonucleotide with kinase and then anneal the two strands.[13,14] We also use slightly different PCR cycling conditions to those published.[13] The PCR conditions are as follows: one cycle of denaturing at 94° for 5 min; followed by 39 cycles of denaturing at 93° for 1 min, annealing at 65° for 1 min, and polymerization at 72° for 3 min; followed by a final polymerization at 72° for 5 min.

Generation of Probes by Interspersed Repetitive Element Polymerase Chain Reaction

The interspersed repetitive element polymerase chain reaction (IRS–PCR) is a technique that was first described in humans using the Alu repetitive element,[15] and it has also been described in mice.[7,16–19] This technique uses primers homologous to mouse repeats, which if close

[13] J. Riley, R. Butler, D. Ogilivie, R. Finniear, D. Jenner, S. Powell, R. Anand, J. C. Smith, and A. F. Markham, *Nucleic Acids Res.* **18,** 2887 (1990).

[14] R. W. Wu, T. Wu, and A. Ray, this series, Vol. 152 [38].

[15] D. L. Nelson, S. A. Ledbetter, L. Corbo, M. F. Victoria, R. Ramirez-Solis, T. D. Webster, D. H. Ledbetter, and T. Caskey, *Proc. Natl. Acad. Sci. U.S.A.* **86,** 6686 (1989).

[16] R. D. Cox, N. G. Copeland, N. A. Jenkins, and H. Lehrach, *Genomics* **10,** 375 (1991).

enough together and in the appropriate orientation will allow amplification of the intervening (usualy unique) DNA. We routinely use the following primers: B1, GTG CGG CCG CCT GGA ACT CAC TCT GAA GAC; B2A, TAG ACG CGG CCG CTC TTC TGG AGT GTC TGA AGA; B2B, TAG ACG CGG CCG CGA CTG CTC TTC CGA AGG TCC; and R2, AGC ATT TGA AAT GTA AAT GA. Full descriptions of these primers can be found in the references. Primers are used singly or in pairs.[18] Under appropriate conditions these primers easily distinguish mouse DNA from yeast DNA. The advantage of IRS–PCR is that probes can be generated from YACs at random without having to digest or clone DNA. Sufficient product can be generated to use directly for labeling as a probe. Probes are easily made by purifying the products on low melting point agarose gels, excising the desired bands, and labeling directly.[20,21] The competition of repeat sequence DNA should be carried out as described above.

Comparison of products from different YACs can provide "fingerprint" information, especially if combined with hybridization. This allows a quick assessment of the similarity of YACs isolated using a particular probe.

This method can be combined with the vectorette method described earlier to generate end (combining vector and IRS primers) and internal YAC probes (combining vectorette and IRS primers), although higher annealing temperatures need to be used to ensure specificity of the vector and vectorette primers. As with any probes generated from YACs they should be mapped back to the YAC and on chromosome mapping panels.

1. Equilibrate one plug in TE.

2. Melt at 65° for 15 min and then dilute in 400 μl of sterile distilled water. At this point, the sample can be stored frozen at −20° and thawed when required.

3. Set up PCR reactions. For a 100 μl reaction, use 5–10 μl of YAC DNA (from Step 2 above), 10 μl of 10× PCR buffer [100 mM Tris-HCl, pH 8.3, 500 mM potassium chloride, 15 mM magnesium chloride, 0.1% gelatin: standard Cetus (Norwalk, CT) buffer], 10 μl of 10× dNTPs [2 mM of each of the four deoxynucleotide triphosphates (Pharmacia, Piscataway NJ), ultrapure], 1 μl of oligonucleotide primer (1 μg; when using two primers double this quantity), sterile distilled water to 99.5 μl final volume, and finally 0.5 μl of AmpliTaq polymerase (Cetus; Other thermostable

[17] R. D. Cox, L. Stubbs, T. Evans, and H. Lehrach, *Nucleic Acids Res.* **19**, 2503 (1991).
[18] M. C. Simmler, R. D. Cox, and P. Avner, *Genomics* **10**, 770 (1991).
[19] N. G. Irving and S. D. M. Brown, *Genomics* **11**, 679 (1991).
[20] C. P. Hodgson and R. Z. Fisk, *Nucleic Acids Res.* **15**, 6295 (1987).
[21] F. Cobianchi and S. H. Wilson, this series, Vol. 152 [10].

polymerases can be used. When using cycling systems other than the Cetus 9600 (which uses its own PCR tubes), add 50 μl of light mineral oil (Sigma).

4. Cycling conditions are dependent on the machine being used, and should be optimized accordingly. These reactions will maintain specificity over a range of temperatures. Always run a yeast DNA negative control. For the Cetus 9600 system we use an initial denaturing cycle of 94° for 5 min, followed by 35–40 cycles of denaturing at 94° for 45 sec, annealing at 54° for 1 min, and polymerization at 72° for 3 min; this is followed by a final cycle at 72° for 5 min and then refrigeration at 4°.

5. Analyze the products by electrophoresis (10 μl of each reaction) on a 1.5% agarose gel.

If there is a lack of specificity in the PCR reactions, adjust the annealing temperature upward. The concentration of nucleotides and primers can also be reduced to increase specificity. Conversely, more products can be generated by relaxing the stringency of the PCR reactions.

Other Methods for Isolating Probes and Mapping Yeast Artificial Chromosomes

A disadvantage of the pYAC4 vector is the inability to rescue both ends of a YAC as a plasmid; having only a single choice of restriction enzyme for the one end (left end) that can be rescued is also a drawback. A method has recently been described to replace sequences in the vector arms of YACs, by homologous recombination in yeast, with rescue vector sequences.[22] This modification of the YAC arms allows rescue of the ends of the YAC using a selection of enzymes provided in the rescue vector polylinker. We find this method to be very useful.

One could also consider making cosmid or λ libraries[23,24] from the YAC. A library can be prepared from the whole yeast plus YAC genome and mouse clones from the YAC identified by hybridization with repetitive DNA. End clones can be identified by hybridization with pBR322. Another advantage of doing this is that it provides DNA of a more manageable size for other experiments such as screening zoo blots (Southern blots of restrict enzyme digested DNA from a selection of species; for example chicken, pig, bovine, human, hamster, any marsupial, yeast, dog, rat, and fruit fly) for conserved sequences (a sequence present in several species, especially if the evolutionary distance between them is large, indicating

[22] G. G. Hermanson, M. F. Hoekstra, D. L. McElligott, and G. A. Evans, *Nucleic Acids Res.* **19**, 4943 (1991).
[23] A.-M. Frischauf, this series, Vol. 152 [16, 17].
[24] A. G. DiLella and S. L. C. Woo, this series, Vol. 152 [18].

selective pressure for conservation which is most likely to be associated with functional sequences) and exon trapping procedures.[25] If a cosmid library is available from a cell hybrid containing the chromosome of interest, and the clones containing mouse DNA have been isolated, it is possible to use whole YACs as a probe.[26] Similarly, it may be possible to use a whole YAC to probe a cDNA library.[27]

Future Prospects

Since preparing this chapter our laboratory has made several improvements to arraying YAC clones. Clones are now replicated for spotting into 384 well microtiter plates (Genetix, 'Q'-system), and spotted using a 384 pin device, thus increasing fourfold the number of clones arrayed per robot cycle. A new robot has also been built, capable of spotting 12 membranes simultaneously in about 3 hr. In order to facilitate clone identification each YAC is spotted in duplicate within a 3 × 3 array (total 8 clones, the center position being empty) and each pair of spots is oriented within the array so that they produce and an unique pattern/orientation of signals. These membranes are available on request.

We have also developed library scale PCR technology, opening the way to IRS–PCR based genome wide continuing experiments.[28]

[25] A. J. Buckler, D. D. Chang, S. L. Graw, J. D. Brook, D. A. Haber, P. A. Sharp, and D. E. Housman, *Proc. Natl. Acad. Sci. U.S.A.* **88,** 4005 (1991).
[26] S. Baxendale, G. P. Bates, M. E. MacDonald, J. F. Gusella, and H. Lehrach, *Nucleic Acids Res.* **19,** 6651 (1991).
[27] P. Elvin, G. Slynn, D. Black, A. Graham, R. Butler, J. Riley, R. Anand, and A. F. Markham, *Nucleic Acids Res.* **18,** 3914 (1990).
[28] S. Meier-Ewert, E. Maier, A. Ahmadi, J. Curtis, and H. Lehrach, *Nature (London)* **361,** 375 (1992).

[39] Cloning Developmentally Regulated Gene Families

By Brian J. Gavin and Andrew P. McMahon

Introduction

Over the past few years it has become increasingly clear that the principal means of evolution at the genomic level is through gene duplication and divergence. This has resulted in the realization that many genes once described as "single copy" are actually members of relatively large gene families. The primary method used to search for members of a gene family has been to screen either genomic or cDNA libraries at low

METHODS IN ENZYMOLOGY, VOL. 225

Copyright © 1993 by Academic Press, Inc.
All rights of reproduction in any form reserved.

stringency with a DNA probe encoding an amino acid motif that is believed to be conserved. This approach has been quite successful when the conserved motif is relatively large, as in the case of the homeotic genes.[1-3] However, it does have a number of disadvantages. In particular, one is relying on conservation of the DNA sequence used as a probe when in the majority of cases it is the encoded amino acid sequence that is being evolutionarily conserved. As a result, when the conserved motif within a gene family is relatively small, the degeneracy of the triplet code can make detection of other family members virtually impossible using this method.

In attempts to address this problem and reduce the amount of time required for this type of analysis, a number of groups have developed procedures for the identification of novel members of a gene family based on the polymerase chain reaction (PCR).[4] Typically, the problem of differential codon usage is eliminated by generating mixtures of oligonucleotides representing all possible combinations of codons encoding a conserved amino acid motif. These "degenerate" primers are then used to amplify related sequences from either genomic DNA or cDNA. This approach has been successfully employed to identify members of a variety of gene families including ion channels, pattern formation genes, and G-protein coupled receptors.[5,6] We have used a similar strategy in combination with conventional screening methods to identify six novel members of a family of genes which are related to the protooncogene *Wnt*-1.[7] This chapter focuses on the methodological considerations involved in this type of approach using the *Wnt* gene family as a paradigm.

Primer Selection

The first step in any screening procedure is the identification of a motif(s) which appears to be conserved between different members of a gene family and/or is conserved within a single family member across

[1] W. McGinnis, M. S. Levine, E. Hafen, A. Kuroiwa, and W. J. Gehring, *Nature (London)* **308,** 428 (1984).

[2] W. McGinnis, R. L. Garber, J. Wirz, A. Kuroiwa, and W. J. Gehring, *Cell (Cambridge, Mass.)* **37,** 403 (1984).

[3] W. McGinnis, C. P. Hart, W. J. Gehring, and F. H. Ruddle, *Cell (Cambridge, Mass.)* **38,** 675 (1984).

[4] R. K. Saiki, D. H. Gelfand, S. Stoffel, S. Scharf, R. Higuch, G. T. Horn, K. B. Mullis, and H. A. Erlich, *Science* **239,** 487 (1988).

[5] A. Kamb, M. Weir, B. Rudy, H. Varmus, and C. Kenyon, *Proc. Natl. Acad. Sci. U.S.A.* **86,** 4372 (1989).

[6] F. Libert, M. Parmentier, A. Lefort, C. Dinsart, J. VanSande, C. Maenhaut, M.-J. Simons, J. E. Dumont, and G. Vassart, *Science* **244,** 569 (1989).

[7] B. J. Gavin, J. A. McMahon, and A. P. McMahon, *Genes Dev.* **4,** 2319 (1990).

species. In the case of the *Wnt* gene family, comparison of the amino acid sequences of mouse,[8,9] *Drosophila*,[10] and *Xenopus*[11] *Wnt*-1 with that of mouse *Wnt*-2[12] revealed a number of short regions which are completely conserved both between family members and throughout evolution (Fig. 1). Many of these regions contain conserved cysteine residues believed to be involved in the tertiary structure of these proteins and thus may have a greater likelihood of being evolutionarily conserved in other family members.

Three motifs, denoted I, II, and III (Fig. 1), were selected for the generation of degenerate PCR primers based on both their sequence conservation and the presence of conserved cysteine residues. Four sets of primers were synthesized which corresponded to either the sense (domain I), antisense (domain III), or both (domain II) strands of these motifs (Fig. 2). In addition, the primers incorporate *Eco*RI and *Xba*I restriction enzyme sites at the 5' and 3' ends of the PCR products to allow directional cloning following amplification (Figs. 2, 3A).

There are a number of issues to consider when selecting the amino acid motifs on which degenerate primers will be based. First, the complexity of the degenerate primer mixture can be controlled by selecting conserved regions that consist primarily of amino acids encoded by a small number of codons. Second, it is important to consider the sequence conservation within the amplified region. If this sequence is completely divergent in known members of the family, it may be impossible to distinguish real amplification products from those generated through nonspecific hybridization of the primers. It is therefore helpful to have some degree of conservation in the intervening amino acid sequence. For example, in the *Wnt* gene family conserved cysteine residues within the amplified region were used as a framework for orientation of novel PCR products (see Fig. 4).

Polymerase Chain Reaction Amplification

Conditions for the amplification of sequences using degenerate primers are similar to those described by Kamb.[5] Standard reactions contain 200

[8] A. van Ooyen and R. Nusse, *Cell (Cambridge, Mass.)* **39**, 233 (1984).
[9] Y.-K. T. Fung, G. M. Shackleford, A. M. C. Brown, G. S. Sanders, and H. E. Varmus, *Mol. Cell. Biol.* **5**, 3337 (1985).
[10] F. Rijsewijk, M. Schuermann, E. Wagenaar, P. Parren, D. Weigel, and R. Nusse, *Cell (Cambridge, Mass.)* **50**, 649 (1987).
[11] J. Noordemeer, F. Meijlink, P. Verrijzer, F. Rijsewijk, and O. Destrée, *Nucleic Acids Res.* **17**, 11 (1989).
[12] J. A. McMahon and A. P. McMahon, *Development (Cambridge, UK)* **107**, 643 (1989).

```
                                                                       55
Mm   Wnt -1   MGLWALLPSWVSTTLL-LALTALPAALAANSSGR---WWGIVNIASSTNLLTDS
Xl   Wnt -1       MRI*TFLLGLK**WV**FSS*SNTI*V*N**K---*****V**AG*V*PG*
Dm   Wnt -1   MDISYIFVICLMALCSG-GSSLSQVEGKWWSGR**GSM****AKVGEPN*I-T--
Mm   Wnt -2       MNVPLGGIW*W-*P*LLTWLTPEVS**-----**YMRATGG*SRVMCD-

                                                                       110
Mm   Wnt -1   KSLQ--LVLEPSLQLLSRKQRRLIRQNPGILHSVSGGLQSAVRECKWGFRNRRWN
Xl   Wnt -1   DARPVp***D********-*K*********Q*ITR**H**I*****H*******
Dm   Wnt -1   -PIM--YMDPAIHST*R******V*D***V*GALVK*ANL*IS**QHQ*******
Mm   Wnt -2   NVPG--**---*----RQR*--*CHRH*DVMRAIGL*VAEWTA**QHQ**QH***

                                                                       165
Mm   Wnt -1   CPTAPGPH-L--FGKIVNR-GCRETAFIFAITSAGVTHSVARSCSEGSIESCTCD
Xl   Wnt -1   ***GT*NQ-V--***I**-********V**************************S**
Dm   Wnt -1   *S*RNFSRGKNL****D*-*****S**Y*****A****I**A*****T*******
Mm   Wnt -2   *N*LDRD*S*--**RVLL*SS-**S**VY**S****VFAIT*A**Q*ELK**S**

                                     ___I___                          220
Mm   Wnt -1   YRRRGP------------GGPDWHWGGCSDNIDFGR1FGREFVDSGEK-GRDLRF
Xl   Wnt -1   ******------------*********E***FI*******S*R-****KY
Dm   Wnt -1   *SHQSRSPQANHQAGSVA*VR**E********G**FK*S*****T**R-**N**E
Mm   Wnt -2   PKKKGSAKDSK-------*TF**--*********Y*IK*A*A***AK*RK*K*A*A

                                     ___II___                         275
Mm   Wnt -1   LMNLHNNEAGRTTVFSEMRQECKCHGMSGSCTVRTCWMRLPT1RAVGDVLRDRFD
Xl   Wnt -1   *V*****Q***L**LT************SL*******PF*S***A*K****
Dm   Wnt -1   K*********AH*GA*******************K******ANF*VI**N*KA***
Mm   Wnt -2   *******R***KA*KRFLK*******V*****L****LAMADF*KT*EY*WRKYN

                                                                       330
Mm   Wnt -1   GASRVLYGNRGSNRA-SRAE------------------------------------
Xl   Wnt -1   ***K*T*S*N***WG**SD------------------------------------
Dm   Wnt -1   **TR*QVT*--*L**-TN*LAPVSPNAAGSNSVGSNGLIIPQSGLVYGEEEERML
Mm   Wnt -2   **IQ*VMNQD*TGFT-VANK------------------------------------

                                                                       385
Mm   Wnt -1   -----------------------------------------------------LLRL
Xl   Wnt -1   -----------------------------------------------------PPH*
Dm   Wnt -1   NNDHMPDILLENSHPISKIHHPNMPSPNSLPQAGQRGGRNGRRQGRKHNRYHFQ*
Mm   Wnt -2   -----------------------------------------------------RFKK

                                                                       440
Mm   Wnt -1   EPEDPAHKPPSPHDLVYFEKSPNFCTYSQRLGTAGTAGRACNSSSPALDGCELLC
Xl   Wnt -1   ***N*T*AL**SQ***********SP*EKN**P**T**I***T*LG********
Dm   Wnt -1   N*HN*E****GSK***L*P**S**EKNL*Q*IL**H**Q**ET*LGV***G*M*
Mm   Wnt -2   -*T--------KN******N**DY*IRDREA*SL****V**LT*RGM*S**VM*

                              ___III___                                492
Mm   Wnt -1   CGRGHRTRTQRVTERCNCTFHWCCHVSCRNCTHTRVLHECL-----------
Xl   Wnt -1   ****Y*SLAEK****H***N*****T*L***SSQIV***-----------
Dm   Wnt -1   ****YD*SHVTRMTK*E*K*****A*R*QD*LEALDV*T*KAPKSADWATPT
Mm   Wnt -2   ****YD*SHVTRMTK*E*K*****A*R*QD*LEALDV*T*KAPKSADWATPT
```

FIG. 1. Comparison of the amino acid sequences of *Drosophila, Xenopus,* and mouse *Wnt*-1 with mouse *Wnt*-2. Identical amino acids are indicated by asterisks, gaps by dashes. Conserved motifs used for the generation of degenerate primers are denoted I, II, and III.

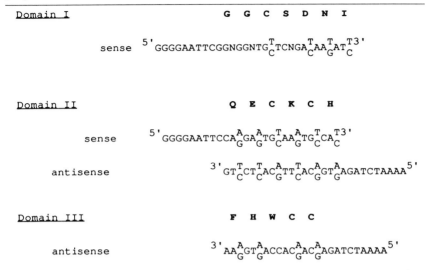

FIG. 2. Amino acid sequences of domains I, II, and III and nucleotide sequences of the corresponding degenerate primers.

ng of each primer in a final volume of 20 µl of 50 mM KCl, 10 mM Tris-HCl (pH 8.4), 1.5 mM MgCl$_2$, 20 µg/ml gelatin, 0.2 mM dNTPs (Pharmacia, Piscataway, NJ) and 0.6 units Taq polymerase (Perkin-Elmer/Cetus, Norwalk, CT). Reactions are overlaid with 25 µl of mineral oil and amplified using a thermal cycler (M-J Research, Watertown, MA) under the following conditions: 2 min at 94°; 30 cycles of 90 sec at 55°, 1 min at 72°, and 1 min at 94°; 30 min at 37°. Following amplification, the reactions are analyzed by gel electrophoresis of 5 µl of the reaction mix on 2% agarose gels. Templates consist of either 1 ng of control Wnt-1 or Wnt-2 containing plasmid DNA or 1/40 of a cDNA reaction. First-strand cDNA is synthesized in a total volume of 50 µl from 10 µg of total 9.5-day fetal mouse RNA using an oligo(dT) priming kit (Invitrogen, San Diego, CA), as recommended by the manufacturer.

The degenerate primers corresponding to domains I–III were tested in pairwise combinations for their ability to amplify sequences from plasmid DNA containing either Wnt-1 or Wnt-2 cDNA sequences. These experiments revealed clear differences in the efficiency of amplification by each primer set (data not shown) and emphasize the need to test multiple primers in this type of analysis. Of the primers tested, the combination of the domain II sense-strand and domain III primers gave the best signal-to-noise ratio and is thus used for subsequent analyses.

Amplification of plasmids containing Wnt-1 and Wnt-2 cDNAs with the II/III primer mixture generated the expected products of 434 and 406

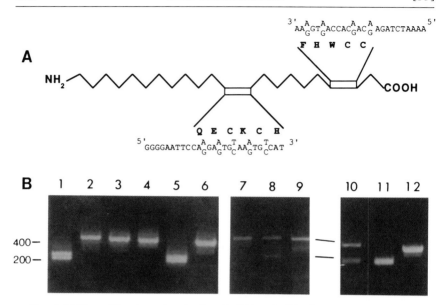

FIG. 3. PCR amplification using the domain II/domain III primer mixture. (A) Position of domains II and III within the *Wnt*-1 sequence and degenerate primers encoding these regions. (B) Restriction enzyme analyses of amplification products. Templates were as follows: lanes 1–3, *Wnt*-1-containing plasmid; lanes 4–6, *Wnt*-2-containing plasmid; lanes 7–10, 9.5-day fetal cDNA; lanes 11 and 12, *Wnt*-1 and *Wnt*-2 plasmid mix. Amplified DNAs were digested with the following restriction enzymes: *Acc*I (lanes 1, 4, and 7); *Eco*RI (lanes 2, 5, and 8); *Acc*I and *Eco*RI (lanes 10 and 11); undigested (lanes 3, 6, 9, and 12). (Reproduced by permission of Cold Spring Harbor Laboratory Press from Gavin *et al.*[7])

bp, respectively (Fig. 3B, lanes 3 and 6). The fidelity of the PCR reactions were assessed by digesting the amplification products with restriction enzymes specific for each gene. As predicted from the restriction maps of these cDNAs, digestion of the *Wnt*-1 amplification product with AccI yields two bands of 233 and 201 bp (Fig. 3B, lane 1), whereas in the case of *Wnt*-2 digestion with *Eco*RI reduces the amplification product to two bands of 216 and 190 bp (Fig. 3B, lane 5). In each case, the reciprocal reaction shows no effect (Fig. 3B, lanes 2 and 4).

In an attempt to identify novel *Wnt*-1-related genes that might play a role in development, 9.5-day mouse fetal cDNA was used as a template for amplification by the II/III primer mixture. *Wnt*-1 and *Wnt*-2 are both expressed at this stage of mouse development and thus serve as internal controls for the reaction. Amplification of the cDNA template generates a major band similar in size to *Wnt*-1 and *Wnt*-2 and a variable minor artifactual band. To determine quickly whether novel sequences had been

```
          II
Wnt - 4   VECKCHGVSG SCEVKTCWRA VPPFRQVGHA LKEKFDGATE VEPRRVGSSR
Wnt - 5a  VACKCHGVSG SCSLKTCWLQ LADFRKVGDA LKEKYDSAAA MRLNSRG---
Wnt - 6   TECKCHGLSG SCALSTCWQK LPPFREVGAR LLERFHGASR VMGTNDGKA-
Wnt - 7a  LECKCHGVSG SCTTKTCWTT LPQFRELGYV LKDRYNEAVH V--EPVRASR


Wnt - 4   A-------LV PRNAQFKPHT DEDLVYDEPS PDFCEQDIRS GVLGTRGRTC
Wnt - 5a  -------KLV QVNSRFNSPT TQDLVYIDPS PDYCVRNEST GSLGTQGRLC
Wnt - 6   --------LL PAVRTLKPPG RADLLYAADS PDFCAPNRRT GSPGTRGRAC
Wnt - 7a  NKRPTFLKIK KPLSYRKPMD T-DLVYIELS PNYCEEDPVT GSVGTQGRAC

                                           III
Wnt - 4   NKTSKAIDGC ELLCCGRGFH TAQVELAERC GCRFHWCC
Wnt - 5a  NKTSEGMDGC ELMCCGRGYD QFKTVQTERC HCKFHWCC
Wnt - 6   NSSAPDLSGC DLLCCGRGHR QESVQLEENC LCRFHWCC
Wnt - 7a  NKTAPQASGC DLMCCGRGYN THQYARVWQC NCKFHWCC
```

FIG. 4. Comparison of the amino acid sequence encoded by the *Wnt*-4, *Wnt*-5a, *Wnt*-6, and *Wnt*-7a amplification products. Conserved cysteine residues are indicated in boldface type, and dashes denote gaps introduced to maximize alignments.

amplified, the reaction products were digested with *Eco*RI and *Acc*I. Under conditions that result in complete digestion of a control reaction (Fig. 3B, lane 11), a fraction of the amplified product remains undigested (Fig. 3B, lane 10), suggesting the primers had indeed amplified sequences other than *Wnt*-1 and *Wnt*-2.

Cloning and Analysis of Amplified Products

The cDNA amplification reactions are brought to 100 μl with water, extracted once with aqueous phenol, extracted once with chloroform/ isoamyl alcohol (24 : 1), and precipitated by the addition of 1/10 volume of 3 M sodium acetate and 2 volumes of 100% ethanol. Ethanol precipitates were recovered by centrifugation at 13,000 g for 5 min, washed with 100 μl of 70% (v/v) ethanol, dried under vacuum, and dissolved in 10 μl water. The reaction products were next digested for 4 hr at 37° with 20 units each of *Xba*I and *Eco*RI, purified and precipitated as described above, and dissolved in 15 μl water. Then 7.5 μl of the digested DNA was cloned into *Xba*I/*Eco*RI-cut pGEM3Zf(+) (Promega, Madison, WI) and recombinant clones identified by restriction analysis of alkaline lysis mini-preparation[13] DNA. Clones that contained inserts were then analyzed by double-stranded dideoxy sequencing using a T7 polymerase kit (Pharmacia) as recommended by the manufacturer.

[13] T. Maniatis, E. F. Fritsch, and J. Sambrook, "Molecular Cloning: A Laboratory Manual." Cold Spring Harbor Laboratory, Cold Spring Harbor, New York, 1982.

TABLE I
INSERT DISTRIBUTION OF *Wnt*-1-RELATED
PCR CLONES[a]

Gene	Number of isolates[b]
Wnt-1/*int*-1	1
Wnt-2/*irp*	13
Wnt-4	1
Wnt-5a	2
Wnt-6	5
Wnt-7a	3

[a] Reproduced by Permission of Cold Spring Harbor Laboratory Press from Gavin *et al.*[7])
[b] Of 80 colonies screened, 28 contained inserts, 25 of which were *Wnt*-1-related.

Colonies generated from cloning of the cDNA amplification products were analyzed for the presence of insert by digestion with *Eco*RI and *Xba*I. Of 80 colonies examined, 28 contained inserts greater than 100 bp. At this stage of the analysis it is important to sequence all clones that contain inserts of reasonable size and not just those with inserts of the expected size. This takes into account the possibility of genetic changes within the amplified region such as insertions and deletions or the generation of an internal cleavage site(s) corresponding to one of the cloning sites introduced by the primers. Sequence analysis revealed that in addition to *Wnt*-1 and *Wnt*-2 the clones contained sequences representing four novel Wnt-related genes (see Table I), which were named *Wnt*-4, -5a, -6, and -7a. All of the novel clones contained a single long open reading frame and preserve the characteristic pattern of cysteine residues seen in *Wnt*-1 and *Wnt*-2 (Fig. 4). It is interesting to note that the sequence of the domain III motif, FHWCC, is absolutely conserved in all family members. In contrast, the domain II sequence is relatively divergent, showing conservation of only the 4 terminal amino acids, CKCH. Examination of the DNA sequences encoding this motif reveals that cDNA containing as little as 12 nucleotides of contiguous homology to the primer (Fig. 5) can be amplified using this procedure.

Isolation of Clones Containing Complete Coding Sequence

To isolate the remainder of the coding sequence for the *Wnt* genes identified by cDNA amplification, the PCR products were used to screen

	Q	E	C	K	C	H
CONSENSUS	C A ^G_A G A ^G_A T G ^C_T A A ^A_G T G ^T_C C A ^T_C

	V	E	C	K	C	H
Wnt - 4	g t G G A G T G C A A G T G T C A C

	V	A	C	K	C	H
Wnt - 5a	g t A G c c T G T A A G T G T C A T

	L	E	C	K	C	H
Wnt - 7a	C t G G A G T G T A A G T G C C A T

	T	E	C	K	C	H
Wnt - 6	a c c G A G T G T A A G T G C C A T

FIG. 5. Comparison of the amino acid and nucleotide sequences within domain II in the *Wnt* family members amplified from 9.5-day mouse fetal cDNA. The top line shows the conserved amino acid sequence noted in Fig. 1 and the nucleotide sequence of the corresponding degenerate primer. Lowercase letters in the amplified sequences indicate regions of mismatch with the primer.

an 8.5-day mouse fetal cDNA library.[14] Probes for screening the library were generated by amplification of 1 ng of plasmid DNA containing one of the four novel genes using the degenerate primers and reaction conditions described above. Following amplification, the reactions are extracted and precipitated as previously described and then dissolved in 20 μl LoTE (1 mM Tris-HCl, pH 8.0; 0.1 mM EDTA). One microliter of each reaction is then labeled with [α-^{32}P]dCTP using a random priming kit (Multiprime, Amersham, Arlington Heights, IL).

Approximately 6 × 10^5 recombinants from the λgt10 8.5-day mouse fetal cDNA library were plated and duplicate plaque lifts made using standard conditions.[13] Filters were prehybridized for 2–3 hr at 42° in 50% formamide, 0.2% polyvinylpyrrolidone, 0.2% bovine serum albumin (BSA), 0.2% Ficoll, 50 mM Tris-HCl (pH 7.5), 1.0 M NaCl, 0.1% sodium pyrophosphate, 1.0% sodium dodecyl sulfate (SDS), 10% dextran sulfate, and 100 μg/ml yeast RNA. Then 5 × 10^5 counts/min (cpm)/ml of each probe was added and the hybridization continued overnight. Following hybridization the filters were washed three times for 15 min each at room temperature in 2× SSC/0.2% SDS, washed three times for 30 min each at 68° in 0.2× SSC/0.1% SDS, air-dried, and autoradiographed at − 70° with an intensifying screen. To identify plaques corresponding to individual family members, quadruplicate filter lifts were made at the final round of screening and hybridized with individual probes.

[14] K. Fahrner, B. L. M. Hogan, and R. A. Flavell, *EMBO J.* **6,** 1265 (1987).

To subclone the λ phage inserts, individual plaques were stabbed with a toothpick and spun into 100 μl of TE (10 mM Tris-HCl, pH 8.0; 1 mM EDTA). Two microliters of a 1 : 10 dilution of this stock was used as a template for PCR amplification using primers complementary to the EcoRI linker which flanks all library inserts. Conditions for PCR reactions, analysis, and purification of the reaction products were identical to those described above. The amplification products were then digested with EcoRI and cloned into EcoRI-cut pGEM7Z (Promega). A subset of the positive plaques did not produce amplification products, apparently owing to concentration of the stab stocks. DNA from these clones was isolated using lambda-sorb (Promega, Madison, WI) as recommended by the manufacturer, digested with EcoRI, and the inserts subcloned into EcoRI-cut pGEM7Z. The DNA sequence of the inserts was then determined using a T7 polymerase dideoxy-sequencing kit (Pharmacia).

The initial round of library screening resulted in the isolation of clones containing the majority of the coding sequence for each of the novel Wnt family members. To obtain clones encoding the extreme 5' and 3' ends, probes outside of the amplification region were generated from isolated λ clones and used to rescreen up to 5 × 10^6 recombinants from the same library. In the process of screening, clones encoding two additional family members were isolated by virtue of their homology to Wnt-5a and Wnt-7a. These genes, named Wnt-5b and Wnt-7b, are approximately 90% identical to their cognate pair at the amino acid level and presumably represent recent gene duplication events. Thus, the combined approach of PCR amplification and library screening resulted in the identification of 6 novel members of the mouse Wnt gene family.

Concluding Remarks

The advent of PCR has brought about a revolution in molecular biology. What once required complex subcloning or construction strategies can now be reduced to a single amplification reaction. The isolation of a specific cDNA used to require many rounds of library screening (if the correct library even existed) and analysis of multiple clones. Now, if one has enough sequence information to generate two primers, the sequence can be amplified from a cDNA library or directly from first-strand cDNA in a matter of hours. The degenerate primer technology discussed in this chapter provides yet another example of the power of PCR to streamline existing protocols. Moreover, in the case of the Wnt gene family, attempts to identify related genes using previously available methods had been fruitless. Indeed, the only family member known prior to the studies described above had been isolated serendipitously.[12,15] Thus, without development of degenerate PCR, systematic identification of these genes would have been virtually impossible.

As with any approach there are, however, limits to the power of this technique. Analysis of the expression patterns of the six *Wnt* gene family members isolated in our study revealed that all are expressed at day 9.5 of mouse development. In addition, Roelink[16] has reported the isolation of another mouse *Wnt*-1 homolog (*Wnt*-3) which is also expressed at day 9.5. Therefore, a question can be raised as to why only four of the seven novel homologs expressed at this time were amplified using the degenerate primers. One simple explanation is that our primer for domain II was not completely degenerate. It provided for only a CAT codon at the penultimate histidine residue, and all of the nonamplified cDNAs contain a CAC codon at this position. However, mismatch at this codon did not completely prevent priming from genes encoding a CAC, as both *Wnt*-1 and *Wnt*-4 were amplified in our studies (Fig. 5). Nonetheless, to determine whether lack of degeneracy at the histidine in domain II had biased our screen we have repeated the 9.5-day cDNA amplification experiments using a CAC encoding primer. To date, these experiments have resulted in the isolation of clones for some but not all of the missing genes.[17] Thus, there appear to be some unexplained biases within the system, which may reflect both message levels and efficiency of reverse transcription. One possible way to circumvent these biases is to use the degenerate primers to amplify related sequences from genomic DNA. However, if this approach is attempted, it is advisable to pick primers that are contained within a single exon, and this limits the combination of motifs on which the primers can be based.

In summary, the use of degenerate PCR primers provides a powerful new approach for the identification of families of related genes based on conservation of relatively small amino acid motifs. Owing to the exponential amplification of signal provided by PCR, this approach should be particularly useful in developmental systems where one is often limited with regard to experimental material.

[15] B. J. Wainwright, P. J. Scambler, P. Stanier, E. K. Watson, G. Bell, C. Wicking, X. Estivill, M. Courtney, A. Bowe, P. S. Pedersen, R. Williamson, and M. Farrall, *EMBO J.* **7,** 1743 (1988).

[16] H. Roelink, E. Wagenaar, S. Lopes da Silva, and R. Nusse, *Proc. Natl. Acad. Sci. U.S.A.* **87,** 4519 (1990).

[17] J. A. McMahon and A. P. McMahon, unpublished results (1990).

[40] Screening for Novel Pattern Formation Genes Using Gene Trap Approaches

By DAVID P. HILL and WOLFGANG WURST

Introduction

Genetic screens have been used to identify a large number of developmental mutations in organisms such as *Drosophila melanogaster* and *Caenorhabditis elegans*. Mutations in these organisms have been generated by chemical mutagens, radiation, or insertion of foreign DNA into the genome. Originally, the mutations were identified by their phenotypic effects on the process of normal development. Later, screens using reporter genes such as β-galactosidase contained within transposable elements were used to screen for insertional mutations into developmentally regulated genes.[1–3] These mutations are recovered regardless of the phenotype. Using this strategy, genes displaying both novel and familiar patterns of expression during embryogenesis have been identified, and mutations in these genes have been generated.

Analogous "trapping" strategies have been applied to the mouse to identify enhancers and the genes they drive that are important in mammalian development.[4,5] Enhancer trap vectors have been introduced into the mouse genome either by zygote injection[4] or by transfection of embryonic stem (ES) cells.[5] Reporter gene activity has been assayed during different stages of embryonic development in the offspring of founder animals or in chimeric embryos generated from ES cells. The major drawback of zygote injection is that transgenic mouse lines have to be established for each integration event before expression patterns can be analyzed. Maintenance of these animals becomes both costly and time consuming.

The use of ES cell technology is a more efficient and straightforward way of performing trapping screens in mice.[6] Embryonic stem cell lines

[1] C. J. O'Kane and W. J. Gehring, *Proc. Natl. Acad. Sci. U.S.A.* **84,** 9123 (1987).
[2] E. Bier, H. Vaessin, S. Shepherd, K. Lee, K. McCall, S. Barber, L. Ackerman, R. Carretto, T. Uemura, E. Grell, L. Y. Jan, and Y. N. Yan, *Genes Dev.* **3,** 1273 (1989).
[3] H. J. Bellen, C. J. O'Kane, C. Wilson, U. Grossniklaus, R. K. Pearson, and W. J. Gehring, *Genes Dev.* **3,** 1288 (1987).
[4] N. D. Allen, D. G. Gran, S. C. Barton, S. Hettle, W. Reik, and M. A. Surani, *Nature (London)* **333,** 852 (1988).
[5] A. Gossler, A. L. Joyner, J. Rossant, and W. C. Skarnes, *Science* **244,** 463 (1989).
[6] A. L. Joyner, *BioEssays* **13,** 649 (1991).

Copyright © 1993 by Academic Press, Inc.
All rights of reproduction in any form reserved.

have been isolated from the mouse inner cell mass, and after genetic manipulations and a period of *in vitro* culture they remain pluripotent and can contribute to all tissues of the mouse including the germ line.[7-9] This technology has allowed for the establishment of thousands of random reporter gene insertions that can be used to screen for embryonic expression patterns in chimeras and/or phenotypic effects after germ line transmission.

The prototype of promoter trapping vectors contained coding sequences of a reporter gene whose activity could only be seen after integration in frame into an exon. More sophisticated vectors, termed gene trap vectors, have been designed containing a splice-acceptor site upstream of the β-galactosidase gene.[5,10,11] Integration of these vectors into introns results in the generation of a fusion transcript between the endogenous gene and the β-galactosidase gene. Because introns rather than exons are the targets of this type of vector, the number of target sites in the genome is much greater than for the original vectors. Fusion transcripts from insertion of these types of vectors mimic endogenous gene expression at the insertion locus, and this expression can be monitored by visualizing β-galactosidase expression.[12] Furthermore, the gene trap vector can also act as an insertional mutagen, disrupting endogenous gene function.[10,12] Recent evidence suggests that these types of vectors can be further improved by alleviating the reading frame requirement between the coding regions of the β-galactosidase and the endogenous gene.[13]

The rich source of potentially mutated genes can be explored in at least two different ways. The mutation can be transmitted through the germ line and the offspring screened for a recessive mutant phenotype,[10] or the expression pattern of tagged genes can be prescreened in chimeric embryos.[5] Mutations that exhibit "interesting" expression patterns can be transmitted through the germ line, and their phenotype can be studied. Prescreening has the advantage that more integrations can be assayed without having a large animal facility. It also can lead to the discovery of novel patterns of developmental gene expression, and it can uncover genes that are involved in developmental processes but do not display an obvious

[7] M. J. Evans and M. H. Kaufman, *Nature (London)* **292**, 154 (1981).
[8] G. R. Martin, *Proc. Natl. Acad. Sci. U.S.A.* **78**, 7634 (1981).
[9] A. Gossler, T. Doetschman, R. Korn, E. Scherfling, and R. Kemler, *Proc. Natl. Acad. Sci. U.S.A.* **83**, 9065 (1986).
[10] G. Friedrich and P. Soriano, *Genes Dev.* **5**, 1513 (1991).
[11] W. G. Kerr, G. P. Nolan, A. T. Serafin, and L. A. Herzenberg, *Cold Spring Harbor Symp. Quant. Biol.* **59**, 767 (1989).
[12] W. C. Skarnes, B. A. Auerbach, and A. L. Joyner, *Genes Dev.* **6**, 903 (1992).
[13] D. P. Hill and J. Rossant, unpublished results (1993).

phenotype when mutated. Of course, by biasing oneself to genes with striking patterns of expression, important genes whose expression is ubiquitous will be overlooked.

In addition to the approaches described above, where genes that are active in undifferentiated ES cells are identified, gene trap vectors can also be used to screen for genes that are expressed in ES cells as they are induced to differentiate *in vitro*. This will allow investigators to use ES cells as an "*in vitro* mouse," and gene trap screens can be performed on the cells as they are placed in a variety of experimental environments. One exciting prospect for this type of screen would be to look for genes that are expressed in ES cells after they have been exposed to a variety of growth factors. This would allow investigators to identify genes that are downstream in the pathway of pattern formation after growth factor stimulation.

Gene Trap Vectors

A variety of gene trap vectors have been described in detail elsewhere.[5,10,14] The gene trap vectors we are using contain a En-2 splice-acceptor site, which has been shown to work in the predicted manner (Fig. 1).[12] To increase the number of neomycin-resistant colonies, we replaced the human β-actin promoter driving neomycin with the phosphoglycerate kinase (PGK-1) promoter, which results in about a five-fold increase of neomycin-resistant colonies.[13] The ratio of *lacZ*-expressing to neomycin-resistant colonies remained unchanged. Introduction of a translational start signal (ATG) in the *lacZ* gene allows one also to detect integrations into an untranslated first exon. This modification increases the number of *lacZ*-expressing clones about 3-fold (1/30 neomycin-resistant colonies express *lacZ*)[15] (Fig. 2).

Large-Scale Gene Trap Screen in Mouse Embryonic Stem Cells

We have undertaken a large-scale screen in ES cells by prescreening the expression patterns of the reporter gene integrations in chimeric embryos. From approximately 289 integrations tested, about 12% showed spatially restricted expression patterns in 8.5 days postcoitum (dpc) chi-

[14] W. C. Skarnes, *Bio/Technology* **8**, 827 (1990).
[15] W. Wurst and A. L. Joyner, unpublished results (1992).

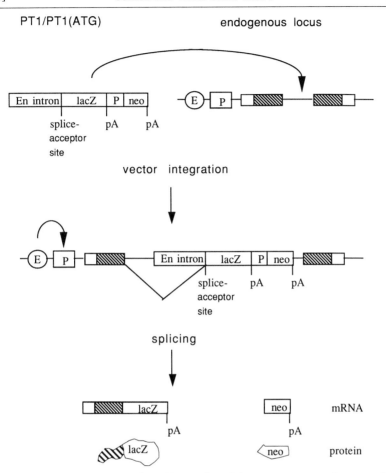

FIG. 1. Schematic representation of integration of the gene trap vector into an endogenous locus.

meric embryos.[16] Of those that have exhibited interesting expression patterns, 20% showed expression patterns restricted to extraembryonic and 80% to embryonic and extraembryonic tissues. Central nervous system (CNS) and heart staining was predominant in the embryonic patterns, which might reflect the complexity of CNS development and the stage at which the embryos were analyzed. Three integrations showed gradient-

[16] W. Wurst, J. Rossant, V. Prideaux, M. Kownacka, A. Joyner, D. P. Hill, F. Guillemont, S. Gasca, D. Cado, A. Auerbach, and S. L. Ang, manuscript in preparation (1993).

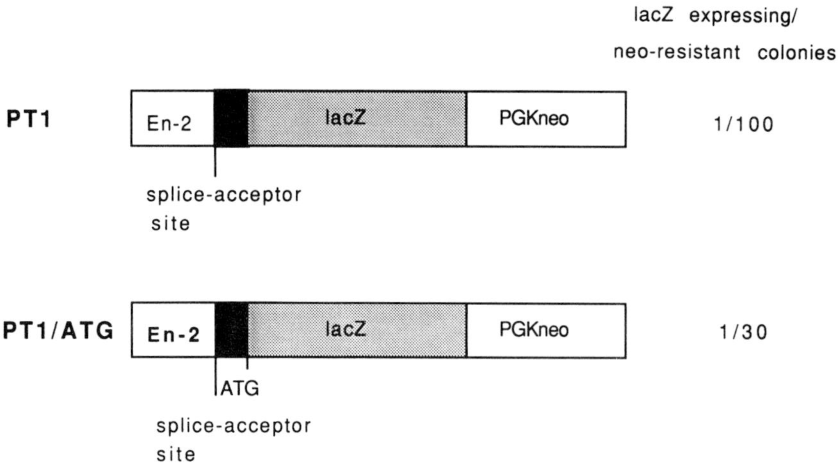

FIG. 2. Schematic presentation of gene trap vectors PT1 and PT1(ATG). Both vectors contain En-2 intron sequences including a splice-acceptor site in front of the bacterial *lacZ* gene and a neomycin gene driven by the PGK-1 promoter. PT1(ATG) includes a translational start signal (ATG) in the *lacZ* gene. The ratio of *lacZ*-expressing to neomycin-resistant colonies is shown for each vector.

like expression patterns with strong staining in the posterior and/or anterior end of the embryo. All the integrations exhibiting specific expression patterns will be transmitted through the germ line, and the mutated gene will be characterized molecularly.

Directed Gene Trap Approach

We have designed a gene trap screen in ES cells to identify specifically genes that act downstream of inducing factors. This allows us to prescreen *in vitro* for genes that are inducible and which might be involved in cell differentiation, cell proliferation, or pattern formation.

We are currently exploring this system using retinoic acid (RA). From 12,000 gene trap vector integrations screened, 90 colonies expressed β-galactosidase after RA application. Three of the colonies showed repression of the reporter gene activity after RA treatment, whereas all the other clones expressed β-galactosidase independent of the RA induction.[17] This preliminary screen suggests that gene trap integrations responding to certain inductive stimuli can be identified using the direct gene trap approach.

[17] W. Wurst and A. Joyner, unpublished results (1992).

Propagation and Maintenance of Embryonic Stem Cells

Many different ES cell lines such as CP-1,[18] CC1.2,[18] D3,[19] CCE,[20] E14,[21] AB-1,[22] and R1[23] have been established from 129 mouse inbred stains. The conditions required to maintain the pluripotency of the cells are slightly different, and it is advisable to follow the recommended growth conditions. The conditions we describe here are based on our experience using the D3 line. Under these conditions we obtained very good results with respect to tissue contribution in chimeras as well as germ line transmission.

Culture Media and Solutions

To culture ES cells we use Dulbecco's modified Eagle's medium (DMEM) with high glucose and glutamine (Flow Labs, McLean, VA, powder, No. 430-1600). The DMEM powder is dissolved in distilled water, buffered with bicarbonate, adjusted to pH 7.2, and filter-sterilized. Immediately prior to use we add 0.1 mM nonessential amino acids (100× stock, GIBCO, Grand Island, NY, No. 320-1140AG), 1 mM sodium pyruvate (100× stock, GIBCO, No. 320-1360AG), 10^{-6} mM 2-mercaptoethanol (100× stock stored at −20°; Sigma, St. Louis, MO, No. M-6250), 2 mM L-glutamine (100× stock, GIBCO, No. 320-5030), and 15% (v/v) fetal calf serum (FCS). Penicillin and streptomycin can be added to the medium but are not essential. The quality of the FCS is very important for the growth of ES cells, and we recommended testing different batches from various suppliers on the ES cell line that will be used.

Primary Embryonic Fibroblast Feeder Layer

The D3 ES line was established on primary embryonic fibroblast feeder layers (EMFI), and it is advisable to maintain it under these conditions. However, it should be noted that batches of EMFI cells will vary, and each should be thoroughly tested for ES cell doubling time and morphology.

[18] A. Bradley, M. Evans, M. H. Kaufman, and E. Robertson, *Nature (London)* **309**, 255 (1984).

[19] T. C. Doetschman, H. Eichstetter, M. Kertz, W. Schmidt, and R. Kemler, *J. Embryol. Exp. Morphol.* **87**, 27 (1985).

[20] E. Robertson, A. Bradley, M. Kuehn, and M. Evans, *Nature (London)* **323**, 445 (1986).

[21] A. H. Handyside, G. T. O'Neil, M. Jones, and M. L. Hooper, *Roux's Arch. Dev. Biol.* **198**, 48 (1989).

[22] A. P. McMahon and A. Bradley, *Cell (Cambridge, Mass.)* **63**, 1073 (1990).

[23] A. Nagy, personal communication (1992).

Preparation of Feeder Cells

1. Dissect 14–16 dpc mouse embryos from the uterus and extraembryonic membranes into phosphate-buffered saline (PBS).

2. Remove heads and internal organs and wash 10 carcasses at least twice in 50 ml PBS.

3. Mince the carcasses with a pair of watchmaker's scissors into cubes about 2–3 mm in diameter in 2 ml of PBS.

4. Using a syringe plunger, press the cubes from embryos through a screen (1 mm in diameter) into an Erlenmeyer flask that contains 20 ml of sterile beads and a stir bar.

5. Flush 50 ml of trypsin/EDTA over the remaining clumps on the screen [1 liter of $5\times$ trypsin/Tris–saline stock contains 8 g NaCl, 0.4 g KCl, 0.1 g Na_2HPO_4, 1.0 g glucose, 3.0 g Trizma base, 0.06 g penicillin G, 0.1 g streptomycin, and 2.5 g trypsin (Difco, Detroit, MI, No. 0153-59); the stock is diluted 1:4 with saline/EDTA (saline/EDTA is 0.02 g EDTA, 0.8 g NaCl, 0.02 g KCl, 0.115 g Na_2HPO_4, 0.02 g KH_2PO_4, and 0.01 g/phenol red, pH 7.2)].

6. To digest released DNA, add 200 μl of DNase I (10 mg/ml) and incubate at 37° for 30 min with stirring.

7. Add another 50 ml of trypsin/EDTA and incubate at 37° for 30 min.

8. Repeat Step 7.

9. Decant the cell suspension and pellet cells (1000 rpm, 5 min).

10. Wash the cell pellet twice in DMEM plus 10% (v/v) FCS. If the solution is too viscous, add 200 μl of DNase I and incubate at 37° for 30 min.

11. Pellet the cells and resuspend the pellet in 10 ml DMEM plus 10% FCS and count the viable cells using trypan blue (Flow Labs, No. 16-910-49).

12. Plate 5×10^6 viable cells per 150-mm tissue culture dish in 25 ml DMEM plus 10% FCS and culture overnight at 37°, 5% CO_2.

13. Change the medium after 24 hr.

14. After 2–3 days of culture the cells should form a confluent monolayer.

15. Trypsinize each plate and replate onto five 150-mm dishes.

16. When the plates are confluent (after 2–3 days), freeze all the cells from each plate in one freezing vial in 1 ml of freezing medium [DMEM, 25% (v/v) FCS, 10% (v/v) dimethyl sulfoxide (DMSO)] and store at $-70°$ for 1 day before transferring to liquid nitrogen.

Preparation of Feeder Layers

1. Thaw one frozen vial of EMFI cells quickly at 37°, add 10 ml DMEM plus 10% FCS, and centrifuge at 1000 rpm for 5 min at room temperature.

2. Resuspend the cell pellet in 10 ml DMEM plus 10% FCS, seed onto five 150-mm dishes, and grow the cells until they form a confluent monolayer.

3. Remove the medium from the confluent plates and add 10 ml DMEM plus 10% FCS containing 100 μl mitomycin C (1 mg/ml PBS, Sigma, M-0503).

4. Incubate plates at 37°, 5% CO_2 for 2–2.5 hr.

5. Wash the monolayer of cells twice with PBS and add 10 ml trypsin/EDTA to each plate.

6. Incubate the plate for 5–10 min (37°, 5% CO_2), then add 10 ml DMEM plus 10% FCS.

7. Centrifuge the cells (1000 rpm, 5 min) and resuspend the pellet in DMEM plus 10% FCS. Split cells onto suitable size plates.

8. Allow the feeder cells to attach and use plates within 1 week.

Growth of Embryonic Stem Cells on Feeder Layers

The parameters important for keeping ES cells pluripotent are as follows: growth medium, length of growth time between passages, number of cells plated at each passage, extent of typsinisation, CO_2 concentration in the incubator, and the source of the feeder layer. However, even after using stringent tissue culture conditions, abnormal cell clones will arise with time in culture. Some of these variants might not contribute to the germ line and also might have a growth advantage by having a shorter doubling time. Therefore, we recommend freezing as many vials as possible from an early passage of the ES lines.

Culture Conditions

1. Quickly thaw one vial of frozen ES cells and transfer the cells to a 10-ml tube.

2. Add 10 ml ES cell medium and centrifuge at 1000 rpm for 5 min.

3. Resuspend the cell pellet in 5 ml ES cell medium and plate onto a 60-mm plate containing a feeder layer.

4. Change the medium the following day.

5. On the second day, wash cells with PBS and add 2 ml trypsin/EDTA.

6. Incubate for 3–5 min or until the cells begin to come off the plate. The cells may be pipetted gently to break up clumps.

7. Add 5 ml ES cell medium and centrifuge for 5 min at 1000 rpm.

8. Aspirate the supernatant and resuspend the pelleted cells in 5–7 ml ES cell medium.

9. Pass 1 ml of the cell suspension ($\sim$0.2 $\times$ 10^6 cells) to a 60-mm feeder layer dish containing 4 ml ES cell medium.

In general, ES cells should be passed every second day and the medium changed every day.

Freezing and Storage of Embryonic Stem Cell Lines

ES cells can be frozen and stored in DMEM plus 25% (v/v) FCS and 10% (v/v) DMSO. As a general rule, freeze slowly and thaw quickly. For long-term storage cells should be kept under liquid nitrogen, but for short-term storage they can be kept at $-70°$.

Freezing

1. After the second day of ES cell culture, trypsinize and centrifuge the cells.
2. Resuspend the cell pellet from a 60-mm dish in 1 ml freezing medium and transfer to a freezing ampoule on ice (4°).
3. Quickly transfer the ampoule to a precooled Styrofoam box in a $-70°$ freezer and transfer them the next day to liquid nitrogen.

Electroporation of Embryonic Stem Cells and Selection Procedure

DNA can be introduced into ES cells by the application of a high electrical pulse to a suspension of cells and DNA.[24] This procedure is relatively simple but has the disadvantage that about 50% of the cells die under conditions giving optimal transformation efficiency. The parameters that influence transformation efficiency are capacitance, voltage, ionic strength of the electroporation buffer, and DNA concentration. The following protocol is based on our experience using the D3 cell line.

Electroporation

1. On the second day after passing, trypsinize ES cells to form a single cell suspension and stop the trypsinization with 7 ml ES cell medium.
2. "Preplate" the cells by incubating the dish for 30 min at 37° to allow feeder cells to attach to the plate.
3. Harvest the ES cells by centrifugation at 1000 rpm for 5 min at room temperature.
4. Aspirate the supernatant and resuspend the cells in cold PBS.
5. Count the cells and adjust the cell concentration to 7×10^6 cells/ml. One can expect about $2-3 \times 10^7$ cells from a 100-mm dish.
6. Mix 0.8 ml of the cell suspension with 40 μg of linearized vector DNA and transfer it to an electroporation cuvette (Bio-Rad, Richmond, CA, No. 165-2088).

[24] A. Potter, L. Weir, and P. Leder, *Proc. Natl. Acad. Sci. U.S.A.* **81**, 7161 (1984).

7. Set up the electroporation conditions (240 V, 500 μF for the Bio-Rad gene pulser).

8. Transfer the cuvettes into the cuvette holder and deliver the pulse.

9. Remove the cuvette and place it on ice for 20 min.

10. Transfer the cells into 20 ml ES cell medium containing 500 units/ml leukemia inhibitory factor (LIF) (GIBCO, No. 3275) and plate onto two 100-mm gelatinized dishes (to gelatinize dishes, add 10 ml of 0.1% gelatin to each plate, aspirate the gelatin solution, and air-dry).

11. Change the medium the next day.

12. On the second day after the electroporation, add medium containing G418 for drug selection. For each batch of G418 we set up killing curves to determine the optimal drug concentration [range between 150 and 250 μg/ml active G418 (GIBCO, No. 860-1811)].

13. Change the medium every day or every second day for 8 days. At this point clearly visible G418-resistant colonies should be seen.

Replica Plating Technique

Cells expressing β-galactosidase can be identified using a "replica plating technique,"[5,25] fluorescence-activated cell sorting (FACS),[26] or directly selectable reporter genes such as βgeo.[10] We routinely use the replica plating technique (Fig. 3).

1. After selection, aspirate the medium from the plates and carefully place autoclaved marked (cut by notches) polyester filters (1 μm pore size) on top of the colonies.

2. Cover 80% of the filter with sterile glass beads (3 mm in diameter) and add 10 ml ES cell medium including G418.

3. After 2–2.5 days of incubation, aspirate the medium, pour off glass beads, mark the orientation of the filter on the back of the plate, and carefully pull off the filter with sterile forceps.

4. Place the filter cell side up in a petri dish containing 10 ml PBS and add ES cell medium to the "master" plate. The filters are now ready to use for the *LacZ* expression assay.

Assaying for LacZ Expression

ES cells and chimeric embryos can be assayed for *lacZ* gene expression using the following protocol.

[25] S. Gal, this series, Vol. 151, p. 104.

[26] S. Reddy, H. Rayburn, H. von Melchner, and H. E. Ruley, *Proc. Natl. Acad. Sci. U.S.A.* **89**, 6721 (1992).

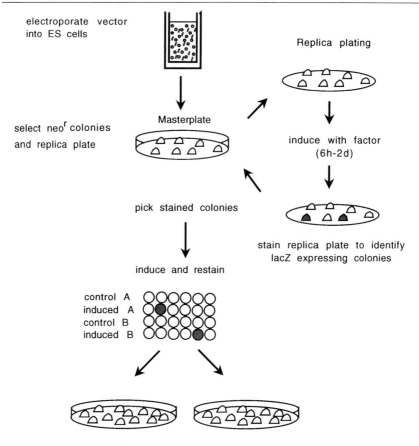

FIG. 3. Schematic representation of electroporation, selection procedure, induction by factors, and replica plating technique to isolate *lacZ*-expressing ES cell colonies. The induction step can be omitted.

1. Fix the cells/embryos by incubating in fix solution (fix solution is 0.40 ml of 25% glutaraldehyde, 3.00 ml of 0.1 M EDTA, pH 7.3, 0.12 ml of 1 M MgCl$_2$, 46.4 ml of 0.1 M phosphate buffer, pH 7.3) at room temperature for 5 min for cells and 7.5 dpc embryos, 10 min for 8.5–11.5 dpc embryos, and 20 min for 12.5–14.5 dpc embryos.

2. Wash the cells/embryos three times for 5 min at room temperature in wash solution [wash solution is 0.4 ml of 1 M MgCl$_2$, 2.0 ml of 2% (v/v) Nonidet P-40 (NP-40), 195.6 ml of 0.1 M phosphate buffer, pH 7.3].

3. Incubate the cells/embryos in fresh staining solution [staining solution is 2.0 ml 4-chloro-5-bromo-3-indolyl-β-galactoside (X-Gal) (25 mg/ml in dimethylformamide), 48.0 ml wash solution, 0.106 g potassium ferrocyanide (Sigma, P-9387), 0.082 g potassium ferricyanide (Sigma, P-8131)] at 37° until the color develops (usually 1 hr to overnight).

Identification of β-Galactosidase-Expressing Cells on Master Plates

1. After identification of β-galactosidase-expressing colonies on the filters and the "master" plate, pick them under a dissecting microscope with a drawn-out Pasteur pipette.

2. Transfer picked colonies to a 24-well plate containing feeder cells and ES cell medium. After two days "tryplate" ES cell colonies by aspirating the medium, washing with PBS, and adding 75 μl of trypsin.

3. Incubate for 5 min until the colonies disaggregate to a single cell suspension and add 500 μl of medium. Break cell clumps by gently pipetting up and down.

4. Incubate at 37° in 5% CO_2 for 2 more days.

5. If the colonies are confluent, trypsinize them and establish two identical plates, one to restain and verify the β-galactosidase-expressing colonies and one to expand and freeze. If the colonies are not confluent, tryplate again and grow the colonies for another 2 days. For assaying *lacZ* expression, use the same protocol as described earlier.

Replica Plating and Screening for Inducible Genes

In principle the screening procedure is the same as described above with the addition of an induction step after replica plating. After removing the filter from the plate, place the filter cell side up side in a petri dish containing DMEM medium (5% FCS) plus the inductive stimuli. In our case, we used retinoic acid (10^{-6} M) as an inductive reagent. Incubate the filters for an appropriate time (6 hr to 2 days) to screen for early or late inducible genes. β-Galactosidase expression is assayed, and all *lacZ*-expressing colonies are picked and expanded. To distinguish between constitutively expressed genes and genes that are specifically inducible, duplicate 24-well plates should be prepared with one plate being used for induction and the other as a control (Fig. 3).

Chimera Production

One of the distinct advantages of using embryonic stem cells for genetic manipulation is the ability to introduce them back into mouse embryos, where they can contribute to all tissue types.[27,28] There are currently two procedures used to introduce embryonic stem cells into mouse embryos: aggregation chimera production and blastocyst injection. Although both procedures can be used to generate chimeras, each has its own advantages and drawbacks. Production of aggregation chimeras requires no special equipment and leads to a high contribution of embryonic stem cells in the embryo. If chimeric embryos are to be used to assay for gene expression of reporter genes in the embryonic environment, this method is quite satisfactory. However, a large contribution of ES cells to the embryo often leads to decreased embryonic survival, particularly at later stages of development depending on the quality of ES cell lines used. Another disadvantage in this procedure is that two embryos are used to generate each chimera. Thus, twice the number of embryos must be collected than the number of chimeras generated. Recently, however, we were able to generate aggregation chimeras using a single embryo and a clump of ES cells. Injection chimeras generally lead to a lower contribution of ES cells to the embryo, but these embryos are very healthy and usually survive to term.

All solutions for embryo culture and manipulation should use water of the highest purity possible. All glassware must be free of detergents and residue. When possible, it is advantageous to use disposable plasticware or to have a set of glassware specifically dedicated to embryo culture and manipulation.

Diploid Aggregation Chimeras

For generating aggregation chimeras we use the method described by Nagy et al.[29]

1. Prepare microwells in the bottom of a plastic petri dish by grinding a smooth darning needle through a drop of M16 (M16 culture medium is 94.66 mM NaCl, 4.78 mM KCl, 1.71 mM CaCl$_2$, 1.19 mM KH$_2$PO$_4$, 1.19 mM MgSO$_4$, 25 mM NaHCO$_3$, 23.28 mM sodium lactate, 0.33 mM sodium pyruvate, 5.56 mM glucose, 4.0 g/liter BSA, 100 units/ml penicillin G potassium salt, 50 μg/ml streptomycin sulfate, 10 mg/liter phenol red) that has been overlayed with paraffin oil into the plastic at the bottom of the dish. Care must be taken at this step to ensure that the sides of the

[27] A. Bradley, in "Teratocarcinomas and Embryonic Stem Cells: A Practical Approach" (E. J. Robertson, ed.), p. 113. IRL Press, Oxford and Washington, D.C., 1987.

[28] B. Hogan, F. Costantini, and E. Lacy, "Manipulating the Mouse Embryo." Cold Spring Harbor Laboratory, Cold Spring Harbor, New York, 1986.

[29] A. Nagy, E. Gocza, E. Merentes Diaz, V. R. Prideaux, E. Ivany, H. Markkula, and J. Rossant, Development (Cambridge, UK) 110, 815 (1990).

well are smooth and that the bottom of the well is rounded so that embryos and cells are not damaged or retained by protruding edges at the sides of the well. Many needles may need to be tried before a suitable brand is found that will generate a smooth well.

2. Overlay the microwells with paraffin oil and place them into a 37° incubator with 5% CO_2 for preequilibration.

3. Flush 2.5 dpc mouse embryos from the oviduct of a pregnant female. Embryos should be flushed into M2 medium (M2 culture medium has the same composition as M16 only HEPES is used to buffer the solution instead of bicarbonate, 20.85 mM HEPES and only 4.15 mM NaHCO$_3$).

4. Wash the embryos once through a drop of preequilibrated M16 medium and then incubate the embryos for a few minutes in a drop of acid Tyrode's solution [8.0 g NaCl, 0.2 g KCl, 0.24 g CaCl$_2 \cdot$ H$_2$O, 0.1 g MgCl$_2 \cdot$ 6H$_2$O, 1.0 g glucose, 4.0 g polyvinylpyrrolidone (PVP); adjust to pH 2.5 with concentrated HCl, filter-sterilize, and store at $-20°$] until the zona pellucida is dissolved. Monitor the procedure and quickly transfer the embryos back into M2 medium to neutralize the acid when the zona pellucida begins to disappear. If embryos remain in acid too long they will be damaged. It is also important to transfer as little solution as possible with the embryos to maintain the correct pH in each drop.

5. Briefly trypsinize the embryonic stem cells using 1.5 ml of trypsin for a 30-mm plate until the cells form small clumps of 5–20 cells. When this is accomplished, add 3.5 ml of ES cell culture medium to the plate and place the plate in the incubator to allow the clumps of cells to recover.

6. Remove the plate containing the microwells from the incubator and, using a drawn-out, flame-polished Pasteur pipette, place two embryos in each drop of embryonic culture medium. Place one embryo at the bottom of the well and one embryo next to the well.

7. Remove clumps of embryonic stem cells that contain 10–15 cells from the trypsinized plate and wash these clumps through two changes of M16. Next, place a single clump of ES cells into the microwell such that the cells land on top of the embryo at the bottom of the well.

8. Drop the remaining embryo onto the cells so that a sandwich is made between the two embryos and the cells. Aggregation chimeras can also be generated by placing a single embryo on top of a clump of 10–15 ES cells in the microwell. The sandwiches are then incubated overnight in M16 at 37° with 5% CO_2.

9. The next morning remove the chimeras from the incubator. The embryos should have compacted and formed small blastocoels. If the embryos have only compacted but have not yet formed blastocoels, they can be transferred as is or incubated for another day. We transfer the embryos as soon as possible because longer *in vitro* culture leads to a higher risk of the embryos developing abnormally. Embryos should be

transferred to a 2.5 dpc pseudopregnant host as described in detail by Hogan et al.[28] The developmental timing of the embryo will now coincide with that of the host mother.

10. Embryos can be dissected at any stage of development or allowed to continue until term.

Production of Tetraploid Embryos

Nagy et al.[29] showed that chimeric embryos can be generated that contain almost entirely ES cell-derived tissue in their embryonic lineages. The strategy behind this approach is to place the host embryo at a disadvantage so that its cells divide more slowly. This allows the ES cells to completely outgrow the tetraploid cells in the embryonic lineages of the chimera. Chimeric embryos are produced essentially as described above except that tetraploid host embryos are used.

1. Flush embryos at the 2-cell stage (1.5 dpc) from the oviducts of pregnant mice.

2. Place the embryos between the electrodes of a fusion apparatus so that the plane between the two cells is parallel to the plane between the two electrodes. We usually leave the zona pellucida intact at this stage; however, if it is very difficult to orient the embryos correctly, the zona may be removed so that the embryo is easier to orient. If the zona is removed, it is imperative that the embryos be cultured individually in the next step or they will aggregate to form one giant embryo.

3. Deliver a square-wave pulse of 100 msec at 100 V/cm to the embryos.

4. Allow the embryos to recover for 15 min. By this time the two cells should have fused. If they have not, another pulse can be given to the embryos.

5. Allow the embryos to develop overnight. The next day use the healthy 4-cell embryos to generate chimeras.

Injection

A large variety of microinjection apparatuses are currently available commercially, and the choice of which to use is up to the individual investigator. The procedure has been described in detail by Hogan et al.[28] We currently use two different procedures for microinjection: microinjection into a hanging drop using a conventional compound microscope and microinjection into a drop on the surface of a coverslip using an inverted microscope. We briefly describe these two methods but stress that microinjection technique varies from individual to individual; personal comfort is the most crucial factor in choosing a microinjection system.

Preparation of Needles

1. Holding pipettes are pulled by hand using a Bunsen burner until the diameter of the pipette is about three-fourths that of a blastocyst. The pipette is broken using a microforge. The pipette is slowly brought into close proximity of the heating element. Just as the pipette begins to melt, the heat is lowered quickly. As the element cools and contracts it will break the pipette cleanly. This technique requires some practice at first, as it requires good coordination between pipette movement and heat application. Once a cleanly broken pipette is obtained, the end is polished and smoothed by moving it close to the heating element of the microforge. The holding pipette can now be back-filled with light oil and mounted into a micromanipulator.

2. Injection pipettes are pulled on a mechanical pipette puller set so that the thin end of the pipette has an internal diameter that is slightly smaller than that of an embryonic stem cell. After pulling, the pipette is rebroken under a dissecting microscope by laying it across a piece of Tygon tubing and applying pressure with a scalpel blade. This creates a sharp end on the pipette, which facilitates entry into the blastocyst. The finished pipette is then back-filled with light oil and mounted on a micromanipulator.

Preparation of Injection Chambers

Hanging Drop Chamber. Hanging drop chambers are constructed using a slotted microscope slide, vacuum grease, and a coverslip. The edges of the coverslip are coated with vacuum grease and pressed onto the edges of the notched slide to create a "tunnel." Using a pulled Pasteur pipette, place a series of drops of embryonic culture medium along the coverslip such that they hang from the coverslip by surface tension. It helps if the drops are slightly elongated rather than perfectly round. Next, fill the "tunnel" with light oil. This must be done quickly so that the surface tension on the ends of the tunnel holds the oil in place. Cells and blastocysts can now be added to the drops, and the chamber can be placed on the microscope. The cells and blastocysts collect at the bottom of the drops, and the holding and injection pipettes enter the drop horizontally from opposite ends of the tunnel. The pipettes are used to orient a blastocyst onto the holding pipette such that its inner cell mass is at the bore of the pipette. Next, approximately 10 cells are picked up in the injection pipette and are injected into the blastocoel. It is easiest to enter the blastocyst between two trophectoderm cells if possible. The injected blastocysts are then transferred to pseudopregnant host mice.

Surface Injection Chamber. To make a surface injection chamber, an aluminum microscope slide with an oval hole through its center is used as a chamber. The chamber is constructed by coating the edges of a coverslip with vacuum grease and pressing it against the aluminum slide to form a well. It is crucial that this well does not leak! A drop of embryonic culture medium is placed on the coverslip and the drop is overlayed with light paraffin oil. The chamber is then placed on an inverted microscope, and the holding and injection pipettes are lowered into the drop. The blastocyst and cells are manipulated in the same way as is described above and transferred to the uterus of 2.5 dpc pseudopregnant hosts.

Creation of Germ Line Chimeras

Germ line chimeras are generated by creating chimeric embryos as described above and allowing the embryos to continue to term. The chimeric mice can then be bred with tester strains to determine if they receive a dominant marker that is characteristic of the ES cell line used. It is important that neither the host embryo nor the tester strain carry this marker. One good marker is the coat color marker dominant agouti, which is contained in a variety of ES cell lines. The mice showing the coat color marker are analyzed by Southern blot to check for vector integration. Once germ line chimeras are obtained, they can be used as a source of heterozygous mice carrying the inserted trapping vector. It is a good idea to check mice that are to be used in breeding protocols by Southern blot to be sure that the gene trap vector has inserted as a single integration. If it has not, other integrations may be bred away by continuous backcrossing. This is a time-consuming task.

Rapid Cloning of *lacZ* Fusion Transcripts

Using gene trap vectors containing a splice-acceptor site allows rapid cloning of the fusion transcript taking advantage of the rapid amplification of cDNA ends (RACE) protocol.[12,30] The partial sequence of the fusion transcript will provide information about the nature of the mutated gene and can be used in the isolation of full-length cDNAs. These techniques bring together the power of insertional mutagenesis and molecular biology

---

[30] M. A. Frohman, M. K. Dush, and G. R. Martin, *Proc. Natl. Acad. Sci. U.S.A.* **85,** 8998 (1988).

and will allow researchers to dissect the molecular basis of mammalian development.

Acknowledgments

The authors thank Alex Joyner and Janet Rossant for constant support, Lesley Forrester, Francois Guillemot, and Monica McAndrews-Hill for critical reading of the manuscript, and Andras Nagy for his help in making aggregation chimeras. D.P.H. was a Postdoctoral Fellow of the Medical Research Council of Canada. W.W. was partly supported by the Deutscher Akkademischer Austausch Dienst (DAAD) and Deutsche Forschungsgemeinschaft (DFG). Our work using gene trap vectors is supported by the National Institutes of Health.

[41] Insertional Mutagenesis by Retroviruses and Promoter Traps in Embryonic Stem Cells

By GLENN FRIEDRICH and PHILIPPE SORIANO

Introduction

A powerful and versatile tool to study embryonic development is genetic manipulation. The effect that mutations have on development facilitates an understanding of the underlying mechanisms in the same way that throwing a wrench into an engine can reveal the importance of the crankshaft. Although the connection between the part (or gene) affected and the eventual result is not always obvious, persistent tampering and observing the effect of combining one mutant with another will reveal the elegant workings of the "machinery." The evidence for the success of applying random mutagenesis to dissecting developmental pathways is ample, and to realize this one needs only to read through a review describing *Drosophila* or *C. elegans* development.[1,2] However, using traditional genetic methodologies with mammals is difficult due to their long generation time and the complexity of the genome. Therefore, those wishing to study genes important in mammalian development have had to rely mainly on the "reverse genetic" approach to isolate genes by sequence similarity and subsequent genetic analysis via homologous recombination.[3] Although these methodologies have certainly not been exhausted, an approach directed toward random mutagenesis will expand the base of genes to study.

[1] L. Cooley, R. Kelley, and A. Spradling, *Science* **239**, 121 (1988).
[2] I. A. Hope, *Development (Cambridge, UK)* **113**, 399 (1991).
[3] A. Bradley, *Curr. Opin. Biotechnol.* **2**, 823 (1991).

Copyright © 1993 by Academic Press, Inc.
All rights of reproduction in any form reserved.

Fortunately, the variety of methods to create random mutants in a model mammal, the mouse, has expanded.[4]

This chapter describes current methods used to mutate genes at random within the germ line of the mouse. This discussion is limited to insertional mutagenesis either directly into the germ cell lineage of an embryo or into the germ line by the use of embryonic stem (ES) cells. Insertional mutation in this chapter does not refer to targeted mutation by homologous recombination as described elsewhere in this volume.[5]

These methods were designed to address three major concerns: (1) the mutagen should simplify cloning of the mutated locus, (2) the mutation should not be associated with any rearrangements of the genome, and (3) the frequency of mutation should be high. Because genetic analysis of mice can involve a major commitment of resources, the relative efficiencies of each method are addressed.

This chapter is organized into two sections. The first describes how to generate proviral insertions by retroviral infection. The second describes promoter traps, which are a means to select or screen for insertions that occur in transcribed regions of the ES cell genome.[6] A brief discussion of mouse strains and genotyping concludes the chapter.

Retroviral vectors based on Moloney murine leukemia virus (MoMLV) can be used to infect the mouse embryo *in vitro* before implantation or *in utero* after implantation. Approximately 5% of the insertions that occur in the germ line of infected embryos create a recessive mutation with an easily observable phenotype.[7] Examples of this are the *Mov*13 mutation in the collagen $\alpha 1$(I) gene[8,9] and the *Mov*34 mutation, which results in recessive embryonic lethality.[10] A frequency of 5% is quite low, even though there may be preferred sites for retroviral insertion, for example, areas of open chromatin or actively transcribed areas.[11,12]

The dominant emerging technology for mouse genetics has been the *in vitro* manipulation of totipotent ES cells. As described in other chapters of this volume, targeted mutations to specific loci can be made in ES cell lines and then transferred to the germ line of the mouse. Retroviral

[4] J. Rossant and N. Hopkins, *Genes Dev.* **6**, 1 (1992).

[5] R. Ramírez-Solis, A. C. Davis, and A. Bradley, this volume [51].

[6] Note that some laboratories refer to promoter traps as "gene traps."

[7] G. Friedrich and P. Soriano, *Genes Dev.* **5**, 1513 (1991).

[8] R. Jaenisch, K. Harbers, A. Schnieke, J. Löhler, I. Chumakov, D. Jähner, D. Grotkopp, and E. Hoffman, *Cell (Cambridge, Mass.)* **32**, 209 (1983).

[9] J. Löhler, R. Timpl, and R. Jaenisch, *Cell (Cambridge, Mass.)* **38**, 597 (1984).

[10] P. Soriano, T. Gridley, and R. Jaenisch, *Genes Dev.* **1**, 366 (1987).

[11] E. Barklis, R. C. Mulligan, and R. Jaenisch, *Cell (Cambridge, Mass.)* **47**, 391 (1986).

[12] C.-C. Shih, J. P. Stoye, and J. M. Coffin, *Cell (Cambridge, Mass.)* **53**, 531 (1988).

insertions can also be delivered into the germ line this way; however, one can also take an approach directed toward disrupting transcribed loci. Promoter and enhancer traps, widely used in prokaryotes and eukaryotes, are DNA constructions designed to detect insertions into transcribed loci. Insertions into active loci in ES cells increase the frequency of mutagenesis and to date represent the most efficient means to perform insertional mutagenesis in the mouse.

Because proviral insertion and promoter traps depend on the insertion of foreign DNA to create a mutation, the disrupted locus is accessible to standard cloning methodologies. It is important to note that retroviral infection, in contrast to DNA transfection or microinjection, does not cause rearrangement of the genome. This should be a concern when planning to undertake insertional mutagenesis on any scale.

Retroviral Infection of Mouse Embryos

Retroviral vectors are ideal as insertional mutagens. They are easy to use, infect a wide variety of cell types, are stable through multiple generations, and do not cause rearrangement of the host genome when integrated. For these reasons, a number of mouse lines with proviral integrations have been generated. The frequency with which these insertions have caused mutations detectable by an overt phenotype is, based on over 100 lines, approximately 5%.[7]

Retroviral Vectors

Although infection of embryos places few restrictions on vector design, the following points should be considered. The vector should include elements to aid cloning of the integration site such as a bacterial *supF* gene[13] or rescuable plasmid sequences.[14] The resulting provirus should be easily detected on a Southern blot to aid genotyping, and the viral titer of the producer cell line should be relatively high. Whereas initial experiments of this type used replication-competent viruses,[15] now disabled viruses generated by a packaging line such as $\psi 2$[16] or GP + E86[17] are used.

Retroviral particles are harvested from medium [Dulbecco's modified Eagle's medium (DMEM) plus 10% fetal calf serum (FCS) and antibiotics]

[13] W. Reik, H. Weiher, and R. Jaenisch, *Proc. Natl. Acad. Sci. U.S.A.* **82**, 1141 (1985).
[14] C. L. Cepko, B. E. Roberts, and R. C. Mulligan, *Cell (Cambridge, Mass.)* **37**, 1053 (1984).
[15] R. Jaenisch, H. Fan, and B. Croker, *Proc. Natl. Acad. Sci. U.S.A.* **72**, 4008 (1975).
[16] R. Mann, R. C. Mulligan, and D. Baltimore, *Cell (Cambridge, Mass.)* **33**, 153 (1983).
[17] D. Markowitz, S. Goff, and A. Bank, *J. Virol.* **62**, 1120 (1988).

overlaying producing cells. Virus stocks can be prepared in advance and kept frozen at $-70°$ until use. Virus-producing cells are seeded into 100-mm or 150-mm culture dishes and left to divide until just subconfluent. A minimal amount of fresh medium (7 ml for 100-mm dish, 20 ml for 150-mm dish) is placed on the cells and left for 14–16 hr. Virus-containing medium is collected directly from the dish into a sterile syringe and expelled through a syringe-fitted 0.22-μm filter to remove cells and cell debris. The viral titer should be determined at this time by an appropriate means (e.g., transduction of a selectable marker). If a sufficiently high titer is expected, aliquots of 200 μl should be frozen as soon as possible in sterile 500-μl Eppendorf tubes.[18] Before freezing, Polybrene is added to a final concentration of 4 μg/ml for infection of preimplantation embryos or 50 μg/ml for infection of postimplantation embryos.

Viral particles can be concentrated by centrifugation if the titer is low [e.g., $<5 \times 10^5$ colony-forming units (cfu)/ml]. Virus-producing cells are grown in several 150-mm dishes and virus-containing medium is harvested every 12 hr for a 48-hr period. Store the media at $4°$ until the entire amount can be pooled, and keep media on ice for all subsequent manipulations. Remove cells by passing pooled media through a bottle-top 0.22-μm filter. Viral particles are pelleted in a sterile, plastic 250-ml centrifuge bottle at 10,000 g in an angled rotor for 16 hr at $4°$. Before spinning media, it is useful to mark the bottle with a felt-tip pen at the expected position of the pellet. Discard the supernatant; the pellet will likely be invisible or small and translucent. Suspend pellet in 1/100 of the initial volume using fresh medium at $4°$. Be gentle with the pellet; swirl slowly or pipette up and down carefully without introducing any bubbles. Add Polybrene as described above, store in 50-μl aliquots at $-70°$, and determine the resulting titer.

Infection of Preimplantation Embryos

Four- to eight-cell preimplantation embryos are infected either by incubation in the presence of virus-producing cells or by incubation in microdrops of virus-containing medium.[19,20] Embryos are collected from the oviduct at the 4- to 8-cell stage (i.e., before compaction) in M2 medium and the zona pellucida removed with acid Tyrode's solution as previously described.[21] It is necessary to infect embryos at this stage since it is

---

[18] The half-life of retroviral particles at room temperature is approximately 6 hr.

[19] R. Jaenisch, D. Jähner, P. Nobis, I. Simon, J. Löhler, K. Harbers, and D. Grotkopp, *Cell (Cambridge, Mass.)* **24**, 519 (1981).

[20] P. Soriano and R. Jaenisch, *Cell (Cambridge, Mass.)* **46**, 19 (1986).

[21] B. Hogan, F. Costantini, and E. Lacy, "Manipulating the Mouse Embryo: A Laboratory Manual." Cold Spring Harbor Laboratory, Cold Spring Harbor, New York, 1986.

difficult to culture the embryos of some strains from the zygote up to the 8-cell stage. Infection should occur prior to compaction since cells destined to compose the embryo proper will become sequestered into the inner cell mass and thus become inaccessible to viral particles. Injection of virus particles or virus-producing cells into the blastocoel is possible, but this method is less efficient than coculture.[19]

After the zona is removed, embryos are transferred to a 60-mm culture dish containing subconfluent virus-producing cells covered with fresh DMEM plus 10% FCS, 4 μg/ml Polybrene, 23 mM sodium lactate, and 0.33 mM sodium pyruvate. Avoid having embryos contact one another since they will aggregate readily without the zona pellucida. Be gentle when moving a culture dish containing embryos from the microscope to incubator since even a little agitation can cause the embryos to clump together. It is neither practical nor necessary to do any of these manipulations in a sterile laminar flow hood.

Embryos are cultured in the presence of virus-producing cells for 16–20 hr in a humidified incubator at 37° and 5% CO_2. During this time, the embryos should have compacted, and some will have begun to form a blastocoel cavity. The embryos are then removed from the culture dish into fresh M2 before transferring them into the uterus of 2.5-day pseudopregnant females.[21]

Embryos can also be cultured in microdrops of virus-containing medium.[19] Medium prepared as described above is placed in small drops (30–50 μl) on the bottom of a 35-mm petri dish made of plastic which has not been treated for tissue culture or glass which has been siliconized. Ideally, the titer of virus should be above 1×10^6 cfu/ml. After light paraffin oil or silicone oil is carefully layered over the drops, the medium is allowed to equilibrate in an incubator for at least 30 min. It may be necessary to screen paraffin oil from different sources since some lots have contaminants that diffuse into the microdrops and affect the viability of the embryos (we use EM Science, Gibbstown, NJ, No. PX0047-1).

Precompaction 4- to 8-cell embryos are transferred into the equilibrated microdrops in as little volume as possible to avoid diluting the virus. The culture is returned to the incubator for 5–6 hr. Infected embryos are then transferred into a dish of warm DMEM or M16 medium[21] and cultured overnight. Resulting morulae are transferred into the uterus of 2.5-day pseudopregnant females.

In animals that show proviral integrations in somatic tissue, only a proportion will transmit these to the next generation. In one set of published experiments, 50% of all surviving embryos had somatic mosaicism for provirus integrations. Of these mosaic animals, 70% transmitted at least one provirus to subsequent generations.[20]

The results from preimplantation infections can vary greatly. However, in general the higher the titer, the higher the proportion of embryos infected. The study cited above showed that MoMLV could infect over one-half of the embryos. Another study showed that defective retroviruses with titers above 10^6/ml were able to infect embryos with high efficiency, but the percentage of infected embryos did not depend strictly on the titer.[22] Although one could imagine that the viral sequence should not interfere with the infection process, some sequences might be less efficiently inserted into the host genome owing to factors restricted to early embryonic cells. Nevertheless, it should be possible to infect embryos with a wide variety of retroviral vectors, with titer playing a predominant role in determining success.

Infection of Postimplantation Embryos

Infection of postimplantation embryos is the least efficient method for transferring proviruses into the germ line. In total, only four lines have been derived out of 1500 gametes tested.[10] However, the technique is useful for lineage tracing (if the retrovirus transduces a histochemically detectable marker) or for transducing a gain-of-function marker. For these purposes, the reader is directed to a description of the method by Hogan et al.[21]

Promoter Traps in Embryonic Stem Cells

Considerations for Construct Design

Promoter traps are primarily a means of insertional mutagenesis. They allow insertions into transcribed loci to be detected from among a large group of random insertions. Promoter traps are designed so that the only possibility of activating a reporter gene is if the trap integrates into the proper position to be transcribed under control of a genomic locus. Therefore, promoter trap insertions cause a mutation and allow the regulation of the locus to be studied by assaying the activity of the reporter gene.

There is a distinction between promoter traps and enhancer traps. Promoter traps use a reporter gene that lacks any transcriptional control elements, whereas enhancer traps use a reporter gene driven by a basal promoter element (ideally only a TATA box and a transcriptional initiation

---

[22] H. van der Putten, F. M. Botteri, A. D. Miller, M. G. Rosenfeld, H. Fan, R. M. Evans, and I. M. Verma, *Proc. Natl. Acad. Sci. U.S.A.* **82,** 6148 (1985).

site).[23,24] Enhancer traps therefore detect an increase of transcription resulting from integration near an active genomic enhancer element and any tissue specificity conferred by that element. Because integration of enhancer traps does not necessarily disrupt the normal transcription of the locus, they are likely to be less mutagenic. In addition, because enhancer effects can be exerted over large physical distances, cloning of genes identified by this method could be laborious. Enhancer traps are not discussed further here; however, the techniques described here can be readily adapted for their use.

There have been a number of promoter trap constructs described,[7,25-29] a selection of which are represented schematically in Fig. 1. Commonly used reporter genes are β-galactosidase (βgal) and neomycin phosphotransferase (neo); only the latter is selectable, but βgal is a useful reporter since its activity is easily detectable using a chromogenic substrate, X-Gal. In promoter traps β-galactosidase coding sequences can be configured as an individual terminal exon followed by a polyadenylation signal. If a mammalian initiator codon is included, βgal activity would result if the reporter gene integrated into coding sequences so translation can initiate at the βgal ATG codon (Fig. 1A). Insertion into an intron would not result in translation due to RNA processing. Alternatively, if insertion occurred immediately proximal to an endogenous promoter, the transgene would become the first exon (Fig. 1B). Removing the initiator codon of βgal identifies integration events into transcribed coding regions in the proper translational frame so that a fusion protein is produced. Only if the fusion does not affect βgal activity will a promoter trap event become detectable (Fig. 1C). βgal activity is not selectable so all these constructs require the addition of a selectable marker, usually neo under control of separately engineered promoter elements. Therefore, actual promoter trap events occur among a background of total random insertions.

Another major type of promoter trap construct uses splice recognition sequences. The reporter gene, either with or without an initiator codon, is placed downstream from a splice-acceptor sequence, thereby forcing

[23] N. D. Allen, D. G. Cran, S. C. Barton, S. Hettle, W. Reik, and M. A. Surani, *Nature (London)* **333**, 852 (1988).

[24] R. Kothary, S. Clapoff, A. Brown, R. Campbell, A. Peterson, and J. Rossant, *Nature (London)* **335**, 435 (1988).

[25] D. G. Brenner, S. Lin-Chao, and S. N. Cohen, *Proc. Natl. Acad. Sci. U.S.A.* **86**, 5517 (1989).

[26] A. Gossler, A. L. Joyner, J. Rossant, and W. C. Skarnes, *Science* **244**, 463 (1989).

[27] H. von Melchner and H. E. Ruley, *J. Virol.* **63**, 3227 (1989).

[28] S. Reddy, J. V. DeGregori, H. von Melchner, and H. E. Ruley, *J. Virol.* **65**, 1507 (1991).

[29] W. G. Kerr and L. A. Herzenberg, *Methods Companion Methods Enzymol.* **2**, 261 (1991).

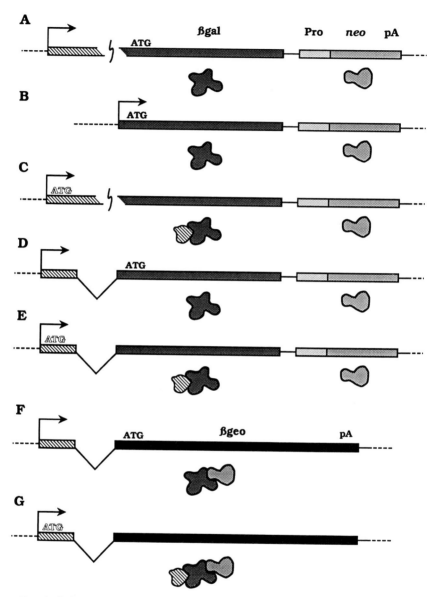

Fig. 1. Schematic diagram of generic promoter trap constructs; see text for details. Arrows represent transcription initiation sites; striped boxes, genomic exons; dark shaded boxes, β-galactosidase gene sequences; light shaded boxes, neomycin-resistance gene sequences with the promoter (Pro) and polyadenylation signal (pA) indicated; dashed lines, genomic DNA sequences; and solid lines, construct sequences. The solid ATG is supplied by the promoter trap construct, whereas the outlined ATG is supplied by the genomic coding sequence.

the construct to be processed as a separate exon (Fig. 1D,E). If the site of integration is into the first intron of a gene, a fusion mRNA is produced: for example, a short 5′ untranslated exon spliced onto the promoter trap exon. In such a case the first initiator codon is that of βgal (Fig. 1D). When an initiator codon is not included with the reporter gene, an insertion downstream from the endogenous initiator ATG would result in a translational fusion if the initiation codon is in frame with βgal after a splicing event (Fig. 1E). Therefore, it is important to ensure that there are no stop codons 5′ to and in the same frame as βgal.

The reason that βgal is of such utility as a reporter gene in promoter traps is that its activity is easily assayed. Using a chromogenic substrate (X-Gal), tissues expressing βgal can be identified by a blue precipitate. In a strain of mice carrying a promoter trap using βgal, the spatial and temporal activity of the trapped locus can be easily detected. For example, the tissue specificity of a promoter during embryonic development can be visualized by simply staining a whole embryo or sections of an embryo with X-Gal (described below). However, it would be convenient to not screen numerous random integrants for those which activate βgal but rather to design a trap which activates a selectable gene. neo can be used in this way if it is engineered in the contexts described above for βgal. However, the advantage of X-Gal staining is lost, and in situ techniques would have to be used to determine patterns of expression. We have previously described construction of a reporter gene that is a translational fusion between βgal and neo and codes for a protein with both activities.[7] Using this gene (βgeo) as a reporter gene thus allows direct selection of promoter trap events while maintaining the ability to detect patterns of expression with X-Gal staining (Fig. 1F,G).

Transfection Promoter of Traps by Electroporation

The primary concern for all work with ES cells is to maintain impeccable culture conditions so that totipotency of the cell line is unaltered after several in vitro manipulations. Therefore, when undertaking random mutagenesis via ES cells, careful adherence to the procedures described in this volume is crucial.[30] In our experience, the most important variables include using low passage ES cells, testing serum lots for highest ES cell plating efficiency, and using feeder cells such as STO SNL 76/7 cells.[31] In addition, the following points concerning preparation of media should be considered. Use either glass containers in which there is absolutely no trace of detergents or disposable tissue culture ware. Use freshly drawn

---

[30] S. J. Abbondanzo, I. Gadi, and C. L. Stewart, this volume [49].

[31] A. P. McMahon and A. Bradley, *Cell (Cambridge, Mass.)* **62,** 1073 (1990).

Milli-Q-purified (Millipore, Bedford, MA) water to make DMEM (we use GIBCO, Grand Island, NY, No. 430-2100ED). Buffer DMEM with 2.2 g/liter of sodium bicarbonate, not 3.7 g/liter as recommended by the manufacturer. This will allow proper pH equilibration in a 5% CO_2 incubator. The osmolality of the completed medium (without serum) should be tested for each batch and should be approximately 290–300 mmol/kg. Once medium has been stored for 1 month, it is necessary to replenish the glutamine component to a final concentration of 2 mM from a 100× stock (e.g., GIBCO, No. 320-5030AG).

Plasmid DNA containing the promoter trap construction should be prepared by standard techniques; preferably, CsCl equilibrium density gradient purification is used. The plasmid DNA is made linear by digestion with a restriction enzyme. Promoter traps employing a splice-acceptor sequence should be cut in the plasmid backbone several hundred base pairs away from the plasmid insert. This ensures that any degradation from the DNA ends will not affect efficiency. In the case of promoter traps acting as exon fusions, care should be taken when choosing the site for digestion since sequences at the ends of the integration will act either as exon untranslated sequences or as coding sequence. After complete digestion, the DNA is purified by organic extraction and ethanol precipitation. Cleaned DNA is dissolved in sterile TE [10 mM Tris-Cl (pH 7.4), 1 mM EDTA (pH 8.0)] at about 1 μg/μl.

Embryonic stem cells (e.g., AB1) are grown on mitomycin-inactivated feeders (e.g., SNL 76/7) until just subconfluent, at which time the cells are trypsinized and counted. Approximately 1×10^7 cells are transferred to a 15-ml culture tube, pelleted gently (250 g for 5 min at room temperature), and suspended in 0.8 ml of phosphate-buffered saline (PBS). Twenty-five micrograms of linear DNA is placed into an electroporation cuvette, the cells are added, and the combination is mixed gently by pipette. The cells should be electroporated as soon as possible at room temperature. Optimal conditions for electroporation should be determined empirically; we use a Bio-Rad (Richmond, CA) Gene Pulser unit set at 230 V and equipped with a capacitance extender set at 500 μF. Time constants should not deviate greatly between experiments; our values range between 5 and 7 msec. Unusual time constants can indicate improperly made PBS, an incorrect number of cells, or other problems. Treated cells are transferred immediately to a 100-mm culture dish with feeders. A large amount of cell death resulting in dramatically increased viscosity is normal, and so the cuvette should be rinsed with medium to ensure transfer of all viable cells.

The cells are cultured overnight to allow attachment and recovery. About 20 hr after electroporation, the medium is replaced with medium used for selection. In the case of *neo* or *βgeo* markers, the medium is supplemented with 300 μg/ml (150 μg/ml active ingredient) of G418 (Geneticin, GIBCO, No. 860-1811). Cells are kept under selection for 10 to 12 days. The medium should be changed every day for the first 5 or 6 days and when needed thereafter. For rapidly growing ES cells, the medium should be replaced whenever an obvious change in color of the medium to a golden yellow is observed.

G418-resistant colonies are visible macroscopically after 6 or 7 days of culture and should be harvested after 10 days. At this point, various procedures can be followed, depending on the mutagenic strategy employed. If random insertions need to be screened for promoter trap events, the colonies should be either replica plated or analyzed individually. Replica plating of ES cell colonies has been described elsewhere.[32] To screen colonies individually, one can either stain the colonies on the dish after removing a portion for expansion or simply clone and expand individual colonies for later analysis.

ES cell clones grow in tight, bulging clumps. This morphology makes it easy to transfer cells by mouth pipette. Well-separated colonies are identified under magnification using a stereo microscope, and clumps of cells are drawn up into a small pipette. A tissue culture hood equipped with a recessed dissection microscope serves to maintain sterility while providing sufficient magnification. Glass capillaries (0.7–1.0 mm) are pulled by hand into pipettes, the tip of which is made by breaking the glass at the taper. Experience is the best guide for determining the optimal width of pipette to use (e.g., 100–200 μm). The medium is removed from the plate and the cells washed once with PBS and kept in a second wash of PBS for subsequent manipulations. The following technique should be performed quickly to ensure cell viability. Part or all of a ES colony is removed by applying gentle suction by mouth (Fig. 2). If the colony is to be expanded, the cells are transferred in a minimal volume into a small (~10 μl) drop of trypsin. After about 5 min at room temperature, the clump of cells is broken up by adding an equal volume of medium and pipetting up and down with a small tip on an Eppendorf micropipettor (0.5–10 μl). The homogeneous suspension of cells is transferred into a 4-well dish (10-mm diameter, e.g., Nunc, Roskilde, Denmark, No. 176740) with feeders and fresh, warm medium. When expanding the colony from this point onward, it is important to maintain a relatively high cell density. This usually means passaging all of the cells from the 4-well dish into a 35-mm dish and then into a 100-mm dish every 3–4 days.

As mentioned, it may be necessary to identify positive promoter trap events among a background of random insertions, for example, traps which use *βgal* as a reporter gene and the *neo* gene for detecting stable

[32] S. Gal, this series, Vol. 151, p. 104.

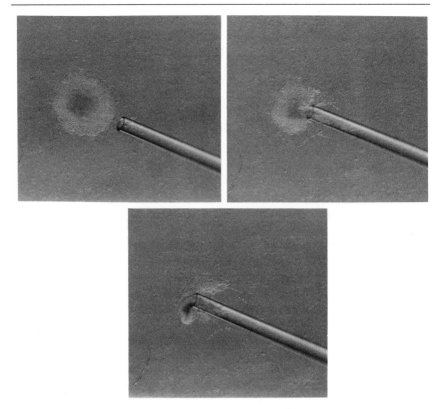

Fig. 2. Procedure for picking individual ES cell colonies using a glass mouth pipette. See text for details.

integrants. A large number of individual colonies can be isolated and X-Gal-positive clones identified during expansion. A less labor-intensive method is to remove only a portion of the colony for expansion and leave the remainder for staining with X-Gal. Colonies can be identified by a circle and number printed on the bottom of the dish with colored markers. One-half of each clone is transferred to numbered wells as described above. In this way 100 or more colonies can be analyzed from one 100-mm plate. Once all colonies are transferred, the dish is fixed and stained with X-Gal as described below. X-Gal-positive colonies are expanded, and negative colonies can be discarded.

These methods are designed to detect genes which are expressed in ES cells. Alternatively, it is possible to select for promoter trap events that insert into genes which are not active in ES cells but which are activated after differentiation. Differentiated tissues can be derived from ES cells either by inducing cystic embryoid bodies[33] or, more simply, by creating primary embryoid bodies and allowing them to attach and differentiate on a plastic dish. Part of an ES cell colony (numbered on the bottom of the dish) is removed with a glass mouth pipette as described above and placed in a 24-well dish (16 mm diameter wells, e.g., Costar, Cambridge, MA, No. 3424) in which the bottom of each well has been covered with 1% (w/v) agarose (0.5 ml) previously equilibrated in culture medium (DMEM + 15% FCS). The plate of colonies is returned to the incubator and can be maintained for the following week with frequent changes of medium. The sample of each ES cell clone is prevented from attaching to the bottom of the well by the agarose and will thus form a primary embryoid body within 3 days. At this time, the entire cell mass is transferred gently into a new 24-well dish coated with 0.1% (w/v) gelatin. During the next 3 days, the embryoid body will attach and begin to differentiate. Some of the resulting types of tissue can include endoderm, neural derivatives, and beating myocardium. The differentiated clones are fixed and stained with X-Gal as described below. Clones with interesting staining patterns can be recovered from the original plate with a mouth pipette and expanded for blastocyst injections and further analysis.

Experiments based on activation of a selectable reporter gene such as *neo* or *βgeo* result in 100% of colonies being promoter trap events. Because there is no need to screen for activity, all G418-resistant colonies are used for making chimeric animals. Each colony can be expanded and injected into blastocysts individually, or a number of colonies can be pooled for injection into blastocysts. The former approach has the advantage of having ES cells on hand for each potential mutant strain generated, whereas the later involves much less tissue culture. These approaches can be combined, that is, individual colonies are maintained in culture and are pooled at the time of blastocyst injection. In this way the major advantage of injecting pools of ES clones is realized: up to four different integration events can be transmitted by one male "germ line" chimera.[7,34]

Transduction of Promoter Traps by Retroviral Vectors

Electroporation is an efficient and simple means to deliver DNA into the cell. However, integrants are often concatemers of two or more unit lengths with small deletions or rearrangements. It is unknown whether rearrangements of the genome occur as in traditional transgenic animals

---

[33] E. J. Robertson, *in* "Teratocarcinomas and Embryonic Stem Cells: A Practical Approach" (E. J. Robertson, ed.), p. 71. IRL Press, Oxford and Washington, D.C., 1987.

[34] E. Robertson, A. Bradley, M. Kuehn, and M. Evans, *Nature (London)* **323**, 445 (1986).

A

B

FIG. 3. Retroviral promoter traps. (A) Promoter trap vector based on a splice acceptor and the *βgeo* gene with an initiator ATG. (B) Promoter trap vector with the reporter gene inserted into the LTR. Each construct is shown in proviral form, with the wavy lines representing genomic sequences. Unique features are indicated. HIS-D is the histidinol dehydrogenase gene, a selectable marker, and Δenh indicates deletion of the LTR enhancers.

produced by pronuclear injection. To avoid these problems, we and others have designed retroviral vectors that deliver promoter traps into ES cells.[7,25,27,29] Retroviral vectors integrate cleanly into the genome without deletion of host sequences and are simple to use. Each of the basic promoter trap designs depicted in Fig. 1 can be used in the context of a retroviral vector. Vectors that delete the long terminal repeat (LTR) enhancer when integrating into the genome are used to avoid the possible interaction of the viral enhancer with the genomic promoter. Such an interaction could potentially alter the spatial and temporal pattern of expression of the mutated allele *in vivo*.

At least two basic vector designs have been used to date. Vectors can be constructed with *βgal* and *βgeo* reporter genes like those diagrammed in Fig. 1 inserted between the LTRs[7] or inserted directly into the LTR.[27] Vectors with an internal reporter gene depend on a splice acceptor to create promoter traps that integrate into introns (e.g., the ROSA series of vectors). Because the polyadenylation signal at the end of the reporter gene should not interfere with transcription of the full-length viral transcript in the packaging cell line, the promoter trap cassettes are inserted in the opposite transcriptional orientation relative to the LTRs (Fig. 3A). By placing a reporter gene into the LTR, promoter traps are required to integrate into coding sequences since the reporter gene sequence is separated from the genomic sequences by only a few base pairs and no splicing signals are present (Fig. 3B).[27]

Retroviral vectors should be constructed by transfection of the retroviral vector construct into an appropriate packaging cell line (e.g., GP + E86) using standard procedures.[35] Note that a selectable marker may need to

[35] P. Soriano, G. Friedrich, and P. Lawinger, *J. Virol.* **65**, 2314 (1991).

be included in the transfection if the retroviral vector lacks a continuously active marker of its own. We find that a 10:1 molar ratio of vector to marker DNA results in over 90% of selected cells containing both DNAs.

The titer of promoter trap vectors should be determined using ES cells. For retroviral vectors with a selectable marker under control of its own promoter, infect ES cells with 1 ml of serially diluted medium in the presence of 4 μg/ml Polybrene. For this purpose, 6-well plates (35 mm diameter) are convenient. Seed 1×10^5 ES cells per well (previously prepared with inactivated feeders) the evening before virus-containing medium is to be collected. At the same time, overlay virus-producing cells with fresh medium. After 16 hr, collect and filter virus and make appropriate dilutions in warm medium containing Polybrene. Infect ES cells with 1 ml of each dilution for 6 hr. Replace virus-containing medium with fresh medium and allow cells to recover overnight. Start selection the following morning and continue until colonies can be counted macroscopically (10 days) by staining with 0.5% (w/v) methylene blue in 50% (w/v) methanol.

For vectors that rely on a promoter trap event to activate a selectable marker, it is necessary to determine the approximate titer by extrapolating the observed titer (determined as described above) by a factor based on the trapping efficiency. For example, when the construct depicted in Fig. 1D is transduced by a retrovirus, 10% of G418-resistant colonies are βgal-positive. Therefore, when a selectable marker (e.g., βgeo as in Fig. 1F) is placed behind the same splice-acceptor sequence, the actual viral titer is 10 times the value determined by counting the number of G418-resistant colony-forming units.

The method to generate a large number of infected colonies is simply a scaled-up version of determining the titer. On a 100-mm plate prepared with inactivated feeders, seed 1×10^6 ES cells. The following day, overlay these cells with medium containing 4 μg/ml Polybrene and sufficient virus to obtain approximately 200–500 drug-resistant colonies. This represents a multiplicity of infection significantly lower than one virus per cell and ensures that only one provirus is present in a clone. This is important to avoid the complication of separating multiple proviruses by breeding to isolate the one which is a promoter trap. Infect cells for 6 hr and leave to recover overnight before starting drug selection. Resulting colonies are manipulated as described above.

Assessing Patterns of Expression

In promoter trap events, the regulatory elements of the mutated gene drive transcription of the reporter gene. If the level of transcription and the spatial and temporal specificity of this promoter is unaffected by the insertion, then following the activity of the reporter gene is a direct measure of these various properties. However, the insertion may affect some regulatory regions found in introns, or the stability of the reporter gene transcript may be different than that of the endogenous gene; both situations would likely result in misleading reporter gene expression patterns. Also, any unique translational or posttranslational control of gene expression may not be reproduced by the reporter gene.

Tissues expressing βgal or βgeo are identified with X-Gal (5-bromo-4-chloro-3-indolyl-β-D-galactopyranoside). *In situ* localization of mRNA or antibody staining is required to detect expression of most other reporter genes. These techniques are well documented and are not discussed here.[36,37]

The following technique is used to stain whole embryos. One to four wild-type females are left overnight with a male who is heterozygous for a promoter trap allele. The females are checked for a vaginal plug the next morning before 10 AM. Midnight of the night of the plug is time 0, so that noon of the plug day is embryonic day 0.5 (i.e., E0.5) and so on.

The activity of βgal (or βgeo) can be easily detected in whole embryos from fertilization to embryonic day 13 (E13). The procedure is the same for any size embryo, varying only the length of time for fixation and the volume of fixative and staining solution (refer to the Appendix for the composition of these solutions). All dissections are carried out in PBS; for preimplantation embryos, PBS is supplemented with 0.4% polyvinylpyrrolidone (PVP) to reduce stickiness.

Preimplantation embryos are flushed from the oviduct or uterus using standard procedures.[21] A hand-drawn mouth pipette is used to transfer preimplantation embryos between solutions. Incubations can be carried out conveniently either in a small spot plate (glass or plastic) or in microdrops covered with paraffin oil. It is important to avoid having the embryos stick to any surface; therefore, use siliconized spot plates or bacterial-grade plastic. It may also be necessary to siliconize transfer pipettes if embryos stick to the inner surface. Fix preimplantation embryos for no more than 10 min at room temperature. Transfer fixed embryos through two changes of PBS (with PVP) to rinse away fixative, leaving each wash about 5 min. Finally transfer embryos into X-Gal staining solution and incubate at room temperature for several hours to overnight. When the male used for plugs is heterozygous, one-half of the embryos are wild-type and thus serve as an internal control for any background X-Gal staining.

[36] C. Bonnerot and J.-F. Nicolas, this volume [27].
[37] D. G. Wilkinson, J. A. Bailes, and A. P. McMahon, *Cell (Cambridge, Mass.)* **50,** 79 (1987).

TABLE I
CONDITIONS FOR X-GAL STAINING OF MOUSE EMBRYOS

Age of embryo (days)	Volume of fixative and stain[a] (ml)	Time for fixation (min)
E7.5–E8.5	2	15
E9.5–E10.5	5	30
E11.5–E12.5	10	60
E13.5–E14.5	15	90

[a] Volumes are suitable for a litter of up to 12 embryos.

Implanted embryos are dissected away from the uterus and removed from the decidua.[21] Fixation times and appropriate volumes are given in Table I. Incubations can be carried out either in small dishes or in test tubes. Wash embryos for at least 20 min between fixation and staining with two changes of PBS. Embryos are stained overnight at room temperature to avoid detecting endogenous βgal activity found at higher temperatures. After staining, embryos are transferred through several changes of 70% ethanol to inactivate βgal. Embryos can be stored indefinitely in glass vials filled completely with 70% ethanol. Wild-type embryos serve as a control for any background staining; note that the yolk sac of embryos older than E10 often stain nonspecifically with X-Gal.

Positive X-Gal staining tissues can be enhanced and internal structures visualized by clearing the embyros with methyl salicylate (oil of wintergreen).[38] Completely dehydrate embryos by several 1-hr changes of 100% ethanol and transfer into a minimal volume of methyl salicylate. Clearing should occur quite quickly, in about 2 hr for E12.5 embryos. Record results by photomicroscopy immediately since the X-Gal precipitate is soluble in methyl salicylate and will slowly vanish. Use of this technique may reveal a limitation of staining whole embryos, namely, that the X-Gal stain does not completely penetrate larger embryos (older than about E11.5).

The same solutions are used to fix and stain ES cell colonies. Aspirate the medium and replace it with enough fixative to cover the colonies completely. Leave fixative at room temperature for 10 min, wash twice with PBS, and add staining solution. Stain colonies overnight at room temperature.

[38] R. D. Lillie, "Histopathologic Technic and Practical Histochemistry." Blakiston, New York, 1954.

βgal activity can also be detected in frozen cryostat sections of tissue.[39] Embryos dissected away from the decidua are frozen in OCT compound in an appropriately sized mold by slowly immersing into liquid nitrogen. Cryostat sections about 10 μm thick are left to dry and firmly attach to glass slides treated with poly(L-lysine). The section is fixed for no longer than 1 min. After two brief washes in PBS, the slide is transferred into staining solution and left to incubate overnight at 37°. After staining, rinse twice in PBS and mount the section in glycerol; document the results by photomicroscopy.

Genetic Analysis of Mutant Alleles

Mice

Many of the ES cells in use today are derived from the 129/Sv mouse strain. The 129/Sv strain is wild-type at the agouti locus (A/A), which is dominant over the black, nonagouti C57BL/6J strain (a/a) from which blastocysts are obtained for injections. Male chimeras are bred to wild-type C57BL/6J females, and germ line transmission is indicated by agouti offspring (A/a). These F_1 animals are hybrid between 129/Sv and C57BL/6J and are screened for the presence of the promoter trap allele as described below.

For genetic studies of any sort it is best to isolate a mutant on a homogeneous, inbred background. Crossing a chimera to a C57BL/6J mouse mixes two separate inbred strains. Therefore, once germ line transmission is demonstrated, the chimera is bred to a wild-type 129/Sv female to maintain an inbred background. Unfortunately, in our experience, this particular mouse strain is ill-suited for laboratory breeding since they litter infrequently and have small litter sizes.[40] This low fecundity can make analysis of mutant alleles in a 129/Sv inbred background impractical. However, the 129/Sv × C57BL/6J hybrid animals are much better breeders. We have limited most studies of mutant alleles in our laboratory to hybrid animals while maintaining a minimal number of 129/Sv inbred breeding pairs for each strain. However, this results in a significant increase in the size of the colony. It is important to note that the phenotype may have a different penetrance or vary significantly depending on the background. These problems indicate a need for additional ES cell lines derived from mouse strains other than 129/Sv.

[39] J. R. Sanes, J. L. R. Rubenstein, and J.-F. Nicolas, *EMBO J.* **5,** 3133 (1986).
[40] A significant improvement in the fecundity of 129/Sv breeding pairs is possible using a high-fat diet (e.g., Purina 5015).

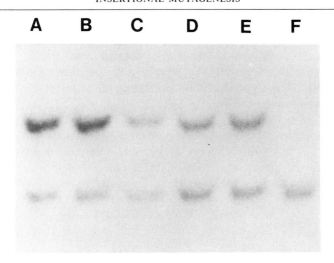

FIG. 4. Southern blot of tail DNAs from six mice. The top band is specific to the promoter trap integration, and the lower band is an unrelated single-copy control. Animals A and B are homozygous for the promoter trap insert, animals C, D, and E are heterozygous, and animal F is wild type.

Genotyping

Initial breeding experiments determine the mutant phenotype of animals homozygous at the promoter trap locus. Heterozygous breeding pairs are set up and their progeny screened for homozygous animals at weaning. If no homozygotes are found among at least 30 progeny, the mutation likely causes embryonic or perinatal lethality ($p > 0.99$ using a χ^2 test). Viable homozygotes can be screened for a mutant phenotype, such as infertility or morphological or metabolic abnormalities. Fertile homozygotes can be maintained by inbreeding homozygous animals through several generations.

Genotypes are determined by Southern blotting, dot hybridization, or polymerase chain reaction (PCR). Distinguishing heterozygotes from wild type is simply a matter of detecting the promoter trap construct by an appropriate probe or set of PCR primers. Genomic DNA is isolated from tail biopsies when the animals are about 8–14 days old. Distinguishing heterozygotes from homozygotes is best done by Southern blotting and comparing band intensities. For example, Fig. 4 shows a Southern blot of DNA from six animals digested with an enzyme that cuts within the LTRs of a promoter trap vector. The upper band is promoter trap-specific (a *neo* probe in this case) and the lower band is an unrelated single-copy gene that serves as an internal control for the amount of DNA loaded in each lane. The promoter trap-specific band in lanes A and B (Fig. 4) is twice as intense as the same band in lanes D and E, especially considering the equivalent intensities of the internal control (lower) band. This indicates that animals A and B are homozygous and animals D and E are heterozygous. The amount of DNA loaded in lane C (Fig. 4) is less than the other lanes since the control band is less intense. However, animal C is likely heterozygous, considering that the ratio of the promoter trap band to the control band is the same as the heterozygous animals D and E. Animal F is wild type. These results can then be confirmed by breeding the homozygous animals to wild-type animals and determining that all progeny in two or three litters are heterozygous.

Concluding Remarks

Promoter trap insertions are a highly efficient means of creating new mutant mouse strains. To date, we have generated 50 mouse lines, each with a promoter trap insertion of the type depicted in Fig. 1F,G and transduced into ES cells by retroviral vectors. Of 33 mutations tested, 14 are recessive embryonic lethal and at least 1 is homozygous male sterile (G. Friedrich and P. Soriano, unpublished data, and Ref. 7). Thus the frequency of obtaining mutant phenotypes with this system is 45%. Whether the preponderance of embryonic lethality is due to the bias of trapping genes which are expressed in ES cells or whether these data reflect the proportion of all genes which are required for embryonic development is unclear at the moment. Limiting oneself only to genes expressed in ES cells may be a problem if one wishes to screen for a specific phenotype. However, it should be possible to generate many different types of mutant phenotypes given the appropriate promoter trap vector and screening protocol.

Creating numerous promoter trap ES cell lines is relatively simple and requires only basic tissue culture and molecular biology skills. The rate-limiting step to creation of large collections of mutant mouse strains is to generate reproducibly chimeric animals whose germ line is populated by ES cell derivatives. With better defined techniques, creating germ line chimeras will become routine, and it is important therefore to begin to consider whether some mechanism should exist for centralized maintenance of the impending deluge of new mutant strains. Considering the high cost of mouse husbandry, it is unlikely that any individual investigator will want to keep all of the "peripheral" mutant strains generated in the search for the few strains with a phenotype or expression pattern of interest. It is a waste of a valuable resource, however, if these strains are terminated. A mutant repository that would maintain and distribute strains would ultimately save granting agencies considerable funds and provide an invaluable scientific collection—"A Book of the Mouse," with book-marks.

Appendix: Media and Solutions

Polybrene stock (1000×): 4000 $\mu g/ml$ in water, filter sterilized

Phosphate-buffered saline (PBS): 8.0 g/liter NaCl, 0.2 g/liter KCl, 1.15 g/liter Na_2HPO_4, 0.2 g/liter KH_2PO_4, pH 7.2

Fixative solution for embryos, embryo sections, and ES cells: 2% formaldehyde (from a 37% stock), and 0.2% glutaraldehyde (grade I, from a 25% stock) in PBS; make fresh for each use

Staining solution for embryos, embryo sections, and ES cells: 1 mg/ml X-Gal, 5 mM potassium ferricyanide, 5 mM potassium ferrocyanide, 2 mM $MgCl_2$, 0.01% sodium deoxycholate, and 0.02% Nonidet P-40 (NP-40) in PBS. This solution, without X-Gal, can be prepared in advance and stored at room temperature in the dark. Add X-Gal from a stock solution just before use.

X-Gal (5-Bromo-4-chloro-3-indolyl-β-D-galactopyranoside) stock: 40 mg/ml in dimethylformamide; store at $-20°$ in small aliquots and ensure that X-Gal is well dissolved before use otherwise crystals may form in the staining solution

[42] Two-Dimensional Gel Analysis of Repetitive DNA Families

By RACHEL D. SHEPPARD and LEE M. SILVER

Introduction

The use of DNA probes that detect restriction fragment length variants or polymorphisms (commonly referred to as RFLPs) as means for establishing whole genome linkage maps has revolutionized the field of mammalian genetics. The power of this approach has been seen most keenly in the field of human genetics, where the analysis of RFLPs has led to the localization and cloning of numerous disease loci including those for cystic fibrosis and retinoblastoma. Although effective, the traditional RFLP approach is tedious. To construct a 95% complete, whole human genome

METHODS IN ENZYMOLOGY, VOL. 225

Copyright © 1993 by Academic Press, Inc.
All rights of reproduction in any form reserved.

linkage map, Donis-Keller and colleagues[1] analyzed the segregation of loci identified by 359 independent probes.

A more efficient approach would be to couple the use of a smaller number of probes, which each detect a larger number of polymorphic loci spread throughout the genome, with a fractionation system that allows the simultaneous resolution of 100 or more individual loci. Probes to several classes of DNA families have been described that would be ideal for such an analysis. These probes include different retroviral-like genes[2] and some of the minisatellite sequences.[3] Although studies with these probes have been performed using standard agarose gel electrophoresis protocols, it is difficult to resolve more than 15 to 20 bands, and, as the copy number of a DNA family increases, resolution of individual bands goes down.

It should be possible to increase the resolution of complex DNA patterns by a power of 2 through the separation of individual fragments in two dimensions rather than one. This can be accomplished by extending the standard one-dimensional gel separation of restriction fragments into a second dimension by digesting these fragments *in situ* with a second restriction enzyme, followed by a second round of electrophoresis in an orthogonal direction. In theory, such a protocol should provide one with the ability to visualize hundreds of individual loci. A number of reports have appeared describing the use of two-dimensional DNA fractionation systems of various kinds.[4–17] These reports have clearly demonstrated the

[1] H. Donis-Keller, P. Green, C. Helms, S. Cartinhour, B. Weiffenbach, K. Stephens, T. Keith, D. Bowden, D. Smith, E. Lander, D. Botstein, G. Akots, K. Rediker, T. Gravius, V. Brown, M. Rising, C. Parker, J. Powers, D. Watt, E. Kauffman, A. Bricker, P. Phipps, H. Muller-Kahle, T. Fulton, S. Ng, J. Schumm, J. Braman, R. Knowlton, D. Barker, S. Crooks, S. Lincoln, M. Daly, and J. Abrahamson, *Cell* (*Cambridge, Mass.*) **51**, 319 (1987).

[2] G. L. C. Shen-Ong and M. D. Cole, *J. Virol.* **42**, 411 (1982).

[3] A. J. Jeffreys, V. Wilson, and S. L. Thein, *Nature* (*London*) **314**, 67 (1985).

[4] E. C. Rosenvold and A. Honigman, *Gene* **2**, 273 (1977).

[5] S. G. Fischer and L. S. Lerman, this series, Vol. 68, 183.

[6] R. C. Parker and B. Seed, this series, Vol. 65, 358.

[7] R. DeWachter, J. Maniloff, and W. Fiers, "Gel Electrophoresis of Nucleic Acids: A Practical Approach," p. 1. IRL Press, Oxford and Washington, D.C., 1982.

[8] T. Yee and M. Inouye, *J. Mol. Biol.* **154**, 181 (1982).

[9] T. L. J. Boehm and D. Drahovsky, *J. Biochem. Biophys. Methods* **9**, 153 (1984).

[10] D. A. Jackson and P. R. Cook, *EMBO J.* **4**, 919 (1985).

[11] T. G. Fanning, W. S. Hu, and R. D. Cardiff, *J. Virol.* **54**, 726 (1985).

[12] T. Woolf, E. Lai, M. Kronenberg, and L. Hood, *Nucleic Acids Res.* **9**, 3863 (1988).

[13] K. A. Nawotka and J. A. Huberman, *Mol. Cell. Biol.* **8**, 1408 (1988).

[14] M. A. Walter and D. W. Cox, *Genomics* **5**, 157 (1989).

[15] J. A. Mietz and E. L. Kuff, *Proc. Natl. Acad. Sci. U.S.A.* **87**, 2269 (1990).

[16] M. Yi, L.-C. Au, N. Ichikawa, and P. O. Ts'O, *Proc. Natl. Acad. Sci. U.S.A.* **87**, 3919 (1990).

general feasibility of using two-dimensional gel electrophoresis to enhance the resolution of complex gene families. Here we provide a detailed description of an optimized methodology and apparatus to accomplish this task in a reproducible fashion. These methods have been previously discussed.[17]

Apparatus

A combination of gel tray–gel chamber–tank has been designed for use in both the first and second dimensions of electrophoresis with two different types of companion combs, as shown in Fig. 1A. This apparatus is required to ensure the success of the protocol on a reproducible basis, and it can also be used for traditional one-dimensional agarose gel protocols. The gel chamber–tray combination (Fig. 1A) has been designed to eliminate the possibility of leakage during pouring of the gel.

Methodology

DNA Preparation and Restriction Enzyme Digestion

Genomic DNA is prepared according to standard protocols that yield fragments in the size range of 100–150 kb or greater. Aliquots of DNA (25–40 μg) are digested in volumes of 400–800 μl with *Xba*I in a buffer provided by the vendor. (Other restriction enzymes with 6–8 bp recognition sites can also be used for this first digestion.) Aliquots of enzyme (2.5 units of enzyme/μg of DNA) are added in two steps. On addition of the first aliquot, homogenization of the reaction mixture is performed by pipetting up and down 10 times with a Gilson Pipetman (Rainin, Woburn, MA). The reaction proceeds for at least 3 hr at 37° prior to the addition of the second aliquot of enzyme. This is followed directly by simple vortex-mixing and incubation for an additional 3 to 16 hr.

Confirmation of complete digestion is accomplished by direct minigel analysis of a 20-μl aliquot of the reaction mixture. The ethidium bromide pattern of the experimental sample should be compared to a control sample, known to be completely digested, in an adjacent lane. The sample is extracted once by adding an equal volume of phenol–chloroform–isoamyl alcohol (25 : 24 : 1, v/v), followed by vigorous vortex-mixing and centrifugation for 5 min. The aqueous phase is obtained and adjusted to 0.3 *M* sodium acetate followed by the addition of 2 volumes of ethanol and

[17] R. Sheppard, X. Montagutelli, W. Jean, J.-Y. Tsai, J.-L. Guenet, M. D. Cole, and L. M. Silver, *Mammal. Genome* **1**, 104 (1991).

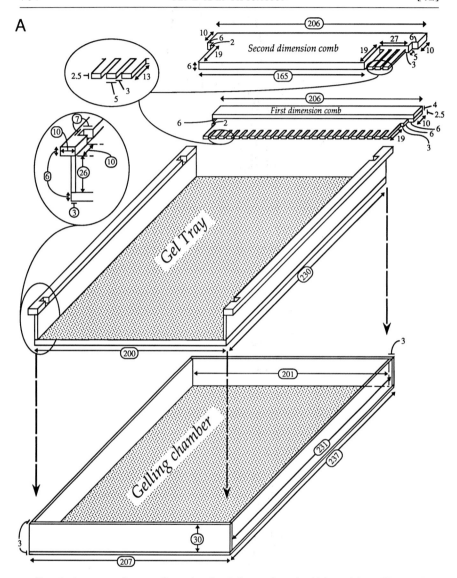

FIG. 1. Apparatus for two-dimensional gel electrophoresis. (Adapted from Sheppard *et al.*[17]) (A) Gel tray, gelling chamber, and combs. The gel tray, gelling chamber, and first-dimension comb are constructed from clear acrylic plastic sheets that are sold under various trade names (Acrylite GP, Plexiglas) and supplied from the manufacturer in 4 foot × 8 foot × 1/4 inch and 4 foot × 8 foot × 1/8 inch sizes. The stated 1/4 inch (6.37 mm) thickness is actually 5.9 mm and is indicated as 6 mm in the diagram; 1/8 inch thickness is actually 2.95 mm and is indicated as 3 mm in the diagram. Clear plastic sheets were obtained from CYRO Industries, Mt. Arlington, NJ 07856, and Rohm & Haas, Philadelphia, PA 19105.

B

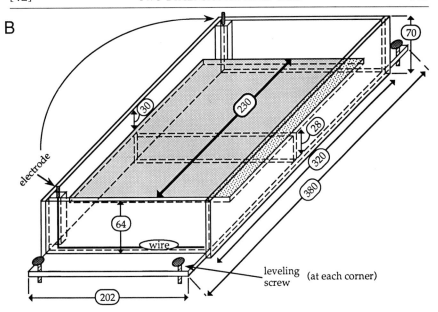

Sections were machine-cut according to the schematic diagrams shown and glued together with acrylic cement, also available from the companies noted above. Combs having teeth with other dimensions can also be constructed for use of the apparatus with routine agarose gels. (B) Electrophoresis tank. The basic construction is entirely of clear acrylic sheets of 5.9 mm thickness. Nylon leveling screws (1/4 inch diameter, 20 threads/inch) were obtained from McMaster-Carr Supply Co., New Brunswick, NJ, and are compatible with screw holes made at each corner of the tank. The electrode wires are 0.25-mm diameter platinum soldered to nickel-plated brass electrodes held in place in an acrylic block as shown. A safety cover (not shown) sits on top of the tank and has two sides that extend approximately 3 cm along the length of the tank. Acrylic sheets of 2.6 and 3.7 mm thickness are used to create the first-dimension comb. The second-dimension comb is constructed from a nylon sheet of 6.3 mm thickness.

vortex-mixing. Precipitation is carried out in an alcohol/dry ice bath for 20 min, or at $-20°$ for at least 1 hr. Precipitated DNA is pelleted by centrifugation for 30 min, and ethanol is removed by pipetting. (Tubes can be recentrifuged briefly to concentrate the remaining liquid for more efficient removal.) Wet pellets are resuspended in 30 μl of 10 mM Tris, 1 mM EDTA. Resuspension is facilitated by pipetting up and down with a Gilson Pipetman at intervals over a period of at least 1 hr at room temperature (or at 65° to facilitate solubilization), followed by the addition of 3 μl of 10× loading buffer (50% (v/v) glycerol, 100 mM EDTA, 2% (v/v) bromphenol blue). Samples can be stored indefinitely at 4° prior to, or after, the addition of loading buffer.

First-Dimension Gel Preparation

To begin the protocol, the gel tray is placed into the gelling chamber, and identical first dimension combs are inserted at both ends; the second comb is used to produce wells for aligning lanes to be excised subsequent to electrophoresis as described in the next section. The comb will sit at a distance of 1 mm from the tray surface. The tray/chamber must be placed on a level surface (crucial because of the small gel volume used) in a room having an ambient temperature of 22° or higher. A gel solution containing 150 ml of 0.5–0.7% (w/v) GTG grade SeaKem agarose (FMC, Rockland, ME) in 1× TBE buffer (90 mM Tris base, 90 mM boric acid, 2 mM EDTA) is prepared and cooled to 55° in a water bath before it is poured into the tray to produce a gel having a theoretical depth of 3.36 mm; owing to evaporation, the actual depth of the solidified gel is usually at or below 3.0 mm.

First-Dimension Sample Loading

Once the gel has solidified, the tray is removed from the gelling chamber and placed into the companion electrophoresis tank (Fig. 1B). TBE buffer is added to a height that covers the gel by 0.5–1 cm, and both combs are removed. One portion of the gel is used for loading 25 μl of each sample, separated from each other by at least one intervening well. The other side of the gel is used for loading small test aliquots (1–2 μl) of each sample along with appropriate size markers. It is useful to retain an additional 2–3 μl aliquot of each sample to be run on the corresponding second dimension gel. Bromphenol blue will readily leach out of this thin gel during the course of the run, so at least one sample should have a higher concentration of the dye (0.4%) than usual to visualize the front for a determination of when to end the run.

First-Dimension Electrophoresis

Electrophoresis is performed at 0.8–1.6 V/cm (distance measured between electrodes) until the bromphenol blue dye front has proceeded to the opposite end of the gel (~24 hr). The area of the gel containing the 25 μl samples is excised as a single unit, notched in one corner for orientation, and placed into a tray containing at least 20 volumes of sterile distilled water. The remaining portion of the gel is stained with ethidium bromide and photographed according to standard procedures.

DNA Digestion within Gel Strips

The gel portion containing the 25 μl samples is rocked gently for 1 hr (fast enough to allow the gel to float back and forth in the tray), followed by a change of water (at least 20 volumes) and additional rocking for at least 1 hr more. At this stage, individual sample-containing lanes are excised for further processing steps as follows. (Clean gloves must be worn during the procedure.) The gel is placed on a dry glass plate, and all excess liquid is eliminated from the sides and surface with paper towels, which are maintained along the sides to prevent lateral motion. A flexible plastic ruler is placed along sample-containing lanes (one side at a time), aligned according to wells at both ends, and a scalpel blade is drawn along the ruler across the length of the gel. The actual excision line should fall halfway between adjacent wells. The portion of the lane between the head of the gel and the well is cut away, as is the portion between the well and the limiting mobility line of DNA estimated from ethidium bromide staining of the control lanes. Further trimming is performed at the foot of the lane to produce a gel strip with a maximum length of 16 cm. Finally, a small cut is made at a 45° angle across the top left-hand corner of the gel strip, for future orientation during placement into the second-dimension gel. The gel strip is pushed from the glass plate onto the flexible plastic ruler and transported to a 7 × 20 cm plastic freezer storage bag sealed on two adjacent sides. [An alternative possibility is to use quart size Ziploc storage bags (Dow Chemical Company, Midland, MI) directly for all incubation steps. It is not necessary to heat-seal the bags, and the internal volume can be adjusted by folding the bag over multiple times and securing with clips.] The same process is repeated for each sample lane in the gel.

The second 7 cm side of each bag is sealed, 30 ml of *Taq* buffer (100 mM NaCl, 10 mM MgCl$_2$, 6 mM 2-mercaptoethanol, 10 mM Tris-HCl, pH 8.4) is added, and the fourth side of the bag is sealed to produce an incubation chamber measuring approximately 19 × 4.5 cm. Bags are rocked at a speed fast enough to allow the gel strip to float gently back and forth at ambient temperature for 2 hr. Subsequently, each bag is opened at one end, buffer is poured off, and the gel is situated as close as possible to two edges of the bag. A new seal is now formed along the length of the gel to reduce the width of the incubation chamber to approximately 1.5 cm. Five milliliters of *Taq* buffer supplemented with nuclease-free bovine serum albumin (BSA) (100 μg/ml) and *Taq*I enzyme (600 units, from New England Biolabs, Beverly, MA) is added to the bag by Pasteur pipette, bubbles are eliminated, and a final seal is made along the fourth edge. Bags containing gel strips are incubated at 65°–68° for 12–16 hr rocking rapidly in a water bath or oven. Some success has been achieved using alternative restriction enzymes for second-dimension digestions when the initial incubation is carried out at 4° for 1–2 hr followed by incubation at the optimal temperature for enzyme activity.

Following *Taq*I digestion, bags are cut open on three sides, and gel strips are released into individual containers with at least 200 ml of 0.1× TBE supplemented with bromphenol blue to 0.005%. Equilibration in this buffer is allowed to continue for 1–2 hr at ambient temperature with gentle rocking, during which time the second-dimension gels are prepared as described below.

Second-Dimension Gel Electrophoresis

The second-dimension gels are formed using the same gel tray and chamber described above, with a special nylon comb shown in Fig. 1A. This comb has a 16.5 cm long tooth, which is 6 mm (1/4 inch) wide, and three additional teeth of normal size (5 × 2.5 mm). The comb will sit at a distance of 2 mm from the bottom of the gel tray. We use 500 ml of 0.8% (w/v) agarose (type II, medium EEO grade, Sigma, St. Louis, MO, or LE grade SeaKem, FMC) in 1× TBE to prepare a gel having a theoretical depth of 11 mm. After the gel is poured and the comb is put in, bubbles that may form along the front side are eliminated. The gel is allowed to solidify for at least 1 hr.

After removing the comb from the second-dimension gel, the first-dimension gel strip is transferred from the equilibration buffer onto a flexible plastic ruler (cut to a length of 16.5 cm), which is used to transfer the strip into the long well. The strip should be oriented such that the notched corner is positioned upward at the left side of the well; this allows the DNA-containing side of the strip to abut the front of the well. Using a spatula, one should press the gel strip along the front wall of the well, taking care to eliminate all bubbles between the strip and the wall. The well is now filled with molten (55°) 0.8% agarose in 0.25× TBE, which is allowed to harden for a maximum of 10–30 min. The gel tray is transferred to the electrophoresis tank, and 1× TBE is added to a height that covers the gel by approximately 1 cm. Appropriate control samples and size markers are loaded into the small wells, and electrophoresis is carried out at 0.8–1.6 V/cm until the bromphenol blue dye front has proceeded to the edge of the gel.

Gel Processing and Blotting

The gel is stained with ethidium bromide and photographed. Denaturation is accomplished by rocking in 1 liter of 0.6 *N* NaOH, 1.5 *M* NaCl for at least 2–3 hr (the increased concentrations of NaCl and NaOH, as well as the extended time of equilibration are due to the large gel volume), followed by blotting, from the backside of the gel, in 0.2 *N* NaOH, 0.5 *M* NaCl, onto GeneScreen filter membranes (New England Nuclear, Boston,

MA) for at least 24 hr, as originally described by Southern.[18] The blotting protocol is easily carried out within the gel tank used for electrophoresis, and the empty gel chamber can be placed on top of the paper towels to provide a uniform base for weighting. Filters are neutralized by immersion in 40 mM sodium phosphate (pH 7) for 1 min, patted dry between Whatman (Clifton, NJ) 3MM paper, transferred to a fresh Whatman 3MM paper sandwich, baked for 30 min at 80°–85°, and exposed to UV light to cross-link DNA according to standard procedures. Probes are labeled according to the procedure of Feinberg and Vogelstein,[19] and multiple hybridizations are performed according to the method of Church and Gilbert.[20] Two probes were used in the examples presented. The pH2IIa probe was used to detect members of the *H-2* class I gene family.[21] The pIII/3 probe was used to detect members of the VL30 family of endogenous retrovirus-like sequences.[22]

Time Schedule for Complete Protocol

Day 1: Start with digested DNA samples suspended in sample loading buffer. Build first-dimension gel, load samples, and begin electrophoresis.

Day 2: End first-dimension run. Stain and process gel. Excise and process gel strips. Begin overnight *Taq*I incubation.

Day 3: Equilibrate gel strips. Build second-dimension gels and begin electrophoresis.

Day 4: End second-dimension runs. Stain and denature gels. Begin overnight blotting onto filter membranes.

Day 5: Neutralize, bake, and cross-link filter membranes. Prehybridize and hybridize according to standard procedures.

Rationale and Choice of Reagents

Apparatus and Protocol. It is important to consider several factors in the design of the apparatus and the protocol. First, the first-dimension gels must be sufficiently shallow to permit complete diffusion of *Taq*I enzyme throughout each gel strip in a reasonable amount of time. Another critical optimization factor is the equilibration of the *Taq*I-treated first-dimension gel strips in 0.1 × TBE prior to second-dimension electrophore-

[18] E. Southern, *J. Mol. Biol.* **98,** 503 (1975).
[19] A. P. Feinberg and B. Vogelstein, *Anal. Biochem.* **137,** 266 (1984).
[20] G. M. Church and W. Gilbert, *Proc. Natl. Acad. Sci. U.S.A.* **81,** 1991 (1984).
[21] M. Steinmetz, J. G. Frelinger, D. Fusher, T. Hunkapiller, D. Pereira, S. M. Weissman, H. Uehara, S. Nathenson, and L. Hood, *Cell (Cambridge, Mass.)* **24,** 125 (1981).
[22] G. Rotman, A. Itin, and E. Keshet, *Nucleic Acids Res.* **14,** 645 (1986).

M 1 2   X,T

FIG. 2. Example of an ethidium bromide-stained two-dimensional gel. Adapted from Sheppard *et al.*[17]) Lane M contains λ/*Hin*dIII size markers, and lanes 1 and 2 represent two mouse DNA samples that have been doubly digested with *Xba*I and *Taq*I. The triangular pattern of fractionated DNA represents a two-dimensionally fractionated sample that was digested by *Xba*I before electrophoresis in the first dimension and by *Taq*I before electrophoresis in the second dimension.

sis in a 0.8% agarose–1× TBE gel. The salt and agarose concentration differentials created between the gel strip and the surrounding gel within which it is embedded work together to focus DNA fragments into well-defined spots.

Choice of Restriction Enzymes. Two factors are important to consider in choosing the first restriction enzyme. First, the restriction site recognized should be present sufficiently less frequently in the genome than the site recognized by the second-dimension enzyme so that independent separation in two dimensions actually occurs for a majority of genomic

A/J

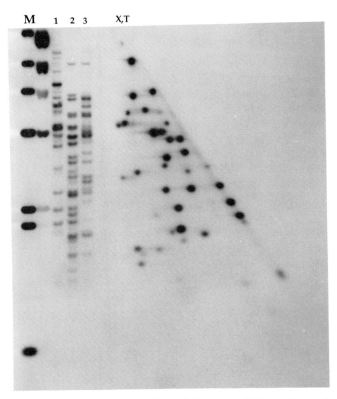

FIG. 3. One- and two-dimensional analysis of the mouse *H-2* class I gene family from strain C57BL/10I. (Adapted from Sheppard *et al.*[17]) Lane M contains markers as described for Fig. 2; lane 1 represents mouse C57BL/10J DNA digested with *Xba*I; lane 2 contains double-digested DNA (*Xba*I and *Taq*I); and lane 3 represents digestion with *Taq*I. The two-dimensional pattern was generated by digestion with *Xba*I for first-dimension fractionation and *Taq*I for second-dimension fractionation. Hybridization was performed with the pH2IIa probe.[21]

fragments. This condition is met by all 6-bp recognition enzymes. However, enzymes that detect sites with a CpG sequence are particularly useful since this dinucleotide is underrepresented in the mammalian genome. With the use of field inversion electrophoresis in the first dimension,[23] one can alter the size window of fractionation to accommodate

[23] G. F. Carle, M. Frank, and M. V. Olson, *Science* **232,** 65 (1986).

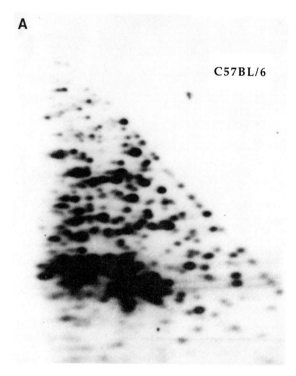

FIG. 4. Two-dimensional analysis of the VL30 gene family. (Adapted from Sheppard *et al.*[17]) Digestion and fractionation conditions are the same as described in Fig. 3. (A) C57BL/6 DNA; (B) *M. spretus* DNA; (C) C57BL/6 × *M. spretus* F_1 hybrid DNA. Hybridization was performed with the pIII/3 probe.[22]

samples digested with moderately rare-cutting enzymes such as *Xho*I, *Sma*I, *Xma*I, *Spl*I, *Sal*I, *Sac*II, and *Sfi*I.

The second factor concerns the presence of the restriction site of interest within the repetitive element being studied. Ideally, one would choose an enzyme that does not cut within any of the members of the DNA family; this would provide for a one-to-one relationship between each resolvable DNA spot and a particular genomic locus. Although in practice this will be difficult, if not impossible, to accomplish, one should begin with an enzyme that does not cut within the probe used for analysis.

For the second-dimension fractionation, *Taq*I is the enzyme of choice for several reasons. First, and most importantly, it works consistently within agarose strips. This is not true for a number of other enzymes that

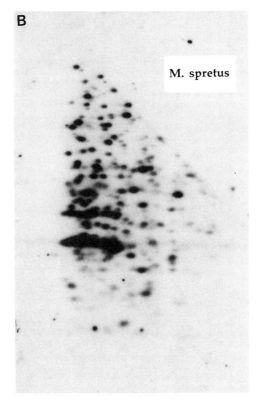

FIG. 4B.

were tested. Second, the required incubation temperature of 65° acts to inhibit potential nuclease activities. Finally, the frequency of TaqI sites within mammalian genomes is greater than that of all restriction enzymes with 6-bp recognition sites; this means that most first-dimension-fractionated fragments will be independently fractionated in a second dimension as well.

Results and Discussion

Ethidium Bromide Staining Pattern

Figure 2 shows an example of the ethidium bromide staining pattern obtained with mouse DNA digested with XbaI for first-dimension separation and TaqI for second-dimension separation. Fractionated DNA is

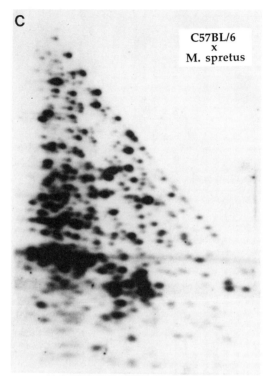

FIG. 4C.

observed within a triangular pattern. The intensely stained diagonal represents the subset of XbaI fragments that do not contain any TaqI sites. Incomplete TaqI digestion would be observed as a more intensely stained diagonal relative to the smear below it. To confirm that complete digestion has occurred during both enzyme incubation steps, one can hybridize blots with any single-copy genomic probe known to detect only a single band in both XbaI and TaqI one-dimensional patterns.

H-2 Class I Gene Family

Twenty H-2 class I-related DNA fragments can be resolved with a traditional one-dimensional gel separation of DNA that has been doubly digested with XbaI and TaqI. In contrast, 42 spots can be resolved readily within the two-dimensional pattern generated with the use of the same two enzymes in different dimensions (Fig. 3). In several instances where a single band appears in one dimension, two to four spots can be clearly observed in two dimensions. Band F (see Ref. 17) corresponds to spots 8, 9, 10, and 13; band G corresponds to spots 11 and 12; band H corresponds to spots 14 and 15; the doublet D/E corresponds to spots 5, 6, and 7; and the doublet N/O corresponds to spots 24, 25, and 26. The two-dimensional pattern shown is likely to represent the maximum resolution possible with this gene family, which is known to have approximately 30 cross-hybridizing members. Because the number of observed spots is greater than the number of loci, it is clear that some loci have become fragmented into two or more spots on digestion with either or both *Xba*I and *Taq*I.

VL30 Family of Retrovirus-like Sequences

Two-dimensional patterns obtained with a more highly repeated, dispersed, endogenous retrovirus-like gene family called VL30 are shown in Fig. 4. The pattern in Fig. 4B is from an inbred *Mus spretus* mouse strain (SEG), and that in Fig. 4C is from an F_1 hybrid between SEG and the inbred laboratory strain C57BL/6J (B6; cf. Fig. 4A). As expected from the fact that the two genomes have a 50% identity, a large number of the same spots are present in both patterns (examples are indicated with lowercase letters; see Ref. 17). This result serves to establish the reproducibility of the protocol, and it also demonstrates the ease with which one can compare and align complex patterns obtained with the same probe but different genotypes. It is possible to count approximately 135 spots in the SEG × B6 pattern shown in Fig. 4C. Similar results were obtained with the use of a probe that detect the type II subfamily if the intracisternal type A particle (IAP) elements.[17]

Potential Applications

Two-dimensional analysis of dispersed, moderately repetitive gene families has applications to a number of different areas of research. These include the following: (1) the use of individual complex family loci as genetic markers in high-resolution mapping experiments; (2) monitoring of the genetic purity of inbred strains of mice; (3) whole-genome fingerprinting of somatic cell hybrid lines; (4) restricted cloning of individual loci from complex families via partial transfer methodologies; and (5) correlations of phenotypic loss or gain of function with large-scale deletions or chromosome loss.

Acknowledgments

This research was supported by grants from the National Institutes of Health to L.M.S. In addition, L.M.S. thanks the American Cancer Society for an ACS Scholar award used to conduct a portion of this work at the Pasteur Institute. We thank Don Peoples, at Princeton University, for construction and advice on the electrophoresis apparatus.

Section IX

Nuclear Transplantation

[43] Transplantation of Nuclei to Oocytes and Embryos

By KEITH E. LATHAM and DAVOR SOLTER

Introduction

Nuclear transplantation is a valuable technique for investigating the mechanisms that control embryogenesis and is broadly applicable to two basic types of analyses. First, it allows the investigator to evaluate the genetic equivalence and developmental potential of diverse types of nuclei. For example, nuclear transplantation experiments in amphibians indicate that nuclei from cells of early stages of embryogenesis can support development of fertile adult individuals.[1] Additionally, nuclei from several differentiated somatic cell types and from primordial germ cells can support development to advanced stages.[1-3] The low frequency with which this occurs, however, indicates significant developmental restrictions in somatic cell nuclei, but less severe restrictions in germ-lineage nuclei.[3] In mammals, nuclear transplantation experiments revealed the existence of gamete-of-origin dependent epigenetic modifications of male and female pronuclei (i.e., genomic imprinting).[4-6] The second type of analysis for which nuclear transplantation is well suited is in the study of nuclear–cytoplasmic interactions. Nuclear reprogramming and the acquisition of transcriptional competence[7-9] have been documented using this technique, as have interactions between pronuclei and the egg cytoplasm.[10,11] Although nuclear transplantation has been applied primarily to the study of amphibian embryos, these recent experiments with mouse embryos demonstrate that the technique is likely to find additional applications in mammalian embryology as the methods for molecular analysis of gene expression become more sensitive. This chapter summarizes the procedures used in

[1] M. A. DiBerardino, N. J. Hoffner, and L. D. Etkin, *Science* **224**, 946 (1984).
[2] J. B. Gurdon and V. Uehlinger, *Nature (London)* **210**, 1240 (1966).
[3] M. A. DiBerardino and N.J. Hoffner, *Exp. Zool.* **176**, 61 (1971).
[4] J. McGrath and D. Solter, *Cell (Cambridge, Mass.)* **37**, 179 (1984).
[5] S. C. Barton, M. A. Surani, and M. L. Norris, *Nature (London)* **311**, 374 (1984).
[6] M. A. Surani, S. C. Barton, and M. L. Norris, *Nature (London)* **308**, 548 (1984).
[7] K. E. Latham, D. Solter, and R. M. Schultz, *Mol. Reprod. Dev.* **30**, 182 (1991).
[8] K. E. Latham, D. Solter, and R. M. Schultz, *Dev. Biol.* **149**, 457 (1992).
[9] S. K. Howlett, S. C. Barton, and M. A. Surani, *Development (Cambridge, UK)* **101**, 915 (1987).
[10] K. E. Latham and D. Solter, *Development (Cambridge, UK)* **113**, 561 (1991).
[11] C. Babinet, V. Richoux, J.-L. Guenet, and J.-P. Renard, *Development (Cambridge, UK)* **81**(Suppl.), (1990).

Copyright © 1993 by Academic Press, Inc.
All rights of reproduction in any form reserved.

the transfer of pronuclei, blastomere nuclei, somatic cell nuclei, and oocyte germinal vesicles between embryos and for the removal of metaphase spindles from mature oocytes.

Equipment

The procedure for transferring pronuclei has been described in detail.[4,12,13] A brief summary of the basic technique follows. The following materials are required:

Manipulation chamber

Holding pipette attached to a 25- to 50-ml plastic syringe by a Leitz instrument tube and plastic tubing

Enucleation/transfer pipette

Oil-filled syringe equipped with a mechanism for fine movement (e.g., Beaudouin, Paris, France; Nikon IM-6; or Hamilton, Reno, NV) with attached Leitz instrument tube

Inactivated Sendai virus

Embryo culture medium (e.g., Whitten's medium plus 100 μM EDTA),[14,15] with and without 5 μg/ml cytochalasin B and 0.1 μg/ml Colcemid (N-deacetyl-N-methylcochicine)

Embryo manipulation medium (e.g., HEPES-buffered Whitten's), with 5 μg/ml cytochalasin B and 0.1 μg/ml Colcemid

Compound microscope with 10–15× eyepieces and 40× objective with long working distance

Two micromanipulators (e.g., Leitz)

Manipulation Chamber

In most cases, it will be necessary to segregate embryos according to the manipulations that are performed on them. For this purpose, embryos are placed individually in separate droplets of medium under oil. For conventional microscopes, embryos are manipulated in hanging drops. A siliconized (Prosil-treated) coverslip with droplets is supported on two glass strips (2.5–3 cm apart) attached to a glass slide (4 × 4.5 × 0.3 cm), leaving two sides open for access to the embryos. The space between the coverslip and the glass slide is filled with silicone oil (100 centistokes viscosity). Pipettes are inserted horizontally between the coverslip and

[12] J. McGrath and D. Solter, *Science* **220**, 1300 (1983).
[13] B. Hogan, F. Costantini, and E. Lacy, "Manipulating the Mouse Embryo: A Laboratory Manual," p. 197. Cold Spring Harbor Laboratory, Cold Spring Harbor, New York, 1986.
[14] W. K. Whitten, *Acta Biosci.* **6**, 129 (1971).
[15] J. Abramczuk and D. Solter, *Dev. Biol.* **61**, 378 (1977).

A

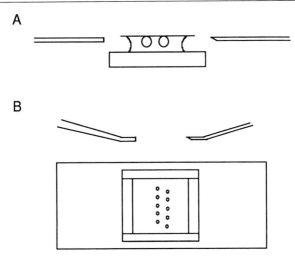

B

FIG. 1. Diagram showing chambers for performing nuclear transplantations using either (A) conventional upright optical configuration or (B) inverted microscope. For the upright microscope, single embryos are placed in hanging drops suspended from a Prosil-treated coverslip that is supported at front and back by two glass strips. The space between the coverslip and glass slide is filled with oil. Pipettes are inserted into this space from either side. For an inverted microscope, single embryos are placed in droplets under oil within a reservoir created by attaching a glass frame to a Prosil-treated glass slide with wax. The pipettes are angled near the tips to provide a horizontal approach to the embryos.

the slide (Fig. 1A). For inverted microscopes, a convenient chamber can be made by attaching a glass frame to a Prosil-treated glass slide (5 × 7.5 cm) with wax to create a reservoir. Droplets are made on the glass slide inside the frame and the reservoir filled with oil (Fig. 1B). The tips of the pipettes can be angled slightly to provide a horizontal approach to the embryos.

Pipettes

Holding and transfer pipettes are prepared as described.[13,16] Briefly, holding pipettes are prepared from capillaries drawn to an outside diameter of approximately 100 μm. The holding pipette is broken at a convenient point using a microforge (e.g., DeFonbrune) with a thick filament (0.12–0.3 mm). After breaking to create a flat tip, the tip is polished, leaving an opening of approximately 20 μm.

[16] B. Hogan, F. Costantini, and E. Lacy, "Manipulating the Mouse Embryo: A Laboratory Manual," p. 164. Cold Spring Harbor Laboratory, Cold Spring Harbor, New York, 1986.

Transfer pipettes are prepared from soft flint glass (e.g., type R-6, 1.0 mm OD, 0.65 mm ID; Drummond Scientific, Broomall, PA). The pipette is pulled on a vertical or horizontal puller (e.g., Sutter Instruments, Novato, CA, Model P-87), and the tip is broken to an outside diameter of approximately 20 μm on a microforge using a thin filament (0.1 mm). The tip is beveled by grinding with diamond paste or by using a specially designed grinding wheel (Narishige Scientific Instruments or Sutter Instruments). After beveling, the tip is rinsed once with dilute (10%, v/v) hydrofluoric acid and 5–6 times with deionized water. The wall of the tip is thinned on its outer surface by dipping into the dilute acid while expelling air. The tip is rinsed again with water and then with 95% (v/v) ethanol. Using the microforge, a small point is pulled on the leading edge of the bevel. This is accomplished by touching the tip of the pipette to the thin filament heated just enough to fuse the glass to the filament and then drawing the pipette away to form a spike. If the tip is too long after being drawn out, the excess length can be broken off to leave a shorter tip that can pierce the zona pellucida without damaging the embryo. After sharpening, the pipette is rinsed 5–6 times with 100% Nonidet P-40 (NP-40) and 10–12 times with deionized water. Excess liquid is removed from the tip with a Kimwipe, and the pipette is stored overnight in a dustproof container. It is helpful to repeat the NP-40 treatment on pipettes that are not used within a few weeks of preparation. The enucleation pipette is filled with silicone oil (50 centistokes viscosity) and attached to an oil-filled syringe with plastic tubing (e.g., Tygon tubine), avoiding bubbles.

Pronuclear Transplantation

Embryos are cultured at 37° for at least 20 min in Whitten's medium containing cytochalasin B and Colcemid prior to manipulation in HEPES-buffered Whitten's medium with the same cytoskeletal inhibitors. Micromanipulations can typically be performed at approximately 20–22 hr following injection with human chorionic gonadotropin (hCG). Male and female pronuclei remain identifiable until approximately 27 hr post-hCG. Embryos are placed singly in separate droplets of medium. In addition, a small droplet of inactivated Sendai virus diluted (2500–3000 hemagglutinating units/ml) in HEPES-buffered medium is placed on the slide. The dilution of the Sendai virus should be determined for each batch of inactivated virus that is prepared and should be sufficient to avoid adhesion between the karyoplast and pipette while still providing for efficient fusion of karyoplasts to recipient embryos. The tip of the transfer pipette is gently inserted under the zone pellucida of the recipient embryo without penetrating the plasma membrane. Lysis of the embryo most frequently occurs when the transfer pipette is inserted too far and contact is made with the opposite surface of the embryo. For ease in relocating the original opening in the zona, it is convenient to hold the embryo near the polar body and pierce the zona at a point that is opposite to the polar body.

The beveled opening of the transfer pipette tip is placed adjacent to a pronucleus and, with suction, the overlying membrane, a minimal amount of intervening cytoplasm, and the pronucleus are drawn into the pipette. Suction is applied with a Beaudouin syringe or similar device. Once the desired pronucleus is inside the pipette, the pipette is slowly withdrawn to remove the pronucleus within a membrane-bound karyoplast. One or both pronuclei can be extracted. An appropriate pronucleus is taken from a donor embryo and then a small volume (approximately equal to the karyoplast volume) of inactivated Sendai virus is drawn into the pipette. The tip of the pipette containing the pronucleus and inactivated Sendai virus is placed under the zona of the recipient embryo through the original opening, and the pronucleus and virus are slowly expelled from the pipette. After all of the embryos on the slide have been manipulated, they are cultured at 37° in Whitten's medium without cytoskeletal inhibitors. Fusion of the karyoplast should occur within 20 min and can be verified visually. Intermediate stages of a successful transfer are shown in Fig. 2. Male and female pronuclei are distinguishable on the basis of location and size, with the smaller female pronucleus located closer to the polar body (Fig. 2).

Electrofusion

Electrofusion offers an alternative that is useful in cases where Sendai virus-mediated fusion cannot be employed.[17,18] The electrofusion technique is described in detail by Tsunoda and co-workers.[18] These investigators optimized electrofusion conditions based on the fusion of 2-cell stage blastomeres and then applied these conditions to the fusion of karyoplasts with enucleated 1-cell embryos. The embryos are placed in a small glass petri dish containing phosphate-buffered saline (PBS) and fused with two pulses of either 15 V for 15 μsec or 10 V for 200 μsec. The two pulses are administered 3 sec apart from two platinum electrodes mounted on micromanipulators and separated by a distance of 100 μm.

Electrofusion under these conditions is less efficient than Sendai virus-mediated fusion with respect to both the efficiency of fusion (78 versus 94%, respectively) and the efficiency of development to the blastocyst stage (67 versus 94%, respectively). The overall efficiency of 52% of

[17] J. Kubiak and A. K. Tarkowski, *Exp. Cell Res.* **157**, 561 (1985).
[18] Y. Tsunoda, Y. Kato, and Y. Shioda, *Gamete Res.* **17**, 15 (1987).

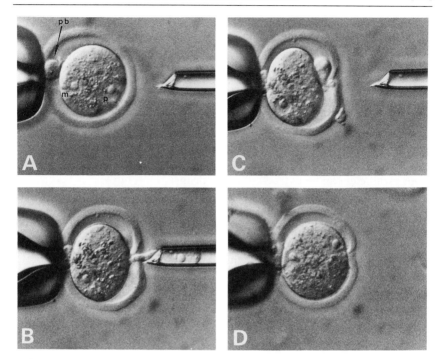

FIG. 2. Transplantation of pronuclei. (A) An embryo is held in the region of the polar body (pb), with maternal (m) and paternal (p) pronuclei visible. (B) The male pronucleus is removed. (C) The donor karyoplast containing the female pronucleus from another egg is placed in the perivitelline space along with inactivated Sendai virus. (D) The karyoplast fuses with the recipient embryo to generate a gynogenone containing two female pronuclei.

manipulated embryos developing to the blastocyst stage with electrofusion compares with 88% blastocyst formation achieved with Sendai virus-mediated fusion, which in other reports was in excess of 90%.[12] Thus, although electrofusion can produce embryos that develop at a high frequency, the overall efficiency is less than that achieved with Sendai virus-mediated fusion. An additional limitation of electrofusion is that large and small cells do not fuse with each other as readily as with Sendai virus.[19]

Cutting of Zona Pellucida

An alternative method of nuclear transplantation involves cutting the zona pellucida with a fine glass needle and then transplanting nuclei with

[19] R. S. Prather, F. L. Barnes, M. M. Sims, J. M. Robl, W. H. Eyestone, and N. L. First, *Biol. Reprod.* **37,** 859 (1987).

a pipette that is not beveled.[20] The glass needle is inserted into the perivitelline space of both donors and recipients and rubbed against the wall of the holding pipette to tear the zona. The blunt transfer pipette is pushed through the tear in the zona, and the nuclei are transferred as described above. Although this method incorporates an additional step into the procedure, it eliminates the need for beveling the transfer pipette and then pulling a point at the tip. The efficiency of transfer and subsequent embryonic development is comparable for the two methods.

Other Manipulations

Transfer of Cleavage-Stage Nuclei

The procedure described above can be used to transfer nuclei between embryos of different ages. Development to term has been obtained in mice following transplantation of nuclei from 4- and 8-cell stage blastomeres into 2-cell stage recipient cytoplasms, although the frequency of development decreases with the age of the donor.[21] Additionally, identical twin and triplet mice have been produced by transplanting nuclei from 2- and 4-cell stage blastomeres into enucleated 2-cell stage blastomeres.[22] Development following transplantation of blastomere nuclei into enucleated 1-cell stage cytoplasms, however, fails to produce embryos that develop to term, although limited development can occur.[23]

The fraction of recipient embryos that develop after transferring cleavage-stage nuclei to enucleated 1-cell recipients is affected by the position of the donor and recipient embryos in the cell cycle at the time of manipulation.[24,25] For experiments in which nuclei from 1- or 2-cell stage donors are transplanted to enucleated 1-cell stage recipients, development is improved by combining cytoplasts and karyoplasts of the same cell cycle phase and by the use of late, as opposed to early, 1-cell stage recipient cytoplasms.[24] Furthermore, the efficiency of cleavage following transplantation to enucleated 1-cell stage recipients is greater when early 1- or 2-cell stage nuclei are transplanted to late 1-cell stage cytoplasms than when late nuclei are transplanted to early cytoplasms.[25] This indicates that it is easier for a nucleus from an early cell cycle phase to synchronize

[20] Y. Tsunoda, T. Yasui, K. Nakamura, T. Uchida, and T. Sugie, *J. Exp. Zool.* **240**, 119 (1986).
[21] Y. Tsunoda, T. Yasui, Y. Shioda, K. Nakamura, T. Uchida, and T. Sugie, *J. Exp. Zool.* **242**, 147 (1987).
[22] T. Kono, Y. Tsunoda, and T. Nakahara, *J. Exp. Zool.* **257**, 214 (1991).
[23] J. McGrath and D. Solter, *Science* **226**, 1317 (1984).
[24] L. C. Smith, I. Wilmut, and R. H. F. Hunter, *J. Reprod. Fertil.* **84**, 619 (1988).
[25] L. C. Smith, I. Wilmut, and J. D. West, *J. Reprod. Fertil.* **88**, 655 (1990).

with a later cytoplasm than it is for a later cell cycle phase nucleus to synchronize with an earlier cytoplasm.[25] The reason for this is not known but may be related to a need for completion of certain events in the recipient embryo prior to enucleation. This is an important consideration for experiments that seek to evaluate the degree to which blastomere nuclei can be reprogrammed. Although cleavage is apparently not essential for some of the events that occur during the second cell cycle,[7,9] the possibility that cleavage may affect other events remains. Selection of donor and recipient embryos of ages that allow cleavage to occur may simplify such experimental analyses.

For experiments involving the transplantation of nuclei from 8- or 16-cell stage embryos, the removal of donor nuclei is facilitated by incubating the donors in calcium-free Whitten's medium containing cytoskeletal inhibitors. Incubation in calcium-free medium disrupts the calcium-dependent cell–cell adhesion system and facilitates the removal of karyoplasts without rupturing the blastomeres.

Transfer of Inner Cell Mass and Cultured Cell Nuclei

Nuclei from isolated inner cell mass (ICM) and cultured cells can also be transferred to enucleated embryos, provided that the donor cell type has receptors for Sendai virus and is of a sufficient size to provide good surface contact. ICM cells can be obtained directly from blastocyst stage embryos by immunosurgery.[26] Embryos containing nuclei from ICM cells, embryonic stem (ES) cells, and parietal endoderm-like PYS-2 cells have been constructed by fusing whole ICM, ES, or PYS-2 cells with enucleated 1-cell stage recipients.[7] Because of the comparatively small size of the ICM cells and some cultured cells, the amount of cell contact and efficiency of fusion can be reduced. The efficiency of fusion can be improved by culturing manipulated embryos briefly (3–5 min, 37°) in hypotonic Whitten's medium diluted by 20% with deionized water. During this incubation, the embryo swells and presses against the transferred cell. The embryos should be watched carefully and removed from the hypotonic medium before lysis or extrusion of the transferred cell from the perivitelline space occurs. In some experiments where nuclear–cytoplasmic interactions were investigated, up to five ES or PYS-2 cells were placed in the perivitelline space to improve the probability of introducing at least one ES or PYS-2 cell nucleus into the recipient embryo.[7] Fusion can usually be verified by using Nomarski optics to visualize the small somatic cell nuclei.

[26] D. Solter and B. B. Knowles, *Proc. Natl. Acad. Sci. U.S.A.* **72,** 5099 (1975).

Transplantation of Nuclei into Metaphase II Oocytes

For some experimental purposes, it is useful to transfer nuclei to unfertilized metaphase II oocytes. Both primordial germ cell and ICM nuclei have been transplanted to unfertilized eggs.[27] This can be done by first removing the meiotic spindle and chromosomes from the unfertilized egg. The chromosomes can be visualized by pretreatment with the fluorescent DNA intercalating agent Hoechst 33342 (1 μg/ml) for 5 min at 37°.[27] Alternatively, to avoid the use of Hoechst 33342 and UV illumination, the spindle can be visualized by polarized light microscopy; mouse eggs are relatively free of birefringent material, and the spindle birefringence can be readily detected (Fig. 3). Unfertilized eggs are collected from superovulated females and incubated for 20 min at 37° in Whitten's medium containing 5 μg/ml cytochalasin B but no Colcemid. After incubating in cytochalasin B, the eggs are placed on the slide for manipulation in HEPES-buffered medium containing both 5 μg/ml cytochalasin B and 0.1 μg/ml Colcemid. Typically, six eggs are placed on the slide at one time, and spindles are removed when an approximately 50% reduction in spindle birefringence has been induced by the presence of Colcemid.

An egg is held on the holding pipette and examined while rotating the compensator (maximum extinction of $\lambda/2$) slightly off extinction in both directions. The spindle, which appears as a bright/dark signal in the cytoplasm (Fig. 3), is removed as described above for pronuclear removal. If the birefringence is not immediately apparent, the egg should be reoriented; the strength of birefringence will be greatest when the spindle is horizontal and oriented 45° to the polarizer axis. Following the introduction of nuclei, the egg is activated with medium containing 7% (v/v) ethanol for 7 min at room temperature.

Transfer of Germinal Vesicles

Intact germinal vesicles (GVs) from fully grown oocytes can also be transferred to enucleated embryos.[7] GV-intact oocytes are isolated from mice 46–48 hr following injection with pregnant mare serum gonadotropin (PMSG). The oocytes are isolated, cultured, and manipulated in the presence of 0.2 mM 3-isobutyl-1-methylxanthine (IBMX) to prevent GV breakdown. A slightly higher concentration of cytoskeletal inhibitors (7 μg/ml cytochalasin B and 0.2 μg/ml Colcemid) is used. It is difficult to penetrate the zona pellucida of a GV-intact oocyte using the larger-diameter transfer pipettes required to accommodate the GV. If the oocyte cytoplasm is not

[27] Y. Tsunoda, T. Tokmuna, H. Imai, and T. Uchida, Development (Cambridge, UK) 107, 407 (1989).

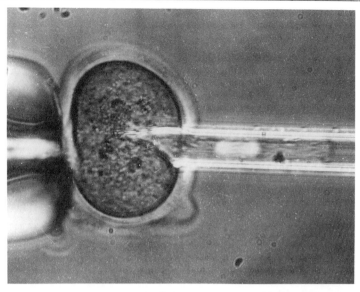

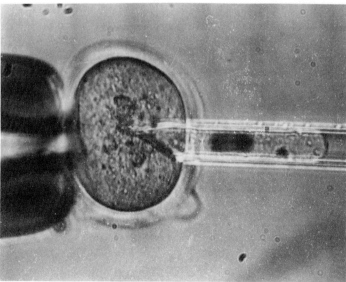

FIG. 3. Removal of chromosomes and spindle from a mature metaphase II arrested oocyte. The spindle with attached chromosomes (in pipette) appears bright with additive compensation (top) and dark with subtractive compensation (bottom) under polarized light.

required as a recipient for other nuclei, the zona can be removed with acidified Tyrode's buffer (0.137 M NaCl, 2.7 mM KCl, 1.6 mM CaCl$_2$, 0.5 mM MgCl$_2$, 0.1% glucose, 0.4% polyvinylpyrrolidone, pH 2.5). Recipients are enucleated in advance as described for pronuclear removal. The zona-free oocyte is held by gentle suction with a holding pipette that has an opening of only 7–10 μm. A transfer pipette approximately 25 μm in diameter is made as described above except that the beveled tip is left without the sharp point normally required to pierce the zona. The tip of the transfer pipette is placed adjacent to the GV, and the GV is drawn into the pipette slowly. Once the GV is inside, the transfer pipette is moved away slowly while applying slight suction with the holding pipette until the cytoplasm divides near the tip of the transfer pipette (Fig. 4). The GV is placed under the zona of the enucleated recipient embryo through the original opening along with inactivated Sendai virus.

Because of its size, an intact GV is more difficult to transfer than pronuclei, and a somewhat lower efficiency is to be expected. Enucleated fully grown oocytes can be used as recipients for other nuclei by removing the GV with the zona intact and following the procedures outlined for pronuclear transfer. Alternatively, cutting the zona (see above)[20] should allow karyoplasts or whole cells to be fused to the enucleated oocyte by placing them in the perivitelline space. Other investigators have fused zona-free oocyte cytoplasms with cells or blastomeres by aspirating them into a pipette with a narrow lumen.[28]

Nuclear Transplantations in Other Species

Nuclear transplantation has been applied quite successfully to embryos of other species such as sheep, pigs, cows, and rabbits.[29–38] Unlike the mouse embryo, embryos of these species have proved amenable to cloning by transplantation of nuclei from blastomeres to the metaphase II oocyte.

[28] A. K. Tarkowski and H. J. Balakier, *J. Embryol. Exp. Morphol.* **55,** 319 (1980).

[29] R. S. Prather and N. L. First, *Int. Rev. Cytol.* **120,** 169 (1990).

[30] R. S. Prather and N. L. First, *J. Reprod. Fertil. Suppl.* **40,** 227 (1990).

[31] S. M. Willadsen, *Nature (London)* **320,** 63 (1986).

[32] M. E. Westhusin, J. H. Pryor, and K. R. Bondioli, *Mol. Reprod. Dev.* **28,** 119 (1991).

[33] J. M. Robl, R. Prather, F. Barnes, W. Eyestone, D. Northey, B. Gilligan, and N. L. First, *J. Anim. Sci.* **64,** 642 (1987).

[34] R. S. Prather, M. M. Sims, and N. L. First, *Biol. Reprod.* **41,** 414 (1989).

[35] R. S. Prather, M. M. Sims, and N. L. First, *J. Exp. Zool.* **255,** 355 (1990).

[36] P. Collas and J. M. Robl, *Biol. Reprod.* **43,** 877 (1990).

[37] S. L. Stice and J. M. Robl, *Biol. Reprod.* **39,** 657 (1988).

[38] X. Yang, L. Zhang, A. Kovacs, C. Tobback, and R. H. Foote, *Mol. Reprod. Dev.* **27,** 118 (1990).

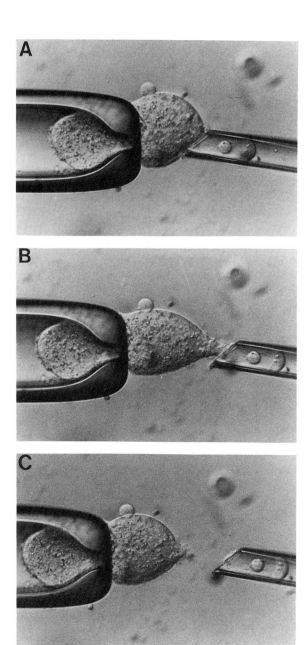

FIG. 4. Removal of the germinal vesicle from a fully grown oocyte. After removal of the zona pellucida with acidified Tyrode's buffer, the oocyte is held by gentle suction while the tip of the transfer pipette is placed adjacent to the germinal vesicle. After drawing the GV into the pipette by suction (A), the transfer pipette is slowly withdrawn (B) while maintaining suction with the holding pipette until the karyoplast containing the GV is pulled free (C). The GV can then be transferred to a recipient embryo that has previously had an opening made in its zona.

TABLE I
EFFICIENCY OF NUCLEAR TRANSPLANTATION IN MAMMALIAN EMBRYOS

Species	Donor stage	Recipient stage	Sendai virus fusion (%)	Electrofusion (%)	Activation (%)	Blastocysts (%)[a]	Refs.
Sheep	8-Cell	Oocyte	50	90	n.e.[b]	33–48	31
Cow	5-Day	Oocyte	8	77	90	20.7	29, 32, 33
Pig	1-Cell	1-Cell	—	76	69	38	34
Rabbit	32-Cell	Oocyte	—	94	85	48	36
	16-Cell	Oocyte	—	56	68	45	38
Mouse	1-Cell	1-Cell	99	—	n.a.[c]	>90	12, 23
	1-Cell	1-Cell	94	—	n.a.	94	18
	1-Cell	1-Cell	—	78	n.a.	67	18

[a] Fraction of successfully enucleated and manipulated eggs.
[b] n.e., Not evaluated.
[c] n.a., Not applicable.

As described above, cloning in the mouse can only be achieved by transplanting nuclei to 2-cell stage recipients or by separating blastomeres to generate separate individuals.

Nuclear transplantation in species other than the mouse requires (1) removal of the chromosomes from the recipient oocyte, (2) introduction of the donor nucleus, and (3) activation of the egg. The efficiency with which these steps are accomplished varies with the species (reviewed by Prather and First),[29] as does the developmental potential of the manipulated embryos (Table I). One important technical consideration for these manipulations appears to be the method used to induce fusion and activation. Sendai virus induces fusion in the other species at a reduced frequency, as compared with mouse embryos.[29–31] Consequently, electrofusion is used to induce fusion of eggs and karyoplasts and to achieve egg activation as well. The efficiency with which activation occurs is dependent on the species, age of oocyte, and number and duration of electrical pulses. A recent study reports improvements in the efficiency of nuclear transplantation in the rabbit embryo,[36] which in earlier reports exhibited a lower efficiency of egg activation than other species.[37] These improvements include the use of multiple pulses during electrofusion, which increased the efficiency of activation and subsequent development. The use of multiple pulses also allows a greater efficiency of activation of recently ovulated oocytes, which can be enucleated more efficiently.[36] Multiple pulses have also been used for nuclear transplantations in sheep and mouse embryos.[18,31]

Concluding Remarks

Using the techniques described and referenced above, nuclear transplantation in the mouse embryo, and in other species as well, offers a useful approach to the study of the mechanisms that control early embryogenesis. Specifically, this approach offers the means to examine such processes as nuclear remodeling and reprogramming, genome activation, and genomic imprinting. Additionally, interspecies nuclear transplantation offers a potential means of investigating the evolutionary conservation of critical controlling molecules and processes that mediate these early events. The technique of nuclear transplantation should increase in applicability with further refinement in molecular techniques for analyzing one or a few preimplantation-stage embryos.

Acknowledgments

This work was supported in part by U.S. Public Health Service Grants HD-17720, HD-23291, and HD-21355 from the National Institute of Child Health and Human Development and CA-10815 from the National Cancer Institute. K.E.L. was supported in part by a training grant from the NCI (CA 09171-14).

[44] Manipulations of Genetic Constitution by Nuclear Transplantation

By SHEILA C. BARTON and M. AZIM SURANI

Introduction

Techniques of nuclear transplantation have been developed which make it possible to vary the genetic constitution of mammalian eggs in ways not possible by natural fertilization. For instance, heterozygous diploid eggs whose genome is wholly of paternal (androgenetic) or maternal (gynogenetic) origin can be made; similarly, eggs whose maternal/paternal genome is normal but asynchronous or whose female pronucleus is of a different genetic background to the cytoplasm of the egg can be constructed, and nuclei from later stage embryos can be introduced into the cytoplasm of 1- or 2-cell stage embryos. The technique has become a basic tool in the study of imprinting and the vital differences in the maternal and paternal contributions to the

Copyright © 1993 by Academic Press, Inc.
All rights of reproduction in any form reserved.

developing embryo.[1] The technique also makes it possible to address the question of genomic totipotency and the role of oocyte cytoplasmic factors in reprogramming the genome.

Injecting a naked pronucleus directly through the membrane of mammalian eggs is not practicable because, unlike in amphibians where this method has been successful, the pronuclei are large relative to the size of the egg. In the mouse the diameter of each pronucleus (female 14–19 μm, male 17–22 μm) is between 20 and 30% that of the diameter of the whole egg (about 73 μm), and piercing the egg membrane with a pipette large enough to carry a pronucleus usually leads to irreparable damage to the membrane. To overcome this, a method was devised by McGrath and Solter[2] whereby the pronucleus (or both pronuclei), surrounded by an envelope of egg membrane, can be withdrawn from the egg without the pipette penetrating the membrane. The resulting karyoplast can then be injected under the zona pellucida of a recipient egg, and fusion of the two membranes is induced by exposure to inactivated Sendai virus.

Other methods of inducing fusion of the egg membranes include electro-fusion[3,4] and incubation in polyethylene glycol (PEG),[5,6] but with both of these methods the fusion efficiency is greatly reduced when there is a substantial difference in size between the two cells to be fused, as is necessarily the case if nuclear transfer is the object of the exercise. Furthermore, PEG may have some deleterious effects on the development of reconstituted zygotes. However, further advances in the technique of electrofusion may be possible and desirable in order to overcome possible batch variations with Sendai virus and the potential risk to mouse colonies in the event of incomplete inactivation of the virus.

The methods and equipment used to achieve nuclear transfer with Sendai virus-assisted fusion, or with other fusogens, vary from one laboratory to another and are evolving all the time. This chapter describes the regime presently used in our laboratory and refers to other methods only when they differ significantly from ours. It is hoped that the method described here will provide a basis on which to build. As an example we describe preparing heterozygous androgenetic embryos.

[1] M. A. Surani, R. Kothary, N. D. Allen, P. B. Singh, R. Fundele, A. C. Ferguson-Smith, and S. C. Barton, *Development* (*Cambridge, UK*) (*Suppl.*) 89 (1990).
[2] J. McGrath and D. Solter, *Science* **220**, 1300 (1983).
[3] J. R. Robl, P. Collas, R. Fissore, and J. Dobrinsky, *in* "Guide to Electroporation and Electrofusion" (D. C. Chang, B. M. Chassy, J. A. Saunders, and A. E. Sowers, eds.), Chap. 34. Academic Press, San Diego, 1992.
[4] V. Zimmerman and J. Vienken, *J. Membr. Biol.* **67**, 165 (1982).
[5] M. A. Eglitis, *J. Exp. Zool.* **213**, 309 (1980).
[6] A. Spindle, *Exp. Cell Res.* **131**, 465 (1981).

Basic Requirements

A laboratory set up for routine mouse embryo work, such as the production of transgenic mice by gene injection into pronuclear eggs, would include the following: (1) animal facilities to provide fertilized eggs of suitable strains, in sufficient numbers (usually by superovulation), and at the right stage for operation; (2) facilities for harvesting, handling, and culturing eggs, including a good dissecting microscope (e.g., Wild M 5) and an incubator set up to provide a humidified atmosphere of 5% CO_2 in air for culturing fertilized eggs, androgenetic eggs, and parthenogenetic eggs at a temperature of 37.8°–38.0°; and (3) equipment for micromanipulation, including a fixed-stage microscope with suitable optics (preferably Nomarski), manipulators for both holding and operating instruments, micrometer syringes attached to oil lines to operate the instruments, and a pipette puller and microforge for making the instruments. In addition, for nuclear transfer a machine for beveling the operating instruments is needed, and also a supply of inactivated Sendai virus.

We use a Leitz micromanipulator assembly with a fixed-stage upright microscope with image-erected optics and Nomarski attachments and work in hanging drops in Puliv manipulation chambers (internal height 3 mm), which are entered horizontally from the sides for all manipulations, including blastocyst injection and reconstruction, gene injection, as well as nuclear transfer. However, most of the manipulations can be carried out with the more commonly used (and more easily available) inverted fixed-stage microscopes, with which the operations are usually performed on the floor of a petri dish, manipulation chamber, or depression slide, with the instruments being introduced at an angle from above.[7]

To control the operating instrument we use a de Fonbrune suction/ force pump (Beaudouin, Paris, France) and for the holding instrument a Wellcome Agla micrometer syringe fitted with a return spring. The syringes are attached to the instruments with clear, firm plastic tubing filled with heavy liquid paraffin oil to transmit positive and negative pressure to the tips. When setting up the system great care should be taken to exclude any air bubbles large enough to occupy the whole diameter of the tubing or instrument holders because such bubbles will dampen responses to the micrometer controls.

Preparation of Instruments

The glass instruments are made from 15 cm lengths of borosilicate capillary tubing, outside diameter 1.0 mm. The holding pipettes are made

[7] B. Hogan, F. Constantini, and E. Lacy, "Manipulating the Mouse Embryo: A Laboratory Manual" Cold Spring Harbor Laboratory, Cold Spring Harbor, New York, 1986.

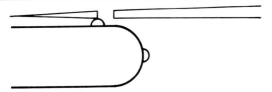

FIG. 1. Tip of filament, seen from the side, with beads of solid glass fused to the tip and upper surface, with a drawn thin-walled capillary fused to one of the glass beads and broken by switching off the heat. (Not to scale.)

from thick-walled capillary (inside diameter 0.58 mm), and the nuclear transfer pipettes are made from thin-walled capillary (inside diameter 0.78 mm) and are drawn on a Brown–Flaming pipette puller (Sutter Instruments, Novato, CA) and beveled on a grinding machine with a rotating plate whose surface is of very fine diamond (such machines are available from Sutter Instruments and Narishige Co. Ltd). To fashion the instruments we use a de Fonbrune (Beaudouin) microforge with a 10× objective, a filament of platinum/iridium wire, diameter 0.25 mm, and a 1 mm graticule marked in one-hundredths fitted in the eyepiece. The V-shaped filament is mounted horizontally in the microforge, and two small beads of solid glass are fused to it, one at the tip and one on the upper surface near to the tip (Fig. 1).

Nuclear Transfer Pipettes

A thin-walled capillary is drawn on the pipette puller at a setting that will produce a taper of 10–15 mm from shoulder to tip. The pipette is mounted horizontally in the microforge and broken off squarely at an outside diameter of about 28 μm. To achieve this the glass bead near the tip of the filament is gently fused to the capillary at the chosen diameter and the heat is switched off. Contraction of the filament should break the pipette cleanly (Fig. 1). The resulting tip may curve a little but should not be narrowed nor the walls be thickened. The pipette is transferred to the beveling machine and the tip ground to about 45° on the rotating diamond surface, which has been moistened with distilled water. If the tip is asymmetrical, the shoulder of the pipette can be marked with a pen before removing it from the forge so that the orientation of the curve is known and it can be ground so that the bevel is directly into the curve rather than away from it or askew (see Fig. 2).

The pipette is returned to the microforge and the bevel checked for completeness (Fig. 3A). The ground edge is then smoothed by bringing the bevel close to the glass bead, which has been heated to a very dull glow, care being taken not to narrow the opening. The tip of the bevel is

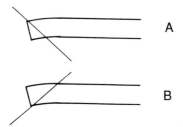

FIG. 2. Acceptable (A) and unacceptable (B) orientation of bevel on a broken-off capillary with a slight curve at the tip.

sharpened again by bringing it into contact with the bead of glass on the tip of the filament, again at a fairly low temperature (dull glow), and drawing out a small sharp point of glass (Fig. 3B), which is knocked off against the cold glass bead to make a sharp edge (Fig. 3C). The resulting pipette has a smooth-edged bevel with a tip that is sharp enough to penetrate the zona pellucida but does not damage the egg membrane.

Finally, a bend of about 5° is put in the shaft of the pipette near the shoulder in such a way that the vertical plane of the orifice bevel is the same as the plane of the bend. This will ensure that, when mounted in the manipulator with the taper sloping upward toward the tip (Fig. 4A), the orientation of the opening will be convenient for nuclear transfer (Fig. 4B). For other kinds of manipulation chambers the arrangements of bends in the shaft will be different to achieve the same final orientation of the tip. The tip of the finished pipette is then coated with undiluted Nonidet P-40 (NP-40), rinsed a few times through two changes of distilled water, and dried in a warm oven. Transfer pipettes for the smaller female pronu-

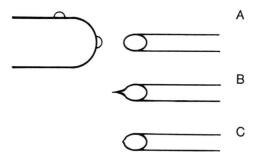

FIG. 3. (A) Orientation of tip to check completeness of the bevel. (B) Sharp point drawn on the tip of the smoothed bevel. (C) Point knocked off against a cold glass bead to make a sharp edge. (Not to scale.)

A

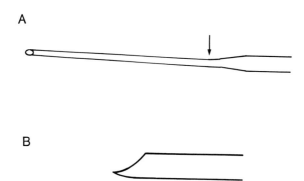

B

FIG. 4. (A) Nuclear transfer needle seen from side; arrow marks position of the 5° bend. (B) Tip of nuclear transfer needle seen from above, as it will appear in the microscope. (Not to scale.)

clei or later stage nuclei would be made by making the original break in the capillary at a smaller outside diameter, between 20 and 24 μm.

Holding Pipettes

Thick-walled capillary tubing is drawn by hand using a small Bunsen gas flame and broken off absolutely square by hand at between 90 and 120 μm outside diameter (the squareness of the break is more important than the precise outside diameter of the tip). The remaining length of taper should be between 2 and 3 cm, and the relative thickness of the wall should not be increased. The pipette is mounted on the microforge, the tip checked for squareness and size, then polished with the filament at a moderate heat (red glow) so that the opening is narrowed to an apparent internal diameter of 15–20 μm, as seen though the thickness of the glass. Again, the pipette needs a bend of about 5° near its shoulder which, if not introduced during drawing, is introduced with the forge.

Embryos for Operation

When preparing androgenetic or gynogenetic embryos, it is advisable to have the egg of a genotype identifiably different from the sperm genotype, for instance, at the glucose-phosphate isomerase-1 (*Gpi-1*) locus, so that the provenance of the pronuclei can be checked later by simple enzyme analysis. 129/Sv is often the preferred genotype for androgenetic embryos because a relatively large proportion of androgenones with this genetic background have been found to develop to the blastocyst stage

(25–40%). These mice are GPI-1A, so that a mating of 129 males with C57BL/6 or (C57BL/6 × CBA)F₁ females, which are GPI-1B and whose eggs develop reliably in culture, will provide suitable eggs for operation.

To obtain eggs at the right stage for pronuclear transfer, the superovulated and mated females are sacrificed about 20 hr after administration of human chorionic gonadotropin (hCG). The cumulus masses are removed from the oviducts and incubated briefly (2–3 min) in hyaluronidase (300 μg/ml) in phosphate-buffered medium (PB1) plus 0.4% bovine serum albumin (BSA) to disperse the cumulus cells. The eggs are washed thoroughly and cultured under oil in a CO_2 incubator in carbonate-buffered medium (T6) plus 0.4% BSA for 1–2 hr before operation. Fifteen minutes before operation enough eggs for an operating session of about 1.5 hr are selected (pronuclei expanding and migrating toward the center of the egg) and put into a drop of operating medium, PNC (PB1 + BSA to which have been added the cytoskeletal inhibitors cytochalasin B and nocodazole). The composition of the media used in our laboratory (modified from Whittingham[8,9]) is presented in Table I. Both T6 and PB1 media have a relatively low osmolarity of about 240 mOsM. Many laboratories use other formulations of egg culture medium, HEPES-buffered medium in place of phosphate-buffered medium, and other forms of cytoskeletal inhibitors, such as Colcemid instead of Nocodazole and cytochalasin D rather than cytochalasin B.

Sendai Virus

Sendai virus is prepared in embryonated chicken eggs and inactivated with β-propiolactone, as described previously.[10–12] After inactivation of a batch, it is stored at $-70°$ in aliquots of about 20 μl, and the final dilution for the batch is determined empirically. Immediately before use an aliquot is thawed, the dilution is adjusted, and the aliquot centrifuged very gently for 2 min.

Nuclear Transfer

A manipulation chamber is prepared with one large drop of Sendai virus and several smaller drops of PNC. With the hanging drop method

[8] D. G. Whittingham, *J. Reprod. Fertil. Suppl.* **14**, 7 (1971).
[9] D. G. Whittingham and R. G. Wales, *Aust. J. Biol. Sci.* **22**, 1065 (1969).
[10] R. E. Giles and R. H. Ruddle, *In Vitro* **9**, 103 (1973).
[11] J. M. Neff and J. F. Enders, *Proc. Soc. Exp. Biol. Med.* **127**, 260 (1968).
[12] S. C. Barton, M. L. Norris, and M. A. H. Surani, in "Mammalian Development, A Practical Approach" (M. Monk, ed.), Chap. 12. IRL Press, Oxford and Washington, D.C., 1987.

TABLE I
MEDIA FOR EGG CULTURE
AND BENCH MANIPULATION[a]

| Component | Concentration (g/liter) | |
	T6[b]	PB1[c]
NaCl	4.72	5.97
KCl	0.11	0.2
$NaH_2PO_4 \cdot 2H_2O$	0.06	—
KH_2PO_4	—	0.19
$CaCl_2 \cdot 2H_2O$	0.26	0.14
$MgCl_2 \cdot 6H_2O$	0.1	0.1
Glucose	1.0	1.0
Sodium pyruvate	0.03	0.04
$NaHCO_3$[d]	2.1	—
Na_2HPO_4 (anhydrous)[d]	—	1.14
Penicillin G	0.06	0.06
Streptomycin sulfate	0.05	0.05
Phenol red	0.01	0.01
Sodium lactate (60% syrup)	3.4 ml	—

[a] Both media are Millipore-filtered (Millipore, Bedford, MA) and stored in 5-ml aliquots at 4°. Bovine serum albumin (BSA) (fraction V, Sigma, St. Louis, MO, A4503) is added to both media before use at 4 mg/ml. Hyaluronidase for dispersion of cumulus cells is dissolved in distilled water at 6 mg/ml, Millipore-filtered, and stored at −20° in 50-μl aliquots; when required an aliquot is made up to 1 ml with PB1 + BSA to give a working concentration of 300 μg/ml. PNC, the operating medium, is PB1 + BSA to which have been added cytochalasin B (CCB) at 5 μg/ml and nocodazole {methyl [5-(2-thienylcarbonyl)-1H-benzimidazol-2-yl]carbamate} at 1.5 μg/ml; both are dissolved in dimethyl sulfoxide (DMSO), CCB at 5 mg/ml and Nocodazole at 3 mg/ml, and stored in small aliquots at −20°. All chemicals used in the preparation of the media are obtained from Sigma or British Drug House (BDH, Poole, UK). Media are made up in Analar water from BDH.
[b] T6, Egg culture medium.
[c] PB1, Bench preparation medium.
[d] These are dissolved separately and added last.

A

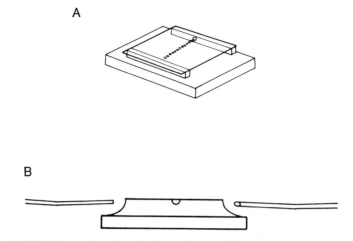

B

FIG. 5. (A) Puliv manipulation chamber with coverslip attached and drops of virus and operating medium place on the underside of the coverslip, prior to introduction of heavy liquid paraffin. (B) Chamber filled with oil, seen from the front, with holding and operating instruments set ready for introduction. (Not to scale.)

of micromanipulation, these can be arranged in a line across the chamber, first the large drop of virus (~3 mm diameter), then about 12 drops of PNC (~1.5 mm diameter) placed on the underside of a siliconized coverslip that has been secured to the Puliv chamber with petroleum jelly stiffened with 5% hard paraffin wax (Fig. 5A). The chamber is then filled with heavy liquid paraffin oil (Fig. 5B). After about 15 min of incubation in PNC, the eggs are put into the hanging drops in groups of convenient numbers; as a precaution against confusion, the donor eggs can be, for instance, in groups of four, the recipient eggs in pairs.

The chamber is transferred to the manipulator microscope stage and the microscope focused on some of the eggs to determine the height of the base of the operating drops; the chamber is moved clear of the microscope. The focusing of the microscope is then left untouched while the instruments are being set up, focusing of the tips under the microscope being achieved by using the vertical adjustment on the manipulators. The two manipulators are set to move horizontally (mark 0 on the dial), the angles of the instruments being adjusted by the ball joints of the instrument holding heads. Holding and nuclear transfer pipettes are mounted in the oil-operated instrument holders, oil brought to the tips of the instruments (care being taken to exclude air bubbles), and the tip of the transfer pipette oriented so that the plane of the orifice is vertical. The tip will appear from above, down the microscope, as in Fig. 4B. The chamber can now

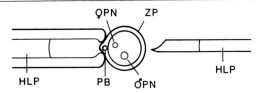

FIG. 6. Egg held ready for extraction of the female pronucleus. HLP, Heavy liquid paraffin; PN, pronucleus; PB, polar body; ZP, zona pellucida.

be replaced (Fig. 5B) and the instruments taken into the operating drops at approximately the right height.

Enucleation and nuclear transfer can be performed either quite separately or as a single two-step operation. There are advantages and disadvantages to both approaches. In the first case all the recipient eggs are prepared by removing the female pronuclei, and the eggs have to be picked up a second time in such a way that the same hole in the zona pellucida is entered to deposit the second paternal karyoplast; otherwise, the egg can be squeezed out of the first hole in the zona when making the second hole. In the second case the paternal karyoplast is introduced under the zona of the intact recipient egg, then deposited together with some Sendai virus and the female pronucleus removed immediately after; there is a danger that the maneuvering of the transfer pipette will cause the karyoplast to break. We find that the first strategy is more expeditious in the long run despite the extra time taken to position the egg carefully.

A recipient egg is prepared by holding it firmly by the area of the zona pellucida over the second polar body with the egg oriented so that both pronuclei can be seen clearly and compared for size and position. The female pronucleus will usually be closer to the second polar body than the male pronucleus and it will always be the smaller of the two, although the difference is sometimes too slight to be obvious. If there is any doubt about the identity of the pronuclei the egg should be discarded. With polar body, target pronucleus, and transfer pipette tip all in the same optical plane (Fig. 6) the pipette tip is pressed through the zona pellucida into the perivitelline space. The consistency of the zona pellucida is variable in mouse eggs, and penetration may require considerable force; if this is the case several little bouncing jabs are less likely to cause damage to the egg than one hard push. Once through the zona the pipette is pressed into the egg without breaking the egg membrane, making an indentation as deep as necessary to reach the female pronucleus. The cytoskeletal inhibitors in the operating medium render the membrane sufficiently elastic to withstand this. Keeping the tip of the pipette and the pronuclear membrane at its widest diameter in the same optical plane, gentle suction is

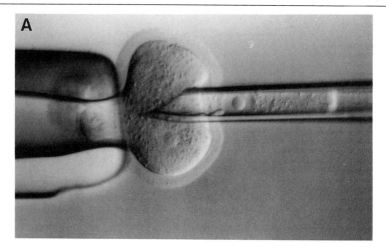

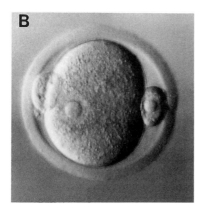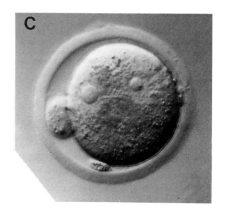

FIG. 7. Extraction of female pronucleus, in this case with a large amount of cytoplasm to show the great elasticity of the egg membrane. (B) Egg with female pronucleus removed and a second male pronuclear karyoplast introduced under the zona pellucida, with a small amount of Sendai virus. (C) Fusion of the pronuclear karyoplast with the egg.

applied to the enucleation pipette, and egg membrane, pronucleus, and a small amount of cytoplasm are drawn into it (Fig. 7A). The pipette is gently withdrawn through the zona and the membrane continues to stretch until it pinches off, healing on both sides of the break. The enucleation process is repeated until all the recipient eggs are ready. The female pronuclei can then be discarded into an empty drop.

In the second stage the male pronucleus is withdrawn from a donor egg in essentially the same way. It can be helpful to take a small separate cytoplast into the operating pipette before taking the karyoplast to act as a buffer later on, and it is not necessary to hold the egg by the polar body region. The pipette with karyoplast in its tip is moved to the drop of Sendai virus, and the karyoplast is protruded a little from the tip to bathe it in the virus and then drawn back into the pipette about 75 μm so that a small volume of virus is held in the tip of the pipette. The pipette, carrying buffer cytoplast, karyoplast, and virus, is taken to a prepared recipient egg, which is picked up by the polar body region again (the zona pellucida will often retain its deformation from the first operation, and the combination of polar body and deformation helps in finding the original orientation). The operating pipette is eased through the original hole in the zona, virus and karyoplast are gently expelled into the perivitelline space and the pipette withdrawn (Fig. 7B). Fusion of egg and karyoplast membranes will probably follow sometime during the next hour, perhaps within a few minutes (Fig. 7C).

As soon after the operation as conveniently possible, whether or not fusion has yet taken place, the eggs should be washed and returned to culture in medium without cytoskeletal inhibitors, since long exposure to these drugs (more than about 2 hr) may render them oversensitive to the virus and some of the successfully operated eggs may later lyse. Sensitivity to the drugs and virus may also be to some extent dependent on the mouse strain.

At the end of the session the operated eggs can be washed again and checked for fusion before delivering them to their longer term culture place. A proportion of androgenetic embryos will start to cavitate after about 4 days in culture (embryonic day 5, counting the 1-cell stage as day 1), 1 day later than normal fertilized eggs, and after 1 or 2 more days (day 6–7) some will be well-expanded blastocysts with a substantial inner cell mass. Hatching from the zona pellucida tends to be premature because of the hole made during the nuclear transplantation operation, and therefore is usually incomplete. About 30% will form substantial blastocysts and about 30% of these (i.e., 10% of the androgenetic embryos made) will have inner cell masses capable of further development, as judged both by culture and immunosurgery retrieval[13] and by transfer at the 4- or 8-cell or morula stage (day 3–4 embryos) to the oviducts of day 1 pseudopregnant foster mothers.

The process described above can be adapted to produce gynogenones with two maternal pronuclei. Similarly, the procedure for transplantation of nuclei from advanced preimplantation embryos (e.g., 8-cell morulae)

---

[13] S. C. Barton, A. C. Ferguson-Smith, R. Fundele, and M. A. Surani, *Development (Cambridge, UK)* **113,** 679 (1991).

into enucleated oocytes is similar to the above with minor variations in detail. Transfer of nuclei or pronuclei is also possible into enucleated 2-cell embryos, although inducing fusion of karyoplasts to each cell without also fusing the two enucleated cells can sometimes be a problem. A compromise occasionally employed if the end point of the experiment is fairly limited (e.g., to preimplantation development) and a large number of experimental embryos is required, is to enucleate both cells of the 2-cell embryos but to transfer nuclei to only one of the enucleated cells.

Section X

Transgenic Animals: Pronuclear Injection

[45] Production of Transgenic Mice

By Jon W. Gordon

Introduction

Historical Background

The ability to introduce foreign genes into the mammalian germ line has proved to be one of the most powerful experimental tools in mammalian developmental genetics. This technology arose from the knowledge, attained through gene transfer experiments in tissue culture, that new DNA could be inserted into the genome of a host cell and subsequently efficiently expressed (reviewed by Scangos and Ruddle[1]). The key unknown factors at the time efforts were begun to insert DNA into an intact mammal related to the frequency of gene insertion and the question of whether an embryo, which, in contrast to cultured cells, must complete a complex developmental program, could tolerate insertion of new genes without suffering fatal consequences.

In typical tissue culture gene transfer experiments, selectable markers were transferred into cells, after which the selective agent was employed to identify transformants.[2,3] The frequency of success for such experiments was prohibitively low for insertion of genes into embryos[1]; however, subsequent studies in which genes were injected directly into cell nuclei gave gene transfer frequencies approaching 20%.[4,5] Concerning the ability of developing embryos to tolerate integration and expression of new genetic material, little was known. However, in 1976, Jaenisch demonstrated that a murine retrovirus, Moloney murine leukemia virus (MMLV), could infect mouse embryos, insert a proviral copy of its DNA into the genome, and have that DNA transmitted as a Mendelian trait to offspring.[6] Because successful MMLV-mediated gene insertion depended on biological activity of a virus that might seek selective sites in the genome for DNA integration, and because retroviral genes have highly restricted patterns of expression, it was not at all apparent from these experiments that

[1] G. A. Scangos and F. H. Ruddle, *Gene* **14,** 1 (1981).
[2] H. Maitland and J. McDougall, *Cell (Cambridge, Mass.)* **11,** 233 (1977).
[3] M. Wigler, S. Silverstein, L. S. Lee, A. Pellicer, Y. C. Cheng, and R. Axel, *Cell (Cambridge, Mass.)* **11,** 223 (1977).
[4] A. Graessmann, M. Graessmann, W. C. Topp, and M. Botchan, *J. Virol.* **32,** 989 (1979).
[5] M. R. Capecchi, *Cell (Cambridge, Mass.)* **22,** 479 (1980).
[6] R. Jaenisch, *Proc. Natl. Acad. Sci. U.S.A.* **73,** 1260 (1976).

Copyright © 1993 by Academic Press, Inc.
All rights of reproduction in any form reserved.

recombinant DNA could be inserted into the genome of the mouse embryo. Nonetheless, these experiments demonstrated that, at least under highly specialized conditions, some new DNA could be tolerated by the developing mouse. In the mid 1980s, these findings were exploited to develop a gene transfer system that employs retroviral vectors.[7-9]

The feasibility of inserting DNA from any source directly into the mouse was first demonstrated by Gordon *et al.*,[10] who devised the strategy of pronuclear microinjection. Mice transformed in this way were subsequently called "transgenic" by Gordon and Ruddle,[11] a term which has since been applied to all forms of gene transfer into intact organisms, both animal and plant. The pronuclear microinjection technique, to be described here, has been used without significant modification since its inception, and it has been proved applicable to a wide variety of mammalian species.[12] The procedure is illustrated in Fig. 1. Before describing this technique and the biological factors that impact on its successful use, the process of gene integration as it is presently understood, factors influencing gene expression, and some applications of transgenic technology are briefly described. This information is useful for successful design of transgenic experiments.

Fundamentals of Gene Regulation

It was Brinster *et al.*[13] who first convincingly demonstrated efficient expression of "transgenes" in transgenic mice. After ligating the 5' regulatory region of the mouse metallothionein-1 (MT-1) gene to the coding sequence for herpes virus thymidine kinase (*tk*), transgenic mice were produced by pronuclear microinjection. Adult mice expressed the herpes *tk* gene in a pattern typical of MT-1, with highest expression in liver and kidney and low expression in brain. These experiments and many which followed (reviewed by Palmiter and Brinster[14] and Gordon[15]) demonstrated

---

[7] R. Mann, R. C. Mulligan, and D. Baltimore, *Cell* (*Cambridge, Mass.*) **33**, 153 (1983).

[8] D. Jahner, K. Haase, R. Mulligan, and R. Jaenisch, *Proc. Natl. Acad. Sci. U.S.A.* **82**, 6927 (1985).

[9] H. van der Putten, H. Boteri, F. M. Miller, A. D. Rosenfeld, M. G. Fan, R. Evans, and I. Verma, *Proc. Natl. Acad. Sci. U.S.A.* **82**, 6148 (1985).

[10] J. W. Gordon, G. A. Scangos, D. J. Plotkin, J. A. Barbosa, and F. H. Ruddle, *Proc. Natl. Acad. Sci. U.S.A.* **77**, 7380 (1980).

[11] J. W. Gordon and F. H. Ruddle, *Science* **214**, 1244 (1981).

[12] R. E. Hammer, V. G. Pursel, C. E. Rexsroad, Jr., R. J. Wall, D. J. Bolt, K. M. Ebert, R. D. Palmiter, and R. L. Brinster, *Nature* (*London*) **315**, 680 (1985).

[13] R. L. Brinster, H. Y. Chen, M. E. Trumbauer, A. W. Senear, R. Warren, and R. D. Palmiter, *Cell* (*Cambridge, Mass.*) **27**, 223 (1981).

[14] R. D. Palmiter and R. L. Brinster, *Annu. Rev. Genet.* **20**, 465 (1986).

[15] J. W. Gordon, *Int. Rev. Cytol.* **115**, 171 (1989).

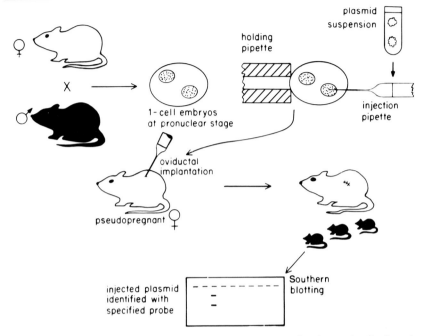

FIG. 1. Production of transgenic mice. Females are superovulated, mated to fertile males, and sacrificed the next day (top left). Zygotes with two pronuclei are recovered (top center), and one of the pronuclei is microinjected (top right) with DNA. Surviving embryos are reimplanted into pseudopregnant foster females (center), and DNA samples from newborns are evaluated (bottom) for the presence of foreign genes.

that cis-acting regulatory elements in and around genes are the factors most important to determining spatial and temporal patterns of mammalian gene expression, with the chromosomal integration site playing a relatively minor role.

Although these cis-acting "promoter/enhancer" elements were sometimes found to be relatively simple, as in the case of pancreatic elastase I,[16] they were demonstrated to be far more complex in the case of many other genes, perhaps best exemplified by β-globin. Even though the globin gene is active in only a single cell type, multiple enhancers are dedicated to its regulation.[17] Important to the present discussion is that in the absence of the full complement of enhancer elements, β-globin expression, while

[16] G. H. Swift, R. E. Hammer, R. J. MacDonald, and R. L. Brinster, Cell (Cambridge, Mass.) 38, 639 (1984).
[17] F. Grosveld, G. B. van Assendelft, D. R. Greaves, and G. Kollias, Cell (Cambridge, Mass.) 51, 975 (1987).

still tissue-specific, is far lower and more subject to the influence of integration site.[18,19] Thus, it is important when designing transgenes to recognize that the absence of key enhancer elements can significantly affect the level of gene expression. It should also be realized that when the promoter region of a gene is grafted to a heterologous coding element, enhancers which may reside within the gene from which the promoter was taken will be lost, and thus the level of tissue-specific expression dictated by the promoter may be less than that of the genomic sequence from which the promoter was obtained.

Genes may also have multiple enhancers, each devoted to a region of a single organ. For example, in the case of Thy-1, an immunoglobulin-like gene expressed throughout the entire brain in all mammals studied, separate enhancers for different cerebellar layers clearly exist.[20]

Trans-acting factors which interact with transgenes to stimulate transcription may also affect the pattern of expression. For example, trans-activators of human fetal globin gene expression are apparently elaborated by the mouse embryonic yolk sac; thus, insertion of the γ-globin gene into transgenic mice results in embryonic expression of this gene, which, in humans, is activated in fetal liver.[21,22] In the case of human Thy-1, the *absence* of interspecific differences in trans-activator production between human and mouse results in expression of the human transgene in organs that do not normally express mouse Thy-1 but which are sites of human Thy-1 expression.[20] Thus, unexpected patterns of expression may be encountered when transgenes from heterologous species are used to produce transgenic mice.

Integration Mechanism

How microinjected genes become integrated into the recipient genome has never been experimentally established. However, several features of integration are important. Transgenes most commonly insert as head-to-tail concatamers into a single site in the genome[23]; rarely, two sites of integration are obtained in a single founder animal. It is not unusual to find 50 or more copies of the transgene at the integration site. In our

[18] K. Chada, J. Magram, K. Raphael, G. Radice, E. Lacy, and F. Costantini, *Nature (London)* **314**, 377 (1985).
[19] T. M. Townes, J. B. Lilngrel, H. Y. Chen, R. L. Brinster, and R. D. Palmiter, *EMBO J.* **4**, 1715 (1985).
[20] J. W. Gordon, P. G. Chesa, H. Nishimura, W. Rettig, J. E. Maccari, T. Endo, E. Seravilli, T. Seki, and J. Silver, *Cell (Cambridge, Mass.)* **50**, 445 (1987).
[21] K. Chada, J. Magram, and F. Costantini, *Nature (London)* **319**, 685 (1986).
[22] G. Kollias, N. Wrighton, J. Hurst, and F. Grosveld, *Cell (Cambridge, Mass.)* **42**, 89 (1986).
[23] F. Costantini and E. Lacy, *Nature (London)* **294**, 92 (1981).

experience, no integrated concatamer has failed to contain at least one complete copy of the transgene. However, we have additionally observed aberrant restriction fragments repeated many times within a single concatamer.[15] These findings cumulatively suggest that homologous recombination, probably between microinjected fragments prior to integration, occurs, but also that pieces of DNA may be reduplicated during the integration process. As a consequence of these observations, we are generally confident that an intact, expressible transgene fragment is present whenever a concatamer is seen. We accordingly do not conduct exhaustive Southern analyses to demonstrate the presence of the coding element. Typically, we screen mice by Southern blotting using a restriction enzyme that cuts once within the microinjected sequence. When concatamers are present, this digest yields a major band of a size equal to the microinjected fragment, and possibly other minor bands. We have observed that microinjection of fragments with a 3' overhang on one end and a 5' overhang on the other favors insertion of short concatamers.

Another reasonably common feature of integration is that it may be delayed, causing the resultant transgenic founder to be a genetic mosaic for the integrated transgene. It is important to recognize this situation, since transmission of the foreign gene to offspring may occur less frequently than the predicted 50%. When Southern analysis is performed after digestion of the DNA with an enzyme that cuts once within the transgene, a hybridization band fragment equal in size to the intact transgene but with an intensity approximating that of a unique endogenous sequence suggests mosaicism. This is true because a "unit length" restriction fragment can be produced after such a digest only when more than one copy of the fragment is present in the head-to-tail array. Therefore, in the absence of mosaicism the blotting intensity should be higher than that of a unique sequence. Of course, intense hybridization can still occur in mosaics with large integrated concatamers.

A final feature of foreign genes in transgenic animals which can be significant is that concatamers may be unstable. This instability is most commonly manifest as deletion of one or members of the concatamer, presumably as a result of recombination. This phenomenon can lead to a reduction in blotting intensity and/or the appearance of new minor bands as the transgene is propagated through the germ line.

It is not uncommon that, when separating a DNA fragment from cloning vector material by restriction enzyme digestion, a small amount of contamination by the vector DNA occurs and leads to its cotransfer with the transgene fragment. It is remarkable how little of such contamination is required in order to obtain cotransfer of unlinked DNA into the mouse. Accordingly, care should be taken to assure purity of the

DNA to be microinjected. For example, we do not load molecular weight markers on preparative gels, as this creates the risk of cotransfer into the transgenic mouse genome of prokaryotic sequences derived from the size markers.

The phenomenon of cotransfer can be employed to the advantage of the experimenter. When linkage of independent DNA markers into the same site in the genome is desired, but the restriction maps do not make feasible the ligation of the independent fragments prior to microinjection, independent DNAs can be simply mixed and microinjected. This frequently leads to cotransfer of the unlinked fragments.[24,25]

Unless the full complement of enhancer sequences is present on the donor fragment, larger concatamers with more transgenes do not necessarily express at higher levels. However, when all enhancers of a gene are included in the microinjected fragment, an approximately linear relationship exists between the number of integrated copies and the level of expression.[16,17]

Another phenomenon that should not be overlooked is that of host DNA modification in association with integration. Bizarre structural alterations of host DNA sequences, including deletions, duplications, interspersion of islands of genomic DNA within transgene sequences,[26] and even translocations,[27,28] are not unusual. These aberrations can be mutagenic, causing abnormal phenotypes, usually in homozygous transgenic animals. In addition, such changes can interfere with the production of fertile homozygotes (see section on colony management).

Applications of Transgenic Technology

We have undoubtedly not yet exhausted the many potential applications of transgenic technology. Not only has this technique been enormously useful in elucidating mechanisms of gene regulation as alluded to above, it has been exploited to investigate many of the most complex developmental and homeostatic functions of mammals. A few examples,

[24] T. M. Ryan, T. M. Townes, M. P. Reilly, T. Asakura, R. D. Palmiter, R. L. Brinster, and R. R. Behringer, *Science* **247,** 566 (1990).
[25] E. A. Robey, B. J. Fowlkes, J. W. Gordon, D. Kioussis, H. von Boehmer, F. Ramsdell, and R. Axel, *Cell* (*Cambridge, Mass.*) **64,** 99 (1991).
[26] L. Covarrubias, Y. Hishida, and B. Mintz, *Proc. Natl. Acad. Sci. U.S.A.* **83,** 6020 (1986).
[27] K. A. Mahon, P. A. Overbeek, and H. Wesphal, *Proc. Natl. Acad. Sci. U.S.A.* **85,** 1165 (1988).
[28] J. W. Gordon, D. Pravtcheva, P. A. Poorman, M. J. Moses, W. A. Brock, and F. H. Ruddle, *Somatic Cell Mol. Genet.* **15,** 569 (1989).

not at all intended to represent a thorough review of the literature, are mentioned briefly below.

In the field of immunology, transgenic mice have been exploited to elucidate mechanisms of allelic exclusion by insertion of rearranged, expressible immunoglobulin genes.[29-31] Fascinating studies illustrating mechanisms of self-tolerance in the B-cell lineage also illustrate the power of this experimental approach.[32] In addition, it has been possible to modify the immune response profile of inbred mouse strains by transfer of expressible histocompatibility genes.[33,34] In the area of oncology, transgenic mice have been used to demonstrate that overexpression of protooncogenes can predispose to the development of malignancies,[35] to show that aberrant expression of oncogenes can transform tissues normally resistant to neoplastic degeneration,[36] and to reveal the oncogenic potential of genes within viruses, infection by which is epidemiologically associated with development of cancers.[37] Pronuclear microinjection has been used to demonstrate the feasibility of using transgenic livestock as factories for production within breast milk of complex, biologically active proteins that cannot be readily produced in bacteria by standard recombinant DNA techniques.[38]

These are just a few of the uses of transgenic animals, the vast majority of which are mice. Moreover, use of transgenic technology in animals larger than mice necessitates establishment of a transgenic mouse colony for rapid testing of recombinant constructs. Thus, it is hoped that whatever the intentions of the reader regarding employment of transgenic technology, this chapter will be of assistance.

[29] K. A. Ritchie, R. L. Brinster, and U. Storb, *Nature (London)* **312**, 517 (1984).
[30] U. Storb, C. Pinkert, B. Arp, P. Engler, K. Gollahon, J. Manz, W. Brady, and R. L. Brinster, *J. Exp. Med.* **164**, 627 (1986).
[31] M. C. Nussenzweig, A. C. Shaw, E. Sinn, D. B. Danner, K. I. L. Holmes, H. C. Morse III, and P. Leder, *Science* **236**, 816 (1987).
[32] C. C. Goodnow, J. Crosbie, H. Jorgensen, R. A. Brin, and A. Basten, *Nature (London)* **342**, 385 (1989).
[33] M. Le Meur, P. Gerlinger, C. Benoist, and D. Mathis, *Nature (London)* **316**, 38 (1985).
[34] K. Yammamura, H. Kikutani, V. Folsom, L. K. Clayton, M. Kimoto, S. Adira, S. Kashiwamura, S. Tonegawa, and T. Kishimoto, *Nature (London)* **316**, 67 (1985).
[35] T. A. Stewart, P. K. Pattengale, and P. Leder, *Cell (Cambridge, Mass.)* **38**, 627 (1984).
[36] K. A. Mahon, A. B. Cheplinsky, J. S. Khillan, P. A. Overbeek, J. Piatigorsky, and H. Westphal, *Science* **235**, 1622 (1987).
[37] M. Nerenberg, S. H. Hinrichs, R. K. Reynolds, G. Khoury, and G. Jay, *Science* **237**, 1324 (1987).
[38] K. Gordon, E. Lee, J. Vitale, A. Smith, H. Westphal, and L. Henninghausen, *Biotechniques* **5**, 1183 (1987).

TABLE I
SCHEDULE FOR TRANSGENIC MOUSE PRODUCTION

Step	Schedule
Light cycle for animal room	Lights off, 10 PM
	Lights on, 8 AM
Injection of PMSG (2.5 IU)	Day −3: 1300 hr
Injection of hCG (5 IU)	Day −1: 1200–1400 hr
	Mating of foster females to vasectomized males: 1300–1600 Hr
Sacrifice and manipulation	Day 0
	Sacrifice: 1230–1400 hr
	DNA microinjection: 1300–1600 hr
	Embryo transfer: 1400–1700 hr
Examination for birth of newborns	Days 18–21

Use of Procedures

Many papers describing the method of producing transgenic animals by pronuclear microinjection already exist, including a previous one in this series.[39] The strategy taken for this chapter is to provide equipment, reagent, and supply lists for rapid assembly of material needed for production of transgenic mice, a description of the technique itself, advice on colony management, and some discussion of the minor aspects of the procedure that can significantly affect success. This chapter does not address methods other than pronuclear microinjection, as these are described by other contributors.

This chapter includes all procedures required for production of transgenic mice and isolation of DNA for genetic analysis. Reagents, supplies, and equipment lists include all items needed and preferred in our laboratory. The symbol "Tg" in the text and tables substitutes for "transgenic."

A schedule for producing transgenic mice is shown in Table I. Obviously, the timing of various steps can be altered to suit the needs of the individual investigator. However, significant changes in schedule will require compensatory alteration of the light–dark cycle in the animal room. Equipment for producing transgenic mice is listed in Table II. Suggestions concerning optional equipment are made in another chapter in this volume, which describes a variety of other micromanipulation procedures.[40] Reagents for transgenic production excluding those required for basic culture media are shown in Table III, and supplies are listed in

[39] J. W. Gordon and F. H. Ruddle, this series, Vol. 101, p. 411.
[40] J. W. Gordon, this volume [12].

TABLE II

EQUIPMENT FOR TRANSGENIC MOUSE PRODUCTION

Microscope, preferably inverted, equipped with interference
  optics (Nomarski, Hoffmann, or phase contrast)
Two micromanipulators
Marble table for manipulation[a]
One syringe micrometer for applying suction or positive pressure
  through the micromanipulation system[b]
Video system for micromanipulator[c]
Microforge
Pipette puller
Dissecting microscope equipped with transmitted light and
  overhead illumination
Microburner or alcohol lamp
Bench-top microcentrifuge
Rocking shaker for DNA extraction.

[a] A marble table is rarely needed. See this volume [12] Ref. 40.
[b] We use a single syringe micrometer for the holding pipette and
  a standard syringe for the microneedle (see text).
[c] Optional, but recommended.

Table IV. Formulas for culture media stocks along with instructions for
mixing the stocks for production of media are listed in the Appendix
(Tables A.I–A.VII). Other specialized media and solutions are also listed
in the Appendix.

Transgenic Mouse Production

DNA Preparation

Because cloning vector sequences (bacteriophage and plasmids) can
interfere with transgene expression, most laboratories purify the coding
elements from such sequences as much as possible prior to microinjection.
Such purification is normally accomplished by restriction enzyme digestion and either sucrose gradient centrifugation or agarose gel electroelution. Although agarose is believed by many to contain toxic contaminants
that can interfere with production of transgenic mice, we have found that
careful cleaning of DNA after agarose gel electroelution gives a high rate of
gene transfer. In addition, agarose is superior for separation of fragments
similar in size, and it is faster and easier to use. Therefore, we use gel
electroelution exclusively for fragment isolation, according to the following protocol.

TABLE III
SUPPLIES FOR TRANSGENIC MOUSE PRODUCTION

Rubber tubing equipped with mouthpiece for pipetting small volumes (Baxter, Edison, NJ)

Microcentrifuge tubes, 1.5-ml (Baxter)

Sterile plastic tissue culture dishes, of suitable size for micromanipulation and other procedures (Fisher, Springfield, NJ)

Disposable filters for sterile filtration of volumes of 100 to 1000 ml (Fisher)

Watchmaker's forceps, 2 pair (Fullam, Latham, NY)

Small iris scissors, 1 pair (Fullam)

Hemostat

4-O silk thread for closing wounds after mouse surgery (MSMC, New York)

Small needles with cutting edge for closing mouse wounds (e.g., Baxter, Anchor No. 1834-8)

Microcapillary tubing for producing microneedles used for holding pipettes (e.g., Mertex, 100 × 0.5 mm ID, No. MX-999, Mercer Glass Works, NY)

Self-filling microcapillary tubing for zona drilling or DNA microinjection (e.g., 12111 Omega dot tubing, Glass Co. of America, Millville, NJ)

Spinal needle for filling barrel of microcapillary tubing used for DNA injection (e.g., Popper & Sons, Quinke Babcock 26-gauge spinal needle)

Polyethylene tubing for connecting syringe to microneedle (e.g., Intramedic, PE60, PE90, PE160, PE190) (Clay Adams, Parsippany, NJ)

Plastic tubing adaptor with Luer-lock for PE tubing (e.g., Clay Adams, A-1026)
  Size B fits PE60 and PE90
  Size C fits PE160 and PE190

Glass depression slides for harvesting mouse oocytes or embryos, or, if Nomarski optics are used for micromanipulation, for use on this microscope (e.g., Fisher, Pittsburgh, PA, 12-565A)

Anesthesia for mice (e.g., sodium pentobarbital, 6 mg/ml, see Table A.IX for formulation) (Ft. Dodge, Ft. Dodge, IA)

Sealing wax

Automatic pipettors capable of delivering 0–200 and 0–20 μl (Rainin, Woburn, MA)

Pasteur pipettes, 9 inch (Kimble is best) (Thomas Scientific, Swedesboro, NJ)

Diamond pen for scoring and breaking glass (Fisher)

Dialysis tubing for DNA fragment electroelution (Baxter)

Sterile conical centrifuge tubes for medium stock preparation, 15 and 50 ml (Baxter)

Single-edge razor blades for cutting pipette tips, polyethylene tubing, etc. (VWR, Philadelphia, PA)

After digestion of sufficient material to yield at least 30 μg of transgene fragment and electroelution of that fragment into a dialysis bag, we extract the DNA with DNA-grade phenol alone until, after centrifugation, the interface is free of agarose (usually one to four extractions). Note that the interface never becomes completely clean, but extractions with phenol alone should continue until a clear layer of white flocculent material between the phenol and aqueous phases disappears. Only then do we proceed to phenol–chloroform and then chloroform extraction. After extraction, the DNA is ethanol precipitated and redissolved in 10 mM Tris-HCl, pH

TABLE IV
REAGENTS FOR TRANSGENIC MOUSE PRODUCTION[a]

Concentrated HCl (Fisher)
NaOH (Sigma, St. Louis, MO)
Fluorinert (3M, FC-77) (Industrial Chemical Products,
  St. Paul, MN)
Light mineral oil (Sigma)
Pregnant mare serum gonadotropin for superovulation
  (Sigma)
Human chorionic gonadotropin for superovulation
  (Sigma)
Sodium pentobarbital stock solution (50–60 mg/ml) (Ft.
  Dodge)
100% Ethanol (MSMC)
Propylene glycol (Sigma)
Hyaluronidase (Sigma, H-3606)
Bovine serum albumin (fraction V) (Sigma)
Proteinase K (Sigma)
DNA-grade phenol (BRL, Grand Island, NY)
Chloroform–isoamyl alcohol (24 : 1, v/v) (Sigma, Fisher)
Sodium dodecyl sulfate (SDS) (Sigma)
EDTA (Sigma)
NaCl (Sigma)
Trizma base (Sigma)

[a] Not including those contained in basic media.

7.5, 0.2 mM EDTA, pH 7.5, to a final concentration of about 2000 copies/pl. This concentration is about 10-fold higher than in many published protocols,[41] but the higher concentration is useful because we find that it does not reduce the frequency of DNA integration, and because a "near miss" of the pronucleus will still result in insertion of adequate quantities of DNA. We then dialyze the DNA (100–300 μl) against 2000 ml of the same Tris-HCl solution for 3 days with two exchanges per day. After dialysis, 5 μl of the DNA is visualized on an agarose gel to confirm adequate recovery and absence of degradation.

DNA to be microinjected, or stock solutions of the fragment, are maintained at 4°. Freezing is not recommended. Immediately prior to the first two sessions of microinjection, the DNA is centrifuged in a microcentrifuge at maximum speed (12,000–14,000 rpm) for 15 min at 4°. The solution is then carefully harvested with an automatic pipettor, using a tip washed 3 times in distilled water, into a clean microcentrifuge tube which has also been washed 3 times in water. Only the top 80–90% of the solution is harvested.

We prefer to microinject a fragment for three or fours experiments as

[41] R. L. Brinster, H. Y. Chen, M. W. Trumbauer, M. K. Yagle, and R. D. Palmiter, *Proc. Natl. Acad. Sci. U.S.A.* **82**, 4438 (1985).

soon as possible after the dialysis. If the DNA is stored for more than 2 weeks, it is rechecked on a gel before use to assure that no degradation has taken place. After long periods of storage, we again centrifuge the DNA immediately before use. Special caution should be taken with DNA fragments larger than 10 kb. High rates of gene transfer can be obtained with large fragments, but DNA degradation produces a disproportionately large molar contribution of small pieces of DNA, and functionally increases the DNA concentration. Under these circumstances, the rate of transgenic mouse production can be significantly reduced.

Microtool Production

For successful microinjection, it is important to appreciate the factors that most affect the ease of handling embryos. General principles covering the production of holding pipettes are covered in detail elsewhere in this volume.[40] It is advisable to fashion the holding pipette such that its outside diameter is equal to or slightly greater than the zygote with its surrounding zona pellucida. This allows for placement of the holding pipette against the floor of the culture dish with the embryo affixed, a position which provides support to the embryo and reduces trauma during the microinjection process. If the shaft of the holding pipette is narrower than the embryo, lowering it to the floor of the culture dish will result in the zona pellucida making first contact with the dish, with subsequent "prying off" of the embryo from the holding pipette aperture occurring as the pipette is lowered further.

When we conduct microinjections, microneedles are pulled and used directly. If this is to be done, a gradual taper is preferable. Generally, resistance to flow across the tip of a microneedle that has not been beveled on a needle grinder is exceedingly high, and it is necessary to open the tip slightly to initiate flow of the DNA solution (see below).

Embryo Retrieval and Microinjection

Effective transgenic mouse production is dependent on successful superovulation of the female donors. Although protocols vary for this procedure, we find that 2.5 international units (IU) of pregnant mare serum gonadotropin (PMSG), followed 48–52 hr later by 5 IU of human chorionic gonadotropin (hCG) is effective (Table I). In our experience, mice under 4 weeks of age, despite giving a high frequency of mating to males, produce fewer normal oocytes. This problem is partly due to the fact that, in our experience, higher body weight tends to be correlated with a good response to hormones. Consequently, we order 5- to 6-week-old females for superovulation.

Mice are sacrificed at the appropriate time relative to the light cycle (Table I), and the oviducts are removed to 2 ml of M16 medium (Tables A.I–A.VII in a 30-mm tissue culture dish. When the tissue is excised, we generally remove the ovary, oviduct, and a small portion of the uterine horn so as to avoid damage to the oviduct, where the zygotes are located. Each oviduct is then placed in M16 medium supplemented with hyaluronidase (Table A.VIII), about 75 μl of which as been placed in a depression slide and preequilibrated for temperature and pH in the incubator. The ampulla of the oviduct can be recognized under transmitted light as a swollen, translucent region with longitudinal striations. Usually the ampulla is especially distended because of the presence of the fertilized eggs and associated cumulus cells. This area is isolated with forceps and torn open. If the cumulus mass is still intact, the zygotes will be extruded spontaneously into the hyaluronidase. If the cumulus mass has been dispersed by hyaluronidase from spermatozoa, which is more likely if the animals are sacrificed later in the day, then it may be necessary to express the zygotes from the oviduct. After several washes in M16 medium to remove debris and abnormal eggs, the fertilized ova are then loaded into the microdrop of M2 medium (Tables A.I–A.VII) for microinjection. At least one pronucleus is almost always visible with the dissecting microscope using transmitted light.

Microneedle tubing that can be filled from the base (Table III) is pulled on the pipette puller, and the tip is then filled by placing the base in the DNA solution. During this procedure rubber gloves are not used, as the particulate matter on the inside of the glove can contaminate the DNA and obstruct flow during injection. However, the base of the microneedle, to be inserted into the DNA, is never handled directly. After the tip is filled with DNA, a spinal needle (Table III) is used to fill the barrel of the microneedle with Fluorinert or mineral oil (Table IV).

To initiate flow of DNA we brush the microneedle tip against the holding pipette to break off the end. We do not bevel the needle. Then with the zygote positioned such that a pronucleus is accessible to the microneedle (Fig. 2A), the zona is traversed. Positive pressure is then applied and flow of DNA is observed as a slight expansion of the perivitelline space. The microneedle is then inserted into the pronucleus (Fig. 2A), and after swelling of the pronucleus is observed, it is withdrawn (Fig. 2B). Because many zygotes are usually manipulated, it is advisable to place them in an elongated microdrop, such that it is possible to separate cells that have been injected from those yet to be manipulated. We routinely draw a line on the underside of the manipulation dish in order to divide the microdrop into areas from embryos pre- and postmicroinjection.

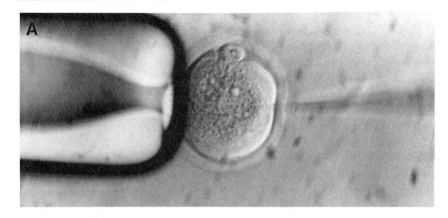

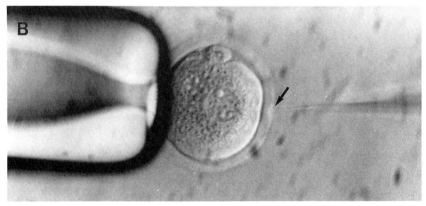

FIG. 2. Pronuclear microinjection. (A) Microneedle in the male pronucleus, injecting DNA. (B) Microinjection complete, with a viable zygote and an intact male pronucleus. The point of entry into the zona pellucida is still seen as a dimple (arrow).

It is frequently necessary to pass the microneedle completely through the pronucleus in order to achieve penetration of the membrane. If the needle is sharp this maneuver often does not kill the zygote. However, probably because pronuclear DNA is associated with the membrane, this maneuver can lead to adherence of embryo DNA to the microneedle. Subsequently, DNA is withdrawn from the cell with the microneedle, and the embryo dies. Such trauma is best avoided if DNA is flowing slowly but continuously from the microneedle. Continuous positive flow, although it runs the risk of destroying the zygote by adding excessive fluid volume, reaps dividends by preventing pronuclear DNA from adhering to the microneedle and being removed from the embryo when the microneedle is withdrawn.

Inadvertent removal of pronuclear DNA on withdrawal of the microneedle is by far the most frequent cause of embryo death with microinjection. As noted above, continuous positive flow from the microneedle helps alleviate this problem. However, when this difficulty becomes severe, with more than 40% of embryos dying, the most likely cause is contamination of the microinjected DNA with particulate matter. For reasons we do not understand, such contamination greatly encourages adherence of pronuclear DNA to the microneedle. Thus, when we encounter this problem, the DNA solution is centrifuged and harvested as before the first two microinjection sessions.

We find that microinjection is best controlled if the microneedle is connected to an empty 10-ml syringe fitted to the manipulator tubing with a tubing adaptor that contains a Luer-lock. It is easy to establish a desired flow rate by progressively increasing pressure on the syringe. When it is necessary to terminate flow to open the microneedle tip, the Luer-lock is simply opened, and pressure within the system is relieved. When particles clog the microneedle to the point where the tip must be broken to reestablish flow, it is important to relieve accumulated pressure within the microneedle system before opening the tip. If this is not done, a sudden increase in flow through the microneedle will carry particles to the tip and immediately replug it. Thus, before servicing the microneedle we always open the Luer-lock.

Another common problem is dullness of the microneedle. After many embryos are microinjected, accumulation of material on the microneedle tip can make penetration of the plasma membrane more difficult. Under these circumstances it is also not unusual for flow of DNA to be impeded or halted altogether. We have recently devised a procedure for "rescuing" such microneedles. Although not always successful, this procedure can clear the microneedle while at the same time producing a very sharp tip.

The holding pipette is checked to assure contact between the barrel and the floor of the petri dish. The microneedle is then lowered until the tip also touches the dish. The tip is then advanced under the holding pipette until it becomes trapped between the holding pipette and the culture dish. A slight, forward movement of the microneedle frequently produces a short, longitudinal cleavage to yield a sharp, open tip. If this procedure fails, the microneedle must be changed. If DNA is prepared properly, however, it should be possible to microinject as many as 50–100 embryos with a single microneedle.

After microinjection, we place surviving zygotes in M16 medium for 0.5–1 hr, after which the survivors are implanted into pseudopregnant foster mice. Several anesthetics for this surgery, as well as for the vasectomy required for generating sterile males used for production of pseudopregnant females, are acceptable. The formula for sodium pentobarbital anesthesia is given in Table A.IX.

Embryo Transfer

We do not culture zygotes overnight to the 2-cell stage; in our view, culture only reduces the likelihood of survival, and it is not required to identify the embryos that have survived microinjection. The latter is true because cell death, recognized by a loss of membrane integrity, disappearance of the perivitelline space, and a pale appearance, is usually seen within 30 min of microinjection, and because implantation of inviable embryos has no known inhibitory effect on the embryos that do survive. We prefer to implant at least 30 embryos into one mouse. However, if only a single foster mouse is available, we have successfully inserted as many as 70 embryos into a single animal. We generally insert embryos into both oviducts, though it is possible to obtain success by transferring into only one. If one side is chosen, it is technically easiest to operate on the side opposite the spleen. Principles governing the transfer are that minimal fluid and air should accompany embryos into the oviduct, and minimal trauma from the glass pipette should be incurred by the wall of the oviduct. A method for producing a suitable transfer pipette for this procedure, discussed in detail in a previous publication,[39] is described briefly here.

A Pasteur pipette is heated and drawn out to yield a thin-walled shaft of glass with an inside diameter of about 120 μm. This is scored with the diamond pen, broken to give a clean break with no protruding shards of glass, and inserted into the truncated barrel of a Pasteur pipette such that about 3 cm of the shaft protrudes from the end of the pipette. The air space between the shaft and the pipette barrel is then sealed with sealing wax. When embryos are loaded, medium is first drawn to fill 0.5 cm of the protruding tip. This is followed by a small air bubble marker, an additional 0.5–1 cm of medium, and then the embryos. The embryos should not be separated by significant amounts of medium, and they should be positioned just at the aperture of the pipette for transfer into the oviduct.

Successful embryo transfer depends on rapid location of the ostium and avoidance of trauma to the oviduct. The ovary is approached through a dorsal incision about 1 cm lateral to the midline, 1–2 cm below the costal margin. The ostium lies immediately adjacent to the ovary, and the opening is always oriented toward the tail of the animal. Unlike humans, the mouse ovary is surrounded by a bursa which must be opened to gain access to the oviduct. When the bursa is opened, trauma to large blood vessels should be avoided. The bursa is attached to the neck of the ostium. Often, after the bursa is opened, strands of tissue can cross the oviduct at the point of bursal attachment and pinch the neck of the ostium. To assure successful embryo transfer, the pipette should be advanced beyond the point of bursal attachment. The best way to accomplish this is to gently grasp the bursa immediately adjacent to the oviduct and lift up. This relieves pressure against the oviduct and allows the pipette to be advanced. Embryos are then expelled until the air bubble marker enters the oviductal lumen. Resistance to flow through the transfer pipette during embryo expulsion is usually due to contact between the pipette opening and the wall of the oviduct. To overcome this problem, the pipette is withdrawn slightly while pressure is applied and is held in position as soon as flow is appreciable.

Pups born 19–21 days after transfer are then reared to weaning age and evaluated by the polymerase chain reaction (PCR) or Southern blotting for integration of the transgene. Rates of integration should be 15–90% of newborn pups.

Vasectomy

Vasectomy for production of breeders for rendering foster mice pseudopregnant is a simple procedure. We find that if 6-week-old mice are vasectomized, sperm will have not yet reached the vas deferens, and it is not necessary to breed the animals in order to clear the ducts of spermatozoa.

The vasectomy can be done without ligation or suturing of the vas deferens. A ventral midline incision about 1 cm in length is made above the position of the preputial gland, which is situated under the skin and which can be appreciated as a bulging of the skin in the lower abdomen. The scrotal sac is squeezed to force the testes toward the incision, and the testes are both exposed by grasping the associated fat pads with forceps. The vas deferens is identified and isolated by blunt dissection. A large loop of the vas deferens is then isolated in the jaws of a hemostat, crushed, and torn away. The animal can then be closed with a single suture through both the skin and peritoneum. After about 1 week the animals can be used as breeders. The strain of animals should of course be chosen such that failure to successfully sterilize these males will lead to newborns that are distinguishable by coat color markers.[39]

Tail Biopsy and DNA Analysis

When mice are 3–5 weeks old they can be screened by DNA analysis. A protocol for biopsy and DNA isolation is given in Table V. This protocol yields sufficient amounts of DNA for several Southern blots or PCR analy-

TABLE V
TAIL BIOPSY AND DNA EXTRACTION

1. Remove 1/3 to 1/2 of tail (depending on size of animal). For tail biopsy, the animal is held over an open cage by the end of the tail and the tail is cut, allowing the animal to drop into the cage. Anesthesia is not used because it is more traumatic than the biopsy itself and endangers the life of the animal
2. Cut the tail, including the bone, into 5–10 pieces and place it in 500 μl of the following solution in a 1.5-ml centrifuge tube: 10 mM NaCl, 10 mM Tris-HCl, pH 8.0, 10 mM EDTA, pH 8, and 2% SDS
3. Add 50 μl of proteinase K (20 mg/ml in water), and incubate at 37° on a rocking shaker for 2–2.5 hr
4. Add 300 μl DNA-grade phenol, shake, and microcentrifuge. Harvest the viscous supernatant with a 200 μl automatic pipettor equipped with a tip that has had the end cut off with a razor to produce a wide opening
5. Extract the supernatant with 150 μl phenol, harvest again, and extract with 100 μl chloroform–isoamyl alcohol (24 : 1, v/v)
6. Precipitate the DNA with 2.5 volumes of ice-cold 100% ethanol
7. Redissolve the DNA pellet in 100–200 μl of 10 mM Tris-HCl, pH 8, 0.1 mM EDTA, pH 8. When redissolved in 100 μl of buffer the DNA concentration is usually 2–3 μg/μl

ses. Although the extraction procedure is costly because of the large quantity of proteinase K used, it is complete within 3 hr and consistently gives high yields of DNA that is readily digestible with restriction enzymes.

When large concatamers are integrated, it is sufficient to employ dot-blot hybridization for identification of transgenic progeny. When fewer than 5 copies of the transgene per cell are present, we prefer to follow inheritance with Southern blots. PCR analysis, using transgene-specific primers, can also be employed, though this technique examines only a small portion of the transgene and cannot detect structure changes.

Selection of Strains for Transgenic Mouse Production

The vast majority of Tg mice have been produced on outbred genetic backgrounds because such mice produce large numbers of embryos in response to superovulation. In many instances the genetic background is irrelevant to the experiment, but in some circumstances it is crucial that the recipient strain be inbred (e.g., immunological studies). In other cases, the penetrance of a desired phenotype may be variable because of modifying alleles in an outbred line.

Most inbred strains can be superovulated and used for Tg mouse production. Usually, inbred animals produce zygotes in which pronuclei are more difficult to visualize. One exception is the FVB/N strain, which is highly inbred but which produces large numbers of healthy embryos with large pronuclei.[42] We have used this strain extensively and found no difference in gene transfer frequency. Moreover, we have compared transgene constructs for expression on the FVB/N background and outbred backgrounds and found no significant differences. Therefore, we find the FVB/N mouse highly useful for Tg mouse production.

Transgenic Colony Management

Although not directly related to micromanipulation technique, transgenic colony management, especially at a time when animal rights activism is increasing animal care costs, is a very important skill. In this section, the production colony and maintenance procedures used in our laboratory are described, along with suggestions for strain preservation.

Production Colony

For the production colony we maintain 30 males as breeders to be mated to superovulated females every other day. Five- to six-week-old females are imported weekly for superovulation, and 7–8 such females are superovulated for each experiment (when FVB mice are used; for inbred strains it may be necessary to superovulate more animals per experiment). Mature CD-1 females are imported for use as pseudopregnant foster mothers and are ordered at a weight of 30 gs. We maintain 50 of these animals, housed in five groups of 10.

It is important that each group of 10 foster mothers always remain together in order to avoid synchronization of estrus cycles among the entire foster population. When the animals in each cage (labeled cages A–E) remain together, the cage may become synchronized, but the estrus day will differ between cages. The foster mothers are mated without superovulation to vasectomized males, which are maintained as five groups of 5. Foster mothers are not superovulated because mating may not result in pregnancy after embryo transfer. The "A" foster mothers are always mated with the five "A" vasectomized males, etc. Since mice reach estrus every 5 days, this program gives a high probability of mating on each day. Both fertile and vasectomized males are changed every 6 to 8 months. When several foster mothers are available but only one is needed, we sacrifice the additional animals and replace them. If the animals are to be reused, they must be housed

[42] M. Taketo, A. C. Shroeder, L. E. Mobraaten, K. B. Gunning, G. Hanten, R. R. Rox, T. H. Roderick, C. L. Stewart, F. Lilly, C. T. Hansen, and P. A. Overbeek, *Proc. Natl. Acad. Sci. U.S.A.* **88**, 2065 (1991).

for 14 days until estrus cycling resumes. Such prolonged housing is costly and can predispose to the establishment of infection in conventional colonies. It should be noted that our laboratory conducts microinjections almost every day; the numbers of animals can be scaled down for more moderate production efforts.

Establishing Lines

When transgenic founders are first identified, it is important to breed them aggressively until the line is established. This is important because transgene integration and/or expression can compromise fertility and because the founder may be a mosaic, with the transgene present in only a subset of germ cells. Once positive progeny are identified, efforts can begin to establish homozygous lines if desired (see below) and to evaluate the animals. Transgenic animals should always be observed for health problems relating to transgene integration. Such problems can lead to loss of valuable lines and can frequently occur when harmful genes such as oncogenes are inserted.

Embryo Freezing

When freezing embryos for transgenic strain preservation, it is important to realize that large numbers of embryos are required to assure rapid reconstitution of the strain. These numbers may be augmented further if the transgenic line cannot be bred to homozygosity. The inability to produce fertile homozygotes of both sexes is not uncommon, and it is due either to insertional mutagenesis or to chromosomal rearrangment associated with transgene integration. Embryo freezing also requires a backup system to secure against freezer breakdown and the availability of skilled labor. Given these factors, establishment of a freezing program is not recommended unless a large colony is developed. Commercial freezing is available but can be costly. For these reasons, it is often desirable to maintain small samples of breeding nuclei for various strains "on the shelf." When homozygotes can be produced this strategy is simple and cost effective, as DNA analysis in the line becomes unnecessary.

Tests for Transgene Homozygosity

It is necessary to make sure that putative homozygotes are indeed homozygous. This is done both by Southern blotting using a reference probe to track the amount of DNA loaded[43] and by genetic testing. Although it is theoretically possible to identify homozygotes by quantitative PCR, variability in PCR efficiency between samples usually makes Southern hybridization more reliable. When reduced litter sizes are seen on inbreeding a transgenic line, one should be alerted to the possibility of recessive insertional mutagenesis leading to embryo lethality. Once putative homozygotes are produced and crossed with each other, they should be watched carefully for reproductive competence, as homozygotes of either sex may be infertile. When homozygotes cannot be produced, we cross heterozygotes, screen small litters, and replace breeders with newly reared transgenic pups.

Cost Containment

As animal purchase and maintenance costs steadily increase while grant funds decrease, cost containment of transgenic colonies becomes very important. Perhaps the most important cost-saving measure is effective design of transgenes. When the sequence to be microinjected is poorly designed, mice are produced that give less than the expected result. This situation frequently leads to new rounds of microinjection with new constructs, but few investigators discard the previous mice. Therefore, because the effort to produce transgenic mice is substantial and because animal care costs are high, it is advisable to plan experiments carefully.

When several lines are produced with a single construct, it may be advisable to discontinue some of the lines. This is psychologically difficult because of the potential loss of data but may prove to be unavoidable. Obviously it is important to characterize the phenotypes of the various lines before making such a decision.

A third measure toward cost containment involves reproductive management. Since male mice can be maintained for a significant period of time without losing fertility, it is possible to maintain some lines by simply keeping 1–3 males on the shelf for up to 6 months without breeding them, provided that the males are known to transmit the transgene effectively. The inclination to expand every line to large numbers should be avoided; instead, a rational plan should be developed concerning the number of animals, both positive and negative, that will actually be needed for each experiment. Although establishment of homozygous transgenic mice can be difficult and is labor intensive, it

[43] T. F. Krulewski, P. E. Neumann, and J. W. Gordon, *Proc. Natl. Acad. Sci. U.S.A.* **86**, 3709 (1989).

is, in the long term, the most efficient cost-saving measure. When every animal in a line is transgenic, DNA screening can be halted, and very few animals are needed to assure preservation of the line. However, it is of paramount importance that homozygosity be unequivocally established before DNA screening is halted.

Appendix

This appendix contains the formulations for stock solutions and other media.

TABLE A.I

M2 AND M16 CONCENTRATED STOCKS: STOCK A[a]

Compound	Amount (g)
NaCl	5.534
KCl	0.356
KH_2PO_4	0.162
$MgSO_4 \cdot 7H_2O$	0.293
Sodium lactate, 60% syrup	4.439 (or 2.610 ml)
Glucose	1.000
Penicillin	0.060
Streptomycin	0.050

[a] Stock A (100 ml), 10× solution. Storage time: 3 months.

TABLE A.II

M2 AND M16 CONCENTRATED STOCKS: STOCK B[a]

Compound	Amount
$NaHCO_3$	2.101 g
Phenol red, 0.5% solution	0.010 ml

[a] Stock B (100 ml), 10× solution. Storage time: 2 weeks.

TABLE A.III

M2 AND M16 CONCENTRATED STOCKS: STOCK C[a]

Compound	Amount
Sodium pyruvate	0.180 g

[a] Stock C (50 ml), 100× solution. Storage time: 2 weeks.

TABLE A.IV

M2 AND M16 CONCENTRATED STOCKS: STOCK D[a]

Compound	Amount
$CaCl_2 \cdot 2H_2O$	0.252 g

[a] Stock D (10 ml), 100× solution. Storage time: 3 months.

TABLE A.V

M2 AND M16 CONCENTRATED STOCKS: STOCK E[a]

Compound	Amount
HEPES solid	5.958 g
Phenol red, 0.5% solution	0.010 ml

[a] Stock E (100 ml), 10× solution. Storage time: 3 months.

TABLE A.VI

M2 AND M16 CONCENTRATED STOCKS: MAKING UP STOCKS

Stock	Procedure
Stocks A–D	Weigh solids into a clean flask or appropriate sized disposable conical centrifuge tube and add appropriate quantity of distilled water (18 MΩ). If sodium lactate syrup is weighed (stock A), use a small weighing boat and decant contents into flask, then rinse boat 2–3 times with distilled water and decant rinse into flask
Stock E	Weigh out HEPES and add Phenol red. Add 50 ml distilled water, then adjust pH to 7.4 with 5 N NaOH. Add distilled water to 100 ml

TABLE A.VII
M16 MEDIUM AND M2 MEDIUM FROM
CONCENTRATED STOCKS

M16 (500 ml) Stock	Amount
A	50 ml
B	50 ml
C	5 ml
D	5 ml
Water	390 ml
BSA (fraction V)	2 g
Rinse all pipettes, etc., thoroughly into final flask, then add water to 500 ml. Add 2 g BSA, dissolve without excessive frothing, and sterile filter into 50-ml sterile disposable conical centrifuge tubes	

M2 (100 ml) Stock	Amount
A	10.0 ml
B	1.6 ml
C	1.0 ml
D	1.0 ml
E	8.4 ml
Water	78.0 ml
Make up as for M16, but recheck pH prior to sterile filtration (pH should be 7.3–7.4)	

TABLE A.VIII
MEDIUM SUPPLEMENTED WITH HYALURONIDASE
FOR CUMULUS CELL REMOVAL

Add hyaluronidase to a final concentration of 1–2
  mg/ml to M2 or M16
Filter sterilize before use

TABLE A.IX
PREPARATION OF PENTOBARBITAL ANESTHESIA
FOR MOUSE SURGERY[a]

Compound	Amount (ml)
100% Ethanol	10
Propylene glycol	20
Water	70

[a] To make 100 ml, mix the above components, then add sodium pentobarbital stock solution (see Table IV in text) to a final concentration of 6 mg/ml. Administer intraperitoneally to mice at 60 mg/kg, or 0.1 ml/10 g body weight.

Acknowledgments

Support for work toward preparation of this chapter was supported by National Institutes of Health Grants A124460, HD20484, and CA42103.

[46] Factors Influencing Frequency Production of Transgenic Mice

By Jeffrey R. Mann and Andrew P. McMahon

Introduction

Certain parameters have been defined that are essential for the efficient production of transgenic mice utilizing pronuclear injection of DNA. These include the concentration of DNA injected, composition of injection buffer, form of DNA, and mouse strain.[1] Nevertheless, in the literature it is often revealed that good efficiency is not obtained. Therefore, in this discussion, we attempt to define some additional parameters that may influence success.

Each step involved in the production of transgenic mice is discussed with respect to what factors may influence efficiency. It is emphasized that not all comments are necessarily backed up by a wealth of experimental evidence. Some are based on limited observation by the authors.

[1] R. L. Brinster, H. Y. Chen, M. E. Trumbauer, M. K. Yagle, and R. D. Palmiter, *Proc. Natl. Acad. Sci. U.S.A.* **88**, 4438 (1985).

Copyright © 1993 by Academic Press, Inc.
All rights of reproduction in any form reserved.

Choice of Mice

In general, fertilized eggs for injection obtained from hybrid F_1 females give better results than eggs obtained from inbred females, although the success with inbred strains has been observed to vary according to the strain used, and probably varies with the parental inbred strains used to produce the F_1 hybrids as well. Good results are obtained when one of the parental inbred strains is C57BL/6. Traditionally, this is the female. CBA, C3H, or SJL/J males are often used as the other parent in the cross, but males of other strains can also be used. The better results obtained with eggs of F_1 females is due to hybrid vigor, and it is a maternal or egg cytoplasmic effect. This means that the eggs can be fertilized by males of any strain with no alteration in the efficiency of transgenesis. The authors have obtained the same efficiencies with eggs of (C57BL/6J × CBA/J)F_1 females fertilized by (C57BL/6J × CBA/J)F_1 hybrid or C57BL/6J or 129/Sv inbred males.

Eggs of inbred females can be injected as easily as eggs of F_1 females, with no increase in the frequency of lysis. However, in most inbred strains, the ability of the eggs to recover from the injection relative to eggs of F_1 females is reduced. Fewer injected 1-cell eggs of inbred mice undergo cleavage, and from those that do cleave, less mice are obtained. However, of the mice born, the frequency of transgenic individuals is equivalent to that obtained with eggs from F_1 hybrids. With eggs of C57BL/6J inbred females, one-sixth overall efficiency of transgenesis in comparison to eggs of (C57BL/6J × SJL/J)F_1 females was obtained, although when related to the number of mice born, the frequency of transgenesis was similar.[1] The present authors obtained even lower overall efficiency using eggs from 129/Sv females fertilized by 129/Sv males. In at least one inbred strain, FVB/N, efficiencies comparable to those obtained with F_1 hybrids have been reported.[2] Indeed, one may wish to use such inbred strains in the place of F_1 hybrids, as it is often advisable to study the effects of a transgene when the genetic background is uniform between individual founders.

Outbred females may be used for economy. Albino Swiss types (e.g., CD-1, Charles River, Wilmington, MA) are available from many commercial suppliers and are one-quarter the cost of inbreds and hybrids. However, their eggs perform more like those of inbred than eggs of F_1 strains following microinjection.

---

[2] M. Taketo, A. C. Schroeder, L. E. Mobraaten, K. B. Gunning, G. Harten, R. R. Fox, T. H. Roderick, C. L. Stewart, F. Lilly, C. T. Hansen, and P. A. Overbeek, *Proc. Natl. Acad. Sci. U.S.A.* **88**, 2065 (1991).

Production of Fertilized Eggs for Microinjection

It is most convenient to superovulate females for obtaining fertilized eggs, although again the success of the procedure varies with mouse strain. The standard regime is to administer by intraperitoneal injection 5 IU of pregnant mare serum gonadotropin (PMSG) followed about 48 hr later by 5 IU of human chorionic gonadotropin (hCG).[3,4] F_1 females respond very well. Usually, prepubertal females 3 to 5 weeks of age are used, but good results are also obtained with mature females of 8 weeks of age or more. In the latter case, fewer eggs may be obtained than with prepubertal mice, but a higher proportion are morphologically normal. The frequency of mating should be at least 75% following the hormone regime, provided the stud males are not used more than about twice a week, and are replaced when they get to about 6 months of age. A record should be kept of the mating performance of studs, and males that do not produce plugs consistently (e.g., three times out of four) should be replaced.

When using inbred females, it is usually better to obtain eggs by natural mating. Females of many strains will superovulate well, but they are unpredictable with regard to whether they will subsequently mate and, if they do, whether the eggs will be fertilized. Nevertheless, if working with an untried strain, in the first instance an attempt could be made to obtain fertilized eggs by superovulation.

Egg Culture Conditions

Before attempting to make transgenic mice, the egg culture conditions must be optimized. Media and incubation conditions are acceptable if approximately 90% of noninjected 1-cell eggs of F_1 females reach the blastocyst stage by the middle of the fourth day of culture in medium CZB,[5] or about one-quarter of a day later in medium M16.[6]

Microinjection of Eggs

It is possible to inject eggs from F_1 females such that at least 90% will be viable and undergo cleavage following injection (Table I). It is our

[3] A. H. Gates, in "Methods in Mammalian Embryology" (J. C. Daniel, ed.), p. 64. Freeman, San Francisco, 1971.
[4] B. Hogan, F. Costantini, and E. Lacy, "Manipulating the Mouse Embryo," p. 92. Cold Spring Harbor Laboratory, Cold Spring Harbor, New York, 1986.
[5] C. L. Chatot, C. A. Ziomek, B. D. Bavister, J. L. Lewis, and I. Torres, J. Reprod. Fertil. 86, 679 (1989).
[6] D. G. Whittingham, J. Reprod. Fertil. Suppl. 14, 7 (1971).

experience that an injected egg that does not lyse during subsequent culture will almost always undergo cleavage. Exceptions could occur; for example, transient expression of a particular construct might affect cleavage. The three factors that will excessively damage eggs and lead to lysis are the inherent vibration in the micromanipulator, the pipette design, and the injection technique of the operator.

Vibration

The tip of the injection pipette should appear completely stable when present in the injection chamber. Vibration transmitted from the floor to the micromanipulator and pipette tip can cause eggs to lyse after the tip has punctured the membrane. Vibration may be eliminated by supporting the micromanipulator on a metal plate, under which is placed a semiinflated tire inner tube or squash balls, and a heavy table. Gas-cushion tables (e.g., Vibraplane, Kinematic Systems, Roslindale, MA) are effective in eliminating vibration.

Pipette Design

Tip openings less than 0.5 μm will clog quickly, and openings larger than 1.5 μm will begin to cause irreparable damage. It can be difficult to calibrate pipette pullers to pull pipettes with a defined tip opening. Therefore, the tips can be chipped to size on a glass bead with the aid of a microforge. One can get an idea of the tip size created by microscopic observation under a 100× objective using an eyepiece fitted with a graticule. A small piece of modeling clay fixed to the mechanical stage will immobilize the pipette for inspection. Open tips can also be created by chipping against the holding pipette in the injection chamber. It is easiest to chip a pipette tip by bringing it perpendicular to the glass bead or holding pipette.

The angle of the taper of a pipette should not be too large, as this will force larger holes in the egg membrane and increase the frequency of lysis. The approximate shape and dimensions of a desirable pipette are given in Fig. 1.

Injection Technique

Egg penetration should be smooth and direct. Ideally, a sharp pipette will puncture the membrane before or as the tip reaches the pronucleus. Otherwise, it is often necessary to continue to push the tip into and past the pronucleus before puncture occurs. Such variations should not

a

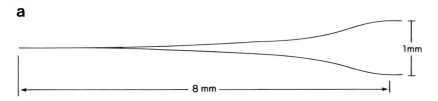

b

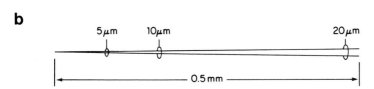

FIG. 1. Dimensions of pipette for injection of egg pronuclei. (a) Overall dimensions of pulled portion. The glass used is thin-walled borosilicate with filament. (b) Dimensions of last 0.5 mm.

affect the frequency of lysis. Puncture is evidenced by sudden relaxation of the plasma membrane. The tip is pushed into the pronucleus, or retracted if it is not already inside, and then the solution is ejected. The authors routinely swell the pronucleus to its maximum extent, with swelling being rapid but not sudden. Occasionally, solution is seen to burst through the nuclear membrane into the egg cytoplasm. Some variation in these swelling characteristics should not influence the frequency of DNA integration. Retraction of the tip from the egg following swelling of the pronucleus should be done very quickly. Although the injection pressure is released at the same time as retraction occurs, some time must elapse before flow is reduced, and flow of solution may perturb the plasma membrane if the tip is retracted slowly. If leakage of egg cytoplasm into the perivitelline space is observed on retraction of the tip, the egg will eventually lyse.

The pipette tip should be designed and arranged in the injection chamber such that it enters the egg perpendicularly and not obliquely. This will facilitate membrane puncture and ensure that the membrane does not tear on tip entry. If the injection design is such that the capillary must enter at a larger angle to the floor of the chamber (e.g., with concavity slide chambers), the pipette tip should first be bent with the microforge such that the first 500 μm or so is horizontal to the viewing plane.

Transfer of Injected Eggs to Pseudopregnant Recipients

Strain of Recipient

Again, F_1 females are excellent foster mothers. The less expensive outbred albino Swiss type mice are very good, but their infundibuli tend to be less accessable.

Stage of Transfer

Eggs can be transferred to oviducts of recipients on the same day as they are injected, when they will be at the 1-cell stage. Alternatively, injected eggs can be cultured overnight, then transferred to oviducts at the 2-cell stage. Eggs that do not cleave after overnight culture are inviable. The data in Table I show that the efficiency of transgenesis is similar when eggs are transferred at these two stages.

If one has technical difficulty in performing oviduct transfers, injected eggs could be cultured to the blastocyst stage, then transferred to the uterus. Although the authors have not used this approach, it seems unlikely that this alternative would markedly affect overall efficiency.

Success of Transfer

At least 80% of noninjected control 2-cell eggs of F_1 females transferred to the oviducts of F_1 pseudopregnant recipients on day 1 of pseudopregnancy (day 1 = day of plug) can develop into fetuses of normal morphology. This is regardless of whether eggs have been fertilized naturally or in vitro.[7] Given these frequencies, approximately 30% of injected eggs should give rise to mice (Table I). Higher frequencies often, but not always, indicate low DNA integration frequency. Transfer approximately 13 eggs to each oviduct of a recipient to give an expected litter size of about 8. It is rare for an F_1 recipient not to become pregnant following transfer of 1-cell or 2-cell eggs to oviducts. In the experiments presented in Table I, where a record was kept, all of 44 recipients became pregnant.

DNA Integration Frequency

The data in Table I show that, for every construct injected, at least 25% of embryos obtained or mice born were transgenic. The constructs varied in size between 4600 to 18,000 bp, many being essentially unrelated

---

[7] J. R. Mann, *Biol. Reprod.* **38**, 1077 (1988).

in sequence. The DNA concentration of the solution injected was 2.5 μg/ ml in all cases. If the frequency of transgenic mice per embryos obtained or per mice born is less than 25%, then four explanations are possible.

Improper Injection. It is possible that not every transferred egg was injected. Successful injection is best judged by a visible swelling of the pronucleus. However, to familiarize oneself with the flow characteristics from the pipette tip, and to observe what happens to the flow when the tip becomes blocked by particulates from the solution or capillary, load it with an aqueous solution of dye (e.g., trypan blue), to visualize the injected solution in a practice run (it should not be used in a real experiment).

Dilution of DNA. The DNA solution at the pipette tip may have been diluted. It is important not to allow medium in the injection chamber to be drawn into the pipette tip at any stage during the procedure, as this will dilute the DNA solution. Commercial or laboratory-made gas injection systems that provide a back pressure when not injecting safeguard against this possibility. If injecting by hand using a 50-ml syringe,[8] the DNA solution should be expelled at various intervals, including just before injection, to guard against dilution of the DNA solution at the pipette tip.

Lethality. The construct may have been lethal to embryos on integration. Lethality of transgenic embryos could have occurred if no or few mice born were positive for the construct. If this is suspected, probably the best stage at which to perform an initial analysis *in utero* is at day 10 of gestation. Early embryonic death will be apparent at this stage, and positive mice can be identified by polymerase chain reaction (PCR) analysis using the yolk sac, or even by Southern blotting. Approximately 5 μg of DNA can be recovered from the yolk sac at this stage (A. McMahon, unpublished observations, 1991).

Insufficient Purity of DNA. The procedure used in purifying DNA for injection may be the most important consideration in obtaining a high frequency of transgenesis. When separating the insert from the plasmid, the use of agarose might drastically lower the integration frequency. This has been our experience when using a variety of procedures for isolating DNA from agarose gels. Therefore, if possible, it is advisable to perform the separation by rate zonal sucrose gradient centrifugation, which has been the principal method used in the experiments documented in Table I. Otherwise, if agarose is used, load as much DNA onto as low as a percentage gel as possible for electrophoresis (e.g., 20–30 μg into one 6 × 1.5 mm slot in a 0.5% gel). Good integration frequencies have been

---

[8] B. Hogan, F. Costantini, and E. Lacy, "Manipulating the Mouse Embryo," p. 169. Cold Spring Harbor Laboratory, Cold Spring Harbor, New York, 1986.

obtained with a number of methods for purifying DNA from agarose, for example, sodium perchlorate dissolution followed by glass matrix adsorption (Prep-A-Gene, Bio-Rad, Richmond, CA), Gelase (Epicentre Technologies, Madison, WI) digestion followed by Elutip (Schleicher & Schuell, Keens, NH) adsorption, and phenol extraction from low-melt agarose after melting the gel. Ensure that the phenol used in extraction is buffered at pH 8.0 or higher.

If one is not experienced in producing transgenic animals, and obtains only low integration frequency per mouse born in the first attempts, it is advisable to check intrinsic operator efficiency by injecting a DNA solution that can give a high frequency of integration. The DNA used should be (1) innocuous (e.g., a construct lacking a promoter or one in which the product of the bacterial gene *lacZ* would be expressed), and (2) an insert purified using a sucrose gradient, or a linearized construct purified by phenol/chloroform extraction and precipitation only.

Explanation of Tabular Data

The data in Table I represent all experiments conducted over a period of 1.5 years, after experimental conditions had been optimized. Each experiment (Expt) was conducted during 1 day. The capital letter (Table I, column 1) indicates one DNA preparation, and the numeral indicates an individual experiment conducted with this DNA preparation, usually with a fresh aliquot. The constructs injected were as follows: C, human neurofilament promoter–mouse *wnt-1* cDNA; D, human β-actin promoter–bacterial β-galactosidase (*lacZ*) gene; E, human neurofilament promoter–mouse *wnt-1* cDNA with a small fragment of the *myc* gene inserted; I, human β-actin promoter–mouse *wnt-1* cDNA; J, mouse *wnt-1* promoter–*lacZ* gene; N, mouse *wnt-1* gDNA–*lacZ* gene (A. McMahon, unpublished data, 1990); K, rat nestin promoter–*lacZ* gene (L. Zimmerman and R. McKay, unpublished data, 1991); M, mouse gutamic acid decarboxylase promoter–*lacZ* gene (Z. Katarova and G. Szabo, unpublished data, 1991); O, mouse phosphoglycerate kinase-1 (*Pgk-1*) short promoter–*lacZ*–*Pgk-1* short promoter–bacterial *neo* gene (J. Mann, unpublished data, 1990). The size of constructs is given in base pairs. DNA in experiments D and E were linearized or double-digested to release the insert, then purified by phenol–chloroform extraction and precipitation. In the remainder, inserts for injection were separated from plasmids by sucrose gradient centrifugation (see below for methods).

TABLE I
PRODUCTION FREQUENCY OF TRANSGENIC MICE[a]

Expt.	Size of construct (bp)	Number of eggs						Number of	
		Injected	Lysed	Did not cleave	Transferred	Implanted	Embryos obtained	Mice born	Transgenic
C2	7200	76	1	—	75	—	21 (28%)	—	9 (43%)
C4	7200	46	3	0	43	—	24 (56%)	—	12 (50%)
E2	4600	100	2	0	98	89 (91%)	52 (53%)	—	22 (42%)
E3	4600	96	2	6	88	48 (55%)	26 (30%)	—	13 (50%)
E4[b]	4600	110	—	—	—	—	41	—	17 (41%)
I1	6600	83	1	—	68	49 (72%)	24 (35%)	—	14 (58%)
I2	6600	79	4	0	75	45 (60%)	33 (28%)	—	9 (27%)
I4	6600	149	0	—	149	77 (52%)	26 (34%)	—	26 (27%)
I5	6600	167	12	1	153	—	38 (25%)	—	13 (34%)
I6	6600	131	8	—	120	—	39 (32%)	—	10 (26%)
J1	13,000	123	1	—	106	67 (63%)	40 (38%)	—	10 (25%)
J2	13,000	63	1	—	62	46 (74%)	17 (27%)	—	11 (65%)
K1	8200	90	0	1	89	52 (58%)	12 (13%)	—	7 (58%)
K2	8200	90	3	—	86	39 (45%)	21 (24%)	—	9 (43%)
N1		205	25	—	109	—	40 (37%)	—	8/25 (32%)
O1	6000	103	5	—	97	—	37 (38%)	—	15 (41%)
Total		1601	68 (4%)	—	1418	—	450 (32%)	—	188/435 (43%)
C3	7200	136	1	—	135	—	—	37 (27%)	13 (35%)
D2	6500	99	2	0	97	—	—	34 (35%)	11 (32%)
M2	18,000	105	2	—	97	—	—	29 (30%)	10 (34%)
Total		340	5 (1.5%)	—	329	—	—	100 (30%)	34 (34%)

[a] See text for explanation.
[b] Data in this row are excluded from the totals.

Regarding the number of eggs (Table I, columns 3–7), lysed refers to the number of eggs that underwent lysis as a consequence of the injection procedure before and after subsequent culture. The figure in parentheses is the percentage of injected eggs that lysed. Did not cleave refers to the number of eggs that remained at the 1-cell stage after overnight culture; in experiments marked with a dash, eggs were transferred to recipients on the same day of injection. Implanted refers to the total number of implantation sites, including resorbing deciduoma and embryos. Embryos obtained and mice born (Table I, columns 8 and 9) indicate that recipients were sacrificed at or between days 9 to 13 of gestation for analysis. The figure in parentheses represents the percent embryos or newborns that were obtained from the total number of eggs transferred. For the last column of Table I, labeled transgenic, the figure in parentheses is the percentage of transgenic individuals that were obtained from all embryos obtained or mice born. They were identified by PCR analysis of digested yolk sacs and/or Southern blot of DNA purified from the yolk sac or placenta of embryos or tails of postnatal mice.

Methods

For production and injection of eggs, (C57BL/6J × CBA/J)F$_1$ (B6CBF$_1$) females greater than 8 weeks of age are superovulated as previously described, and usually mated to B6CBF$_1$ males. The light cycle of the mouse room dictates lights on from 0630 to 1830 hr. Injection of fertilized eggs begins at approximately 1400 hr, about 6 hr before they begin to cleave. Eggs are injected usually in two to three batches, with about 35 eggs being injected in each batch with usually two or three pipettes in approximately 40 min. Eggs are then transferred at the 1-cell stage to B6CBF$_1$ pseudopregnant recipients later the same afternoon, or, more often, cultured overnight to the 2-cell stage and then transferred at any time during the second day of culture. Injection and transfer of eggs to recipients are carried out in medium M2 without cytoskeletal inhibitors,[9] and eggs are cultured in medium M16 at 37° in 5% CO$_2$ in air.

Microinjection of DNA is carried out under differential interference contrast optics (25× objective, 10× eyepieces) with a Labovert inverted microscope and mechanical micromanipulators (Leitz, Germany). A siliconized (Sigmacote; Sigma Chemical Co., St. Louis, MO) coverslip "sandwich" injection chamber is used.[10] Glass capillaries used for manufacture

[9] M. J. Wood, D. G. Whittingham, and W. F. Rall, *in* "Mammalian Development: A Practical Approach" (M. Monk, ed.), p. 259. IRL Press, Oxford and Washington, D.C., 1987.
[10] E. G. Diacumakos, *in* "Methods in Cell Biology" (D. M. Prescott, ed.), Vol. VII, p. 287. Academic Press, New York and London, 1973.

of pipettes are thin-walled with filament (1.0 mm OD, 0.78 mm ID, Cat. No. 30-30-0, Frederick Haer Co., Brunswick, ME). Capillaries of equivalent dimensions available from other suppliers give the same results. Capillaries are acid-washed prior to use[11] to remove possible particulate matter, but are not siliconized. The DNA concentration is always 2.5 μg/ml, measured accurately with a Beckman DU-65 spectrophotometer, in filtered (0.2-μm filter) 10 mM Tris-HCl, pH 7.5, 0.1 mM EDTA (TE buffer),[1] and is stored at $-20°$ as 100-μl aliquotes in Eppendorf tubes. An aliquot is centrifuged at 13,000 rpm for 10 min just prior to use for additional removal of particulates. A small volume of the solution, about 5 μl, is deposited into the injection pipette close to the tip by backfilling with a pulled Pasteur pipette. The DNA solution is injected into eggs utilizing an Eppendorf gas injector system (Eppendorf microinjector 5242, Carl Zeiss, Germany). The injection pressure is approximately 100 psi, with the back pressure always set at one-third the injection pressure.

DNA Purification

For rate zonal sucrose gradient centrifugation to separate the insert from plasmid (M. Nerenberg, personal communication), 30 μg of DNA is digested with restriction enzyme, sodium acetate/ethanol precipitated, and resuspended in 50 μl of TE buffer. This is layered onto a continuous linear gradient of 10 to 40% sucrose in 1.0 M NaCl, 10.0 mM Tris-HCl, pH 8.0, 1 mM EDTA, and centrifuged at 35,000 rpm at 15° for 16 hr (average conditions) in a swinging-bucket rotor with an individual tube capacity of 5 to 10 ml. Following centrifugation, a hole is pushed through the bottom of the tube with a 25-gauge needle for gravity collection of single-drop fractions. Small aliquots of each fraction are electrophoresed to determine which contain only insert DNA. These are pooled, and gradient medium is removed by five successive rinses with injection buffer (10 mM Tris-HCl, pH 7.5, 0.1 mM EDTA) in a Centricon 100 microconcentrator (Amicon, Danvers, MA). Much lower recovery of DNA is obtained when dialysis is used to remove the gradient medium.

For linearized and double-digested constructs, 30 μg of DNA is digested, extracted twice with phenol/chloroform, four times with chloroform alone, then sodium acetate/ethanol precipitated, and then the pellet washed once with 70% ethanol. The pellet is allowed to dry for a few minutes, then is resuspended in injection buffer at the desired concentration for injection. No dialysis of the DNA preparation is performed.

[11] M. L. De Pamphilis, S. A. Herman, E. Martinez-Salas, L. E. Chalifour, D. O. Wirak, D. Y. Cupo, and M. Miranda, *BioTechniques* **6**, 662 (1990).

[47] Surgical Techniques in Production of Transgenic Mice

By JEFFREY R. MANN

Methods for transferring eggs to the oviduct and uterus, cesarian section, and vasectomy are described in detail.

Anesthesia

To carry out the surgical procedures described, one of the following injectable anesthetics are recommended. (1) For the Ketaset/Rompun mixture, add 1.0 ml of Ketaset (Aveco, Ft. Dodge, IA; contains 100 mg/ml ketamine hydrochloride) to 0.32 ml of Rompun (Haver, Mobey Corp., Shawnee, KA; contains 20 mg/ml xylazine) and make up to a total volume of 10 ml with sterile phosphate-buffered saline (PBS). Store at room temperature. The dose is approximately 0.3 ml per 25 g body weight by intraperitoneal injection. (2) For the Hypnorm/Hypnovel mixture[1] (a neuroleptanalgesic), mix one part Hypnorm (Janssen Pharmaceutical Ltd., Oxford, UK; contains 0.315 mg/ml fentanyl citrate and 10 mg/ml fluanisone), one part Hypnovel or Versed (Hoffman-La Roche, Nutley, NJ; contains 5.0 mg/ml midazolam), and six parts of sterile water. Store at room temperature. The dose is approximately 0.3 ml per 25 g body weight by intraperitoneal injection. This dosage can be increased up to 2-fold. (Note: Hypnorm is not available in the United States, and fentanyl and midazolam are controlled substances.) Avertin (2,2,2-tribromoethanol dissolved in tertiary amyl alcohol) has been used extensively, but it is losing favor because of its possible lethal side effects.

Following the surgical procedure, the mouse should be placed in a warm environment to aid recovery. It is convenient to place it onto a slide warming tray set at approximately 35°. Alternatively, a closed plastic lunch box filled with warm to hot water will provide a warm surface. Wrapping the mouse in a Kimwipe provides some insulation. A good indication that the mouse has recovered is if it rights itself on rolling it onto its back.

Transfer of Eggs to Reproductive Tract

Following injection of 1-cell eggs with DNA, the eggs must be transferred back to the reproductive tract of pseudopregnant recipients to

[1] P. A. Flecknell, *Vet. Rec.* **113,** 574 (1983).

Copyright © 1993 by Academic Press, Inc.
All rights of reproduction in any form reserved.

enable further development to postimplantation stages or to term. As the efficiency of obtaining transgenic animals is dependent on this step, it is best to become proficient at transferring eggs before attempting to inject them with DNA. Pseudopregnancy is induced in females when they copulate with a sterile male. Males are made sterile by vasectomy, which is described in the last section.

Stage of Eggs and Pseudopregnant Recipients at Transfer

It is standard procedure to transfer 1-cell and cleavage stage eggs to the oviduct at any time during day 1 of pseudopregnancy (day 1 = day of vaginal plug, indicative of mating the previous night), and compacted morulae and blastocysts to the uterus late in the afternoon of day 3, or blastocysts only to the uterus in the morning of day 4 of pseudopregnancy. Morulae and blastocysts can also be transferred to recipients on day 1 of pseudopregnancy with success.[2] These procedures ensure that the eggs will be at the blastocyst stage at the time the uterus is receptive for implantation, which is at day 5.

The stage of development of postimplantation embryos and fetuses developing from transferred eggs corresponds to the stage of pregnancy of the recipient. Therefore, eggs 1 day apart in preimplantation development will be at the same stage of postimplantation development if transferred to recipients which mated on the same night.

Transfer of 1-Cell and Cleavage Stage Eggs to Oviduct

Equipment Required

Stereo microscope with transmitted light base
Stereo microscope with reflected light base (optional)
Fiber optic light source with ring light (preferable) or gooseneck accessory
Anesthetic
Syringes, 1-ml, and 25-gauge needles
70% Ethanol in a squeeze bottle
Kimwipes
Iris scissors, 10 cm long, one pair
Iris forceps, 10 cm long, one pair straight, one pair curved
Watchmaker's forceps, No. 5, in good condition, two pair
Dieffenbach clip, 2.5 cm long, one
Pasteur pipettes or 1.5 mm outer diameter glass capillaries

[2] R. A. Bronson and A. McLaren, *J. Reprod. Fertil.* **22**, 129 (1970).

Mouth-controlled suction/expulsion device (Fig. 1)
Small Bunsen burner or alcohol burner
Petri dishes, 3.5 cm
Medium M2[3]
Wound clips and applicator, or surgical needles and 4.0 surgical thread
Warm surface

The steps described are given in the sequence in which they are carried out. (They are written for a right-handed person and should be appropriately altered if the operator is left-handed.)

1. Manufacture transfer pipettes. These are made from Pasteur pipettes. Alternatively, 1.5 mm outer diameter glass capillaries can be used. Hold the pipette from underneath between the thumbs and forefingers. Using a small Bunsen burner with a blue flame, heat the Pasteur pipette until the glass is soft at the region where the thin portion begins to widen. Take the pipette out from the flame, then draw it out at moderate speed so that the drawn portion is about 10 cm long. Pause briefly to allow the glass to cool and harden, then continue to pull the hands in opposing directions so as to snap the glass within the drawn region. The aim is to produce a clean, perpendicular break, with the drawn out region being about 3.0 cm long and with an internal diameter twice that of a cleavage stage egg [Fig. 1 (g)]. To some extent, this relies on chance. Check the shape of the break and the diameter of the pipette with a stage micrometer and stereomicroscope. The outer diameter is ideally about 0.2 mm. Alternatively, the break can be made by scoring the pipette at the desired region with a diamond pencil, then snapping the pipette by bending. This often results in a good perpendicular break. The orifice of the pipette must be flame-polished, as sharp edges readily collect blocking debris when inserted into the oviduct. To flame-polish the pipette, pass the tip through the edge of the blue flame at moderate speed, then check the degree of polishing under the stereo microscope. Only a slight amount of polishing is desirable. Make two or three transfer pipettes before proceeding.

2. Prepare a transfer pipette with medium. Place 2 ml of medium M2 into a 3.5-cm petri dish. Select a transfer pipette, and load it with approximately five alternating regions of medium and air [Fig. 1 (f)]. This is achieved by dipping the pipette tip in and out of the medium while applying mouth-controlled suction (the suction/expulsion device is detailed in Fig. 1). The purpose of the bubbles is to inhibit capillary action.

[3] M. J. Wood, D. G. Whittingham, and W. F. Rall, *in* "Mammalian Development: A Practical Approach" (M. Monk, ed.), p. 259. IRL Press, Oxford and Washington, D.C., 1987.

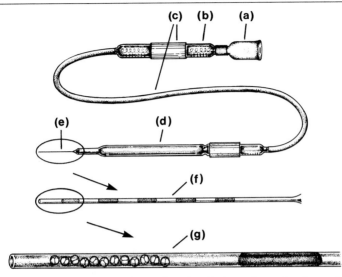

Fig. 1. Mouth-controlled expulsion and suction device and transfer pipette. (a) Mouthpiece; (b) barrel of plastic disposable 1-ml syringe, cut at the 0.6 ml mark, containing a wad of cotton wool; (c) silicon tubing; (d) Pasteur pipette; (e) drawn region of Pasteur pipette; (f) drawn region of pipette containing air bubbles; (g) pipette tip showing loading of eggs to be transferred.

Also, the bubbles serve as a guide for the transfer of eggs into the oviduct as described below. Stand the pipette base down in a rack before proceeding.

3. Anesthetize the recipient. While awaiting for it to succumb, proceed to the next step.

4. Prepare the eggs for transfer. Using a separate drawn Pasteur pipette, transfer the eggs from the incubator to the petri dish containing the 2 ml of medium M2 prepared previously, and place the dish on the transmitted light base of the stereomicroscope. The medium should be at room temperature. Deposit only those eggs that will be transferred to the recipient that has just been injected with anesthetic.

5. Expose the reproductive tract. When the mouse is anesthetized, place it belly down onto a folded Kimwipe with its head facing away from you. Wet the back with a small amount of 70% (v/v) ethanol, then make a dorsal midline incision approximately 1 cm in length through the skin using the scissors and straight iris forceps (Fig. 2A). As the skin of the mouse is quite loose, the incision can be moved around with forceps, as shown in Fig. 2B, to overlie the ovary. In lean mice, the ovary will be visible through the muscle wall. Otherwise, it takes some experience until one becomes familiar with the position of the ovary. Grasp the overlying

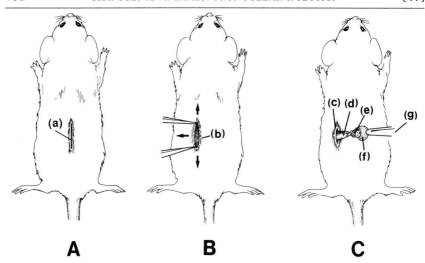

A   B   C

FIG. 2. Exposing the reproductive tract. (A) Dorsal midline incision (a) of 0.5 to 1 cm. (B) Using forceps, the incision serves as a movable window (b) with which to find the position of the ovary. (C) Incision (c) in body wall; (d) uterus; (e) oviduct; (f) ovarian fat pad; (g) Dieffenbach clip.

muscle wall with the straight forceps, and make a very small incision of a few millimeters with the scissors. This is made easier by first cutting away some of the loose overlying membranes. While holding the edge of the incision in the body wall with the straight iris forceps, take the curved forceps, insert them into the body cavity through the incision, and grasp the ovarian fat pad. Then pull out from the body cavity the ovary, the oviduct, and a small portion of the uterus, clasp the fat pad with the Dieffenbach clip, and lay the exposed portion of the reproductive tract across the back as shown in Fig. 2C.

6. Expose the infundibulum of the oviduct. Lift the mouse by the Kimwipe, and place it onto the reflected light stand of the stereo microscope, and position it such that its head is to the left and the oviduct is in the field of view. If one does not have a second stereo microscope, then place the mouse onto the transmitted light stand next to the petri dish of eggs awaiting transfer. Focus on the oviduct. Illumination should be accomplished with a fiber optic light source to minimize desiccation. A ring light accessory attached to the objective is the most convenient means for illumination; otherwise, a goose-neck arrangement is satisfactory. The oviduct is comprised of a series of twists and turns, which are consistent between mice, and from which it can be ascertained where the infundibulum will lie beneath the ovarian bursa, if it is not already visible.

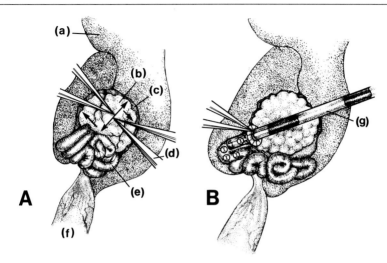

FIG. 3. Transfer of eggs to the oviduct. (A) Tearing the bursa open with the watchmaker's forceps avoiding blood vessels, as indicated by arrows. (a) Ovarian fat pad; (b) ovary; (c) bursa; (d) watchmaker's forceps; (e) oviduct; (f) uterus. (B) Expulsion of eggs into the oviduct. (g) Transfer pipette containing air bubbles.

Using the No. 5 watchmakers' forceps, tear the ovarian bursa longitudinally and latitudinally, and tuck it back out of the way beneath the ovary (Fig. 3A). To avoid undesirable bleeding during this procedure, which can obscure vision, care must be taken not to (1) tear through large blood vessels within the ovarian bursa, which are not always present or in the way, (2) prick the ovary with the tips of the forceps, or (3) tear up into the ovarian fat pad or down into the oviduct, wherein lie blood vessels. Once the bursa has been tucked back, locate the infundibulum. It is helpful to remember that, for the right and left sides, the infundibular opening faces the posterior end of the mouse.

7. Load the transfer pipette with eggs. Once familiarized with the position of the infundibulum, take the transfer pipette, which already contains medium and air bubbles, and collect the desired number of eggs into the pipette tip using the stereo microscope with the transmitted light base. If this is the only microscope available, then the mouse will have to be pushed to one side on the base so that the microscope can be used to collect eggs. The eggs should be collected so that they are a good distance in front of the first bubble–medium interface, and such that there is no air bubble in front of the eggs [Fig. 1 (g)]. The pipette is then rested between the base of the middle and forefingers, and the two watchmaker's

forceps are picked up, one in each hand. The mouthpiece is kept resting inside the mouth.

8. Transfer the eggs to the oviduct. The pipette tip must now be inserted into the oviduct by way of the infundibulum. The distance it is inserted should be such that the tip always reaches the first kink in the oviductal tube. This can be accomplished in two ways, depending on the accessibility of the infundibulum, which varies between recipients. (1) If the infundibulum is readily accessible, it is sufficient to simply straddle the infundibulum with the tips of the forceps, and then slowly open them so as to push the overlying ovary and underlying oviductal coils clear. When this is achieved, the forceps held in the right hand are put down, the transfer pipette is brought into a position as if holding a pencil, and then the pipette tip is inserted. (2) If the infundibulum is buried deeply, or at an awkward angle for pipette insertion, it can be gripped at the rim to stabilize it while inserting the pipette as shown in Fig. 3B. When the pipette is inserted, the eggs are expelled by applying gentle air pressure with the mouthpiece. At this point, the passage of air bubbles from the pipette and down the oviductal lumen indicates that the eggs are being successfully transferred. Usually, two or three bubbles are passed into the oviduct, with the leading bubble being close to, or within, the ampulla before the pipette is withdrawn. The ampulla is a swelling in the oviduct, a short distance down from the infundibulum, where the eggs congregate for fertilization after they have been ovulated. At the stage of transfer, the naturally ovulated unfertilized eggs of the recipient, surrounded in cumulus cells, should be visible through the translucent wall of the ampulla.

Some operators transfer eggs to the lumen of the oviduct by way of a tear made in the oviductal wall, which obviates the need to tear open the ovarian bursa to access the infundibulum. Although in some hands this technique is successful, it is generally not recommended, as there is a chance that transferred eggs will escape from the oviduct through the tear.

9. Following transfer of the eggs, the instruments are put down and the mouse is removed from the microscope stage with the Kimwipe. The edge of the incision in the muscle wall is then located and gripped with the straight iris forceps, and the reproductive tract is gradually coaxed back into the body cavity using the curved iris forceps. No suturing of this incision is required.

10. The right side of the mouse is now approached. With the head of the mouse facing toward you, exteriorize the reproductive tract as described in Step 5. Expose the infundibulum as described in Step 6, then turn the mouse around 180° such that its head faces toward the left. The infundibulum will now face the right-hand side, which will allow the pipette to be

inserted in the same way as already described for the left-hand side. Proceed as described in Steps 7, 8, and 9.

11. Seal the body wall with sutures, or by application of one or two wound clips.

12. Place the mouse on a warm surface to aid recovery.

Transfer of Eggs to Uterus

The equipment required is essentially the same as described for oviduct transfer, except that the Dieffenbach clip and watchmaker's forceps are not required.

1. Follow Steps 1 to 5 as described for oviduct transfer with the following modifications. The transfer pipette (Step 1, above) ideally has an outer diameter of about 0.3 mm. Once the ovary and first part of the uterus are pulled out from the body cavity, allow this to rest outside the body cavity by its own accord. Do not be concerned if it tends to move back inside, as it is necessary for only the fat pad to be exteriorized at this stage.

2. Collect the eggs as described in Step 7 above, except that there should be a very large distance (about 1 cm) between the first medium–air interface and the tip of the pipette containing the eggs to be transferred. Place the transfer pipette with eggs between the base of the middle and forefingers, and pick up the iris forceps, one in each hand.

3. Transfer the eggs to the uterus. Under the stereo microscope, pull out and manipulate the reproductive tract with the iris forceps so as to expose the upper region of the uterus. The blood vessels supplying it should be out of the way and lying underneath [Fig. 4A (a)]. Grip the tip of the uterus just below the uterotubal junction with iris forceps held in the left hand [Fig. 4A (b)]. Put the forceps held in the right hand down, and pick up a 1-ml syringe barrel fitted with a 25-gauge needle. Pierce the uterus just below the position of the forceps such that the needle enters the uterine lumen (Fig. 4A). This is best accomplished with the bevel of the needle facing up. Put the syringe down, then adjust the position of the transfer pipette so as to hold it like a pencil. Insert the pipette tip through the hole in the uterine wall into the lumen of the uterus (Fig. 4B). Retract it slightly then expel the eggs into the lumen. Expulsion will be indicated by movement of bubbles down the pipette. The pipette tip should be removed when the first bubble approaches the uterus. Therefore, no bubbles are transferred. Again, as with oviduct transfer, the flow of bubbles should occur readily. If not, do not force the eggs into the uterus. Instead, remove the pipette and inspect under the stereo microscope for blockage.

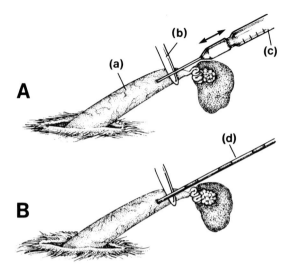

FIG. 4. Transfer of eggs to the uterus. (A) Making a hole in the uterus. (a) Barrel of a 1-ml disposable syringe fitted with a 25-gauge needle; (b) straight iris forceps; (c) uterus. (B) Expulsion of eggs into the uterine lumen. (d) Transfer pipette loaded with eggs.

Success of Transfer

It is possible to perform the procedures for oviduct transfer such that at least 80% of transferred noninjected eggs of F_1 mice will develop into fetuses.[4] The frequency of normal development that can be obtained following transfer of blastocysts to the uterus is similar (J. R. Mann, unpublished observations). In addition, virtually all hybrid pseudopregnant recipients are able to become pregnant following transfer of eggs[4] (see also [46] in this volume). Given these frequencies, a mean of 30% of eggs that have undergone pronuclear injection, and which remain viable, will give rise to mice (see [46], this volume). Therefore, transfer approximately 13 viable eggs to each oviduct to obtain a litter size of approximately 8. Viable eggs are those that have the potential to reach the 2-cell stage after overnight culture.

Very large numbers of fetuses in a recipient (greater than 16) may cause difficulties at parturition, with mice being encapsulated in extraembryonic membranes and dying at birth. In any event, much larger than expected litter sizes are generally indicative of low DNA integration frequency. With very small numbers of fetuses (1 or 2), parturition is often delayed,

[4] J. R. Mann, *Biol. Reprod.* **38**, 1077 (1988).

and as a consequence, survivability of newborns is reduced.[5] Also, in such pregnancies, lethality of the recipients might occur if they do not give birth. The presence of one or two fetuses in otherwise apparently nonpregnant recipients can be checked by palpating females. If pups are not born by the morning of day 21 (day 1 = day of plug), it is a good idea to rescue them by cesarian section if foster mothers are available for nursing.

In transferring eggs to oviducts on day 1 of pseudopregnancy, it should be noted that intraperitoneal injection of anesthetics can result in parthenogenetic activation of the unfertilized eggs of the recipient. If these reach the blastocyst stage, they can implant, resulting in additional decidual responses than expected.[6] This is not a problem when transferring eggs to the uterus, as by day 3 of pseudopregnancy, the eggs of the recipient have degenerated or are no longer responsive to activating stimuli.

Cesarian Section

Equipment Required

Iris scissors, 10 cm long, one pair
Iris forceps, 10 cm long, one pair straight, one pair curved
Kimwipes
Petri dish, 10 cm
Normal saline
Warm surface

Sacrifice the recipient by cervical dislocation, and dissect out the fetuses, being careful not to damage them when cutting off the yolk sac. Wipe them down with a tissue wetted with normal saline or PBS to remove yolk sac fluid, especially around the nasal areas. At this stage, the most important thing is to keep the pups warm as they attempt to initiate breathing. This can be accomplished by placing them onto prewarmed damp tissue paper inside a petri dish. The lid of the dish can be placed on top but should be left well ajar. Warming can be achieved with the same devices as described to aid recovery from anesthesia. Initially, the pups may be a litle blue and will gulp, but within about 10 min they should become pink and breathe in a more regular fashion. If any pup does not appear to be attempting to breathe, it can be stimulated by pinching the skin with forceps. When the pups are pink and breathing well, they can be placed with the foster mother. Usually foster mothers will accept

[5] A. McLaren, *J. Endocrinol.* **47,** 87 (1970).
[6] M. H. Kaufman, *J. Embryol. Exp. Morphol.* **33,** 941 (1975).

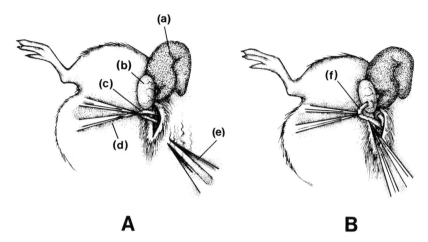

Fɪɢ. 5. Vasectomy. (A) Immobilization of vas deferens for cauterization. (a) Testicular fat pad; (b) left testis; (c) loop of vas deferens; (d) straight iris forceps; (e) red hot watchmaker's forceps. (B) Cauterization of vas deferens. (f) Vas deferens with one side cauterized.

foreign pups without trouble. However, it is a good idea (1) to leave some of the foster mother's own pups in the nest and (2) to induce the foster mother to urinate onto a bench top by picking it up, then to roll the cesarian-derived pups around in the drops of urine before placement in the nest.

Vasectomy

Equipment Required

Anesthetic
Syringe, 1-ml, and 25-gauge needles
Iris scissors, 10 cm long, one pair
Iris forceps, 10 cm long, one pair straight, one pair curved
Watchmaker's forceps, No. 5 one old pair
Bunsen burner
70% (v/v) Ethanol in a squeeze bottle
Kimwipes
Surgical needles and size 4.0 surgical thread
Warm surface

Choose vigorous males to act as vasectomized studs (e.g., outbred Swiss mice or hybrids). Before performing vasectomy for the first time, it is helpful to dissect a cadaver and familiarize oneself with the male reproductive tract.

1. Anesthetize the mouse, lay it on its back, and wet the posterior half of the ventral surface with 70% ethanol.

2. Just anterior to the preputial glands, which lie in the subcutaneous space around the penis, make a lateral incision of the skin of approximately 1.0 cm. Then grip the underlying muscle with the straight iris forceps and make a lateral incision of approximately 0.5 cm through the muscle of the body wall.

3. Grasp the cut edge of the body wall with the straight iris forceps and, by moving the incision, locate the testicular fat pad lying along the left side. Grip the fat pad with the curved iris forceps, and pull it out from the body cavity. This procedure may also exteriorize the left testis. The purpose is to exteriorize the vas deferens, which will appear as a loop. Some maneuvering of the testis and fat pad may be required before the vas deferens is located.

4. Grip the vas deferens with the straight iris forceps and, with the points of the watchmaker's forceps, strip away the transparent membrane inside the loop.

5. While still holding the vas deferens, heat the old pair of watchmaker's forceps in the flame of the Bunsen burner until red hot, then cauterize one side of the loop, then the other, such that a whole section of the vas deferens is removed (Fig. 5).

6. Grasp the cut edge of the body wall with one pair of iris forceps, and with the other pair, push the testis and fat pad back into the body cavity.

7. Repeat steps 3 through 6 for the right-hand side.

8. Suture the body wall with three to four individual stitches, then suture the skin in the same manner. Do not use metallic wound clips for sealing the skin, as the vasectomized males will be used for a number of months. With the fingers, stroke the testicular regions in an anterior to posterior direction a couple of times to encourage the testes back into place.

9. Place the mouse onto a warm surface to aid recovery.

The newly vasectomized male should be left for a few days before it is used for mating. If properly vasectomized, there is no chance that the male will ever regain fertility. The most important thing is to be sure to vasectomize both sides of the mouse. Place the sections of vas deferens to one side as they are removed, and check that the correct number are present when all males have been completed.

[48] Identification of Transgenic Mice

By MAUREEN GENDRON-MAGUIRE and THOMAS GRIDLEY

Introduction

The production of transgenic mice by any of a number of procedures (e.g., pronuclear injection of DNA, retroviral infection, and embryonic stem cell injection) requires a simple assay for the identification of mice that contain the transgene of interest. In the past, identification of transgenic mice was generally performed by transferring DNA to membranes by various blotting procedures (dot, slot, or Southern blotting), followed by hybridization with a labeled probe for the gene of interest. The advent of the polymerase chain reaction (PCR)[1,2] has permitted an alternative method for the identification of transgenic founder animals and their transgenic progeny. A number of protocols have been published that describe the use of PCR techniques to identify transgenic animals.[3-7] We describe here the methods used in our laboratory for preparing genomic DNA and performing the PCR analyses. We also discuss situations where the identification of transgenic animals by the PCR is inappropriate and where more traditional blotting and hybridization techniques should be used.

Preparation of Genomic DNA

Genomic DNA is prepared from tail fragments of the mice. Tails can be cut at approximately 10 days after birth, or alternatively at the time of weaning (~3 to 4 weeks of age). At the time the tail fragment is taken, the mice must also be permanently marked to permit proper identification after the DNA analysis is complete. Figure 1 shows a scheme to assign numbers up to 9999 continuously by toe clipping. We prefer toe clipping to methods such as ear punching, as ear punches are easily torn and proper identification can be more difficult. Alternative numbering schemes

[1] R. K. Saiki, D. H. Gelfand, S. Stoffel, S. J. Scharf, R. Higuchi, G. T. Horn, K. B. Mullis, and H. A. Erlich, *Science* **239**, 487 (1988).

[2] M. A. Innis, D. H. Gelfand, J. J. Sninsky, and T. J. White, "PCR Protocols: A Guide to Methods and Applications." Academic Press, San Diego, 1990.

[3] C. Abbot, S. Povey, N. Vivian, and R. Lovell-Badge, *Trends Genet.* **4**, 325 (1988).

[4] C. A. Walter, D. Nasr-Schirf, and V. J. Luna, *BioTechniques* **7**, 1066 (1989).

[5] C. S. Lin, T. Magnuson, and D. Samols, *DNA* **8**, 297 (1989).

[6] S. Chen and G. A. Evans, *BioTechniques* **8**, 32 (1990).

[7] D. G. Skalnik and S. Orkin, *BioTechniques* **8**, 34 (1990).

Copyright © 1993 by Academic Press, Inc.
All rights of reproduction in any form reserved.

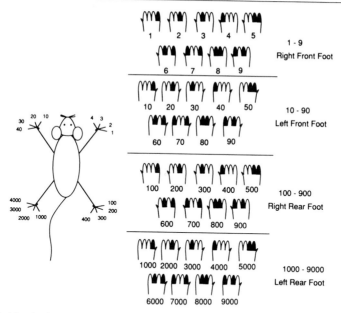

FIG. 1. Numbering system for identification of transgenic mice. Numbers from 1 to 9999 can be assigned using the toe clipping system illustrated. Black toes indicate toes to be cut.

have also been described.[8] Another possible alternative to toe clipping or ear punching is the subcutaneous implantation of microchips (BioMedic Data Systems, Maywood, NJ). As with all animal procedures, local and national guidelines for working with mice should be consulted and followed.

To isolate tail fragments from 10-day-old mice, approximately one third of the tail is cut and placed into a screw-capped 1.5-ml microcentrifuge tube containing 500 μl of TB buffer (see below). The animal is assigned a number by clipping the toes with fine-pointed surgical scissors, and is then placed back into the litter. It is advisable when weaning the animals to check the toe clips, as the clipping may need to be redone after it is performed on very young animals. If tail fragments are taken from older mice (3 weeks or older) the animal should be anesthetized with chloroform vapor before cutting a 2 cm section of tail and clipping the toes. In addition, to prevent bleeding, the cut tail should be cauterized with a soldering iron while the animal is still anesthetized.

The tubes containing the tail fragments are incubated overnight at 55° in a horizontal position on a rocking platform, or with periodic agitation.

[8] B. Hogan, F. Costantini, and E. Lacy, "Manipulating the Mouse Embryo." Cold Spring Harbor Laboratory, Cold Spring Harbor, New York, 1986.

They are extracted once with 500 μl of 1 : 1 (v/v) equilibrated phenol–chloroform, and precipitated with 2 volumes of ethanol. After centrifugation, pellets are resuspended in 500 μl water. A typical yield from the tail of a 10-day-old mouse is 250 μg. DNA prepared by this procedure is pure enough for uses other than PCR analysis (e.g., Southern blotting).

If a simpler and quicker DNA isolation protocol is desired, a small piece of the ear of a 10-day-old or older mouse can be cut into a microcentrifuge tube containing 20 μl PK buffer (see below). The tubes are incubated overnight at 55°, then boiled for 10 min to inactivate the proteinase K. Then 380 μl water is added, and 5 μl of this preparation is used in a 50-μl PCR analysis. We have found, however, that DNA isolated by this procedure is somewhat less reliable for PCR analysis. It is a good idea when using this DNA isolation protocol to include positive control primers (discussed below) in the PCR amplifications to minimize the problem of false negatives. Other investigators have described the analysis of transgenic mice by PCR techniques using DNA isolated from small quantities of blood.[6,7]

Materials

TB buffer: 50 mM Tris-HCl, pH 8, 100 mM EDTA, 100 mM NaCl, 1% sodium dodecyl sulfate (SDS), 600 μg/ml proteinase K (Boehringer-Mannheim, Indianapolis, IN)

PK buffer: 20 mM Tris-HCl, pH 8, 10 mM EDTA, 0.5% SDS, 400 μg/ml proteinase K

Water: double-distilled or other high-quality water

Polymerase Chain Reaction Amplification

The PCR amplification reactions are performed using standard procedures.[2] In general, oligonucleotides used as primers do not require special purification procedures. In our laboratory, after deprotection the oligonucleotides are dried in a SpeedVac concentrator (Savant Instruments) and resuspended in water. The oligonucleotide solutions are then centrifuged at 12,000 g for 10 min in a microcentrifuge to pellet debris. The supernatant is transferred to a new tube, and the concentration of the oligonucleotide is determined spectrophotometrically.

One particular advantage to the identification of transgenic mice by PCR techniques is the ability to use specific primer sets in situations where the same transgene (or part of the same transgene) is being used repeatedly [e.g., *lacZ* reporter transgenics or embryonic stem cell-derived mice containing the neomycin phosphotransferase (*neo*) gene]. In addition to primer sets specific for individual transgenes, we routinely use primer sets specific for the *neo* and *lacZ* genes. We also use a primer set for the thyroid-stimulating hormone β (TSHβ) gene as an internal control for the quality of the DNA preparation and the efficiency of the PCR.[4] Inclusion of the internal control primer set is particularly useful when using rapid DNA preparation methods which include minimal purification of the DNA (e.g., the protocol described above for isolating DNA from ear fragments). We have observed that DNA prepared by these rapid preparations is somewhat less reliable in the PCR amplifications. Inclusion of the internal control primers can minimize the problem of false negatives by indicating which DNA preparations may not have amplified well. DNA can then be prepared again from these mice, and they can be retested.

Procedure

The PCR amplification conditions we use are those recommended by Perkin-Elmer (Norwalk, CT) for use with AmpliTaq DNA polymerase. For a 50-μl reaction, mix the following components:

Template DNA, ~1 μg genomic DNA	50.0	μl
Primer 1, 20 μM (gene-specific primer)	2.5	μl
Primer 2, 20 μM (gene-specific primer)	2.5	μl
Primer 3, 20 μM (control primer)	2.5	μl
Primer 4, 20 μM (control primer)	2.5	μl
10× PCR buffer	5.0	μl
Sterile water	29.75	μl
AmpliTaq DNA polymerase	0.5	μl
	50.0	μl

(*Note:* If internal control primers are not included in the amplification reaction, the additional volume is made up with sterile water.)

Cycling Parameters. For most primer sets, we use the following cycling parameters: initial denaturation at 94° for 3 min; 35 cycles of 94° for 30 sec, 55° for 1 min, and 72° for 1 min; and final extension at 72° for 7 min.

Oligonucleotide Sequences. Shown below are the sequences of oligonucleotide primers we use for amplification of the *lacZ*, *neo*, and TSHβ[4] genes.

lacZ (825-bp amplified fragment):
  Upstream: GACACCAGACCAACTGGTAATGG
  Downstream: GCATCGAGCTGGGTAATAAGCG
neo (150-bp amplified fragment):
  Upstream: TGAATGAACTGCAGGACGAGG
  Downstream: AAGGTGAGATGACAGGAGATC

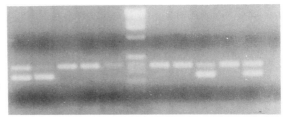

Internal Control
Transgene

Fɪɢ. 2. Identification of transgenic animals by PCR. Nine animals were tested for the presence of the transgene using both a transgene-specific primer set and the TSHβ internal control primer set. PCR amplifications were performed as described in the text. The reaction products were run on a 1% agarose gel, which was stained with ethidium bromide and photographed. Animals 1, 2, and 8 had inherited the transgene. The lanes marked + and − contained positive and negative controls, respectively.

TSHβ (386-bp amplified fragment):
   Upstream: TCCTCAAAGATGCTCATTAG
   Downstream: GTAACTCACTCATGCAAAGT

Materials

   10× PCR buffer: 100 mM Tris-HCl, pH 8.3, 500 mM KCl, 15 mM MgCl$_2$, 0.01% gelatin, 200 μM dATP, 200 μM dTTP, 200 μM dCTP, 200 μM dGTP
   AmpliTaq DNA polymerase (Perkin-Elmer)

Comments

   When choosing oligonucleotides for transgene-specific primer sets, we generally try sequences approximately 21 bp in length with a GC content of approximately 50%. Primer sets are tested by diluting the linearized transgene construct with genomic DNA prepared in the same way that DNA from the transgenic animals will be prepared. For a 10-kb construct, you should be able to detect 1 pg or less of the transgene construct diluted into 1 μg of genomic DNA. As with all PCR analyses, conditions may need to be optimized for individual primer sets.[2] An example of the identification of transgenic progeny using a transgene-specific primer set and the TSHβ internal control primer set is shown in Fig. 2.

Other Methods for Identifying Transgenic Mice

   Detection of transgenic mice by PCR analysis provides a simple and quick method for the identification of transgenic founder animals and their progeny. It is particularly well-suited to the analysis of large numbers of animals and to situations where the same transgene (or part of the same transgene) is being used repeatedly. However, PCR analysis is not appropriate in all situations. For example, there are instances where transgenes must be detected by Southern blots. Certain experiments using replication-competent retroviruses as insertional mutagens have required the detection and monitoring of a particular band on a Southern blot.[9]

PCR analysis is also not particularly well-suited for determining whether a particular transgenic animal is heterozygous or homozygous for the transgene of interest. In instances where we are identifying progeny homozygous for a transgene we generally use a slot-blotting procedure. A series of dilutions of DNA from the progeny to be tested (along with a known heterozygote and a known nontransgenic mouse for positive and negative controls) are transferred to two membrane filters with a dot- or slot-blotting apparatus. One filter is hybridized with a probe specific for the transgene of interest, while the other filter is hybridized with a probe for an endogenous control gene. Generally, any single-copy gene will suffice for the endogenous control. The signal from each slot of the two filters is quantitated with an isotope detector and imager (e.g., Betascope 603, Betagen, Waltham, MA) or by scanning densitometry. The signal from the filter hybridized with the probe for the transgene is then normalized to the signal from the control gene. The normalized signal from homozygous mice should be twice that of heterozygous mice. Mistakes can be made, however, and the safest procedure to follow before setting up matings to establish a homozygous line is to test-breed each parent with wild-type mice in order to detect transmittance of the transgene to 100% of the progeny.

Acknowledgments

We thank Sergio Lira and Maobin Zhang for helpful discussions and Rudolf Jaenisch for the transgenic numbering system.

[9] P. Soriano, T. Gridley, and R. Jaenisch, *Genes Dev.* **1**, 366 (1987).

Section **XI**

Transgenic Animals: Embryonic Stem Cells
and Gene Targeting

[49] Derivation of Embryonic Stem Cell Lines

By Susan J. Abbondanzo, Inder Gadi, and Colin L. Stewart

Introduction

Embryonic stem (ES) cells are the pluripotent derivatives of the inner cell mass (ICM) of the blastocyst. They have many features in common with embryonal carcinoma (EC) cell lines; morphologically they are indistinguishable, have the same growth requirements, and their differentiation *iv vitro* results in the formation of similar cell types.[1-3] However, unlike EC cells, ES cells are derived directly from embryos explanted in culture.[2-7] EC cells are always derived from teratocarcinomas that either arise spontaneously or are induced in adult mice.[1] The derivation of EC cells is a comparatively complicated and involved process, and the reliance on first establishing cells from tumors (teratocarcinomas) may contribute to the fact that a number of established EC lines behave in a diverse manner. In contrast, ES cells are derived directly from the ICM of blastocysts explanted *in vitro*. A variety of procedures have been employed to obtain ES cells, including using blastocysts that have undergone delayed implantation as well as culturing cells directly from ICMs isolated from blastocysts following immunosurgery.[2-7]

The *in vitro* growth of ES cells is dependent on the cytokine leukemia inhibitory factor (LIF).[8,9] This protein is essential for maintaining the growth of ES cells *in vitro* since, in its absence, ES cells differentiate and eventually will cease to proliferate. Other factors, such as human oncostatin M (OSM) and rat ciliary neurotrophic factor (CNTF), can

---

[1] C. F. Graham, *in* "Concepts in Mammalian Embryogenesis" (M. I. Sherman, ed.), p. 315. MIT Press, Cambridge, Massachusetts, 1977.

[2] M. J. Evans and M. H. Kaufman, *Nature (London)* **292,** 154 (1981).

[3] G. R. Martin, *Proc. Natl. Acad. Sci. U.S.A.* **78,** 7634 (1981).

[4] H. R. Axelrod, *Dev. Biol.* **101,** 225 (1981).

[5] A. M. Wobus, H. Holzhauser, P. Jäkel, and J. Schöneich, *Exp. Cell Res.* **212,** 152 (1984).

[6] E. F. Wagner, E. Keller, E. Gilboa, U. Rüther, and C. L. Stewart, *Cold Spring Harbor 50th Anniversary Symp.* 691 (1985).

[7] T. C. Doetschman, H. Eistetter, M. Katz, W. Schmidt, and R. Kemler, *J. Embryol. Exp. Morphol.* **87,** 27 (1985).

[8] R. L. Williams, D. J. Hilton, S. Pease, T. A. Willson, C. L. Stewart, D. P. Gearing, E. F. Wagner, D. Metcalf, N. A. Nicola, and N. M. Gough, *Nature (London)* **336,** 684 (1988).

[9] A. G. Smith, J. K. Heath, D. D. Donaldson, G. G. Wong, J. Moreau, M. Stahl, and D. Rogers, *Nature (London)* **336,** 688 (1988).

Copyright © 1993 by Academic Press, Inc.
All rights of reproduction in any form reserved.

substitute for LIF since they are capable of binding to the same receptor complex,[10–12] although preliminary observations suggest that OSM and CNTF are not as effective as LIF at supporting ES cell proliferation.[10]

Leukemia inhibitory factor can be supplied to ES cells in different ways. Currently the best approach, and still the most effective one for long-term culture, is to grow the ES cells on a feeder layer of fibroblasts. The feeder layers synthesize and secrete LIF into the culture medium, and, in addition, an alternative form of LIF is also produced that remains closely associated with the extracellular matrix deposited by the fibroblasts.[13] LIF is the only factor produced by the feeder layers that is essential for ES cell growth, since fibroblasts isolated from mutant embryos that are deficient in LIF production do not support ES cell growth and proliferation.[14]

Embryonic stem cell lines can also be established and maintained from embryos in the absence of a feeder layer. Under these conditions the culture medium is supplemented with recombinant LIF, which is available from commercial suppliers (GIBCO–BRL, Grand Island, NY; R and D Systems, Minneapolis, MN). It is also possible to use regular culture medium supplemented with medium "conditioned" by growing certain cell lines (see below) that secrete relatively large quantities of LIF into the culture medium. The medium can be collected and used at an appropriate dilution as a source of LIF.

Culture Requirements

Equipment

To establish and culture ES cells, a laboratory equipped with standard tissue culture facilities is required, namely, a sterile/filtered air culture hood, a 37°, CO_2-gassed incubator, and a tissue-culture microscope equipped with phase-contrast optics for viewing cells. In addition, a good stereo dissection microscope is required with ×40 magnification, along with a mouth-controlled pipette that is used for transferring blastocysts

---

[10] C. L. Stewart, unpublished observations.

[11] N. Y. Ip, S. H. Nye, T. G. Boulton, S. Davis, T. Taga, Y. Li, S. J. Birren, K. Yasukawa, T. Kishimoto, D. J. Anderson, N. Stahl, and G. D. Yancopoulos, *Cell (Cambridge, Mass.)* **69**, 1171 (1991).

[12] D. P. Gearing and A. G. Bruce, *New Biol.* **4**, 61 (1992).

[13] P. D. Rathgen, S. Toth, A. Willis, J. K. Heath, and A. G. Smith, *Cell (Cambridge, Mass.)* **62**, 1105 (1990).

[14] C. L. Stewart, P. Kaspar, L. J. Brunet, H. Bhatt, I. Gadi, F. Köntgen, and S. J. Abbondanzo, *Nature (London)* **358**, 76 (1992).

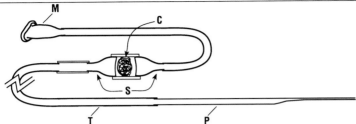

FIG. 1. Mouth-controlled pipette. A plastic mouthpiece (M) that is gripped between the teeth is connected via a rubber tube to two syringe ends (S) from 1-ml syringes. These are cut from the syringes and joined back-to-back with a wider diameter piece of Teflon tubing. Between the connected ends, a small wad of sterile cotton wool (C) is inserted to act as a filter and saliva trap. The other end is connected to a stretch of Teflon tubing 2 feet long with an outer diameter of 1/8 inch and inner diameter of 1/16 inch. The pulled capillary (P) made from hard glass capillary tubing is inserted at the end.

and for picking the ICMs or ES colonies. An example of such a pipette is shown in Fig. 1. Although these can be made from Pasteur pipettes, we prefer to use hard glass capillary tubes (BDH Ltd., Poole, UK, or Gallard-Schlesinger, New York, NY). To pull the capillary, the center of the tube is rotated over a small gas or alcohol burner flame until it begins to buckle or melt. The tube is then withdrawn from the flame and the ends pulled in opposite directions, thus stretching out the center of the tube into one with a finer, narrower diameter. With practice, the degree of force applied to the pull can be varied to obtain the desired internal diameter of the pipette. The pipette is then snapped in two by bending. The hard glass capillaries usually give a clean vertical break without any jagged edges. The opening at the tip can be flame polished and narrowed by holding it at the base (the cooler part) of the flame.

Culture Media

The effective maintenance of ES cells requires that all culture media be made with very pure water. The Millipore (Bedford, MA) Five-bowl Milli-Q purification system provides water that is of satisfactory quality. A variety of different media have been used to culture embryos and ES cells: Dulbecco's modified Eagle's medium (DMEM), Glasgow modified Eagle's medium, and a DMEM/Ham's F12 mixture. We use DMEM with high glucose (4.5 g/liter), L-glutamine, and no sodium pyruvate. The medium is purchased in powdered form, although $1\times$ to $10\times$ concentrated liquid forms are available. It is made up according to the manufacturer's instructions and buffered with 2.2 g/liter sodium bicarbonate. It is supplemented with MEM nonessential amino acids to a final concentration of

0.1 mM [these can be obtained from GIBCO–BRL as a 100× (10 mM) solution]. In addition, L-glutamine to a final concentration of 2 mM is added together with 2-mercaptoethanol at a final concentration of 0.1 mM [a stock 0.1 M solution is made by adding 70 μl of the standard 14 M solution (Sigma, St. Louis, MO) to 10 ml of phosphate-buffered saline (PBS)]. Penicillin (50 IU/ml) and streptomycin (50 IU/ml) are also included in the final formulation, and 100× solutions can be obtained from GIBCO–BRL. This formulation is referred to as ES-DMEM.

The formulations for M2 and CZB plus glucose media are given elsewhere (see [9] and [10] in this volume). These media are used for the isolation of the blastocysts from the uteri of pregnant females and for the short-term culture of embryos.

Serum Requirements for Embryonic Stem Cell Cultures

Embryonic stem cells are very sensitive to the type and quality of serum in which they are grown. Because the quality of fetal calf serum (FCS) can vary from batch to batch, it is essential to test each batch of serum to ensure that it is optimal for ES cell growth. This is particularly important in the establishment of ES lines from ICMs, where the quality of serum can make all the difference between success and failure.

We routinely use fetal calf (bovine) serum (FCS) for all our cultures. However, a 50:50 mixture of FCS and newborn calf serum (NCS) can also be used,[2] since NCS is considerably less expensive than FCS. There are a number of companies that supply serum, and test samples can be readily ordered. Once the best sample has been determined, the batch can be purchased. Serum can be stored frozen at −20° for 2–3 years.

The procedure to determine the optimal batch of FCS is based on a colony-forming assay in which each batch is tested to determine the maximum number of ES colonies that grow in the serum following their plating as a single cell suspension. Three thousand cells are plated onto already prepared petri dishes (60 mm) containing a layer of feeder cells (see below). The single cells attach and over the next 5 days will proliferate to form colonies of ES cells, which are counted. This assay is duplicated for each batch, and the serum is tested at two different concentrations: 15%, which is the usual concentration in which ES cells are grown, and 30%, to determine if there is any toxicity associated with the batch. The latter concentration is frequently useful in determining which batches, giving similar results at the 15% concentration, are better. A good batch of serum should give a colony-forming efficiency of about 30% or more of the plated cells. We have found that FCS from HyClone Labs (Logan, UT) is consistently of sufficient quality for the isolation and maintenance of ES cell lines.

Contamination of Cultures

Occasionally, cell cultures can become contaminated, and this can be particularly frustrating if it is a newly isolated ES line or clone derived from a cell line. The most common contaminants are yeast or fungal infections.

The best method to rid the culture of contaminating organisms is to wash the culture vessel 3 times in PBS to remove most of the organisms. The cells are then refed with fresh medium to which nystatin (Sigma) has been added. Nystatin is a fungal inhibitor that is very effective at killing yeast and fungi, and it does not appear to be toxic to ES cells at the doses required. A $200 \times$ stock suspension (the reagent is supplied as a yellow crystalline powder that goes partly into solution) is made by adding 10 mg of powder to 10 ml of PBS. Shake to suspend and add the appropriate volume to give the final dilution in the culture vessel. Change and refeed the cells with fresh medium containing nystatin for the next 4–5 days of culture. This should be effective at killing all yeast/fungal contaminants. In the culture dish, the nystatin suspension forms fine, long crystals, which can be seen under a microscope.

Preparation of Feeder Layers

Embryonic stem cells are dependent on the cytokine LIF to maintain them as an undifferentiated proliferating population. The cytokine is usually supplied by growing the cells on feeder layer fibroblasts that produce LIF.[14] Recombinant LIF is commercially available but is expensive. ES cells have been derived from blastocyst cultures in the absence of feeders, but with the medium supplemented with recombinant LIF.[15,16] However, the majority of these lines contain a significant percentage of aneuploid karyotypes, rendering them unsuitable for the generation of germ line chimeras.[16] Only in a few instances have germ line chimeras been produced with ES cells established in feeder-free LIF-containing medium.[15,16] It remains to be determined, however, if ES cells derived and maintained in LIF alone can in the long term be an efficient substitute for those maintained on feeder layers. Consequently, we still prefer to establish our ES cells on a feeder layer of mouse fibroblasts.

[15] S. Pease, P. Braghetta, G. D. Gearing, D. Grail, and R. L. Williams, *Dev. Biol.* **141**, 344 (1990).

[16] J. Nichols, E. P. Evans, and A. G. Smith, *Development (Cambridge, UK)* **110**, 1341 (1990).

As yet it is unclear as to whether feeders are providing, in addition to LIF, other factors that help to establish and maintain ES cells. Possibly, the matrix-associated form of LIF, along with the extracellular matrix deposited by the feeders, is more effective in maintaining ES cells than the soluble form alone. We have found that the maintenance of feeder-dependent ES cells, under feeder-free conditions in the presence of LIF, is more effective (in inhibiting ES differentiation) when the ES cells are grown on extracellular matrix deposited by fibroblasts (see below) rather than on gelatine alone, which is the standard procedure.

The feeders, which are fibroblasts, can either be permanently growing lines (e.g., STO fibroblasts) or primary mouse embryo fibroblasts (PMEFs). The advantage of STO cells is that they are continuously proliferating, so they do not need to be repeatedly derived. The disadvantage with STO cells is that there is variation between different sublines, with some being more effective than others at sustaining ES cells. Furthermore, there is evidence that even though STO fibroblasts are as effective as PMEFs as a substrate for establishing ES cell lines, they appear to be less effective at maintaining the stem cell phenotype over the long term in culture. Furthermore, ES cells maintained on PMEFs have a more stable euploid karyotype.[17]

Procedure for Preparing Primary Mouse Embryonic Fibroblasts. The PMEFs can be made from any strain of mouse.

1. Set up females with males. The following day, check for those that mated by looking for females that have a copulation plug. This is called day 1 of pregnancy (day 1 = day of plug). Others have used the day of plug as being equivalent to day 0 or day 1/2. Remove the mated females to a separate cage.

2. On day 13 of pregnancy, sacrifice the pregnant females by cervical dislocation. Open the peritoneal cavity and remove the uteri carrying the embryos. Dissect the embryos from the uteri in PBS, and remove the yolk sac, amnion, and placenta. The embryos and are then washed twice in fresh PBS to remove any blood.

3. Using a pair of fine forceps, pinch off the head (Fig. 2). Pinch out and remove the liver.

4. Place 5–8 of the carcasses in the barrel of a sterile 3-ml syringe to which a sterile 18-gauge hypodermic needle has been attached.

5. Add 2 ml of sterile PBS and replace the syringe plunger. With the tip of the needle in a 150-mm tissue culture dish, expell and draw up the carcasses through the needle, 4–5 times, breaking them up into small

[17] H. Suemori and N. Nakatsuji, *Dev. Growth Differ.* **29**, 133 (1987).

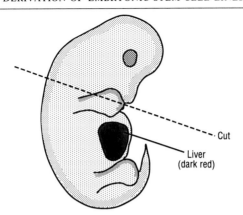

FIG. 2. Diagram of a day 13 embryo from which PMEFs will be prepared. The cut where the head is removed is shown, as is the liver.

clumps of cells. Add 20 ml of DMEM plus 10% FCS with penicillin/ streptomycin to the dish and transfer to a 37° gassed (5% CO_2) incubator. The clumps of cells begin to attach almost immediately and will start to give rise to fibroblasts over the next 2–3 days. By the third day, the dish should be confluent.

6. Wash the petri dish twice in Ca^{2+}/Mg^{2+}-free PBS and add 3 ml of 0.25% trypsin and 1 mM EDTA in PBS containing 1% (v/v) chicken serum. Trypsinize the culture and transfer the cells to three other 150-mm dishes. After further culture, the fibroblasts can be split at no more than a 1 : 3 ratio to expand their numbers. However, PMEFs are only good as a feeder layer for 5–6 passages, since they then will begin to cease proliferating, and fresh cultures will need to be established.

7. Once sufficient numbers of PMEFs are established, they can either be prepared as a feeder layer for ES cells, or they can be frozen in liquid nitrogen for storage in anticipation of future use.

Mitomycin C Treatment of Primary Mouse Embryonic Fibroblasts. To use PMEFs as feeders, it is essential to stop their proliferation so that they do not overgrow any ES cultures, but still support the growth of the ES cells. This can be achieved in two ways. If available, a source of γ radiation can be used to inactivate the cells with 600 rads, which is sufficient to inhibit their proliferation. A more common method is to treat a proliferating population of cells with mitomycin C.

1. Mitomycin C can be purchased from Sigma in 2-mg vials.
2. Dissolve the mitomycin C by injecting 5 ml of sterile PBS into a vial.

3. Add the dissolved mitomycin C to 195 ml of DMEM medium plus 5% FCS, so that the final concentration is 10 μg/ml. Aliquot into 10-ml volumes. Store frozen at $-20°$ in the dark (mitomycin C is light-sensitive and its activity decreases with exposure to light).

4. Aspirate the culture medium from an almost confluent 150-mm dish of PMEFs that are still proliferating and replace with 10 ml of the mitomycin C solution. Return the dish to the incubator and culture for 3 hr. This is sufficient time for the mitomycin C to permanently arrest PMEF proliferation.

5. After 3 hr, remove the mitomycin C-containing medium. The PMEFs can be trypsinized for replating onto small tissue culture dishes as feeders, or they can be washed twice with fresh medium (to remove any traces of mitomycin C), refed with 20–25 ml of DMEM plus 5% FCS, and kept for up to 1 week in the incubator prior to use.

6. For preparation prior to seeding into smaller culture dishes, the treated PMEFs are washed twice in PBS and trypsinized. The trypsinized cells are harvested, pelleted in a centrifuge (a speed of 1000 rpm for 5 min is sufficient), resuspended in 10 ml of DMEM plus 10% FCS, and counted using a hemocytometer or Coulter counter.

7. The dishes to which the PMEFs are to be added at this stage are pretreated with a 0.1% (w/v) gelatin solution. The gelatin, which is denatured collagen, attaches to the surface of the tissue culture dish and enhances the attachment of the feeders and the ES cells to the culture dish surface. A 0.1% solution is made by adding 0.5 g of swine skin type III collagen (Sigma) to 500 ml of PBS. The PBS–gelatin mixture is dissolved and sterilized by autoclaving.

8. To gelatinize a culture dish, sufficient autoclaved gelatin–PBS solution is added to the dish so that the culture surface is covered with a thin layer. The dishes are then kept at $4°$ for a minimum of 30 min, and excess solution is pipetted off prior to use. The dishes containing the gelatin solution can be kept indefinitely (under sterile conditions) at $4°$ and can even be allowed to dry since the gelatin will remain attached to the surface.

9. To the gelatinized tissue culture dishes, and adequate number of mitomycin C-treated PMEFs are added so that they will form a uniform monolayer when they attach. The number of cells per dish varies according to the surface area and, hence, the diameter of the culture dish. The numbers we use, depending on the dish (diameter) are as follows: 100-mm dish, $1.7–2 \times 10^6$ cells; 60-mm dish, 1×10^6 cells; 35-mm dish, $5–6 \times 10^5$ cells; and 16-mm well, 1×10^5 cells.

10. The PMEF cells are allowed to attach for 6–12 hr prior to the addition of the ES cells. The medium in which they are cultured is ES-

DMEM (see above). PMEF feeder layers are good for up to 2 weeks before they should be discarded.

Preparation of Conditioned Medium

Leukemia inhibitory factor is the cytokine that ES cells require in order to remain as an undifferentiated proliferating population of cells. The recombinant protein is commercially available, but an alternative source of this factor is from cells that synthesize and secrete LIF into the culture medium. Such conditioned medium can be concentrated or used directly (at the appropriate dilution) as a supplement to the feeder cells or as a substitute, when it is preferable not to have any feeders present in the culture. Two cell types have been used to produce conditioned media, the buffalo rat liver (BRL)[9] cell line and the human bladder carcinoma line 5637.[8] Both produce LIF in approximately equal amounts, although different BRL sublines produce varying amounts of active LIF. Line 5637 can be obtained from the American Type Culture Collection (ATCC, Rockville, MD).

1. The 5637 cells are grown in standard tissue culture flasks/dishes at 37° in DMEM plus 5% FCS.

2. Grow the cells to confluency. Large cultures may be grown in roller bottles or cell factories, T-75 flasks, or 175 flasks depending on the volume required.

3. Once the cells have reached confluency, fresh medium is added, and the cells are cultured for 3–4 days before the medium is collected. Fresh medium is immediately added to the cells, and repeated collections can be made twice a week for up to 6 weeks.

4. The collected medium is centrifuged at 12,000–16,000 rpm for 1 hr to remove cells and any insoluble material.

5. Filter through a 0.45-μm sterile filter and store at $-20°$ until required.

6. Conditioned medium (100 ml) for growing ES cells under feeder-free conditions is made up as follows:

65 ml of sterile-filtered 5637-conditioned medium

23 ml of DMEM

12 ml of FCS

MEM nonessential amino acids to final concentration of 0.1 mM [from 100× (10 mM) stock solution, GIBCO-BRL]

Penicillin/streptomycin (100× liquid, GIBCO–BRL)

L-Glutamine (200 mM from 100× stock, GIBCO–BRL)

2-Mercaptoethanol at 0.1 mM from 0.1 M stock

Development of Embryonic Stem Cell Lines

Isolation of Embryonic Stem Cells from Blastocysts

A variety of different procedures have been used to establish ES cells from the ICM of blastocysts. Most lines have been derived from the 129/ Sv strain, but ES lines have also been derived from the C57BL/6J[18,19] and C3H/He strains.[19] The simplest procedure (and the one we prefer), which has been effective for producing ES cell lines from wild-type blastocysts as well as from parthenogenetic and androgenetic embryos,[20] is as follows:

1. Day − 1: Set up matings between male and female mice of the strain of choice. 129/Sv mice do not breed well, so if this strain is being used, it may be necessary to set up at least 12–15 stud males with 3 females each to obtain 20–30 embryos.

2. Day 1: Check females for those that have mated by looking for vaginal plugs. Remove mated females to a separate cage.

3. Day 2: Prepare three or four 60-mm dishes with mitomycin C-treated PMEFs. Culture the cells in 5 ml of ES–DMEM.

4. Day 4: The fertilized eggs will, by day 4, have developed to the blastocyst stage and will be found in the uterine lumen. Sacrifice the pregnant females by cervical disslocation. Dissect open the peritoneal cavity and isolate the uteri by cutting the uterus at the uterine–oviduct junction on both sides. The mesentery containing the blood vessels that supply the uterus is then cut away along the length of the uteri, being careful to avoid cutting into or through the uteri. The two uterine horns are removed as one piece by cutting through the vulva just at the point where the two uteri join. The uteri are placed in a dish containing Ca^{2+}/Mg^{2+}-free PBS, and any remaining blood vessels and mesentery are removed with a pair of fine scissors. The cleaned uteri are given one more wash in PBS to remove any blood and are then flushed with M2 medium to isolate the blastocysts as described below.

5. Fill a 3-ml syringe with M2 medium and attach a sterile 25-gauge hypodermic needle. Looking through the low power of a dissection microscope, pick up the uterus using watchmaker's forceps and insert the needle into the uterine lumen at the uterine–oviduct junction. Clamp the uterus around the tip of the needle using the forceps and inject between 0.2 and 0.5 ml of M2 medium into the uterine lumen. The vulval end of the uterus

[18] B. Ledermann and K. Burke, *Exp. Cell Res.* **197**, 254 (1991).
[19] C. L. Stewart, unpublished observations.
[20] J. R. Mann, I. Gadi, M. L. Harbison, S. J. Abbondanzo, and C. L. Stewart, *Cell (Cambridge, Mass.)* **62**, 251 (1990).

should be lying in a 35-mm culture dish. The 0.2–0.5 ml of M2 should flush the uterine lumen and expel the blastocysts into the 35-mm dish. Repeat the procedure with other uterine horns and discard the flushed uteri. The blastocysts should now be sitting in the M2 medium on the surface of the dish and can be located by using a stereo dissection microscope with × 20 or × 40 magnification. Once a blastocyst is identified, it is removed from the dish using a mouth-controlled pipette. The embryos are then transferred to fresh dishes of M2 to wash away any contaminating blood cells or uterine tissue, and any undeveloped eggs/embryos are discarded. The embryos can then be cultured in microdrops of CZB plus glucose medium under paraffin oil.

6. The D4 blastocysts are transferred to 60-mm dishes containing prepared feeders, adding no more than 20 to each dish. The ES–DMEM medium is supplemented with 1000 IU of recombinant LIF (murine or human is equally effective). If LIF is unavailable, the embryos and feeders can be grown in 70% 5637-conditioned medium (see previous section for preparation). The dishes with the embryos are returned to a 37° incubator and left undisturbed for 2 days.

7. Over this period, embryos will hatch from the zona pellucida and attach to the surface of the dish. The trophoblast spreads out to form a monolayer of cells on which the ICM can be seen [Fig. 3 (1)]. Over the next 2 days (i.e., up to day 4 from the time of explanting the blastocysts), the ICM grows and forms a distinct mound of cells on the trophoblast monolayer [Fig. 3 (4)]. At the end of 4 days and in the first half of the fifth day of culture, the ICMs should be picked for disaggregation. There appears to be an optimal window in time when the ICM is best suited for producing ES lines. Generally, blastocysts are too far developed if picked any period after 5 days of explanting, and the frequency of forming ES lines declines. This point can often be recognized by the formation of an endoderm layer around the core of ICM cells [Fig. 3 (2)]. These explants rarely, if ever, give rise to ES lines.

8. To pick the ICMs, the culture medium is aspirated and the dish washed twice in $Ca2+/Mg^{2+}$-free PBS, with embryos remaining covered by the PBS. Microdrops of 0.25% trypsin and 1 mM EDTA plus 1% chicken serum are set up under paraffin oil (see [50] in this volume). Chicken serum is included in the trypsin–EDTA solution because, unlike FCS, it does not contain a trypsin inhibitor, and the added protein protects the cells from lysis.

The ICMs are picked off the trophoblast by gently dislodging them using a mouth-controlled pipette (Fig. 3). Each ICM is then transferred into a single microdrop of trypsin–EDTA solution plus 1% chicken serum and left for approximately 3–5 min. The cells in the ICM clump start to

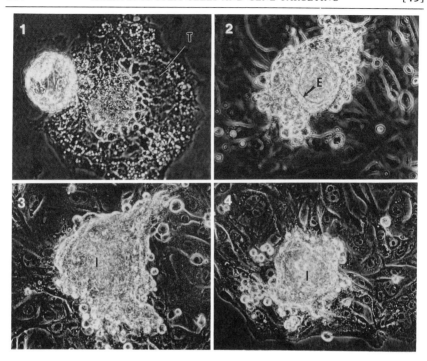

FIG. 3. Blastocyst attachment and outgrowth. (PMEFs have been omitted for the sake of clarity, although they would normally have been included.) (1) Two blastocysts explanted into a culture dish 2 days previously. One has attached, with the trophoblast (T) forming a monolayer. At its center the cells of the inner cell mass (ICM) are visible. To the left is a blastocyst that has not attached, possibly because it is sitting on top of the trophectoderm of the other embryo. (2) Two or three days later the ICM has proliferated to form a distinct outgrowth of cells on top of the trophoblast. The ICM shown has already developed a layer of endoderm (E) and thus is unlikely to give rise to ES lines when picked and disaggregated. (3, 4) ICMs (I) that are ideal for picking and disaggregation.

lose contact with each other. Using another mouth-controlled pipette, whose tip has been flame-polished to remove any sharp edges and whose diameter is between 50 and 100 μm, the clumps are broken up into smaller clusters of cells and single cells by pipetting up and down a few times (Fig. 4). The entire cell suspension is transferred to a single well of a 16-mm tissue culture dish which already contains a fibroblast feeder layer. The culture medium (1 ml) is ES-DMEM supplemented with 1000 IU of LIF or with 70% 5637-conditioned medium. We use Nunclone 4 × 16 mm well multidishes (Nunc) as the culture vessel for the dissaggregated ICMs, allowing one well per ICM. When all the ICMs have been disaggre-

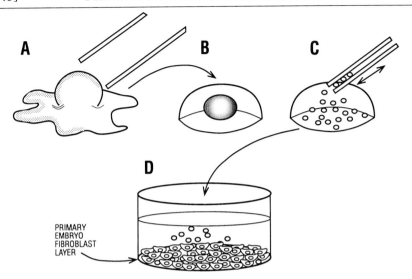

Fɪɢ. 4. Procedure for picking and disaggregation of the ICM. (A) The ICM, sitting on top of the trophoblast outgrowth, is removed with a mouth pipette. (B) The ICM is transferred to a microdrop of trypsin–EDTA containing 1% chicken serum, where it is incubated for about 5 min. (C) Using a narrower diameter mouth-controlled pipette, the trypsinized ICM is disaggregated to individual cells and clumps of cells. (D) Cells of the disaggregated ICM are transferred to a 16-mm (diameter) culture well that already contains a layer of mitomycin C-treated PMEFs. The ICM cells settle onto the feeder layer, where they should give rise to ES colonies.

gated and each one has been transferred to a well, the culture dishes are returned to the incubator.

9. Between 3 and 4 days after explanting the ICMs, the wells should be inspected to check that ICM cells are present and have started to form colonies. The explanted ICM cells do not just give rise to ES cells. In many instances, other cell types appear with the continued culture of the primary explants. These colonies, some of which are shown in Fig. 5, may at first resemble ES colonies. However, over time they differentiate and cease to proliferate. ES cell colonies, which have a characteristic morphology [Fig. 5 (1)], continue to proliferate, usually as tight round colonies that have smooth edges. It is difficult to distinguish the individual cells in the colony, although their nuclei can be recognized and contain one or two prominent nucleoli. By observing the well on a daily basis, it is possible to see whether a colony continues to increase in size as it proliferates without differentiation. These colonies are most often found at the perimeter of the well, which is sometimes difficult to view with a tissue culture microscope. Careful inspection should therefore be made

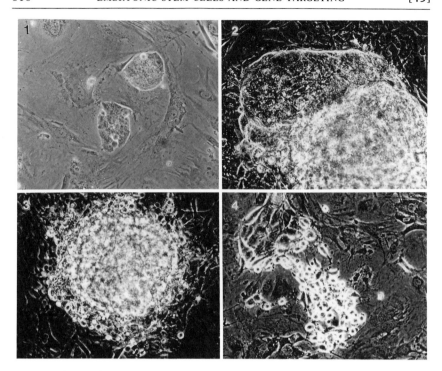

FIG. 5. Some of the cell types that grow after plating a disaggregated ICM. (1) Two primary ES colonies. The colonies, with PMEFs around them, have a typical morphology, with the individual cells comprising the colony being difficult to distinguish. The pale nuclei with the prominent dark nucleoli can be seen. (2) Large colony of what superficially looks like an ES colony. However, the edges of the colony are not as smooth as for the ES cells, the prominent nucleoli are absent, and, with continued observation, the colony did not proliferate after it was trypsinized and replated into a well with a fresh feeder layer. (3) ICM clump in which a layer of endoderm has formed on the periphery of the clump. (4) Group of parietal cells, which can proliferate. However, they have a distinctive morphology (they are highly refractile, and the individual cells can also be recognized).

of the perimeter to ensure that no colonies are missed. ES colonies should be apparent within 7–10 days after picking and disaggregating the ICM.

The process of producing new ES cell lines can be frustrating, with no success after weeks or even months. Then, for no apparent reason, the process works and lines are established. It appears that using early passage (P2–3) PMEFs and including either recombinant LIF or 5637-conditioned medium in the culture medium can help in the establishment of ES cells from the disaggregated ICMs. Overall, ES lines can be established at a frequency of 10–30% from the picked ICMs.

Expansion of Embryonic Stem Cells

When colonies of ES cells have been identified in the primary explants, their numbers can be expanded. It is not necessary to isolate the ES cells in the primary cultures from other differentiated cell types that may be present, since one of the characteristics of ES cells is rapid and continuous proliferation.

The entire well containing the ES colonies is washed 2 times in PBS, and the PBS is aspirated. To each well, 0.2 ml of trypsin solution plus 1% chick serum is added, and the well is left to trypsinize for 5 min. Then 0.5 ml of ES-DMEM is added, and all clumps of cells are broken up by gently pipetting the suspension, with care being taken to ensure that no bubbles are introduced into the well. If only one or two ES colonies are present in the well, the cell suspension is left in the well to reattach. The medium is replaced, the next day, with 1 ml of ES-DMEM plus 1000 IU/ ml LIF or 5637-conditioned medium. Over the next 3–5 days, if ES colonies were correctly identified, many new colonies of ES cells should become visible. The well can then be trypsinized again and the contents transferred to a 60-mm dish containing a PMEF feeder layer. The colonies of ES cells should continue to proliferate without differentiation. At this point, it is no longer necessary to include LIF or 5637-conditioned medium, and the cells can be maintained on feeder layers in ES–DMEM.

Procedure for Karyotyping Embryonic Stem Cell Lines

A significant proportion (30–50%) of newly established ES lines do not contain a normal diploid (euploid) karyotype. If the intention is to use ES cells for genetic manipulation of the mouse germ line, it is essential to identify lines with a euploid karyotype that preferably are male. Once the sex and karyotype has been determined, and if it is normal, the line should be tested for its ability to form functional gametes by determining whether it can form germ line chimeras (see [50] in this volume).

The karyotype and sex of the line can be determined by making *in situ* metaphase chromosome spreads from the ES cells by the C-banding technique so that the Y chromosome can be identified.[21] The procedure is based on the *in situ* method for preparing chromosome spreads. The cells are grown in microscope coverslips. They are swollen with a hypotonic solution while still attached and are fixed to the coverslip. The coverslips are processed for C banding and stained. The mouse has 40 chromosomes, 19 pairs of autosomes and the 2 sex chromosomes. Unlike human chromosomes, mouse chromosomes do not exhibit a great variation in size, and

[21] P. M. Iannaccone, E. P. Evans, and M. D. Burtenshaw, *Exp. Cell Res.* **156**, 471 (1985).

so identification of specific autosomes is more difficult. However, for the purpose of determining whether a newly derived line is suitable for generating germ line chimeras, counting the chromosomes is sufficient to give a good indication of the potential of the line. Any Robertsonian translocation will be easily detected, and any loss or gain of a chromosome(s) can also be determined. Determination of the sex is relatively straightforward using the C-banding technique, which highlights the regions of constitutive heterochromatin, principally the centromeres. All autosomes in the mouse have a darkly stained centromere, as does the X chromosome. The centromere of the Y chromosome does not stain as darkly, and thus a male karyotype can be recognized by lack of staining on one of the three smallest chromosomes visible in a typical spread (Fig. 6). The other two chromosomes that stain are both copies of chromosome 19, which are about the same size as the Y chromosome.

1. Sterilize standard 22 × 22 mm microscope coverslips by soaking in 100% ethanol and dry by flaming over a burner in the tissue culture hood.

2. Place a coverslip into a 35-mm culture dish.

3. Trypsinize a 60-mm dish of ES cells in the log phase of growth, that is, at the stage in which the ES colonies are easily visible and covering about one-third to one-half of the dish surface.

4. Seed about 200–400 ES cells (as a single cell suspension) onto the coverslip and culture for 48 hr in ES–DMEM plus 1000 IU LIF. This should allow ample time for the ES cells to attach to the surface of the coverslip. On the third day, to arrest cells in mitosis, add 50 μl Colcemid (5 μl/ml) to each dish. Incubate at 37° for 45 min.

5. Aspirate the medium. Gently add 2 ml of 0.8% sodium citrate hypotonic solution. Incubate for 20 min at room temperature. This causes the cells and nuclei to swell, thus releasing the mitotic chromosomes onto the surface of the coverslip. Gently add 1 ml of freshly prepared fixative (3 parts methanol to 1 part glacial acetic acid) to each dish.

6. Aspirate the hypotonic/fixative solution from the edge of each dish. Add 2 ml of fixative to fix chromosomes on the coverslip. Leave for 30 min at room temperature. Repeat the fixation twice, leaving for 20 min each time at room temperature.

7. Remove the fixative by aspiration and dry the coverslip immediately over a spirit lamp, with a hair dryer, or with a convection oven set at 70°–80°. Carefully remove the coverslip from the petri dish, keeping it cell side up. Label with a permanent marker and leave on a slide warmer set to 60° for 18–24 hr. They can now be processed for cytogenetic analysis.

Staining Chromosomes for C-Banding Analysis

1. Bring 100 ml of distilled water to a boil.

2. Add 0.5 g of barium hydroxide Ba(OH)$_2$ and dissolve. Let the solu-

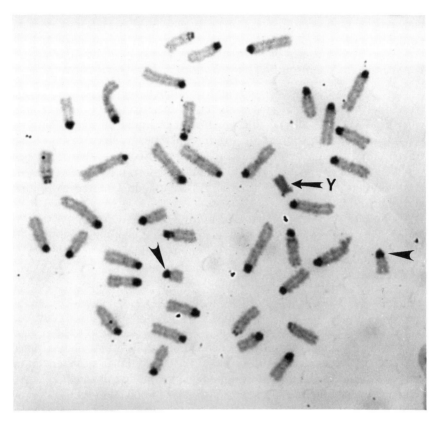

FIG. 6. Chromosome spread of 40 chromosomes from a male ES line. The chromosomes have been stained by C banding, resulting in intense staining of the centromere in all chromosomes except the Y. The two smallest chromosomes (number 19) are also indicated by arrowheads.

tion cool to 55° and remove any surface scum with filter paper (always prepare fresh before use).

3. Immerse the coverslip for 2–3 min in the Ba(OH)$_2$ solution.

4. Rinse the coverslip 3 times with distilled water to remove excess Ba(OH)$_2$.

5. Incubate the coverslip for 2 hr at 50° in 2X SSC (30 mM NaCl, 30 mM trisodium citrate).

6. Rinse the coverslip 2 times in washes of distilled water.

7. Stain with 5% Giemsa for 5–10 min. Rinse in water to remove excess stain and allow coverslips to dry.

8. Mount coverslips in DPX and examine for chromosome spreads using a microscope with ×200 magnification. Count and photograph the

chromosomes, examining between 30–40 spreads. This should be sufficient to give a good description of the karyotype of the ES line, i.e. what percentage of cells have a normal diploid chromosome number.

Expansion, Freezing, and Routine Culture of Embryonic Stem Cells

Once an ES line has been found to contain a high percentage of cells with a normal diploid karyotype, it should be expanded so that as many early passage cells as possible are frozen in liquid nitrogen. This will provide sufficient resources for future experiments, since early passage ES cells tend to make better chimeras at a higher frequency than if passages 15–20 and later are used. However, there is no absolute correlation, since relatively late passage lines such as D3 have been reported to produce germ line chimeras.

The ES cells can be maintained as an undifferentiated population by trypsinizing and replating the cells onto dishes containing fresh feeders, every 5–6 days if the cells are plated out at a sufficiently low density. A 60-mm dish at maximum density will contain about $1–2 \times 10^7$ ES cells, and a 150-mm dish can contain up to $2–3 \times 10^8$ cells at maximal density. The cells will start to differentiate or die if they are maintained beyond the maximum density level, and thus the optimal period of time they can be maintained before they have to be passaged (Fig. 7) is about 5–7 days. To maintain a line, trypsinizing a semiconfluent dish and plating out of the single cell suspension with 1 : 100 to 1 : 500 dilution is sufficient. If the cells are replated at reasonably low density, the culture medium needs changing every other day to keep cells under optimal conditions. If more cells and higher densities are required, then the cells should be refed every day. Under optimal conditions, the ES cells should grow as small clusters or mounds [Fig. 7 (1)]. If the conditions are suboptimal, differentiated derivatives will appear, and the mounds of ES cells will start to flatten out, with individual cells becoming more distinct [Fig. 7 (2)]. Under extreme conditions the majority of the cells will have differentiated [Fig. 7 (3)].

Freezing of Embryonic Stem Cells

1. A culture of ES cells should be in the log phase of growth, that is, not at maximal density. Wash the dish 2 times in PBS and trypsinize.

2. Harvest the cells, resuspened in medium, and count with a hamocytometer.

3. The medium for freezing the cells consists of a 50 : 50 mixture of DMEM and FCS containing a final concentration of 10% (v/v) dimethyl sulfoxide (DMSO) (Sigma).

4. One milliliter of medium containing $1–5 \times 10^6$ ES cells is aliquoted into a 1-ml sterile freezing vial (Nunc) that has a screw cap and rubber seal.

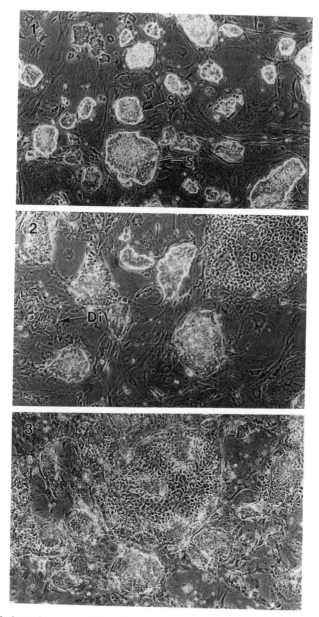

FIG. 7. Embryonic stem cells in culture. (1) Culture growing under optimal conditions. The ES clumps (S) are growing on a feeder layer of PMEFs. None of the clumps show any overt differentiation. (2) Culture in which the conditions are suboptimal. Colonies of differentiated cells (D) are visible, and other ES colonies are flattened out (Di), which is the first overt sign of their starting to differentiate. (3) Culture in which the majority of cells have differentiated, with few stem cell colonies remaining.

5. The vials are labeled with the ES line and passage number, placed in a holding rack, and left overnight in a −70° freezer.

6. The following day the frozen vials should be transferred to a liquid nitrogen container for long-term storage.

7. To thaw ES cells, a 60-mm tissue culture dish containing a feeder layer in ES-DMEM medium should be prepared in advance. Remove the vial of ES cells and place in a beaker of sterile distilled water prewarmed to 37° until the contents of the vial have melted. Remove the vial, swab with 100% ethanol to sterilize the outside, and remove the cell suspension with a sterile Pasteur pipette. The cells can be immediately plated out in the 60-mm dish. The next day the culture medium is replaced with fresh ES–DMEM to remove all the DMSO and any dead cells. If freezing and thawing of the ES cells were performed correctly, then ES colonies should already be visible in the culture dish.

Feeder-Free Culture of Embryonic Stem Cells

Under certain circumstances (e.g., metabolic labeling of proteins in ES cells or preparation of mRNA for cDNA libraries), it is preferable to obtain ES cells without any fibroblast feeder layer contaminants. To do this, it is necessary to culture the ES cells in the absence of any feeders for 2–3 passages in order to dilute out residual fibroblasts. ES cells can be grown in absence of feeders on gelatinized dishes, provided that LIF has been included in the medium. The substitution of feeder cells by LIF alone appears to work reasonably well for some ES cells that were originally maintained on feeder layers. For other lines, it does not work as well, and extensive differentiation of the stem cells occurs.

An effective way to reduce the extent of differentiation and grow almost 100% pure populations of stem cells is to use the extracellular matrix deposited by the feeder layer as a substitute for gelatin. ES cells grown on such a matrix in the presence of exogenous LIF retain their stem cell phenotype, and much less differentiation occurs. Furthermore, the ES cells can be put through multiple rounds of passaging, and, provided that they are replated onto such matrix with exogenous LIF, they do not differentiate.

Preparation of Extracellular Matrix for Feeder-Free Growth of Embryonic Stem Cells

1. Plate out non-mitomycin C-treated proliferating fibroblasts at the appropriate numbers depending on the dish size. The dishes should have been gelatinized (with 0.1% gelatin solution), and the cells are then cultured in DMEM plus 10% FCS.

2. The cells are allowed to grow to confluency over 3–4 days. As they grow, they secrete and deposit an extracellular matrix.

3. To remove the cells without disrupting the matrix, wash the fibroblast cultures 2 times with PBS. Add an amount of lysis buffer just sufficient to cover the surface of the cells (~0.5–1 ml for a 60-mm dish). The lysis buffer consists of 0.5% (v/v) of Triton X-100 (nonionic detergent) in PBS and 35 μl of amonium hydroxide solution (NH_4OH) per 100 ml of PBS–Triton X-100.

4. Incubate the dish at room temperature for 5–10 min. The cell membranes lyse and release the cytoplasmic contents, but the nuclear membranes remain intact. The extracellular matrix, which can sometimes be seen as fibrous strands on the surface of the dish, remains attached to the surface.

5. Wash off the cellular contents, nuclei, detergent, etc., with three changes of PBS plus Ca^{2+}/Mg^{2+}. The dish can now be used to culture ES cells in ES-DMEM plus 1000 IU LIF. The plates should be refed every 2 days. The cells can be passaged onto freshly prepared plates of lysed feeder cells to dilute out remaining feeder contaminants.

Acknowledgments

We thank Alisoun Carey for critical reading of the manuscript and Sharon Perry for efficient typing.

[50] Production of Chimeras between Embryonic Stem Cells and Embryos

By Colin L. Stewart

Introduction

The principal uses of embryonic stem (ES) cells are screening for new genes of potential developmental interest (enhancer/promoter traps)[1-3] (see [40] and [41] in this volume and introducing a specifically mutated gene or other form of new genetic information into the mouse germ line[4-9]

[1] G. Friedrich and P. Soriano, Genes Dev. 5, 1513 (1991).
[2] W. C. Skarnes, B. A. Auerbach, and A. L. Joyner, Genes Dev. 6, 903 (1992).
[3] H. Von Melchner, J. V. DeGregori, H. Rayburn, S. Reddy, C. Fredel, and E. H. Ruby, Genes Dev. 6, 19 (1992).
[4] S. L. Mansour, K. R. Thomas, and M. R. Capecchi, Nature (London) 336, 348 (1988).

Copyright © 1993 by Academic Press, Inc.
All rights of reproduction in any form reserved.

[45]–[47] in this volume). The generation of chimeras between embryonic stem (ES) cell lines or clones and embryos is an essential step in these processes, which when successful leads to the derivation of new strains of mice with an altered genome.

Most ES lines that are currently in use have an XY or male genotype. This has two advantages. The first is that male XY ES lines, when injected into female XX blastocysts, will tend to bias the development of the resulting chimera toward a male phenotype.[10,11] In phenotypically male chimeras, only XY-bearing germ cells (i.e., those derived from the ES cells) will form functional gametes. XX primordial germ cells (i.e., those derived from the host blastocyst) will not form functional gametes and are lost. This will, therefore, favor the development of gametes derived from the ES cells. Second, a male chimera can produce more offspring over its reproductive life span than a female, so that even chimeras with a relatively low percentage contribution of the ES cells to the germ line can be detected.

Another important consideration is the strain of mouse from which the ES cells are derived and the strain of embryo into which they will be introduced. This is particularly relevant for producing germ line chimeras since certain strains have, in combination with others, a competitive advantage in their development. Thus, many of the chimeras will be predominantly composed of cells derived from the dominant strain,[12] affecting the likelihood that the ES cells will contribute to the germ line. Most ES cell lines have been derived from the 129/J and 129/Sv strains of mice, and, when injected into embryos of the C57BL/6J strain (which is readily available commercially and breeds relatively well), the 129 ES cells tend to predominate in the chimeras. The 129 ES cells have an agouti coat color genotype, whereas that of C57BL6 embryos is black. Therefore, the chimeric mice produced from this combination will be a color mix of agouti/black. Frequently, if the ES line is particularly "good" at making chimeras, extreme individuals composed almost entirely of the 129/Sv

[5] M. Zijlstra, E. Li, S. Fereydown, S. Subramani, and R. Jaenisch, *Nature (London)* **342**, 435 (1989).

[6] A. P. McMahon and A. Bradley, *Cell (Cambridge, Mass.)* **62**, 1073 (1990).

[7] C. L. Stewart, M. Vanek, and E. F. Wagner, *EMBO J.* **4**, 3701 (1985).

[8] R. L. Williams, S. A. Courtneidge, and E. F. Wagner, *Cell (Cambridge, Mass.)* **52**, 121 (1988).

[9] C. L. Stewart, P. Kaspar, L. Brunet, H. Bhatt, I. Gadi, F. Köentgen, and S. Abbondanzo, *Nature (London)* **358**, 76 (1992).

[10] A. McLaren, "Mammalian Chimeras." Cambridge Univ. Press, Cambridge, 1976.

[11] P. M. Iannaccone, E. P. Evans, and M. D. Burtenshaw, *Exp. Cell Res.* **156**, 471 (1985).

[12] P. L. Schwartzberg, S. P. Goff, and E. J. Robertson, *Science* **246**, 799 (1989).

genotype can be produced. Germ line chimeras, using 129 ES cells, can also be produced at a reasonable frequency with the outbred albino MFI strain.[12] Some ES lines have been made from C57BL/6 embryos.[13-15] The best host embryo strain into which they can be injected has not yet been identified, although reasonable chimeras, some of which are germ line, have been produced using the BALB/c strain.[13-15] Finally, the length of time that ES cells have spent in culture since their derivation can also affect their ability to make germ line chimeras. Chimeras that are the strongest and of the highest frequency are usually those derived with early passage clones (i.e., up to 10–15 passages); thereafter, it has been noted that the extent and frequency of chimerism may often, but not always, start to decline.

To generate germ line chimeras efficiently it is essential that the ES line be tested, prior to any manipulation or selection, for its capability of generating chimeras at a high frequency. The criterion is that more than 50% of the offspring born should be chimeric, with the majority of these being able to transmit the ES genotype through the germ line. It is also recommended to determine the karyotype of any subsequent clones isolated by selection, prior to injection into blastocysts, thereby avoiding any clones having aneuploid karyotypes that may not produce germ line chimeras. This procedure will result in considerable savings in time and effort and need only involve counting of the chromosomes, using the C-banding staining technique (see [49] in this volume), if the ES cell line used has already been assessed as to its ability to produce germ line chimeras. Any deviation from a mean number of 40 chromosomes will almost inevitably result in weak chimeras being produced, with little possibility of the ES cells contributing to the germ line. The exception, however, is loss of the Y chromosome from a male ES line, resulting in a 39X0 karyotype. Such clones can produce very good chimeras, resulting in germ line transmission by the females.[9]

Presently, there are three methods of producing ES cell chimeras: (1) blastocyst injection, (2) morula injection, and (3) morula aggregation.

Materials for Blastocyst and Morula Injection

The methods for blastocyst and morula injection are essentially the same, involving the injection of cells into preimplantation embryos at different developmental stages. Both procedures require the same type of microinjection apparatus.

[13] C. L. Stewart, unpublished observations.
[14] B. Ledermann and K. Bürki, *Exp. Cell Res.* **197,** 254 (1991).
[15] F. Köntgen, Ph.D. Thesis, University of Freiburg, Germany (1992).

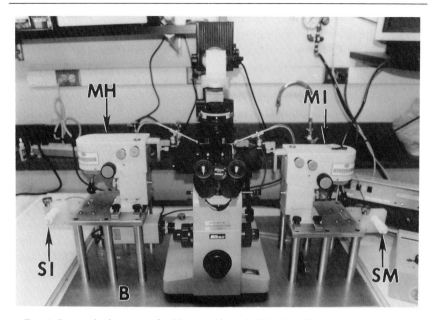

Fig. 1. Inverted microscope for blastocyst/morula injection. The lens turret is obscured by the binocular eyepieces at the front of the microscope. MH, Leitz manipulator that controls the holding pipette; MI is the right-hand manipulator that holds and controls the injection pipette. The joysticks (for x and y axis control) can be seen hanging underneath the manipulators. SI is the micrometer syringe that controls the suction to the injection pipette through a plastic tube, filled with light paraffin/mineral oil. The pipette holder and tube connecting it to the syringe can be seen attached to the right-hand manipulator (MI). SM is the syringe that controls the holding pipette on the left-hand manipulator. B is the customized base plate on which the manipulators are attached.

Microinjection Apparatus

Embryo injections are best performed on a fixed-stage microscope where the object lens is moved in order to focus. A variety of fixed-stage microscopes are available from Zeiss, Nikon, Leitz, and Olympus. The objective lenses should be $10 \times$ and $20 \times$ magnification with phase-contrast optics. The magnification of the eye pieces should be $10 \times$. Phase objectives are essential because only with these can one distinguish between viable and dead ES cells, although Nomarski optics give very clear images of embryos and cells during injection. The microscope should be mounted on an air table (e.g., Vibraplane made by Kinetic Systems, Boston, MA) in order to suppress any vibrations that would make observation of the embryos difficult (Fig. 1).

In addition to the microscope, two micromanipulators are essential. Two types of manual (mechanical) micromanipulators are readily available (Leitz and Narishige). One of the micromanipulators is required for immobilizing the embryo for injection (the holding manipulator); the other is used for picking up the ES cells and injecting them into the embryo (the injection manipulator). The manipulators are capable of moving the injection/holding pipettes in all three axes, with the movements in the horizontal axes being controlled by a joystick. Manual or mechanical manipulators are preferable to those that are electrically driven, since they offer greater flexibility and speed of control.

If the Leitz micromanipulators are used, they need to be clamped to a base plate. The type of base plate depends on the type of microscope purchased. Leitz offers a base plate in conjunction with the Labovet FS microscope. The Nikon, Olympus, and Zeiss microscopes all require a customized base plate[16] that can be purchased from H. I. Instrumentation (293 Carlton Terrace, Teaneck, NJ 07666). No base plate is needed for the Narishige manipulators since they can sit on or be fixed to shelves placed on the air table next to the microscope.

Preparation of Injection and Holding Pipettes

The introduction of ES cells into the blastocyst or morula is performed by injecting groups of individual cells into the embryo. Control of the embryos/cells is accomplished by exerting positive/negative suction pressure through fine glass pipettes. The pipettes are inserted into the holding instruments (which, in turn, are attached to the micromanipulators). The holding instruments (Fig. 1) are connected by flexible plastic tubing to two micrometer syringes (Gilmont makes an excellently designed syringe that is also relatively inexpensive). It is important that the micrometer syringe controlling the suction to the particular pipette be placed opposite to the manipulator/pipette it is controlling. This allows the operator to control the movement of the pipette with one hand while the other hand controls the micrometer and, hence, the suction applied to the pipette. The whole setup is filled with light paraffin oil (Mallinkrodt), with care being taken to ensure that no air bubbles are present in the plastic tubing, syringes, or pipettes, since these will disrupt the suction pressure applied to the pipettes.

Two types of pipettes are required, namely, holding and injection pipettes, both of which can be made from the same glass capillary tubing,

---

[16] B. Hogan, F. Costantini, and E. Lacy, in "Manipulating the Mouse Embryo," p. 279. Cold Spring Harbor, Laboratory, Cold Spring Harbor, New York, 1986.

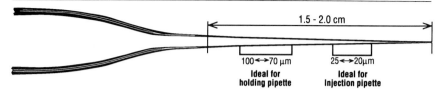

1.5 - 2.0 cm

100←→70 μm   25←→20μm
Ideal for   Ideal for
holding pipette   Injection pipette

FIG. 2. Dimensions of a pipette pulled on a pipette puller. The regions for breaking the pipette to make either an injection or holding pipette are shown.

which should have an external diameter of 1 mm and an internal diameter of 0.8 mm. Capillary tubing can be obtained from World Precision Instruments (175 Sarasota Center Blvd., Sarasota, FL 34240), in either 4- or 6-inch lengths (Cat. No. 1B100-4/6). The capillaries have to be heated and pulled in order to produce a pipette of the appropriate diameter. This is done by using an electromagnetic electrode/pipette puller. A variety of pullers are available such as the flying magnet (Narishige), flying coil (Campden Instruments), or twin pulling system (Sutter Instrument Co., Novato, CA). All of these are capable of producing pulled pipettes with the appropriate dimensions. The pulled pipettes should taper to a point of less than 1 μm over a distance of at least 1.5–2.0 cm from the shoulder of the pipette (Fig. 2). Both the holding and injection pipettes can be made from the same pulled capillary tubing.

Holding Pipettes. Holding pipettes are used to immobilize the blastocyst/morula that is be injected. The embryo is clamped to the opening of the pipette by suction. The holding pipettes have a wider diameter than the injection pipettes, with an opening of between 20 and 40 μm, and are made using a DeFonbrune microforge. The steps in preparing a holding pipette are shown in Fig. 3.

1. A pulled pipette is placed over the glass bead that is fused to the heating element of the microforge, at a point where the external diameter of the pipette is between 70 and 100 μm [Fig. 3A, B (1)].

2. When the heating element is switched on, it causes the element and glass bead to expand slightly and move along the length of the pipette.The amount of heat applied should be just sufficient to result in the pipette fusing to the glass bead when brought in contact with the bead, without the pipette bending or melting to any great extent.

3. When the heat is switched off the heating element/bead cools and contracts rapidly, and the pipette fused to the bead snaps. The break should be (and usually is) a vertical one [Fig. 3A, B (2)], relative to the long axis of the pipette.

4. The tip of the broken pipette, which usually remains attached to the glass bead, is removed. The opening of the pipette is then positioned

in front of the glass bead [Fig. 3A, B (3)]. It is important that the opening be situated at least 100–200 μm away from the bead. This is because when the power to the heating element/bead is switched on, it causes the bead to move toward the opening of the holding pipette; if the bead contacts the pipette, the two will fuse and the holding pipette will be ruined.

5. Switch on the power to the heating element so that it glows yellow. The holding pipette is then brought close to the bead such that the edges will start to melt, so narrowing (and polishing) the opening [Fig. 3A, B (4)]. The opening of the pipette should have a final diameter between 20 and 40 μm, which is sufficient to hold a blastocyst/morula in position for injection.

6. The tip of the holding pipette then has to be bent. Most injections are performed in shallow depression slides or injection chambers (see below). To avoid continually having to adjust the height of the holding pipette during embryo manipulation, a bend is placed near the tip of the pipette. The pipette tip can then be kept parallel to the bottom of the injection chamber in which the blastocysts are lying. Only minimal adjustments will then be necessary to pick up and move the blastocyst into position for injection. Usually a 30° bend is introduced (2.5–3 mm) near the tip of the pipette.

7. The procedure for introducing the bend is shown in Fig. 3A, B (5) and (6). The holding pipette is brought at a 30° angle near to the filament. The heating element should be switched on to a level where the glass bead glows orange and the temperature adjusted so that the pipette starts to bend toward the filament. It is important to ensure that the filament and pipette do not touch; otherwise they will fuse. The bending is allowed to proceed until the desired angle is achieved, and the heat is switched off. The holding pipettes can be stored in large petri dishes, resting on a strip of plasticine so that the tips are not in contact with any surface, or they can be stored upright in metal blocks that have been drilled with a 1.2-mm drill to a depth of 2 cm.

Injection Pipettes. The injection pipette is basically an extremely fine, sharp-pointed hypodermic needle, which is used to introduce individual ES cells either into the blastocoel cavity of the blastocyst or between blastomeres of the morula. The internal diameter of injection pipettes should be at least slightly larger than that of the cells (ES cells have a diameter of 15–20 μm). The pipettes are pulled under the same conditions used to make holding pipettes. However, because pipettes with a small opening are required, the pipette is broken at a narrower point (see Fig. 2).

The injection pipettes can be broken by using a sharp scalpel blade. The pipette is placed on a transparent rubber sheet and viewed through

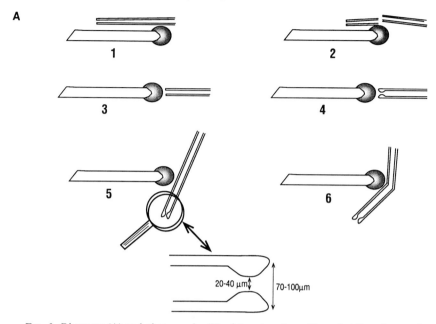

FIG. 3. Diagrams (A) and photographs (B) of the steps in making a holding pipette. At the appropriate diameter the pipette is placed over a glass bead melted onto the heating element of the microforge (1). The pipette is then fused to the glass bead. The power to the heating element is switched off, resulting in cooling and contracting of the heating element/bead, so snapping the pipette fused to the bead (2). The tip of the broken pipette is placed in front of the bead (3). The power is switched on to a level that results in the tip melting, producing an opening with a narrower diameter and with the edges being smoothed by flame polishing (4). A bend is introduced in the pipette at a distance of about 1–3 mm from the tip (5, 6). The bend is induced by heating the glass bead, which causes the pipette tip to bend toward the bead.

a dissecting microscope. The diameter of the pipette is then gauged with the help of a micrometer scale, and the scalpel blade is used to break the pipette at the region of the pipette where the internal diameter is about 20 μm. The break at that region frequently results in the tip breaking with a sharp point that is sufficient for immediate use in blastocyst injections. Usually, an adequate number of pipettes can be made in this way, but we have found that this is not very reproducible and sometimes requires breaking 5–10 pulled pipettes before a satisfactory one is generated.

B

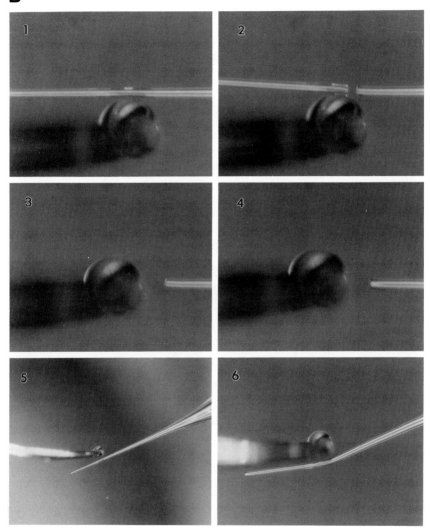

FIG. 3B.

A highly reproducible alternative procedure is to use a microforge and pipette grinder. The pulled pipette is mounted on a holder and broken at the appropriate diameter in the same manner as the holding pipette was made [Fig. 4A (1–4) and Fig. 4B (1–3)]. The tip of the pipette is then

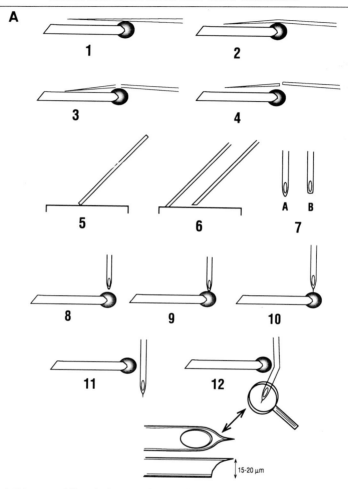

FIG. 4. Diagrams (A) and photographs (B) of the steps in making an injection pipette. The pipette is placed over the glass bead at the appropriate diameter and is fused to the heated bead; then the power is switched off, causing the bead simultaneously to cool and contract, so breaking the pipette attached to the bead [A (1–4), B (1–3)]. The pipette tip is beveled at a 40°–45° angle on a rotating grinding wheel [A (5–7)]. The beveling should result in the tip being ground to a fine point (7A). If insufficient grinding occurs, the tip will have a squared end (7B). To generate a tip that is sharp enough for blastocyst injections, an even finer point is pulled using the microforge. The beveled tip is placed vertically over the glass bead [A (8), B (4)]. The bead is heated, with the air blower keeping the pipette cool. In one smooth motion, the tip should briefly touch the heated bead and then be pulled away [A (9, 10), B (5, 6)]. This causes the very tip to melt, resulting in the pulling of a very fine, sharp point. The tip (3–4 mm) of the injection pipette is then bent by placing it close to the heated bead, with the tip bending toward the bead [A (11, 12), B (7, 8)]. *Note*: In diagrams 11 and 12 (a), the opening to the pipette faces the operator so that when the bent pipette is placed in the injection chamber, the opening will be at the side.

B

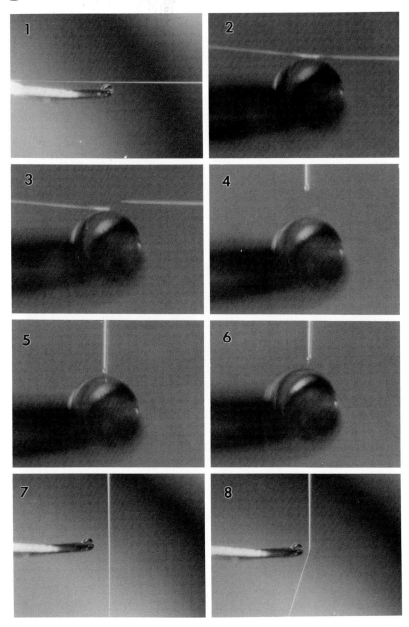

FIG. 4B.

FIG. 5. Pipette beveling device from Bachofer. (Top) An injection pipette is placed at an angle over the wheel in the center of the picture. The grinding process is viewed using the stereo microscope and light at left. (Bottom) Close-up of the pipette (P) being beveled on the rotating grinding wheel. The water supply that lubricates the surface of the wheel is at top left.

beveled to a 35°–40° angle using a grinding wheel [Fig. 4A (5, 6)]. Grinding wheels can be purchased from Narishige; however, the simplest and easiest to use is one made by Bachofer Laboratoriumsgeräte (D-7410 Reutlingen, Germany). This is a rotating wheel mounted at a horizontal plane on an electric motor (Fig. 5). Lubrication is provided by dripping filtered water onto the wheel as it rotates. The tip of the pipette is brought down into contact, at the appropriate angle, with the spinning wheel. Beveling will take about 1 min in order to prepare a tip with a suitable, fine point [Fig. 4A (7A)]. If the beveling is not sufficient, then the tip will have a square edge and will be unsuitable for injection purposes [Fig. 4A (7B)].

Beveled pipettes can be used for morula injection. For blastocyst injection, however, a finer point is required, and this can be pulled on the microforge as shown in Fig. 4A (8–10) and Fig. 4B (4–6).

1. The pipette is positioned in the vertical plane over the glass bead with the opening of the pipette facing the operator.

2. The glass bead, which must be a sufficient distance from the pipette tip so that the tip does not melt and fuse, is heated to a point where the bead is barely glowing a dull orange (it is at this step that cooling jets of air, a feature of the DeFonbrune microforge, will regulate the temperature of the bead and keep the pipette cool, except at its very tip).

3. The tip of the pipette needs to be slowly lowered toward the bead to a point where it is just above the bead. Then, in one smooth motion, the tip is lowered, briefly allowed to touch the surface of the bead, and is then pulled away. This should result in a very fine, short and sharp point [Fig. 4A (8–10) and Fig. 4B (4–6)]. If the temperature of the bead is too hot, or if the pipette is kept in contact with the bead for too long a time, a longer point may be drawn at the tip and the opening to the pipette will start to melt, resulting in a narrower diameter, too small to allow drawing up of cells. With a minimal amount of practice, the optimal conditions can be found that reproducibly result in very fine, short points being produced.

4. The last step, as with the holding pipette, is to introduce a bend within 3–4 mm of the pipette tip so that it is parallel to the bottom of the injection chamber, which simplifies the process of drawing up individual ES cells into the pipette. The bending is performed exactly as with the holding pipettes, except that the orientation of the bend, relative to the pipette opening, must be such that the opening remains to the side of the pipette [see Fig. 4A (11, 12) and Fig. 4B (7, 8)].

5. A pipette prepared in such a fashion should last for at least 50–70 blastocyst injections, and frequently the same pipette can be used on successive days. Care should be taken to not blunt the tip, which can occur by breaking it against the bottom of the injection chamber or the tip of the holding pipette.

Injection Chamber

Embryo injections are performed in microdrops of medium that have been placed in an injection chamber. The injection chambers can be made from pieces of plastic or aluminum that have the same dimensions as a standard microscope slide (i.e., 75 × 25 × 1 mm). We favor aluminum since it has good heat conductance and the slides can be washed, sterilized, and reused. As an alternative disposable 35- or 60-mm petri dishes can also be used.

The injection chamber consists of an aluminum slide with a hole, 1.8 cm (3/4 inch) in diameter, drilled through its center (Figs. 6 and 7). Over the hole is fixed a standard glass histology coverslip, which is stuck into position using paraffin wax.

1. To attach the coverslip, two small beads of molten paraffin wax (mp 60°) are placed diametrically opposite each other near to the edge of the hole. A coverslip is balanced on top of the beads, with care being taken to ensure that it completely covers the hole.

2. The aluminum slide is placed on a preheated hot plate, which causes the beads of wax to melt. The coverslip drops down over the hole, with the molten wax spreading between the slide and the coverslip. When the slide is removed from the hot plate, the wax cools and the coverslip adheres to the slide.

3. Ensure that the wax has spread all around the edge of the coverslip–slide; otherwise, there will be gaps present, and this will result in leakage. The injection medium consists of Dulbecco's modified Eagles' medium (DMEM) plus 10% fetal calf serum (FCS), buffered with 20 mM HEPES. In addition we add 300 IU of DNase I. The enzyme is included because cells in the injection chamber inevitably lyse during the injection process and release their DNA, which is the principal cause of "stickiness," resulting in adherence of debris, cells, etc., to the tips of the pipettes. The enzyme does not affect the viability of the cells, and its presence greatly reduces problems associated with stickiness. To set up the injection chamber, a drop of medium (~20 μl) is pipetted onto the surface of the coverslip, and the chamber is then flooded with paraffin oil (Figs. 6 and 7).

Embryo injections can be performed either at room temperature or on a cooling stage. Although it is not absolutely essential, a cooling stage is recommended because it makes the injection of blastocysts easier. With cooling of the embryos to between 4° and 8°, (1) the blastocysts become more rigid, and hence easier to penetrate with the injection pipette, and (2) the lower temperatures reduce the extent of ES cell death (some of which is inevitable), which results in lysis of the cells and accumulation of cell debris that can stick to the tips of the pipettes and reduce their effectiveness. Such a cooler (Fig. 7), which is available either with a microscope slide cooler (ideal for the chambers described above) or for

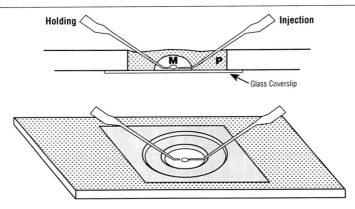

Fig. 6. Injection chamber. A glass coverslip is attached by paraffin wax over a hole drilled through the center of an aluminum slide. A drop of approximately 10–20 μl of injection medium (M) is placed on the coverslip, and the well is flooded with paraffin oil (P). The holding pipete and injection pipettes are placed in the drop with the bent tips parallel to the surface of the coverslip. Blastocysts and ES cells are placed in the medium.

a 60-mm petri dish, is the KT Stage Controller supplied by 20/20 Technology Inc. (Whitehouse Station, NJ).

Embryo Injection

Preparation of Embryonic Stem Cells

Embryonic stem cells for injection should be in the log phase of growth. ES cells that are nearing their maximum colony size and are reaching confluency, or those exhibiting overt and extensive differentiation, generally do not produce good levels of chimerism. A 35- or 60-mm petri dish of cells is usually adequate, providing that a reasonable number of stem colonies are visible. Three to four hours prior to trypsinization, the cells should be refed with fresh culture medium. The dish should be washed 2 times in phosphate-buffered saline (PBS) (Ca^{2+}/Mg^{2+}-free) and then trypsinized with 0.25% trypsin and 1 mM EDTA for about 3 min at 4°. The colonies are broken up into a single cell suspension in 3–5 ml of DMEM plus 10% FCS. The suspension is then plated onto two gelatinized dishes (see [49] for routine ES culture methods). The dishes are returned to the incubator at 37° for 1 hr, which will allow the feeder fibroblasts to attach and viable ES cells to begin to adhere to the surface of the dish. After 1 hr, the nonadherent cells are gently aspirated off, and the loosely adhering ES cells (which are viable) are gently blown off with a pipette, using DMEM plus 10% FCS. These are then transferred, in fresh medium,

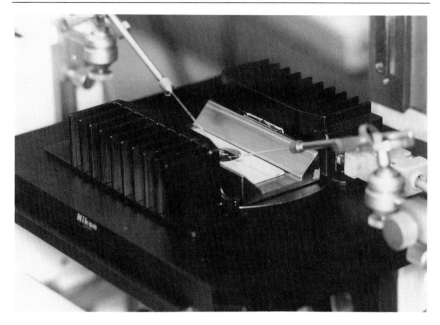

FIG. 7. Injection chamber on the KT cooling stage. The injection pipette at right and holding pipette at left can be seen; the tips are in the microdrop of the injection medium in the chamber.

to a 35-mm dish, where the cells are allowed to settle. Aliquots and cells can be removed, using a mouth-controlled pipette, and transferred to the injection chamber. This procedure greatly increases the probability of having viable ES cells for injection and maximizes the efficiency of producing chimeras.

Embryo Preparation

Blastocysts are flushed, using M2 medium (see [9] and [10] in this volume), from the uteri of day 4 pregnant females (day of plug = day 1 of pregnancy), whereas 8-cell stage morulae are removed from the oviducts of D3 pregnant females. The blastocysts to be used for injection should be surrounded by an intact zona pellucida (those lacking a zona are difficult to inject), and a blastocoel cavity should be visible. Between 10 and 20 blastocysts are introduced into the microdrop of medium in the injection chamber, followed by the introduction of 1–3 μl of medium containing the trypsinized but recovering ES cells, which should still be in a single cell suspension. The injection chamber containing the embryos and ES cells can be kept at 4° for up to 30 min in a refrigerator to precool the embryos/ES cells.

Before the injection and holding pipettes are mounted onto the instrument holder of the micromanipulators, the bottom 2–3 mm of each pipette is broken off to prevent blockage in the pipettes, caused by dust, etc., that may have accumulated at the bottom, from being pushed down into the pipette when it is backfilled with paraffin oil. Once mounted onto the instrument holders, the pipettes are backfilled with oil until a drop appears at the tip of each pipette.

The pipettes are positioned so that their movement is directly in line to the x axis movement of the manipulators. Care is taken to ensure that no air bubbles have been introduced into the pipettes or the tubing that connects them to the micrometer syringes that regulate the movement of the paraffin oil in the pipettes. Once the pipettes have been positioned, the cooling chamber is switched on. The temperature is allowed to fall to 4°–6°, and the injection chamber is then placed on the cooling stage.

The injection and holding pipettes are lowered into the drop of injection medium, and adjustments are made to ensure that the terminal 1–2 mm of the tips are parallel to the bottom of the chamber (Fig. 6). In other words, the holding pipette should readily be able to pick up and hold a blastocyst, by suction, without any extensive movements in the vertical plane, and the injection pipette should be able to draw up individual ES cells easily. A common problem at this stage is that the side of the pipettes are in contact with the edge of the injection chamber. This prevents the tips of the pipette from being aligned so that they are parallel and lying above the bottom of the chamber. This can be remedied by increasing the angle at which pipettes are inserted into the injection chamber.

Blastocyst Injection

1. Individual ES cells are selected and drawn up by slowly increasing the negative pressure in the injection pipette. It is important that cells appear viable and healthy. Under phase-contrast optics, the cells are round and refractile, with a yellow color. Any cell that appears brown or gray is dying or dead. Up to 100 cells can be picked up using the injection pipettes described above, with up to 10 embryos being serially injected.

2. Having picked up the ES cells, a blastocyst is selected for injection. The blastocyst is maneuvered into the correct orientation by gentle prodding, using the holding and injection pipettes. The blastocyst is immobilized at the tip of the holding pipette by suction. It is important that the blastocyst be held either by the side on which the inner cell mass is lying or at

the side of the trophectoderm, such that the inner cell mass (ICM) is facing the operator [Fig. 8 (1)]. Once the blastocyst is fixed in position, the holding pipette can be lowered until the blastocyst is in contact with the bottom of the injection chamber, which further restricts its movement.

3. To facilitate injection, the wall of the trophectoderm should be perpendicular to the point of the injection pipette, that is, at the equatorial plane [Fig. 8 (2)]. To locate this point, the equatorial plane of the blastocyst is determined by focusing. The injection pipette is positioned such that it just touches the wall of the trophectoderm. It is at this point that it is either above or below the plane of focus and should be raised or lowered, accordingly, until it is in focus. It is best to select an area on the blastocyst at a boundary between trophectoderm cells, since it is easier to penetrate into the blastocoel cavity by passing the injection pipette between two cells rather than through a cell. This will also prevent the collapse of the blastocyst prior to injection.

4. Once the optimal point of entry has been determined, the injection pipette should be pushed with one, smooth movement into the blastocoel cavity [Fig. 8 (3)]. If done too fast, there is a risk of hitting the tip of the injection pipette against the holding pipette and blunting the injection tip. If done too slowly, the blastocoel may collapse and prevent the needle from penetrating the trophectoderm.

5. As soon as the opening of the injection pipette is clearly in the blastocoel cavity, the ES cells are slowly expelled by applying positive pressure [Fig. 8 (4)]. Between 10 and 20 ES cells should be released into the blastocoel cavity. Once a sufficient number have been injected, a slight amount of negative pressure should be applied to draw up any medium injected along with the cells. This will also reduce the pressure in the blastocoel, and the blastocoel will start to collapse. When this happens, the injection pipette should be slowly withdrawn. Some cells may also escape from the hole left by the injection pipette in the trophectoderm; however, the majority should remain in the blastocoel cavity [Fig. 8 (5)]. The holding pipette is then m⋯ed toward the operator and the injected blastocyst released [Fig. 8 (6)]. A new blastocyst is then picked up and the process repeated.

It is important to ensure that the tip of the injection pipette remains sharp, that it not be blunted by touching the holding pipette. The tip must also be free of all accumulated debris (if this occurs, it can be remedied by pulling the tip out through the medium–paraffin oil interface of the drop). With practice, 30–40 blastocysts can be injected per hour. If cultured at 37° for 1.5–2 hr, they will start to reexpand their blastocoels, and the injected ES cells should become visible in the blastocoel cavity, many

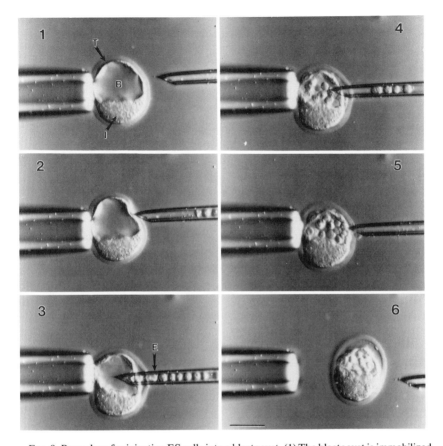

FIG. 8. Procedure for injecting ES cells into a blastocyst. (1) The blastocyst is immobilized on the tip of the holding pipette at the meural trophectoderm (T). The ICM (I) is facing the bottom. Injections can also be made by holding the blastocyst at the ICM. The blastocoel cavity (B) is expanded. (2) The injection pipette is brought into focus by adjusting its vertical position to a point at which it will preferably be inserted between two trophectoderm cells, in the mid or equatorial plane of the blastocyst. If the injection pipette is above or below this point, it may be more difficult to penetrate the trophectoderm. (3) Once a point has been selected, the injection pipette is pushed into the blastocoel cavity. The force used to push the injection pipette should not be so strong that the tip hits the holding pipette nor too gentle so that it results in the trophoblast being only partly penetrated and starting to collapse. (4) ES cells already loaded in the injection pipette are then slowly expelled into the blastocoel cavity, with an average of between 10 and 15 being injected into each blastocyst. (5) Once a sufficient number of ES cells have been injected, a slight negative pressure is applied to the injection pipette, resulting in some of the medium injected with the cells being drawn into the pipette. This helps to prevent the injected ES cells from escaping out through the hole in the trophoblast made by the injection pipette. A few cells may do so, but this is not important. The blastocoel should also start to collapse, trapping the cells against the ICM. (6) The injection pipette is withdrawn clear of the blastocyst. The blastocyst, with the blastocoel starting to collapse, is released from the holding pipette and is moved to a region of the microdrop where it will not interfere with subsequent injections.

of them adhering to the ICM. The injected blastocysts can then be transferred to the uteri of pseudopregnant recipients.

Alternative Routes to Producing Embryonic Stem Chimeras

Chimeras can also be produced by injecting 8-cell stage morulae with ES cells, in which the zona pellucida remains intact.[17,18] Alternatively, a small group of ES cells can be aggregated with 8-cell stage embryos in which the zona is removed prior to aggregation.[19–21] The advantage of the latter technique is that it is relatively simple and requires no sophisticated injection apparatus.

The disadvantage of the aggregation technique is that derivation of viable chimeras requires the use of two embryos per chimera. To circumvent the requirement for additional embryos, attempts have been made to inject cells into a single 8-cell stage embryo. This is relatively easier than blastocyst injections, since no penetration of the trophectoderm is required and fewer (3–6) ES cells are injected. Any increase in the number of ES cells will result in abnormal development. The cells are deposited close to, or between, blastomeres, where they will aggregate with the blastomeres as the embryos compact and end up being located in the ICM of the chimeric blastocyst.

The results from 8-cell (morula) injection experiments are preliminary because fewer embryos have developed to term (compared to an average of 50% with blastocyst injections). However, initial results have indicated that a high percentage of those that implant, following injection, are extreme chimeras, that is, with most of the embryo being derived from the ES cells[17,18] (see Fig. 15).

Procedure for Injection of 8-Cell Embryos

The basic procedure in terms of the microscope, pipettes, and injection chamber is identical to that for blastocyst injection, except that 8-cell stage embryos are recovered from the oviducts of day 3 pregnant females.

1. In the morning of the third day of pregnancy (day of plug = day 1 pregnancy), pregnant females, from either maternal or superovulated

[17] C. L. Stewart, unpublished observation, 1990.
[18] Y. Lallemand and P. Brulet, *Development (Cambridge, UK)* **110**, 1241 (1990).
[19] C. L. Stewart, *J. Embryol. Exp. Morphol.* **67**, 167 (1982).
[20] E. F. Wagner, G. Keller, E. Gilboa, U. Rüther, and C. L. Stewart, *Cold Spring Harbor 50th Anniversary Symp.* 691 (1985).
[21] J. T. Fuji and G. R. Martin, *J. Embryol. Exp. Morphol.* **54**, 79 (1983).

matings, are sacrificed by cervical dislocation. The oviducts are exposed and excised by cutting between the ovary and the uterine–tubal junction. In dishes of M2 medium (see [9] and [10] in this volume), the oviducts can either be squeezed between forceps or flushed with medium using a syringe and 30-gauge needle to expel the embryos. The embryos are harvested and washed through two changes of M2, then cultured in microdrops of CZB plus glucose medium under paraffin oil in a 37° incubator gassed with 5% CO_2 in air until they are needed. Only embryos either undergoing cleavage from the 4- to 8-cell stage or already at the uncompacted 8-cell stage should be collected.

2. The 8-cell stage embryos and ES cells are both placed in a microdrop of medium under paraffin oil in an injection chamber, as described for blastocyst injections. It is not necessary to precool the chamber or to use a cooling stage for injection. However, it is essential to include cytochalasin D (Sigma, St. Louis, MO) at a final concentration of 1 μg/ml in the injection medium. The blastomeres are very susceptible to lysis if touched by the tip of an injection pipette, which is almost an inevitable occurrence with injection of the morula. Inclusion of cytochalasin D in the medium will depolymerize (in a reversible manner) the cytoskeleton of the blastomere. The cell membranes thus are more distortable, making the blastomeres less susceptible to lysis.

3. A single morula is drawn onto the end of the holding pipette. The injection pipette containing ES cells is pushed through the zona pellucida, and 3–6 ES cells are deposited between, or close to, the blastomeres. The injection pipette is withdrawn, and the injected cells are moved to a remote part of the microdrop prior to injection of the next morula (Fig. 9).

4. After 20–30 morulae have been injected, they are removed from the injection chamber, washed 3 times in CZB plus glucose to remove residual cytochalasin D, and incubated overnight in a microdrop of medium made up of a 50:50 mixture of CZB plus glucose and DMEM plus 10% FCS. The next day the embryos should have developed to the early blastocyst stage, with formation of a blastocoel cavity. They are then transferred, surgically, to the uteri of day 3 pseudopregnant recipients for further development.

Aggregation of 8-Cell Stage Embryos with Embryonic Stem Cells

All that is necessary for the aggregation procedure is a good stereo dissection microscope with magnification to 40× and a mouth-controlled micropipette. This procedure has also been modified to produce embryos/mice that are entirely derived from the ES cells. This involves the aggrega-

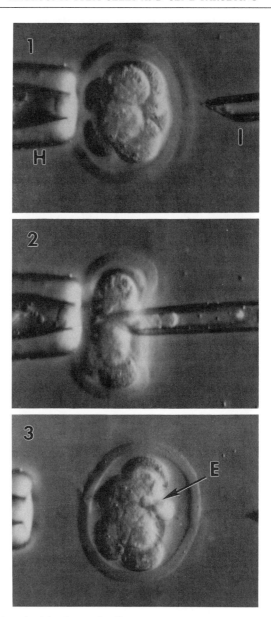

FIG. 9. Procedure for injecting an 8-cell stage morula. (1) An embryo is drawn onto the tip of the holding pipette (H). The injection pipette (I) containing ES cells is positioned close to the embryo. (2) The injection pipette is pushed through the zona pellucida, and between 3 and 8 ES cells are injected between two of the blastomeres. (3) The injection pipette is withdrawn, leaving the ES cells (E) between the blastomeres. The injected embryos are

tion of ES cells with two tetraploid 4-cell stage embryos (Fig. 10).[22] Tetraploid embryos are routinely produced by electrofusion of diploid blastomeres at the 2-cell stage (see [55] in this volume). Aggregating the diploid ES cells with tetraploid blastomeres results in the ES cells forming most of the ICM, whereas derivatives of the tetraploid embryos tend to form the extraembryonic membranes such as the trophectoderm and yolk sac endoderm. Thus, at birth, the embryo derived from the ICM will be largely or entirely derived from the ES cells. The extraembryonic membranes derived from the tetraploid embryos, in the form of the placenta and yolk sac, are lost at birth.[22]

Preparation of 8-Cell Stage Embryos for Aggregation

1. On the morning of the third day of pregnancy, the oviducts are removed from the females. They are placed in dishes of prewarmed M2 medium, where the 8-cell stage embryos are isolated either by flushing the oviduct using a syringe and 28-gauge needle or by squeezing them out using fine forceps. The embryos are washed twice in M2 to remove any cellular debris, blood cells, etc., and are cultured in drops of CZB plus glucose medium under paraffin oil.

2. To aggregate ES cells with the embryos, it is necessary to remove the zona pellucida. This is done by incubating the embryos for 20–40 sec in dishes of prewarmed (37°) acidified Tyrode's solution.[23] In batches of 10, the 8-cell stage embryos should be introduced into a 35-mm dish containing acidified Tyrode's solution. The low pH of the Tyrode's solution results in the zona pellucida dissolving in the saline solution. The acidified Tyrode's solution should be between pH 2 and 3, if the embryos are to be completely freed of their zonae. As soon as the zona has disappeared, the embryos are removed from the Tyrode's solution and washed 3 times in M2 medium.

3. In a 60-mm bacteriological grade petri dish, set up three 20-μl drops of medium containing a 50 : 50 mixture of DMEM plus 10% FCS and CZB plus glucose. In addition, set up 20 1-μl drops of the same medium. Cover with light paraffin oil (Fig. 11). The three 20-μl drops will hold the ES

---

[22] A. Nagy, E. Gocza, E. Merentes Diaz, V. R. Prideaux, E. Ivanyi, M. Markkula, and J. Rossant, *Development* (Cambridge, UK) **110**, 815 (1990).
[23] G. M. Nicolson, R. Yamagimachi, and H. Yamagimachi, *J. Cell Biol.* **66**, 263 (1975).

---

moved to an area of the injection microdrop where they will not be confused with uninjected embryos. After washing in fresh medium to remove excess cytochalasin D, the embryos are cultured overnight, at which time the ES cells will aggregate with the blastomeres, subsequently forming a blastocyst.

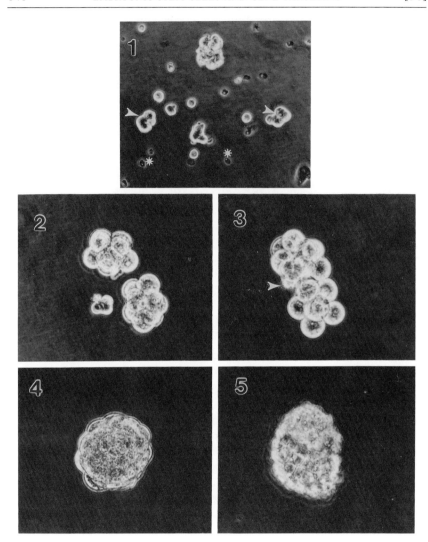

FIG. 10. Procedure for aggregating ES cells with two 8-cell stage embryos. (1) Colonies of ES cells are disaggregated into clumps of cells using 0.5 mM EGTA in PBS. Clumps containing 4–8 ES cells are removed for aggregation (arrows). These are ideal for aggregation since larger clumps contain too many cells and will not develop properly with the embryos. Dead/dying ES cells are indicated by white stars. (2) Two uncompacted 8-cell stage embryos are in a microdrop of aggregation medium with a single clump of ES cells. (3) The two embryos and ES cells are pushed/blown together with a mouth-controlled pipette so that they all adhere to each other. The ES cell clump is indicated by an arrow. (4) Following overnight culture, the cells and embryos have aggregated to form a single embryo. (5) The single embryo forms a blastocyst (this one started to collapse while being photographed). The ES cells are located in the ICM.

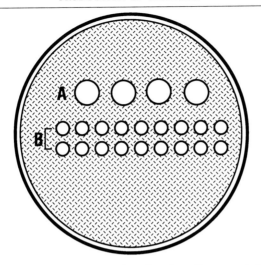

FIG. 11. Arrangement of microdrops of medium in a 60-mm petri dish. Larger drops contain the disaggregated ES cell clumps (A). The smaller drops (1–2 μl) contain two 8-cell stage embryos each (B), and in these drops the aggregation is performed. The whole dish is covered with light paraffin oil.

clumps (see below) that will be aggregated with the embryos. Into each 1-μl drop of medium, transfer two 8-cell stage embryos. The benefit of the small drops is that they not only provide sufficient nutrients for overnight culture, but also physically confine the embryos. When 20 pairs have been set up, the dish is returned to the incubator.

Preparation of Embryonic Stem Cells for Aggregation

1. The ES cells are prepared as small aggregates of between 5 and 10 cells each rather than single cells (which would be difficult to manipulate).

2. A 35- or 60-mm dish of ES cells, in which the cells are growing (in the log phase) as colonies on feeders, is washed twice in Ca^{2+}/Mg^{2+}-free PBS. The cells are then covered in Ca^{2+}/Mg^{2+}-free PBS containing 0.5 mM EGTA and left for 5 min. This causes the cells in the colonies to loosen their attachment to each other. The loosened colonies of ES cells are drawn up using a mouth-controlled pipette having an internal opening diameter of about 50–75 μm with the edges of the tip smoothed by flame polishing. The colonies are then transferred to 20-μl microdrops of 50 : 50 DMEM plus 10% FCS and CZB medium. By gently blowing the colonies back and forth between the pipette and microdrops, the colonies will fall apart into clumps of ES cells [Fig. 10 (1)]. The clumps are allowed to settle onto the surface of the dish. Individual clumps of 5–10 cells are

selected and then introduced into the 1-μl drops containing the two 8-cell stage embryos [Fig. 10 (2)].

3. The aggregation procedure consists of using a mouth-controlled pipette to push the clump of ES cells into a crevice between two blastomeres. It is important to ensure that the embryos have not started to compact because aggregation with uncompacted embryos is easier and usually results in the clump of cells adhering to the blastomeres. The second embryo is then maneuvered by the pushing/gentle blowing of medium into a position so that it sandwiches the ES clump that is attached to the first embryo. Both embryos must be in contact with each other [Fig. 10 (3)]. Adherence and subsequent aggregation of the ES cells to the embryos are temperature-dependent, and the whole process is more difficult if the dish and embryos are allowed to cool substantially. When all the embryos have been aggregated, the dish is returned to the incubator. Fifteen to twenty minutes later, each aggregate should be checked to ensure that the embryos are still attached to each other and to a clump of ES cells. If a clump of ES cells is not adhering to the embryo (this can be determined by gently blowing the whole aggregate around the microdrop to ensure that all components are sticking to each other), replace the cells with another group. The aggregated ES cells/embryos are then cultured overnight. The following morning, the majority of aggregates [Fig. 10 (4, 5)] should have formed blastocysts. These are then surgically transferred to the uteri of pseudopregnant recipients.

Preparation of Pseudopregnant Recipients

Vasectomy of Male Mice

For manipulated embryos to develop to term, they have to be returned to the uterus for proper implantation and development. Female mice must be mated with males for them to initiate the physiological changes associated with pregnancy. If females are mated to normal males, they would contain viable embryos resulting from that mating. The presence of these embryos would compete with any experimentally manipulated embryos transferred to the uteri of the pregnant female. To avoid this but to still induce pregnancy, female recipients are mated with vasectomized males, which can mate with females but cannot fertilize eggs.

Vasectomizing Male Mice

1. Anesthetize a 6-week-old male by a single injection of Avertin. To make Avertin add 0.5 g of 2,2,2-tribromoethanol to 0.63 ml of tert-amyl alcohol prepared in a 1-ml Eppendorf tube. Vortex to dissolve the tribro-

moethanol. Add 0.5 ml of this solution to 19.5 ml of prewarmed 0.9% saline solution, in which the anesthetic will dissolve after shaking, and allow to cool. The dose injected is 0.012 ml/g body weight.

2. The anesthetized male is laid on its back, the belly is swabbed with 70% ethanol solution, and a horizontal incision using scissors is made through the skin at a position shown in Fig. 12 (1). All surgical procedures should be performed under a stereo dissection microscope with an incident light source.

3. Expose the underlying peritoneum and make a horizontal incision. This should expose two fat pads.

4. Using a pair of blunt forceps, grasp one of the fat pads and pull it out of the body cavity. This results in the testis also being pulled out with it. Beneath the fat pad and connected to the testis is a muscular tube, the vas deferens. This can be recognized by the single blood vessel that runs along its side [Fig. 12 (3)].

5. Using a pair of fine forceps, a loop is made in the vas deferens. With a pair of forceps, the tips having been preheated, the loop of vas deferens is cauterized and severed. This results in a section of the tissue being removed, with the remaining ends being sealed [Fig. 12 (4–6)].

6. The testis/fat pad is then gently moved back into the peritoneal cavity, and the process is repeated for the other testis.

7. Once the procedure is completed, the peritoneal incision is ligated together using a surgical needle and thread. The skin cut is then clamped together using wound clips.

8. The male is allowed to recover. The animal should be set up and test-mated with females to ensure sterility. The wound clips should be removed 10–14 days after the operation.

Transfer of Manipulated Embryos to Pseudopregnant Recipients

For the injected/aggregated embryos to develop to term, they have to be transferred to the uteri of pseudopregnant recipients (i.e., females mated with vasectomized males). Blastocysts injected with ES cells are transferred to recipients on the same day as they were injected. For Morula injection/aggregation, transfer occurs the following day, that is, once they have developed to the blastocyst stage, which follows overnight culture *in vitro*.

It is best to transfer the blastocysts to pseudopregnant recipients whose stage of pregnancy is 1 day behind that of the blastocyst. In normal pregnancy, blastocysts are found in the uteri of day 4 pregnant mice, so the manipulated embryos are transferred to the uteri of day 3 pseudopregnant recipients. This apparently gives blastocysts time to recover *in vivo* from

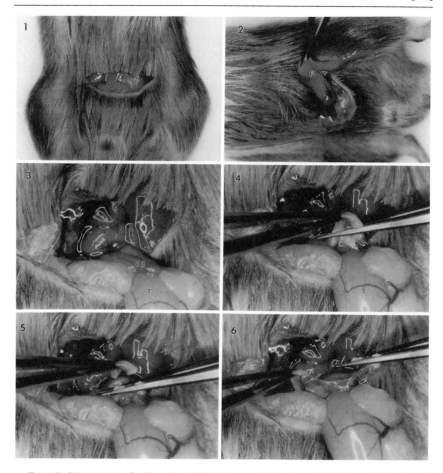

Fig. 12. Vasectomy of male mice. (1, 2) The anesthetized male is laid on its back, and a horizontal incision is made with scissors through the skin at the inguinal region. A second incision is made through the peritoneum, with care being taken to ensure that no blood vessels are cut. Directly beneath the peritoneum lies a fat pad (F). Grasp this with a pair of blunt forceps (2) and gently pull it out through the incision (3). This will result in one of the testes (T) also being pulled out of the peritoneum. The vas deferens (VD) lies beneath the fat pad, connected to the testis (T), and has a distinctive blood vessel running down its side. (4) A loop is made in the vas deferens using a pair of fine forceps. (5) Using another pair of preheated forceps (tips held over a spirit flame), the loop is cauterized and removed from the rest of the vas deferens. (6) The two ends of the epididymis are checked to ensure that they have been sealed by cauterization and are separated. The testis and fat pad are gently pushed back into the peritoneum, and the process is repeated on the other epididymis (vas deferens). Once surgery is completed, the peritoneum wall is sutured together using a surgical thread and needle. The skin is then clamped together using wound clips.

the *in vitro* manipulations. Transfer to day 3 recipients also results in a higher incidence of implantation than when blastocysts are transferred to synchronized recipients (i.e., day 4 pregnant females).

The best recipients are F_1 hybrids, with (C57BL/6 × CBA) F_1 hybrids being the standard choice. These females make good mothers because they do not neglect or cannibalize their progeny. No more than 6–7 embryos should be transferred to each uterine horn. If fewer are available, then transferring to only 1 horn is satisfactory.

Procedure for Transferring Embryos to Pseudopregnant Recipients

1. Female mice that were mated 3 days previously with vasectomized males are anesthetized by an injection of Avertin (females should be between 6 and 12 weeks in age).

2. After weighing, the female is injected intraperitoneally with the appropriate volume of Avertine (see section on vasectomizing male mice). The animal should be fully anesthetized within 2–3 min, which is determined by gently squeezing one of the rear paws. If the animal responds by rapidly shaking back and forth, the animal is not anesthetized and needs to be left longer for the anesthetic to take its full effect or be given an additional injection of about one-third the original dose.

3. Once fully anesthetized, the female is laid on the ventral side with the head facing to the left [Fig. 13 (1)]. Swab the back with a 70% ethanol solution, and, using fine forceps, part a line of hairs along the spine [Fig. 13 (2)] about 0.5–1 cm long, just below the bottom of the rib cage [Fig. 13 (1)]. With a pair of scissors, an incision is made through the exposed skin. The incision is opened, and some of the transparent mesentery attaching the skin to the peritoneum lying immediately beneath the skin is cut or pulled away. The skin incision is moved over the peritoneum to the point where the right ovary is seen to be lying just beneath the peritoneum [Fig. 13 (3)] [the position of the ovaries is shown by stars in Fig. 13 (1)]. The ovary is recognized by its bright cherry red color (owing to the numerous copora lutea). An incision of no more than 0.5 cm is made through the peritoneum, with care being taken to avoid cutting any of the blood vessels visible in the peritoneum. The ovary is attached to a fat pad and to the oviduct and uterus. By grasping the fat pad, the ovary, oviduct, and uterus are pulled out of the peritoneal cavity with a pair of blunt forceps, exposing the ovarian end of the uterus. To keep the uterus from sliding back into the peritoneal cavity, the fat pad is clamped with a small pair of aneurism clips, which is of sufficient weight to prevent the organ from sliding back. It is important that the uterus not be touched during the surgical procedure, since trauma may result in failure of the embryos to implant.

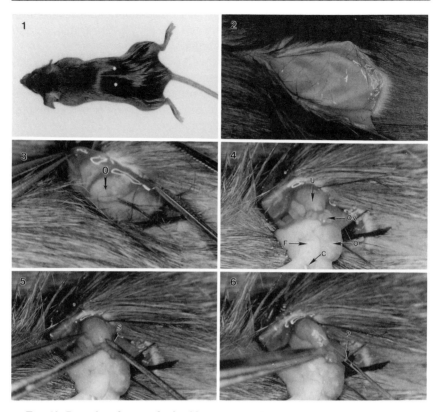

FIG. 13. Procedure for transferring blastocysts to the uterus of a day 3 pseudopregnant recipient. (1) The anesthetized recipient is shown lying on its front. Positions of the ovaries are shown by white stars. A parting of the hairs on the back (after swabbing with 70% ethanol) is also present. (2) An incision (using scissors) is made through the skin along the hair parting at the point midway between where the left and right ovaries lie. (3) The skin incision is moved over the approximate region where an ovary lies, and a second incision is made through the peritoneal wall. The ovary (O) is seen lying beneath the incision. (4) A fat pad (F) attached to the ovary (O) is grasped using blunt forceps and pulled out through the incision. The fat pad is clamped with a pair of aneurism clips (C) to prevent the ovary (O), oviduct (Ov), and uterus (U) from sliding back into the peritoneal cavity. (5) A sterile 25-gauge needle (S) is used to pierce a hole through the top of the uterus. (6) A transfer pipette containing the embryos is inserted through the hole into the uterine lumen. The embryos are expelled by blowing through the pipette, with the air bubbles (P) being used to follow their position in the pipette.

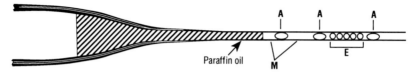

FIG. 14. Transfer pipette pulled such that the tip has an inner diameter of approximately 100 μm. The capillary is first filled with paraffin oil and then culture medium (M). Air bubbles (A) act as markers indicating where the embryos (E) are located in the pipette.

4. With the ovarian end of the uterus lying on the peritoneum wall, a hole is made in the uterus just above the uterine–oviduct junction, using a new (sterile) 25-gauge syringe needle. It is only necessary to penetrate the wall of the uterus using the tip of these extremely sharp needle, which should be inserted no more than 1–2 mm [Fig. 13 (5)].

5. The blastocysts to be transferred have, at this point, already been picked up and are lying in the transfer pipette (Fig. 14). These pipettes can be readily pulled on a gas or alcohol burner flame. The internal diameter should be about 100 μm, and the tip should be no longer than 2–4 cm. Light paraffin oil is drawn into the barrel of the pipette using mouth control (see Fig. 1 in [49], this volume). The viscosity of the paraffin oil gives a much finer level of control in pipetting medium, which is required for picking up and transferring the blastocysts into the uterine lumen. The embryos to be used for transfer are sitting in a 35-mm dish of prewarmed M2 medium with no paraffin oil covering the medium. The transfer pipette, with the tip filled with paraffin oil, is introduced into the M2 medium. A small amount of medium is drawn up into the tip, followed by a small air bubble. More medium is taken up at about 0.5–1 cm, and then a second small air bubble. This is followed by drawing up 6–7 blastocysts in as small a volume of M2 medium as possible, followed by a third air bubble. This arrangement can be seen in Fig. 14. The air bubbles act as markers for determining where the embryos are lying, since they are more visible in the pipette than the embryos. The two lowermost bubbles, which sandwich the embryos, indicate where the embryos are lying in the pipette. The first, uppermost bubble acts as a marker to indicate when all the embryos have been transferred into the uterus.

6. Using a pair of fine forceps, grasp the oviduct to steady the uterus. The tip of the transfer pipette is inserted into the hole in the uterine wall and is pushed about 3–5 mm into the uterine lumen [Fig. 13 (6)]. This should be done gently; any resistance indicates that the tip is in contact with the uterine endometrium. Once the transfer pipette has been inserted sufficiently deep into the uterus, it is withdrawn about 1–2 mm to ensure that the opening at the tip (still within the lumen) is not in contact with

FIG. 15. Three chimeras produced by injecting C57BL6 ES cells into BALB/c embryos. The two individuals at left and center were derived by blastocyst injection, and they have an agouti color on a white background. The individual at right was derived by injection of an 8-cell stage BALB/c embryo with a C57BL6 ES line. The individual is completely black, indicating that it was entirely derived from the ES line.

the endometrium, which would block the exit of embryos into the uterine lumen. The embryos are expelled into the lumen, with the transfer being followed by watching the air bubbles. When the last air bubble (i.e., the one nearest the paraffin oil) is seen to enter the uterus, the pipette is withdrawn. The tip is immediately placed into the dish containing the remaining blastocysts, and medium is gently drawn back and forth through the tip. This cleans any blood that may be adhering to the tip which, if clotted, will block the tip. This washing also ensures that all the embryos were transferred to the uterus. The next set of blastocysts can then be picked up in the transfer pipette using the same arrangement of medium and air bubbles.

7. The uterus into which the embryos were transferred is gently pushed back into the peritoneal cavity after the aneurism clip is removed from the fat pad. The wall is pinched together and can be sutured, although this is not usually necessary. The process is repeated for the remaining uterine horn. When the operation is completed, the edges of the skin where the incisions were made are stapled together by two or three 0.9-

mm wound clips (Clay Adams, Becton-Dickinson and Co., Parsippany, NJ). The recipients are placed on a 37° slide warmer to keep the mice warm until they regain consciousness. They are then returned to their cages. The manipulated embryos should be born within 16–18 days of the day of transfer.

If 129/Sv agouti ES cells have been used for injection into C57BL/6 embryos, the chimeras should be apparent by day 5–7 after birth when the coat colors become visible. Chimeras will be recognized by the presence of agouti stripes/patches intermingled with areas of black fur. Extensive chimeras will be predominantly agouti. If C57BL/6 ES lines have been injected into BALB/c embryos, then the chimeras will have an agouti or black color on a white (albino) background (Fig. 15). The chimeric distribution of the ES cells to the internal tissues is assessed by electrophoretic analysis of the glucose-6-phosphate isomerase (GPI) enzyme.

To determine whether a chimera has a contribution to its germ line from the ES cells, the 129/Sv↔C57BL/6 chimeras should be mated with C57BL/6 mice. Germ line chimeras are detected by the presence of agouti offspring in the resulting litters, since agouti is dominant to black. With C57BL/6↔BALB/c chimeras it will be necessary to cross with BALB/c mice. All ES-derived offspring will also be agouti.

Acknowledgments

We thank Jeff Mann for fruitful discussions and suggestions, Alisoun Carey for critical reading of the manuscript, and Sharon Perry for efficient typing.

[51] Gene Targeting in Embryonic Stem Cells

By RAMIRO RAMÍREZ-SOLIS, ANN C. DAVIS, and ALLAN BRADLEY

I. Introduction

Analysis of classical mouse mutants has been a useful tool for discovering the functions of gene products. Cloning of some of the involved loci has provided dramatic insights into mammalian development.[1] However, there are only a limited number of mutant mice where the gene involved has been identified and cloned. In contrast, there are a large number of

[1] A. Gossler and R. Balling, *Eur. J. Biochem.* **204**, 5 (1992).

Copyright © 1993 by Academic Press, Inc.
All rights of reproduction in any form reserved.

cloned genes whose map positions do not correlate with any of the classical mouse mutants. It has become possible to introduce mutations in this kind of loci and then generate mutant mice for the study of gene function *in vivo*. Two key technologies have facilitated this experimental system: the isolation of embryonic stem cells as permanent *in vitro* cell lines that can repopulate the blastocyst stage embryo[2,3] and the discovery that mammalian cells could recombine introduced vector DNA with a homologous chromosomal target, a process known as gene targeting.[4-7]. Gene targeting in embryonic stem cells allows specific mutations to be introduced into the mouse germ line in virtually any gene so that the gene function can be studied by mutational analysis *in vivo*. This technology has produced a number of valuable mouse mutants that have proved to be useful in many different genetic systems *in vivo*.[8-16]

Embryonic stem (ES) cells are derived from 3.5 days postcoitum (dpc) mouse embryos and arise from the inner cell mass (ICM) of the blastocyst.[2,3] The ES cells can grow *in vitro* and retain the potential to contribute extensively to all of the tissues of an animal when injected back into a host blastocyst, which is allowed to develop in a foster mother. The animal formed from the injected blastocyst is called a chimera since it is formed by cells from two different individuals.[17] The injected cells can form all or part of the functional germ cells of the chimera and in doing so establish

---

[2] M. J. Evans and M. H. Kaufman, *Nature (London)* **292**, 154 (1981).

[3] G. R. Martin, *Proc. Natl. Acad. Sci. U.S.A.* **78**, 7634 (1981).

[4] F.-L. Lin, K. Sperle, and N. Steernberg, *Proc. Natl. Acad. Sci. U.S.A.* **82**, 1391 (1985).

[5] O. Smithies, R. G. Gregg, S. S. Boggs, M. A. Koralewski, R. S. Kucherlapati, *Nature (London)* **317**, 230 (1985).

[6] T. Doetschman, R. G. Gregg, N. Maeda, M. L. Hooper, D. W. Melton, S. Thompson, and O. Smithies, *Nature (London)* **330**, 576 (1987).

[7] K. R. Thomas and M. R. Capecchi, *Cell (Cambridge, Mass.)* **51**, 503 (1987).

[8] H. Schorle, T. Holtschke, T. Hünig, A. Schimpl, and I. Horak, *Nature (London)* **352**, 621 (1991).

[9] O. Chisaka and M. R. Capecchi, *Nature (London)* **350**, 473 (1991).

[10] D. Kitamura, J. Roes, R. Kühn, and K. Rajewsky, *Nature (London)* **350**, 423 (1991).

[11] T. M. DeChiara, A. Efstradiatis, and E. J. Robertson, *Nature (London)* **345**, 78 (1990).

[12] T. Lufkin, A. Dierich, M. LeMeur, M. Mark, and P. Chambon, *Cell (Cambridge, Mass.)* **66**, 1105 (1991).

[13] A. P. McMahon and A. Bradley, *Cell (Cambridge, Mass.)* **62**, 1073 (1990).

[14] V. L. J. Tybulewicz, C. E. Crawford, P. K. Jackson, R. T. Bronson, and R. C. Mulligan, *Cell (Cambridge, Mass.)* **65**, 1153 (1991).

[15] P. L. Schwartzberg, A. M. Stall, J. D. Hardin, K. S. Bowdish, T. Humaran, S. Boast, M. L. Harbison, E. J. Robertson, and S. P. Goff, *Cell (Cambridge, Mass.)* **65**, 1165 (1991).

[16] P. Soriano, C. Montgomery, R. Geske, and A. Bradley, *Cell (Cambridge, Mass.)* **64**, 693 (1991).

[17] R. L. Gardner, *Nature (London)* **220**, 596 (1968).

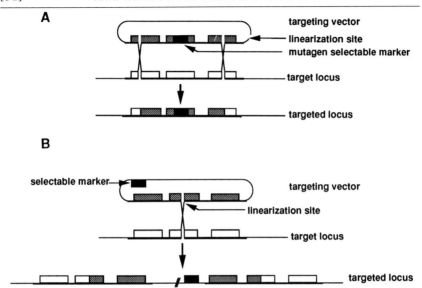

FIG. 1. Replacement and insertion type vectors. (A) Replacement vectors target the locus by double reciprocal recombination or by gene conversion of genomic to targeting vector sequences, and they insert only the information contained inside the homologous region. (B) Insertion vectors target the locus by single reciprocal recombination and insert the whole vector sequence.

themselves in the mouse germ line.[18] This can occur even after the ES cells have been modified by the introduction of exogenous DNA.[19-21]

Gene targeting by homologous recombination in mammalian cells was first reported in mouse L cells[4] and in human or mouse–human hybrid cells.[5] The first examples of gene targeting in ES cells were at the hypoxanthine–guanine phosphoribosyltransferase gene (*HPRT*), a selectable locus located on the X chromosome.[6,7] Both kinds of vectors, insertion and replacement, were used to target the *HPRT* locus (Fig. 1). The first report of germ line transmission of a targeted correction of a mutation in ES cells was also achieved in the *HPRT* locus.[22]

[18] A. Bradley, M. Evans, M. H. Kaufman, and E. J. Robertson, *Nature (London)* **309**, 255 (1984).
[19] A. Gossler, T. Doetschman, R. Korn, E. Serfling, and R. Kemler, *Proc. Natl. Acad. Sci. U.S.A.* **83**, 9065 (1986).
[20] E. Robertson, A. Bradley, M. Kuehn, and M. Evans, *Nature (London)* **323**, 445 (1986).
[21] M. R. Khuen, A. Bradley, and E. Robertson, *Nature (London)* **326**, 295 (1987).
[22] S. Thompson, A. R. Clarke, A. M. Pow, M. L. Hooper, and D. W. Melton, *Cell (Cambridge, Mass.)* **56**, 313 (1989).

The invention of methods that lower the background of nontargeted events, namely, positive–negative selection,[23] promoterless resistance markers[24,25] (expressed only if integrated properly downstream of a promoter active in ES cells), and polyadenylation signal-less markers[26] (produce stable transcripts only if inserted upstream of a genomic polyadenylation signal), made targeted mutations at many nonselectable loci easier to detect. Equally important has been the development of methods that improve the screening of clones, namely, the polymerase chain reaction (PCR) on pools of clones.[26] Germ line transmission of a nonselectable gene disrupted by gene targeting was first reported at the c-abl[25] and at the β_2-microglobulin[27] loci in 1989.

The first kind of modifications achieved with gene targeting technology was the gross disruption of targeted loci. Usually, it was caused by the insertion of a piece of DNA that carried a selectable gene necessary to separate the cells that incorporated exogenous DNA from those that did not in the original transfection. A second generation of targeting strategies has been developed to introduce specific small mutations without the concomitant presence of a marker gene (*in vivo* site-directed mutagenesis). Subtle specific mutations of nonselectable genes have been reported at the *Hox-1.1* gene by direct DNA microinjection into ES cells[28] and at the *Hox-2.6* locus by the "hit and run" technique.[29] Introduction of small mutations on selectable genes has been reported for the RPII215 gene, which encodes the largest subunit of the RNA polymerase II,[30] and for the *HPRT* gene.[29,31,32]

In this chapter, we describe the basic protocols to generate targeted clones, with special emphasis on the important technical details to minimize the risk of destroying the capacity of the ES cells to form part of the mouse germ line. A general description of ES cell culture conditions is followed by methodology for gene targeting and blastocyst injection.

[23] S. L. Mansour, K. R. Thomas, and M. R. Capecchi, *Nature (London)* **336**, 348 (1988).

[24] J. Charron, B. A. Malynn, E. J. Robertson, S. P. Goff, and F. W. Alt, *Mol. Cell. Biol.* **10**, 1799 (1990).

[25] P. L. Schwartzberg, S. P. Goff, and E. J. Robertson, *Science* **246**, 799 (1989).

[26] A. L. Joyner, W. C. Skarnes, and J. Rossant, *Nature (London)* **338**, 153 (1989).

[27] M. Zijlstra, E. Li, F. Sajjadi, S. Subramani, and R. Jaenisch, *Nature (London)* **342**, 435 (1989).

[28] A. Zimmer and P. Gruss, *Nature (London)* **338**, 150 (1989).

[29] P. Hasty, R. Ramírez-Solis, R. Krumlauf, and A. Bradley, *Nature (London)* **350**, 243 (1991).

[30] C. M. Steeg, J. Ellis, and A. Bernstein, *Proc. Natl. Acad. Sci. U.S.A.* **87**, 4680 (1990).

[31] V. Valancius and O. Smithies, *Mol. Cell. Biol.* **11**, 1402 (1991).

[32] A. C. Davis, M. Wims, and A. Bradley, *Mol. Cell. Biol.* **12**, 2769 (1992).

Reagents and Materials

The following list includes only items not commonly found in a molecular biology or tissue culture laboratory.

Fetal calf serum (FCS) tested for toxicity and cloning efficiency on ES cells[33]

Fibroblast feeder SNL76/7 cell line[13] (resistant to G418)

Gelatin solution: 1% (w/v) tissue culture grade gelatin mixed in water and sterilized by autoclaving; the working solution is 0.1% and is made by diluting the 1% stock solution in sterile water. Store at room temperature

Dulbecco's modified Eagle's medium (DMEM): DMEM (GIBCO, Grand Island, NY, Cat. No. 430-2100) plus 2.2 g/liter of tissue culture grade sodium bicarbonate, adjusted to pH 7.4 with HCl and filter sterilized. Store at 4°

Glutamine–penicillin–streptomycin (GPS): For the 100× stock solution, 300 mg (1650 U/mg) of sodium penicillin G and 500 mg of streptomycin sulfate are added to 100 ml of 200 mM glutamine (GIBCO), and filter sterilized. Store at 4°

2-Mercaptoethanol 100× stock solution: 10 mM 2-mercaptoethanol in PBS, filter sterilized. Store at 4°

Phosphate-buffered saline (PBS): For 1 liter, mix 8 g NaCl, 0.2 g KCl, 0.2 g KH_2PO_4, and 1.072 g $Na_2HPO_4 \cdot 7H_2O$ in water and adjust to pH 7.2 using a saturated solution of $Na_2HPO_4 \cdot 7H_2O$. Filter sterilize and store at room temperature

M15 medium: DMEM, 15% FCS, 1× GPS, 1× 2-mercaptoethanol. Store at 4°

Trypsin solution: For 1 liter, mix 7 g NaCl, 1 g D-glucose, 0.18 g $Na_2HPO_4 \cdot 7H_2O$, 0.37 g KCl, 0.24 g KH_2PO_4, 0.4 g EDTA, 2.5 g trypsin (1 : 250, GIBCO), and 3 g Trizma base, dissolve in water, and adjust to pH 7.6 using HCl. Filter sterilize and store at $-20°$

Freezing medium (2×): 60% (v/v) DMEM, 20% (v/v) FCS, and 20% (v/v) dimethyl sulfoxide (DMSO). Store at 4°

Lysis buffer for 96-well DNA microextraction: 10 mM Tris, pH 7.5, 10 mM EDTA, 10 mM NaCl, 0.5% (w/v) sarcosyl. Add proteinase K to a final concentration of 1 mg/ml just prior to use

G418 (Geneticin, GIBCO) 100× stock solution: Dissolve 180 mg active ingredient in 10 ml PBS and filter sterilize. Store at 4°

---

[33] E. J. Robertson, *In* "Teratocarcinomas and Embryonic Stem Cells: A Practical Approach" (E. J. Robertson, ed.), p. 74. IRL Press, Oxford and Washington, D.C., 1987.

FIAU 1000× stock solution (200 μM): Add 388 mg of FIAU [1-(2'-deoxy-2'-fluoro-β-D-arabinofuranosyl)-5-iodouracil] to 9 ml PBS and add NaOH until the FIAU has dissolved. Make total volume to 10 ml. This solution is 100 mM and must be diluted 1:500 in alkaline PBS to give the 1000× stock solution. Store at −20°

Mitomycin C 50× stock solution: Dissolve 2 mg in 4 ml of PBS. Store at 4° for 1 week and then discard (Note: Toxic by inhalation, skin contact, or if swallowed)

Light paraffin oil: Sterilize by filtration

Avertin solution: To make stock solution, dissolve 1 g of 2,2,2-tribromoethanol in 0.5 ml of tertiary amyl alcohol. To prepare the working solution, dilute 1.2 ml of 2,2,2-tribromoethanol stock in 100 ml of boiling PBS. Mix well and store at room temperature

12-Channel multichannel pipettor: ICN (Irvine, CA)

Multichannel tissue culture aspirator system: Inotech (Lansing, MI)

Electroporation apparatus: Bio-Rad (Richmond, CA) Gene Pulser with a capacitance extender

Bio-Rad electroporation cuvettes

Sterile transfer pipettes

II. Gene Targeting

A. Culture of Embryonic Stem Cells

The purpose of using ES cells for gene targeting is to transfer the mutation generated in culture into the mouse germ line. For this reason, culture conditions that prevent the overgrowth of abnormal cells are critical.[34] ES cells should be grown on mitotically inactivated feeder cell layers (Section II,A, Step 1). In addition, the cells should be grown at high density and passaged frequently at 1 : 3 to 1 : 6; this usually means replacing the medium daily. ES cells should be fed 4 hr before passage. To passage, the cells should be washed twice with PBS and trypsinized for 10 min; there is no need to prewarm the trypsin solution. M15 medium is added, and the cell clumps are mechanically disrupted by vigorous pipetting. It is important to generate a single cell suspension before passage as clumps have a tendency to differentiate. The passage number of the cell line should be recorded to give an estimate of the time the cells have been in culture. If the cells are not to be used immediately, they should be frozen and then recovered when needed.

---

[34] A. Bradley, *Curr. Opin. Cell Biol.* **2,** 1013 (1990).

The cultured ES cell population includes totipotent cells, as well as cells with limited potential to contribute to all tissues of the mouse. Because targeted events are usually rare and single cell cloning is necessary, it is advisable to optimize targeting vectors and conditions such that several targeted clones can be recovered. Also, cloning involves culture at low cell concentrations and potentially for a prolonged period while screening for the desired clone. The following protocols have been designed to reduce the overall time in culture of targeted clones prior to blastocyst injection and, in particular, the length of time the cells are grown at low density. This is done by picking colonies directly into 96-well feeder plates and by freezing the colonies directly in these plates.

Preparation of Feeder Layers. ES cells are grown on mitotically inactive feeder layers of G418^r SNL76/7 fibroblasts. The factors secreted from these cells are unknown but appear to be important for maintaining totipotency *in vitro*. All feeder cells are grown on gelatin-coated plates to aid in adhesion.

1. Coat tissue culture plates with gelatin by covering the bottom of the plate with a 0.1% gelatin solution and incubating at room temperature for 2 hr. Aspirate the gelatin before plating the inactivated feeder cells.

2. Grow G418^r SNL76/7 cells to confluence on 15 cm gelatinized tissue culture plates in DMEM plus 7% FCS and 1X GPS. To inactivate the cells, mitomycin C stock solution (0.5 mg/ml) is added to the medium to give a final concentration of 10 μg/ml, and the plate is incubated at 37°, 5% (v/v) CO_2, for 2 hr.

3. Aspirate the mitomycin-containing medium and wash the plate twice with PBS.

4. Add 2 ml of trypsin solution and incubate at 37°, 5% CO_2, for 5 min.

5. Add 5 ml of medium and suspend the cells by vigorous pipetting. Transfer the cells to a 50-ml sterile centrifuge tube. Wash the plate with medium once again. Pool all the mitomycin-treated cells and centrifuge at 1000 rpm for 5 min at room temperature.

6. Aspirate the supernatant and resuspend the pellet in 5–10 ml of medium. Count the cells and add medium to give a concentration of 3.5 × 10^5 cells/ml.

7. Transfer aliquots of feeders onto gelatinized plates, 12 ml per 10-cm plate (4.2 × 10^6 cells/plate), 4 ml per 6-cm plate (1.4 × 10^6 cells/plate), etc. Leave plates in the incubator overnight before use to give cells time to attach to the plate. Feeder plates can be stored for 3–4 weeks in the incubator, but they should be checked under the microscope before use to confirm that the layer is intact.

B. Electroporation

The first step of any targeting experiment is the introduction of DNA into the recipient cells. For ES cells, DNA microinjection[28] and electroporation[8-16] have been shown to be useful to permit gene targeting. DNA microinjection is technically difficult and has the potential to cause gross chromosomal disruption,[35] which may lower the potential of the ES cells to populate the germ line of chimeras. Electroporation, on the other hand, has been used extensively to generate targeted clones that have gone through the germ line. The electroporation protocol used in our laboratory is basically similar to those used for other cell types, but some things are particularly important for the specific case of electroporation of ES cells. The cells should be growing actively at the time of the electroporation; this can be achieved by passaging the ES cells 1 day before the electroporation and adding fresh medium a few hours before harvesting the cells. The trypsin treatment should be long enough to allow mechanical disaggregation of the cell clumps to avoid differentiation. The electroporated cells should be plated on feeder cells with M15 medium within 5–10 min.

1. Prepare targeting vector DNA by the CsCl banding technique.

2. Cut 200 μg of targeting vector DNA with the appropriate restriction enzyme to linearize it. Assess the completion of the restriction digest by agarose gel electrophoresis.

3. Clean the DNA with phenol–chloroform, chloroform, and precipitate it with NaCl and ethanol. Resuspend the DNA in sterile $0.1 \times$ Tris–EDTA buffer (TE) and adjust the concentration to 1 mg/ml.

4. One day before the electroporation, passage the actively growing ES cells (~80% confluent) 1 : 2.

5. Feed the cells with fresh M15 medium 4 hr before harvesting them for the electroporation.

6. Wash the plates twice with PBS and detach the cells by treatment with trypsin solution for 10 min at 37° (1 ml trypsin solution for a 10-cm plate).

7. Stop the action of the trypsin solution by adding 1 volume of M15 medium and dissociate the cell clumps by moving the cell suspension up and down with the transfer pipette.

8. Centrifuge the cells at 1000 rpm for 5 min in a clinical centrifuge and discard the supernatant. Resuspend the cells in 10 ml of PBS and determine the total number of cells.

[35] L. Covarrubias, Y. Nishida, M. Terao, P. D'Eustachio, and B. Mintz, *Mol. Cell. Biol.* **7**, 2243 (1987).

9. Recentrifuge the cells, aspirate the supernatant, and resuspend the cells in PBS at a final density of 1.1×10^7 cells/ml.

10. Mix 25 μg of the linearized targeting vector with 0.9 ml of the cell suspension in an electroporation cuvette. Incubate for 5 min at room temperature.

11. Electroporate in the Bio-Rad Gene Pulser at 230 V, 500 μF. Incubate for 5 min at room temperature.

12. Plate the entire contents of the cuvette on a 10-cm tissue culture plate with feeder cells. The medium on the feeder plate should be changed to M15 prior to plating the cells.

13. Apply G418 selection 24 hr after the electroporation. FIAU selection can also be applied if a positive–negative selection protocol using the herpes simplex virus-1 thymidine kinase (HSV-1tk) gene is being followed.

14. Refeed the cells when the medium starts turning yellow, usually daily for the first 5 days.

15. Ten days after the electroporation, the colonies are ready to be picked.

C. Picking and Expansion of Colonies after Electroporation

After electroporation, the ES cell colonies take 8–12 days of growth to become visible to the naked eye and can be picked at this time (Fig. 2A). Care should be taken that only a single colony is seeded per well to avoid a further cloning step.

1. Wash the plate containing the colonies twice with PBS and add PBS to cover the plate.

2. Prepare a 96-well U-bottomed plate by adding 25 μl of trypsin solution per well.

3. Place the original 10-cm plate on an inverted microscope and pick individual colonies with a micropipettor and disposable sterile tips in a maximum volume of 10 μl. Each colony is transferred to the trypsin solution in a well of the plate prepared in Step 2.

4. After 96 colonies have been picked, place the 96-well plate in the 37°, 5% CO_2 incubator for 10 min.

5. During the incubation, take a previously prepared 96-well feeder plate (flat-bottomed wells), aspirate the medium, and add 150 μl of M15 per well. Use a multichannel pipettor (12 channels) for all following steps.

6. Retrieve the trypsinized colonies from the incubator and add 25 μl of M15 per well. Break up the clumps of cells by moving the cell suspension up and down with the multichannel pipettor about 5–10 times.

7. Transfer the entire contents of each well to a well in a 96-well plate prepared in Step 5. Change tips each time.

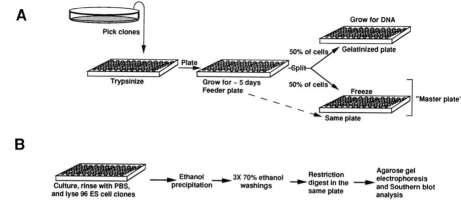

FIG. 2. DNA extraction protocol for fast and simple mass screening of ES cell clones. (A) The ES cell clones are seeded into a 96-well plate and grown for 5 days. Then, one-half of the cells are frozen in the original plate while the remainder are grown in a replica plate for Southern blot analysis. (B) For the DNA extraction, the cells are rinsed and lysed in the plate. The rest of the steps are carried out in the same plate, discarding the solutions by gentle inversion while the nucleic acids remain attached to the plate. The samples are removed from the plate to be loaded in the agarose gel for electrophoresis.

8. Put the plate in the incubator and grow for 3–5 days, changing the medium as necessary.

9. When the wells are approaching confluence, wash twice with PBS and trypsinize using 50 μl of trypsin solution per well during 10 min. Add 50 μl of M15 and break up cell clumps by vigorous pipetting. Replate 50 μl onto a gelatinized 96-well plate without feeder cells. The remaining cells in the original 96-well plate may be frozen by adding 50 μl of 2× freezing medium and proceeding through the next protocol from Step 4.

The gelatinized plate can be grown to confluence for DNA preparation and analysis by "mini-Southern" blotting (Section II,E). Once the targeted clones have been identified, the appropriate wells can be retrieved from the freezer and expanded for blastocyst injection and further DNA analysis (Section II,D).

D. Freezing and Thawing ES Cells in 96-Well Plates

Freezing ES clones in individual vials while screening for targeted clones is laborious and time-consuming work, especially if the number of clones to be screened is very large. We have devised a strategy to freeze ES cells in 96-well tissue culture dishes that consistently allows us to recover 100% of the thawed clones.

1. Change the medium on the cells 4 hr before freezing.
2. Discard the M15 medium by aspiration and rinse the cells twice with PBS.
3. Add 50 μl of trypsin solution per well with the multichannel pipettor and incubate the plate for 10 min at 37°, 5% CO_2.
4. Add 50 μl of 2× freezing medium per well and dissociate the colonies.
5. Add 100 μl of sterile light paraffin oil per well to prevent degassing and evaporation during storage at $-70°$.
6. Seal the 96-well plate with Parafilm and put it into a Styrofoam box; close the box and store it at $-70°$ for at least 24 hr. For long-term storage, transfer the plate to a $-135°$ freezer.
7. To thaw, take the 96-well plate out of the freezer and place it into the 37° incubator for 10–15 min.
8. Identify the selected clones and put the entire contents of the well into a 1-cm plate (24-well) with feeder cells containing 2 ml of M15 medium. Change the medium the next day to remove the DMSO and the oil.

E. Southern Blot Analysis Using DNA Prepared Directly on Multiwell Plates

Screening by Southern blotting necessitates that the colonies be expanded *in vitro* to provide enough DNA to carry out such an analysis. In this context, it is very important to increase the efficiency of DNA recovery during the extraction process, which will consequently diminish the time that the cells have to be expanded. A replica of the clones may be frozen while carrying out the analysis. A protocol to freeze cells directly in a 96-well plate has been given (Section II,D). To further improve the efficiency of the gene targeting protocol, we designed a DNA extraction technique that provides a fast, simple, and reliable way to screen a large number of clones by Southern analysis[36] (Fig. 2B). After the cell suspensions have been divided into halves and one-half has been frozen, the other is plated on a gelatin-coated 96-well replica plate (Section II,C, Step 9). This last plate provides the initial material for the DNA microextraction procedure. Lysis of the cells is carried out in the plate by adding lysis buffer and incubating overnight at 60° in a humid atmosphere. The nucleic acids are precipitated in the plate and remained attached to it while the solution is discarded by simply inverting the plate; the nucleic acids are then rinsed, dried, and the DNA cut with restriction enzymes in the plate. All 96 samples

[36] R. Ramírez-Solis, J. Rivera-Pérez, M. Wims, J. D. Wallace, H. Zheng, and A. Bradley, *Anal. Biochem.* **201,** 331 (1992).

can be separated by electrophoresis in a single gel. This greatly accelerates the rate at which screening can be done by Southern blotting. This protocol has been tested for several restriction enzymes, and all give complete DNA restriction using this procedure. However, a pilot reaction with the enzyme of choice should be performed before starting a large screen. When handling a large number of plates, label bottoms and lids to avoid confusion.

1. Allow the cells on the gelatin-coated plates to grow until they turn the medium yellow every day (4–5 days).

2. When the cells are ready for the DNA extraction procedure, rinse the wells twice with PBS and add 50 μl of lysis buffer per well.

3. Incubate the plates overnight at 60° in a humid atmosphere. This is easily achieved by incubating the plates inside a closed container (Tupperware) with wet paper towels in a conventional 60° oven.

4. The next day, add 100 μl per well of a mix of NaCl and ethanol (150 μl of 5 M NaCl to 10 ml of cold absolute ethanol) using a multichannel pipettor.

5. Allow the 96-well plate to stand on the bench for 30 min at room temperature without mixing. The nucleic acids precipitate as a filamentous network.

6. Invert the plate carefully to discard the solution; the nucleic acids remain attached to the plate. Blot the excess liquid on paper towels.

7. Rinse the nucleic acid 3 times by dripping 150 μl of 70% ethanol per well using the multichannel pipettor. Discard the alcohol by inversion of the plate each time.

8. After the final wash, invert the plate and allow it to dry on the bench. The DNA is ready to be cut with restriction enzymes.

9. Prepare a restriction digestion mix containing the following: 1× restriction buffer, 1 mM spermidine, bovine serum albumin (BSA, 100 μg/ml), RNase (100 μg/ml), and 10 units of each restriction enzyme per sample.

10. Add 30 μl of restriction digest mix per well with a multichannel pipettor; mix the contents of the well using the pipette tip and incubate the reaction at 37° overnight in a humid atmosphere.

11. Add gel electrophoresis loading buffer to the samples and proceed to conventional electrophoresis and DNA transfer to blotting membranes. We use a 6 by 10 inch 1% (w/v) agarose gel with three 33-tooth combs spaced 3.3 inches apart. This gives enough space for 96 samples plus one molecular weight marker lane for every comb. Gel electrophoresis in 1× TAE at 80 V for 4–5 hr gives a good separation in the 1–10 kb range.

F. Freezing and Thawing Embryonic Stem Cells in Vials

Clones that appear to have the desired mutation should be expanded and frozen in vials.

1. Dissociate the cells that have been expanded in the 1-cm plate (Section II,D, Step 8) with 0.2 ml of trypsin solution for 10 min at 37°, then stop the action of the trypsin by adding 1 volume of M15 and disaggregate the cell clumps as mentioned before.

2. Take the necessary cells for blastocyst injection and for expansion for further DNA analysis, and freeze the rest as follows.

3. Slowly add 1 volume of 2× freezing medium and mix the cell suspension gently.

4. Distribute the cell suspension into aliquots in sterile freezing vials. Place the vials in a Styrofoam container, close it, and store it at −70° overnight. The next day, transfer the vials to a −135° freezer, or to liquid nitrogen.

5. To thaw, transfer the vial containing the frozen cells to a 37° water bath.

6. When the cell suspension has thawed, transfer it to a sterile 15-ml tube. Add M15 medium slowly, while shaking the tube; fill the tube with M15 medium and collect the cells by centrifugation at 1000 rpm for 5 min at room temperature.

7. Discard the supernatant by aspiration, resuspend the cell pellet in 2 ml of M15 medium, ensure the absence of cell clumps, and plate the cell suspension onto a 1-cm plate with feeder cells. Incubate at 37°.

G. Vector Design

We outline some general principles about the design of the targeting vectors that should be kept in mind when attempting to target a new gene. A detailed description of vector design has been published elsewhere.[37]

1. Selectable Mutations. Generally, gene targeting by homologous recombination occurs at a low frequency in comparison to random integration events. For most genes, vectors can be designed to reduce the frequency of random integration events surviving selection (Fig. 3). A gene that is expressed in ES cells can be targeted using a selectable marker with no promoter. The selectable marker can either have its own translation initiation signal or form a fusion protein with the targeted gene. Alternatively, the selectable marker can be placed within the gene so that the polyadenylation signal must be supplied by the genomic integration site.

---

[37] P. Hasty and A. Bradley, *in* "Gene Targeting: A Practical Approach" (A. Joyner, ed.), in press. IRL Press, Oxford and Washington, D.C., 1993.

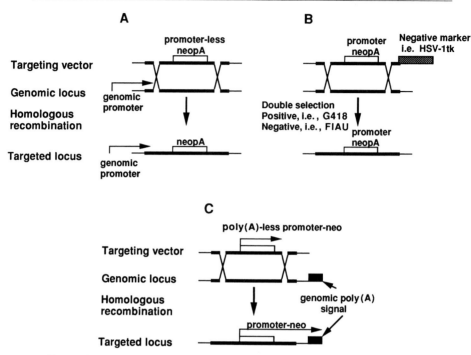

Fig. 3. Three common ways to increase the relative gene targeting frequency. (A) Promoterless genetic markers (i.e., *neo*) require integration next to a genomic promoter to be expressed and confer resistance to a drug. This approach requires that the target gene be expressed in the ES cells. (B) Positive–negative selection is based on the fact that random insertions of the targeting vector will incorporate a genetic marker that can be selected against (i.e., HSV-1*tk*), whereas targeted insertions will not integrate this marker. (C) Polyadenylation signalless genetic markers require that the vector integrates in the appropriate way with respect to a genomic polyadenylation signal to produce stable transcripts of the positive genetic marker.

For any gene, a negative selectable marker (i.e., HSV-1*tk*) can be used outside the homologous region in the targeting vector. In a correct targeting event, the negative selectable marker will be excised and the cells will be resistant to FIAU, but in the random events, the negative marker will generally be integrated and expressed, causing cell death via metabolism of the toxic nucleoside analog. These strategies can be used alone or in combination to help increase the relative gene targeting frequency. The number of clones with random integration events that survive selection will be reduced which will make the targeted event easier to detect.

    The factors that determine the frequency with which a genomic locus will be targeted have not as yet been determined completely. Factors which do affect the targeting frequency include the length of perfect homology between the targeting vector and the genomic locus, the placement of the selectable marker within the homologous stretch, and the site of linearization of the vector.[38,39] The standard replacement vector using positive–negative selection has shown targeting frequencies of 1/10 to 1/1000 G418- and FIAU-resistant colonies for many genes. Regarding the length of homologous sequences in the targeting vector, a convenient compromise between vector construction, diagnosis of targeted events, and targeting frequency is 6 kb with at least 1 kb on either side of the selectable marker.[38] It is best to construct the targeting vector with DNA from the same inbred mouse strain as the ES cell line since polymorphisms could disrupt the length of perfect homology and result in a lower targeting frequency. Careful consideration should be given to the structure of the locus after the desired recombination event, especially if a null allele is desired. For small genes, replacement vectors can be designed in which the coding sequence is replaced by the selectable marker. For larger genes, disruption of the first coding exon is most likely to give a null allele.

We routinely screen 500 colonies by "mini-Southern" analysis (Section II,E) after the first round of targeting. If targeted clones are found, they should be examined by several digests on Southern analysis using probes and enzymes specific for both the 5' and the 3' ends of the homologous sequences, to ensure that the desired recombination event has occurred. If clones are not identified, it is best to redesign the vector rather than continue further screening.

Insertion vectors have been shown to target between 5- and 12-fold more frequently than replacement vectors and could be used for subsequent attempts at targeting.[39] Depending on the design of the original replacement vector, it may be possible to linearize the same vector within the area of homology to take advantage of the higher targeting frequency of insertion events.

2. Creation of Small Nonselectable Mutations in the Mouse Genome. Conventional targeting vectors result in the inclusion of a selectable transcription cassette within the gene. Most cassettes include a promoter, enhancer, and polyadenylation signal. These could all have unpredictable effects on the transcriptional control of the targeted gene or other nearby genes. To use the maximum potential of the ES cell system to investigate molecular changes in a physiological context, the ability to create any mutation, not just a selectable one, will be useful. For instance, more

---

[38] P. Hasty, J. Rivera-Perez, and A. Bradley, *Mol. Cell. Biol.* **11,** 5586 (1991).
[39] P. Hasty, J. Rivera-Perez, C. Chang, and A. Bradley, *Mol. Cell. Biol.* **11,** 4509 (1991).

exact models of human disease could be created in the mouse; regulatory regions could be altered and analyzed in their normal genomic context; truncated proteins or proteins with a single amino acid change could be analyzed *in vivo*; and deletion of regulatory regions could allow extinction of expression only in certain tissues. For this reason, we, and others, have designed protocols which allow the inclusion of small nonselectable mutations in the genomic locus, namely, coelectroporation and "hit and run."

a. *Coelectroporation*. Coelectroporation results in the introduction of two separate DNA fragments into cells during a single electroporation event. When using coelectroporation for gene targeting, the desired event is homologous recombination at the targeted locus and a random insertion event of a selectable marker in a different genomic location. Coelectroporation disrupts the targeted locus at a similar frequency to conventional replacement vectors without negative selection as tested in the *HPRT* locus.[32] Of those targeted events, 1 in 12 had the desired subtle mutation, with no other alteration of the locus. The other targeted events usually resulted from the integration of a concatemer. Thus, this method is most useful for introducing alternative alleles into genes that have been shown to target at high frequency using conventional vectors. The nature of the targeted subtle mutation can be quite diverse, although it is advisable to introduce a new restriction site so the targeted event can be identified on Southern analysis. If the PCR is to be used for screening recombinant clones, the mutation must be asymmetrically positioned with respect to the homologous sequences of the vector so that the expected PCR product does not become too large to amplify reliably. One PCR primer should span the mutation. Since the selectable marker is located at a different chromosomal site in the targeted ES cells, it can be segregated from the desired mutation simply by breeding the mice, once it is in the germ line. The protocol is similar to the electroporation procedure outlined above (Section II,B), so only the differences are noted here.

1. The targeting vector and a different plasmid containing a selectable marker expressed in ES cells (PGKneopA[16]) are prepared separately by the CsCl banding method, linearized, and cleaned as previously described. The DNA is resuspended individually at 10 mg/ml in $0.1 \times$ TE.

2. A total of 200 μg of DNA at a 1 : 1 molar ratio of targeting vector to selectable marker is added to PBS to give a final volume of 400 μl and mixed thoroughly. A control with the same amount of DNA and of selectable marker, but no targeting vector, is prepared in a similar manner to check for toxicity of the DNA preparation.

3. The ES cells are harvested and resuspended at 2×10^7 cells/ml.

Four hundred microliters of DNA solution is mixed with 500 μl of the cell suspension and incubated at room temperature for 5 min. The final cell concentration is 1.1×10^7/ml.

4. Nine hundred microliters of the DNA/cell mixture is placed in an electroporation cuvette, and electroporation is carried out. Selection, etc., is as previously described.

Targeted colonies can be identified by "mini-Southern" analysis or by PCR amplification. Owing to the frequent integration of concatemers at the locus, it is important to do an extensive Southern analysis on targeted clones. Because targeting vectors also integrate in other locations, it is important to use probes that are diagnostic for targeting but are not included within the vector and to check the size of the targeted locus by restriction digests with enzymes that do not cut the targeting vector or the selectable marker.

b. "Hit and Run." The hit and run procedure leads to the generation of ES cells with subtle site-specific mutations. The strategy was originally tested at the HPRT locus[29,31] and then applied to introduce a premature stop codon in the homeodomain of the Hox-2.6 gene.[29] The "hit and run" (H&R) design is based on a two-step recombination process (Fig. 4). The first step consists of the insertion of the targeting vector into the target locus by single reciprocal recombination. The vector contains sequences homologous to the target gene which have been modified to include a desired mutation. It also includes both positive and negative selectable cassettes. Gene targeting results in a duplication of homology separated by the plasmid and the selection cassettes. The targeted clones must be identified at this time. The targeted locus can have the mutation in any, both, or none of the duplicates. The screening can be done by PCR amplification or by Southern blot analysis. When the primary clone has been found, it is used as the starting material for the second step of the procedure. The second step of the H&R protocol is based on the resolution of the duplication by single reciprocal intrachromosomal recombination between the direct repeats that arose as a product of the first step. The second step deletes the selectable cassettes, the plasmid backbone, and one repeat of the targeted locus. The cells that have spontaneously resolved the duplicate can be selected from the rest by their capacity to survive a drug treatment which kills cells containing the marker that can be selected against [i.e., FIAU kills cells that express the herpes simplex virus-1 thymidine kinase (HSV-1tk) gene]. Depending on where the crossover occurs, the second step will produce clones that retain the small mutation and clones whose sequence is restored to wild type.

Although, the in vitro culture time is longer than for traditional one-

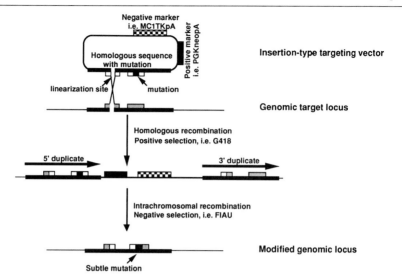

FIG. 4. "Hit and run" technique to introduce subtle mutations. The two-step "hit and run" targeting strategy requires the use of an insertion type vector. The vector contains a piece of sequence homologous to the target gene (bold line), which has been modified with the desired mutation. It also contains a genetic system that can be selected both for and against its presence. The positive selection is used during the first step; the targeted event creates a duplication of direct repeats, which is resolved in the second step. The negative selection is used to obtain the cells that excised the markers and one of the repeats by intrachromosomal recombination. The site of the crossover determines whether the mutation is left in the genome. The bold lines represent the homologous sequences in the targeting vector and in the genomic locus.

step targeting techniques, cells targeted by this method have been proved to contribute to the mouse germ line.[39a]

Selection schemes for "hit and run." For the "hit and run" strategy to work, markers that can be selected for and against need to be used. Either a single selectable marker (i.e., *gpt*) or a set of markers (i.e., *neo* and HSV1-*tk*) can be used. An artificial fusion protein formed by *neo* and HSV-1*tk* has been reported that could potentially be useful for the hit and run strategy.[40] The neomycin phosphotransferase (*neo*) gene provides positive selection by resistance to G418. The herpes simplex virus-1 thymidine kinase gene provides negative selection by making the cells sensitive to FIAU, a nucleoside analog that is specifically recognized

[39a] R. Ramirez-Solis, H. Zheng, J. Whiting, R. Krumlauf, and A. Bradley, *Cell (Cambridge, Mass.)* **73**, 279 (1993).

[40] F. Schwartz, N. Maeda, O. Smithies, R. Hickey, W. Edelman, A. Skoultchi, and R. Kucherlapati, *Proc. Natl. Acad. Sci. U.S.A.* **88**, 10416 (1991).

TABLE I
TYPICAL PARAMETERS IN BLASTOCYST INJECTION EXPERIMENT [a]

ES cell line	Number of injected blastocysts	Number born	Males (%)	Agouti coat color (%)	Germ line chimeras
Good	16	8–10	80	50–100	6–8
Bad	16	8–10	50	<30	0

[a] Numbers are intended to provide a general guide of how many blastocysts to inject per ES cell clone and what can be expected in the case of a totipotent clone (good) compared to a clone unable to contribute extensively to the chimeras (bad).

by the viral thymidine kinase but not by the cellular enzyme. For the H&R technique, we believe that it is important to maintain the G418 selection until immediately prior to FIAU selection in order to minimize the generation of early revertants that can be amplified and that would increase the background. Also, making replica plates to add FIAU at different times after the G418 selection has been stopped may help to get an estimate of the reversion rate at the targeted locus.

1. Thaw the cells as indicated above, plate them on feeder cells, and expand them under G418 selection.

2. Make eight replica plates containing 1×10^5 G418[r] targeted cells per 10-cm tissue culture plate with feeder cells.

3. Add FIAU to a final concentration of 0.2 μM to the first plate and label it day 0.

4. The next day, add FIAU to the second plate and label it day 1. Continue this routine for 6 more days to complete the eight replica plates.

5. FIAU-resistant colonies are conspicuous by 10 days and are ready to be picked and screened for the reversion event as described above for the colonies resulting from an electroporation event (Sections II,C–II,E).

III. Getting Mutations into the Germ Line

The protocols described to date have all had the aim of generating a mutation in ES cells in such a way that the cells remain totipotent and can thus contribute both to somatic tissues and, most importantly, to the germ line of mice. Thus, it is important always to grow ES cells on feeder layers, to keep the time in culture to a minimum (particularly at low density), and to dissociate clumps of cells at each passage. To test the pluripotency of each targeted clone, sufficient blastocysts should be injected to give two litters (Table I). The sex of the offspring should be determined and the animals evaluated for the extent of agouti coat color (derived from the ES cells). Agouti and black are distinguishable approximately 1 week after birth.

The ES cell lines are usually derived from male blastocysts, and extensive contribution to the injected embryo will convert a female blastocyst to a male animal. This gives a disproportionate number of males in the litter. In addition, males that are converted female blastocysts are desirable, as they transmit only ES cell-derived genes to their offspring. They often have reduced fertility,[41] but this disadvantage is more than offset by the efficient transmission of the mutation by the fertile animal. This becomes particularly important when the chimeras are bred to females of the 129 strain to establish the mutation on an inbred background, since no coat color-based selection is possible in this case. Our experience indicates that if a clone does not give high ES cell contribution chimeras or a good sex distortion in 10–12 offspring, then repeated injections of that clone are unlikely to result in germ line transmission. Time is better spent testing several clones to find those that contribute extensively to the coat color and give a disproportionate number of males. Male chimeras from those clones should be test bred. Ideally, for any mutation, two clones should be established in the germ line to confirm that the phenotype is the result of the engineered change. Under ideal conditions, 80–90% of injected clones should be transmitted through the germ line.

Here we only outline the basic procedures to inject ES cells into blastocysts and to reintroduce the injected blastocysts into a foster mother. A more detailed description has been published elsewhere.[42]

A. Setting up Mice to Produce Blastocysts

C57BL/6 (B6) mice are used to obtain the recipient blastocysts. Mating pairs are set up 4 days before the injection, and vaginal plugs are checked the next morning.

1. In the afternoon, select B6 females in estrus (vulva is swollen, pink, and slightly moist). Mate these females to fertile B6 males.

2. The next morning, check every female for the presence of a vaginal plug. A plug is a white, semisolid condensation of semen. Usually, the plug is very obvious in the vaginal opening, but sometimes it is necessary to probe a little inside to detect it.

[41] C. E. Patek, J. B. Kerr, R. G. Goslen, K. W. Jones, K. Hardy, A. L. Muggleton-Harris, A. H. Handyside, D. G. Wittingham, and M. L. Hooper, *Development (Cambridge, UK)* **113**, 311 (1991).

[42] A. Bradley, *in* "Teratocarcinomas and Embryonic Stem Cells: A Practical Approach" (E. J. Robertson, ed.), p. 113. IRL Press, Oxford and Washington, D.C., 1987.

3. Collect the plugged females and keep them in a separate cage. They will be sacrificed 3 days later to obtain the 3.5 dpc blastocysts. Conventionally, the day of the plug is embryonic Day 0.5.

B. Preparing Foster Mothers for Injected Blastocysts

F_1 hybrid females (B6 × CBA) are used as foster mothers for the chimeras as they have a high pseudopregnancy rate and usually provide good care for their litters. Foster mothers are used at 2.5 days of pseudopregnancy. Therefore they should be set up for mating with sterile males on the day that the plugs are checked in the B6 females.

1. Select F_1 hybrid females (B6 × CBA) in estrus the same way as for the B6 females and mate them to vasectomized F_1 hybrid males (B6 × CBA). Use two females per male.

2. The next day, check for the presence of plugs. Collect the females that have plugs, and put them in a separate cage. These will serve as foster mothers to and recipients of the injected blastocysts 2 days later.

C. Obtaining Blastocysts

On the morning of the injection day the blastocysts are collected from the B6 females and are stored in the incubator until ready for injection. Follow the approved protocols for animal research.

1. Sacrifice the B6 females by cervical dislocation. Open the abdomen and excise the genital system, including the uterus, oviducts, and ovaries.

2. Cut the fat pad and mesenteric blood vessels away from the uteri. Separate the two horns of the uterus by cutting between the ovary and the oviduct at the proximal end and at the bifurcation of the uterine horns at the distal end.

3. Flush the blastocysts out of the uterus with a 1-cm syringe equipped with a 25-gauge needle inserted at the oviduct end of the uterus. Use DMEM plus 10% FCS and 20 mM HEPES for the flushing. Collect the blastocysts in a 6-cm tissue culture plate. Do each horn separately.

4. When finished with all the females, collect the blastocysts from the plate with a finely drawn Pasteur pipette and put them together in a 1-cm tissue culture plate with the medium described in Step 3. Store them in the 37° incubator until fully expanded for injection.

D. Injection Needles and Holding Pipettes

The needle will carry the ES cells into the blastocoel. To work properly, it should have a diameter of 20 μm at the end. Special care should be taken to produce a smooth surface in the tip of the needle (Fig. 5A). A good needle is the key for a successful injection day; if a needle is not working properly (i.e., flow of the cells cannot be controlled properly or the point is not sharp enough to penetrate the blastocyst), it is best to prepare a new one.

The holding pipette will keep the blastocyst steady while the needle penetrates it and injects the cells. A holding pipette should be prepared with an external diameter of 100 μm and an internal diameter of 20 μm. The very end of the pipette should be fire-polished to permit good holding and to avoid damaging the blastocyst (Fig. 5A).

E. Preparing Injection Chamber

When the needle and holding pipette are ready, they should be mounted in the micromanipulator, taking special care to avoid any air bubbles in the hydraulic system, as this will cause problems with the handling of the flow of the cells. The injection chamber is filled with cold medium (DMEM, 10% FCS, 20 mM HEPES), and it should be kept at 10° to prevent stickiness of the cells and the necessity of changing the needle prematurely because of clogging. The needle and pipette should be parallel to the bottom of the injection chamber.

F. Blastocyst Injection

1. The cells should be fed 3 hr before the injection, then washed and trypsinized for 10 min at 37°. Add M15 medium and disaggregate the cells by vigorous pipetting. Separate some cells for injection in a sterile 15-ml tube, and put the tube on ice to avoid clumping of the cells. Transfer some cells and 10–16 blastocysts to the injection chamber, and proceed to inject.

2. Collect 10–15 ES cells with the needle.

3. Pick up the blastocyst with gentle vacuum on the holding pipette and lower it until it is seen touching the bottom of the plate. The blastocyst will be kept in place by the holding pipette and the bottom of the chamber.

4. Focus the microscope on an intercellular junction at the equatorial plane of the blastocyst, and adjust the height of the needle to bring it into focus in the same plane (Fig. 5B).

5. Insert the needle through the intercellular space and start injecting the cells slowly (Fig. 5B). When finished, withdraw the needle (Fig. 5C), then separate the blastocyst from the uninjected group.

6. Repeat this procedure with each blastocyst in turn.

A

B

C

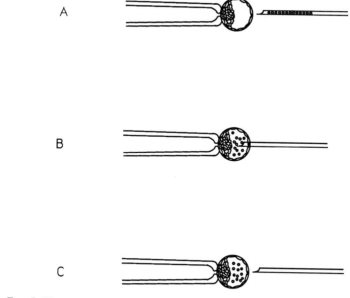

FIG. 5. Blastocyst injection. (A) Lateral view of the injection chamber. It is important that the tips of the injection needle and holding pipette are smooth. (B) The penetration site is at an intercellular junction at about the equatorial plane of the blastocyst. The movement to penetrate the blastocoel must be done as smoothly as possible, and the cells are injected slowly. (C) When injection is finished, the needle is pulled out carefully.

G. Transferring Blastocysts Back into Foster Mother

When the injections have been completed, the blastocysts must be put in a prepared female that can carry them to term. For this, use the F_1 hybrid females that were mated with vasectomized males.

1. Anesthetize the females by intraperitoneal injection of Avertin (0.02 ml/g, ~0.5 ml/mouse).

2. Rinse the back of the mouse with ethanol and proceed to make a small incision in the back at approximately one-third the length of the mouse from the base of the tail. Move the skin with the incision ventrally until the ovary can be seen through the body wall as a pink structure surrounded by fat.

3. Cut a small incision in the body wall, and carefully pull out the ovary, oviduct, and uterus, handling the reproductive tract by the attached fat.

4. To transfer the embryos, make a hole in the uterus with a 25-gauge needle close to the oviduct end.

5. Using a finely drawn Pasteur pipette, transfer the injected blastocysts into the uterus. Six to seven blastocysts per uterine horn per female is an appropriate number. If possible, use only one horn per female.

6. Put the uterus back inside the abdominal cavity, and close the body wall with one suture.

7. Close the skin with a metal clip, and put the female on a warm surface to allow recovery. When the female is awake, put the animal in a new cage. The chimeras should be born in 17 days.

IV. Conclusion

In this chapter, we have outlined the basic tools required to introduce directed mutations into the mouse germ line. The gene targeting field is moving quickly, and every advance will permit the study of new aspects of mammalian biology as represented in the mouse. New techniques that can increase the efficiency of introducing mutations into the germ line may increase the potential scope of the gene targeting experiments, and consequently lead to a better understanding of the relevant and fascinating genotype–phenotype relationships that occur during the life of a mammal.

Acknowledgments

This work was supported by grants from the National Institutes of Health, the Cystic Fibrosis Foundation, and the Searle Scholars Program. A.B. is a Leukemia Society Fellow. A.C.D. has been a fellow of the Medical Research Council of Canada. R.R.S. has been partially supported by the Centro Internacional de Biología Molecular y Celular, A.C. (CIBMYC), Monterrey, Mexico. We acknowledge Paul Hasty, Jaime Rivera, and Richard Behringer for critical reading of the manuscript and valuable suggestions.

[52] Simple Screening Procedure to Detect Gene Targeting Events in Embryonic Stem Cells

By FRANK KÖNTGEN and COLIN L. STEWART

Introduction

Gene Targeting

The mouse has been used for many years as an experimental system to study gene function. The latest developments have made it possible to

Copyright © 1993 by Academic Press, Inc.
All rights of reproduction in any form reserved.

mutate a gene of interest in the mouse by gene targeting.[1] Gene targeting is the exchange of a specific endogenous gene or gene segment for an *in vitro* manipulated DNA fragment by homologous recombination. This technology offers for the first time the possibility of specifically mutating genes in the mouse germ line and studying their function, for example, in early mouse ontogeny[2] or in the developing immune system.[3,4]

Murine Embryonic Stem Cells

Gene targeting is achieved in murine embryonic stem (ES) cells.[5] Totipotent ES cells are derived from the inner cell mass (ICM), the part of the mouse blastocyst that usually gives rise to the embryo proper. Several ES cell lines have been derived from the 129/Sv strain to date that are capable of introducing mutations into the mouse germ line,[6–8] but only one has been described for the C57BL/6 mouse strain.[3] ES cells are maintained *in vitro* and kept in their totipotent state by addition of leukemia inhibitory factor (LIF) to the culture medium and/or by growth on primary mouse embryonic fibroblasts (PMEFs) which produce and secrete LIF.[9] After injection into host blastocysts and transfer to foster mothers, ES cells are able to participate in the development of every tissue of the mouse, including the germ line (germ line chimeras). Mutated ES cells are used in this way to create new mouse strains.

DNA Transfer

DNA transfer into ES cells is accomplished by electroporation. Although this method shows a considerably low transfection efficiency ($\leq 10^{-3}$), it is relatively easy to reproduce. With the use of positive and negative selectable markers, such as the neomycin resistance (*neo*) and

[1] S. L. Mansour, K. R. Thomas, and M. R. Capecchi, *Nature (London)* **336**, 348 (1988).

[2] C. L. Stewart, P. Kaspar, L. J. Brunet, H. Bhatt, I. Gadi, F. Köntgen, and S. J. Abbondanzo, *Nature (London)* **359**, 76 (1992).

[3] F. Köntgen, "Generation and Analysis of MHC Class II Deficient Mice." Albert-Ludwigs-Universität, Freiburg, Germany, 1992.

[4] J. Travis, *Science* **256**, 1392 (1992).

[5] E. J. Robertson, *Biol. Reprod.* **44**, 238 (1991).

[6] T. C. Doetschman, H. Eistetter, M. Katz, W. Schmidt, and R. Kemler, *J. Exp. Med.* **87**, 27 (1985).

[7] P. L. Schwartzberg, S. P. Goff, and E. J. Robertson, *Science* **246**, 799 (1989).

[8] S. Thompson, A. R. Clarke, A. M. Pow, M. L. Hooper, and D. W. Melton, *Cell (Cambridge, Mass.)*, **56**, 313 (1989).

[9] R. L. Williams, D. J. Hilton, S. Pease, T. A. Willson, C. L. Stewart, D. P. Gearing, E. F. Wagner, D. Metcalf, N. A. Nicola, and N. M. Gough, *Nature (London)* **336**, 684 (1988).

thymidine kinase (*tk*) genes, it has become feasible to screen large numbers of transfectants for targeted integrations.[10]

Positive–Negative Selection

Positive–negative selection takes advantage of the fact that in a homologous recombination event heterologous sequences at either end of the targeting vector are eliminated. They are not integrated into the genome, whereas in the case of nonhomologous integration usually the complete construct is integrated. Having the *tk* gene cloned at either or both ends of the targeting vector provides a means for negative selection of virtually all clones with a random integration. The *neo* gene, if placed internally into the targeting vector, serves as a marker for positive selection as well as a potential inactivator of the targeted gene.[11]

Neomycin and Thymidine Kinase Genes

The *neo* gene, encoding a bacterial aminoglycoside phosphotransferase, renders a given cell resistant to the antibiotic G418, so that it can be positively selected.[12] The *tk* gene is of viral origin (herpes simplex virus, HSV) and facilitates the incorporation of antiviral drugs such as ganciclovir and FIAU [1-(2'-deoxy-2'-fluoro-β-D-arabinofuranosyl)-5-iodouracil] into DNA during replication. These base analogs block the incorporation of additional bases and thereby induce a termination of replication that eventually will lead to cell death. Thus, *tk*-transfected cells can be negatively selected.[13,14]

Detection of Homologous Recombinants

Screening for homologous recombinants is first undertaken with pools of ES cell clones and subsequently with single clones. The screening method of choice is the polymerase chain reaction (PCR).[15] The PCR is very sensitive and rapid; it can also be performed on crude cell lysates. For the detection of a gene targeting event, the homologously recombined

[10] A. L. Joyner, W. C. Skarnes, and J. Rossant, *Nature (London)* **338**, 153 (1989).

[11] T. DeChiara, A. Efstratiadis, and E. J. Robertson, *Nature (London)* **345**, 78 (1990).

[12] F. Colbere-Garapin, F. Horodniceanu, P. Kourilsky, and A. Garapin, *J. Mol. Biol.* **150**, 1 (1981).

[13] E. Borrelli, R. Heyman, M. Hsi, and R. M. Evans, *Proc. Natl. Acad. Sci. U.S.A.* **85**, 7572 (1988).

[14] M. H. St. Clair, C. U. Lambe, and P. A. Furman, *Antimicrob. Agents Chemother.* **31**, 844 (1987).

[15] R. K. Saiki, D. H. Gelfand, S. Stoffel, S. J. Scharf, G. Higuchi, G. T. Horn, K. B. Mullis, and H. A. Erlich, *Science* **239**, 487 (1988).

TABLE I
SCREENING FOR GENE TARGETING EVENTS IN EMBRYONIC STEM CELLS[a]

Day	Experimental procedures
1	Electroporate ES cells in PBS or DMEM and plate onto five 60-mm dishes
2	Start positive selection with G418
4	Start negative selection with ganciclovir or FIAU
10	Pick single colonies into 48-well plates (6 rows/8 columns)
13	From each of the 6 rows per 48-well plates prepare a pooled lysate for PCR analysis
14	Analyze each "row lysate" by PCR and then prepare from each PCR-positive row eight individual well lysates
15	Analyze single well lysates by PCR
17	Expand clones for Southern blot analysis, karyotyping, and/or injection of blastocysts

[a] Time schedule given may vary slightly from experiment to experiment. Times were taken from a successful MHC class II targeting experiment.

DNA region is amplified with two primers, one specific for the mutation (targeting vector) and the other for the endogenous (nonvector) DNA. It is possible to detect one homologous recombinant in a background of several hundred thousand nontargeted ES cells.[16] Homologous recombinants are then further characterized for correct integration by Southern blot analysis.

Materials and Methods

This chapter describes a straightforward, relatively easy, and reproducible method to screen for gene targeting events in ES cell clones by PCR amplification. It has been applied to detect the disruption of genes for the major histocompatibility complex (MHC) class II antigens,[3,17,18] LIF,[2] tumor necrosis factor receptor (TNF-R),[19] ZP3 (sperm receptor on the zona pellucida of the mouse oocyte),[20] and Thy-1 (a T-lymphocyte marker)[21] in the mouse. A general outline of the experimental procedures is given in Table I.

[16] H. S. Kim and O. Smithies, *Nucleic Acids Res.* **16**, 8887 (1988).
[17] Deleted in proof.
[18] Deleted in proof.
[19] J. Rothe, personal communication, 1992.
[20] R. Kinloch, personal communication, 1991.
[21] C. L. Stewart and J. Silver, personal communication, 1991.

Embryonic Stem Cell Culture

Embryonic stem cell lines are always cultured on an irradiated or mitomycin C-inactivated[2] feeder layer of PMEFs at 37°, 95% relative humidity, 10% (v/v) CO_2 in supplemented Dulbecco's modified Eagle's medium (DMEM) on gelatinized tissue culture dishes. Supplemented DMEM contains 15% fetal calf serum (FCS), 2 mM glutamine, 1000 units/ml LIF, 1× nonessential amino acids (NEAA), 0.1 mM 2-mercaptoethanol, and 100 units/ml penicillin–streptomycin.

The PMEFs are irradiated in a cesium source with 3000 rads either as a confluent cell layer or as a cell suspension. PMEF medium contains 10% FCS only and no LIF. ES cells are passaged 1 : 2 or 1 : 4 at approximately 75% confluency every 2–3 days. Trypsinizations are done in 1× trypsin–EDTA, 1% chicken serum (CS) for 5–7 min at 37°, reactions are stopped with culture medium, and cells are plated onto PMEFs. The CS prevents cells from becoming sticky and keeps them in a protein-buffered environment (CS does not contain a trypsin inhibitor). For freezing, trypsinized cells are pelleted in a cytocentrifuge (1000 rpm, 3 min), taken up in culture medium with 5–10% dimethyl sulfoxide (DMSO), frozen at −80° (≥12 hr) in a 1.0-ml freeze vial, and transferred to liquid nitrogen. ES cells are thawed at 37° (water bath or incubator), centrifuged through 10 ml medium, and plated. PMEFs are thawed and plated directly.

Notes. The 0.1% (w/v) gelatin solution is prepared at 1 g/liter in phosphate-buffered saline (PBS). Autoclave to dissolve and sterilize. Incubate the cell culture dishes with 0.1% gelatin solution for at least 30 min, withdraw the solution, and plate the cells; drying of the surface is not necessary. All PBS used is Mg^{2+}- and Ca^{2+}-free. PBS, trypsin, and DMEM are used straight out of the refrigerator (4°). Finally, ES cells are frozen from a 75% confluent dish (60 mm) in 3× aliquots or from a flask (150 cm²) in 25× aliquots.

Preparation of Feeder Fibroblasts. PMEFs are routinely made by dissection of 13-day-old ICR-M-TK*neo*2 embryos[22] under a stereo microscope in a 100-mm petri dish containing 10 ml PBS. After removal of extraembryonal tissue, liver, heart, and head, the bodies are transferred into a 5-ml syringe barrel (take out plunger and load bodies from the top end). They are disaggregated by repeated passage through an 18-gauge needle in 5 ml culture medium containing 350 μg/ml G418 (see below). Suspensions of about four embryos are then plated into gelatinized culture flasks (175 or 150 cm²) and grown for 3 days. The medium is changed daily. PMEFs are passaged 1 : 5 on day 3 and frozen 1 : 3 on day 6. PMEFs are only used for up to 7 passages.

[22] C. L. Stewart, S. Schütze, M. Vanek, and E. F. Wagner, *EMBO J.* **6,** 383 (1987).

The amount of PMEFs needed for a confluent feeder layer is calculated in terms of surface area. A confluent 175-cm^2 flask will provide enough cells for eight or nine 60-mm dishes (8 × 21 cm^2 = 168 cm^2); the area of a 100-mm dish is 55 cm^2, and the total area of a 48-well plate is 48 cm^2.

Materials for Cell Culture

Chicken serum, GIBCO (Grand Island, NY), Cat No. 033-6110 H
DMEM, GIBCO, Cat. No. 074-02200 P
DMSO, Merck (Darmstadt, Germany), Cat. No. 027-IA26178
LIF (ESGRO), Amrad (Victoria, Australia)
FCS, HyClone (Logan, UT), GIBCO–BRL, etc.; the FCS is tested to give the best plating efficiency and lowest cytotoxicity at 30% FCS
L-Glutamine, GIBCO, Cat. No. 043-05030 H
2-Mercaptoethanol, Merck, Cat. No. 15433
NEAA, GIBCO, Cat. No. 043-01140 H
PBS Penicillin–streptomycin (Pen–Strep), GIBCO, Cat. No. 043-05140 H
Porcine skin gelatin type III, Sigma (St. Louis, MO), Cat. No. G-1890
Trypsin–EDTA, GIBCO, Cat. No. 043-05300 H

Transfection of Embryonic Stem Cells

For electroporation, ES cells are thawed and plated onto 60-mm dishes. After 2 days they are passaged onto a 100-mm dish and after another 2 days seeded into a 175-cm^2 flask. After an additional 2 days, approximately 1.5 × 10^8 ES cells are harvested and split into three aliquots.

Day 1: Electroporation. Each aliquot is resuspended in 800 μl PBS or DMEM (4°) containing 25 μg linearized targeting vector and pulsed with 250 V/500 μF or 320 V/500 μF at 0°–4° in a 0.4-cm cuvette. Electroporated ES cells are then resuspended in 25 ml medium and immediately plated onto five 60-mm dishes containing PMEFs.

Days 2 and 4: Positive and Negative Selection. Positive selection is started 1 day after electroporation with medium containing 350 μg/ml G418 (~175 μg active substance). Negative selection with 2 μM ganciclovir or 0.2 μM FIAU is started either on day 2 or day 4. The culture medium is changed on a daily basis in order to wash away dead cells. Resistant colonies become visible around day 7 after electroporation.

Notes. G418 is prepared as a 100× stock stolution (50 ml) in PBS, filter sterilized, and stored at 4°. Ganciclovir or FIAU is prepared as a 10,000× stock solution in water, filter sterilized, and stored at −20° in 50-μl aliquots. As a control for the efficiency of negative selection, one dish with electroporated ES cells may be treated with G418 only.

Materials for Transfection

Ganciclovir (Cymevene), Syntex
FIAU, Oclassen
G418 (Geneticin), GIBCO, Cat. No. 066-01811 B
Gene Pulser, Bio-Rad (Richmond, CA)

Screening Procedure

Day 10: Pick Single Colonies. Ten days after elecroporation (9 days
of selection), the surviving colonies are picked (up to 960 colonies) in the
following manner. The medium is aspirated, culture dishes are put upside
down under a stereo microscope, and colonies are marked with a circle
and counted. ES cell colonies are then washed twice in PBS, covered
with 2 ml trypsin–EDTA, 1% CS, and put back under the stereo microscope. Single colonies are drawn into a mouth-controlled, drawn-out glass
capillary and blown into one well of a 48-well plate (up to 20 plates)
containing PMEFs and 250 μl Geneticin-containing medium. Bubbles are
introduced into the medium so that one can distinguish filled from empty
wells. Each colony is picked with a new capillary, to avoid cross-contamination between clones. About 20–30 min is required to fill a 48-well plate
with clones. Cultures are returned to the incubator and grown for 2 or 3
days without changes of medium.

Notes. Colonies can be counted in the morning or 1 day in advance
so that one can prepare the appropriate number of 48-well plates
required. Do not shake culture dishes during trypsinization, as colonies
will detach and mix. The diameter of the capillary tip should be smaller
than the size of the colonies so that a suspension of single cells is
generated during the pipetting process. In addition, do not pipette
colonies up and down, since this may cause cross-contamination of
individual ES cell clones.

Day 13: Prepare Pooled Lysates from Each Row. Pooled lysates are
prepared from each row (8 wells) of a 48-well plate. Cultures are washed
in 250 μl PBS and trypsinized with 100 μl trypsin–EDTA, 1% CS for
5–7 min at 37°. Trypsinizations are stopped with 250 μl G418-containing
medium. Cells are resuspended by pipetting up and down approximately
20 times with a 12-channel pipettor, set at 180 μl volume. However, only
8 of the 12 channels contain pipette tips, to fit the 8 wells of a row. Pipette
tip columns 2, 5, 8, and 11 in each pipette tip rack are therefore emptied.
These tips are transferred into an empty, sterile rack (columns 1, 3, 4, 6,
7, 9, 10, and 12) with an 8-channel pipettor and are used for the following
48-well plates.

From each row, eight cell suspensions (180 μl each) are pooled in a reagent basin tilted on its end to an angle of 45°, so that the suspension will collect at the bottom (reagent basins are small plastic containers usually used to load tips of 8- or 12-channel pipettors). After addition of 250 μl G418-containing medium including inactivated PMEFs to each well, cultures are returned to the incubator. Trypsinization of the next plate is then started. During this trypsinization time, collected pools are transferred into 1.5-ml Eppendorf tubes (safe lock).

After preparing all pools, cells are pelleted in a bench-top centrifuge (3000 rpm, 10 min at room temperature). After the supernatants are aspirated, the pellets are vortexed, resuspended in 50 μl PCR–lysis buffer containing 60 μg/ml proteinase K, again vortexed, covered with 50 μl paraffin oil, and incubated overnight at 56°. For further handling, see the section on PCR analysis.

Notes. Medium, PBS, and trypsin are added with a multistep pipette. Reagent basins are washed with PBS, sprayed with 70% ethanol, and dried with a tissue after each use. Prepare PBS (5 liters) for washing of the reagent basin in advance. The pool preparation time per 48-well plate is approximately 15 min.

Day 14: Prepare Lysates from Single Wells. After identification of targeted ES cells in a "row lysate," the corresponding wells are screened for homologous recombinants. Supernatants and one wash with PBS or medium (contains floating and/or dead ES cells) are taken from each well and transferred to an Eppendorf tube. The ES cells are pelleted, lysed, and analyzed by the PCR as described below (pellet is mostly not visible). Cultures are fed with G418-containing medium and returned to the incubator.

Notes. Because of the low number of floating/dead cells per well a pellet is mostly not visible, however, there is a sufficient number to produce reliable PCR data. If one well is contaminated with yeast or bacteria, add 1 or 2 tablets of sodium hydroxide to prevent spreading.

Day 17: Expansion and Verification of Targeted ES Cells. Confluent 48-well cultures of targeted ES cells are either expanded in a 60-mm dish for Southern blot analysis and karyotyping, and/or 50% are immediately injected into blastocysts. Clones are frozen as early as possible (see above). Verification of a targeted mutation by Southern blot analysis is performed in accordance with standard techniques. ES cells are maintained in G418-medium for four or more passages.

Equipment for Screening

8-channel pipettor, Flow Labs (McLean, VA)
12-channel pipettor, Flow Labs

48-well plates, Costar (Cambridge, MA)
Hard glass capillaries, British Drug House, (Poole, UK), Cat. No. 32124
Multistep pipette, Eppendorf
Reagent basins, Flow Labs, Cat. No. 77-824-01

Polymerase Chain Reaction Analyses

Mimicking Gene Targeting Event. The method described is based on PCR-dependent detection of the homologous recombinants. Therefore the PCR has to be optimized so that 10–100 targeted alleles can be detected in a background of approximately 100,000 nontargeted alleles. The conditions may be adjusted with lysates from a cell line which has been transfected with a "mock knockout construct" that contains the mutation-specific *neo* primer sequence as well as the endogenous primer sequence. However, the generation of such a plasmid and cell line is time consuming and increases the risk of contamination (e.g., the subsequent detection of false positives in a gene targeting experiment).

Therefore, a safer and less labor-intensive procedure, not needing a transfected cell line or a "mock knockout construct," has been established and shown to give reliable results. For this purpose crude cell lysate or genomic DNA from about 10,000 ES cells is mixed with approximately 1 μg linearized targeting vector and used as a calibration substrate. The positive signal obtained on an ethidium bromide-stained agarose gel shows the same intensity as the one obtained from about 1000 homologous recombinants or control cells as used above. In such a PCR the targeting vector does not serve as template DNA but rather as a very large primer (Fig. 1). This relatively easy and straightforward PCR optimization procedure has been successfully used to calibrate PCRs for knock out experiments of the MHC class II, LIF, TNF-R, Thy-1, and ZP3 genes.

Primers. It is recommended that the PCR primers be about 24 base pairs in length with 55% GC content. The GC residues should be equally distributed throughout the primer, and obvious secondary structures are to be avoided. It is preferable to have a G or C at the 3' end of the primers to provide a more stable extension start site for the Taq polymerase. In the beginning several different primer combinations (six pairs) are tested for the most efficient and most specific amplification result. The choice of primers is one of the most crucial aspects in achieving the desired sensitivity. Some primer sequences that have been shown to work efficiently are shown in Table II.

Optimizing Conditions. The first amplifications are carried out at an annealing temperature of 65° with varying amounts of $MgCl_2$ (1.0, 1.5,

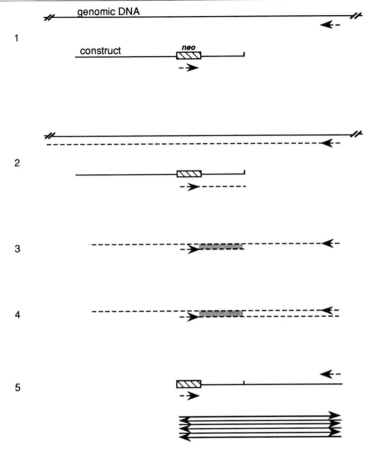

FIG. 1. Mimicking homologous recombination. (1) First annealing of *neo* primer and endogenous primer on physically separate templates, after the first denaturation. (2) Extension of both primers. (3) Annealing of extended *Neo* primer to homologous sequences on extended endogenous primer, after the second denaturation. (4) Extension of *neo* primer strand and thereby generation of a template containing both *neo* and endogenous primer sequences. (5) Standard PCR amplification as it also occurs in a gene targeting event.

2.0, 5.0, 7.5, and 10.0 mM) and 1 μg linearized targeting vector in a lysate of approximately 10,000 ES cells or equivalent genomic DNA. If no signal is obtained the annealing temperature is lowered to 60° and 55°. Further optimization can be gained by varying the concentration of deoxynucleoside triphosphates (dNTPs, $\leq$800 μM) and primer ($\leq$1 μM). Doing so, one has to remember that increasing or decreasing the molarity of dNTPs also raises or lowers the free magnesium concentration proportionally.

TABLE II
POLYMERASE CHAIN REACTION PRIMERS

Gene	GC content (%)	Number of base pairs	Sequence 5' → 3'
neo	52	23-mer	ATA TTG CTG AAG AGC TTG GCG GC
PGK	57	21-mer	GCA CGA GAC TAG TGA GAC GTG
MHC class II	54	24-mer	CAC AGT CTC TGT CAG CTC TGT GAC
LIF	57	21-mer	GGA TTG TGC CCT TAC TGC TGC

Problems can also be caused by GC stretches longer than about 100 bp in the target sequence, which may interfere with amplification or in the worst case abolish it completely. This problem can sometimes be overcome by the addition of small amounts of DMSO (≥0.5%) to the reaction mixture.

Furthermore, the order in which the reaction components are pipetted seems to be critical. All lysates (25 µl) are distributed into reaction tubes first, and then the master mix is prepared (*Taq* polymerase goes in last). This mix is then quickly pipetted into the lysates (multistep pipette, vortexing is not required) and covered with paraffin oil (multistep pipette or dropping out of a 1 ml Pipetman), and thermocycling (see below) is started. Predenaturation without *Taq* polymerase is not necessary. Nevertheless, the PCR machine should be prerun and restarted before loading the samples. In this way nonspecific amplification in the first cycle is minimized since the heating blocks have at least reached the annealing temperature, preferrably 95°. In the beginning one may want to set up the polymerase chain reactions on ice, since this has been found to increase the amplification yield.

Nevertheless, in some cases the PCR will still not be sensitive enough for the detection of recombinants (e.g., TNF-R). Under these circumstances a second PCR can be performed after the first 25 cycles, using nested primers. The nested primers will amplify a target within the first amplification product and thereby increase the sensitivity of the detection assay about 10-fold.

Days 14 and 15: Template Preparations. PCR amplifications are done on 25-µl samples of crude lysates of ES cells without further purification. The template material is incubated at 56° overnight in PCR–lysis buffer

60 μg/ml proteinase K and heated at 95° for 30 min on the following day to inactivate proteinase K.

Reaction Setup. A 50-μl reaction mix contains 50% PCR–lysis buffer (lysate/template) and 50% PCR buffer. If a smaller volume of lysed template or purified DNA is used, the remainder is made up with PCR–lysis buffer. A master mix from concentrated stock solutions is prepared as follows:

Calculation for master mix:

1× Lysis buffer or lysate	10 × 25 μl
2 mM each dNTP	50 μl
10 μM primer I and II	50 μl
10× PCR buffer	25 μl
10 mM MgCl$_2$	25 μl
5 units/μl *Taq* polymerse	5 μl
Doubly distilled water	95 μl
Total (10 reactions)	500 μl

Notes. All stock solutions are made in bulk and stored in 500-μl aliquots at −20°, except the PCR–lysis buffer which is stored in 50-ml tubes at −20°; however, the aliquot which is in use is kept at 4°. Do not leave the PCR–lysis buffer at room temperature for longer periods of time (e.g., overnight) as this will decrease detection sensitivity. Take *Taq* polymerase from the −20° freezer only for the removal of a required aliquot. All other solutions may be kept at room temperature during setup; there is no need to keep the dNTPs on ice.

Materials for Polymerase Chain Reaction

PCR–lysis buffer (1×): 50 mM KCl (Merck, Cat. No. TA 573936), 10 mM Tris-HCl, pH 8.3 (BRL, Gaithersburg, MD, Cat. No. 5504 UA), 2 mM MgCl$_2$ (Merck, Cat. No. A 192833), 0.45% Nonidet P-40 (NP-40, Sigma, Cat. No. N-3516), and 0.45% Tween 20 (Sigma, Cat. No. P-1379)

PCR buffer (10×): 166 mM (NH$_4$)$_2$SO$_4$ (Fluka, Ronkonkoma, NY, Cat. No. 09980), 670 mM Tris-HCl, pH 8.8, and 1 mg/ml bovine serum albumin (BSA, Serva, Cat. No. 11930; use of "pure" BSA is recommended, as ultrapure or nuclease-free BSA could decrease sensitivity)

dNTP set (Pharmacia, Piscatasay, NJ, Cat. No. 27-2035-01)

Paraffin oil (Fluka, Cat. No. 76235)

Proteinase K (Merck, Cat. No. E 123668)

Taq polymerase (Perkin-Elmer Cetus, Norwalk, CT, or Boehringer-Mannheim, Mannheim, Germany)

TRIOtherm (Biometra)

DNA thermal cycler Perkin-Elmer Cetus

Cycling Conditions. The PCR amplification cycles at 95° for 20 sec, 65° for 30 sec, 72° for 30 sec; after 40 cycles are complete, keep the reactions at 72° for 5 min. Samples are stable at room temperature indefinitely.

Notes. Preincubation of template mix (with or without *Taq* polymerases) is not necessary. The 30-sec extension time is appropriate for up to a 1.5-kb target size; for larger sizes 1 min is used. Do not increase the denaturation time, since this will decrease the half-life of the *Taq* polymerase and therefore reduce signal intensity. Increasing the annealing time (to 60 sec) might be helpful in some cases.

Amplification Product Analysis. After amplification, a 10-μl sample of each reaction is separated on an ethidium bromide-containing 1.0% agarose gel. Polaroid photos are taken under UV light (302 nm).

Acknowledgments

We thank Horst Blüthmann and Michael Steinmetz as well as many other people in the laboratory for help and critical support. Thanks also to A. Carey for critically reading the manuscript. Furthermore, F.K. is thankful to Gabrielle Süss for generous support during writing of his Ph.D. thesis.

[53] Manipulation of Transgenes by Site-Specific Recombination: Use of Cre Recombinase

By BRIAN SAUER

Introduction

Transgenic mice are now routinely generated either by direct pronuclear injection of exogenous DNA into fertilized zygotes[1] or by injection of genetically engineered embryonal stem (ES) cells into the mouse blastocyst.[2] Direct pronuclear injection results in random integration of the injected DNA into the genome and relies on the dominant nature of the transgene to give a useful phenotype (e.g., the expression of a foreign protein or of a mutated version of an endogenous gene). Manipulation of endogenous genes (gene targeting) has primarily been performed with ES

[1] R. D. Palmiter and R. L. Brinster, *Annu. Rev. Genet.* **20**, 465 (1986).
[2] P. I. Schwartzberg, S. P. Goff, and E. J. Robertson, *Science* **246**, 799 (1989).

Copyright © 1993 by Academic Press, Inc.
All rights of reproduction in any form reserved.

cells because they can be manipulated and grown *in vitro* as a permanent cell line and then reintroduced into the mouse blastocyst to contribute to the mouse germ line. Gene targeting uses homologous DNA recombination by the cell to substitute a manipulated transgene for the endogenous chromosomal allele. The technique has been used primarily to generate null alleles or gene "knockouts."

Further manipulation of the transgene can be achieved *in vivo* by use of a site-specific recombination system such as the Cre/*lox* system of bacteriophage P1.[3] Cre (causes recombination) is a member of the Int family of recombinases[4] and has been shown to perform efficient recombination at *lox* sites (locus of X-ing-over) not only in bacteria but also in eukaryotic cells.[5,6] The Cre recombinase can efficiently excise DNA bracketed by *lox* sites from the chromosome. Cre-mediated excision is useful both for removing unwanted DNA segments from the genome and for designing recombination-dependent switches to control gene expression.[7] In addition, the 38-kDa Cre protein can direct integration of *lox* vectors specifically to a *lox* site previously placed into the genome.[8] This chapter discusses the use of Cre-mediated recombination both in precisely removing DNA from the genome and in site-specific insertion of DNA into the genome.

Cre-Mediated Recombination

Cre-mediated recombination and its use in recombination-activated gene expression (discussed below) is illustrated in Fig. 1. Two components are required for recombination: expression of the Cre recombinase and an appropriate *lox*-containing substate DNA(s). Recombination between two directly oriented *loxP* sites excises the intervening DNA as a circular molecule having a single *lox* site. As indicated in Fig. 1, intermolecular recombination results in integrative recombination but is less efficient than the intramolecular recombination event. The 34-bp *loxP* site (Fig. 1) consists of two 13-bp inverted repeats, binding sites for the Cre protein, and an 8-bp asymmetric core region in which recombination occurs and

[3] R. H. Hoess and K. Abremski, *in* "Nucleic Acids and Molecular Biology" (F. Eckstein and D. M. J. Lilley, eds.), Vol. 4, p. 99. Springer-Verlag, Berlin and Heidelberg, 1990.

[4] P. Argos, A. Landy, K. Abremski, J. B. Egan, E. H. Ljungquist, R. H. Hoess, M. L. Kahn, B. Kalionis, S. V. L. Narayana, L. S. Pierson, N. Sternberg, and J. M. Leong, *EMBO J.* **5**, 433 (1986).

[5] B. Sauer, *Mol. Cell. Biol.* **7**, 2087 (1987).

[6] B. Sauer and N. Henderson, *Proc. Natl. Acad. Sci. U.S.A.* **85**, 5166 (1988).

[7] B. Sauer and N. Henderson, *Nucleic Acids Res.* **17**, 147 (1989).

[8] B. Sauer and N. Henderson, *New Biol.* **2**, 441 (1990).

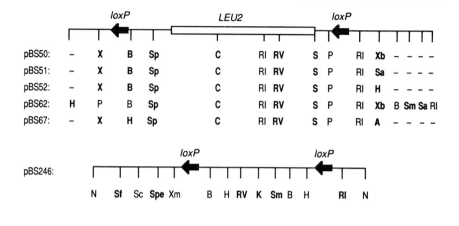

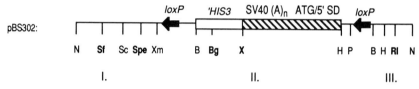

FIG. 3. *lox²* excision cassette vectors. Short, stubby black arrows represent *loxP* sites. All cassettes, except for that in pBS62, are in a 2.2-kb pBR322 backbone having the Apʳ gene and plasmid *ori*. The pBS62 cassette is in plasmid pSP64 (Promega, Madison, WI). The 2.2-kb *LEU2* and 0.45-kb *HIS3* yeast DNA fragments are shown as white boxes. The 0.93-kb SV40 DNA fragment and synthetic aberrant ATG start is represented by the striped box. Maps are not drawn to scale in order to emphasize available restriction sites. Boldface type indicates that the restriction site is unique in the plasmid. Restriction sites are as follows: A, *Aat*II; B, *Bam*HI; Bg, *Bgl*II; C, *Cla*I; H, *Hind*III; K, *Kpn*I; N, *Not*I; P, *Pst*I; RI, *Eco*RI; RV, *Eco*RV; S, *Sal*I; Sa, *Sac*II; Sc, *Sca*I; Sf, *Sfi*I; Sm, *Sma*I; Sp, *Sph*I; Spe, *Spe*I; X, *Xho*I; Xm, *Xmn*I.

of the selectable marker after the first round of gene targeting would allow targeting of the second gene copy using the same selectable marker.

Recombination-activated gene expression (RAGE) is shown in Fig. 1. This strategy uses an intervening DNA sequence ("STOP") to prevent expression of a downstream gene, in this example *neo*, by the SV40 or other desired promoter. Excision of STOP results in productive expression of the *neo* gene and the generation of G418-resistant cells. We have characterized a mouse L cell line containing such an activatable *neo* gene to assess the efficiency of Cre-mediated excision from a chromosome.[7] Typically 10% of the surviving cells become G418 resistant after electroporation with pBS118 (Table I), a frequency that is sufficiently high to permit screening for recombinant colonies by the polymerase chain reaction

TABLE I
Cre-MEDIATED DNA EXCISION IN CULTURED CELLS[a]

Cre vector	Cell viability (%)	Fraction of viable cells resistant to G418 (%)
None	10.2	0.1
pBS118	8.7	9.7

[a] Cell line 12HG-1 [3 × 10⁶ cells in 0.8 ml HeBS buffer; G. Chu, H. Hayakawa, and P. Berg, *Nucleic Acids Res.* **15,** 1311 (1987)] was electroporated with 15 µg of the indicated DNA using a single pulse of 1800 V at 25 µF with the Bio-Rad (Richmond, CA) Gene Pulser. Selection for resistance to 800 µg/ml G418 was imposed 48 hr after electroporation to allow Cre expression. Results shown are the average of two experiments. A low incidence of non-Cre-mediated G418 resistance occurs from gene amplification events in this cell line.[7]

(PCR). Higher frequencies can be obtained by using pBS185 and by optimizing the electroporation protocol for the particular cell line being used.

Plasmid pBS302 (Fig. 3) contains a *lox*[2] STOP cassette that we have used in transgenic mice[14] and which is useful in the design of recombination-activated gene expression. Although the yeast *LEU2* gene can act as a weak STOP element,[7] the STOP cassette on plasmid pBS302 is much more efficient. The cassette consists of a spacer DNA (from the yeast *HIS3* gene), the small intron and polyadenylation signal from SV40, followed by a gratuitous ATG translation start and 5' splice donor signal to quash correct expression from any residual transcription of the desired downstream gene. The STOP sequence is flanked by directly repeated *lox* sites. The entire cassette is on a *Not*I fragment that can be inserted between a promoter and the desired reporter gene.

Although we have used the *lox*[2] STOP cassette as a switch to control SV40 T antigen expression in the lens,[14] and hence lens tumor formation, RAGE strategies can be used to control expression of any transgene and may be particularly useful in marking cells for cell lineage analysis. For example, a reporter gene such as *lacZ* can be inserted into region III of pBS302 (Fig. 3) and a promoter of interest placed into region I. The resulting construct is contained on a *Not*I fragment that can be used for generation of transgenic mice. The STOP sequence in region II prevents expression of the *lacZ* reporter gene. Mating of the *lox*[2]–STOP–reporter mice with transgenic mice having the *cre* gene under either developmental or inducible control will generate doubly transgenic mice suitable for the desired lineage experiment. To verify that the lox^2–STOP–reporter mice exhibit the desired pattern of expression, it may be convenient to also mate them with a constitutive Cre expression mouse to activate the *lacZ* gene.

There are two interesting corollaries to this strategy. First, toxic genes will be tolerated in an inactive state in a lox^2–STOP mouse and can then be activated in progeny by mating to a Cre expression mouse. Because Cre protein is required only transiently to activate expression from the dormant gene (just a single recombination event per cell is needed), continual high-level expression of Cre in the target tissue is not required. Second, temporal or tissue-specific gene expression may be more precisely targeted by using a combination of Cre and lox^2–STOP mice. For example, if *cre* were designed to be expressed in tissues A and B, but not C, and the lox^2–STOP–reporter designed to express (after recombination) in A and C, but not B, then in the double transgenic animal expression would be expected to occur only in tissue A (because recombination would not have occurred in C).

Integration

Random integration of the transgene into the genome results in variable gene expression owing to position effects and variation in copy number of the transgene.[15] These effects can be circumvented either by gene targeting strategies or by incorporating the appropriate locus control region,[16] if known, on the transgene vector. A third strategy is to use Cre-mediated integrative recombination to target the transgene to a predetermined locus in the genome in ES cells. This strategy is likely to be most convenient when a series of alleles of a particular transgene need to be compared. A set of vectors that can be adapted to this purpose is shown in Fig. 4.

The target to be placed into the chromosome consists of both a *lox* site and a defective *neo* gene that lacks a promoter and the first five codons of the *neo* gene.[17] The *lox* site is designed to be in frame with the *neo* structural gene, but it does not provide a translational start. A selectable marker can be incorporated at one of the unique sites outside of the *lox–neo* fusion gene in pSF1 and the target introduced into the genome either randomly or by homologous gene targeting. The chromosomal *lox* site in the resulting cell line is targeted by Cre-

[15] E. Lacy, S. Roberts, E. P. Evans, M. D. Burtenshaw, and F. D. Constantini, *Cell* (*Cambridge, Mass.*) **34,** 343 (1983).

[16] F. Grosveld, G. B. van Assendelft, D. R. Greaves, and G. Kollias, *Cell* (*Cambridge, Mass.*) **51,** 975 (1987).

[17] S. Fukushige and B. Sauer, *Proc. Natl. Acad. Sci. U.S.A.* **89,** 7905 (1992).

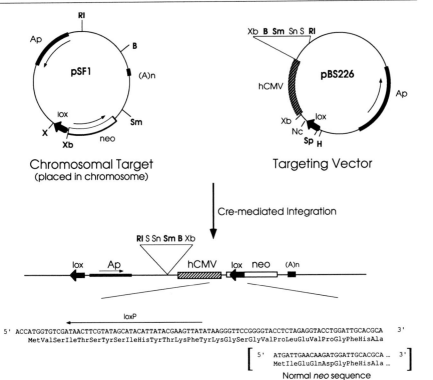

Chromosomal Target
(placed in chromosome)

Targeting Vector

Cre-mediated Integration

FIG. 4. *lox–neo* fusion shuttle vectors. Short, stubby black arrows represent *loxP* sites. Restriction sites in boldface type are unique. B, *Bam*HI; H, *Hin*dIII; Nc, *Nco*I; RI, *Eco*RI; S, *Sal*I; Sm, *Sma*I; Sn, *Sna*BI; Sp, *Sph*I; X, *Xho*I; Xb, *Xba*I.

mediated recombination by using a targeting vector such as pBS226 (Fig. 4). This targeting vector is essentially a site-specific integrating shuttle vector. Site-specific integration both restores a translational start to the defective *lox–neo* fusion gene and provides a promoter for expression. The resulting integrated DNA is flanked by *lox* sites and, hence, can be removed as discussed above.

To illustrate use of these vectors, we constructed a Chinese hamster ovary (CHO) cell line derivative in which a single copy of the *lox* target in pSF1 had been integrated into the genome.[17] In these cells Cre-mediated targeting with pBS226 gives G418-resistant colonies that are all single-copy site-specific integrants of the targeting vector. Figure 5 shows a typical cotransformation experiment of a Cre expression vector (pBS185) with increasing amounts of the targeting vector pBS226. Because of the design of the defective *lox–neo* fusion target, the occurrence of spontane-

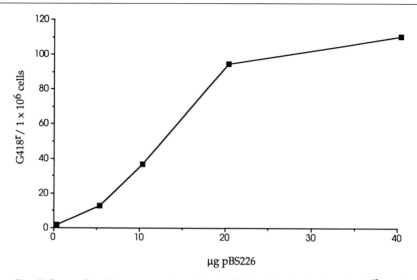

Fig. 5. Cre-mediated chromosome targeting. A CHO cell line derivative, 14-1-2,[17] containing a single randomly integrated copy of the defective *lox–neo* fusion gene was electroporated with the indicated amount of the pBS226 targeting vector and 10 μg of the Cre expression vector pBS185. Selection for G418 resistance was performed 48 hr later. The number of G418-resistant colonies per 1 × 10⁶ viable cells is shown. Electroporation conditions were as follows: 1 × 10⁷ cells in 0.8 ml HeBS buffer [G. Chu, H. Hayakawa, and P. Berg, *Nucleic Acids Res.* **15,** 1311 (1987)] were given a single pulse of 450 V at 500 μF with the Bio-Rad Gene Pulser.

ous G418 resistance is rare. To facilitate use of the targeting vector in ES cells, the hCMV promoter of pBS226 can be easily replaced with a promoter active in ES cells by digestion with *Xba*I.

In Vitro Cre Recombination

To verify that *lox²* constructs are behaving properly, either with respect to recombination or to activation of gene expression, it is useful to make the excised version of the *lox²* construct. This can be done by simply transforming the *lox²* construct into a Cre⁺ *Escherichia coli* strain like BS591,[18] a *recA* strain harboring a λ *cre* prophage. Alternatively, the recombination can be performed *in vitro* with commercially available Cre protein. Integrative recombination *in vitro* may also facilitate some plasmid constructions.

[18] B. Sauer and N. Henderson, *Gene* **70,** 331 (1988).

Reagents

	Amount	
Component	Cre dilution buffer	3× Cre reaction mix
Water	8.2 ml	216 μl
1 M Tris-HCl (pH 7.5)	0.25 ml	45 μl
1 M NaCl	0.5 ml	30 μl
0.25 M EDTA	0.04 ml	—
1 M MgCl$_2$	—	9 μl
50% Glycerol	1.0 ml	—
Bovine serum albumin (BSA, 20 mg/ml)	0.02 ml	Optional
	10 ml	300 μl

Procedure

1. Mix the following: 10 μl of 3× Cre reaction mix, 3 μl of 17% poly(vinyl alcohol) (stimulatory for integrative recombination), 0.20 μg of plasmid DNA, 3 μl of Cre protein (diluted to 40 μg/ml), available from Du Pont–NEN (Boston, MA), and water to 30 μl.

2. Incubate for 30 min at 30° followed by 5 min at 70°. The DNA can be used directly for transformation. For gel analysis, add electrophoresis sample buffer and extract once with chloroform–isoamyl alcohol (24 : 1, v/v).

Further Considerations

Site-specific recombination provides a powerful methodology for manipulating the genome in living cells and transgenic animals[14,19] and is likely to be useful in a wide variety of circumstances. Other site-specific recombinases related to the Cre protein will also be useful for these purposes.[20–23] These uses include removal of the selectable marker used in generating gene knockouts, cell lineage analysis using RAGE, and construction of transgenic animals harboring a developmental lethal gene that is activated only after mating to a Cre-expressing mouse. Because knockouts in certain important genes may have both an embryonal lethal

[19] P. C. Orban, D. Chui, and J. D. Marth, *Proc. Natl. Acad. Sci. U.S.A.* **89,** 6861 (1992).
[20] K. G. Golic and S. Lindquist, *Cell (Cambridge, Mass.)* **59,** 499 (1989).
[21] H. Matsuzaki, R. Nakajima, J. Nishiyama, H. Araki, and Y. Oshima, *J. Bacteriol.* **172,** 610 (1990).
[22] S. O'Gorman, D. T. Fox, and G. M. Wahl, *Science* **251,** 1351 (1991).
[23] S. Maesner and R. Kahmann, *Mol. Gen. Genet.* **230,** 170 (1991).

phenotype as well as a second role in the adult animal, it may be useful to complement a particular gene knockout with a lox^2 construct of that gene so that it can be specifically removed at a defined developmental time using an inducible or developmentally controlled Cre expression vector. Moreover, precise chromosome rearrangements, such as inversions and large deletions, should be possible by placing correctly oriented *lox* sites at the desired chromosome locations by gene targeting and then allowing Cre to perform the desired recombination. Mice containing precise chromosome rearrangements would be quite useful in both mapping and strain construction.

Acknowledgments

Many thanks are due to Wendy Baubonis, Shinichi Fukushige, and other members of my laboratory for help and critical support. Thanks also to Heiner Westphal and Merja Lasko (National Institutes of Health) for their unwavering encouragement.

[54] Embryonic Stem Cell Differentiation *in Vitro*

By MICHAEL V. WILES

Introduction

In this chapter we describe an *in vitro* system that allows access to the stages of embryogenesis where mesoderm commitment occurs followed by the development of the primary embryonic hematopoietic system. *In vitro* approaches to mouse embryonic development allow rapid and easy access to a system where controlled intervention can often be difficult. Once such systems are established experimental conditions can be readily altered at will. In addition, the amount of material available for *in vitro* studies is neither as limited nor as expensive as with *in vivo* derived sources. However, *in vitro* approaches are only pathfinders, a rapid way to approach many ideas and questions before going on to the more complex and often technically difficult *in vivo* world. As such, experiments conducted *in vitro* cannot be regarded as a substitute for *in vivo* analysis; by using the various advantages of each experiment as needed, complex life systems can be most effectively analyzed.

Copyright © 1993 by Academic Press, Inc.
All rights of reproduction in any form reserved.

Embryonic Stem Cells and What They Can Do

Mouse embryonic stem (ES) cells are totipotent, nontransformed, primitive ectoderm-like cells derived from the inner cell mass of 3.5 day blastocysts.[1,2] They are thought to resemble most closely the primitive ectoderm of the very early postimplantation embryo.[3] ES cells can be maintained in culture for many passages in an undifferentiated state either on "feeder" cells or in the presence of an inhibitor of ES cell differentiation, namely, leukemia inhibitory factor (LIF).[4,5]

Embryonic stem cells have two main characteristics: (1) They can be reintroduced back into a 3.5 day blastocyst which, when placed into a suitable host (a pseudopregnant mouse) can colonize all tissues of the developing embryo, giving rise to a chimeric animal (i.e., part host derived, part ES derived). As such they provide a fundamental tool for genetic engineering.[6-9] (2) If ES cells are allowed to reach a high density in monolayers or to form aggregates in suspension (embryoid bodies, EBs), they will spontaneously differentiate to various cell types.[1,2,10] For example, Doetschman and colleagues were able to differentiate ES cells to parietal and visceral endoderm, cardiac muscle, and other unidentified cell types.[10] A further dimension to this *in vitro* differentiation is its use where genes have been selectively inactivated in ES cells by homologous recombination. As both alleles can be relatively easily disrupted, a null mutation can be generated.[11] As such, it should be possible to define rapidly *in vitro* the mode of action of

[1] M. J. Evans and M. H. Kaufman, *Nature (London)* **292,** 154 (1981).

[2] G. R. Martin, *Proc. Natl. Acad. Sci. U.S.A.* **78,** 7634 (1981).

[3] M. J. Evans and M. Kaufman, *Cancer Surv.* **2,** 185 (1981).

[4] R. L. Williams, D. J. Hilton, S. Pease, T. A. Willson, C. L. Stewart, D. P. Gearing, E. F. Wagner, D. Metcalf, N. A. Nicola, and N. M. Gough, *Nature (London)* **336,** 684 (1988).

[5] A. G. Smith, J. K. Heath, D. D. Donaldson, G. G. Wong, J. Moreau, M. Stahl, and D. Rogers, *Nature (London)* **336,** 688 (1988).

[6] A. Bradley, M. Evans, M. H. Kaufman, and E. Robertson, *Nature (London)* **309,** 255 (1984).

[7] E. Robertson, A. Bradley, M. Kuehn, and M. Evans, *Nature (London)* **323,** 445 (1986).

[8] T. Doetschman, R. G. Gregg, N. Maeda, M. L. Hooper, D. W. Melton, S. Thompson, and O. Smithies, *Nature (London)* **330,** 576 (1987).

[9] S. Thompson, A. R. Clarke, A. M. Pow, M. L. Hooper, and D. W. Melton, *Cell (Cambridge, Mass.)* **56,** 313 (1989).

[10] T. C. Doetschman, H. Eistetter, M. Katz, W. Schmidt, and R. Kemler, *J. Embryol. Exp. Morphol.* **87,** 27 (1985).

[11] R. M. Mortensen, D. A. Conner, S. Chao, A. A. T. Geisterfer-Lowrance, and J. G. Seidman, *Mol. Cell. Biol.* **12,** 2391 (1992).

any gene involved in early development. In certain cases, gene inactivation will prevent or reduce the contribution of ES cells to the developing embryo or lead to poor chimera formation and/or embryo death; such problems can be circumvented by an *in vitro* approach.[12,13]

Differentiation of ES cells to hematopoietic cells *in vitro* was first observed by Doetschman and colleagues, who found that under certain conditions a low percentage of EBs developed small islands of primitive erythrocytes.[10] Using a similar protocol, the range of hematopoietic cells observed has been extended.[14] However, in both cases the protocols relied on the presence of human cord serum during the differentiation process. This is a significant drawback since human cord serum is not readily available, batches are small, and the composition is subject to extreme variation. As such the lack of reproducibility limits the usefulness of this approach.

The protocols described in this chapter use commercially available reagents and are simple, reproducible, and efficient. Following these protocols, greater than 95% of all ES-derived EBs generated will contain yolk sac-like hematopoietic progenitors within 6 days of differentiation (G. Keller, personal communication; Wiles, personal observations).[15] To obtain ES cell differentiation to other cell types other approaches can be used; for example, in the case of cardiac muscle or endothelial cells, the reader is referred to studies by the Doetschman and Bautch groups, respectively.[16,17]

Basic Considerations of Embryonic Stem Cell Maintenance and Differentiation

Embryonic Stem Cells

It is beyond the scope of this chapter to give a complete outline of how to isolate and maintain ES cells; instead, the reader is referred to the excellent review by Robertson.[18] Briefly, to achieve efficient and repro-

---

[12] L. Pevny, M. C. Simon, E. Robertson, W. H. Klein, S. F. Tsai, V. D'Agati, S. H. Orkin, and F. Costantini, *Nature (London)* **349,** 257 (1991).

[13] M. C. Simon, L. Pevny, M. V. Wiles, G. Keller, F. Costantini, and S. H. Orkin, *Nat. Genet.* in press (1992).

[14] R. M. Schmitt, E. Bruyns, and H. R. Snodgrass, *Genes Dev.* **5,** 728 (1991).

[15] M. V. Wiles and G. Keller, *Development (Cambridge, UK)* **111,** 259 (1991).

[16] J. Robbins, J. Gulick, A. Sanchez, P. Howles, and T. Doetschman, *J. Biol. Chem.* **265,** 11905 (1990).

[17] R. Wang, R. Clark, and V. L. Bautch, *Development (Cambridge, UK)* **114,** 303 (1992).

[18] E. J. Robertson, *in* "Teratocarcinomas and Embryonic Stem Cells: A Practical Approach" (E. J. Robertson, ed.), p. 71. IRL Press, Oxford and Washington, D.C., 1987.

ducible *in vitro* differentiation of ES cells, it is necessary that the cells be maintained in a healthy state. This is best achieved by learning how to maintain ES cells in an established ES cell laboratory. If this is not possible, then the absolute minimum requirements are good tissue culture skills, ultrapure water [Millipore (Bedford, MA) Milli-Q system or equivalent] for preparing media (if necessary, buy ready-made medium from a reputable dealer), an inverted phase-contrast microscope, and an experienced "eye" for cells.

Embryonic stem cells can be maintained in an undifferentiated state on gelatin-treated tissue culture grade plastic plates or flasks in the presence of feeder cells. However, for *in vitro* differentiation experiments we tend to use ES lines that have been adapted to remain undifferentiated in the presence of LIF and in absence of feeders cells. Such cultures will not have the possibility of interference from "contaminating" feeder cells during the differentiation process. Additionally, ES cells grown in the absence of feeders show a higher plating efficiency when seeded in suspension or semisolid methyl cellulose-containing media.

Differentiation Conditions: Suspension versus Matrix

There are many ways to allow ES cells to differentiate spontaneously. The basic scenario is to allow the formation of aggregates, either within a monolayer by crowding or by preventing cell attachment to the substrate. The choice of differentiation system will depend on which cell types are wanted and the questions being asked. The variations described here use methyl cellulose media (MCM) and allow reproducible and efficient hematopoietic differentiation.

Methyl cellulose is wood pulp and has been in use for some time as an inert support matrix for hematopoietic precursor cell development.[19] At the concentrations used here the material handles like a thick "soup" or "semijelly." When cells are placed into it, they tend to remain stationary and will develop from single cells into distinct colonies.

Advantages of Methyl Cellulose Media. (1) When ES cells are plated under conditions where they cannot attach to tissue culture plastic, they will develop into aggregates, the EBs. When standard medium lacking methyl cellulose is used, these aggregates will form rapidly by cell–cell collision and adhesion. As these cultures develop, cells and cell aggregates coalesce, ultimately forming very large structures that can develop necrotic regions. These heterogeneously sized EBs will also rapidly lose any synchrony in differentiation. If ES cells are seeded into a semisolid

---

[19] N. N. Iscove and H. Schreier, *in* "Immunological Methods" (I. Lefkovits and B. Pernis, eds.), Vol. 1, p. 379. Academic Press, New York, 1979.

matrix of MCM, EBs will develop in isolation from single cells, with collision aggregation occurring only rarely. This leads to the development of a synchronized culture which undergoes reproducible temporal development. (2) By using methyl cellulose, individual EBs can be monitored as they develop. (3) Under a number of conditions, for example, in the presence of factors such as erythropoietin and/or interleukin 1α (IL-1α) plus interleukin 3, there is a massive expansion of mature hematopoietic cells for rupturing EBs. In MCM these will remain in close proximity to the originating EB, allowing easy scoring, whereas in standard liquid media the differentiated progeny cells disperse.[15] (4) Even at low initial cell seeding densities using non-tissue culture (bacterial grade) plastic dishes, aggregating ES cells/EBs in standard media have a strong tendency to attach to the plate and then spread; however, this occurs only rarely in MCM. If the three-dimensional structure of the EBs is lost, hematopoietic differentiation is greatly reduced. In addition, once EBs begin to attach and spread there is rapid cell growth, leading to exhaustion of the medium.

Disadvantages of Methyl Cellulose Media. (1) The C57BL/6 (SQ1.2) and DBA2 (BAM3) ES lines we have used have a very low plating efficiency when placed directly into methyl cellulose. This can be overcome by allowing cells to aggregate for 24–48 hr in conventional medium before seeding into MCM, although this will lead to a loss of differentiation synchrony. (2) Monitoring the effect of factors added during an experiment, for example, 2 to 3 days after initiation, is difficult, since they disperse slowly through the MCM gel. (3) In some experiments, for example, in attempted hematopoietic reconstitution of irradiated animals with differentiated ES cells, the presence of methyl cellulose itself is an additional complication. (4) Any additional reagent can always be a source of problems.

Reagents/Supplies

Medium. ES cells are generally maintained in Dulbecco's modified Eagle's medium (DME–high glucose; e.g., BRL-GIBCO, Grand Island, NY, Cat. No. 041-019695) supplemented with 15% fetal calf serum (FCS) and 150 μM (12.6 μl/liter) monothioglycerol (MTG; Sigma, St. Louis, MO, Cat. No. M-1753). Efficient hematopoietic ES cell differentiation is obtained in Iscove's modified Dulbecco's medium (IMD; e.g., GIBCO–BRL, Cat. No. 041-01980) supplemented as described below. If the medium is prepared in house it must be from ultrapure water, as ES cells are very sensitive to impurities, especially at clonal densities (i.e., 200 to 1000 cells/ml).

Plastics. ES cells are routinely grown, plus or minus feeder cells, on gelatin-treated tissue culture plastic.[18] For ES cell hematopoietic differentiation in either suspension or MCM, bacterial grade petri dishes must be used (35 or 90 mm). If normal tissue culture treated plastics are used for EB formation, cells will attach to the plate, spread, and fail to give effective hematopoietic differentiation.

Trypsin/EDTA. On subculturing, ES cells must be dissociated to single cells to prevent aggregate formation. We have found that this is best achieved by a higher than normal concentration of trypsin.[18] Routinely we use 0.25% trypsin (e.g., Sigma, Cat. No. T-8128) in standard phosphate-buffered saline (PBS) with 1 mM Na$_2$EDTA, plus (optional) phenol red, final pH 7.5 (T/V). The main stocks of T/V can be aliquoted and stored at $-20°$. Once a working stock has been thawed it should be used in the next 8 hr or so, as it will self-digest. To produce a single cell suspension of ES cells, the flask/plate of cells is washed once with T/V, then an additional 1 ml/25 cm^2 is added and the plate/flask incubated at 37° for 3–5 min. The cells can then be resuspended in medium plus FCS for subculturing or experimental use.

Leukemia Inhibitory Factor. Many lines of ES cells can be maintained in an undifferentiated state by the addition of LIF to the medium.[4,5] Recombinant LIF is routintely used at approximately 1000 units/ml and can be obtained from Amrad Co Ltd. (Victoria, Australia). Another source is conditioned medium from an LIF-Chinese hamster ovary (CHO) cell line, available as a gift under certain conditions from the Genetics Institute Inc. (Cambridge, MA 02140-2387).

Fetal Calf Serum. FCS batch-related variations on both ES cell growth and differentiation exist. Of the FCS batches we have tested, approximately three-quarters have given reasonable hematopoiesis, with about one-quarter giving excellent hematopoietic differentiation. The FCS should be tested by the following critiera: (1) It is capable of supporting normal ES cell growth at clonal densities (for a full description, see review by Robertson).[18] (2) It can support ES–EB hematopoiesis. This can be easily tested by using a 129/Sv-derived ES cell (e.g., CCE[7]) seeded in MCM in the presence of 15% serum plus 450 μM MTG (see below). If the FCS is good, EBs will be observed after 3 days. After 10 days, at least 40% of all EBs should show overt globinization, that is, will be distinctly red (see scoring section below). (3) If hematopoietic precursor studies are to be undertaken, then the FCS should also be tested for its ability to support hematopoietic colony formation in MCM bone marrow cultures at 10 to 15%.[19] Finally, heat-inactivation of FCS (56° for 30 min) is optional; we have not observed a difference between treated and nontreated serum. Newborn calf serum appears to be toxic and should not be used.

Methyl Cellulose Stock Solution. MCM can be purchased from a number of sources (e.g., Terry Fox Laboratory, 601 West 10th Ave., Vancouver, BC V5Z 1L3); however, it is fairly simple to prepare. The protocol given here is based on that of Iscove and Schreier and will make 500 ml of 2× MCM.[19]

Place 230 ml of ultrapure sterile water plus a *clean* magnetic stirrer into a sterile conical flask with a loose-fitting top, for example, a small beaker. Record the total weight of the assembly. Gently boil the water for 5 to 10 min. While stirring, *slowly* add 10 g methyl cellulose (MC, USP 4000 mPa.s.; Fluka A.G., Buchs, Switzerland, Cat. No. 64630). This should result in a slurry with few or no lumps. Recover the flask and sterilize the contents by boiling for another 10 to 12 min. Allow to cool to below 40°. Add 250 ml of sterile 2× IMD, made by adding half the required amount of ultrapure water to IMD power (available from, e.g., BRL-GIBCO, Cat. No. 074-02200A). Stir rapidly for some minutes to mix and to remove any remaining lumps. Bring the medium–methyl cellulose slurry to a final weight of 506 g (plus beaker, stirrer, etc.) with sterile water. At this stage the methyl cellulose will still appear cloudy and moderately thin. Aliquot and *freeze,* storing at −20°. Only following this initial freezing will the MCM appear as a transparent viscous liquid. MCM should not be refrozen. Working stocks can be maintained at 4° for 1 month.

Growth Factors. The basic differentiation of ES cells to mesoderm, together with the initial wave of embryonic hematopoiesis, appears to be independent of additional exogenous growth factors tested so far. The subsequent expansion of myeloid cells, however, is substantially increased by the presence of certain growth factors.[15] These factors include erythropoietin (Epo), Steel (also known as SCF or KL, the ligand to c-*kit*), IL-1α, IL-3, and macrophage colony-stimulating factor (M-CSF, also known as CFS-1). Other factors that may influence commitment and differentiated cell expansion include the fibroblast growth factor and tumor growth factor β families. In addition, it is recommended that insulin (10 μg/ml; e.g., Sigma, Cat. No. I-6634) be added to the differentiation medium; this will increase plating efficiency and possibly hematopoietic precursor expansion. These factors are available commercially. Under certain conditions some are also available from the Genetics Institute Inc., as gifts. Additionally, many factors are also obtainable in conditioned media from cells transfected with the recombinant gene.

We have used recombinant human erythropoietin (rhEpo), purchased from Cilag A.G. (Schaffhausen, 6300 Zug, Switzerland; Eprex 4000), at

2 units/ml; mouse recombinant IL-3 is provided in a conditioned medium from a transfected X63 Ag8-653 myeloma cell line provided as a gift by Melchers.[20] For M-CSF, L-cell conditioned medium can be used; however, it does contain other factors, for example, Steel, that may complicate its use. Hematopoietic factors are used at concentrations that give at least three times the half-maximal colony response in murine bone marrow colony assays.[19,21]

5637-Conditioned Media. The 5637 cell line was derived from a human bladder carcinoma. These cells produce a number of growth factors including LIF.[22-24] To prepare 5637-conditioned medium, cells are grown to confluence in normal medium (e.g., DME with 5-10% FCS). The medium is removed and replaced with DME supplemented with 2% fetal calf serum (10 ml/75-cm² flask). After 3 to 4 days the conditioned medium is removed, centrifuged to remove cell debris, filter-sterilized, and stored at $-20°$. This procedure can be repeated several times with a single flask of cells.

Reducing Agents. For efficient hematopoietic differentiation from ES cells there is an absolute requirement for the presence of a reducing agent.[15,25] Reducing agents probably act by scavenging free radicals, although the exact mode of action on cells is not currently understood.

We have tested various reducing reagents and have found that monothioglycerol (MTG) gives good, reproducible results in both suspension and MCM cultures. The concentration found to be optimal for ES cell hematopoietic differentiation in the CCE line is three times higher than that normally added to medium namely, 450 μM (equal to 37.8 μl MTG per liter). It is strongly stressed that this reagent must be diluted at the time of setting up each experiment. Diluted stock solutions of MTG deteriorate rapidly, resulting in a sharp decline in both ES cell plating efficiency and hematopoietic cell generation. Other reducing agents such as 2-mercaptoethanol, dithiothreitol, and to a lesser extent vitamin E and ascorbic acid do work, but in our hands are less efficient.

It has also been suggested that the concentration of reducing agents present during the differentiation influences the direction of EB develop-

[20] H. Karasuyama and F. Melchers, *Eur. J. Immunol.* **18,** 97 (1988).
[21] N. N. Iscove, "Methods for Serum-Free Culture of Neuronal and Lymphoid Cells," p. 169. Alan R. Liss, New York, 1984.
[22] H. Gascan, I. Anegon, V. Praloran, J. Naulet, A. Godard, J. P. Soulillou, and Y. Jacques, *J. Immunol.* **144,** 2592 (1990).
[23] R. Sorg, J. Enczmann, U. Sorg, K. Heermeier, E. M. Schneider, and P. Wernet, *Exp. Hematol.* **19,** 882 (1991).
[24] J. G. Kaashoek, R. Mout, J. H. Falkenburg, R. Willemze, W. E. Fibbe, and J. E. Landegent, *Lymphokine Cytokine Res.* **10,** 231 (1991).
[25] R. Oshima, *Differentiation* **11,** 149 (1978).

ment; for example, cardiac muscle is seen rarely in ES differentiations conducted in 450 μM MTG but is abundant in cultures started with 150 μM[15] (M. V. Wiles, unpublished observations). Interestingly, it has also been found that ES cell plating efficiency in both normal medium and MCM is significantly increased if the cells are maintained under low O_2 conditions (e.g., 7% O_2, 5% CO_2, 88% N_2, by volume). Hematopoietic differentiation may also be enhanced by low O_2 (M. V. Wiles, unpublished observation).

Embryonic Stem Cell Lines

Many ES cell lines/strains are available. The protocols given here for hematopoietic differentiation have worked extremely well on all 129/Sv-derived ES cells tested (e.g., CCE and D3).[7,10] The C57BL/6 lines SQI.2[15] and 1E6/2[14] plus the DBA2 ES line BAM3 (M. V. Wiles, unpublished observations) all give reasonable to good hematopoietic differentiation; however, modifications to the basic protocols are required (see below) (M. V. Wiles, personal observations; G. Keller, personal communication).

Methods

Adaptation of Embryonic Stem Cells to Feeder Independence

To adapt ES cells from feeder cell dependence, subculture cells as normal onto gelatin-treated plastic, but in the absence of feeder cells. To minimize trauma, the medium should be made up of 50% 5637-conditioned medium (see above), 35% DME supplemented with (freshly diluted) MTG at 150 μM, 15% FCS, and 1000 units of LIF. As conditioned medium is already partly exhausted, it is necessary to change the medium daily. The cells will require subcultured 1 : 3 to 1 : 8 every 2 to 3 days and should at no time reach more than two-thirds confluent. After two passages the concentration of 5637-conditioned medium can be reduced to 25% while maintaining LIF at approximately 1000 units/ml. By the fourth subculture (~10 to 12 days), the majority of the cells should be in a good (undifferentiated) state, and the 5637-conditioned medium can be left out at further subculture. Once adapted, the majority of the feeder-independent ES cell stock should be divided into a number of ampules and frozen for future use. We have found that continued long-term culture in LIF alone can reduce the differentiation capability of ES cells. As such, it is recommended that ES cells be subcultured for no more than 20 passages in LIF alone when they are to be used for hematopoietic differentiation experiments.

It should be noted that some cell death and differentiation may occur, especially during the primary adaptation passages. The alternative is to use an ES line that was etablished in LIF, in the absence of feeders (e.g., the 129/Sv-derived MBL-1 line).[26] This adaptation protocol has worked well for feeder-dependent 129/Sv-derived ES cells; however, with C57BL/6 and DBA2 feeder-dependent derived lines, a substantial percentage of cells continue to differentiate, even in the presence of high concentrations of LIF (M. V. Wiles, personal observations).

Embryonic Stem Cell Differentiation in Methyl Cellulose Medium

To obtain consistent results, ES cells should be given a medium change 1 to 3 hr before use. Cells that are subconfluent and in an undifferentiated state are dissociated to single cells using T/V (see above), then resuspended in DME or IMD plus 150 μM MTG and 15% FCS but in the absence of LIF. If the cells have been maintained on feeder layers, these can be partly removed by plating the cell suspension onto tissue culture grade plastic for 1 hr, where the feeder cells will preferentially attach. Cells are then counted. The number of cells seeded per milliliter into MCM is moderately low; it is therefore recommended for accurate dilution of cells that the working ES cell suspension be below 5×10^5 cells/ml.

The 1.8% stock methyl cellulose in IMD is extremely thick. To aliquot the required amount either pour directly or, for smaller amounts, use a syringe fitted with a 18-gauge (pink) needle. For normal differentiation, dilute the 1.8% methyl cellulose stock with an equal volume of IMD. Add insulin (optional, at 10 μg/ml), FCS to a final concentration of 15%, and at the *time of use* MTG to a final concentration of 450 μM. The last addition is conveniently done by adding 3 μl of a *freshly* prepared MTG diluted stock per milliliter MCM. (The diluted stock is made by adding 26 μl of neat MTG into 2 ml of medium.) This ready-to-use MCM–cell mix is most easily handled by using a syringe fitted with 18-gauge needle.

The initial ES cell seeding density into MCM depends on a number of variables: (1) the originating mouse strain of the ES cells, (2) whether the cells have been maintained on feeder layers or LIF alone (LIF-maintained cells have a higher plating efficiency), (3) the condition of cells prior to experiment (overgrown cultures have a poor plating efficiency), (4) the duration of the experiment (too many cells/EBs will lead to premature medium exhaustion, necessitating feeding), (5) the presence/absence of feeder or "stroma" cells in the differentiation culture (in some experiments it may be useful to include irradiated non-ES cells in MCM differentiations,

[26] S. Pease, P. Braghetta, D. Gearing, D. Grail, and R. L. Williams, *Dev. Biol.* **141**, 344 (1990).

e.g., stroma cells, as this will lead to an increase in plating efficiency), and (6) the presence of extracellular matrix components, which can lead to an increase in plating efficiency.

To determine the optimal cell density of ES cells at initiation of differentiation, a dilution series covering 100 to 30,000 cells/ml should be set up. In general 300 to 600 ES cells/ml works well. The aim is to have approximately 50 to 300 EBs per milliliter for cultures of 0 to 6 days duration. For cultures beyond 6 days, around 50 EBs/ml is reasonable. Routinely we use 35-mm bacterial dishes, with 1 to 1.25 ml of MCM. If tissue culture treated plastic is used, cells will rapidly attach and spread, and hematopoietic differentiation will not occur. It is also important that the plates not be allowed to dry out appreciably during the experiment. This is conveniently avoided by placing the small dishes of cells/MCM into another, larger plate with a lid which also contains a small open dish of water for additional humidity.

We have found that C57BL/6 and DBA2 ES cell lines exhibit a low plating efficiency when put directly into MCM. Reasonable hematopoietic development can be obtained in MCM by first allowing the cells to form aggregates in liquid culture. This is achieved by seeding cells at $1-3 \times 10^5$ cells/ml in IMD supplemented with 15% FCS and 450 μM MTG for 24 to 48 hr. The aggregates that form are then added to MCM, where they develop into EBs. This approach, however, leads to heterogeneous EB formation and subsequent loss of close synchrony. For a listing of possible problems and possible solutions, the reader is referred to Table I.

What Will Be Seen. After 3 days of differentiation in MCM, ES cells will have developed into tight clusters consisting of 5 to 10 cells (Fig. 1A,B). After 6–8 days these will reach a diameter of 100 to 500 μm. By day 8, overt globinization will be seen with dark-field illumination (i.e., regions of red). The color will appear to be principally at the center of the EB. Over the next 3 to 5 days globinization will increase substantially. If cryostat sections are made of EBs after more than 6 days of culture, cavities, many of which contain pockets of erythroid cells, can be found in most EBs (Fig. 1C,D,F). In the presence of Epo, there will also be a substantial expansion of erythroid cells (Fig. 1E), which will often rupture the EB, resulting in a halo of red cells forming around the EB.[15] After about 12 days, erythropoiesis will begin to fade; however, in the presence of IL-1α and IL-3 there will be a massive, almost explosive expansion of mature macrophages and other hematopoietic lineages (Fig. 1G). If the EB density is not too high, expanding colonies of macrophages and other cell types will exceed 3 mm in diameter.

TABLE I
TROUBLESHOOTING FOR CELL CULTURE IN METHYL CELLULOSE MEDIUM

Effect seen	Possible reason	Possible cure
ES/EBs attach and spread	Tissue grade plates used	Use bacterial grade plates
	"Sticky" bacterial grade plates used	Use bacterial grade plates from another supplier
EBs develop, but stop growing or die after 3–5 days	Water evaporation from medium	Reduce drafts by enclosing plates in a larger dish
	Toxic reagent	Test reagents, if possible individually
	ES line will not plate directly in MCM	Plate cells out in medium without methyl cellulose for 24–48 hr before adding MCM
No sign of EBs forming	No reducing agent added to medium	Repeat this time adding fresh reducing agent
	No FCS added	Repeat this time adding FCS
	No cells added, etc.	Repeat this time adding cells, etc.
Medium becomes yellow by day 6 or so	Plating cell density too high	Titer number of cells to give about 50 to 100 colonies/1.25 ml/35-mm plate
EBs develop, but no or poor hematopoiesis	FCS inhibitory to hematopoiesis	Test batch of FCS
	Residual LIF in medium	Wash cells in LIF-free medium before use
	"Abused" ES line	Return to an earlier passage or change ES line
Few EBs form	Cell line has a low plating efficiency	Repeat using a range of cell densities
Uneven distribution of EBs in MCM	Not completely mixed	Mix MCM/cells thoroughly

Long-Term Culture of Macrophages and Mast Cells Derived from Embryonic Stem Cells

Precursors of both macrophages and mast cells are found by days 6 and 10 of ES/EB development, respectively.[15,27] For some work, it may be required to produce large numbers of ES-derived macrophages and/ or mast cells. This can be conveniently done by replating EBs from MCM into fresh medium and selectively expanding the cells in the presence of hematopoietic factors. Although EBs grown in MCM, 15% FCS, and MTG

[27] U. Burket, T. von Rüden, and E. F. Wagner, *New Biol.* **3,** 698 (1991).

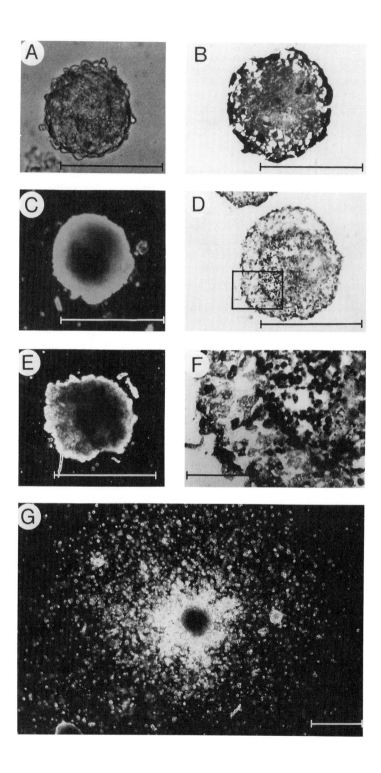

alone can give rise to macrophages and mast cells, those grown with the addition of IL-1α and IL-3 to MCM will give rise to these cells more rapidly. (L-cell-conditioned medium will also greatly increase the numbers of macrophages.)

The basic protocol involves using 10- to 15-day-old EBs grown in MCM plus IL-1α and IL-3. By this time the majority of the EBs will be surrounded by a halo of macrophages and other hematopoietic cells (Fig. 1G). Three to five EBs are transferred from MCM (e.g., by using a Gilson pipette) into 30- to 50-mm tissue culture grade plates with 1 to 4 ml IMD plus 150 μM MTG (12.6 μl MTG per liter of medium) and 10% FCS.

In the presence of IL-3 alone, mast cells will be predominant. For selective expansion of macrophages, a combination of IL-3 and M-CSF (L-cell-conditioned medium at 3–10%) is very effective. Within 48 hr EBs will attach and spread onto the tissue culture plastic. After 3 to 10 days regions of highly refractive, rapidly dividing cells will be seen under phase-contrast microscopy. The medium should be changed every 3 days or as it becomes acidic. If the cultures are maintained well, large numbers of macrophages or mast cells can be harvested twice weekly for at least 10 weeks.

Mast cells can be identified by their characteristic staining with May–Grünwald–Giemsa stain or toluidine blue; both stains will show densely staining cell granules. Macrophages can be identified by their highly vacuolated appearance after staining with May–Grünwald–Giemsa. In addition, macrophages will actively phagocytose cell debris, etc. The monocyte/macrophage-specific antibody F4/80 can be used for final identification of macrophages.[28]

[28] J. M. Austyn and S. Gordon, Eur.J. **11,** 805 (1981).

Fig. 1. Development in MCM from single ES cells of EBs over time, in the presence or absence of additional growth factors. (Note differing scale bars.) (A) Photomicrograph of a day 4 EB. Bar: 100 μm. (B) Cryostat section of a May–Grünwald–Giemsa-stained day 4 EB; note the apparent simple structure. Bar: 100 μm). (C) Dark-field photomicrograph of EB grown for 10 days in MCM (no additional exogenous factors). Although not readily apparent in this black-and-white photograph, the dark center of the EB, is red with globin. Bar: 500 μm. (D) Cryostat section of EB grown for 10 days in MCM with no additional exogenous factors, stained with May–Grünwald–Giemsa; note the clusters of small, densely staining erythropoietic cells. Bar: 500 μm. (E) Dark-field photomicrograph of EB grown for 12 days in MCM supplemented with 2 units/ml rhEpo. The central EB has ruptured and is now surrounded by a halo of red cells. Bar: 1000 μm. (F) Cryostat section showing a close-up of an active hematopoietic region of a day 10 EB. Bar: 100 μm. (G) Dark-field photomicrograph of EB grown in MCM plus IL-1α and IL-3 for 15 days. A central EB has ruptured and is surrounded by a rapidly expanding field of macrophages and other hematopoietic cells. Bar: 1000 μm.

Embryonic Stem Cell Differentiation in Media without Methyl Cellulose

Subconfluent, undifferentiated ES cells are prepared as for MCM culture. MTG (final concentration 450 μM) is added freshly to IMD plus 15% FCS. Various initial cell densities should be used (see above). Under these conditions hematopoietic differentiation appears similar to that seen for MCM cultures except that markers of early differentiation, for example, *Brachyury* (T gene), GATA1, and βH1, appear more rapidly and are not as synchronous as seen with MCM (see below for details on these genes). In addition, under these culture conditions the developing EBs tend to "shed" cells, creating a field of EBs with clusters of floating differentiated progeny and debris. In the presence of IL-1α and IL-3 this sea of differentiated cells can become dominated by macrophages and to a lesser extent mast cells. (Also see comments above about possible problems in the standard medium.)

Scoring of Globinization/Hematopoiesis

There are a number of ways to score hematopoietic development of EBs. These range from simply counting the number of "red" or globinized EBs over time to disrupting EBs into single cells followed by replating under conditions that allow hematopoietic precursor cell development.[27,28a]

In the first instance, the rough-and-ready approach of looking down the microscope and counting EBs that show overt signs of globinization is adequate, if rather subjective. Direct observation scoring is best achieved under dark-field illumination using a white light source (e.g., a halogen light) combined with a pale blue filter. Under these conditions erythropoietic EBs will appear to glow red and can be easily scored. Under conventional yellow light sources it is easy to misidentify EBs having a yellow-brown appearance as globinized cells, which may not be the case. If necessary, direct observation scoring can be verified by picking single EBs, isolating and dot blotting the RNA, followed by probing for mRNA coding for βH1. For such small amounts of material, the acid–guanidinium–phenol–chloroform (AGPC) protocol of RNA extraction and purification is effective.[29]

To monitor accurately the development of hematopoietic development in EBs, disruption of the EBs to single cells is required, followed by plating under conditions that allow precursor development. A procedure for the disruption of EBs is based on a report by Wong *et al.*, who

[28a] G. Keller, M. Kennedy, T. Papayannopoulo, and M. V. Wiles, *Mol. Cell. Biol.* **13**, 473 (1993).

[29] P. Chomczynski and N. Sacchi, *Anal. Biochem.* **162**, 156 (1987).

monitored hematopoietic development in the mouse yolk sac.[30] Briefly, EBs (≥4 days) are washed in PBS without Mg^{2+} or Ca^{2+}, then digested in a solution of 0.25% collagenase (Sigma, Cat. No. C-0130) in PBS plus 20% FCS, at 37° with occasional shaking for about 1 hr. The EBs, which will still appear intact at this stage, are passed three times (6 passages) through a syringe fitted with a 20-gauge needle (yellow). This will cause the majority of EBs to dissociate into single cells. Cells are counted, and 10^4 to 10^5 cells/ml are plated under conditions that will allow myeloid colony formation (see Refs. 19 and 21 for details). This protocol is also effective for single EBs. The digestion is carried out in 50 μl of collagenase, and the disruption is done by pipetting repeatedly in a "yellow tip" of a Gilson pipette. Trypsin/EDTA solutions should not be used, as the survival of hematopoietic precursors is reduced.

Molecular Markers of Embryonic Stem Cell Development

There are a number of molecular markers that can be used to follow the progressive differentiation of ES cells. Gene transcription can be conveniently monitored using the reverse transcription–polymerase chain reaction (RT–PCR) or, where the RNA is abundant, by RNase protection or Northern blot analyses. Here we describe the expression pattern seen for some informative genes. (For others see Ref. 28a.)

As mentioned earlier, ES cells are similar to primtive ectoderm. *In vivo,* primitive ectoderm differentiates to mesoderm soon after embryo implantation. *Brachyury* (T gene) expression is a good marker of presumptive mesoderm and of early mesoderm. It has been detected by *in situ* hybridization in cells at the primitive streak as they develop into mesoderm.[31,32] As such, it would be expected that one of the earliest steps of ES cell differentiation would involve the formation of mesoderm and expression of the T gene; this is indeed the case. Using the ES line CCE differentiated in MCM, T gene mRNA can be found by RT–PCR after 72 hr of differentiation. Expression peaks at 96 hr and then rapidly declines.[28a]

GATA1 is a transcription factor involved in the earliest steps of hematopoiesis. It has been suggested that as mesoderm commits to hematopoietic cells, GATA1 transcription is activated.[33,34] Using the ES line CCE

[30] P. M. Wong, S. W. Chung, D. H. Chui, and C. J. Eaves, *Proc. Natl. Acad. Sci. U.S.A.* **83,** 3851 (1986).

[31] B. G. Herrmann, S. Labeit, A. Poustka, T. R. King, and H. Lehrach, *Nature (London)* **343,** 617 (1990).

[32] D. G. Wilkinson, S. Bhatt, and B. G. Herrmann, *Nature (London)* **343,** 657 (1990).

[33] E. Whitelaw, S. F. Tsai, P. Hogben, and S. H. Orkin, *Mol. Cell. Biol.* **10,** 6596 (1990).

[34] L. I. Zon, C. Mather, S. Burgess, M. E. Bolce, R. M. Harland, and S. H. Orkin, *Proc. Natl. Acad. Sci. U.S.A.* **88,** 10642 (1991).

differentiated in MCM, GATA1 mRNA is first detectable by RT–PCR after 96 hr. GATA1 expression rises rapidly thereafter.[28a,35]

Transcription of the embryonic β-major globin-like gene βH1 is detected *in vivo* in embryonic erythrocytes of the 7.5 to 8 day embryo yolk sac blood islands. Slightly later adult β-major globin is also detected.[30,36,37] Using the cell line CCE in MCM, βH1-specific mRNA is easily detectable after 96 hr of ES/EB development. Thereafter, the abundance rises sharply to rival that of β-actin by day 8 of ES differentiation,[15] whereas β-major globin mRNA is detectable in EBs by day 5.[15,35]

Reverse Transcriptase–Polymerase Chain Reaction "Primer"

As mentioned before, RT-PCR is a convenient method to monitor gene activation, and there are many published protocols.[38] For RNA isolation, the simple AGPC procedure of Chomczynski and Sacchi is recommended; this is especially useful if only small amounts of material are available or required.[29] This procedure has been used successfully with one to hundreds of EBs, and it can also be used to isolate RNA from single hematopoietic colonies. When working with small amounts of RNA, it is suggested that 100 μg/ml of RNase/DNase-free glycogen (e.g., Boehringer-Mannheim, Mannheim, Germany, Cat. No. 901393) be added to aid the 2-propanol RNA precipitation step. For cDNA synthesis, we have found that oligo(dT)-primed first-strand cDNA synthesis of total RNA, following standard protocols, works well.[28a,39,40] Using the PCR primers given in Table II, gene expression can be detected by a diagnostic ethidium band using cDNA derived from approximately 200 differentiated ES cells.

Conclusions and Projections

In this chapter we have illustrated the simplicity of ES cell differentiation *in vitro* and noted that such a system allows one access to cells undergoing the earliest steps involved in the formation of mesoderm and embryonic hematopoietic cells. This is only an outline and can be regarded

[35] M. H. Lindenbaum and F. Grosveld, *Genes Dev.* **4,** 2075 (1990).

[36] P. M. Wong, S. W. Chung, S.M. Reicheld, and D. H. Chui, *Blood* **67,** 716 (1986).

[37] P. M. Wong, S. W. Chung, C. J. Eaves, and D. H. Chui, *Prog. Clin. Biol. Res.* **193,** 17 (1985).

[38] M. A. Innis, D. H. Gelfand, J. J. Sninsky, and T. J. White (eds.), "PCR Protocols: A Guide to Methods and Applications." Academic Press, San Diego, 1990.

[39] J. Sambrook, E. F. Fritsch, and T. Maniatis, "Molecular Cloning: A Laboratory Manual." Cold Spring Harbor Laboratory, Cold Spring Harbor, New York, 1989.

[40] H.-J. Thiesen, G. Casorati, R. Lauster, and M. V. Wiles, *in* "Immunological Methods" (I. Lefkovits and B. Pernis, eds.), p. 35. Academic Press, San Diego, 1991.

TABLE II

MARKER GENES EXPRESSED DURING EARLY DEVELOPMENT OF EMBRYONIC STEM CELLS AND EMBRYOID BODIES[a]

Gene	Base pairs		Annealing temperature (°C)	3' Sequence	5' Sequence
	cDNA	Genomic			
HPRT	249	~1100	50–58	5' CACAGGACTAGAACACCTGC	5' GCTGGTGAAAAGGACCTCT
T gene	947	≥947	50	5' TCCAGGTGCTATATATTGCC	5' TGCTGCCTGTGAGTCATAAC
GATA1	582	Smear	55	5' ATGCCTGTAATCCCAGCACT	5' TCATGGTGGTAGCTGGTAGC
βH1	268	1082	55	5' GCCTAATTCAGTCCCATGG	5' CTCAAGGAGACCTTTGCTCA
β-Major globin	578	1296[b]	55	5' CTGACAGATGCTCTCTTGGG	5' CACAACCCAGAAAACAGACA

[a] Together with PCR primers suitable for their detection. The hypoxanthine phosphoribosyltransferase (HPRT) gene is a useful standard to balance the amount of cDNA used. By using the PCR primers, where a genomic DNA-derived band is indicated, a visible ethidium band can be seen with 1 ng of genomic DNA (30-μl reaction, with one-third loaded onto the "gel").

[b] Some additional bands may occur.

as the starting point. As is the nature of any *in vitro* system, there are many possibilities of modification. Some future avenues of research include the following. (1) Avoidance of FCS. What is required is a system of development where every external component of the system can be controlled. (2) Coculturing of developing EBs with other "inducing" cells. This can be done by mixing non-ES cells, for example, a factor-producing stromal cell line (fetal liver or bone marrow derived) into the MCM. The feeder or stromal cells will condition the local environment, with possible effects on the direction of EB development. (3) The dominant direction of ES differentiation will be manipulated by environmental influences, such as the concentration of reducing agent present and/or the medium used. (4) The current system will be extended to the formation of other cell types. For example, factors that could influence the direction of mesoderm differentiation toward somite formation would be of great interest. (5) Using variations on the given protocol, a number of groups have tried to reconstitute the entire hematopoietic system of the mouse. This has possibly been achieved by the group of Palacios.[41] Once we can determine what controls the generation of hematopoietic stem cells (reconstituting units), an entire field of medically important possibilities will arise. (6) cDNA libraries will be generated, which can be regarded as derived from the primary hematopoietic tissues. Such libraries can be used as a source for noval genes involved in the commitment of hematopoietic cells from the mesoderm.

Acknowledgments

Special thanks go to Dr. Gordon Keller for the exchange of unpublished information and discussions, together with Drs. Marie Kosco, Ulrich Deuschle, Chiara Bertini, and Susan Carson for help in assembling this work. The author also thanks Britt Johansson for excellent help in establishing the protocols. The Basel Institute for Immunology was founded and is supported by Hoffmann-La Roche Ltd. (CH-4005 Basel, Switzerland).

[41] J. C. Gutierrezramos and R. Palacios, *Proc. Natl. Acad. Sci. U.S.A.* **89**, 9171 (1992).

[55] Production of Tetraploid Embryos by Electrofusion

By K. J. McLaughlin

Introduction

The tetraploid embryo has two major facets of interest in mammalian developmental biology. The first is the phenotype of tetraploid development and its relevance to developmental patterns. In nature, the observation of tetraploid development is limited by the low rate of natural occurrence and early arrest of development during pregnancy.[1] However the preimplantation development of experimentally produced murine tetraploids[2] proceeds at high rates despite the increased size of the blastomeres and an overall reduction in cell number.[3] Transferred to foster mothers, these tetraploids can develop to at least midgestation and occasionally to day 15.[4] The surviving embryos have consistent defects in cardiac, craniofacial, and vertebral development[4,5] that may provide information regarding the events controlling normal development of these organs.[6]

Tetraploid embryos are also of interest when used in chimeras with diploid embryos and cells. The resulting fetuses show a very low or undetectable contribution of the tetraploid cells to the embryo, but the cells are able to participate in development of other tissues.[1,7] This provides a system for supporting the development of diploid cells that are incapable of supporting extraembryonic tissue development. It also enables diploid cells of the chimera to dominate the embryonic tissue of the fetus. This approach has shown particular potential for the chimeras of embryonic stem cells with tetraploid blastocysts.[8]

Until the use of electric fields, the production of tetraploid mouse embryos had been achieved using cleavage inhibition with cytoskeletal

---

[1] D. H. Carr, *Annu. Rev. Genet.* **5**, 65 (1971).

[2] A. K. Tarkowski, A. Witkowska, and J. Opas, *J. Embryol. Exp. Morphol.* **41**, 47 (1977).

[3] C. C. Henery and M. H. Kaufman, *J. Exp. Zool.* **259**, 371 (1992).

[4] M. H. Kaufman and S. Webb, *Development (Cambridge, UK)* **110**, 1121 (1990).

[5] C. C. Henery, J. B. L. Bard, and M. H. Kaufman, *Dev. Biol.* **152**, 223 (1992).

[6] M. H. Kaufman, *Hum. Reprod.* **6**, 8 (1991).

[7] T. Lu and C. L. Markert, *Proc. Natl. Acad. Sci. U.S.A.* **77**, 6012 (1980).

[8] A. K. Nagy, E. Gocza, E. Merentes Diaz, V. R. Prideaux, E. Ivanyi, M. Markkula, and J. Rossant, *Development (Cambridge, UK)* **110**, 815 (1990).

METHODS IN ENZYMOLOGY, VOL. 225

Copyright © 1993 by Academic Press, Inc.
All rights of reproduction in any form reserved.

inhibitors[1,9,10] and fusion of 2-cell stage blastomeres with Sendai virus[11,12] or polyethylene glycol.[13] Electrofusion has proved advantageous because it achieves a relatively high frequency of fusion[14,15] and subsequent development[4,16] and avoids the use of cytotoxins and virus. Furthermore the 2-cell mouse embryo is unique in its suitability as an electrofusion candidate. The two blastomeres, with a mutual membrane interface, are more amenable to fusion than cells in most other configurations.

This chapter describes approaches for electrically fusing 2-cell mouse embryos using different combinations and arrangements of electric fields and electrodes.

Electric Fields Generated for Blastomere Fusion

Direct current (dc), creating a single square-shaped dc output, is used to disrupt cell membranes, effectively creating holes that initiate fusion. The current is usually in effect for only a short time period (10–100 μsec) and is often referred to as a pulse. An alternating current (ac) field may be used in conjunction with the dc pulse. When operating in an aqueous solution at particular frequencies, the ac field results in a transmembrane potential difference, which effectively polarizes cells. This has two effects on the cells, the first being the rotation of objects within the field such that the longer axis will be parallel with the field lines (Fig. 1). The second effect is the attraction between proximal membranes, thereby forming a tight junction. As 2-cell mouse embryos already have a tight junction between cells, the ac field is required only to orientate the cells relative to the field lines. Once the membranes are at 90° to the field lines, the dc pulse has its optimal effect in disrupting the membrane. The combination of using an ac field to align the embryo before the dc pulse is referred to as polarized cell fusion. When the embryo is oriented by some other means, the fusion is referred to as nonpolarized.

The actual fusion of the blastomeres is a gradual event initiated by the dc pulse. Holes are created in the cell membrane, which usually quickly reanneal. However, in the region where the blastomeres are directly ap-

[9] R. G. Edwards, *J. Exp. Zool.* **137**, 349 (1958).
[10] M. H. L. Snow, *Nature (London)* **244**, 513 (1973).
[11] C. F. Graham, *Act Endocrinol.* **153**(Suppl.), 154 (1971).
[12] G. T. O'Neill, S. Spiers, and M. H. Kaufman, *Cytogenet. Cell Genet.* **53**, 191 (1990).
[13] M. A. Eglitis, *J. Exp. Zool.* **213**, 309 (1980).
[14] Y. Kato and Y. Tsunoda, *Jpn. J. Anim. Reprod.* **33**, 19 (1978).
[15] J. Z. Kubiak and A. K. Tarkowski, *Exp. Cell Res.* **157**, 561 (1985).
[16] J. Barra and J.-P. Renard, *Development (Cambridge, UK)* **102**, 773 (1988).

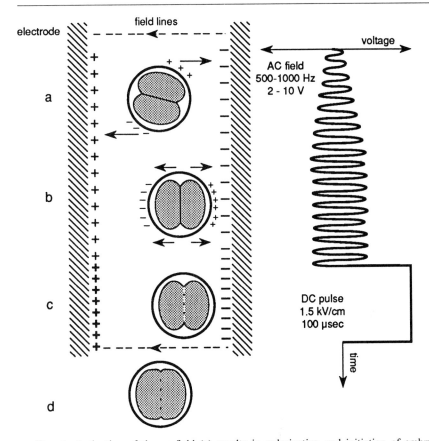

FIG. 1. Activation of the ac field (a) results in polarization and initiation of embryo rotation. The ac field strength is maintained at a constant amplitude (b) until the embryo has finished rotating, with the blastomere interface at 90° to the field lines. The square pulse (c) causes temporary membrane disruption, resulting in permanent cytoplasmic bridges that initiate blastomere deformation toward a single spherical cell (d).

posed, the disrupted regions of one membrane can reform with those of the other, resulting in the formation of cytoplasmic bridges. The cytoplasmic bridges gradually expand until an equilibrium is established (i.e., a spherical object). The number and size of the initial holes formed determine the time required for this event to take place. Extreme disruption arising from an excessive dc pulse will result in the failure of adjacent membranes to reform junctions and subsequently causes cell lysis. This usually occurs at the membrane closest to an electrode and is observed as a blastomere lysing immediately following the dc pulse.

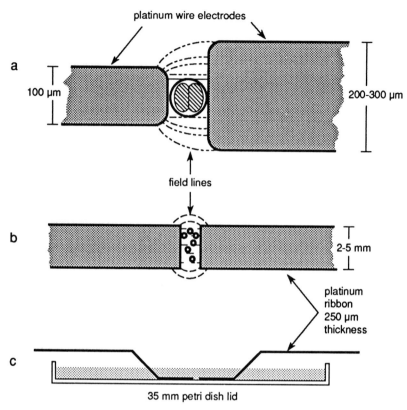

FIG. 2. Two common types of mobile electrodes made from either round platinum wires of different sizes (a) or platinum ribbon electrodes (b) with a larger gap area. Both types of electrodes can be angled to fit into a petri dish lid (c).

Electrofusion Chambers

Mobile Electrode Chambers

The manipulator-mounted wire electrode configuration, first described by Kubiak and Tarkowski,[15] in 1985 uses two round platinum wires of different sizes (Fig. 2a). The larger electrode can be permanently fixed to the base of the petri dish and the smaller one mounted on a manipulator; however, it may be simpler to mount each electrode onto a separate manipulator. Instrument holders can be substituted with 1-ml graduated pipettes, with the wiring mounted internally and the platinum soldered to the wiring at a point external to the medium. Platinum ribbon can be used instead of wire (Fig. 2b). In this case a larger electrode gap is available allowing for the accommodation of several embryos for simultaneous fusion.

Fixed-Electrode Chambers

Fixed-electrode chambers are usually configured to utilize polarized fusion methods and therefore have large electrode gap areas to accommodate large numbers of embryos. Manufacturers of cell-fusion pulse generators (see Appendix) will usually have one or more options for fixed-electrode chambers. Otherwise, several types of chambers can be constructed in the laboratory. The simplest of these consists of two platinum wires mounted in parallel at a predetermined distance onto a microscope slide with adhesive (Fig. 3a,b). A more complex system consists of two plates mounted on clear acrylic with adhesive or screws (Fig. 3c,d). Plates can be made of high-quality stainless steel for economy, but they should be connected to the wiring with platinum wire soldered to the plates. A useful variation is to make one of the plates, and hence the electrode gap, adjustable as shown in Fig. 3c,d.

Direct Current Pulse Generators for Nonpolarized Fusion

Direct current pulse generators can be purchased from suppliers of technical electrical instruments. They are also used for physiology experiments (nerve stimulators) and are often found in teaching laboratories. Most dc pulse generators will produce an output of at least 100 V, which is adequate for a 1.5 kV/cm field strength in chambers with electrode separation over 500 μm. Desirable features and parameters are as follows: dc pulse amplitude, 0–100 V; pulse shape, square; pulse duration, 10–1000 μsec; pulse trigger, manual.

Alternating/Direct Current Cell Fusion Devices for Polarized Fusion

There are several sources for devices capable of producing both a square dc pulse and ac output suitable for cell fusion. Cell fusion devices are produced by several manufacturers, and are primarily designed for fusing large numbers of cells. In general, these devices exceed the requirements for fusing 2-cell embryos and are relatively expensive. Another less expensive option is to have devices constructed by instrument technicians, or to contract a manufacturer of specialist electrical equipment. One other option is to purchase a dc output device and ac generator. As the 2-cell embryos do not require the ac field after orientation of the cells is achieved,

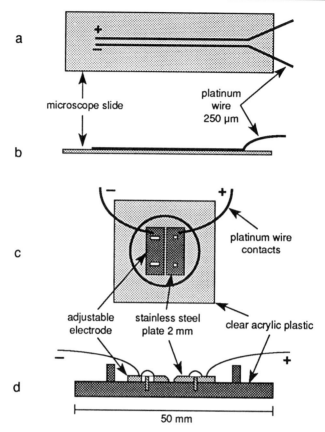

Fig. 3. Fixed-electrode fusion chambers. (a) Top view and (b) side view of a microscope slide chamber. (c) Top view and (d) side view of a fixed-plate fusion chamber with one adjustable electrode.

it is possible to use an ac sine-wave generator to first align the embryos and then to switch the circuit to the dc pulse generator. Specifications required for the ac field generator are as follows: frequency, 500–1000 kHz; waveform, sinusoidal; voltage, 0–10 V. The dc specifications are as for dc pulse generators (see above).

Source of 2-Cell Embryos

Depending on whether the mouse strain being used has a 2-cell culture block, the embryos can be collected at either the 1-cell or 2-cell stage. The collection should be done such that a maximum number of the embryos will be at the 2-cell stage at the desired time to perform fusions. If donors have been superovulated, this will be about 40–44 hr after injection of human chorionic gonadotropin. At this time the embryos will be at the mid 2-cell stage. When polarized fusion is being used, individual embryos should be selected for uniform and symmetrical blastomeres. A group of embryos should be maintained as an experimental control for the fusion manipulations.

Nonpolarized Fusion

The production of tetraploids can be achieved using only a dc pulse. The main advantage of this method is the availability and low cost of the dc pulse generators. The disadvantage is that the embryos need to be individually aligned between the electrodes before the dc pulse is activated. As mentioned previously, the mobile manipulator electrodes are quite suitable for this procedure.

Chamber Preparation

Before embryos are placed in the fusion chamber, several steps should be taken to ensure consistent fusion rates. The electrodes should be totally free of any contaminating material such as dried protein, salts, or electrically induced deposits. These can usually be eliminated with thorough washing in distilled water before and after use. A brief wash in 1 *M* HCl can also remove stubborn residue. Electrodes in good condition should have a shiny appearance at the tips. The chamber is then flooded with a normal electrolyte medium such as medium M2[17] which contains bovine serum albumin (BSA). Verification of an active circuit can be done visually once the electrodes and chamber are flooded and observed under a microscope. A large dc pulse of approximately 30 kV/cm should then be delivered either by minimizing the distance between the electrodes, or by increasing the output voltage. If the circuit is active, then bubbles will form at the electrode tips as a result of electrolysis. The size and number of bubbles are a useful visual reference. Excess bubbles should be removed from the electrodes before they are reset to the desired gap to achieve a field strength of 1.5 kV/cm.

Positioning of Embryos

The embryo(s) can be positioned between the electrodes and aligned using the micromanipulator-mounted electrodes, or manually with a pipett-

[17] P. Quinn, C. Barros, and D. Whittingham, *J. Reprod. Fertil.* **66**, 161 (1982).

ing apparatus. It is imperative that the embryo be placed between the electrodes such that the blastomere interface is perpendicular to the field lines (Fig. 2). If the angle varies much more than 10°, fusion efficiency will be dramatically reduced.

Fusion Pulse

A single, square-shaped dc pulse of 1.5 kV/cm and 100 μsec duration is usually sufficient for fusion. If it is difficult to maintain the orientation of the electrode during the pulse step, the electrode tips can be used to hold the embryo in position. Once a satisfactory pulse has been delivered, the embryo should be removed clear of the electrodes before the next embryo is pulsed.

Postfusion Recovery

Embryos should be inspected 30 min after the dc pulse. At this time, all embryos with the potential to fuse will be undergoing, or will have completed, fusion and can be returned to culture. If less than 90% of the embryos are fusing, then usually there has been (1) a failure to orientate the embryo properly and/or (2) an inadequate pulse strength. The latter may vary with different electrodes and chambers. However, optimal fusion rates for particular setups are usually obtained within a few trials.

Polarized Field Fusion

Use of an ac field to align the embryos before the dc pulse enables the fusion of large numbers of embryos simultaneously. Using a fixed-electrode chamber and a dissecting microscope, the procedure can be done quite rapidly with minimal effort.

Fusion Medium

To create an ac field in aqueous solution, a nonelectrolyte is required as a salt to obtain a normal medium osmolarity. The most commonly used nonelectrolyte is mannitol. Small amounts of calcium and magnesium help in membrane healing.[18] BSA or polyvinyl pyrrolidone acts as a lubricant to minimize adhesion of the embryos to the walls of the chamber and the pipette and to facilitate rotation of the embryos. The contents of the fusion medium are as follows: 0.3 M mannitol, 0.1 mM MgSO$_4$, 50 μM CaCl$_2$, and 3% BSA. The medium is adjusted to pH 7.4, filter-sterilized, and stored at 4° (lasts 3 months).

[18] U. Zimmerman and J. Vienken, *J. Membr. Biol.* **67**, 165 (1982).

Chamber Preparation

As mentioned for nonpolarized fusion, the chamber must be cleaned thoroughly before use, particularly as the fusion medium salt is a nonelectrolyte. Setting the electrode distance depends on the maximum output of the fusion device. Typically, 250–500 μm is convenient for handling the embryos and is within the output limits of most pulse generators. When the distance is set, the voltage should be adjusted for a square dc pulse of 1–1.5 kV/cm. If desired, the chamber can be filled with medium between the electrodes only. However, maintaining a level above the electrodes facilitates handling and visualization of the embryos.

Equilibration

Owing to differences between the composition of fusion medium and mouse embryo culture medium, the embryos require equilibration in the fusion medium before being placed into the electrode gap. This is most easily done by collecting the embryos in a thin drawn Pasteur pipette (100 μm), in a minimum of medium, and placing them into a large drop of fusion medium (e.g., 200 μl). As soon as the embryos equilibrate, they will begin to fall to the bottom of the drop (<60 sec). Then they can be removed and placed between the electrodes.

Positioning of Embryos

Depending on the pipetting skill of the operator, the number of embryos per fusion treatment is limited only by the space between the electrodes. Initially 10–20 embryos is sufficient to verify the operation of the fusion device. The embryos should be taken through the equilibration step and then placed midway between the electrodes. They should be spaced evenly such that embryo–embryo and embryo–electrode contact is minimal.

Alignment of Embryos with Alternating Current Field

The ac field (500–1000 kHz sine wave) can be activated before all the embryos have settled onto the bottom of the chamber so as to facilitate easy rotation. The voltage output required to achieve rotation will depend on the gap distance and other chamber parameters. The best guide is to initially use 2 V ac output, then visually determine if rotation is occurring. Some fusion devices will not allow manual adjustment of ac output during the fusion program, and several programs may therefore have to be tested. This is done by incrementally increasing the ac voltage in conjunction with 0 V dc until satisfactory rotation is observed. Manually controlling ac output allows for a gradual increase in voltage until rotation begins

(Fig. 1). Then the ac output should be maintained until all of the embryos finish rotating, which takes about 5 to 30 sec. For the ac alignment, voltage should not have to exceed 10 V.

As embryos rotate, they will also tend to migrate to the closest electrode (Fig. 1b) and sometimes form chains along the direction of the field lines. When all rotation has ceased, some embryos may not be fully aligned. If the percentage is small, then correction is not warranted. Otherwise, the ac voltage should be stopped temporarily while the nonaligned embryos are relocated in the chamber. Some individual embryos with large cell fragments or poor symmetry will rotate to undesirable angles and will not fuse.

Direct Current Pulse

Once the operator is satisfied with the orientation of the embryos, the dc pulse can be activated at a field strength of 1.5 kV/cm for 100 μsec (with programmable fusion devices, this step may be automatic). Operation of the ac field is not necessary once orientation is achieved, but it may be useful to maintain the orientation of some embryos. Once the dc pulse has been completed, it is necessary to stop the ac field to remove embryos from the chamber.

Postfusion Recovery

The embryos should be washed twice in several milliliters of culture medium before being returned to culture. The frequency of blastomere fusion using polarized cell fusion is expected to be similar to dc fusion.

If problems with fusion occur, a cathode ray oscilloscope can be used to test the signal output of the generator being used as well as the actual voltage output at the electrodes in the chamber. Some fusion generators also include a postfusion ac field, which is used for fusing discrete cells. This is not necessary for 2-cell mouse embryos because of existing junctions between the two blastomeres.

Fusion of 4-Cell Embryos

Tetraploid embryos can also be made from 4-cell embryos by fusing together pairs of blastomeres. To produce these, the best system is manipulator-mounted wire electrodes using only a dc pulse. The success rate of this method is somewhat lower than 2-cell fusion as a result of weaker cell–cell junctions between some blastomeres and variation in the spatial configuration of the blastomeres within the zona pellucida. Owing to the extra manipulation required per embryo, and the lower rates of fusion, this procedure is considerably more time consuming than the 2-cell stage fusion.

Postfusion Development of Embryos

Tetraploid embryos produced by fusion at the 2-cell stage will undergo cleavage at similar cleavage times as unfused two cell embryos. They will typically compact at the 4-cell stage, equivalent to a diploid 8-cell stage, and should form blastocysts at a frequency of greater than 95%.

Appendix: Suppliers of Electrical Equipment for Cell Fusion

Cell fusion devices (ac/dc)

CF-1000 Cell fusion instrument
Biochemical Labor Service
H-1165 Budapest, Zsely Aladar u.31, Hungary
Telephone: 361 271-2896
Telephone/Fax: 361 2712602
Telex: 61 227608

BTX 200 Electro cell manipulator
Biotechnologies and Experimental Research, Inc.
11199 Sorrento Valley Rd.
San Diego, CA 92121-1334
Telephone: (619) 597-6006
Fax: 619 5979594

Biojet CF
B. Braun Biotech Inc.
999 Postal Road
Allentown, PA 18103
Telephone: (800) 258-9000
Fax: 215 266 9319

Waveform generator (ac only)

Wavetek Electronics GmbH
Han-Pinsel-Strasse 9-10
8013 Haar bei Munchen, Germany
Telephone: 0 89 46109-0
Telex: 5 212 996

Square pulse generator (dc only)

S-44 Stimulator
Grass Instruments
101 Old Colony Avenue
P.O. Box 516
Quincy, MA 02269-0516

Telephone: (617) 773-0002
Fax: 617 7730415

Acknowledgment

Preparation of this chapter was supported in part by a fellowship from the Australian Dairy Research and Development Corporation.

Section XII

Lineage Analysis

[56] Lineage Analysis Using Retrovirus Vectors

By Constance L. Cepko, Elizabeth F. Ryder,
Christopher P. Austin, Christopher Walsh,
and Donna M. Fekete

Introduction

The complexity and inaccessibility of the murine embryo has made lineage analysis through direct approaches, such as time-lapse microscopy and injection of tracers, almost impossible. A genetic and clonal solution to lineage mapping in mice is through the use of retrovirus vectors. The basis for this technique is summarized in this chapter, and the strategies and current methods in use in our laboratory are detailed.

Transduction of Genes via Retrovirus Vectors

A retrovirus vector is an infectious virus that transduces a nonviral gene into mitotic cells *in vivo* or *in vitro*.[1] These vectors utilize the same efficient and precise integration machinery of naturally occurring retroviruses to produce a single copy of the viral genome stably integrated into the host chromosome. Vectors that are useful for lineage analysis have been modified so that they are replication-incompetent and thus cannot spread from one infected cell to another. They are, however, faithfully passed on to all daughter cells of the originally infected progenitor cell, making them ideal for lineage analysis.

Retroviruses use RNA as their genome, which is packaged into a membrane-bound protein capsid. They produce a DNA copy of the genome immediately after infection via reverse transcriptase, a product of the viral *pol* gene included in the viral particle. The DNA copy is integrated into the host cell genome and is thereafter referred to as a provirus. Complete synthesis of an integration-competent viral genome requires an S phase, and thus only mitotic cells will serve successfully as hosts for retroviral integration.

Most vectors began as proviruses that were cloned from cells infected with a naturally occurring retrovirus. Although extensive deletions of proviruses were made, vectors retain the cis-acting viral sequences necessary for the viral life cycle. These include the ψ packaging sequence

---

[1] R. Weiss, N. Teich, H. Varmus, and J. Coffin, "RNA Tumor Viruses," Cold Spring Harbor Laboratory, Cold Spring Harbor, New York, 1984 and 1985.

METHODS IN ENZYMOLOGY, VOL. 225

Copyright © 1993 by Academic Press, Inc.
All rights of reproduction in any form reserved.

(necessary for recognition of the viral RNA for encapsidation into the viral particle), reverse transcription signals, integration signals, and viral promoter, enhancer, and polyadenylation sequences. A cDNA can thus be expressed in the vector using the transcription regulatory sequences provided by the virus. Because replication-incompetent retrovirus vectors usually do not encode the structural genes whose products make up the viral particle, these proteins must be supplied through complementation. The structural proteins gag, pol, and env are typically supplied by "packaging" cell lines. These lines are stable mouse fibroblast lines that contain the *gag, pol,* and *env* genes as a result of the introduction of these genes by transfection. However, these lines do not contain the packaging sequence, ψ, on the viral RNA that encodes the structural proteins. Thus, the packaging lines make viral particles that do not contain the genes *gag, pol,* or *env.*

Retrovirus vector particles are essentially identical to naturally occurring retrovirus particles. They enter the host cell via interaction of a viral envelope glycoprotein (a product of the viral *env* gene) with a host cell receptor. The murine viruses have several classes of env glycoprotein which interact with different host cell receptors. The most useful class for lineage analysis of rodents is the ecotropic class. The ecotropic env glycoprotein allows entry only into rat and mouse cells via the ecotropic receptor. It does not allow infection of humans and thus is considered relatively safe for gene transfer experiments. Until 1988, the packaging line most commonly in use was the ψ2 line.[2] It encodes the ecotropic *env* gene and, in our experience, makes the highest titers of vectors, relative to other packaging lines (for unknown reasons). However, it can also lead to the production of helper virus (discussed below). Two newer ecotropic packaging lines, ΨCRE[3] and GP+E-86,[4] have not been reported to lead to production of helper virus to date. A third such "helper-free" packaging line, ΩE,[5] with some improvement in design over the others, has been produced and will probably prove to be very useful. Any of the ecotropic packaging lines can be used to produce vector stocks for lineage analysis. Regardless of the packaging line used, however, all stocks should be assayed for the presence of helper virus.

Production of Virus Stocks for Lineage Analysis

Replication-incompetent vectors that encode a histochemical reporter gene, such as *Escherichia coli lacZ,* are the most useful for lineage studies

[2] R. Mann, R. C. Mulligan, and D. Baltimore, *Cell* (*Cambridge, Mass.*) **33,** 153 (1983).
[3] O. Danos and R. C. Mulligan, *Proc. Natl. Acad. Sci. U.S.A.* **85,** 6460 (1988).
[4] D. Markowitz, S. Goff, and A. Bank, *J. Virol.* **62,** 1120 (1988).
[5] J. P. Morgenstern and H. Land, *Nucleic Acids Res.* **18,** 3587 (1990).

as they allow analysis of individual cells in tissue sections or whole mounts. Stocks of such vectors are typically produced by packaging lines stably transduced with the vector genome. (A detailed description of protocols for making such "producer" lines, titering and concentrating virus stocks, and checking for helper virus contamination has been published[6] and is not given here.)

It is best to obtain lines that make high-titered stocks of lineage vectors from the laboratories that have created them. We have placed Ψ2 and ΨCRE producers of BAG,[7] a *lacZ* virus that we have used for lineage analysis, on deposit at the ATCC (Rockville, MD). They can be obtained by anyone and are listed as ATCC CRL Nos. 1858 (ψCRE BAG) and 9560 (ψ2 BAG). Similarly, ψ2 producers of DAP,[8] a vector encoding human placental alkaline phosphatase (PLAP, described further below), is available as CRL No. 1949. Both of these vectors transcribe the reporter gene from the viral long terminal repeat (LTR) promoter and are generally useful for expression of the reporter gene in most tissues. We compared the expression of *lacZ* driven by several different promoters[9,10] and found that the LTR was generally the most reliable and non-cell type-specific. This is an important consideration as it is desirable to identify all of the cells descended from an infected progenitor, and thus a constitutive promoter is the most useful for lineage studies. However, even with constitutive promoters, it has been noted that some infected cells do not express detectable β-galactoside (β-Gal) protein, even among clones of fibroblasts infected *in vitro*. Thus, it is important to restrict conclusions about lineage relationships to cells that are marked and not to make assumptions about their relationships to cells that are unmarked.

For lineage applications it is usually necessary to concentrate virus in order to achieve sufficient titer. This is typically due to a limitation in the volume that can be injected at any one site. As viruses are macromolecular structures, they can be concentrated fairly easily by a relatively short centrifugation step. Virions also can be precipitated using polyethylene glycol or ammonium sulfate, and the resulting precipitate collected by centrifugation. Finally, the viral supernatant can be concentrated by centrifugation through a filter that allows only small molecules to pass [e.g.,

[6] C. L. Cepko, *in* F. M. Ausubel, R. Brent, R. E. Kingston, D. D. Moore, J. G. Seidman, J. A. Smith, and K. Struhl, "Current Protocols in Molecular Biology," Supplement 17 (1992). Greene, Publishing New York, 1989.

[7] J. Price, D. Turner, and C. L. Cepko, *Proc. Natl. Acad. Sci. U.S.A.* **84**, 154 (1987).

[8] S. C. Fields-Berry, A. L. Halliday, and C. L. Cepko, *Proc. Natl. Acad. Sci. U.S.A.* **89**, 693 (1992).

[9] D. L. Turner, E. Y. Snyder, and C. L. Cepko, *Neuron* **4**, 833 (1990).

[10] C. L. Cepko, C. P. Austin, C. Walsh, E. F. Ryder, A. Halliday, and S. Fields-Berry, *Cold Spring Harbor Symp. Quant. Biol.* **55**, 265 (1990).

Centricon (Amicon, Danvers, MA) filters]. Regardless of the protocols that are used, one must keep in mind that retroviral particles are fragile, with short half-lives even under optimum conditions. To prepare the highest titered stock for multiple experiments, we usually concentrate several hundred milliliters of producer cell supernatant. Concentrated stocks are titered and tested for helper virus contamination. They can be stored indefinitely at $-80°$ in small (10–50 μl) aliquots, although we have noticed reduction in titer upon freeze–thaw for some stocks.

Replication-Competent Helper Virus

Replication-competent virus is sometimes referred to as helper virus as it can complement ("help") a replication-incompetent virus and thus allow it to spread from cell to cell. It can be present in an animal through exogenous infection (e.g., from a viremic animal in the mouse colony), expression of an endogenous retroviral genome (e.g., the *akv* loci in AKR mice), or recombination events between two viral RNAs encapsidated in retroviral virions produced by packaging lines. The presence of helper virus is an issue of concern when using replication-incompetent viruses for lineage analysis as it can lead to horizontal spread of the marker virus, creating false lineage relationships.

The most likely source of helper virus is the viral stock used for lineage analysis. The genome(s) that supplies the *gag, pol,* and *env* genes in packaging lines does not encode the Ψ sequence, but it can still become packaged, although at a low frequency. If it is coencapsidated with a vector genome, recombination in the next cycle of reverse transcription can occur. If the recombination allows the Ψ^- genome to acquire the Ψ sequence from the vector genome, a recombinant that is capable of autonomous replication is the result. This recombinant can spread through the entire culture (although slowly, owing to envelope interference). Once this occurs, it is best to discard the producer clone as there is no convenient way to eliminate the helper virus. As would be expected, recombination giving rise to helper virus occurs with greater frequency in stocks with high titer, and with vectors that have retained more of the wild-type sequences (i.e., the more homology between the vector and packaging genomes, the more opportunity there is for recombination). Note that the helper genome itself will not encode a histochemical marker gene, as apparently there is no room, or flexibility, within murine viruses that allows them to be both replication-competent and capable of expressing another gene like *lacZ*. The way that spread would occur is by a cell being infected with both the *lacZ* virus and a helper virus. Such a doubly infected cell would then produce both viruses.

When performing lineage analysis, there are several signs that can indicate the presence of helper virus within an individual animal. If one allows an animal to survive for long periods of time after inoculation, particularly if embryos or neonates are infected, the animal is likely to acquire a tumor when helper virus is present. Most naturally occurring replication-competent viruses are leukemogenic, with the disease spectrum being at least in part a property of the viral LTR.

If one analyzes either shortly or long after inoculation, the clone size, clone number, and spectrum of labeled cells may be indicative of helper virus. For example, the eye of a newborn rat or mouse has mitotic progenitors for retinal neurons, as well as mitotic progenitors for astrocytes and endothelial cells. By targeting the infection to the area of progenitors for retinal neurons, we only rarely see infection of a few blood vessels or astrocytes, as their progenitors are outside of the immediate area that is inoculated and they become infected only by leakage of the viral inoculum from the targeted area. However, if helper virus were present, we would see infection of a high percentage of astrocytes, blood vessels, and, eventually, other eye tissues since virus spread would eventually lead to infection of cells outside of the targeted area. One would expect to see a correlation between the percentage of such nontargeted cells that are infected and the degree to which their progenitors are mitotically active after inoculation, owing to the fact that infection requires a mitotic target cell. If one were to examine tissues other than ocular tissues, one would similarly see evidence of virus spread to cells whose progenitors would be mitotically active during the period of virus spread. In addition, the size and number of "clones" may also appear to be too large for true "clonal" events if helper virus were present. This interpretation of course relies on some knowledge of the area under study. Finally, if one performs the infections with two different histochemical marker viruses, one can also assess the likelihood of helper virus problems, as discussed below.

Determination of Sibling Relationships

When performing lineage analysis, it is critical to unambiguously define cells as descendents of the same progenitor. This can be relatively straightforward when sibling cells remain rather tightly, and reproducibly, grouped. An example of such a straightforward case is the rodent retina, where the descendents of a single progenitor migrate to form a coherent radial array.[9,11] In such a system, lineage analysis can be performed using one or two distinct histochemical marker viruses, as described below.

[11] D. L. Turner and C. L. Cepko, *Nature (London)* **328,** 131 (1987).

For more complex systems, particularly where cell migration is important, many more markers are needed, as described in a later section (see Clonal Analysis Using Polymerase Chain Reaction/Library Method).

If only a single histochemical marker virus is used, one can perform a standard virological titration, in which a particular viral inoculum is serially diluted and applied to tissue. In the retina, the number of radial arrays, their average size, and their cellular composition were analyzed in a series of animals infected with dilutions that covered a 4-log range. The number of arrays was found to have a first-order relationship to the dose of virus, whereas the size and composition of the radial arrays were independent of the viral dose. Such results indicate that the working definition of a clone, in this case a radial array, fulfilled the statistical criteria expected of a single hit event.

There are several difficulties involved with the use of a single marker virus. First, there must be a wide range of dilutions that can be injected to give countable numbers of events, which is required for determination of a first-order relationship between clone number and viral dose. This type of analysis also relies critically on controlling the exact volume of the injection. These problems may be avoided in cases where it is possible to perform the analysis solely on infections with very small amounts of virus, although generating large amounts of data under these conditions is tedious. Another potential problem with the dilution approach is that aggregates may form during virus concentration which are not separated by dilution. Thus, even at low dilution, a "clone" may be the result of infection of adjacent cells by two or more members of such an aggregate. In addition, it is difficult to calculate an error rate for the assignment of clonal boundaries, which may make interpretation difficult, particularly of events which occur rarely in the data set.

For these reasons, additional viral vectors with histochemically distinguishable marker genes have been developed for use in mixed infections. In addition to addressing the problems mentioned above, use of two marker viruses can often provide a much better initial idea of what clonal boundaries are likely to be, by simple visual inspection of the tissue. By using two markers, one can visualize errors arising from formation of viral aggregates, since some of these aggregates will include two distinguishable virions. Various criteria for clonal boundaries can be tested, and error rates for each can be computed. Extensive dilutions are not needed.

Two viruses that have been used for this approach encode cytoplasmically localized versus nuclear localized β-Gal. This can work when the cytoplasmically localized β-Gal is easily distinguished from the nuclear localized β-Gal.[12,13] We have found that this is not the case in rodent nervous system cells, as the cytoplasmically localized β-Gal quite often is restricted to neuronal cell bodies and is therefore difficult to distinguish from nuclear-localized β-Gal. To overcome this problem, we created the above-mentioned DAP virus,[8] which is distinctive from the *lacZ*-encoding BAG virus.

To perform the two-marker analysis in the rodent retina, a stock containing BAG and DAP was made by growing a Ψ2 producer clone for BAG and a Ψ2 producer clone for DAP on the same dish. The resulting supernatant was concentrated and used to infect rodent retina. The tissue was then analyzed histochemically for the presence of blue (owing to BAG infection) and purple (owing to DAP infection) radial arrays. If radial arrays were truly clonal, then each one should be only one color. Analysis of 1100 arrays indicated that most were clonal. However, 5 comprised blue cells and purple cells, presumably from infection of adjacent progenitor cells with BAG and DAP. (Infection of one cell with both BAG and DAP would not be a problem, since the resulting cells would be clonally related.) The 5/1100 figure will be an underestimate of the true frequency of incorrect assignment of clonal boundaries, as sometimes two BAG or two DAP virions will infect adjacent progenitor cells; the resulting arrays will be a single color, but not clonal. A closer approximation of the true frequency can be obtained by using the following formula (for derivation see Ref. 8):

$$\frac{(\text{\# bicolored arrays})[(a + b)^2/2ab]}{\text{\# total arrays}} = \% \text{ errors}$$

where a and b are the relative titers in the virus stock. The relative titer of BAG and DAP used in the coinfection was 3:1, and thus the value for percent errors in clonal assignments was 1.2%.

Such a low error rate shows that the choice of clonal boundary was reasonable, and that viral aggregation during concentration (or at any other step) was not a problem for this viral stock. To take an extreme counterexample, if every array were the result of infections by two particles, 6/16 of the arrays would be composed of both blue cells and purple cells. (For a stock with relative viral titers of 1:1, half the arrays would have blue cells and purple cells). In analyzing the composition of arrays (e.g., by cell type), one need not worry, as one would if using one marker,

[12] S. M. Hughes and H. M. Blau, *Nature (London)* 345, 350 (1990).
[13] D. S. Galileo, G. E. Gray, G. C. Owens, J. Majors, and J. R. Sanes, *Proc. Natl. Acad. Sci. U.S.A.* **87**, 458 (1990).

that rare cell type combinations may be due to occasional errors in assignment of clonal boundaries. If such rare arrays are always of one color (when the error rate is very low), then they are very likely clones.

The error rate being computed here is the rate of "lumping" errors, that is, the frequency with which the criteria defining a clone lumps together daughters of more than one progenitor. However, this computation does not allow assessment of "splitting" errors, where clones that are more spread out (presumably owing to migration) are split by the definition of a clone into two or more subclones. If the criteria used to define clonal relationships are found to generate too many lumping errors, a more restrictive definition can be made, and then this definition can be tested for percent lumping errors. This process can be done iteratively until the error rate reaches a level that is acceptable relative to the point that is being tested. However, it should be kept in mind that the more restrictive a definition is for clonal relationships, the more it is prone to generate splitting errors. Splitting errors may become obvious in animals injected with very dilute viral stocks; for example, one blue array in a large, otherwise unlabeled area is probably a clone, but it may be divided into several by too restrictive clonal definitions. Thus, combining two-marker analysis with a few dilution experiments may be useful to help balance splitting and lumping errors. If it is necessary to avoid completely both splitting and lumping errors, a much greater number of vectors and more tedious detection method must be used, as detailed below (see Clonal Analysis Using Polymerase Chain Reaction/Library Method).

The value of 1.2% for lumping errors in assignment of clonal boundaries places an upper limit of 1.2% as the frequency of aggregation for this viral stock, since this figure will include errors due to both aggregation and independent virions infecting adjacent progenitors. The presence of helper virus would probably increase the error rate, as it would have the effect of enlarging apparent clone size and creating overlap between clones of different markers. The percentage of errors on injecting this stock in other areas of an animal will depend on the particular circumstances of the injection site and on the multiplicity of infection (MOI, the ratio of infectious virions to target cells). Most of the time the MOI will be quite low (e.g., in the retina it was approximately 0.01 at the highest concentration of virus injected). Concerning the injection site, injection into a lumen, such as the lateral ventricles, should not promote aggregation nor a high local MOI, but injection into solid tissue, in which the majority of the inoculum has access to a limited number of cells at the inoculation site, could present problems. By coinjecting BAG and DAP, one can monitor the frequency of these events and thus determine if clonal analysis is feasible.

The above analysis was performed using viruses that were produced on the same dish and concentrated together. This was done because we felt that the most likely way that two adjacent progenitors might become infected would be through small aggregates of virions. We grew both virions together on the same dish in order for the assay to be sensitive to any aggregation that might occur prior to concentration, but this is probably not necessary in most cases. Aggregation most likely occurs during the concentration step, as one often can see macroscopic aggregates after resuspending pellets of virions. Thus, when the two-marker approach is used to analyze clonal relationships, it is best to coconcentrate the two vectors together in order for the assay to be sensitive to aggregation arising from this aspect of the procedure. [Although aggregation of virions may frequently occur during concentration, it apparently does not frequently lead to problems in lineage analysis, presumably because of the high ratio of noninfectious particles to infectious particles found in most retrovirus stocks. It is estimated that only 0.1–1.0% of the particles will generate a successful infection. Moreover, most aggregates are probably not efficient as infectious units; it must be difficult for the rare infectious particle(s) within such a clump to gain access to the viral receptors on a target cell.]

To determine the ratio of two genomes present in a mixed virus stock (e.g., BAG plus DAP), there are several methods that can be used. The first two methods are performed *in vitro,* and are simply an extension of a titration assay. Any virus stock is normally titered on NIH 3T3 cells to determine the amount of virus to inject. The infected NIH 3T3 cells are then either selected for the expression of a selectable marker when the virus encodes such a gene (e.g., *neo* in BAG and DAP), or are stained directly, histochemically, for β-Gal or PLAP activity without prior selection with drugs. If no selection is used, the relative ratio of the two markers can be scored directly by evaluating the number of clones of each color on a dish. Alternatively, selected G418-resistant colonies can be stained histochemically for both enzyme activities and the relative ratio of blue versus purple G418-resistant colonies computed. A third method of evaluating the ratio of the two genomes is to use the values observed from *in vivo* infections. After animals are infected and processed for both histochemical stains, the ratio of the two genomes can be compared by counting the number of clones, or infected cells, of each color.

When all the above methods were applied to lineage analysis in mouse retina[8] and rat striatum,[14] the value obtained for the ratio of G418-resistant colonies scored histochemically was almost identical to the ratio observed *in vivo.* Directly scoring histochemically stained, non-G418-selected NIH

[14] A. Halliday and C. L. Cepko, *Neuron* **9,** 15 (1992).

3T3 cells led to an underestimate of the number of BAG-infected colonies, presumably because such cells often are only faintly blue, whereas DAP-infected cells are usually an intense purple. *In vivo,* this is not generally the case, as BAG-infected cells are usually deep blue.

Regardless of which method is used to score sibling relationships, one further recommendation to aid in the assignments is to choose an injection site that will allow the inoculum to spread. If one injects into a packed tissue, the viral inoculum will most likely infect cells within the injection tract, and it will be very difficult to sort out sibling relationships (i.e., too many lumping errors). In addition, one must inject such that the virus has clear access to the target population; the virus will blind to cells at the injection site and will not gain access to cells that are not directly adjoining that site.

The procedures described below are those that we have used for infection of rodents, histochemical processing of tissue for β-Gal and PLAP visualization, and preparation and use of a library for the polymerase chain reaction (PCR) method.[15]

Infection of Rodents

Injection of Virus in Utero

The following protocols may be used with rats or mice. Note that clean, but not aseptic, technique is used throughout. We routinely soak instruments in 70% (v/v) ethanol before operations, use the sterile materials noted, and include penicillin/streptomycin (final concentrations of 100 units/ml each) in the lavage solution. We have not had difficulty with infection using these precautions.

Materials

    Ketamine hydrochloride injection (100 mg/ml ketamine)
    Xylazine injection (20 mg/ml)
    Animal support platform
    Depilatory
    Scalpel and disposable sterile blades
    Cotton swabs and balls, sterile
    Tissue retractors
    Tissue scissors
    Lactated Ringer's solution (LR) containing penicillin/streptomycin
    Fiber optic light source
    Virus stock

---

[15] C. Walsh and C. L. Cepko, *Science* **255,** 434 (1992).

Automated microinjector
Micropipettes, 1–5 μl
3-0 Dexon suture
Tissue stapler

1. Mix ketamine and xylazine 1 : 1 in a 1-ml syringe equipped with a 27-gauge needle; lift the tail and hindquarters of the animal with one hand and with the other inject 0.05 ml (mice) or 0.18 ml (rats) of anesthetic mixture intraperitoneally. One or more additional doses of ketamine alone (0.05 ml for mice and 0.10 ml for rats) is usually required to induce or maintain anesthesia, particularly if the procedure takes over 1 hr. Respiratory arrest and spontaneous abortion appear to occur more often if a larger dose of the mixture is given initially, or if any additional doses of xylazine are given.

2. Remove the hair over the entire abdomen using depilatory agent (any commercially available formulation, such as Nair hair remover, works well); shaving of remaining hair with a razor may be necessary. Wash the skin several times with water, then with 70% ethanol, and allow to dry.

3. Place the animal on its back in the support apparatus. For this purpose, we find that a slab of Styrofoam with two additional slabs glued on top to create a trough works well. With the trough appropriately narrow, no additional restraint is needed to hold the anesthetized animal.

4. Make a midline incision in the skin from xyphoid process to pubis using a scalpel, and retract; attaching retractors firmly to the Styrofoam support will create a stable working field. Stop any bleeding with cotton swabs before carefully retracting the fascia and peritoneum and incising them in the midline with scissors (care is required here not to incise the underlying bowel). Continue the incision cephalad along the midline of the fascia (where there are few blood vessels) to expose the entire abdominal contents. If necessary to expose the uterus, gently pack the abdomen with cotton balls or swabs to remove the intestines from the operative field, being careful not to lacerate or obstruct the bowel. Fill the peritoneal cavity with LR, and lavage until clear if the solution turns at all turbid.

Wide exposure is important to allow the later manipulations. During the remainder of the operation, keep the peritoneal cavity moist and free of blood; dehydration or blood around the uterus increases the rate of postoperative abortion.

5. Elevate the embryos one at a time out of the peritoneal cavity, and transilluminate with a fiber optic light source to visualize the structure to be injected. For lateral cerebral ventricular injections, for example, the cerebral venous sinuses serve as landmarks. When deciding on a structure to inject, keep in mind that free diffusion of virus solution through a fluid-

filled structure lined with mitotic cells is best for ensuring even distribution of viral infection events throughout the tissue being labeled. The neural tube is an example of such a structure; when virus is injected into one lateral ventricle, it is observed to quickly diffuse throughout the entire ventricular system.

6. Using a heat-drawn glass micropipette attached to an automatic microinjector, penetrate the uterine wall, extraembryonic membranes, and the structure to be infected in one rapid thrust; this minimizes trauma and improves survival. Once the pipette is in place, inject the desired volume of virus solution, usually 0.1–1.0 μl. Coinjection of a dye such as 0.005% (w/v) trypan blue or 0.025% fast green aids determination of the accuracy of injection and does not appear to impair viral infectivity; coinjection of the polycation Polybrene (80 μg/ml) aids in viral attachment to the cells to be infected.

The type of instrument used to deliver the virus depends on the age of the animal and the tissue to be injected. At early embryonic stages, the small size and easy penetrability of the tissue makes a pneumatic microinjector (such as the Eppendorf 5242) best for delivering a constant amount of virus at a controlled rate with a minimum of trauma. Glass micropipettes should be made empirically to produce a bore size that will allow penetration of the uterine wall and the tissue to be infected. At later ages (late embryonic and postnatal), a Hamilton syringe with a 33-gauge needle works best.

When injecting through the uterine wall, all embryos may potentially be injected except those most proximal to the cervix on each side (injection of these greatly increases the rate of postoperative abortion). In practice, it is often not advisable to inject all possible embryos, if excessive uterine manipulation would be required. At the earliest stages at which this technique is feasible [embryonic day (E) 12 in the mouse or E13 in the rat], virtually any uterine manipulation may cause abortion, so any embryo that cannot be reached easily should not be injected.

7. Once all animals have been injected, lavage the peritoneal cavity until it is clear of all blood and clots, ensure that all cotton balls and swabs have been removed, and move retractors from the abdominal wall/fascia to the skin. Filling the peritoneal cavity with LR with penicillin/streptomycin before closing increases survival significantly, probably by preventing maternal dehydration during recovery from anesthesia as well as preventing infection.

8. Using 3-0 Dexon or silk suture material on a curved needle, suture the peritoneum, abdominal musculature, and fascia from each side together, using a continuous locking stitch. After closing the fascia, again lavage using LR with penicillin/streptomycin.

9. Close the skin using surgical staples (such as the Clay-Adams Autoclip) placed 0.5 cm apart. Sutures may also be used, but these require much more time (often necessitating further anesthesia, which increases abortion risk) and are frequently chewed off by the animal, resulting in evisceration.

10. Place the animal on its back in the cage and allow the anesthesia to wear off. Ideally, the animal will wake up within 1 hr of the end of the operation. Increasing time to awakening results in increasing abortion frequency. Food and water on the floor of the cage should be provided for the immediate postoperative period.

11. Mothers may be allowed to deliver progeny vaginally, or the offspring may be harvested by cesarean section. Maternal and fetal survival are approximately 60% at early embryonic ages of injection and increase with gestational age to virtually 100% after postnatal injections.

Injection of Virus Using exo Utero Surgery

Injections into small or delicate structures (such as the eye) require micropipettes that are too fine to penetrate the uterine wall. In addition, it is impossible to target precisely many structures through the rather opaque uterine wall. These problems can be circumvented, though with a considerable increase in technical difficulty and decrease in survival, by use of the *exo utero* technique.[16] The procedure is similar to that detailed above, with the following modifications to free the embryos from the uterine cavity.

1. The technique works well in our hands only with outbred, virus antigen-free CD-1 and Swiss-Webster mice, but even these strains may have different embryo survival rates when obtained from different suppliers or different colonies of the same supplier. This variability presumably results from subclinical infections, which may render some animals unable to survive the stress of the operation. We have had no success with this technique in rats.

2. After the uterus is exposed and *before* filling the peritoneum with LR, incise the uterus longitudinally along its ventral aspect with sharp microscissors. The uterine muscle will contract away from the embryos, causing them to be fully exposed, surrounded by the extraembryonic membranes.

3. Only two embryos in each uterine horn can be safely injected, apparently because of trauma induced by neighboring embryos touching

---

[16] K. Muneoka, N. Wanek, and S. V. Bryant, *J. Exp. Zool.* **239**, 289 (1986).

each other. Thus, all other embryos must be removed. Using a dry sterile cotton swab, scoop out each embryo to be removed, with its placenta and extraembryonic membranes, and press firmly against the uterine wall where the placenta had been attached for 30–40 sec to achieve hemostasis. It is important to stop all bleeding before proceeding. From this point on, the embryo must be handled extremely gently, as only the placenta is tethering the embryo to the uterus, and it detaches easily.

4. Fill the peritoneal cavity with LR, and cushion each embryo to be injected with sterile cotton swabs soaked in LR. Keeping the embryos submerged throughout the remainder of the procedure is essential for survival.

5. The injection should then be done with a pneumatic microinjector and heat-pulled glass micropipette. This may usually be performed by puncturing the extraembryonic membranes first and then the structure to be injected; for some very delicate injections if may be necessary to make an incision in the extraembryonic membranes, which is then closed with 10-0 nylon suture after the injection.

6. At the time of desired fetal harvest, or at the latest early on the last day of gestation, sacrifice the mother by cervical dislocation, rapidly incise the abdomen, and deliver the fetuses. If survival to a postnatal time point is desired, it is necessary to foster the pups with another lactating female. This is best done with a mouse that has delivered at the same time as the experimental animal, but we have successfully fostered pups with mothers that delivered several days to a week previously. Attempts to reanesthetize the experimental mother for delivery, thus allowing her to survive and obviating the need for fostering, have been unsuccessful owing to poor survival of the pups. This is probably due to the deleterious effect of the anesthesia on the pups, as well as poor milk production by the mother after multiple operations.

Human Placental Alkaline Phosphatase as Histochemical Marker Gene

Human placental alkaline phosphatase (PLAP) has only recently been adapted for use as a histochemical marker for lineage studies. As such, neither its benefits nor its potential drawbacks have been exhaustively analyzed, and this should be kept in mind as the enzyme enjoys wider use. Our laboratory has undertaken a systematic lineage analysis in the postnatal mouse retina[8] and rat striatum[14] comparing β-Gal and PLAP, and little difference between the two has been observed. Thus, for these particular combinations of tissue and vectors (BAG and DAP), there appears to be no effect of ectopic PLAP expression on the choice of cell fate during development. We do not yet know whether this conclusion will hold for other tissues. In the chick retina and cerebellum, for example, clonal analysis does show differences between β-Gal and PLAP in the ease of detection of different cell types (C. L. Cepko, D. Fekete, and E. Ryder, unpublished observations). It is not yet clear whether these differences reflect cell-specific inactivation of the promotor or the enzymes, differential distribution of the two enzymes within intracellular compartments, or (most worrisome) the perturbation of cell fate.

Human PLAP was initially chosen as a potential histochemical marker for several reasons. Among the variety of isoenzymes of alkaline phosphatase that have been studied, human PLAP is by far the most heat stable,[17] by a factor of about 100. Thus, although many tissues express endogenous alkaline phosphatase(s), it is possible in most cases to greatly reduce this background reactivity by preincubating the tissue at 65° for 30 min. In addition, PLAP is resistant to a variety of substances that act to inhibit other isoenzymes of alkaline phosphatase.[18] We have tested the following inhibitors on mouse and chick neural tissue as recommended by Zoellner and Hunter[18]: 0.5 mM levamisole [L(−)-2,3,5,6-tetrahydro-6-phenylimidazo[2,1-b]thiazole], 2 mM mercuric chloride, 5 mM L-leucylglycylglycine, 1 mM EDTA, 1 mM L-phenylalanineglycylglycine, 0.2 M lysine hydrochloride, and 0.3 mM sodium arsenate. Levamisole was the most useful in reducing the background staining in brains, although it also reduced PLAP staining slightly in some cases; it was less effective in retinas. None of the inhibitors was nearly as effective as heat treatment in reducing background in the central nervous system. Nonetheless, it is certainly possible that their use may facilitate staining of other tissues.

Another benefit of PLAP as a marker is the fact that its activity is probably minimal at normal intracellular pH, since this is considerably below the optimal pH for enzyme function. This may suggest that ectopic expression of PLAP during development is not likely to perturb normal physiological processes, but this remains to be tested systematically for different tissues.

Double Staining of Infected Tissues for β-Galactosidase and Alkaline Phosphatase Activities

The following protocol was adopted for the double staining of β-Gal and PLAP in nervous system tissue. Cells expressing β-Gal will be rendered a bright blue, whereas cells expressing PLAP will be rendered purple owing to the presence of nitro blue tetrazolium (NBT) in the reaction. The order

[17] R. B. McComb, G. N. Bowers, Jr., and S. Posen, "Alkaline Phosphatase." Plenum, New York, 1979.
[18] H. F. A. Zoellner and N. Hunter, *J. Histochem. Cytochem.* **37**, 1893 (1989).

in which the staining is done is critical, since β-Gal is inactivated by the heat treatment that is required to inhibit endogenous alkaline phosphatases. Obviously, if only a single marker is needed, the protocol can be minimally adapted for use with either enzyme individually.

Detecting both enzymes in the same cell is not always possible using the following protocols, because the PLAP reaction product is so intense that it usually obscures the β-Gal reaction product. For example, it is difficult to detect cells cotransfected with both genes *in vitro*. This is not a problem when the two enzymatic markers are used in retroviral vectors for lineage mapping studies, because individual cells are unlikely to be infected with both vectors and express both enzymes. (Even if they were, the interpretation of the results would not be affected, since one is still observing a clone.) If detection of cells containing both enzymes is necessary, use of different chromogenic substrates might circumvent this difficulty. PLAP can be reacted with naphthol-AS-BI-phosphate/New Fuchsin (Sigma, St. Louis, MO), which produces a red stain,[19] and NBT can be added to the β-Gal reaction to produce a purple precipitate. Red/blue or red/purple might be distinguishable from the corresponding single stains; however, we have not tested these possible combinations.

We have found that performing immunohistochemistry in conjunction with either β-Gal or PLAP histochemical staining is very difficult. The colored precipitates block fluorescence, thus preventing the use of fluorescently conjugated antibodies. Using horseradish peroxidase (HRP)-conjugated antibodies, it is difficult to distinguish blue/brown double-labeled cells from brown labeled ones, whereas purple and brown/purple cells would probably look identical. Situations in which the subcellular localization of β-Gal and an antigen are significantly nonoverlapping (e.g., a nuclear antigen) may allow for detection of both.[20] In addition, one can try to limit the intensity of the β-Gal or PLAP reaction product by shortening the time of the reactions to allow for more sensitivity in detecting an antigen within the same cell by HRP-conjugated antibodies.[21] Alternatively, antibodies exist for both β-Gal (rabbit polyclonal, Cappel, Malvern, PA; mouse monoclonal, Boehringer-Mannheim, Indianapolis, IN) and PLAP (rabbit polyclonal, Zymed, San Francisco, CA; mouse monoclonal, Medix), although using such reagents is obviously more time consuming than relying on histochemical reactions.

Different species, tissues, or parts of tissues (i.e., brain regions) can

---

[19] E. Harlow and D. Lane (eds.), "Antibodies: A Laboratory Manual." Cold Spring Harbor Laboratory, Cold Spring Harbor, New York, 1988.

[20] E. Y. Snyder, D. L. Deitcher, C. Walsh, S. Arnold-Aldea, E. A. Hartwieg, and C. L. Cepko, *Cell (Cambridge, Mass.)* **68**, 33.

[21] P. J.-J. Vaysse and J. E. Goldman, *Neuron* **5**, 227 (1990).

have varying amounts of background labeling.[22-24] This is especially noticeable with alkaline phosphatase reactions, but is also true for β-Gal. We recommend always including negative controls in order to assess the extent of background for a particular tissue or region of interest. Important variables that affect the signal-to-noise ratio include the following: type of fixative, length of fixation, length of washes, length of heat treatment, prolonged exposure to light, and prolonged storage after staining. For a particularly problematic tissue, it may prove advantageous to try different inhibitors of endogenous alkaline phosphatases (see above), as well as different substrates.

Solutions

Phosphate-buffered saline (PBS) ($10\times$): 80 g NaCl, 2 g KCl, 11.5 g Na_2HPO_4, and 2 g KH_2PO_4 in 1 liter water; adjust to pH 7.2–7.4 and dilute 1 : 10 before using

0.5% Glutaraldhyde: 25% stock (Sigma, St. Louis, MO) can be stored at $-20°$ and frozen/thawed many times; make dilution immediately before use

4% Paraformaldhyde: 4 g solid paraformaldehyde, 2 mM $MgCl_2$, and 1.25 mM EGTA (0.25 ml of a 0.5 M EGTA stock, pH 8.0) in 100 ml PBS, pH 7.2–7.4

Heat ~80 ml water to 60° and add paraformaldehyde; add NaOH to get paraformaldehyde in solution. Cool to room temperature, add 10 ml of $10\times$ PBS, adjust the pH with HCl, add $MgCl_2$ and EGTA, and make up to 100 ml with H_2O. The solution can be stored at 4° for several weeks

X-Gal detection buffer: 35 mM potassium ferrocyanide (can vary from 5 to 35 mM), 35 mM potassium ferricyanide (can vary from 5 to 35 mM), 2 mM $MgCl_2$, 0.02% Nonidet P-40 (NP-40) (diluted from 10% stock solution), and 0.01% sodium deoxycholate (diluted from 10% solution) in PBS; can be stored for at least 1 year at room temperature in a foil-covered container

X-Gal Stock ($40\times$): 40 mg/ml X-Gal (5-bromo-4-chloro-3-indolyl-β-D-galactopyranoside) in dimethylformamide; store at $-20°$ in a glass container covered with foil

X-Gal reaction mix: Make a 1 : 40 dilution of X-Gal stock into X-Gal detection buffer immediately before using; the final concentration of X-Gal is 1 mg/ml

[22] B. Pearson, P. L. Wolf, and J. Vazquez, *Lab. Invest.* **12**, 1249 (1963).
[23] E. Robins, H. E. Hirsch, and S. S. Emmons, *J. Biol. Chem.* **213**, 4246 (1968).
[24] A. M. Rutenburg, S. H. Rutenburg, B. Monis, R. Teague, and A. M. Seligman, *J. Histochem. Cytochem.* **6**, 122 (1958).

X-P detection buffer (Buffer 3, Genius Kit, Boehringer-Mannheim): 100 mM Tris-HCl, pH 9.5, 100 mM NaCl, and 50 mM MgCl$_2$. Store at room temperature. Salts tend to precipitate over several weeks, but this does not seem to markedly affect the staining

X-P stock (100×): 10 mg/ml 5-bromo-4-chloro-3-indolylphosphate (X-P; also referred to as BCIP) in water. Store in the dark as aliquots at −20°; can also be frozen and thawed several times

NBT stock (50×): 50 mg/ml nitro blue tetrazolium in 70% dimethylformamide, 30% water. Store at −20° in a glass container covered with foil (the stock does not freeze at this temperature)

X-P reaction mix: 50 μl X-P stock (100×) 100 μl NBT stock (50×), and 50 μl of 50 mM levamisole (100×) (if desired) in 5 ml X-P detection buffer. Make fresh immediately before using. Final concentrations are as follows: X-P, 0.1 mg/ml; NBT, 1 mg/ml; and levamisole, 0.5 mM. Levamisole was ineffective at reducing background in chick cerebellum, while slightly inhibiting PLAP, and was therefore not used in this tissue. Any other desired inhibitor (see above) would also be added to the X-P reaction mix.

Gelatin subbing solution for slides: 2 g gelatin and 0.1 g chromium potassium sulfate (chrome alum) in 200 ml water. Heat the water to 60°, dissolve the chrome alum, and then gradually dissolve gelatin. Filter before use. The percentage of gelatin can be increased or decreased. Load slides in racks, dip quickly, and air-dry overnight

Gelatin/sucrose embedding medium: 7.5% gelatin (porcine skin, Sigma), 15% sucrose, and 0.05% sodium azide in 1× PBS. Dissolve components gradually at 60°, with stirring. The medium solidifies at room temperature to a transparent gel. Store at room temperature. Liquify in a microwave oven with frequent swirling before embedding samples

Gelvatol mounting medium: Make up according to the protocol given by Rodriguez and Deinhardt.[25] Instead of Elvanol, we use Vinol grade 205 (polyvinyl alcohol) (Air Products and Chemicals, Inc., Allentown, PA).

Whole-Mount Staining Procedure

The primary protocol shown was worked out for staining of intact mouse retinas that had been coinfected with BAG and DAP.[8] The incubation times for X-Gal and X-P reactions are variables that may require adjustment for different tissues. Longer times may be necessary for large

[25] J. Rodriguez and F. Deinhardt, *Virology* 12, 316 (1960).

or dense chunks of tissue, although there is a trade off as the background staining of either enzyme intensifies with increasing reaction time. Whole chick embryos (E7) and chick brains (E10) were found to be incompletely reacted in their centers after 4-hr incubation periods. For tissues that are difficult to stain completely as whole mounts, one can stain as sections (see below). Alternatively, one can do a 3- to 4-hr whole-mount stain to locate cells of interest, dissect out and section only those areas of interest, and restain the sections for 20–30 min to obtain optimal staining. This will only work for recovering more PLAP activity, as the β-Gal activity will be destroyed by the heat treatment.

1. Dissect the tissue into PBS containing 2 mM MgCl$_2$ (PBS + Mg^{2+}) on ice.

2. Fix in 0.5% glutaraldehyde in PBS + Mg^{2+} for not longer than 45 min on ice. The 0.5% glutaraldehyde decreases the alkaline phophatase activity in chick cells stained *in situ* (but not in chicken embryo fibroblasts stained *in vitro*). Therefore, fixation of chick whole mounts is typically done in 4% paraformaldehyde in PBS for 2–4 hr at 4°. In areas where background alkaline phosphatase activity is a problem, increasing the time in 4% paraformaldehyde, even up to several days, can decrease endogenous background without significantly decreasing PLAP. However, such long fixation times may decrease β-Gal activity.

3. Rinse in PBS + Mg^{2+}, 5 times, 5 min each time. Rinsing overnight is fine, but waiting for several days at this step may decrease β-Gal activity.

4. Stain in X-Gal reaction mix for 2–4 hr at 37°. (Tris buffer was also tried in place of PBS, with no success.)

5. Rinse many times in PBS until the solution no longer turns yellow. This usually takes about 5 changes. An overnight rinse is fine. It is important to remove the X-Gal, since residual β-Gal activity in the presence of X-Gal and NBT (added for the following reaction) may enable β-Gal^{+}, PLAP^{-} cells to turn purple. Chick retinas and cerebella have been kept in PBS at 4° for at least 1 month at this point with no appreciable loss of signal in subsequent X-P staining. X-Gal staining can be easier to examine prior to carrying out the X-P reaction, as background alkaline phosphatase staining can obscure the X-Gal signal somewhat.

6. Heat the tissue in PBS at 65° for 30 min. This is usually done by floating the dish containing the tissue in a water bath preset to 65°. For staining of embryonic chick diencephalon (one of the areas of the brain with the highest background), this step has been increased to 1.5 hr.

7. Preincubate in X-P detection buffer for 15 min. Extending the time of this step results in diffusion of the alkaline phosphatase reaction product.

8. Incubate in X-P reaction mix for 3 hr at room temperature. Since background staining increases in light, cover with foil during and after staining.

9. Rinse in 20 mM EDTA in PBS for 2–4 hr. Background can be due to endogenous alkaline phosphatase or other reactions that generate hydride ions and thus reduce NBT to form a purple precipitate. We have noted that background staining appears more slowly in the presence of EDTA. Tissue can be stored in the dark at 4° in PBS plus EDTA or 30% sucrose in PBS plus EDTA for many months, although the background clearly increases over time.

10. Embed in paraffin wax using minimum necessary times for the tissue of interest. For mouse retina, which is approximately 250 μm thick, the following procedure is used. Dehydrate through graded ethanols: 50%, 70%, 95%, 100%, 100% for 20 min each. Clear in xylene, 2 times, 15 min each. Infiltrate with a 1 : 1 mix of xylene and paraffin at 65° for 30 min, then with paraffin, 2 times, 15 min each. Embed in paraffin.

Clearing and paraffin embedding are not recommended for tissues fixed with paraformaldehyde. In paraformaldehyde-fixed chick tissues, both the β-Gal and the PLAP reaction products were found to be very sensitive to xylene treatment; even relatively short exposures to xylene caused the reaction product to diffuse into the surrounding tissue. In some cases, this was true of ethanol as well. Frozen sections are a workable alternative (see below). Other embedding protocols, particularly aqueous-based procedures, may be worth testing. It is worth noting that strong staining in glutaraldehyde-fixed material can even withstand preparation for electron microscopy.[20]

11. Section onto slides coated with gelatin. Silane-treated slides are equally effective.[26]

12. Remove paraffin with xylene and mount with Permount. For frozen sections, fix sections to slides with 4% paraformaldehyde for 15 min. Rinse with PBS, then mount in Gelvatol (plus EDTA if desired). Storing slides at −80° helps prevent background staining from increasing.

Protocol for Staining Frozen Sections

The following protocol was worked out for embryonic rat striatum.[14] Fixation and staining times may need to be altered for other areas of interest.

1. Fix the tissue by perfusion followed by immersion in 4% paraformaldehyde at 4° for 8 hr. Rinse briefly in PBS, then sink in 30% sucrose in

[26] M. Rentrop, B. Knapp, H. Winter, and J. Schweizer, *Histochem. J.* **19**, 271 (1986).

PBS containing 2 mM MgCl$_2$ (PBS + Mg^{2+}) at 4°. Fixation times will vary with the size of the tissue. Perfusion may not be necessary for all tissues, especially in embryonic animals. Shorter fixation times may be preferable, as X-Gal staining may be decreased by lengthy fixation.

2. Embed brain in OCT or gelatin/sucrose mounting medium and freeze using liquid N$_2$. Gelatin/sucrose embedding gives better frozen sections for embryonic tissue than does OCT. Paraffin embedding destroys β-Gal activity. Cells in culture treated as if to embed in paraffin retain PLAP activity.

3. Cut cryostat sections and mount on gelatin-coated slides; air-dry overnight. Sections up to 90 μm thick (the thickest we have tried) have been successfully stained.

4. Fix sections to slides in 4% paraformaldehyde for 10–15 min at 4°.

5. Rinse slides in PBS + Mg^{2+} twice, for 10 min each, at 4°.

6. Stain slides in X-Gal reaction mix for 6 hr at 37°.

7. Rinse slides in PBS 3 times, for 10 min each, or until solution is no longer yellow. Slides can be left in PBS overnight. (See also comments for Whole-Mount Staining Procedure, Step 5.)

8. Transfer slides to preheated PBS at 65° and heat for 30 min.

9. Rinse slides in X-P detection buffer for 10 min.

10. Stain slides in X-P reaction mix for 12 hr. Because background staining increases in light, cover with foil during and after staining.

11. Rinse slides in PBS plus 20 mM EDTA 3 times, 10 min each. Mount in Gelvatol (plus EDTA if desired). (See also comments on background staining for Whole-Mount Staining Procedure, Step 9.) Storing slides at −80° helps prevent background staining from increasing.

Clonal Analysis Using Polymerase Chain Reaction/Library Method

Preparation of Retroviral Library for Polymerase Chain Reaction Analysis

In principle, any retroviral plasmid can be used to make a library. We start with the BAG plasmid,[7] which contains a unique cloning site (*Xho*I) downstream of the reporter gene.[15] The insertion of DNA into this site does not appear to interfere with expression of the upstream genes. The inserted DNA is genomic DNA from *Arabidopsis thaliana* digested with *Mbo*I. The digested DNA is run on an agarose gel, and DNA fragments less than 450 bp in size are used as inserts.[24]

Before making retroviruses with any of these constructs, approximately 100 constructs are identified whose inserts are conveniently distinguishable by size or by their pattern of digestion using restriction enzymes.

PCR products are prepared from bacterial colonies, the products are separated on agarose gels, and the approximate size of each PCR fragment is recorded. PCR products are sorted by size, and each product is digested with a mixture of restriction enzymes with 4-base recognition sites that are chosen because they are inexpensive and compatible within the same digestion buffer (MspI, RsaI, AluI, CfoI, and MseI). The size and restriction pattern of each tag are recorded, and approximately 100 constructs are chosen that are indistinguishable in this standard assay.[15]

Retroviruses are next prepared from the 100 constructs by transfecting "mini-prep" DNA,[27] purified using crushed glass, into a packaging cell line. The 100 DNAs are transfected[6] pairwise into 50 dishes of the ΨCRIP amphotropic packaging line[3] so that each transfected plate contains a mixture of cells producing two viral constructs. The supernatant of each transfected plate is used to infect a dish of the ecotropic packaging cell line Ψ2. Infected Ψ2 producer cells are selected by growth in medium containing G418 for 7 to 10 days, and the population of resistant colonies is raised to confluence.

Viral supernatants are recovered and titered on NIH 3T3 cells in 6-well tissue culture dishes, according to previously described techniques.[6] DNA from the infected NIH 3T3 cells is amplified using the PCR to evaluate the passage of the genetic tags. After the titering reaction, 0.5 ml of a solution of 0.5% Tween 20 and 200 μg/ml proteinase K in 1× PCR buffer is added to each plate of cells. After incubation for 60 min at 65°, the proteinase solution is transferred to a 0.6-ml microcentrifuge tube and incubated at 85° for 20 min, then 95° for 10 min. The undissolved X-Gal precipitate is pelleted by centrifugation at 10,000 rpm for 10 min, and 10 μl samples of the supernatant are used as templates for 50-μl polymerase chain reactions as described below.

The 50 viral supernatants are then mixed to generate approximately equal ratios of each tag, and the mixture is concentrated by centrifugation.[6] The concentrated stock is titered and tested for helper virus in vitro.[6] It is then used to perform experiments in the cerebral cortex.

The method described above can be easily modified in several ways. Any convenient restriction site in a replication-incompetent vector can be used for insertion of tags, so long as it would not interfere with expression of the reporter gene or transmission of the virus. We have successfully created libraries using BAG, DAP, as well as two avian replication-incompetent vectors. Any source of DNA that is not present normally in the host tissue or vector can be used. Any set of enzymes can be chosen for

[27] T. Maniatis, E. F. Fritsch, and J. Sambrook, "Molecular Cloning: A Laboratory Manual." Cold Spring Harbor Laboratory, Cold Spring Harbor, New York, 1982.

the restriction enzyme characterization of the inserts, and any range of insert sizes can be used if the longer ones are not so long that their size would cut down on the efficiency of the PCR. Finally, it may be possible to transfect the entire library as one plasmid preparation into producer cells rather than going through the tedium of making individual stable producers or producer populations. We are currently checking this as a possible way to make more complex libraries.

Tissue Analysis for Polymerase Chain Reaction-Based Clonal Assignments

Animals are infected and processed histochemically for the appropriate reporter gene as described previously. Usually sections are made for this analysis, but in special circumstances whole mounts can be used. Once the histochemistry is complete, the labeled cells are analyzed and their position as well as the outline and landmarks of the tissue are recorded. The goal is to create a permanent record of the morphology and relative location of the labeled cells, since the PCR analysis will destroy the cells and some surrounding tissue. A standard microscope and camera lucida device allow morphological details of cells to be drawn a high magnification, while cell location can be plotted at low magnification. Alternatively, cells can be photographed or plotted on a computerized system (e.g., CARP[28]). The anatomical analysis is often the most time-consuming aspect of the entire analysis.

Polymerase Chain Reaction of Labeled Cells

After anatomical analysis, PCR amplification of the tags that distinguish the vectors allows clonal analysis. For these experiments, start with two nested pairs of oligonucleotides specific for the region of the vector in which the inserts were placed. PCR parameters for each oligonucleotide pair must be optimized, especially the Mg^{2+} concentration in the PCR buffer.[29,30]

Materials

Histological sections prepared and stained as described above, and coverslipped in Gelvatol
Sterile distilled water

[28] C. P. Austin and C. L. Cepko, *Development (Cambridge, UK)* **110,** 713 (1990).
[29] K. B. Mullis and F. A. Faloona, this series, Vol. 155, p. 335.
[30] F. M. Ausubel, R. Brent, R. E. Kingston, D. D. Moore, J. G. Seidman, J. A. Smith, and K. Struhl, "Current Protocols in Molecular Biology." Green, New York, 1989.

1 × PCR buffer (see below)

Two nested pairs of oligonucleotides (as 20 μM solutions)

Deoxyribonucleotide solution (20 mM each in dATP, dTTP, dCTP, and dGTP)

Proteinase K, 10 mg/ml solution, in sterile distilled water

Tween 20, 10% solution, in steriled distilled water

Mineral oil (light), from Sigma

Disposable, breakable razor blades and blade holder (Fine Scientific Tools)

Microfuge tubes, 600 μl (autoclaved and silanized), or 96-well microtiter dishes and lids (e.g., Falcon, Nos. 3911 and 3913)

Automated thermal cycler

Dissecting microscope

Centrifuge tubes, 50 ml

Note. All solutions and containers must be assembled and stored using the most stringent precautions to prevent contamination with DNA that could be amplified in the PCR. A dedicated set of solutions should be handled with dedicated, positive displacement (or other contamination-resistant) pipetting devices in a separate laboratory remote from the thermal cycler. Good discussions of minimizing PCR contamination are available elsewhere.[30]

Dissection of Cells and Digestion of Tissue

1. Prepare a lysis solution. For 50 samples, mix 50 μl, or optimum amount, of 10× PCR buffer; 25 μl of 20 μM solution of each of the outermost oligonucleotides; 10 μl of 10 mg/ml solution of proteinase K (final concentration 0.2 mg/ml); 25 μl of 10% Tween-20 solution (final concentration 0.5%); and 375 μl distilled water.

2. Soak off coverslips in distilled water in a clean, sterile 50-ml centrifuge tube. After the slide has soaked about 30 min, the coverslip can be carefully pried off with a razor blade. After the coverslip comes off, soak the tissue about 5 min more to remove traces of Gelvatol.

3. Pipette 10 μl of the lysis solution into each 600-μl microcentrifuge tube (or each well of a 96-well microtiter dish).

4. Break off a fresh fragment (2–5 mm wide at the edge) of the breakable razor blade in the blade holder.

5. Under the dissecting microscope, locate a labeled cell. Well-stained cells can be seen with a low magnification objective (0.8×), but lighter cells may only be seen with a high magnification objective (5×).

6. Using the razor blade, cut a fragment of tissue that includes the nucleus of the labeled cell. If labeled cells are widely scattered, they can be dissected one at a time, in chunks that contain approximately 1000

unlabeled cells. Chunks are typically less than 500 μm in each dimension, but larger pieces can be used (PCR sensitivity may be less with larger pieces, since they do not dissolve as well). If labeled cells are immediately adjacent, dissect several cells in one chunk.

7. Transfer the tissue piece, on the razor blade, to the lysis solution. Confirm under the microscope that the labeled cell is in the lysis solution. It is important to keep careful notes of which cell goes where (drawings are helpful here).

8. Cover the lysis solution with 100 μl of mineral oil, and cap the tube (or cover the microtiter plate).

9. Replace the blade fragment with a fresh one, and dissect the next cell (return to Step 5). While dissecting cells, prepare negative controls. Intersperse samples that contain no tissue, or unlabeled tissue, among the positive samples.

10. After all cells have been dissected, and all lysis samples covered with mineral oil, cover the tubes (or plate).

11. Transfer samples to a thermal cycler. Digest for 2–3 hr at 65°. Inspect a few samples after this time to confirm that the tissue is totally dissolved. If not, digest longer (e.g., overnight at 37°). The X-Gal precipitate does not dissolve, but it does not interfere with the PCR.

12. Once the tissue is digested, heat it to 85° for 20 min, then heat to 95° for 5 min. This inactivates the proteinase K and denatures the genomic DNA. The samples are now ready for the PCR.

First Polymerase Chain Reaction

1. Prepare the PCR solution. For 50 samples, use 100 μl (or optimum amount) of 10× PCR buffer, 50 μl of 20 μM solution of each of the outermost oligonucleotide primers, 10 μl of the mixed deoxyribonucleotide solution (20 mM in each dNTP), 7.5 μl (37.5 units) of AmpliTaq DNA polymerase (Perkin-Elmer Cetus, Norwalk, CT), and 800 μl of distilled water. The reactions can be scaled up or down proportionately. Taq polymerase from other manufactures can be used but should be tested first.

2. Put the samples in the thermal cycler and start the PCR, which should begin with an initial denaturation at 92°–94° for 3 min. PCR conditions need to be optimized, but will comprise 45 repetitions of denaturation (92°–94° for 30–45 sec), annealing (55°–70° for 1–3 min), and extension (72° for 1 min).

3. Once the samples have reached at least 85°, uncover them and add 20 μl of the PCR solution to each sample. The "hot start" enhances the sensitivity and specificity of the PCR. The added PCR solution does not have to be mixed in. It will sink beneath the oil and join the aqueous phase.

4. Cover the samples and allow the PCR cycling to proceed.

Second Polymerase Chain Reaction. The samples now contain large amounts of amplified target DNA and should not be opened or even transported into the clean laboratory. Because contamination of them is not a concern, they may be handled in the main laboratory. The sensitivity of the second PCR is not critical, and *Taq* polymerase from any manufacturer can be used.

1. Prepare the following solution. For 50 samples, use 10 μl mixed deoxyribonucleotide solution (20 mM in each dNTP), 200 μl (or optimum amount) of 10× PCR buffer, 50 units *Taq* DNA polymerase, 1800 μl sterile distilled water, and 20 μl of each of the internal pair of oligonucleotide primers (final concentration 0.2 μM). The lower concentration of primers and deoxyribonucleotides used in the second reaction does not affect the PCR sensitivity or product yield.

2. Pipette 40 μl of this solution into 50 tubes (or microtiter wells).

3. Transfer 2–4 μl of the product of each first PCR into each tube (or well). Keep samples carefully labeled and take note of any pipetting errors.

4. Run the second PCR for 25–35 cycles. Conditions for the second PCR will also have to be optimized but will likely be similar to those used for the first reaction. If the same thermal cycler is used for both the first and second PCR amplifications, be sure not to open any of the samples of the second reaction near the machine. A ''hot start'' is not necessary for the second reaction.

5. Separate the PCR products on 3% NuSieve/1% SeaKem agarose gels (FMC, Rockland, ME). Use 1× Tris–borate– EDTA (TBE) as the running buffer,[28] and include pBR/*Msp*I or ΦX/*Hae*III DNA size standards. Anticipate that 40–70% of the samples will produce a band. Suspect a problem if yields are consistently below 40%; suspect contamination if many contiguous samples show the same product.

Direct Restriction Enzyme Digestion of Amplified Products

DNA inserts may be distinguished by size or sequence analysis. Restriction mapping offers the most convenient assay, and it is sufficiently specific for libraries with 100–300 different inserts. We chose five restriction enzymes with 4-base recognition sequences (*Cfo*I, *Rsa*I, *Alu*I, *Mse*I, and *Msp*I) that were active in similar buffer conditions (50–100 mM NaCl).[15] The mixture cuts small DNA fragments frequently, allowing them to be easily distinguished. Before loading the diagnostic gel, or prior to restriction digestion, sort the PCR products by size. When samples of similar initial size are run side by side on the gel, it allows the most direct comparison of the restriction fragments.

TABLE I
PROBABILITY OF TWO CLONES HAVING THE SAME TAG IN k
CLONES

Number of tags (n)	Probability			
	$k = 3$	$k = 5$	$k = 10$	$k = 20$
20	0.14	0.40	0.86	0.99
80	0.04	0.12	0.43	0.90
85	0.035	0.11	0.41	0.88
100	0.030	0.10	0.36	0.85
250	0.012	0.04	0.16	0.52

1. Prepare the following solution (for 50 samples): 250 units of each restriction enzyme; 150 μl of 10× restriction enzyme buffer, appropriate for the enzymes chosen; 15 μl of 10 mg/ml bovine serum albumin (BSA); and sterile distilled water to make a total of 500 μl.

2. Pipette 10 μl of this solution into individual wells of a microtiter dish (tubes can also be used).

3. Add 20 μl of one of the PCR products to each well.

4. Cover the microtiter plate and incubate at 37° for 3 hr.

5. Terminate the reactions by adding loading buffer, and separate the products of the reaction on a 3% NuSieve/1% SeaKem agarose gel run in 1× TBE.

6. Record which samples contain the same tags, along with the overall number of tags seen. Compare this information to the original plots of cell location to get clonal information.

Statistical Analysis

The tentative conclusion from the PCR analysis is that cells containing the same tag are members of the same clone. The confidence of this conclusion rests on (1) the number of clones in a given experiment, k, and (2) the number of tags in the library, n. There is a surprisingly large probability that the same tag will appear in two different clones by coincidence, and this probability needs to be considered in the clonal analysis.

The probability of "coincidental double infections" by one tag can be calculated using binomial theory, or by computer, using a Monte Carlo simulation. The simplest assumption for use in either method is that all tags are present in approximately equal ratios. This assumption must be verified experimentally. If it is not the case, the library may still be usable, but more complicated modeling must be performed. Detailed discussions of the statistical analysis of retroviral libraries are presented elsewhere.[31,32]

[31] T. B. L. Kirkwood, J. Price, and E. A. Grove, *Science* **258**, 317 (1992).
[32] C. Walsh, C. L. Cepko, E. F. Ryder, G. M. Church, and C. Tabin, *Science* **258**, 317 (1992).

A computer program (MONTAG) that aids in calculation of expected frequencies of coincidental infections given different distributions and numbers of tags has been written by George Church (Department of Genetics, Harvard Medical School, Boston, MA) and is available through anonymous internet ftp from rascal.med.harvard.edu. It will run on most VMS machines without recompiling. Type run MONTAG and answer the queries. If there are any problems, contact church@gnome.med.harvard.edu. Some typical results, assuming approximately equal ratios, are shown in Table I to provide guidelines. The probability that more than 1 clone show the same tag equals $1 - [n!/(n - k)! \, n^k]$.

To evaluate clonal data, it is important to evaluate the actual complexity of the viral library (i.e., how many tags are seen in total). The best results demand a highly complex library, with very few clones labeled in each experiment (3–4 clones). Under these conditions the probability is maximized that observed patterns faithfully reflect clonal patterns. However, be skeptical of patterns that are seen rarely, as they may only reflect rare coincidental double infections, rather than true clonal events.

Reagents and Solutions for Polymerase Chain Reaction

10× PCR buffer: 10 mM Tris buffer, pH 8.3 (purchase as premixed crystals from Sigma), 50 mM KCl, 0.01% gelatin, and 1.5–2.5 mM MgCl$_2$

Mixed deoxyribonucleotide solution: Deoxyribonucleotides may be purchased from Pharmacia (Piscataway, NJ) as separate 100 mM solutions of dATP, dCTP, dGTP, and dTTP. Mix them 1:1:1:1:1 (by volume) with distilled water to make a working mixture that is 20 mM in each. Store 10-μl aliquots at $-70°$

Proteinase K: Proteinase K can be purchased from many manufacturers, dissolved in sterile distilled water to make a 10 mg/ml solution, and stored as 20-μl aliquots at $-70°$

Oligonucleotide primer solutions: Deprotected oligonucleotides can be passed over a NAP-10 ion-exchange column (Pharmacia) and eluted with sterile distilled water. Adjust the concentration of the effluent to 20 μM by measuring the absorbance at 260 nm. We routinely use oligonucleotides without further purification. Oligonucleotides should be stored as 25- to 50-μl aliquots at $-20°$.

Note. Reagents, instruments, and glass microscope slides should be handled with scrupulous technique and UV-irradiated when needed to destroy contaminating DNA.

Author Index

Numbers in parentheses are footnote reference numbers and indicate that an author's work is referred to although the name is not cited in the text.

A

Aalberg, J., 524, 536(33)
Aarons, D., 145
Abbondanzo, S., 823(9), 824, 825(9)
Abbondanzo, S. J., 689, 804, 812, 879, 882(2)
Abbot, C., 794
Abderrahim, H., 625, 626(16)
Abrahamson, J., 702
Abramczuk, J., 474, 590, 720
Abremski, K., 891, 892
Ackerman, L., 664
Adams, D. A., 305
Adams, N., 195, 196(6)
Adenot, P. G., 492
Adira, S., 753
Agelopoulou, R., 66
Ahmadi, A., 653
Aigner, K., 346
Aitkin, J. R., 154
Akhavan-Tafti, H., 450
Akots, G., 702
Akruk, S. R., 147, 148(37)
Alberti, S., 466
Albertini, D. F., 524, 536(33)
Albertsen, H. M., 625, 626(16)
Alcivar, A. A., 103
Alexander, N., 137
Alfa, C. A., 543
Ali, I., 488
Allen, N. D., 450, 452, 453(7), 454(7), 461(7), 664, 687, 733
Allfrey, V. G., 490, 492
Alliot, F., 186
Almoguera, C., 295
Alt, F. W., 858
Alvarez-Buylla, A., 65, 67
Alwine, J. C., 304
Anad, R., 644
Anakwe, O. O., 104

Anand, R., 625, 626(15), 650, 653
Anderegg, C., 229, 231(16)
Andersen, A., 488
Anderson, D. J., 804
Anderson, D. M., 386
Anderson, E., 87, 114, 116(11), 119(11), 120(11), 122(11), 125(11), 524, 536(33)
Andrulis, I. L., 345, 346(8), 349(8)
Anegon, I., 907
Ang, S. L., 667
Angerer, L. M., 291, 362, 384
Angerer, R. C., 291, 362, 384
Antonarakis, S. E., 613(12), 625, 626(12)
Antoni, F., 273
Appel, R. D., 488
Araki, H., 899
Arboleda, C. E., 104
Arcellana-Panlilio, M. Y., 304
Archibald, A. L., 451
Argos, P., 891
Arimura, A., 502
Arlinghaus, R. B., 112
Arnheim, N., 611
Arnold, H., 398
Arnold-Aldea, S., 189, 948, 952(20)
Aronson, J., 451
Arp, B., 753
Asai, D. J., 526, 533, 534
Asakura, T., 752
Aten, J. A., 103
Attenon, P., 175
Au, L.-C., 702
Auerbach, A., 667
Auerbach, B. A., 665, 666(12), 823
Austin, C. P., 935, 955, 958(28)
Austin, C. R., 137
Austyn, J. M., 913
Ausubel, F. M., 295, 296(1), 297(1), 935, 954(6), 955, 956(30)
Avarameas, S., 509
Aviv, H., 323

Subject Index

ISBN 0-12-182126-9

9 780121 821265

90018

3 5282 00003 9225